APPLIED FINITE MATHEMATICS WITH CALCULUS

APPLIED FINITE MATHEMATICS WITH CALCULUS

HOWARD ANTON
BERNARD KOLMAN
DREXEL UNIVERSITY

NEW YORK
SAN FRANCISCO
LONDON
A Subsidiary of Harcourt Brace Jovanovich, Publishers

Cover art, *Whirl* by Kenneth Noland. Permission granted by The Des Moines Arts Center, Coffin Fine Arts Trust Fund, 1974.

ACADEMIC PRESS, INC.
111 FIFTH AVENUE,
NEW YORK, NEW YORK 10003

UNITED KINGDOM EDITION PUBLISHED BY
ACADEMIC PRESS, INC. (LONDON) LTD.
24/28 OVAL ROAD, LONDON NW1

ISBN: 0-12-059560-55
Library of Congress Catalog Card Number: 77-91305

PRINTED IN THE UNITED STATES OF AMERICA

To our mothers

CONTENTS

4 MATRICES AND LINEAR SYSTEMS

5 LINEAR PROGRAMMING (AN ALGEBRAIC APPROACH)*

6 PROBABILITY

7 STATISTICS

8
APPLICATIONS

9
MATHEMATICS OF FINANCE

10
COMPUTERS

11
FUNCTIONS, LIMITS, AND RATES OF CHANGE

NEW FEATURES IN THIS EDITION

- Five chapters of calculus and its applications have been added. (An alternate version, *Applied Finite Mathematics,* Second Edition, is available without calculus.)
- A new self-contained chapter on the mathematics of finance.
- Stochastic processes are introduced.
- Tree diagrams are used more extensively as a tool in probability problems.
- Additional exercises.
- BASIC replaces FORTRAN in the computer chapter.

PREFACE

This book presents the fundamentals of finite mathematics and basic calculus in a style tailored for beginners, but at the same time covers the subject matter in sufficient depth so that the student can see a rich variety of realistic and relevant applications. Since many students in this course have a minimal mathematics background, we have devoted considerable effort to the pedagogical aspects of this book—examples and illlustrations abound. We have avoided complicated mathematical notation and have painstakingly worked to keep technical difficulties from hiding otherwise simple ideas. Where appropriate, each exercise set begins with basic computational "drill" problems and then progresses to problems with more substance. The writing style, illustrative examples, exercises, and applications have been designed with one goal in mind: *To produce a textbook that the student will find readable and valuable.*

Since there is much more finite mathematics and calculus material available than can be included in a single reasonably sized text, it was necessary for us to be selective in the choice of material. We have tried to select those topics that we believe are *most likely to prove useful* to the majority of readers. Guided by this principle, we chose to omit the traditional symbolic logic material in favor of a chapter on computers and computer programming. Computer programming requires the same kind of logical precision as symbolic logic, but is more likely to prove useful to most students. The computer chapter is optional and *does not* require access to any computer facilities. However, this chapter is extensive enough that the student will be able to run programs on a computer, if desired.

In keeping with the title, *Applied Finite Mathematics with Calculus,* we have included a host of applications. They range from artificial "applications" which are designed to point out situations in which the material might be used, all the way to bona fide relevant applications based on "live" data and actual research papers. We have tried to include

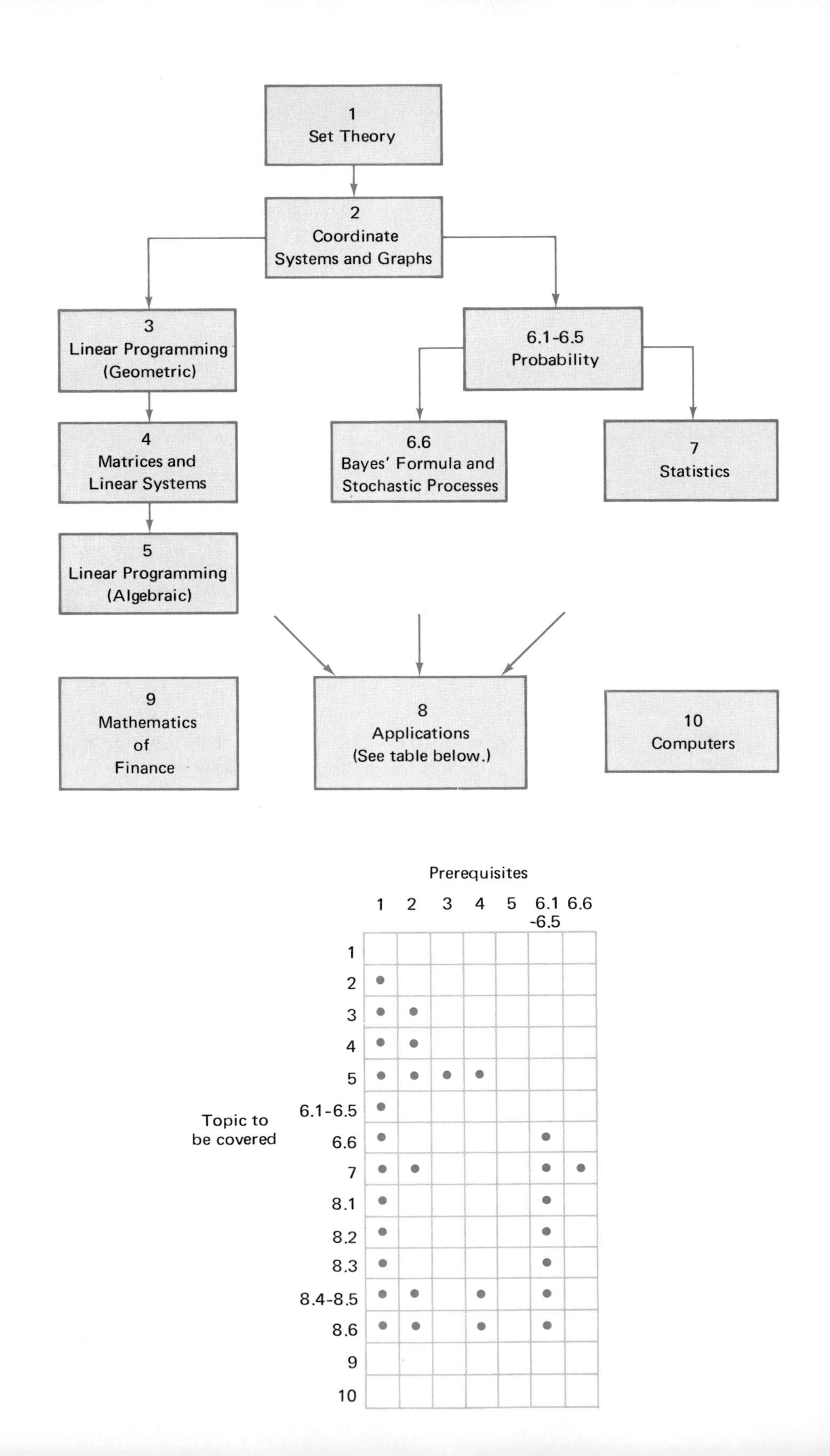

Prerequisites

Topic to be covered	1	2	3	4	5	6.1 -6.5	6.6
1							
2	●						
3	●	●					
4	●	●					
5	●	●	●	●			
6.1-6.5	●						
6.6	●					●	
7	●	●				●	●
8.1	●					●	
8.2	●					●	
8.3	●					●	
8.4-8.5	●	●		●		●	
8.6	●	●		●		●	
9							
10							

a balanced sampling from business, finance, biology, behavioral sciences, and social sciences.

There is enough material in this book so that each instructor can select the topics that best fit the needs of the class. To help in this selection, we have included a discussion of the structure of the book and a flow chart suggesting possible organizations of the material. The prerequisites for each topic are shown in the table below the flow chart.

Chapter 1 discusses the elementary set theory needed in later chapters.

Chapter 2 gives an introduction to cartesian coordinate systems and graphs. Equations of straight lines are discussed and applications are given to problems in simple interest, linear depreciation, and prediction. We also consider the least squares method for fitting a straight line to empirical data, and we discuss material on linear inequalities that will be needed for linear programming.

Portions of this chapter may be familiar to some students, in which case the instructor can review this material quickly.

Chapter 3 is devoted to an elementary introduction to linear programming from a geometric point of view. A more extensive discussion of linear programming, including the simplex method, appears in Chapter 5. Since Chapter 5 is technically more difficult, some instructors may choose to limit their treatment of linear programming entirely to Chapter 3, omitting Chapter 5.

Chapter 4 discusses basic material on matrices, the solution of linear systems, and applications. Many of the ideas here are used in later sections.

Chapter 5 gives an elementary presentation of the simplex method for solving linear programming problems. Although our treatment is an elementary as possible, the material is intrinsically technical, so that some instructors may choose to omit this chapter. For this reason we have labeled this chapter with a star in the table of contents.

Chapter 6 introduces probability for finite sample spaces. This material builds on the set-theory foundation of Chapter 1. We carefully explain the nature of a probability model so that the student understands the relationship between the model and the corresponding real-world problem.

Section 6.6 on Bayes' Formula and stochastic processes is somewhat more difficult than the rest of the chapter and is starred. Instructors who omit this section should also omit Section 8.1 which applies the material to problems in medical diagnosis.

Chapter 7 discusses basic concepts in statistics. Section 7.7 introduces hypothesis testing by means of the chi-square test, thereby exposing the student to some realistic statistical applications. Section 7.4 on Chebyshev's inequality is included because it helps give the student a better feel for the notions of mean and variance. We marked it as a starred section since it can be omitted from the chapter without loss of continuity. An instructor whose students will take a separate statistics course may choose to omit this chapter entirely.

Chapter 8 is intended to give the student some solid, realistic applications of the material. The topics in this chapter are drawn from a variety of fields so that the instructor can select those sections that best fit the needs and interests of the class.

Chapter 9 covers a number of topics in the mathematics of finance. The chapter is self contained and includes an optional review section on exponents and logarithms.

Chapter 10 introduces the student to computers and programming. While there is no need to have access to any computer facilities, the material is presented in sufficient detail that the student will be able to run programs on a computer. It is not the purpose of this chapter to make the student into a computer expert; rather we are concerned with providing an intelligent understanding of what a computer is and how it works. We touch on binary arithmetic and then proceed to some BASIC programming and flow charting. We have starred this chapter since we regard it as optional.

Chapters 11–15 introduce the elements of calculus and its applications. The approach is intuitive with emphasis on applications rather than theory.

ACKNOWLEDGMENTS

We gratefully acknowledge the contributions of the following people whose comments, criticisms and assistance greatly improved the entire manuscript.

Robert E. Beck—*Villanova University*
Elizabeth Berman—*Rockhurst College*
Steve E. Bowser—*Drexel University*
Alan I. Brooks—*Sperry UNIVAC*
Jerry Ferry—*Christopher Newport College*
Mary W. Gray—*American University*
Beryl M. Green—*Oregon College of Education*
Albert J. Herr—*Drexel University*
Robert L. Higgins—QUANTICS
Leo W. Lampone—*Spring Garden College*
Daniel P. Maki—*Indiana University*
Samuel L. Marateck—*New York University*
J. A. Moreno—*San Diego City College*
Donald E. Myers—*University of Arizona*
John Quigg—*Drexel University*
Ellen Reed—*University of Massachusetts at Amherst*
James Snow—*Lane Community College*
Leon Steinberg—*Temple University*
William H. Wheeler—*Indiana State University*

We also thank our typists: Susan R. Gershuni, Judy A. Kummerer, Amelia Maurizio, Kathleen R. McCabe, and Anne Schafer for their skillful work and infinite patience.

We thank: IBM, Sperry UNIVAC, and Teletype Corporation for providing illustrations for the computer material.

Finally, we thank the entire staff of Academic Press for their support, encouragement and imaginative contributions.

I

SET THEORY

A herd of buffalo, a bunch of bananas, the collection of all positive even integers, and the set of all stocks listed on the New York Stock Exchange have something in common; they are all examples of objects that have been grouped together and viewed as a single entity. This idea of grouping objects together gives rise to the mathematical notion of a set, which we shall study in this chapter. We shall use this material in later chapters to help solve a variety of important problems.

1.1 INTRODUCTION TO SETS

A ***set*** is a collection of objects; the objects are called the ***elements*** or ***members*** of the set.

One way of describing a set is to list the elements of the set between braces. Thus, the set of all positive integers that are less than 4 can be written

$$\{1, 2, 3\};$$

the set of all positive integers can be written

$$\{1, 2, 3, \ldots\};$$

and the set of all United States Presidents whose last names begin with the letter T can be written

$$\{\text{Taft, Tyler, Taylor, Truman}\}.$$

We shall denote sets by uppercase letters such as A, B, C, . . . and members of a set by lowercase letters such as a, b, c, ... With this notation, an arbitrary set with five members might be written

$$A = \{a, b, c, d, e\}.$$

To indicate that an element a is a member of the set A, we shall write

$$a \in A,$$

which is read "a is an element of A" or "a belongs to A." To indicate that the element a is *not* a member of the set A, we shall write

$$a \notin A,$$

which is read "a is not an element of A" or "a does not belong to A."

Example 1 Let

$$A = \{1, 2, 3\}.$$

Then $1 \in A$, $2 \in A$, $3 \in A$, but $4 \notin A$.

At times it is inconvenient or impossible to list the elements of a set. In such cases, the set can often be described by specifying a property that the elements of the set have in common. A convenient way of doing this is to use what is sometimes called ***set-builder*** notation. To illustrate, consider the set

$$\text{A} = \{1, 2, 3, 4, 5, 6, 7, 8, 9\},$$

which we have described here by listing its elements. Since A consists precisely of those positive integers that are less than 10, this set can be written in set-builder notation as

$$A = \{x \mid x \text{ is a positive integer less than } 10\},$$

which is read, "A equals the set of all x such that x is a positive integer less than 10."

In this notation x denotes a typical element of the set, the vertical bar $|$ is read "such that," and following the bar are the conditions that x must satisfy to be a member of the set A.

The following examples give further illustrations of this notation.

Example 2 The set of all IBM stockholders can be written in set-builder notation as

$$\{x \mid x \text{ is an IBM stockholder}\},$$

which is read "the set of all x such that x is an IBM stockholder." Note that it would be very inconvenient to list all the members of this set.

Example 3 $\{x \mid x$ is a letter in the word *stock*$\}$ is read "the set of all x such that x is a letter in the word *stock*." This set can also be described by listing its elements as

$$\{s, t, o, c, k\}.$$

Two sets A and B are said to be ***equal*** if they have the same elements, in which case we write

$$A = B.$$

Example 4 If A is the set

$$\{1, 2, 3, 4\}$$

and B is the set

$$\{x \mid x \text{ is a positive integer and } x^2 < 25\},$$

then

$$A = B.$$

Example 5 Consider the sets of stocks

$$A = \{\text{IBM, Du Pont, General Electric}\}$$

and

$$B = \{\text{Du Pont, General Electric, IBM}\}.$$

Even though the members of these sets are listed in different orders, the sets A and B are equal since they have the same members.

It is customary to require that all members of a set be distinct; thus when describing a set by listing its members, all duplications should be deleted.

Example 6 In a study of the effectiveness of antipollution devices attached to the exhaust systems of 11 buses, the following percentage decreases in carbon monoxide emissions were observed:

$$12,\ 15,\ 11,\ 12,\ 14,\ 15,\ 11,\ 16,\ 15,\ 17,\ 12.$$

The set of observations is

$$\{11,\ 12,\ 14,\ 15,\ 16,\ 17\}.$$

Sometimes one sees sets that have no members. For example, the set

$$\{x \mid x \text{ is an integer and } x^2 = -1\}$$

has no members since no integer has a square that is negative. A set with no members is called an ***empty set*** or sometimes a ***null set.*** Such a set is denoted by the symbol $\varnothing$.

Consider the sets

$$B = \{a, e, i, o, u\} \qquad \text{and} \qquad A = \{a, o, u\}.$$

Every member of the set A is also a member of the set B. This suggests the following definition.

If every member of a set A is also a member of a set B, then we say ***A is a subset of B*** and write

$$A \subset B.$$

In addition, it is customary to agree that the empty set $\varnothing$ is a subset of *every* set.

Example 7 Let

$$A = \{a, b, c\}.$$

All possible subsets of A are

$$\varnothing,\quad \{a\},\quad \{b\},\quad \{c\},\quad \{a, b\},\quad \{a, c\},\quad \{b, c\},\quad \{a, b, c\}.$$

Example 8 If

$$A = \{x \mid x \text{ is a positive even integer}\}$$

and

$$B = \{x \mid x \text{ is an integer}\},$$

then

$$A \subset B.$$

Example 9 If S is the set of all stocks listed with the New York Stock Exchange on July 26, 1974 and T is the set of all stocks on the New York Stock Exchange that traded over 100 shares on July 26, 1974, then

$$T \subset S.$$

Example 10 If A is any set, then

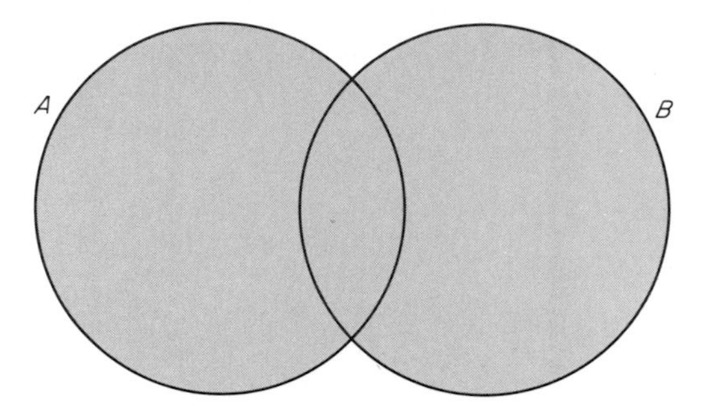

Figure 1.1

If a set A is not a subset of a set B, we write

$$A \not\subset B.$$

Example 11 Let A be the set of points inside the left circle in Figure 1.1 and let B be the set of points inside the right circle. Then

$$A \not\subset B \qquad \text{and} \qquad B \not\subset A.$$

Example 12 Let A be the set of points inside the larger circle in Figure 1.2 and let B be the set of points inside the smaller circle. Then

$$B \subset A \qquad \text{and} \qquad A \not\subset B.$$

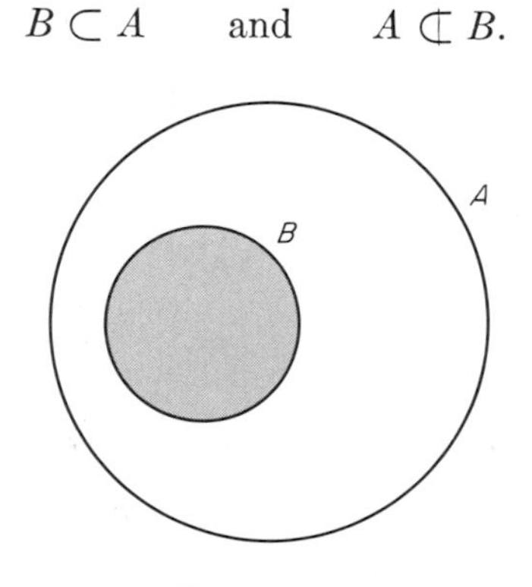

Figure 1.2

EXERCISE SET 1.1

1. Let $A = \{a, b, c, 1, 2, 5, e, f\}$. Answer the following as true or false.

   (a) $4 \in A$ (b) $4 \notin A$ (c) $a \in A$

   (d) $a \notin A$ (e) $\varnothing \notin A$ (f) $A \in \{a\}$.

2. Let $A = \{x \mid x \text{ is a real number satisfying } x \geq 5\}$. Answer the following as true or false.

(a) $2 \in A$ (b) $5.5 \in A$ (c) $7 \notin A$

(d) $5 \in A$ (e) $3 \notin A$ (f) $3 \in A$.

3. Consider the set of water pollutants

$$A = \{\text{sulfur, crude oil, phosphates, mercury}\}.$$

Answer the following as true or false.

(a) sulfur $\in A$ (b) phosphates $\notin A$

(c) arsenic $\notin A$ (d) oil $\in A$.

4. Let $A = \{x \mid x \text{ is a real number and } x^2 = 9\}$. List the elements of A.

5. In each part form a set from the letters in the given words:

(a) *AARDVARK* (b) *MISSISSIPPI* (c) *TABLE.*

6. Write $A = \{1, 2, 3, 4, 5\}$ in set-builder notation.

7. Write the following in set-builder notation:

(a) the set of U.S. citizens;

(b) the set of U.S. citizens over 40 years of age.

8. Let $A = \{1, 2, 3, 4\}$. Which of the following sets are equal to A?

(a) $\{3, 2, 1, 4\}$ (b) $\{1, 2, 3\}$ (c) $\{1, 2, 3, 4, 0\}$

(d) $\{x \mid x \text{ is a positive integer and } x^2 \leq 16\}$

(e) $\{x \mid x \text{ is a positive real number and } x \leq 4\}$

(f) $\{x \mid x \text{ is a positive integer and } x < 4\}$.

9. Consider the set of psychological disorders

$$A = \{\text{schizophrenia, paranoia, depression, megalomania}\}.$$

Which of the following sets are equal to A?

(a) {schizophrenia, paranoia, depression}

(b) {schizophrenia, paranoia, megalomania, depression}.

10. Which of the following sets are empty?

(a) $\{x \mid x \text{ is an integer and } x^2 = 4\}$.

(b) $\{x \mid x \text{ is an integer and } x^2 = -4\}$.

(c) $\{x \mid x \text{ is a real number satisfying } x^2 + 1 = 0\}$.

11. List all subsets of the set $\{2, 5\}$.

12. List all subsets of the set {Roosevelt, Truman, Kennedy}.

13. List all subsets of

(a) $\{a_1, a_2, a_3\}$ (b) $\varnothing$.

14. Let $A = \{3, 5, 7, 9\}$. Answer the following as true or false.

(a) $\{5, 3\} \subset A$ (b) $A \subset A$ (c) $\{3, 5, 9, 7\} \not\subset A$

(d) $A \subset \{3, 5\}$ (e) $A \subset \{3, 5, 7, 9, 0\}$

(f) $\varnothing \subset A$ (g) $\{1\} \not\subset A$.

15. Let $A = \{x \mid x \text{ is an integer and } x^2 \leq 25\}$. Answer the following as true or false:

(a) $\{x \mid x \text{ is a positive integer and } x^2 \leq 16\} \subset A$.

(b) $\{2, -2, 6\} \subset A$.

(c) $A \subset \{1, 2, 3, 4, 5\}$.

(d) $\{x \mid x \text{ is an integer and } x < 5\} \subset A$.

16. In Figure 1.3, let

S = the set of points inside the square

T = the set of points inside the triangle

C = the set of points inside the circle

and let x and y be the indicated points. Answer the following as true or false:

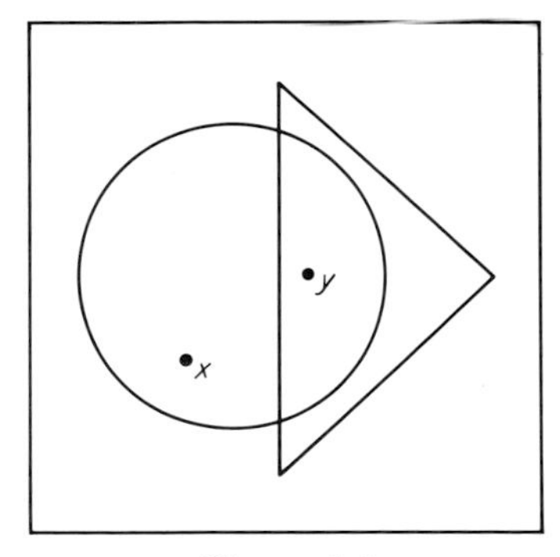

Figure 1.3

(a) $C \subset T$ (b) $C \subset S$ (c) $T \not\subset C$ (d) $x \notin C$

(e) $x \in T$ (f) $y \in C$ and $y \in T$ (g) $x \in C$ or $x \in T$.

17. In each part find the "smallest" possible set that contains the given sets as subsets.

(a) $\{1, 3, 7\}$, $\{3, 5, 9, 2\}$, $\{1, 2, 3, 4, 6\}$, $\{3\}$;

(b) $\{a, b, c\}$, $\varnothing$.

18. In each part find the "smallest" set that contains the given sets as subsets.

(a) {IBM, Du Pont, Xerox}, {Polaroid, Honeywell, Xerox, IBM, Avco}.

(b) $\{1, 3, 5\}$, $\{a, b, 3\}$, $\{a\}$, $\{a, b\}$.

19. Is it true that $\varnothing \in \varnothing$? Is it true that $\varnothing \subset \varnothing$?

20. Is the set of letters in the word *latter* the same as the set of letters in the word *later*?

1.2 UNION AND INTERSECTION OF SETS

We all know that the operations of addition, subtraction, multiplication, and division of real numbers can be used to solve a variety of problems. Analogously, we can introduce on sets operations that can be used to solve many important problems. In this section we shall discuss two such operations and in later sections we shall illustrate their applications.

If A and B are two given sets, then the set of all elements that belong to *both* A and B is a new set, called the ***intersection*** of A and B; it is denoted by the symbol

$$A \cap B.$$

Example 13 If, as in Figure 1.4, A is the set of points inside the circle on the left and B is the set of points inside the circle on the right, then $A \cap B$ is the set of points in the shaded region of the figure.

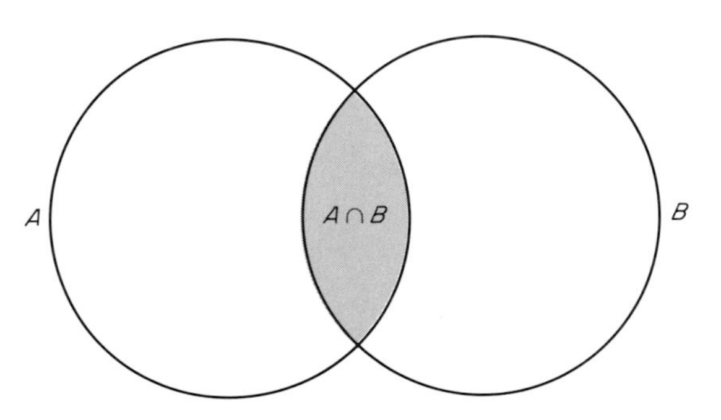

Figure 1.4

Example 14 Let

$$A = \{a, b, c, d, e\}, \qquad B = \{b, d, e, g\}, \qquad C = \{a, h\}.$$

Find $A \cap B$, $A \cap C$, and $B \cap C$.

Solution The only elements that belong to both A and B are b, d, and e. Therefore

$$A \cap B = \{b, d, e\}.$$

Similarly,

$$A \cap C = \{a\}.$$

Since the sets B and C have no elements in common, their intersection is the empty set; that is,

$$B \cap C = \varnothing.$$

Two sets that have no common elements, like B and C in Example 14, are called ***disjoint sets.***

Example 15 In Figure 1.5, let A, B, C, and D be the sets of points inside the indicated circles. Since A and B overlap,

$$A \cap B \neq \varnothing,$$

that is, A and B are not disjoint. On the other hand, C and D are disjoint since

$$C \cap D = \varnothing.$$

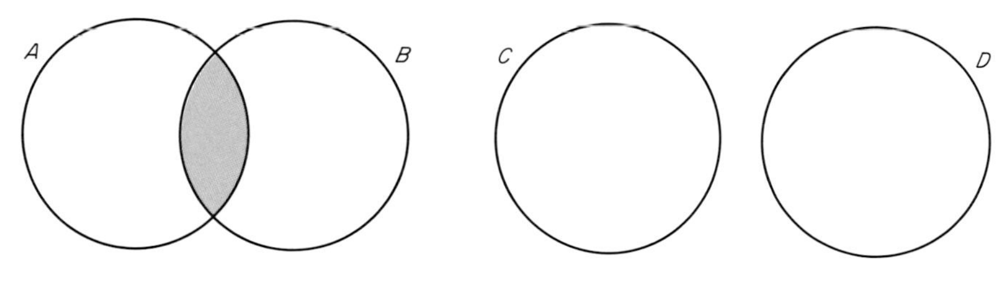

Figure 1.5

We can define intersections of more than two sets as follows:

> The ***intersection*** of any collection of sets is the set of elements that belong to every one of the sets in the collection.

Example 16 If A, B, and C are the sets of points inside the circles indicated in Figure 1.6, then the intersection of these sets, denoted by

$$A \cap B \cap C$$

is the shaded region in the figure.

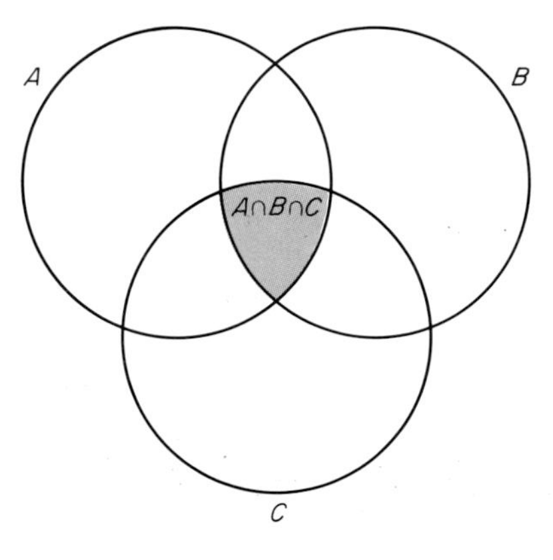

Figure 1.6

Example 17 Let

$$A = \{1, 2, 4, 7, 9, 11\}$$

$$B = \{2, 7, 9, 11, 17, 19\}$$

$$C = \{0, 2, 5, 7, 19, 24\}$$

$$D = \{2, 7, 9\}.$$

Then

$$A \cap B \cap C \cap D = \{2, 7\}.$$

Example 18 Let S be the set of stocks on the New York Stock Exchange that have paid a dividend for each of the past 40 years, and let R be the set of railroad stocks listed on the New York Stock Exchange. Describe the set

$$S \cap R.$$

Solution The members of $S \cap R$ belong to both S and R, so that $S \cap R$ consists of all railroad stocks on the New York Stock Exchange that have paid a dividend for each of the past 40 years.

If A and B are two given sets, then the set of all elements that belong to either A or B or both is a new set, called the ***union*** of A and B; it is denoted by the symbol

$$A \cup B.$$

Example 19 Let

$$A = \{a, b, c, d, e\} \qquad \text{and} \qquad B = \{b, d, e, g\}.$$

Find $A \cup B$.

Solution The elements that belong either to A or to B or to both are a, b, c, d, e, and g. Therefore

$$A \cup B = \{a, b, c, d, e, g\}.$$

Example 20 If A and B are the sets of points inside the indicated circles in Figure 1.7a, then $A \cup B$ is the set of points in the shaded region shown in Figure 1.7b.

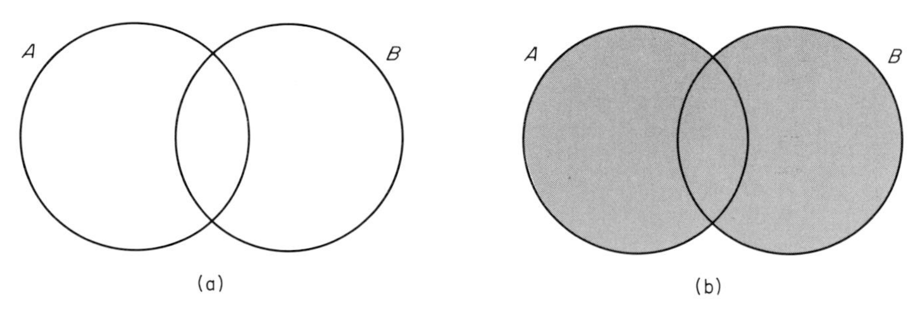

Figure 1.7

Example 21 Let A be any set. Find $A \cup \varnothing$.

Solution The members of $A \cup \varnothing$ are those elements that lie either in A or $\varnothing$ or both. Since $\varnothing$ has no elements, we obtain

$$A \cup \varnothing = A.$$

We can define unions of more than two sets as follows:

> Given any collection of sets, their ***union*** is the set of elements that belong to one or more of the sets in the collection.

Example 22 If A, B, and C are the sets of points inside the circles indicated in Figure 1.8a, then the union of these sets, denoted by

$$A \cup B \cup C$$

is the shaded region in Figure 1.8b.

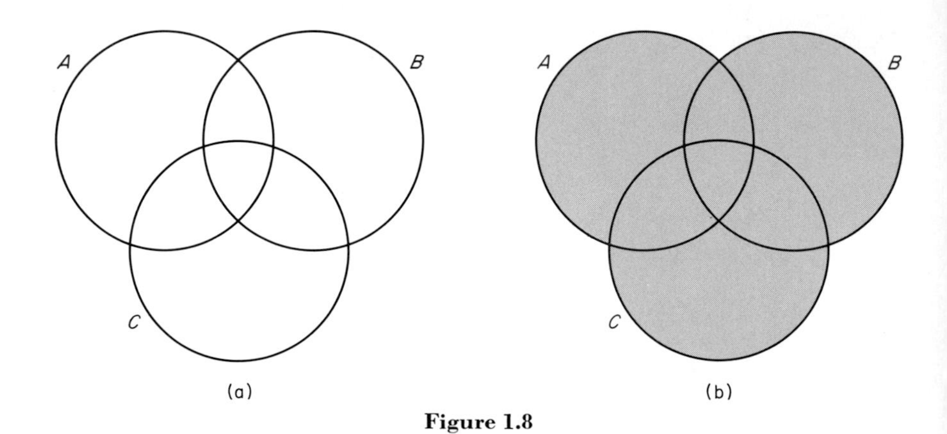

Figure 1.8

Example 23 Let

$$A = \{1, 2, 4, 7, 9, 11\}$$

$$B = \{2, 7, 9, 11, 17\}$$

$$C = \{0, 11\}$$

$$D = \{1, 4\}.$$

Then

$$A \cup B \cup C \cup D = \{0, 1, 2, 4, 7, 9, 11, 17\}.$$

Example 24 Let A, B, and C be the points inside the circles indicated in Figure 1.9. Shade the sets

(a) $A \cup B$ (b) $C \cap (A \cup B)$ (c) $(C \cap A) \cup (C \cap B)$.

Solution (a) See Figure 1.10.

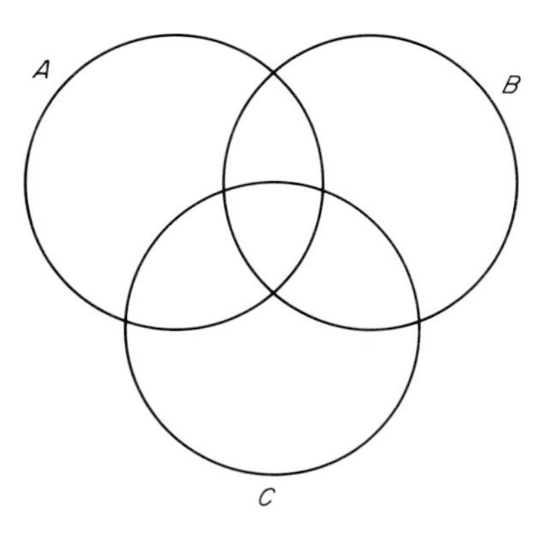

Figure 1.9

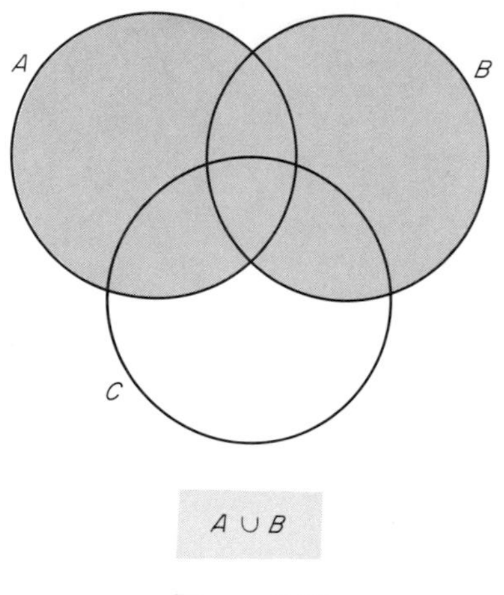

Figure 1.10

Solution (b) To find $C \cap (A \cup B)$ we intersect C with the shaded set $A \cup B$ in Figure 1.10; this yields the shaded set in Figure 1.11.

Solution (c) We begin by shading the sets $C \cap A$ and $C \cap B$; this gives the diagrams in Figure 1.12. To find $(C \cap A) \cup (C \cap B)$, we form the union of the shaded sets $C \cap A$ and $C \cap B$; this yields the shaded set in Figure 1.13.

Observe that the sets $C \cap (A \cup B)$ and $(C \cap A) \cup (C \cap B)$ obtained in parts (b) and (c) are identical. We have thus established the following basic law of sets:

$$C \cap (A \cup B) = (C \cap A) \cup (C \cap B).$$

This is called the ***first distributive law*** for sets. We leave it as an exercise

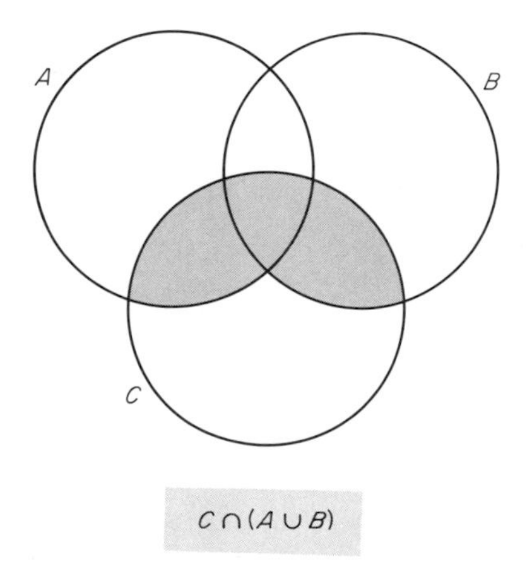

Figure 1.11

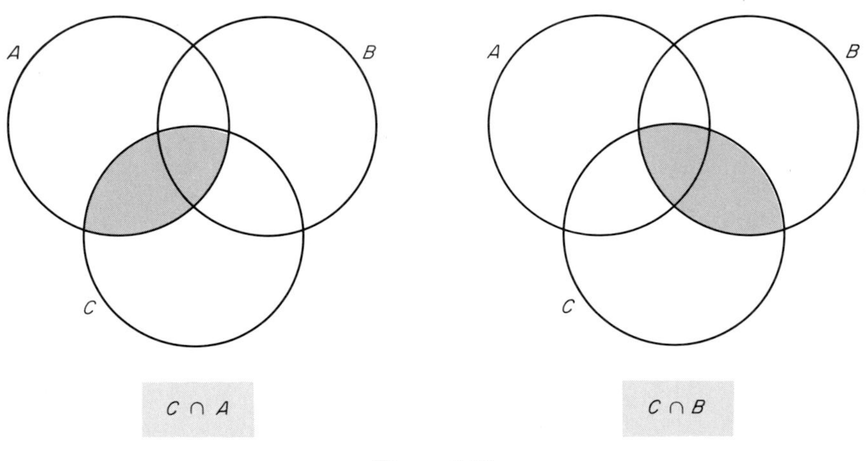

Figure 1.12

to establish the ***second distributive law*** for sets:

$$C \cup (A \cap B) = (C \cup A) \cap (C \cup B).$$

The idea of using shading and circles to establish relationships between sets is attributed to the British logician John Venn.† In his honor, they are called ***Venn diagrams.*** In the exercises we have indicated some other useful set relationships that can be established using Venn diagrams.

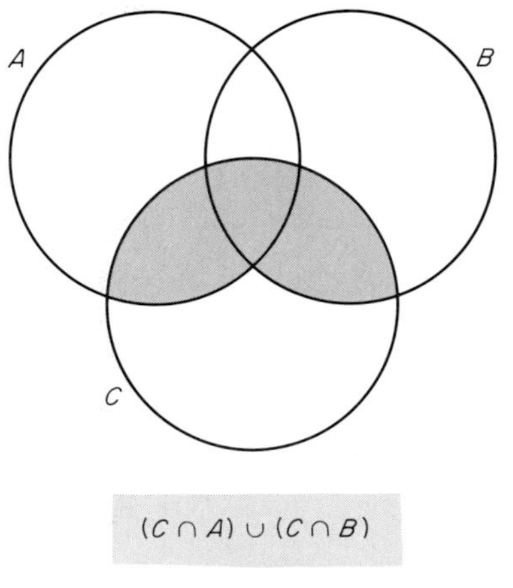

Figure 1.13

† *John Venn* (1834–1923)—Venn was the son of a minister. He graduated from Gonville and Caius College in Cambridge, England in 1853, after which he pursued theological interests as a curate in the parishes of London. As a result of his contact with intellectual agnostics and the works of Augustus DeMorgan, George Boole, and John Stuart Mill, Venn became interested in logic. In addition to his work in logic, he made important contributions to the mathematics of probability. He was an accomplished linguist, a botanist, and a noted mountaineer.

Example 25 In an experiment with hybrid corn, the corn plants were classified into sets as follows:

$$Q = \text{quick-growing} \qquad R = \text{rust resistant}$$
$$W = \text{all white kernels} \qquad Y = \text{all yellow kernels.}$$

Describe the characteristics of the plants in the following sets:

(a) $Q \cap Y$ (b) $R \cup W$

(c) $(R \cap Q) \cup (R \cap Y)$ (d) $Q \cup (W \cap Y)$.

Solution (a) The plants in $Q \cap Y$ are in both Q and Y. Thus $Q \cap Y$ consists of quick-growing, yellow-kerneled plants.

Solution (b) The plants in $R \cup W$ are either in R or W. Thus $R \cup W$ consists of plants that are either rust resistant or white-kerneled.

Solution (c) The plants in $R \cap Q$ are rust resistant and quick-growing. The plants in $R \cap Y$ are rust resistant and yellow-kerneled. Thus $(R \cap Q) \cup (R \cap Y)$ consists of plants that are either rust resistant and quick-growing or rust resistant and yellow-kerneled.

Solution (d) The set $W \cap Y$ is empty since the kernels cannot be both all white and all yellow. Therefore,

$$Q \cup (W \cap Y) = Q \cup \varnothing = Q.$$

Thus $Q \cup (W \cap Y)$ is just the set of quick-growing plants.

EXERCISE SET 1.2

1. Let

$A = \{1, 3, 5, 7\}$, $B = \{2, 3, 4, 7\}$

$C = \{x \mid x \text{ is an integer satisfying } x^2 \leq 9\}$, $D = \{5, 6\}$.

Compute

(a) $A \cap B$ (b) $A \cap C$

(c) $B \cap C$ (d) $B \cap D$

(e) $A \cap B \cap C$ (f) $D \cap \varnothing$.

2. Let A, B, C, and D be the sets in Exercise 1. Compute

(a) $A \cup B$ (b) $A \cup C$

(c) $B \cup C$ (d) $A \cup B \cup D$

(e) $A \cup \varnothing$ (f) $D \cup D$.

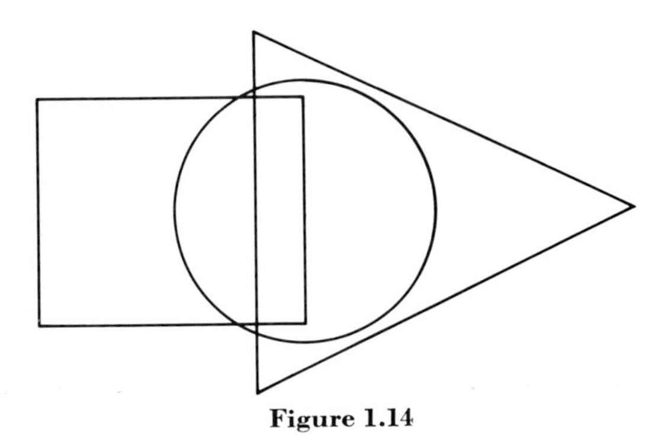

Figure 1.14

3. In Figure 1.14, let

$S =$ the set of points inside the square

$T =$ the set of points inside the triangle

$C =$ the set of points inside the circle.

Shade the following sets:

(a) $S \cap T$ (b) $S \cap C$

(c) $T \cap C$ (d) $S \cap T \cap C$.

4. Let S, T, and C be the sets in Exercise 3. Shade the following sets:

(a) $S \cup T$ (b) $S \cup C$

(c) $T \cup C$ (d) $S \cup T \cup C$.

5. Let S, T, and C be the sets in Exercise 3. Shade the following sets:

(a) $C \cup (S \cap T)$ (b) $C \cap (S \cup T)$

(c) $(C \cap S) \cup T$ (d) $(C \cup S) \cap (C \cap T)$.

6. In each part determine if the given sets are disjoint.

(a) $\{a, b, d\}$, $\{e, f, g\}$

(b) $\{1, 2, 3\}$, $\{3, 7, 9\}$

(c) $\varnothing$, $\{1, 2\}$

(d) {book, candle, bell}, {page, fire, ring}.

For Problems 7–9, refer to Figure 1.15 and let

$S =$ the set of points inside the square

$T =$ the set of points inside the triangle

$C =$ the set of points inside the circle

and let v, w, x, y, z be the indicated points.

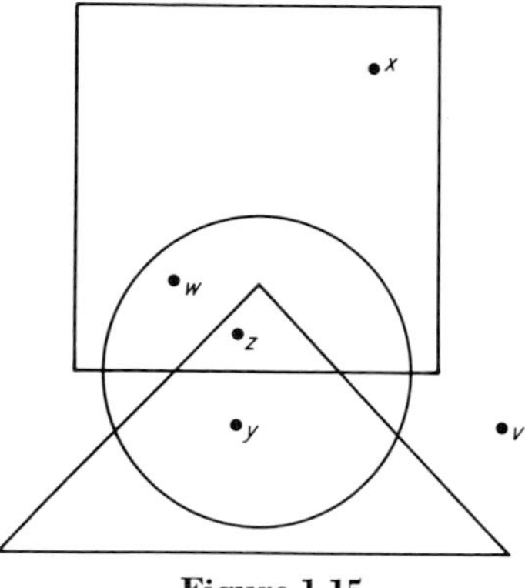

Figure 1.15

7. Answer the following as true or false.

(a) $x \notin S \cap C$ (b) $y \in C \cap T$

(c) $z \in S \cap T \cap C$ (d) $w \in S \cap C \cap T$

(e) $v \notin S \cap C$.

8. Answer the following as true or false.

(a) $x \in C \cup T$ (b) $x \in C \cup S$

(c) $x \in C \cup S \cup T$ (d) $y \in C \cup C$

(e) $v \in C \cup S \cup T$.

9. Answer the following as true or false.

(a) $w \in C \cup (S \cap T)$ (b) $w \in C \cap (S \cup T)$

(c) $x \in (C \cup T) \cap S$ (d) $x \in (C \cup S) \cap (C \cup T)$.

10. Let $A = \{1, 2, 3, 4, 5\}$, $B = \{3, 4, 7\}$, and $C = \{1, 3, 7, 8\}$. Verify the distributive laws

(a) $C \cap (A \cup B) = (C \cap A) \cup (C \cap B)$

(b) $C \cup (A \cap B) = (C \cup A) \cap (C \cup B)$.

11. Use Venn diagrams to establish the second distributive law

$$C \cup (A \cap B) = (C \cup A) \cap (C \cup B).$$

12. Let $A = \{\text{ATT, IBM, GE}\}$, $B = \{\text{Du Pont, Burroughs, GE, Kodak}\}$, $C = \{\text{Avco, Sun Oil, IBM, GE, Du Pont}\}$. Compute

(a) $(A \cap B) \cup C$ (b) $A \cup (B \cap C)$ (c) $A \cup B \cup C$

(d) $A \cap B \cap C$ (e) $A \cap (B \cup C)$ (f) $(A \cup B) \cap C$.

13. Explain why the following are true.

(a) $A \cup B = B \cup A$ (b) $A \cap B = B \cap A$.

14. Use Venn diagrams to establish the following.

(a) $(A \cap B) \cap C = A \cap (B \cap C)$ (b) $(A \cup B) \cup C = A \cup (B \cup C)$.

15. Use Venn diagrams to establish the following.

(a) $A \subset A \cup B$ and $B \subset A \cup B$

(b) $A \cap B \subset A$ and $A \cap B \subset B$.

16. Use Venn diagrams to establish the following.

(a) If $A \subset C$ and $B \subset C$, then $(A \cup B) \subset C$.

(b) If $C \subset A$ and $C \subset B$, then $C \subset (A \cap B)$.

17. The personnel department of a major company classifies its employees into the following categories:

$M =$ the set of all male employees

$F =$ the set of all female employees

$A =$ the set of all administrative employees

$T =$ the set of all technical employees

$S =$ the set of all employees working for the company at least 5 years.

Describe the members of the following sets.

(a) $M \cap A$ (b) $M \cup F$ (c) $A \cap T \cap F$

(d) $A \cup T \cup F$ (e) $M \cap A \cap S$.

18. Let M, F, A, and S be the sets in Exercise 17. Let x designate a male employee who has worked for the company at least 5 years. Answer the following as true or false.

(a) $x \in M \cup F \cup A$ (b) $x \in M \cap S$

(c) $x \in F \cap S$ (d) $x \in (M \cap S) \cup F$.

19. An automobile insurance company classifies its policy holders into the following categories:

$A =$ the set of all policy holders who drive cars with engines that are more than 200 horsepower

$B =$ the set of all policy holders who drive cars with engines that are more than 250 horsepower

$C =$ the set of all policy holders who are over 25 years of age

$D =$ the set of all policy holders who are over 20 years of age

$M =$ the set of all male policy holders

$F =$ the set of all female policy holders.

Describe the policy holders in the following sets:

(a) $A \cap B$ (b) $A \cup B$ (c) $A \cap C \cap M$

(d) $A \cup D \cup F$ (e) $B \cap (D \cup F)$.

20. Let A, B, C, D, M, and F be the sets in Exercise 19. Write the following sets using unions and intersections of A, B, C, D, M, and F:

(a) the set of all female policy holders who are over 25 years of age;

(b) the set of all policy holders who are either male or drive cars with engines that are over 200 horsepower;

(c) the set of all female policy holders over 20 years of age who drive cars with engines that are over 250 horsepower;

(d) the set of all male policy holders who are either over 25 years of age or drive cars with engines that are over 200 horsepower.

1.3 COMPLEMENTATION AND CARTESIAN PRODUCT OF SETS

In this section we introduce some other set operations that will be useful in our later work.

Given a set A, we may want to consider those elements that are not in A. Usually, however, there are elements not in A that are extraneous to the problem being studied. For example, suppose A is the set of all positive numbers and we talk about "an element x that is not in A." Clearly, if x is a negative number, then x is not in A. However, if x is a cabbage or a walrus, then x is also not in A since cabbages and walruses are not positive numbers. If we are concerned with a problem about real numbers, then what we really mean when we talk about

"an element x that is not in A"

is

"a real number x that is not in A"

—cabbages and walruses are extraneous to the problem.

One way of eliminating extraneous elements is to specify in advance that *all* elements in the problem under consideration come from some fixed set U, called the ***universal set*** for the problem. For example, in a problem concerned with real numbers, we might agree to take the universal set U to be the set of all real numbers. Then, if we speak of an "element x that is not in A," we must mean

"a real number x that is not in A"

since we have agreed that all elements come from the universal set U of real numbers.

In many problems, the universal set is intuitively clear from the nature of the problem and is not explicitly stated. For example, in problems con-

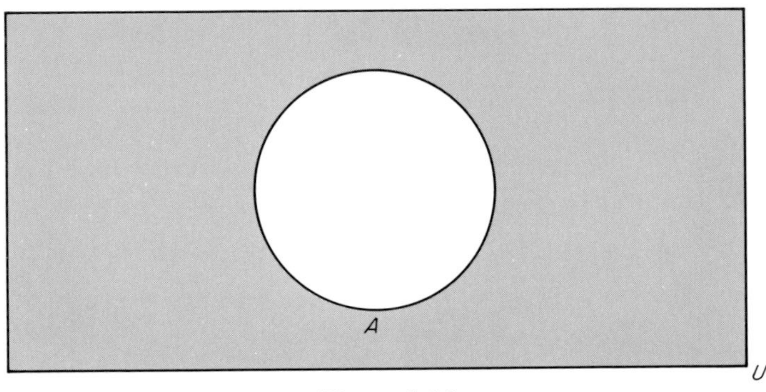

Figure 1.16

cerned with sets of integers, U would be the set of *all* integers; in problems concerned with sets of stocks on the New York Stock Exchange, U would be the set of *all* stocks on the New York Stock Exchange, and so on.

In Figure 1.16, we describe the above ideas using a Venn diagram. The points inside the rectangle form the universal set U. Since all the sets under study must have their members in U, they lie inside the rectangle. In particular, if A is the set of points inside the circle, then the shaded region in Figure 1.16 forms the set of all elements not in A.

If U is the universal set and A is a subset of U, then the set of all elements in U that are not members of A is called the ***complement*** of A. It is denoted by the symbol

$$A'.$$

Example 26 Let A be the set of positive real numbers; that is,

$$A = \{x \mid x \text{ is a real number satisfying } x > 0\}.$$

Find A'.

Solution Since A is a set of real numbers, it is natural to take the universal set U to be the set of all real numbers. Thus A' is the set of real numbers that do *not* satisfy $x > 0$; that is,

$$A' = \{x \mid x \text{ is a real number satisfying } x \leq 0\}.$$

Example 27 Let W be the set of United States citizens who file tax returns listing less than \$10,000 taxable income in 1974. Find the complement of W.

Solution Since W is a set of United States citizens filing tax returns in 1974, it is reasonable to let the universal set U be the set of all U.S. citizens filing tax returns in 1974. Thus W' is the set of all U.S. citizens filing tax returns in 1974 listing at least \$10,000 taxable income.

Example 28 Let

$$U = \{1, 3, 5, 7, 9, 11\}$$

and

$$V = \{1, 5, 7\},$$

then

$$V' = \{3, 9, 11\}.$$

Example 29 Let U be the universal set for a given problem. Find

$$\text{(a)} \quad U' \qquad \text{(b)} \quad \varnothing'.$$

Solution (a) By definition of complement, U' consists of all elements that are not in U. But every element lies in U, so that

$$U' = \varnothing.$$

Solution (b) The set $\varnothing'$ consists of all elements that are not in $\varnothing$. Since $\varnothing$ has no elements, all elements are outside of $\varnothing$; thus

$$\varnothing' = U.$$

Example 30 Let A, B, and C be the sets of points inside the circles indicated in Figure 1.17. Shade the sets:

$$\text{(a)} \quad A' \cap B \cap C \qquad \text{(b)} \quad A \cap B' \cap C'.$$

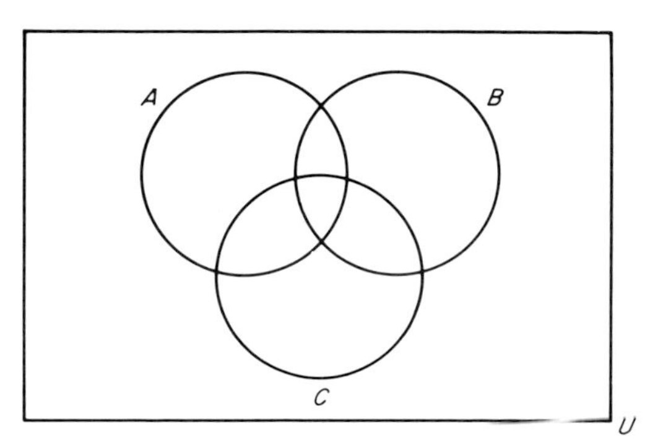

Figure 1.17

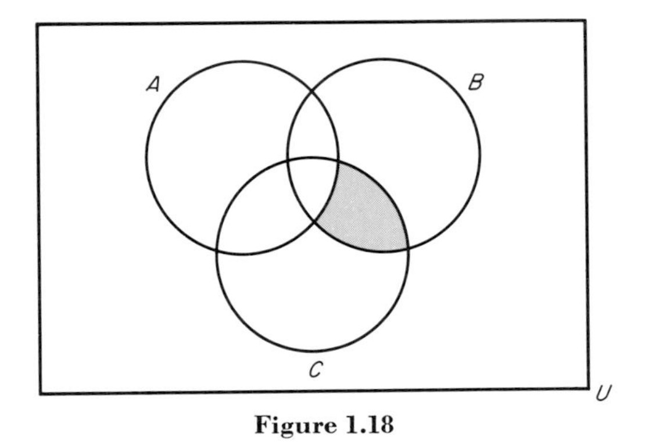

Figure 1.18

Solution (a) The members of $A' \cap B \cap C$ belong to both B and C but not A. This yields the shaded region in Figure 1.18.

Solution (b) The members of $A \cap B' \cap C'$ belong to A but not to B and not to C. This yields the shaded region in Figure 1.19.

There are a number of useful properties of the complementation operation. We shall discuss a few of the more important ones here. We begin with the following two important results, called ***DeMorgan's*† *laws.***

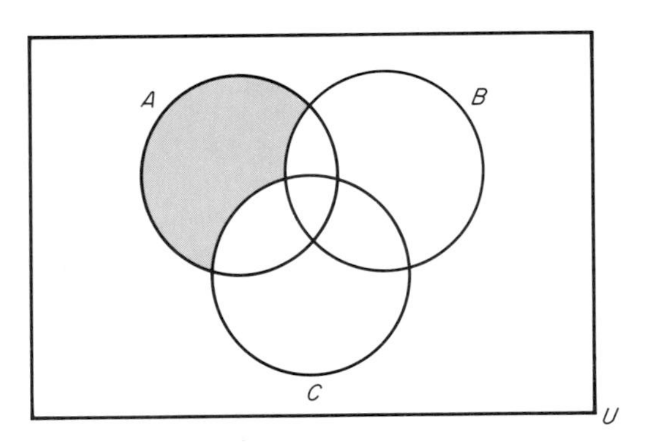

Figure 1.19

† *Augustus DeMorgan* (1806–1871), British mathematician and logician—DeMorgan, the son of a British army officer, was born in Madura, India. He graduated from Trinity College in Cambridge, England in 1827, but was denied a teaching position there for refusing to subscribe to religious tests. He was, however, appointed to a mathematics professorship at the newly opened University of London.

He was a man of firm principles who was described as "indifferent to politics and society and hostile to the animal and vegetable kingdoms." He twice resigned his teaching position on matters of principle (but later returned), refused to accept honorary degrees, and declined memberships in many learned societies.

He is best known for his work *Formal Logic* which appeared in 1847; but he also wrote papers on the foundations of algebra, philosophy of mathematical methods, and probability as well as several successful elementary textbooks.

If A and B are any two sets, then

$$(A \cup B)' = A' \cap B' \qquad \text{first DeMorgan law}$$

$$(A \cap B)' = A' \cup B' \qquad \text{second DeMorgan law}$$

We shall verify the first DeMorgan law using Venn diagrams and leave the second as an exercise. The set

$$(A \cup B)'$$

consists of all elements that are not in

$$A \cup B.$$

This is given by the shaded region in Figure 1.20a.

In Figures 1.20b and 1.20c we have shaded the regions A' and B' and in Figure 1.20d we formed their intersection to obtain

$$A' \cap B'.$$

Comparing Figures 1.20a and 1.20d we see that

$$(A \cup B)' = A' \cap B',$$

which is the first law of DeMorgan.

The next example illustrates DeMorgan's laws.

Example 31 Let

$$U = \{a, b, c, d, e, f, g, h\}$$

$$A = \{a, b, d, e\} \qquad \text{and} \qquad B = \{b, d, f, g\}.$$

Then

$$A \cup B = \{a, b, d, e, f, g\}$$

and

$$(A \cup B)' = \{c, h\}.$$

On the other hand,

$$A' = \{c, f, g, h\} \qquad \text{and} \qquad B' = \{a, c, e, h\},$$

so that

$$A' \cap B' = \{c, h\}.$$

Therefore,

$$(A \cup B)' = A' \cap B',$$

as guaranteed by the first DeMorgan law.

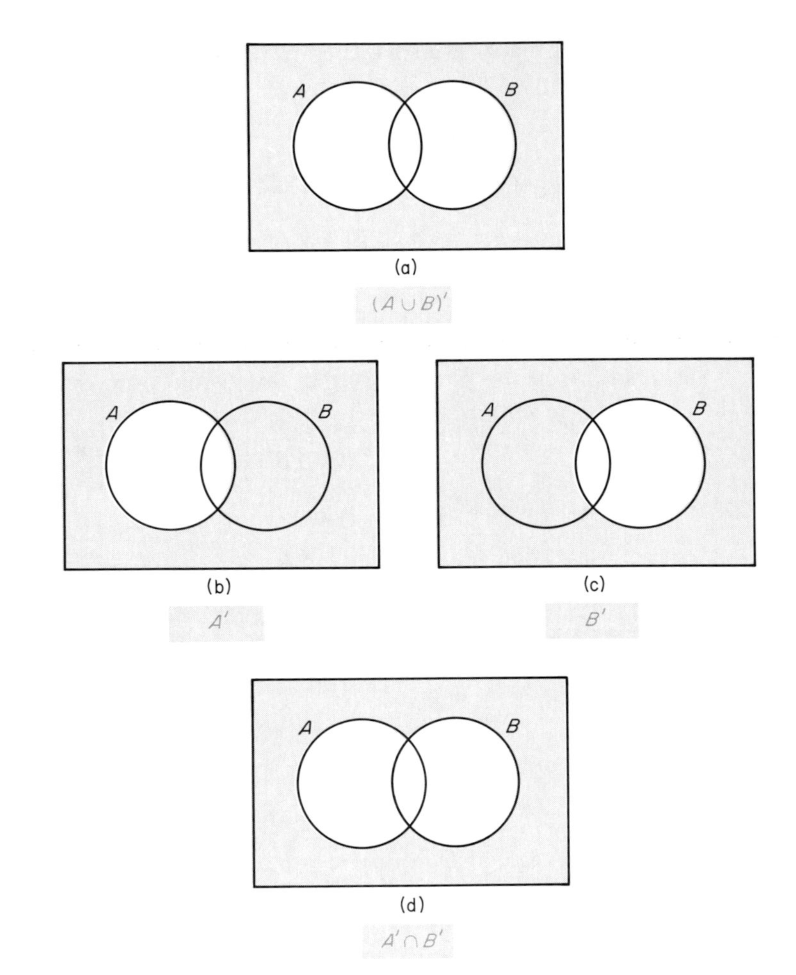

Figure 1.20

The verifications of the following results are left as exercises.

If U is the universal set, and A is a subset of U, then

$$(A')' = A$$

$$A \cup A' = U$$

$$A \cap A' = \varnothing.$$

In many problems we are interested in paired data; for example, the height and weight of an individual, the wind speed and wind direction at a certain time, the total assets and total liabilities of a business firm, and so on. It is often the case that the order in which the information is listed is important. For example, suppose we want to describe the financial status of a business firm by listing a pair of numbers, the first number indicating its total assets in millions of dollars and the second indicating its total liabilities in millions of dollars.

If the pair for Company A is

$$(9, 0)$$

and the pair for Company B is

$$(0, 9),$$

then Company A is in excellent financial position, while Company B is likely on the verge of bankruptcy. Thus, even though the pairs

$$(9, 0) \qquad \text{and} \qquad (0, 9)$$

involve the same numbers, they convey very different information because of the order.

This notion of ordered data gives rise to the following concept.

An ***ordered pair*** (a, b) is a listing of two objects a and b in definite order. The element a is called the ***first entry*** in the ordered pair and the element b is called the ***second entry*** in the ordered pair.

Two ordered pairs are called equal if they list the same objects in the same order. Thus, for (a, b) and (c, d) to be equal ordered pairs, we must have

$$a = c \qquad \text{and} \qquad b = d.$$

We are now in a position to define the last set operation we shall need.

If A and B are two sets, the set of all ordered pairs (a, b), where $a \in A$ and $b \in B$, is called the ***Cartesian*† *product*** of A and B; it is denoted by the symbol

$$A \times B.$$

† René Descartes (1596–1650)—Descartes was the son of a government official. He graduated from the University of Poitiers with a law degree at age 20, after which he went to Paris, where he lived a dissipative life as a man of fashion. In 1618 he joined the army of the Prince of Orange, where he worked as a military engineer. Descartes was a genius of the first magnitude having made major contributions in philosophy, mathematics, physiology, and science in general. His work in mathematics gave new direction to science; along with William Harvey, he is considered a founder of modern physiology.

Example 32 Let

$$A = \{r, s, t\}$$

and

$$B = \{1, 2, 3, 4\}.$$

Then

$$A \times B = \{(r, 1), (r, 2), (r, 3), (r, 4), (s, 1), (s, 2), (s, 3), (s, 4), (t, 1), (t, 2), (t, 3), (t, 4)\}.$$

Note that the members of $A \times B$ can be conveniently arranged in tabular form as in Table 1.1.

Table 1.1

B / A	1	2	3	4
r	$(r, 1)$	$(r, 2)$	$(r, 3)$	$(r, 4)$
s	$(s, 1)$	$(s, 2)$	$(s, 3)$	$(s, 4)$
t	$(t, 1)$	$(t, 2)$	$(t, 3)$	$(t, 4)$

Alternatively, the ordered pairs in $A \times B$ can be systematically constructed using what is called a ***tree diagram.*** This is done as follows. The first entry in each ordered pair in $A \times B$ comes from A. There are three possible choices: r, s, or t. In Figure 1.21 these choices are represented by the three dots labeled r, s, and t at the top of the diagram. The four branches emanating from each of these dots correspond to the possible choices for the second entry in the ordered pair, namely 1, 2, 3, or 4. The various ordered pairs can be listed by tracing out all possible paths or "branches" from the top of the tree to the bottom of the tree. For example, the path in color corresponds to the pair $(s, 2)$ and the heavy black path to the pair $(t, 3)$. The reader will find it instructive to compute the set $A \times B$ by this method.

It is possible to consider Cartesian products of more than two sets. For example, the Cartesian product

$$A \times B \times C$$

of three sets would consist of all ordered *triples*

$$(a, b, c),$$

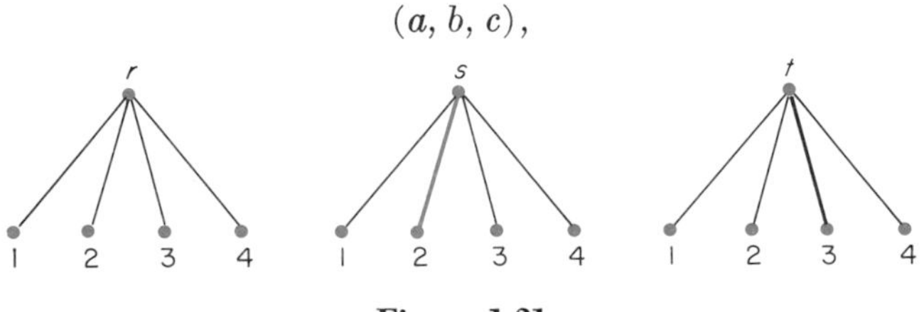

Figure 1.21

where

$$a \in A, \qquad b \in B, \qquad \text{and} \qquad c \in C;$$

and more generally, the Cartesian product

$$A_1 \times A_2 \times \cdots \times A_n$$

of n sets, where n is a positive integer, consists of all ordered n-tuples

$$(a_1, a_2, \ldots, a_n),$$

where

$$a_1 \in A_1, \qquad a_2 \in A_2, \qquad \ldots, \qquad a_n \in A_n.$$

Example 33 A firm that conducts political polls classifies people for its files according to three characteristics

S (sex)	: m = male	f = female	
I (income)	: h = high	a = average	l = low
P (political registration)	: d = Democrat	r = Republican	i = independent.

For example, a person filed under (f, h, r) would be a female with high income who is registered as a Republican.

The Cartesian product

$$S \times I \times P$$

contains all possible classifications. From the tree in Figure 1.22 the different possible classifications are

$$\begin{array}{cccccc} (m, h, d) & (m, h, r) & (m, h, i) & (f, h, d) & (f, h, r) & (f, h, i) \\ (m, a, d) & (m, a, r) & (m, a, i) & (f, a, d) & (f, a, r) & (f, a, i) \\ (m, l, d) & (m, l, r) & (m, l, i) & (f, l, d) & (f, l, r) & (f, l, i). \end{array}$$

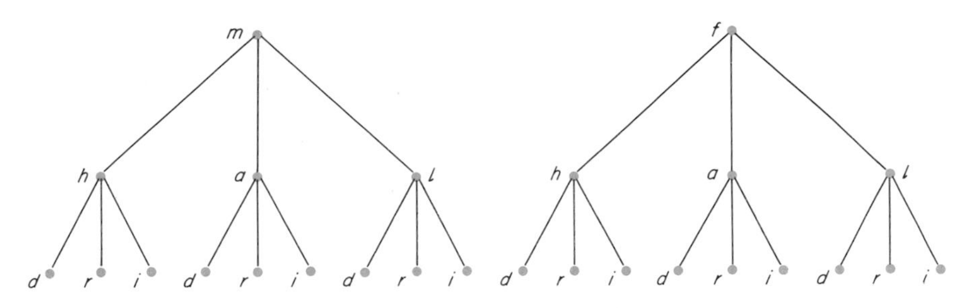

Figure 1.22

EXERCISE SET 1.3

1. Let

$$U = \{a, b, c, d, e, f, g, h\}$$

be the universal set and let

$$A = \{a, d, f, h\}, \qquad B = \{a, d\}, \qquad C = \{e, f\}.$$

Find

(a) A' (b) B'

(c) $(A \cup B)'$ (d) $(B \cap C)'$ (e) U'.

2. Let U, A, B, and C be the sets in Exercise 1. Find

(a) $A' \cap B'$ (b) $B' \cup C'$

(c) $(A \cup A)'$ (d) C' (e) $(C \cap C)'$.

3. Let

$U =$ the set of all stocks traded on the New York Stock Exchange.

$A =$ the set of stocks traded on the New York Stock Exchange that have paid a dividend for the past 10 years without any interruption.

$B =$ the set of stocks traded on the New York Stock Exchange that have a price-to-earnings ratio of no more than 12.

Describe the following sets

(a) A' (b) B'

(c) $(A \cup B)'$ (d) $(A \cap B)'$.

4. Let

$U =$ the set of all real numbers

$A = \{x \mid x \text{ is a solution of } x^2 = 1\}$

$B = \{-1, 2\}$.

Find

(a) A' (b) B'

(c) $(A \cup B)'$ (d) $(A \cap B)'$.

5. Let

$$A = \{x \mid x \text{ is an integer satisfying } x > 0\}.$$

Specify a universal set and compute A'.

6. Let

$$D = \text{the set of all Democratic U.S. senators.}$$

Specify a universal set and compute D'.

7. Let C be the set of all consonants in the English alphabet. Specify a universal set and compute C'.

8. Let A, B, and C be the sets of points in Figure 1.17. Shade the following sets.

(a) $A \cap B' \cap C$ (b) $A \cap B \cap C'$ (c) $A' \cap B \cap C'$ (d) $A' \cap B' \cap C$.

9. Let A, B, and C be the sets of points in Figure 1.17. Shade the following sets.

(a) $A' \cap B' \cap C'$ (b) $(A \cup B \cup C)'$ (c) $(A \cap B \cap C)'$ (d) $[(A \cup B) \cap C]'$.

10. Use a Venn diagram to verify that if $A \subset B$, then $B' \subset A'$.

11. Use a Venn diagram to verify the second DeMorgan law:

$$(A \cap B)' = A' \cup B'.$$

12. Use a Venn diagram to verify the following DeMorgan laws for three sets A, B, and C:

$$(A \cup B \cup C)' = A' \cap B' \cap C'$$

$$(A \cap B \cap C)' = A' \cup B' \cup C'.$$

13. (a) Conjecture DeMorgan laws for four sets A, B, C, and D.

(b) Conjecture DeMorgan laws for n sets $A_1, A_2, \ldots, A_n$.

14. Explain why the following are true.

(a) $(A')' = A$ (b) $A \cup A' = U$ (c) $A \cap A' = \varnothing$.

In Exercises 15 and 16, let A, B, and C be the indicated sets and let x, y, z, and w be the indicated points in Figure 1.23.

15. In each case determine which of the points x, y, z, w belong to the indicated set.

(a) $C' \cap B'$ (b) $A \cap B \cap C'$ (c) $A \cap B' \cap C$ (d) $A' \cap B' \cap C$.

16. Follow the directions of Exercise 15 for the sets

(a) $A' \cup B$ (b) $B' \cup C'$ (c) $A' \cup B' \cup C'$ (d) $A' \cup B \cup C'$.

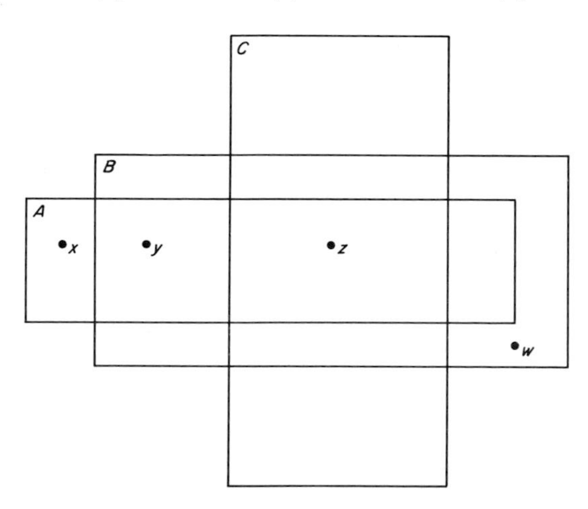

Figure 1.23

17. In each part, find the values of x and y for which the given ordered pairs of integers are equal.

(a) $(x, 7) = (3, 7)$ (b) $(2x, 3) = (6, y)$

(c) $(4, y + 7) = (2x + 2, 14)$ (d) $(x^2, 9) = (16, 9)$.

18. Let $A = \{a, b, c, d\}$ and $B = \{0, 1, 2\}$.

(a) List the elements in $A \times B$. (b) List the elements in $B \times A$.

(c) List the elements in $A \times A$. (d) List the elements in $B \times B$.

19. Let $A = \{a, b\}$, $B = \{1, 2\}$, and $C = \{x, y, z\}$. List the elements of $A \times B \times C$.

20. An air traffic control station supplies the following data to airline pilots:

Traffic: crowded (c), average (a), light (l)

Visibility: poor (p), good (g)

Windspeed: negligible (n), medium (m), high (h).

Flying conditions are described by an ordered triple; for example (c, p, n) means crowded traffic, poor visibility, negligible windspeed. List all the possible flying conditions.

21. In a psychology experiment, a rat is placed in a cage with three doors, a, b, and c (see Figure 1.24). The rat leaves the cage through one of the doors. Upon reaching the intersection he turns either left or right and at the next intersection he turns either left or right again. The rat then proceeds to the reward section.

(a) Show that the various paths that the rat can take can be represented as the elements of the Cartesian product

$$E \times T \times T,$$

where

$$E = \{a, b, c\}, \qquad T = \{l, r\}, \qquad l = \text{left}, \qquad r = \text{right}.$$

(b) List all possible paths.

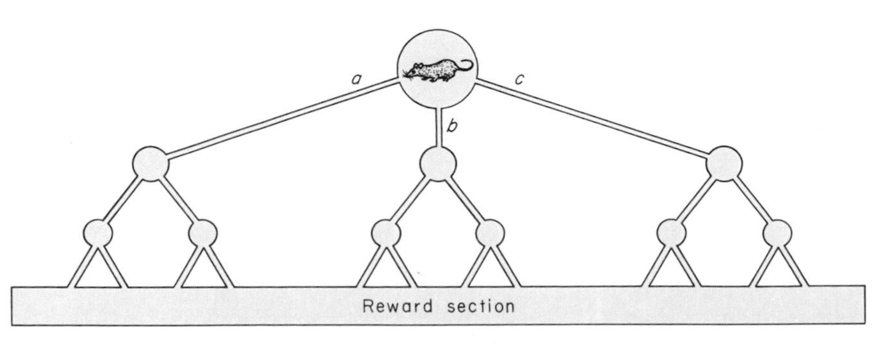

Figure 1.24

22. A menu lists a choice of soup, salad, or juice for an appetizer; a choice of beef, chicken, or fish for the entree; and a choice of pie or fruit for dessert. A complete dinner consists of one choice for each course. Draw a tree for the possible complete dinners.

(a) How many different complete dinners are possible?

(b) If a man refuses to eat chicken or pie, how many different complete dinners can he choose?

(c) A certain customer eats pie for dessert if he has soup as an appetizer; otherwise he chooses fruit for dessert. How many different complete meals are available to him?

1.4 COUNTING ELEMENTS IN SETS

Suppose we know the number of elements in a set A and the number of elements in a set B. What can we say about the number of elements in $A \cup B$ and $A \times B$? In this section we investigate problems like these and illustrate some of their applications.

If S is a set with a *finite* number of elements, then we shall denote the ***number of elements in*** S by the symbol

$$n(S).$$

Example 34 Consider the sets

$$A = \{a, e, g, l\}$$
$$B = \{x \mid x \text{ is a positive integer}\}$$
$$C = \varnothing.$$

Since A has four elements, we have

$$n(A) = 4.$$

Since B contains infinitely many elements and since we defined $n(S)$ only when S has finitely many elements, $n(B)$ is undefined. Finally, since C is empty, it has no elements; thus

$$n(\varnothing) = 0.$$

If, as illustrated in Figure 1.25, A and B are *disjoint* sets with finitely many elements, then the number of elements in $A \cup B$ can be obtained by adding the number of elements in A and the number of elements in B together; in other words

$$n(A \cup B) = n(A) + n(B) \qquad \text{if } A \text{ and } B \text{ are disjoint sets.}$$

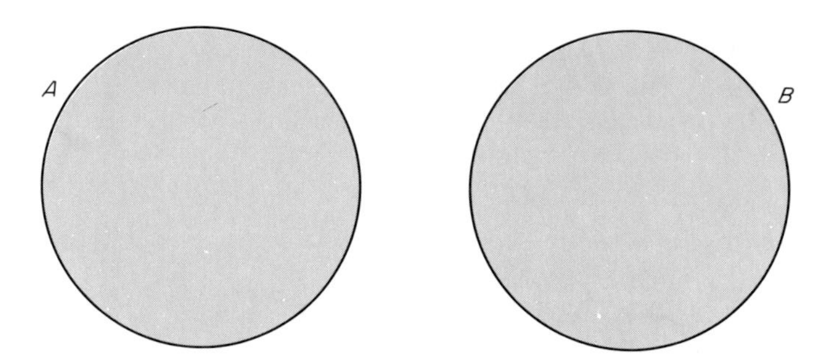

Figure 1.25 The number of elements in the shaded region $A \cup B$ is the number of elements in A plus the number of elements in B.

Example 35 Consider the disjoint sets

$$A = \{1, 2, 5, 7\} \quad \text{and} \quad B = \{3, 9\}.$$

Since

$$A \cup B = \{1, 2, 3, 5, 7, 9\}$$

we have

$$n(A \cup B) = 6.$$

On the other hand,

$$n(A) = 4 \quad \text{and} \quad n(B) = 2,$$

so that

$$n(A \cup B) = n(A) + n(B).$$

In the event that A and B have elements in common, the number of elements in $A \cup B$ is *not* the sum of $n(A)$ and $n(B)$. To see why, consider the sets

$$A = \{a, b, c, d\} \quad \text{and} \quad B = \{d, e, f, g, h\}.$$

Then

$$A \cup B = \{a, b, c, d, e, f, g, h\}.$$

Since

$$n(A) = 4, \quad n(B) = 5, \quad \text{and} \quad n(A \cup B) = 8,$$

we have

$$n(A \cup B) \neq n(A) + n(B).$$

The reason equality does not hold is clear from this example. The element d is counted once in computing $n(A)$ and once again in computing $n(B)$, so that the element d is counted *twice* in computing

$$n(A) + n(B).$$

On the other hand, the element d is only counted *once* in computing

$$n(A \cup B).$$

This example suggests the following idea. Suppose A and B are not disjoint. We know that we cannot simply add $n(A)$ and $n(B)$ to obtain $n(A \cup B)$ because we would be counting the elements in $A \cap B$ twice.

Since we only want to count these elements once, we can correct for this duplication by subtracting $n(A \cap B)$ once. Thus we obtain the formula

$$n(A \cup B) = n(A) + n(B) - n(A \cap B). \tag{1}$$

Example 36 Let

$$A = \{a, b, c, d, e\} \quad \text{and} \quad B = \{d, e, f, g, h, i\},$$

so that

$$A \cup B = \{a, b, c, d, e, f, g, h, i\}$$

and

$$A \cap B = \{d, e\}.$$

Since

$$n(A) = 5, \quad n(B) = 6, \quad n(A \cup B) = 9, \quad n(A \cap B) = 2,$$

we have

$$n(A \cup B) = n(A) + n(B) - n(A \cap B),$$

since

$$9 = 5 + 6 - 2.$$

Example 37 Philadelphia has two newspapers: the *Inquirer* and the *Evening Bulletin*. A survey of 1000 people who subscribe to at least one of the papers revealed the following information:

500 subscribe to the *Inquirer*

100 subscribe to both.

How many of the 1000 surveyed subscribe to the *Bulletin*?

Solution Let A be the set of people surveyed who subscribe to the *Inquirer* and let B be the set of people surveyed who subscribe to the *Bulletin*, so that $A \cup B$ is the entire set of people surveyed. From the above data,

$$n(A \cup B) = 1000, \quad n(A) = 500, \quad n(A \cap B) = 100.$$

Substituting these values in the equation

$$n(A \cup B) = n(A) + n(B) - n(A \cap B)$$

yields

$$1000 = 500 + n(B) - 100$$

so that

$$n(B) = 600.$$

Therefore, 600 of the 1000 surveyed subscribe to the *Bulletin*.

Example 38 A manufacturer of magnetic tape for digital computers finds that an experimental production process can introduce two possible defects in a reel of tape: the tape can contain creases or the magnetic coating can be insufficient. In testing 100 reels, the manufacturer recorded the following information:

15 have creased tape
12 have tape with insufficient magnetic coating
7 have both defects.

(a) How many reels were defective (that is, had at least one defect)?
(b) How many reels were nondefective?

Solution (a) Let A be the set of reels with creased tape and let B be the set of reels containing tape with insufficient coating. Then

$$A \cup B$$

is the set of defective reels, so that we must find $n(A \cup B)$.

From the given data,

$$n(A) = 15, \qquad n(B) = 12, \qquad n(A \cap B) = 7.$$

Substituting these values into Equation (1) yields

$$\begin{aligned} n(A \cup B) &= n(A) + n(B) - n(A \cap B) \\ &= 15 + 12 - 7 \\ &= 20. \end{aligned}$$

Therefore, 20 reels were defective.

Solution (b) From part (a), 20 of the 100 reels tested were defective. Therefore

$$100 - 20 = 80$$

were nondefective.

Formula (1) can be extended to unions of three sets. More precisely, if A, B, and C are sets with finitely many elements, then

$$\begin{aligned} n(A \cup B \cup C) = n(A) &+ n(B) + n(C) \\ &- n(A \cap B) - n(A \cap C) - n(B \cap C) \\ &+ n(A \cap B \cap C). \end{aligned} \tag{2}$$

We omit the proof.

Example 39 Let

$$A = \{a, b, c, d\}, \qquad B = \{d, e, f, g, h, i\}, \qquad C = \{a, g, h, i, j, k\}.$$

Then

$$\begin{aligned} A \cup B \cup C &= \{a, b, c, d, e, f, g, h, i, j, k\} \\ A \cap B &= \{d\} \\ A \cap C &= \{a\} \\ B \cap C &= \{g, h, i\} \\ A \cap B \cap C &= \varnothing. \end{aligned}$$

As guaranteed by formula (2)

$$\begin{aligned} 11 = n(A \cup B \cup C) &= n(A) + n(B) + n(C) \\ &\quad - n(A \cap B) - n(A \cap C) - n(B \cap C) \\ &\quad + n(A \cap B \cap C) \\ &= 4 + 6 + 6 - 1 - 1 - 3 + 0. \end{aligned}$$

Example 40 A new car dealer offers three options to his customers: power steering, a high performance engine, and air conditioning. The dealer listed the following information in his yearly tax records.

200 cars sold
75 without any options
10 with all three options
40 included high performance engine and air conditioning
30 included power steering and air conditioning
20 included power steering and high performance engine
80 included power steering
60 included high performance engine
70 included air conditioning.

Explain why the Bureau of Internal Revenue ordered a complete audit of the dealer's records.

Solution Let

A = the set of all cars sold with power steering

B = the set of all cars sold with a high performance engine

C = the set of all cars sold with air conditioning.

Then

$$A \cup B \cup C$$

is the set of all cars sold with at least one of the three options. From formula (2)

$$\begin{aligned} n(A \cup B \cup C) &= n(A) + n(B) + n(C) \\ &\quad - n(A \cap B) - n(A \cap C) - n(B \cap C) \\ &\quad + n(A \cap B \cap C) \\ &= 80 + 60 + 70 - 20 - 30 - 40 + 10 \\ &= 130. \end{aligned}$$

Thus 130 cars were sold with at least one of the three options. Since the dealer indicated that 75 cars were sold with no options, he should have reported

$$130 + 75 = 205$$

cars sold, rather than 200, as his records showed.

Example 41 In a pollution study of 1500 U.S. rivers the following data were reported:

520 were polluted by sulfur compounds
335 were polluted by phosphates
425 were polluted by crude oil
100 were polluted both by crude oil and sulfur compounds
180 were polluted both by sulfur compounds and phosphates
150 were polluted both by phosphates and crude oil
28 were polluted by sulfur compounds, phosphates, and crude oil.

Use this information to answer the following:

(a) How many of the rivers were polluted by at least one of the three impurities?

(b) How many of the rivers were polluted by exactly one of the three impurities?

Solution Let

S = the set of rivers polluted by sulfur compounds

P = the set of rivers polluted by phosphates

C = the set of rivers polluted by crude oil.

Then the set of rivers polluted by at least one of the three impurities is

$$S \cup P \cup C.$$

To answer (a) we must find

$$n(S \cup P \cup C).$$

We could solve this problem using Equation (2) (see Exercise 18).

Instead we shall illustrate an alternative method using Venn diagrams that will also be helpful in solving part (b) of this problem.

First, we use the fact that 28 of the rivers were polluted by all three of the impurities by placing this number in the region

$$S \cap P \cap C$$

as indicated in the Venn diagram of Figure 1.26a. Since 180 rivers were polluted by both sulfur and phosphates, this leaves

$$180 - 28 = 152$$

that were polluted *only* by sulfur and phosphates. Similarly,

$$150 - 28 = 122$$

were polluted *only* by phosphates and crude oil, while

$$100 - 28 = 72$$

were polluted *only* by sulfur and crude oil. We have placed these numbers in the appropriate regions of the Venn diagram in Figure 1.26a. Next, we fill in the numbers of elements in the remaining portions of the Venn diagram.

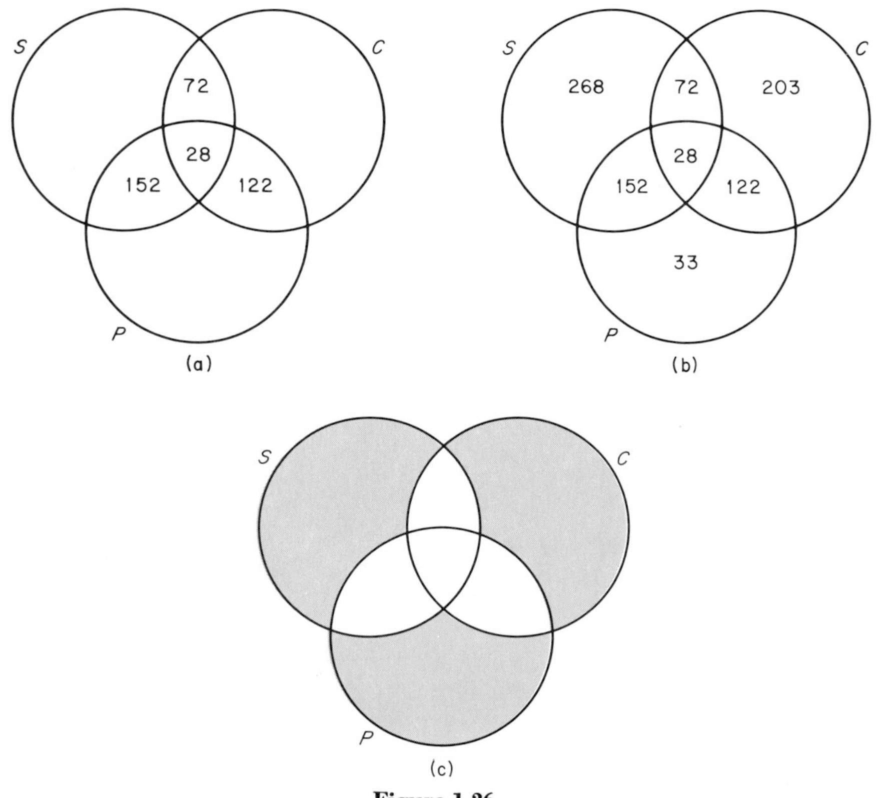

Figure 1.26

From the given data, the number of rivers polluted by sulfur is 520. Therefore, the number of elements in the remaining portion of circle S in Figure 1.26a must be

$$520 - 72 - 28 - 152 = 268.$$

Similarly, the number of elements in the remaining portion of circle C is

$$425 - 122 - 28 - 72 = 203,$$

and the number of elements in the remaining portion of circle P is

$$285 - 152 - 28 - 72 = 33.$$

We have inserted these numbers in the appropriate regions of the Venn diagram in Figure 1.26b. From Figure 1.26b we can now answer questions (a) and (b) posed above.

(a) From Figure 1.26b the number of rivers polluted by at least one of the impurities is

$$n(S \cup P \cup C) = 268 + 72 + 28 + 152 + 203 + 122 + 33 = 878.$$

(b) The number of rivers polluted by exactly one of the three impurities is the number of elements in the shaded portion of Figure 1.26c. Do you see why? From Figure 1.26b this number is

$$268 + 203 + 33 = 504.$$

In Example 32 of Section 1.3 we observed that the members of the Cartesian product $A \times B$ of the sets

$$A = \{r, s, t\} \qquad \text{and} \qquad B = \{1, 2, 3, 4\}$$

could be listed in a table with three rows and four columns (see Table 1.1). With this in mind, it is evident that the number of elements in $A \times B$ is

$$3 \cdot 4 = 12.$$

More generally, if A is a set with m members and B is a set with n members, then the members of $A \times B$ can be listed in a table with m rows and n columns, so that $A \times B$ has

$$m \cdot n$$

members. This yields the formula

$$n(A \times B) = n(A) \cdot n(B). \qquad (3)$$

Example 42 Let

$$A = \{f, a, c, e\} \quad \text{and} \quad B = \{r, o, u, n, d\}.$$

Since

$$n(A) = 4 \quad \text{and} \quad n(B) = 5,$$

the number of elements in $A \times B$ is

$$n(A \times B) = n(A) \cdot n(B) = 4 \cdot 5 = 20.$$

Formula (3) can be extended to Cartesian products of three or more sets. For three sets, the formula is

$$n(A \times B \times C) = n(A) \cdot n(B) \cdot n(C) \tag{4}$$

and more generally

$$n(A_1 \times A_2 \times \cdots \times A_m) = n(A_1) \cdot n(A_2) \cdots n(A_m).$$

Example 43 A coin is tossed four times and the resulting sequence of heads and tails is recorded. Among the possible sequences are

$$(h, h, h, h), \quad (h, t, h, t), \quad (t, h, t, t), \quad \text{and so on.}$$

How many different sequences are possible?

Solution The different sequences can be regarded as the members of the set

$$A \times A \times A \times A,$$

where

$$A = \{h, t\}.$$

Can you see why? Thus, the number of different sequences is

$$\begin{aligned} n(A \times A \times A \times A) &= n(A) \cdot n(A) \cdot n(A) \cdot n(A) \\ &= 2 \cdot 2 \cdot 2 \cdot 2 \\ &= 16. \end{aligned}$$

As a check, the student may wish to list the 16 different possible sequences.

EXERCISE SET 1.4

1. Determine $n(A)$, if possible.

(a) $A = \{x \mid x \text{ is a consonant in the English alphabet}\}$.

(b) $A = \{x \mid x \text{ is a solution of } x^2 = 1\}$.

(c) $A = \{3, -2, 5, 9\}$.

(d) $A = \{f, i, c, k, l, e\}$.

(e) $A = \{x \mid x \text{ is an even integer}\}$.

(f) $A = \{x \mid x \text{ is a real number satisfying } x^2 = -4\}$.

2. Verify the equation

$$n(A \cup B) = n(A) + n(B)$$

for the following disjoint sets

(a) $A = \{a, e, f, z\}$ and $B = \{c, h, k\}$.

(b) $A = \{x \mid x \text{ is a real number satisfying } x^2 < 0\}$ and

$B = \{x \mid x \text{ is a real number satisfying } x^2 = 4\}$.

3. If A and B are disjoint sets such that

$$n(A \cup B) = n(A),$$

what can we say about B?

4. Verify the equation

$$n(A \cup B) = n(A) + n(B) - n(A \cap B)$$

for the following sets.

(a) $A = \{a, c, e, g, i, k\}$, $B = \{e, f, g, h, i, j\}$

(b) $A = \{x \mid \text{is a positive integer} < 10\}$

$B = \{x \mid x \text{ is an integer satisfying } 1 \leq x \leq 7\}$.

(c) $A = \{$Xerox, GE, ATT, IBM, Polaroid$\}$

$B = \{$Sperry Rand, Kodak, Avco, Xerox, GE$\}$.

(d) $A = \{$personal income, gross national product, unemployment$\}$

$B = \{$productivity, unemployment$\}$.

5. Let A and B be sets of integers such that

$$A \cup B = \{1, 2, 3, 4, 5, 6, 7, 8, 9\}.$$

If $A \cap B$ has five elements and A has eight elements, find $n(B)$.

6. Show that the equation

$$n(A \cup B) = n(A) + n(B) - n(A \cap B)$$

reduces to the equation

$$n(A \cup B) = n(A) + n(B)$$

if A and B are disjoint.

7. Verify formula (4) for the sets

$$A = \{1, 2, 3, 5, 7\}, \qquad B = \{2, 4, 8, 9, 11, 12\}, \qquad C = \{1, 5, 8, 9, 13\}.$$

8. In each part of Exercise 4 compute $n(A \times B)$ without enumerating the elements of $A \times B$.

9. In a political poll prior to the 1976 presidential election,

420 people surveyed liked Carter

310 people surveyed liked Mondale

280 people surveyed liked both

If every person who liked either candidate voted the Carter–Mondale ticket, how many votes did the ticket get from the group of people surveyed?

10. An environmental agency needs to hire a total of 50 employees to monitor water pollution and a total of 60 employees to monitor air pollution. Of those hired, 15 will monitor both. How many employees must be hired?

11. A certain kind of item is called defective if it has a major defect, a minor defect, or both. In a batch of 25 defective items, 20 have major defects and 14 have minor defects. How many items in the batch have both major and minor defects?

12. Suppose an ordinary six-sided die is tossed five times and the resulting sequence of numbers is recorded. Among the possible sequences are (3, 2, 5, 1, 4), (1, 3, 2, 6, 6), (1, 1, 2, 2, 2), and so on. How many different sequences are possible?

13. Suppose that a telephone number consists of seven of the digits $\{0, 1, 2, 3, 4, 5, 6, 7, 8, 9\}$ (possibly with repetitions).

(a) Show that a telephone number can be considered as a member of a Cartesian product.

(b) How many telephone numbers are possible?

14. A sales firm wants to use the following system to keep records on the performance of its salesmen. One cabinet will be devoted to the eastern region, a second one to the central region, and a third one to the western region. Each cabinet will be divided into three sections:

I: salesmen selling between \$100,000 and \$199,999 per year

II: salesmen selling between \$200,000 and \$499,999 per year

III: salesmen selling over \$500,000 per year.

In each section there will be a folder for each letter of the alphabet to hold the salesman's file. What is the total number of folders the sales firm will need?

15. In a genetics experiment, the members of a generation of fruit flies were classified as follows

	Short winged	Medium winged	Long winged
Male	26	17	8
Female	21	22	9

(a) Find the number of male fruit flies.
(b) Find the number of long-winged fruit flies.
(c) How many fruit flies were either male or medium-winged?
(d) What percentage of the flies were long-winged males?
(e) How many fruit flies were not short-winged females?
(f) How many fruit flies were either female, long-winged, or short-winged?

16. A survey of 125 people revealed the following information:

27 regularly smoked cigars,
42 regularly smoked cigarettes,
24 regularly smoked a pipe,
9 regularly smoked a pipe and cigars
8 regularly smoked cigars and cigarettes
6 regularly smoked a pipe and cigarettes
2 regularly smoked all three.

(a) How many nonsmokers were surveyed?
(b) How many of those surveyed regularly smoked only cigarettes?

17. A telephone company routes overseas calls in the following manner. The incoming call is routed to one of 200 transmitting stations. Each transmitting station relays the message to one of three satellites. Each satellite in turn relays the call to one of 60 receiving stations, which in turn sends the message directly to the listener. In how many different ways can a message be routed from the sender to the listener?

18. Solve the problem in part (a) of Example 41 using Equation (2).

2
COORDINATE SYSTEMS AND GRAPHS

In many physical problems we are interested in relationships between variable quantities. For example, an astronomer may be interested in the relationship between the size of a star and its brightness, an economist may be interested in the relationship between the cost of a product and the quantity of the product available, and a sociologist may be interested in the relationship between the increase in crime and the increase in population in U.S. cities. In this chapter we discuss two methods for describing such relationships: analytically, by means of formulas, and geometrically, by means of graphs.

Although some of the topics in this chapter may be familiar, others are likely to be new. Because the ideas we shall develop here will be so important in our later work, we shall discuss graphs and functions from the very beginning so we all have a common starting point. The reader may wish to skim the familiar material in this chapter quickly.

2.1 COORDINATE SYSTEMS

In 1637, René Descartes (see p. 25) published a philosophical work called *Discourse on the Method of Rightly Conducting the Reason*. In the back of that book there were three appendices that purported to show how the "method" could be applied to concrete examples. The first two appendices were minor works that endeavored to explain the behavior of lenses and the movement of shooting stars. The third appendix, however, was a landmark stroke of genius. It was described by the nineteenth century British philosopher John Stuart Mill as "the greatest single step ever made in the progress of the exact sciences."

In that appendix René Descartes linked together two branches of mathematics, algebra and geometry. Descartes' work evolved into a new subject called *analytic geometry*; it gave a way of describing algebraic formulas by means of geometric curves and a way of describing geometric curves by algebraic formulas. In this section we shall discuss some of these ideas and in later sections we shall illustrate their application.

We begin with the idea of a ***real number line*** (sometimes simply called the *real line*). The purpose of the real number line is to give a geometric way of describing real numbers. To start, take a line extending infinitely far in both directions. Select a point on this line to serve as a reference point, and call it the ***origin.*** Next, choose one direction from the origin as the ***positive direction*** and let the other direction be called the ***negative direction;*** it is usual to mark the positive direction with an arrowhead as shown in Figure 2.1a. Then, select a unit of length for measuring distances. With each real number, we can now associate a point on the real number line in the following way:

(a) With each positive number r associate the point that is a distance of r units from the origin in the positive direction.
(b) With each negative number $-r$ associate the point that is a distance of r units from the origin in the negative direction.
(c) Finally, associate the origin with the real number 0.

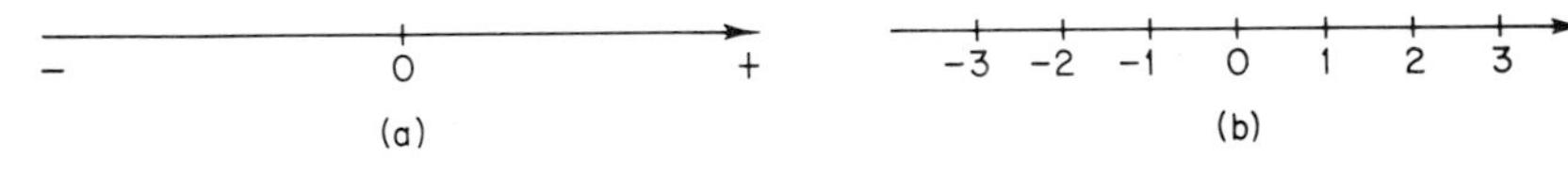

Figure 2.1

In Figure 2.1b we have marked on the real line the points that are associated with some of the integers.

The real number corresponding to a point on the real line is called the ***coordinate*** of the point.

Example 1 In Figure 2.2 we have marked the points whose coordinates are $-\sqrt{2}$, $-\frac{1}{2}$, 1.25, and π.

It is evident from the way in which real numbers and points on the real line are related that each real number corresponds to a single point and each point corresponds to a single real number. This is sometimes described by stating that the set of real numbers and the set of points on the real line are in ***one-to-one correspondence.***

Just as points on a line can be placed in one-to-one correspondence with real numbers, so points in the plane can be placed in one-to-one

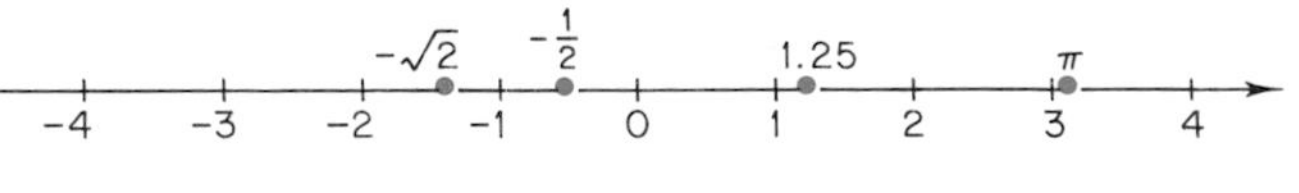

Figure 2.2

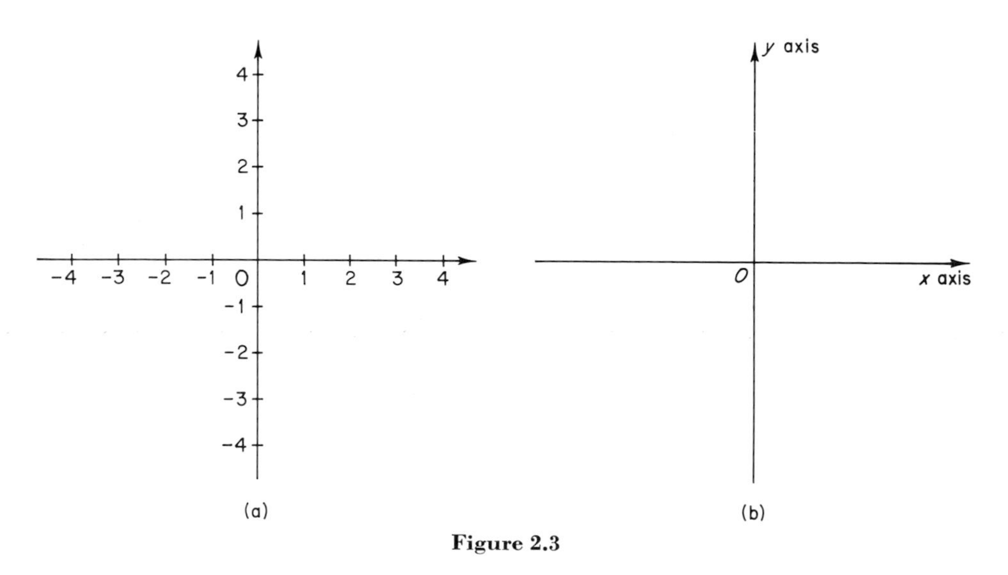

Figure 2.3

correspondence with ordered pairs of real numbers by utilizing two real number lines. This is done as follows. Construct two perpendicular real number lines. For convenience, make one of the lines horizontal, with its positive direction to the right, make the other line vertical with its positive direction upward, and locate the lines so they intersect at their origins (see Figure 2.3a). The two lines are called ***coordinate axes;*** the horizontal line is called the ***x axis,*** the vertical line the ***y axis,*** and the coordinate axes together form what is called a ***Cartesian coordinate system*** or sometimes a ***rectangular coordinate system.*** The point of intersection of the coordinate axes is denoted by O and is called the ***origin*** of the coordinate system (see Figure 2.3b).

We now describe a way of establishing a one-to-one correspondence between the points in the plane and ordered pairs of real numbers. If P is a point in the plane, draw two lines through P, one perpendicular to the x axis and one perpendicular to the y axis. If the first line intersects the x axis at the point whose coordinate is a and the second line intersects the y axis at the point whose coordinate is b, then we associate the ordered pair

$$(a, b)$$

with the point P (Figure 2.4). The number a is called the ***x coordinate*** or ***abscissa*** of P and the number b is called the ***y coordinate*** or ***ordinate*** of P; we say that P is the point with ***coordinates*** (a, b) and denote the point by $P(a, b)$.

Example 2 In Figure 2.5 we have located the points

$$P(4, 3), \qquad Q(-2, 5), \qquad R(-4, -2), \qquad S(2, -5).$$

From the way in which we have formed the construction, each point in

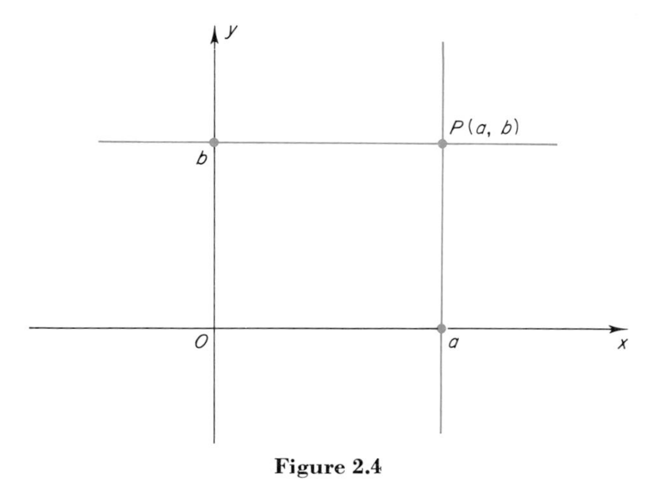

Figure 2.4

the plane determines a unique ordered pair of numbers. Conversely, if we begin with an ordered pair of numbers

$$(a, b)$$

we can construct lines perpendicular to the x axis and y axis at the points with coordinates a and b; the intersection of these lines determines a unique point P in the plane whose coordinates are (a, b) (see Figure 2.4). Thus, we have a one-to-one correspondence between ordered pairs of real numbers and points in the plane.

The coordinate axes divide the plane into four parts, called ***quadrants.*** These quadrants are numbered from one to four as shown in Figure 2.6a. As shown in Figure 2.6b it is easy to determine the quadrant in which a

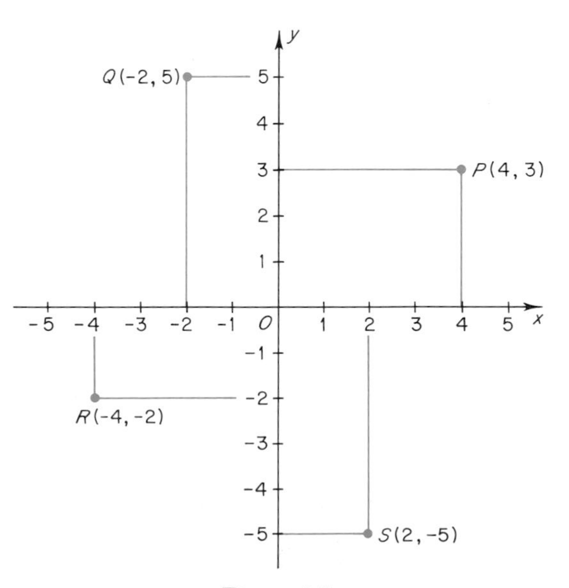

Figure 2.5

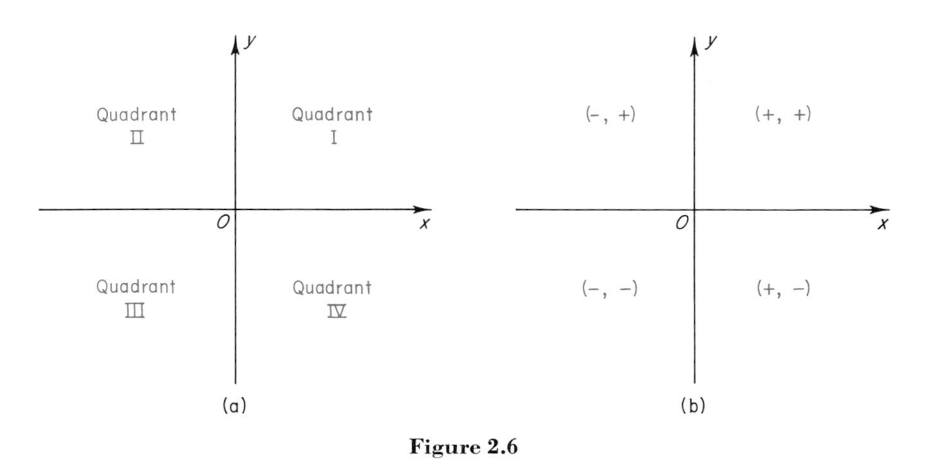

Figure 2.6

point lies from the signs of its coordinates. A point with two positive coordinates $(+, +)$ lies in Quadrant I, a point with a negative x coordinate and a positive y coordinate $(-, +)$ lies in Quadrant II, and so on.

Cartesian coordinate systems are helpful for giving a geometric description of equations involving two variables. To explain how, we shall need some preliminary ideas. We shall assume in our discussion that a Cartesian coordinate system has been constructed and that we are given an equation involving only two variables, x and y, such as

$$3x + 3y = 4, \qquad x^2 + y^2 = 1, \qquad \text{or} \qquad y = \frac{1}{x}.$$

Given an equation involving only the variables x and y, we call an ordered pair of real numbers (a, b) a ***solution of the equation*** if the equation is satisfied when we substitute

$$x = a, \qquad y = b.$$

Example 3 The ordered pair

$$(4, 5)$$

is a solution of the equation

$$3x - 2y = 2$$

since the equation is satisfied when we substitute

$$x = 4, \qquad y = 5.$$

However, the ordered pair

$$(2, 1)$$

is not a solution of the equation since the equation is not satisfied when we substitute

$$x = 2, \qquad y = 1.$$

The set of all solutions of an equation is called its ***solution set.***

Example 4 Like many equations in x and y, the solution set of

$$y = x^2 + 1 \tag{1}$$

has infinitely many members, so that it is impossible to list them all. However, some sample members of the solution set can be obtained by substituting *arbitrary* values for x into the right-hand side of (1) and solving for the associated values of y. Some typical computations are given in Table 2.1.

Table 2.1

x	0	1	2	3	-1	-2	-3
$y = x^2 + 1$	1	2	5	10	2	5	10
solution	(0, 1)	(1, 2)	(2, 5)	(3, 10)	$(-1, 2)$	$(-2, 5)$	$(-3, 10)$

If we plot the points in the solution set of an equation in x and y, we obtain a graphical or geometric picture of the equation. This picture is called the ***graph*** of the equation.

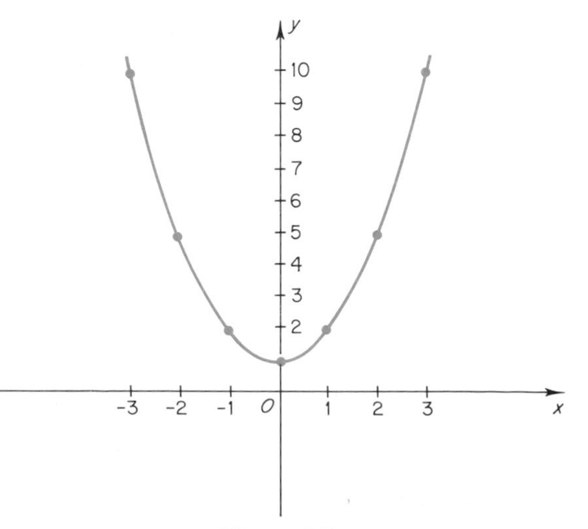

Figure 2.7

Example 5 Sketch the graph of the equation $y = x^2 + 1$ discussed in Example 4.

Solution Since there are infinitely many points in the graph of $y = x^2 + 1$, it is impossible to plot them all. The general procedure for sketching the graph of an equation is to plot enough points in the graph to indicate its general shape. Then the graph is approximated by drawing a smooth curve through the plotted points.

In Figure 2.7 we have sketched the graph of $y = x^2 + 1$ by plotting the points in Table 2.1 and connecting them with a smooth curve.

Example 6 Sketch the graph of the equation

$$y - 3x = 1.$$

Solution By first rewriting this equation as

$$y = 3x + 1$$

and substituting the indicated values of x we obtain Table 2.2. From this table we obtain the graph in Figure 2.8.

Table 2.2

x	0	1	2	-1	-2
$y = 3x + 1$	1	4	7	-2	-5
solution	(0, 1)	(1, 4)	(2, 7)	$(-1, -2)$	$(-2, -5)$

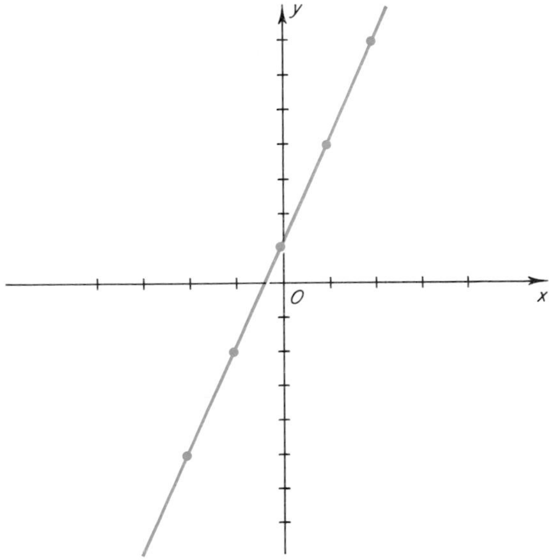

Figure 2.8

There is one flaw in the graphing technique we have just discussed. To see it, consider the graph in Example 6. Although it appears from our sketch that the points in the graph form a straight line, this conclusion is just guesswork based on plotting a few points. It is conceivable that the actual graph oscillates between the points we have plotted or begins to curve off once we pass the points we have plotted. In fact, based on the data given in Table 2.2, it is logically conceivable that the graph of $y = 3x + 1$ might look like the curve in color in Figure 2.9.

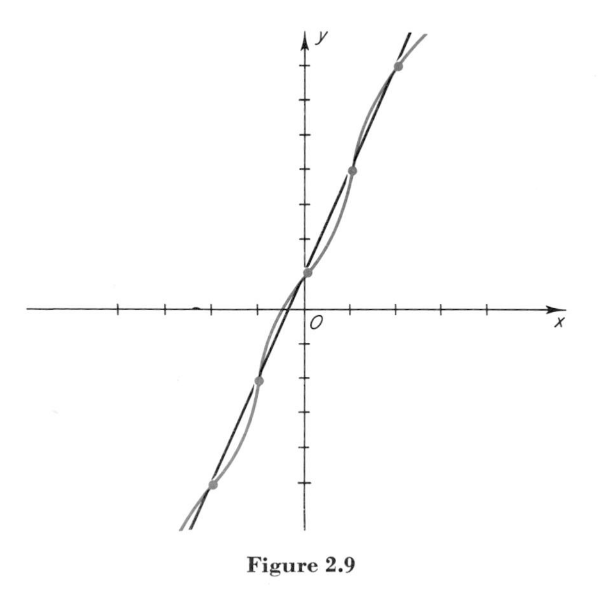

Figure 2.9

Although we shall show in the next section that the graph of $y = 3x + 1$ is actually a straight line, the point is that the graphing technique we have described here provides us only with a reasonable *guess* about the shape of the graph.

EXERCISE SET 2.1

1. Draw a real number line and plot the points whose coordinates are

(a) 6 (b) -4 (c) 2.5 (d) -1.7 (e) $\sqrt{3}$ (f) 0.

2. Draw a Cartesian coordinate system and plot the points whose coordinates are

(a) $(2, 3)$ (b) $(-4, 6)$ (c) $(-3, -5)$ (d) $(7, -3)$.

3. Draw a Cartesian coordinate system and plot the points whose coordinates are

(a) $(3, 0)$ (b) $(0, -3)$ (c) $(-2.5, 3.5)$ (d) $(\frac{1}{2}, -\frac{3}{4})$.

4. Which of the following are solutions of

$$2x + 3y = 1?$$

(a) $(0, \frac{1}{3})$ (b) $(\frac{1}{2}, 0)$ (c) $(-2, 3)$ (d) $(-1, 1)$.

5. Which of the following are solutions of

$$4x^2 + 9y = 36?$$

(a) $(0, 4)$ (b) $(3, 0)$ (c) $(-1, 1)$ (d) $(\frac{1}{2}, \frac{35}{9})$.

6. In each part fill in the blank so that the ordered pair is a solution of the equation

$$3x - 2y = 3.$$

(a) $(\_, 2)$ (b) $(-1, \_)$ (c) $(0, \_)$ (d) $(\_, 3)$.

In Exercises 7–14, sketch the graph of the given equation.

7. $y = 3x + 2$
8. $y = -2x - 3$
9. $2x + 3y = 5$
10. $y = x^2 + 2$
11. $x = y^2 - 4$
12. $4x^2 + y = 3$
13. $y = \dfrac{1}{x^2 + 1}$
14. $xy = 3$.

15. Sketch the set of all points in the plane whose x coordinates are 2.

16. Sketch the set of all points in the plane whose y coordinates are -3.

17. Find the fourth corner of the rectangle, three of whose corners are $(8, 3)$, $(-2, 3)$, and $(8, 9)$.

18. Find the coordinates of the point that lies three units to the left of and four units below the point $(2, -1)$.

19. The formula for changing from Centigrade to Fahrenheit temperature readings is

$$F = \frac{9}{5}C + 32.$$

Draw a Cartesian coordinate system with vertical axis F and horizontal axis C. Sketch the graph of the above equation.

20. It follows from Newton's laws that the distance s traveled after t seconds by a body falling from rest is approximately given by

$$s = 16t^2,$$

where t is in seconds and s is in feet. Draw a Cartesian coordinate system with vertical axis s and horizontal axis t. Sketch the graph of the above equation for $t \geq 0$.

2.2 THE STRAIGHT LINE

In this section we discuss equations and graphs of straight lines. As we shall see in later sections, this material has numerous important applications. Among other things, we shall learn to recognize those equations whose graphs are straight lines. This will eliminate the kind of difficulty that occurred in Example 6 of the previous section; in that example, it appeared, after plotting several points, that the graph of $y = 3x + 1$ was a straight line; however, we could not be absolutely certain.

To begin, consider a line L that is *not parallel to the* y *axis*.

If $P_1(x_1, y_1)$ and $P_2(x_2, y_2)$ are any two distinct points on L, then the number m given by the formula

$$m = \frac{y_2 - y_1}{x_2 - x_1} \tag{1}$$

is called the ***slope of the line*** L.

Example 7 Let L be the line determined by the points $P_1(4, 1)$ and $P_2(6, 9)$. From formula (1), the slope of L is

$$m = \frac{9 - 1}{6 - 4} = 4.$$

Similarly, the slope of the line determined by the points $P_1(1, 11)$ and $P_2(4, 5)$ is

$$m = \frac{5 - 11}{4 - 1} = -2.$$

We emphasize that formula (1) does not apply to lines parallel to the y axis. For such lines we would have $x_2 = x_1$. (Can you see why?) Thus the right-hand side of (1) would have a zero denominator, which is not allowed. Lines parallel to the y axis are said to have *no slope.*

Using plane geometry, it can be shown that the value of m is the same, no matter which two points P_1 and P_2 on L are used (see Exercise 20).

Example 8 Show that a line parallel to the x axis has slope $m = 0$.

Solution If L is parallel to the x axis, then all points on L have equal y coordinates (see Figure 2.10). Thus if $P_1(x_1, y_1)$ and $P_2(x_2, y_2)$ are points on L, we must have $y_2 = y_1$. Therefore $m = 0$ from Equation (1).

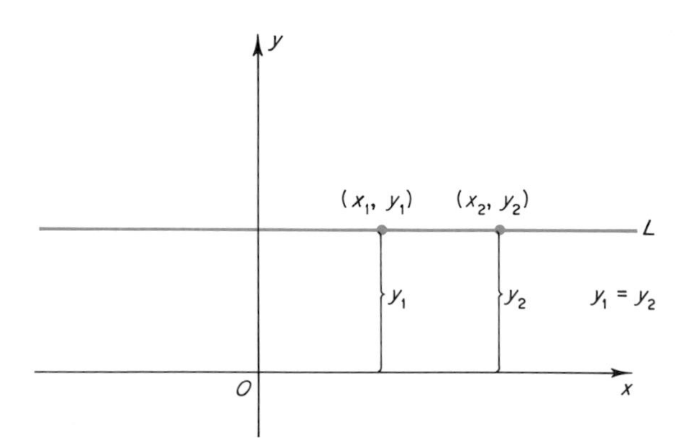

Figure 2.10 All points on L have equal y coordinates.

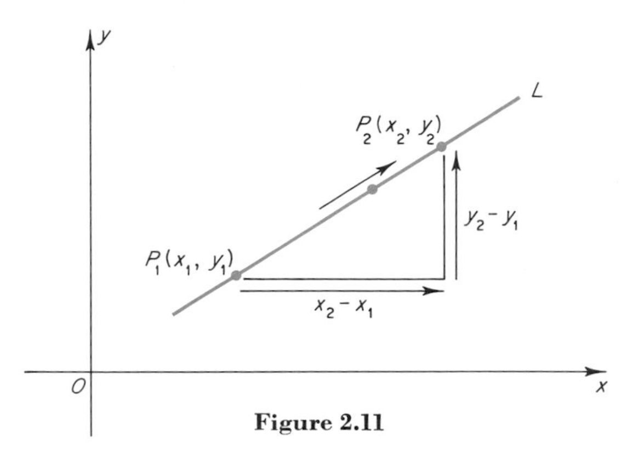

Figure 2.11

The slope of a line has a useful physical interpretation. Imagine a particle moving rightward along a line L from a point $P_1(x_1, y_1)$ to a point $P_2(x_2, y_2)$. As shown in Figure 2.11, the particle moves $y_2 - y_1$ units in the y direction as it travels $x_2 - x_1$ units in the x direction. From (1) these quantities are related by

$$y_2 - y_1 = m(x_2 - x_1), \tag{2}$$

where m is the slope of the line. Equation (2) states that the movement in the y direction is proportional to the movement in the x direction and the slope m is the constant of proportionality. For this reason, m is said to measure the ***rate*** at which y changes with x along the line L.

Example 9 In Example 7, we showed that the line determined by the points $P_1(4, 1)$ and $P_2(6, 9)$ has slope $m = 4$. This means that a particle traveling rightward along this line will gain four units in height for every unit it moves in the positive x direction (see Figure 2.12).

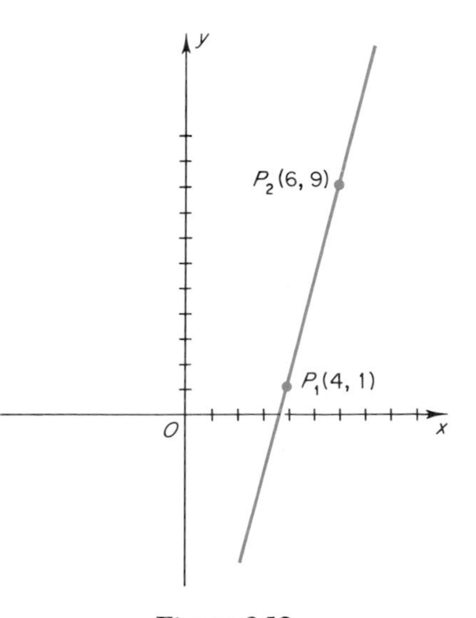

Figure 2.12

It was also shown in Example 7 that the line determined by the points $P_1(1, 11)$ and $P_2(4, 5)$ has slope $m = -2$. This means that a particle traveling rightward along this line will lose two units in height for every unit it moves in the positive x direction (see Figure 2.13).

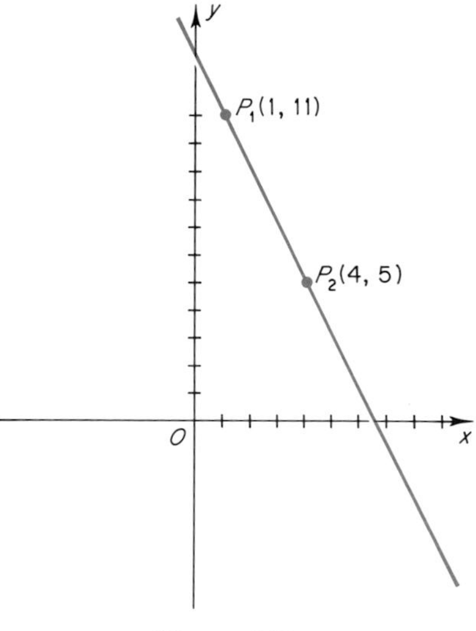

Figure 2.13

In Figure 2.14 we have sketched several lines with varying slopes. Observe that lines with positive slope are inclined upward to the right, while lines with negative slope are inclined downward to the right.

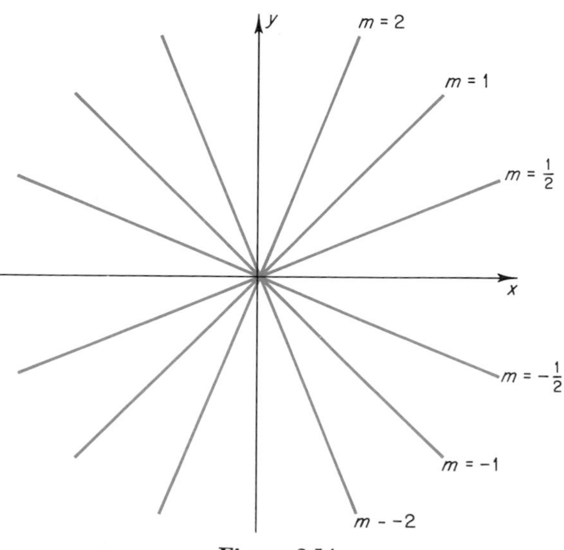

Figure 2.14

Although we shall omit the proof, the following result is evident.

> Two lines with the same slope are parallel, and conversely parallel lines have the same slope.

Example 10 Let L be the line determined by the points $P_1(3, 4)$ and $P_2(4, 1)$ and let L' be the line determined by the points $P_1'(5, 6)$ and $P_2'(7, 0)$. Determine whether L and L' are parallel.

Solution The slope m of the line L is

$$m = \frac{1 - 4}{4 - 3} = -3$$

and the slope m' of the line L' is

$$m' = \frac{0 - 6}{7 - 5} = -3.$$

Since the lines have the same slope, they are parallel. These lines are shown in Figure 2.15.

We are now in a position to find an equation of any line that is not parallel to the y axis. If L is any such line, let m be its slope and suppose that the line intersects the y axis at the point $(0, b)$. If (x, y) is any point on L, then the slope m can be obtained by applying formula (1) to the points $P_1(0, b)$ and $P_2(x, y)$ (see Figure 2.16). This yields

$$m = \frac{y - b}{x - 0}$$

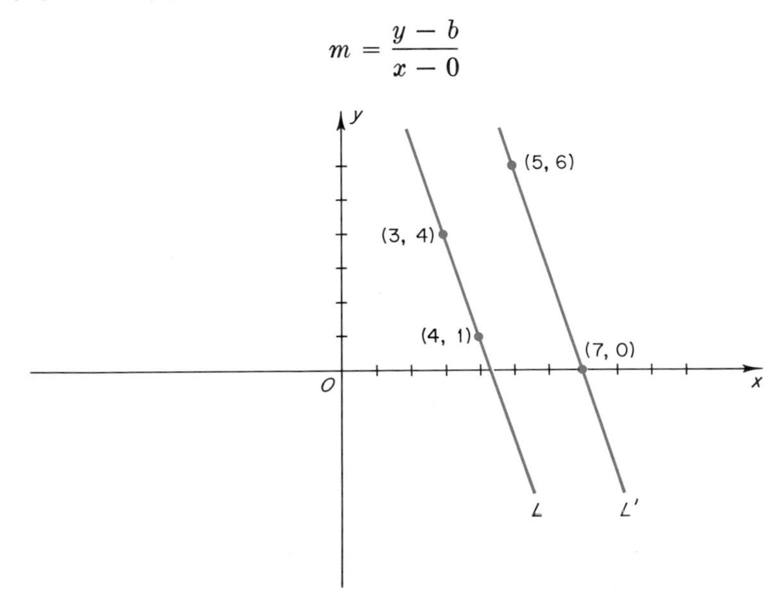

Figure 2.15

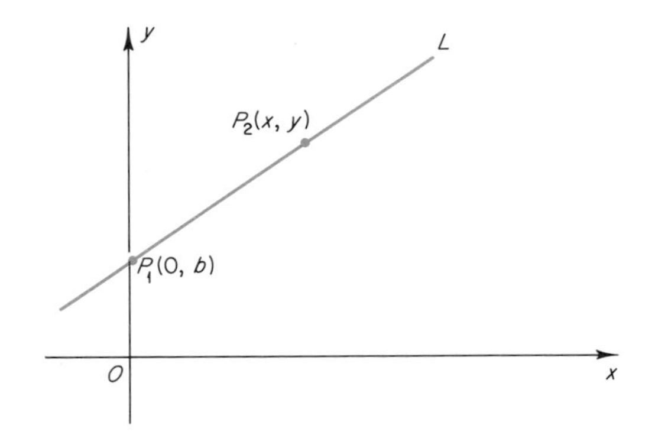

Figure 2.16

or

$$m = \frac{y - b}{x}.$$

Multiplying both sides by x and then adding b to both sides we obtain the following equation, called the ***slope–intercept form*** of the line:

$$y = mx + b. \tag{3}$$

Example 11 Find the slope–intercept form of the line having slope 3 and intersecting the y axis at the point $(0, -2)$.

Solution From the given information, we have $m = 3$ and $b = -2$. Substituting these values in Equation (3) yields

$$y = 3x - 2.$$

Example 12 Find the slope of the line

$$y = 7x + 8.$$

Solution This is the slope–intercept form of a line with $m = 7$ and $b = 8$ (see Equation (3)). Thus the line has slope $m = 7$ and intersects the y axis at the point $(0, 8)$.

Example 13 Find the slope–intercept form of the line passing through the points $P_1(5, 3)$ and $P_2(2, 7)$.

Solution Using the points P_1 and P_2 to determine the slope of the line we obtain

$$m = \frac{7 - 3}{2 - 5} = -\frac{4}{3}. \tag{4}$$

On the other hand, if $P(x, y)$ is any point on the line, then we can also use P_1 and P to determine the slope of the line. This gives

$$m = \frac{y - 3}{x - 5}. \tag{5}$$

Equating (4) and (5) yields

$$\frac{y - 3}{x - 5} = -\frac{4}{3}.$$

This can be rewritten as

$$y = -\frac{4}{3}x + \frac{29}{3},$$

which is the slope–intercept form of the line.

A line can be determined by giving its slope to describe its inclination and specifying a point on the line to pin down its location. Suppose we want the equation of the line passing through the point $P_1(x_1, y_1)$ and having slope m.

If (x, y) is any point on the line, then we can obtain the slope of the line from the points $P(x, y)$ and $P_1(x_1, y_1)$. This gives

$$m = \frac{y - y_1}{x - x_1}.$$

This equation can be rewritten as

$$y - y_1 = m(x - x_1)$$

or

$$y = y_1 + m(x - x_1). \tag{6}$$

This is called the ***point–slope form*** of the line.

Example 14 Find an equation for the line passing through the point $P_1(3, -2)$ and having slope $m = 3$.

Solution Substituting the values $x_1 = 3$, $y_1 = -2$ and $m = 3$ in the point–slope form (6) yields

$$y = -2 + 3(x - 3).$$

Example 15 Find an equation for the line parallel to the x axis and three units above it.

Solution As shown in Figure 2.17, the line intersects the y axis at the point (0, 3). Moreover, as shown in Example 8, a line parallel to the x axis has slope $m = 0$. Thus, substituting the values $x_1 = 0$, $y_1 = 3$, and $m = 0$ in the point–slope form (6) yields

$$y = 3 + 0(x - 0)$$

or

$$y = 3.$$

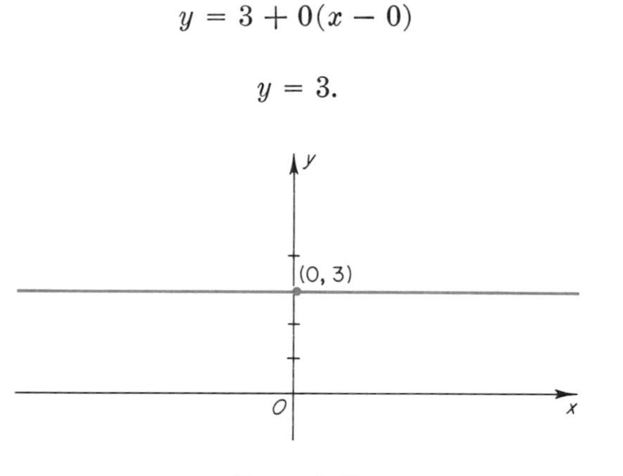

Figure 2.17

As previously mentioned, the slope–intercept form does not apply to lines parallel to the y axis. However, it is particularly easy to obtain equations for such lines. For example, consider Figure 2.18; the line is parallel to the y axis and intersects the x axis at the point (4, 0). This line consists precisely of those points (x, y) whose x coordinate is 4. Thus,

$$x = 4$$

is an equation for this line. More generally,

$$x = c$$

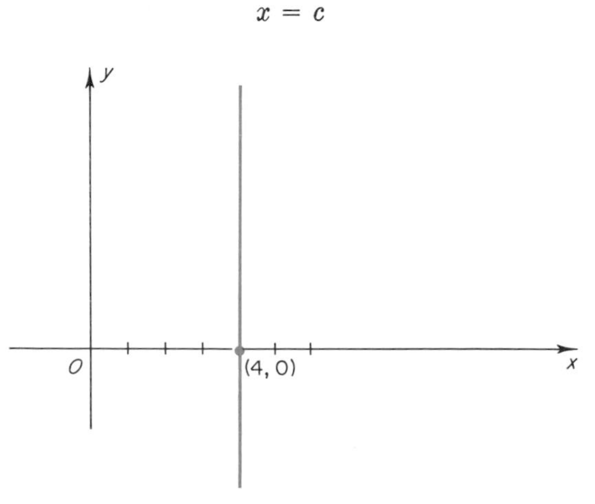

Figure 2.18

is the equation of a line parallel to the y axis and intersecting the x axis at the point $(c, 0)$.

Our next result tells us how to recognize when a given equation will have a straight line as its graph.

Any equation that can be expressed in the form

$$Ax + By = C, \tag{7}$$

where A, B, and C are constants and A and B are not both zero, has a straight line as its graph.

To see this, we distinguish between two cases, $B = 0$ and $B \neq 0$.

If $B \neq 0$, we can solve (7) for y in terms of x, to obtain

$$y = -\frac{A}{B}x + \frac{C}{B}.$$

But this is the slope–intercept form

$$y = mx + b$$

of the line with

$$m = -\frac{A}{B} \quad \text{and} \quad b = \frac{C}{B}.$$

If $B = 0$, then since A and B are not both zero, (7) reduces to

$$Ax = C,$$

which can be rewritten as

$$x = \frac{C}{A}.$$

But this is the equation of a line parallel to the y axis. Thus, in either case (7) has a straight line for its graph. For this reason, an equation that is expressible in the form given in (7) is called a ***linear equation*** in x and y.

Example 16 Since each of the following equations can be rewritten in the form given in (7), they all have straight lines as graphs.

$$x = 7y + 8, \qquad x + 7y - 12 = 0, \qquad x = y.$$

If we are given two linear equations

$$a_1x + b_1y = c_1$$

$$a_2x + b_2y = c_2$$

we can ask for values of x and y that satisfy both equations. The two equations are then said to form a ***system*** of two linear equations and an ordered pair of real numbers (a, b) is called a ***solution of the system*** if both equations are satisfied when we substitute

$$x = a, \qquad y = b.$$

Example 17 Consider the system

$$\begin{aligned} 2x + 3y &= 8 \\ 3x - y &= 1. \end{aligned}$$

The ordered pair $(1, 2)$ is a solution of the system since both equations are satisfied when we substitute

$$x = 1, \qquad y = 2.$$

However, the ordered pair $(-2, 4)$ is not a solution since only the first equation is satisfied when we substitute

$$x = -2, \qquad y = 4.$$

Not all systems of linear equations have solutions. For example, if we multiply the second equation of the system

$$\begin{aligned} x + y &= 2 \\ 3x + 3y &= 9 \end{aligned}$$

by $\frac{1}{3}$, it becomes evident that there are no solutions, since the two equations in the resulting system

$$\begin{aligned} x + y &= 2 \\ x + y &= 3 \end{aligned}$$

contradict each other.

To illustrate the possibilities that can occur in solving systems of linear equations, consider a general system of two linear equations in the unknowns x and y

$$\begin{aligned} a_1x + b_1y &= c_1 \\ a_2x + b_2y &= c_2. \end{aligned}$$

The graphs of these equations are lines, call them l_1 and l_2. Since a point (a, b) lies on a line if and only if the numbers $x = a$, $y = b$ satisfy the equation of the line, the solutions of the system will correspond to points of intersection of l_1 and l_2. There are three possibilities (Figure 2.19).

(a) The lines l_1 and l_2 can be parallel, in which case there is no intersection, and consequently no solution to the system.
(b) The lines l_1 and l_2 can intersect at only one point, in which case the system has exactly one solution.

(c) The lines l_1 and l_2 can coincide, in which case there are infinitely many solutions to the system.

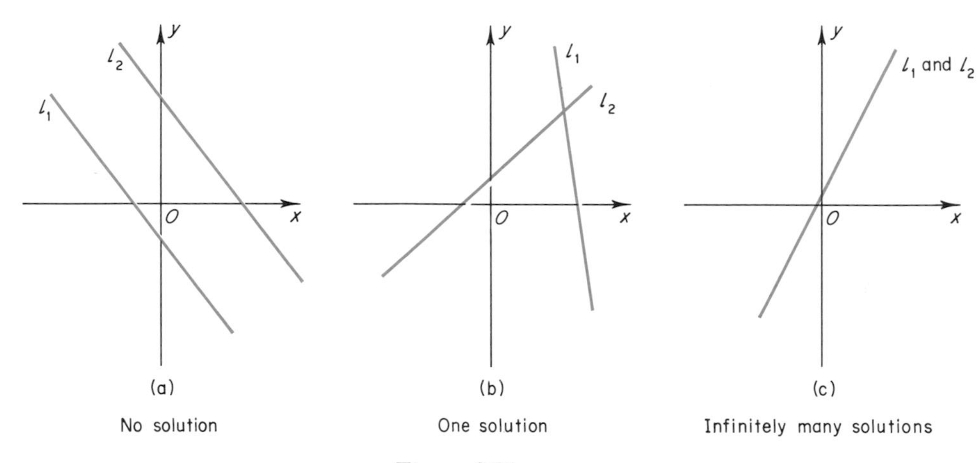

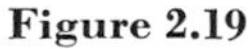

Figure 2.19

To summarize, a system of two linear equations in x and y has either no solutions, exactly one solution, or infinitely many solutions. We shall now develop a method for solving systems of two linear equations.

The idea of the method we shall describe is to replace the given system of equations with a new system that has the same set of solutions, but is easier to solve. This new system will be obtained in a series of steps by applying the following operations:

1. Multiply an equation through by a nonzero constant.
2. Add a multiple of one equation to another.

To see how these operations can be used to solve a system consider the following example.

Example 18 Solve

$$\begin{aligned} 2x - 3y &= 13 \\ 4x + y &= 5. \end{aligned} \tag{8}$$

Solution Our first objective is to add an appropriate multiple of the first equation to the second equation to eliminate the variable x. In this case, this can be done by adding -2 times the first equation to the second; this gives

$$-2(2x - 3y) + (4x + y) = (-2)13 + 5$$

or

$$7y = -21.$$

or

$$y = -3.$$

Replacing the second equation in (8) by this equation we obtain the simpler system

$$\begin{aligned} 2x - 3y &= 13 \\ y &= -3. \end{aligned} \tag{9}$$

Next we eliminate y from the first equation in (9) by adding 3 times the second equation in (9) to the first; this gives

$$3y + (2x - 3y) = 3(-3) + 13$$

or

$$2x = 4$$

or

$$x = 2.$$

Thus the solution is $x = 2$, $y = -3$.

The following examples further illustrate this method.

Example 19 Solve

$$\begin{aligned} 3x + 2y &= -12 \\ 2x - 3y &= 5. \end{aligned}$$

Solution Multiply the first equation by $\frac{1}{3}$ to obtain

$$\begin{aligned} x + \tfrac{2}{3}y &= -4 \\ 2x - 3y &= 5. \end{aligned}$$

Eliminate x from the second equation by adding -2 times the first equation to the second equation. This yields

$$\begin{aligned} x + \tfrac{2}{3}y &= -4 \\ -\tfrac{13}{3}y &= 13. \end{aligned}$$

Multiply the second equation by $-\frac{3}{13}$ to obtain

$$\begin{aligned} x + \tfrac{2}{3}y &= -4 \\ y &= -3. \end{aligned}$$

Eliminate y from the first equation by adding $-\frac{2}{3}$ times the second equation to the first equation. This yields

$$\begin{aligned} x \quad &= -2 \\ y &= -3. \end{aligned}$$

The solution is thus $x = -2$, $y = -3$.

Example 20 Solve the system

$$\begin{aligned} x + 3y &= -1 \\ 2x + 6y &= 6. \end{aligned}$$

Solution Eliminate x from the second equation by adding -2 times the first equation to the second equation. This yields

$$x + 3y = -1$$
$$0x + 0y = 8.$$

Since the second equation is not satisfied by any values of x and y, the system has no solution.

Example 21 Solve the system

$$x + 3y = -1$$
$$2x + 6y = -2.$$

Solution Eliminate x from the second equation by adding -2 times the first equation to the second equation. This yields

$$x + 3y = -1$$
$$0x + 0y = 0.$$

The second equation is satisfied by all values of x and y, so that the solutions of the system consist of all values of x and y that satisfy the first equation

$$x + 3y = -1. \tag{10}$$

The system thus has infinitely many solutions. They can be obtained from (10) by assigning an arbitrary value to y, say

$$y = t$$

and solving for x to obtain

$$x = -1 - 3t.$$

The solutions are thus given by the formulas

$$x = -1 - 3t, \qquad y = t, \tag{11}$$

where t is arbitrary. Particular numerical solutions can be obtained by substituting specific values for t. For example, substituting $t = 3$ in (11) yields the solution

$$x = -10, \qquad y = 3;$$

and substituting $t = -2$ yields the solution

$$x = 5, \qquad y = -2.$$

EXERCISE SET 2.2

1. In each part find the slope of the line determined by the points P_1 and P_2.
   (a) $P_1(2, 4)$, $P_2(\;\;3, 5)$ (b) $P_1(3, 2)$, $P_2(5, 6)$
   (c) $P_1(0, 0)$, $P_2(-3, -2)$ (d) $P_1(4, 2)$, $P_2(2, 2)$
   (e) $P_1(1.846, 2.772)$, $P_2(0.927, 2.253)$.

2. In each part let L be the line determined by the points P_1 and P_2 and let L' be the line determined by the points P_1' and P_2'. Decide whether L and L' are parallel and sketch the lines.

(a) $P_1(4, 3)$, $P_2(1, 4)$, $P_1'(6, 5)$, $P_2'(0, 7)$

(b) $P_1(2, 3)$, $P_2(3, 2)$, $P_1'(5, 5)$, $P_2'(2, 2)$

(c) $P_1(0, 0)$, $P_2(2, 3)$, $P_1'(4, 6)$, $P_2'(0, 0)$.

3. In each part sketch the line with slope m that passes through the point P.

(a) $m = 1$, $P(2, 3)$ (b) $m = -\frac{2}{3}$, $P(0, 2)$

(c) $m = -3$, $P(3, 3)$ (d) $m = 0$, $P(-2, 3)$.

4. In each part use the given information to find the slope–intercept form of the line.

(a) $m = 2$; the point $(0, 4)$ lies on the line.

(b) $m = -\frac{4}{3}$; the point $(0, -2)$ lies on the line.

(c) $m = 0$, the point $(2, 3)$ lies on the line.

5. Find the slope of the given line.

(a) $y = 2x + 3$ (b) $y = -\frac{3}{2}x + 5$

(c) $y = 7$ (d) $x = 2$

(e) $2x + 3y = 2$ (f) $x = \frac{1}{2}y + 5$.

6. In each part find the slope–intercept form of the line passing through the points P_1 and P_2.

(a) $P_1(4, 2)$, $P_2(2, 2)$ (b) $P_1(-3, -2)$, $P_2(0, 0)$

(c) $P_1(-3, 5)$, $P_2(2, 4)$ (d) $P_1(5, 6)$, $P_2(3, 2)$.

7. In each part use the given information to find the point–slope form of the line.

(a) $m = 1$, $P(2, 3)$ (b) $m = -\frac{2}{3}$, $P(0, 2)$

(c) $m = -3$, $P(3, 3)$ (d) $m = 0$, $P(-2, 3)$.

8. Find the point–slope form of the line passing through the points P_1 and P_2.

(a) $P_1(2, 3)$, $P_2(-3, 4)$ (b) $P_1(\frac{3}{2}, 1)$, $P_2(2, -\frac{2}{3})$.

9. (a) Find an equation for the line parallel to the x axis and 2 units below it.
(b) Find an equation for the line parallel to the y axis and 3 units to the left of it.

10. Express each of the following equations in the form $Ax + By = C$ and give the values of A, B, and C.

(a) $x = \frac{1}{3}y - 7$ (b) $y = 2$

(c) $y = 7 + 4(x - 3)$ (d) $x = 3(y - 2)$.

11. Which of the following are linear equations?

(a) $2x + y^2 = 3$ (b) $xy = 1$

(c) $2x - 3y + 2 = 0$ (d) $2(x - 1) + 3(y - 1) = 4$.

12. Which of the following ordered pairs are solutions of the system

$$3x + 2y = 1$$
$$2x - y = -4?$$

(a) $(-1, 2)$ (b) $(-1, -2)$
(c) $(0, \frac{1}{2})$ (d) $(1, -1)$.

13. Solve the following systems:

(a) $2x + 3y = -5$, $3x + 2y = 12$ (b) $4x + 3y = 6$, $3x - 4y = -2$

(c) $3x - 2y = 2$, $2y = -1$ (d) $-2x - 4y = 3$, $5x - 3y = -2$.

14. Solve the following systems geometrically by graphing the lines in the system and determining their point of intersection.

(a) $2x + 3y = 6$, $3x - 4y = -8$ (b) $-3x + 5y = 1$, $2x + 8y = 5$.

15. Solve the following systems, if possible.

(a) $2x + 3y = 4$, $4x + 6y = 2$ (b) $2x - 3y = 5$, $4x - 6y = 10$

(c) $-3x + 4y = 8$, $-\frac{3}{2}x + 2y = 4$ (d) $0.2x + 0.5y = 0.8$, $0.4x + y = 1$.

16. Find the slopes of the sides of the triangle with vertices $(2, 3)$, $(3, -4)$, $(0, 3)$.

17. Find the equation of the line passing through the point $(3, 4)$ and parallel to the line $y = 2x + 3$.

18. In each part use slopes to determine whether the given three points lie on the same line.

(a) $(1, 5)$, $(-2, -1)$, $(\frac{1}{2}, 4)$ (b) $(0, 0)$, $(3, 2)$, $(-4, 3)$.

19. Solve the following systems:

(a) $5s + 8t = -3$, $2s - 3t = 5$ (b) $-V + 4W = -6$, $2V + 3W = 1$.

20. Using plane geometry, show that the value of the slope m of a line L is the same no matter which two points are used in Equation (1). (*Hint:* Use similar triangles.)

2.3 APPLICATIONS OF LINEAR EQUATIONS

In this section we illustrate several applications of linear equations.

Simple Interest

Interest is the fee charged for borrowing money. The fee is called ***simple interest*** if the fee is a fixed percentage of the amount borrowed when it is held for a specified period of time. For example, if one borrows money at a simple interest rate of 6% per year, then for each year the borrowed money is held, the interest fee is 6% of the amount borrowed.

The sum borrowed is called the ***principal.*** Assume the principal is P dollars and the interest rate is $r \times 100\%$ per year. For example, for 6% interest per year, the value of r would be

$$r = 0.06$$

since

$$r \times 100\% = 0.06 \times 100\% = 6\%.$$

By definition, the interest due each year the money is held is

$$Pr, \tag{1a}$$

so that if the money is held for t years, the total interest due is

$$Prt.$$

Consequently, if S is the amount owed after t years, then

$$S = P + Prt \tag{1b}$$

since we must repay the borrowed money (principal) and the interest.

Formula (1b) is often rewritten as

$$S = P(1 + rt)$$

by factoring out the quantity P.

Example 22 If we borrow \$8000 at the simple interest rate of 8% per year, then from (1b) with $P = 8000$ and $r = 0.08$ the amount S owed after t years would be

$$S = 8000 + (8000)(0.08)t$$

or

$$S = 8000 + 640t. \tag{2}$$

Thus, the amount owed after 20 years ($t = 20$) would be

$$S = 8000 + 640(20) = 20{,}800 \quad \text{(dollars)},$$

and the amount owed after three months ($t = \frac{3}{12} = \frac{1}{4}$ years) would be

$$S = 8000 + 640\left(\frac{1}{4}\right) = 8160 \quad \text{(dollars)}.$$

Observe that Equation (2) is a linear equation, where the variables are t and S rather than x and y. Moreover, the slope of this line is 640 (dollars per year) which is the constant annual interest.

Linear Depreciation

Some property loses all or part of its value over a period of time. The decrease in value is called the ***depreciation*** of the property. For example, if a car is purchased for \$2500 and two years later its resale value is \$1400, then the depreciation of the car after two years is

$$\$2500 - \$1400 = \$1100.$$

The depreciation is called ***linear depreciation*** if the loss in value over a specified period of time is a fixed percentage of the original value. For example, if an item depreciates linearly at the rate of 10% per year, then for each year the item is held, its loss in value is 10% of the original value.

Assume the original cost of a certain item is C dollars and that the item depreciates linearly in value at the rate of $r \times 100\%$ per year. For example, if the depreciation is 10% per year, the value of r would be

$$r = 0.1$$

since

$$r \times 100\% = 0.1 \times 100\% = 10\%.$$

By definition, the value lost each year the item is owned is

$$Cr,$$

so that if the item is owned for t years, the total loss in value is

$$Crt.$$

Consequently, if V is the value of the item after t years, then

$$V = C - Crt \tag{3}$$

since V is the original cost less the loss in value.

Formula (3) is often written as

$$V = C(1 - rt)$$

by factoring out the quantity C.

Example 23 If we purchase an item for \$6000 and if the item depreciates linearly at the rate of 10% per year, then from (3) with $C = 6000$ and $r = 0.1$, the value V of this item after t years would be

$$V = 6000 - (6000)(0.1)t$$

or

$$V = 6000 - 600t. \tag{4}$$

Thus, the value of the item after three years ($t = 3$) would be

$$V = 6000 - 600(3) = 4200 \quad \text{(dollars)}$$

and the value of the item after two months ($t = \frac{2}{12} = \frac{1}{6}$ years) would be

$$V = 6000 - 600\left(\frac{1}{6}\right) = 5900 \quad \text{(dollars)}.$$

Observe that Equation (4) is a linear equation, where the variables are t and V rather than x and y. Moreover, the slope of this line is -600 (dollars per year), which is the constant yearly loss in value of the item.

Prediction

Linear equations can sometimes be used to predict future results based on past or current data. The following example illustrates this idea.

Example 24 (Water Pollution) Suppose that tests performed at the beginning of 1970 showed that each 1000 liters of water in Lake Michigan contained 6 milligrams of polluting mercury compounds, while similar tests performed at the beginning of 1972 showed that each 1000 liters of water contained 8 milligrams of polluting mercury compounds. Assuming that the quantity of polluting mercury compounds will continue to increase at a constant rate, predict the number of milligrams of polluting mercury compounds per 1000 liters of water that will be present at the beginning of 1980.

Solution Let y denote the number of milligrams of polluting mercury compounds per 1000 liters of water and let x denote time (in years). We are interested in finding the value of y when $x = 1980$. We know the values of y when $x = 1970$ and when $x = 1972$.

Since we are assuming that y will grow at a constant rate, the graph of y versus x will be a straight line. Moreover, from the given data, this line will pass through the points

$$(1970, 6) \qquad \text{and} \qquad (1972, 8)$$

(see Figure 2.20). From the given data, the slope of the line is

$$m = \frac{8 - 6}{1972 - 1970} = \frac{2}{2} = 1.$$

Using this value of m and the point

$$(1970, 6)$$

together with the point–slope form of the line, we obtain the equation

$$y = 6 + 1(x - 1970).$$

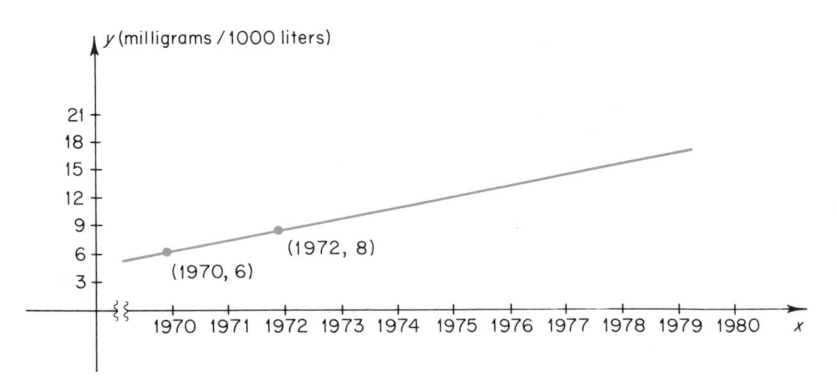

Figure 2.20

or

$$y = x - 1964. \tag{5}$$

The value of y at the end of the year 1980 can now be obtained by substituting the value

$$x = 1980$$

in (5). This yields

$$y = 1980 - 1964 = 16.$$

Thus we would predict that at the beginning of 1980, Lake Michigan will contain 16 milligrams of polluting mercury compounds per 1000 liters.

We emphasize that the result obtained in this example depends on the assumption that the polluting mercury compounds will continue to increase at the same rate. Should it turn out that this assumption is not correct, then the predicted results would, of course, not be accurate.

Trends in Data (Method of Least Squares)

In many physical problems we are interested in determining relationships between two variable quantities x and y. One method for obtaining such relationships consists in gathering data that shows what values of y are associated with various values of x and then plotting the points (x, y) on graph paper. One hopes that after examining the resulting graph, it might then be possible to describe how the variables are related. To illustrate this idea, let us consider the following example.

Example 25 (Astronaut Reaction Time) In an experiment designed to test the effect of weightlessness on an astronaut's reaction time, an astronaut was required to respond to a flashing light by depressing a button. The time elapsed between the light flash and the depression of the button was denoted by y and the length of time that the astronaut was weightless was denoted by x. The resulting data appear in Table 2.3. The points in this table are plotted in Figure 2.21a.

In a problem like this we would not usually try to relate x and y by means of a curve *passing through* the plotted points since the data are really "probabilistic" in nature, that is, we would not expect the astronaut to exhibit exactly the same reaction times if the experiment were repeated. Instead, we observe that the "trend" of the data is upward as x increases and that this upward trend can be described by a straight line (see Figure

Table 2.3

x (hours)	5	6	7	8	9	10	11	12	13	14
y (seconds)	0.70	1.08	1.02	1.26	1.62	1.42	1.47	1.85	2.02	1.93

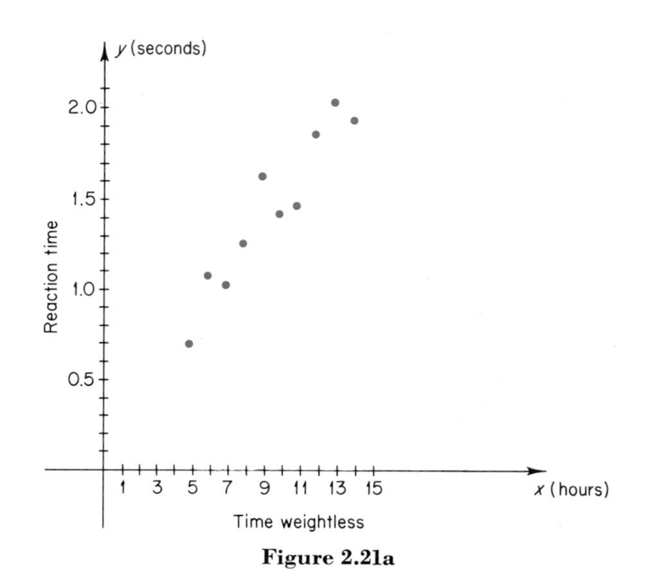

Figure 2.21a

2.21b). We now discuss a technique called the ***method of least squares*** for determining this "trend line." Once this line is obtained it is useful for predicting values of y that are to be expected for values of x not appearing in the data.

The idea of the method of least squares can be illustrated by considering the four points (x_1, y_1), (x_2, y_2), (x_3, y_3), and (x_4, y_4) in Figure 2.22a. Since the four points are not collinear, no line will pass through all four of them. Thus, if we want to describe the upward trend of these points by a

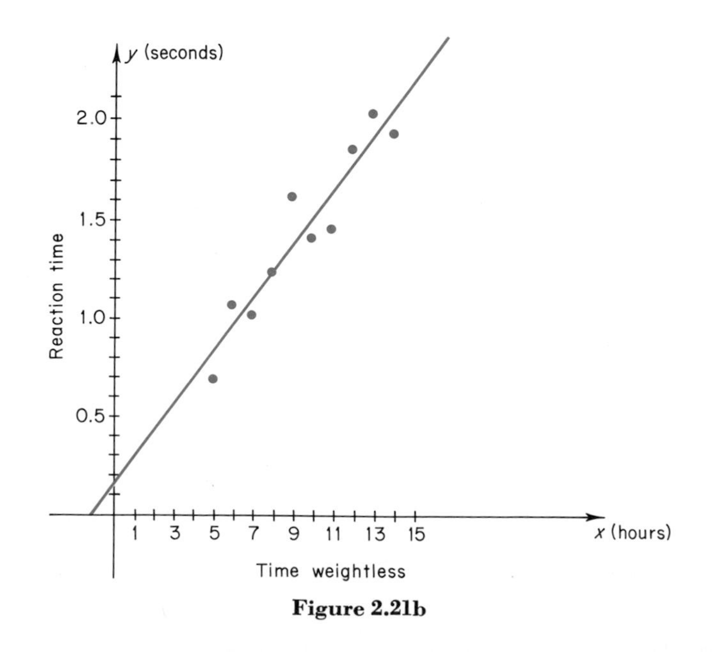

Figure 2.21b

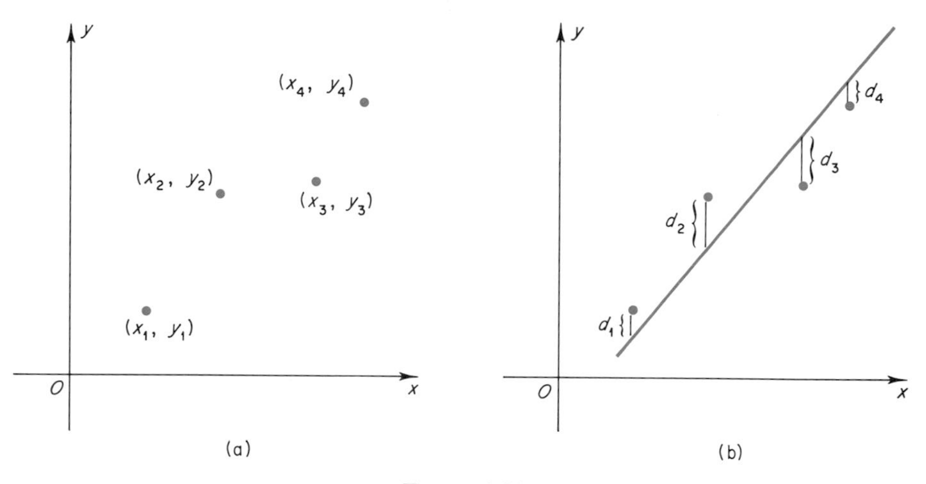

Figure 2.22

line, we must settle for a line that in some sense comes "closest" to passing through all of the points. Although there are various ways of measuring how close a line comes to passing through a set of noncollinear points, the most widely used criterion is the ***principle of least squares*** which states: *The line should be chosen so that the sum of the squares of the "vertical distances" from the data points to the line is as small as possible.* This line is called the line of ***best fit.*** Thus, the line of best fit for the points in (a) of Figure 2.22 is the one that makes

$$d_1^2 + d_2^2 + d_3^2 + d_4^2$$

as small as possible (see Figure 2.22b).

Since the mathematical theory of least squares is most easily worked out with calculus, we shall simply state the main result and give some examples.

Given n data points

$$(x_1, y_1), \quad (x_2, y_2), \quad \ldots, \quad (x_n, y_n)$$

the line of best fit for the data is

$$y = mx + b$$

where m and b satisfy the system of equations

$$\begin{aligned} nb + (x_1 + x_2 + \cdots + x_n)m &= y_1 + y_2 + \cdots + y_n \\ (x_1 + x_2 + \cdots + x_n)b + (x_1^2 + x_2^2 + \cdots + x_n^2)m & \\ &= x_1y_1 + x_2y_2 + \cdots + x_ny_n \end{aligned} \qquad (6)$$

Example 26 Find the line of best fit for the data

$$(1, 3), \quad (2, 1), \quad (3, 4), \quad (4, 3).$$

Solution Denoting the given data by

$$(x_1, y_1) = (1, 3), \quad (x_2, y_2) = (2, 1),$$
$$(x_3, y_3) = (3, 4), \quad (x_4, y_4) = (4, 3),$$

we obtain

$$x_1 = 1, \quad x_2 = 2, \quad x_3 = 3, \quad x_4 = 4,$$
$$y_1 = 3, \quad y_2 = 1, \quad y_3 = 4, \quad y_4 = 3.$$

Thus

$$x_1 + x_2 + x_3 + x_4 = 1 + 2 + 3 + 4 = 10$$
$$x_1^2 + x_2^2 + x_3^2 + x_4^2 = 1 + 4 + 9 + 16 = 30$$
$$y_1 + y_2 + y_3 + y_4 = 3 + 1 + 4 + 3 = 11$$
$$x_1y_1 + x_2y_2 + x_3y_3 + x_4y_4 = 3 + 2 + 12 + 12 = 29.$$

Moreover, since there are four data points, we have $n = 4$. Substituting these values in (6) gives

$$4b + 10m = 11$$
$$10b + 30m = 29.$$

Using the methods described in the previous section, the student should have no trouble showing that the solution to this system is

$$m = 0.3, \quad b = 2.$$

Thus, the line of best fit is

$$y = 0.3x + 2.$$

The graph of this line and the given data are shown in Figure 2.23.

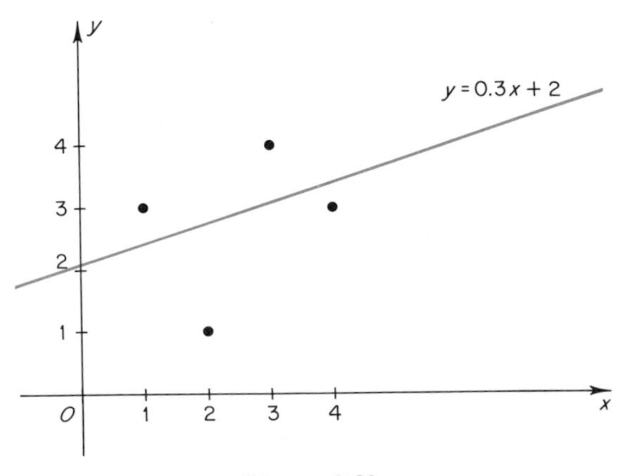

Figure 2.23

Example 27

(a) Find the equation of the line of best fit for the data in Example 25.
(b) Use the equation obtained in part (a) to predict the astronaut's reaction time after 20 hours of weightlessness.

Solution (a) Table 2.4 gives a tabulation of the data needed in Equations (6) to find the line of best fit. Using the sums in this table and $n = 10$, the system of equations in (6) becomes

$$10b + 95m = 14.37$$

$$95b + 985m = 147.63.$$

Solving this system we obtain

$$m = 0.1347 \qquad \text{and} \qquad b = 0.1574.$$

Thus, the line of best fit is

$$y = 0.1347x + 0.1574,$$

which is shown in Figure 2.21b.

(b) To predict the astronaut's reaction time after 20 hours of weightlessness we substitute

$$x = 20$$

in the equation for the line of best fit, obtained in part (a). This yields

$$y = (0.1347)(20) + 0.1574 = 2.8514.$$

Thus, the astronaut's reaction time will be

2.8514 seconds.

Table 2.4

x_i	y_i	x_i^2	x_iy_i
5	0.70	25	3.50
6	1.08	36	6.48
7	1.02	49	7.14
8	1.26	64	10.08
9	1.62	81	14.58
10	1.42	100	14.20
11	1.47	121	16.17
12	1.85	144	22.20
13	2.02	169	26.26
14	1.93	196	27.02
SUM = 95	SUM = 14.37	SUM = 985	SUM = 147.63

EXERCISE SET 2.3

1. Suppose a person borrows $12,000 at the simple interest rate of 7% per year.
   (a) Find a formula for the amount S that he will owe after t years.
   (b) Use this formula to determine how much he will owe after five years.
   (c) Use this formula to determine how much he will owe after 42 months.

2. Draw a Cartesian coordinate system with vertical axis S and horizontal axis t. Sketch the graph of the equation you obtained in Exercise 1a.

3. Suppose a piece of heavy machinery costs $24,000 and depreciates linearly at the rate of 8% per year.
   (a) Find a formula for the value V of the item after t years.
   (b) Use this formula to determine the value V of the item after seven years.
   (c) After how many years will the item be worth $5000?
   (d) After how many years will the item have lost all of its value?

4. Draw a Cartesian coordinate system with vertical axis V and horizontal axis t. Sketch the graph of the equation you obtained in Exercise 3a.

5. Suppose in a biological experiment 64,000 bacteria per unit volume were found in a culture after two hours of growth, and 128,000 were found after five hours of growth. Let N be the number of bacteria per unit volume (in thousands) and let t denote time of growth (in hours). Assuming that N continues to increase with t at a constant rate, find
   (a) an equation relating N and t;
   (b) the graph of the equation in (a) in a Cartesian coordinate system with horizontal axis t and vertical axis N;
   (c) the number of bacteria per unit volume that will be present after nine hours of growth.

6. Suppose that at the present time a communications satellite is handling 1.2 million messages per year. Solve (a) and (b) below assuming that the number N (in millions) of messages will increase at a constant rate of 10% per year. Let t denote time in years and take the present time as $t = 0$.
   (a) Find an equation relating N and t.
   (b) Predict the number of calls that will be handled 12 years from now.

7. There are two common systems for measuring temperature, Centigrade and Fahrenheit. Water freezes at 0° Centigrade and 32° Fahrenheit; it boils at 100° Centigrade and 212° Fahrenheit. Assuming the two are related by a linear equation, show that a temperature of C degrees centigrade corresponds to a temperature of F degrees Fahrenheit by means of the formula

$$F = \frac{9}{5}C + 32.$$

8. Suppose that a taxi fare is 40 cents plus 25 cents for each $\frac{1}{4}$ mile traveled.
   (a) Write an equation for the cost c (in dollars) for traveling x miles.
   (b) Graph the equation in (a) in a Cartesian coordinate system with vertical axis c and horizontal axis x.
   (c) What is the slope of the line in (a).

9. Find the line of best fit for the data points

$$(2, 4), \quad (3, 5), \quad (4, 7), \quad (5, 9).$$

10. Find the line of best fit for the data points

$$(2, 1), \quad (3, 6), \quad (4, 9), \quad (5, 13), \quad (6, 15), \quad (7, 18), \quad (8, 21).$$

11. A department store reports the following sales data over a six year period.

Year	1961	1962	1963	1964	1965	1966
Yearly sales (in hundreds of thousands of dollars)	55.7	68.7	71.3	81.1	82.0	89.0

Let y denote the yearly sales (in hundreds of thousands of dollars) and let the years 1961, 1962, . . ., 1966 be "coded" respectively as 0, 1, 2, . . ., 5. That is, $x = 0$ represents 1961, $x = 1$ represents 1962, and so on.

(a) Find the line of best fit relating x and y.
(b) Use the equation obtained in (a) to estimate the yearly sales in the year 1972 (that is, when $x = 11$).
(c) Explain why it was convenient to "code" the years in this problem.

2.4 LINEAR INEQUALITIES AND THEIR GRAPHS

In the last few sections we studied linear *equations* and their graphs. In this section we shall study linear *inequalities* and their graphs. The ideas we develop in this section will be applied in the next chapter to a variety of important practical problems.

As shown in Figure 2.24, a line divides the plane into two parts called ***half planes.*** Just as the points on the line can be described by means of an equation (or equality), so the points in the half-planes can be described by

Figure 2.24 A line divides the plane into two half-planes.

means of inequalities. To illustrate, consider the line

$$y = x - 1 \tag{1}$$

shown in Figure 2.25a. Call this line L. If $P(x, y)$ is any point on L, then x and y satisfy the equation of L, that is

$$y = x - 1.$$

If A is any point directly above P (see Figure 2.25b), then the y coordinate of A is *larger* than the y coordinate of P so that

$$y > x - 1. \tag{2}$$

Similarly, if B is any point directly below P, then the y coordinate of B is *smaller* than the y coordinate of P so that

$$y < x - 1. \tag{3}$$

Thus the line L consists of all points (x, y) satisfying *equality* (1), while the half-plane above L consists of all points (x, y) satisfying *inequality* (2) and the half-plane below L consists of all points (x, y) satisfying *inequality* (3).

More generally, we can state the following result.

> One half-plane determined by the line
>
> $$ax + by = c$$
>
> is the set of points satisfying
>
> $$ax + by > c$$
>
> and the other half-plane is the set of points satisfying
>
> $$ax + by < c.$$

Example 28 Sketch the set of points satisfying

$$3x + 2y > 6.$$

Solution The set of points we want to sketch is one of the half-planes determined by the line

$$3x + 2y = 6;$$

we have sketched this line in Figure 2.26a.

We must now determine which of the two half-planes is described by the inequality

$$3x + 2y > 6. \tag{4}$$

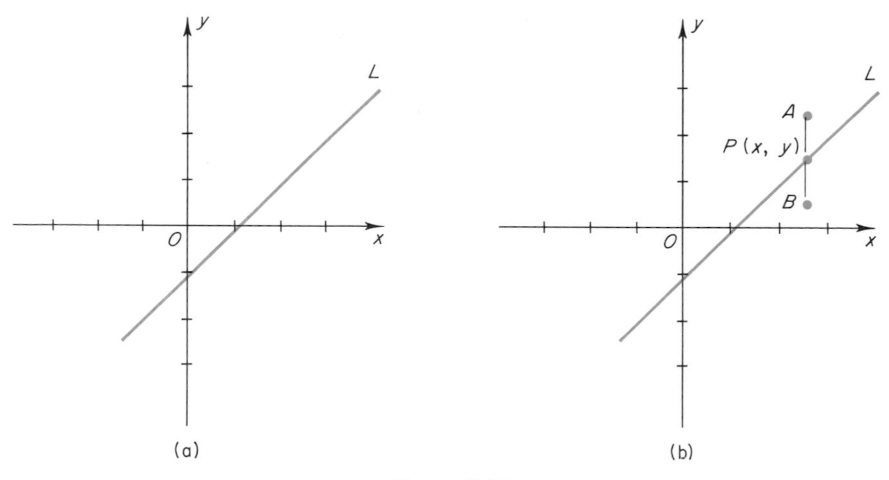

Figure 2.25

To decide, consider the "test point" (0, 0). If we substitute the coordinates $x = 0$, $y = 0$ of the test point into (4), we obtain

$$3(0) + 2(0) > 6$$

or

$$0 > 6,$$

which is false. This tells us that the test point (0, 0) does not lie in the half-plane determined by (4). Thus, the half-plane is the one sketched in Figure 2.26b.

In Figure 2.26b we sketched $3x + 2y = 6$ as a dotted line to emphasize that the points on this line are not included in the shaded region. We also note that we could have used any point not on the line $3x + 2y = 6$ as a test point. We picked (0, 0) merely for convenience.

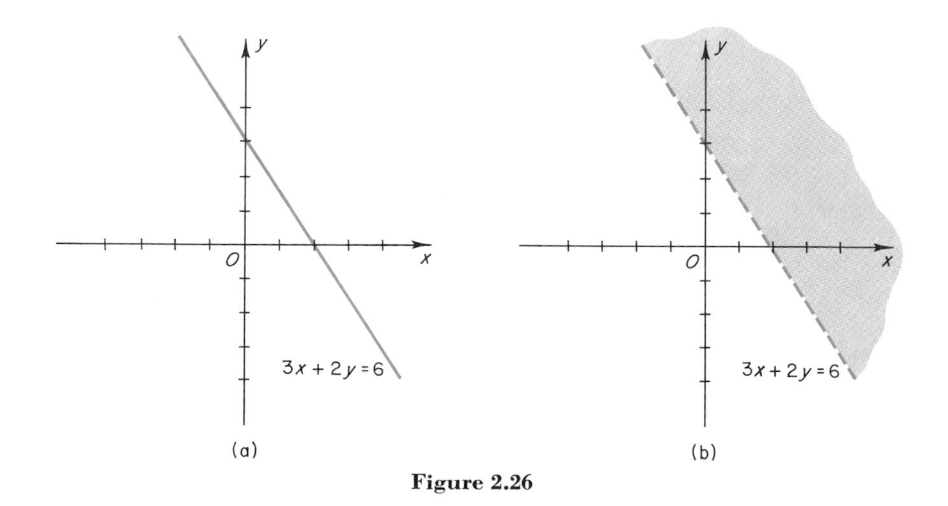

Figure 2.26

Example 29 Sketch the set of points satisfying

$$3x + 2y \geq 6. \tag{5}$$

Solution The points satisying (5) are those points in the half-plane

$$3x + 2y > 6$$

together with those points on the line

$$3x + 2y = 6.$$

The half-plane was sketched in Figure 2.26b and the line in Figure 2.26a, so that the set of points satisfying (5) forms the region shown in Figure 2.27. Observe that the line $3x + 2y = 6$ in Figure 2.27 was sketched as a solid line (rather than dotted) to emphasize that points on this line are part of the shaded region.

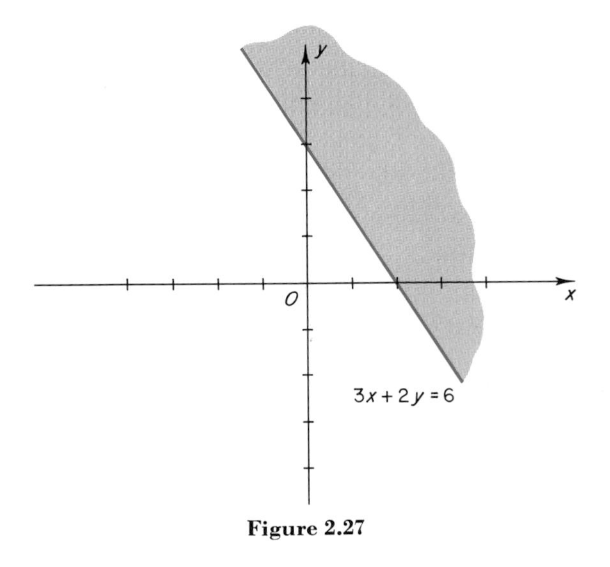

Figure 2.27

Example 30 Sketch the set of points satisfying

$$x + y > 0. \tag{6}$$

Solution The set of points we want to sketch is one of the half-planes determined by the line

$$x + y = 0,$$

which we have sketched in Figure 2.28a. Since (0, 0) lies on this line, we cannot use it as a test point. Instead, we pick any other convenient point, say (1, 1). If we substitute the coordinates $x = 1$, $y = 1$ of the test point into (6) we obtain

$$1 + 1 > 0$$

or

$$2 > 0,$$

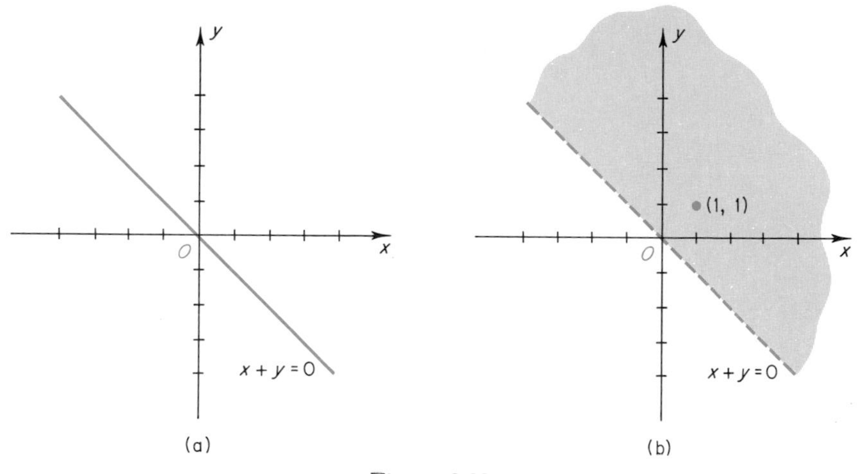

Figure 2.28

which is true. This tells us that the test point lies in the half-plane determined by (6). Thus, the half-plane is the one sketched in Figure 2.28b.

Just as we considered systems of equations of earlier sections, we now consider systems of inequalities.

If we are given two or more inequalities in x and y such as

$$\begin{aligned} 2x + 3y &\geq 6 \\ x + y &\geq 0 \end{aligned} \qquad \text{or} \qquad \begin{aligned} 10x + 15y &\leq 60 \\ x + y &\leq 25 \\ x &\geq 0 \\ y &\geq 0 \end{aligned}$$

we can ask for values of x and y that satisfy all the inequalities. The inequalities are then said to form a ***system*** of inequalities and an ordered pair of real numbers (a, b) is called a ***solution*** of the system if all the inequalities are satisfied when we substitute

$$x = a, \qquad y = b.$$

Example 31 Consider the system

$$\begin{aligned} 2x + 3y &\geq 6 \\ x - y &\leq 0. \end{aligned}$$

The ordered pair $(1, 2)$ is a solution of the system since both inequalities are satisfied when we substitute

$$x = 1, \qquad y = 2.$$

However, the ordered pair $(-3, -2)$ is not a solution since only the second inequality is satisfied when we substitute

$$x = -3, \qquad y = -2.$$

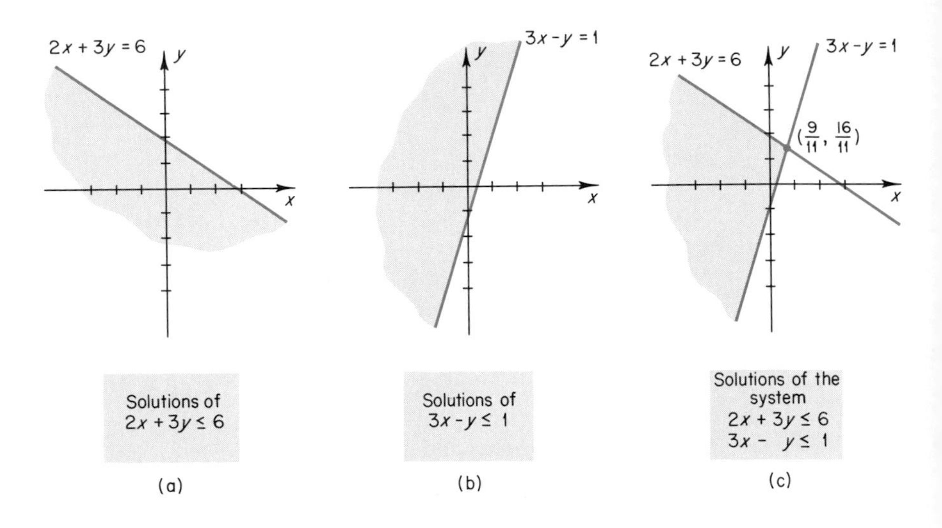

Figure 2.29

Example 32 Sketch the set of solutions of the system

$$\begin{aligned} 2x + 3y &\leq 6 \\ 3x - y &\leq 1. \end{aligned}$$

Solution Using the method illustrated in the preceding examples the student should be able to show that the solutions of the first inequality form the set shaded in Figure 2.29a and the solutions of the second inequality form the set shaded in Figure 2.29b. Since the solutions of the system satisfy both inequalities, the set of solutions of the system is the intersection of the shaded regions in (a) and (b) of Figure 2.29. This set is sketched in Figure 2.29c.

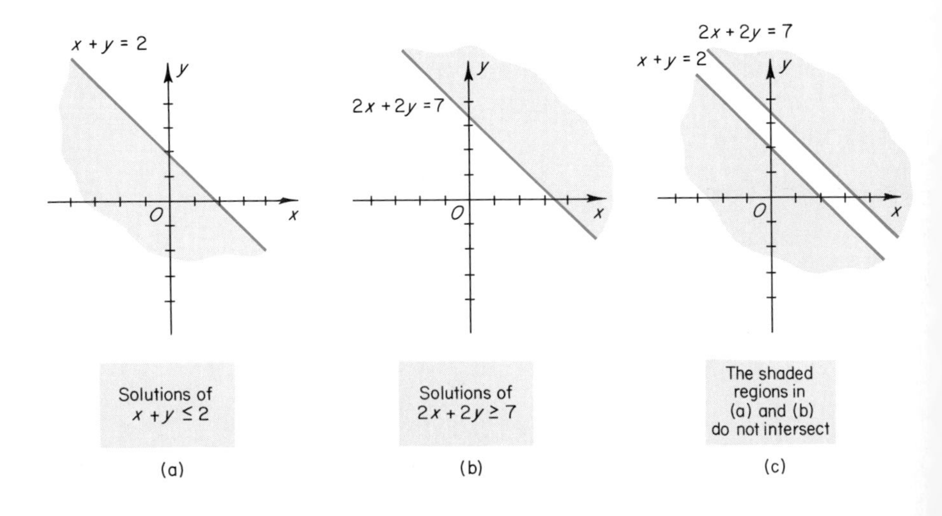

Figure 2.30

Example 33 Sketch the set of solutions of the system

$$x + \ \ y \leq 2$$
$$2x + 2y \geq 7.$$

Solution The solutions of the first inequality are shaded in Figure 2.30a and the solutions of the second inequality are shaded in Figure 2.30b.

Since the solutions of the system satisfy both inequalities, the set of solutions is the intersection of the shaded regions in (a) and (b) of Figure

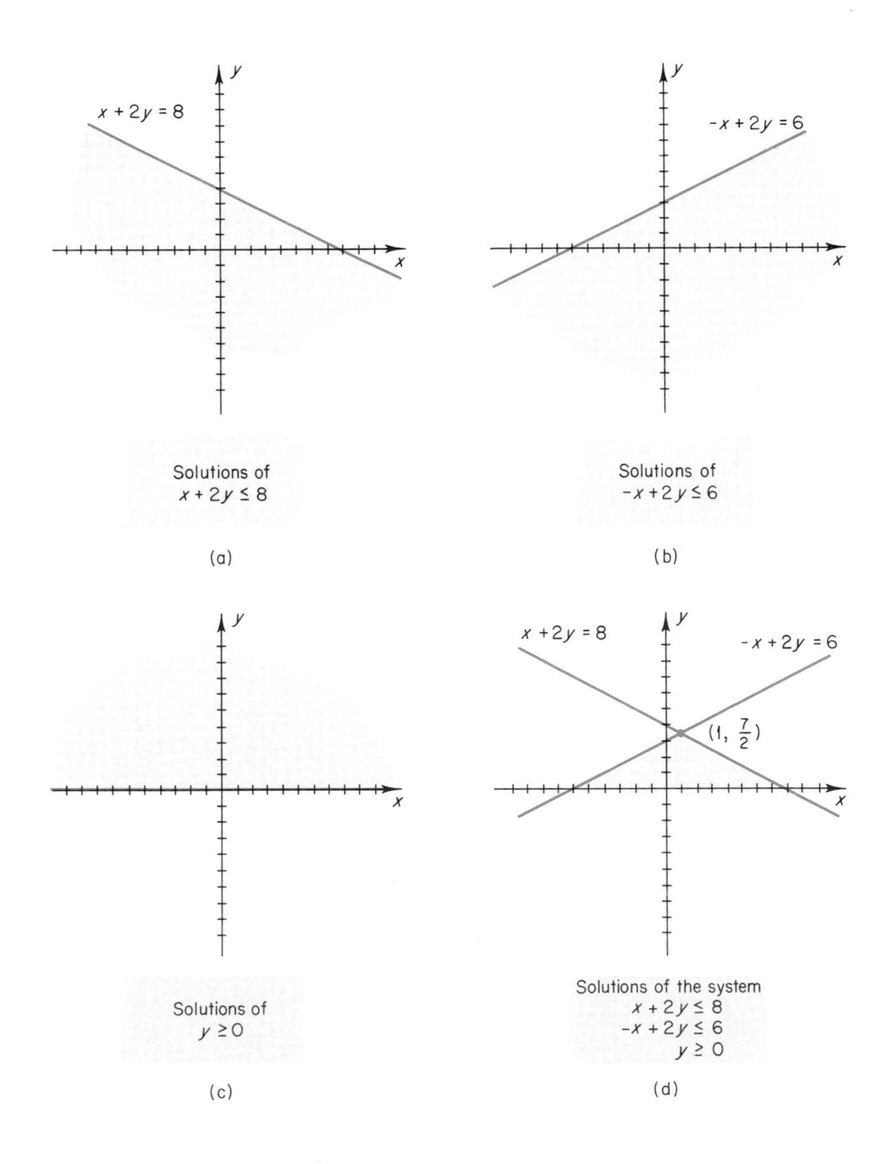

Figure 2.31

2.30. As shown in Figure 2.30c, the intersection of these regions is empty; thus the system has no solutions.

Example 34 Sketch the set of solutions of the system

$$\begin{aligned} x + 2y &\leq 8 \\ -x + 2y &\leq 6 \\ y &\geq 0. \end{aligned}$$

Solution The solutions of each inequality in the system are shown as shaded regions in (a), (b) and (c) of Figure 2.31 on the previous page. The set of solutions of the system is obtained by forming the intersection of these three regions. The set of solutions of the system is the shaded region in Figure 2.31d.

EXERCISE SET 2.4

In each part of Exercises 1–5 sketch the set of points satisfying the given inequality.

1. (a) $3x - 2y < 3$ (b) $3x - 2y \geq 3$.

2. (a) $2x + y \leq 2$ (b) $2x + y \geq 2$.

3. (a) $y > x$ (b) $y \leq x$.

4. (a) $x \geq 2$ (b) $y < -1$.

5. (a) $7x - 8y < 0$ (b) $7x - 8y > 0$.

In Exercises 6–12 sketch the set of points satisfying the given system of inequalities.

6. $$\begin{aligned} x + y &\leq 3 \\ x &\geq 0. \end{aligned}$$

7. $$\begin{aligned} 2x + 3y &< -5 \\ 3x - 2y &> 12. \end{aligned}$$

8. $$\begin{aligned} 2x + 3y &> 4 \\ 4x + 6y &< 4. \end{aligned}$$

9. $$\begin{aligned} 4x + 2y &\leq 5 \\ -x + y &\leq 0 \\ y &\geq 0. \end{aligned}$$

10. $$\begin{aligned} 2x - 7y &\geq 5 \\ 4x + 3y &\leq 3 \\ x - y &\geq 2. \end{aligned}$$

11. $$\begin{aligned} 2x + 3y &< -5 \\ 3x - 2y &> 12 \\ x &\geq 0 \\ y &\geq 0. \end{aligned}$$

12. $$\begin{aligned} x + y &\leq 1 \\ x + y &> 0 \\ 2x + 2y &> 7. \end{aligned}$$

3 LINEAR PROGRAMMING (A GEOMETRIC APPROACH)

In this chapter we give an introduction to a relatively new area of mathematics called linear programming. The importance of this subject was recognized in the late 1940's when it was first used to help solve certain logistics problems for the United States government. Since that time linear programming has been applied to such diverse areas as economics, engineering, biology, agriculture, business, and the social sciences.

Our approach in this introductory chapter will be geometric; as a consequence, the methods we develop will be applicable only to a limited variety of problems. However, in Chapter 5, when we have more machinery to work with, we shall give a more extensive development of linear programming from an algebraic point of view.

3.1 WHAT IS LINEAR PROGRAMMING?

In many problems one is interested in formulating a course of action that will maximize or minimize some quantity of importance. An investor, for example, may want to select investments that will maximize his profits, a nutritionist may want to design a diet with a minimal number of calories, and a businessman may want to select his sources of supply to minimize the shipping distances. It is often the case that the available courses of action are subject to certain constraints. The nutritionist, for example, may have to design a diet to meet certain daily vitamin requirements, the investor may be required to invest 20% of his money in corporate bonds, and the businessman may have to place an order of at least ten truckloads in order to deal with a certain supplier. In problems like these, linear programming helps to select the optimal course of action subject to the given constraints. The following examples illustrate more precisely the kinds of problems to which linear programming is applicable and how these problems can be formulated mathematically.

Production

Example 1 A meat packer wants to blend beef and pork to make two types of frankfurters, regular and deluxe. Suppose that each pound of regular frankfurters contains

0.2 pound of beef
0.2 pound of pork,

while each pound of deluxe frankfurters contains

0.4 pound of beef
0.2 pound of pork.

Suppose also that the profits on the two types of frankfurters are

10¢ per pound on the regular
12¢ per pound on the deluxe.

If the meat packer has 30 pounds of beef in stock and 20 pounds of pork in stock, how many pounds of each kind of frankfurter should he make in order to obtain the largest profit?

As a first step toward finding the solution, we must reformulate the problem in mathematical terms. This can be done as follows. Let x be the number of pounds of regular frankfurters to be made and let y be the number of pounds of deluxe frankfurters. Since the profit on each pound of regular is 10¢ and the profit on each pound of deluxe is 12¢, the total profit z (in cents) will be

$$z = 10x + 12y.$$

Since each pound of regular frankfurters contains 0.2 pound of beef and each pound of deluxe frankfurters contains 0.4 pound of beef, the blend will contain

$$0.2x + 0.4y$$

pounds of beef. Similarly, the blend will contain

$$0.2x + 0.2y$$

pounds of pork. Since the packer can use no more than 30 pounds of beef and no more than 20 pounds of pork, we must have

$$0.2x + 0.4y \leq 30$$

$$0.2x + 0.2y \leq 20.$$

Since x and y cannot be negative, we must also have

$$x \geq 0 \qquad \text{and} \qquad y \geq 0.$$

Therefore, this problem can be reformulated mathematically as follows:

Find those values of x and y that make

$$z = 10x + 12y$$

as large as possible, where x and y must satisfy

$$\begin{aligned} 0.2x + 0.4y &\leq 30 \\ 0.2x + 0.2y &\leq 20 \\ x &\geq 0 \\ y &\geq 0. \end{aligned}$$

In the next section we shall discuss techniques for solving this kind of mathematical problem. For the present, the reader should just concentrate on understanding how the verbal problems are translated into mathematical terms.

Nutrition

Example 2 A nutritionist wants to design a breakfast menu for certain hospital patients. The menu is to include two items, A and B. Suppose that each ounce of A provides

1 unit of iron
2 units of vitamin D,

while each ounce of B provides

2 units of iron
2 units of vitamin D.

Suppose also that the calorie content per ounce for the items is

3 calories per ounce of A
4 calories per ounce of B.

If the breakfast must provide at least

8 units of iron
10 units of vitamin D

how many ounces of each item should be provided in order to meet the iron and vitamin D requirements with the smallest possible intake of calories?

To formulate this problem mathematically, let x be the number of ounces of item A to be provided and let y be the number of ounces of item B. Since there are 3 calories in one ounce of A and 4 calories in one ounce of B the number z of calories provided by the two items will be

$$z = 3x + 4y.$$

Since each ounce of A contains 1 unit of iron and each ounce of B contains 2 units of iron, the two items will provide a total of

$$x + 2y$$

units of iron. Similarly, the two items will provide

$$2x + 2y$$

units of vitamin D.

Thus, to meet iron and vitamin D requirements we must have

$$x + 2y \geq 8$$
$$2x + 2y \geq 10.$$

Since x and y cannot be negative, we must also have

$$x \geq 0 \qquad \text{and} \qquad y \geq 0.$$

Therefore, this problem can be reformulated mathematically as follows:

Find those values of x and y that make

$$z = 3x + 4y$$

as small as possible, where x and y must satisfy

$$x + 2y \geq 8$$
$$2x + 2y \geq 10$$
$$x \geq 0$$
$$y \geq 0.$$

Investment

Example 3 The managers of a pension plan want to invest *up to* \$5000 in two stocks, X and Y. Stock X is considered conservative, while stock Y is considered speculative. The managers agree that the investment in stock X should be

at most \$4000

and that the investment in stock Y should be

at least \$600.

Suppose also that for each \$100 invested

X is expected to return \$8
Y is expected to return \$10.

If the bylaws of the pension plan require that investment in the speculative stock Y can be no greater than one-third of the investment in the conservative stock X, how much should be invested in X and how much in Y to maximize the return on the investment?

To formulate this problem mathematically, let x be the number of dollars to be invested in stock X and let y be the number of dollars to be invested in stock Y. Since stock X will return \$8 for each \$100 invested and stock Y will return \$10 for each \$100 invested, the number z of dollars returned by the two stocks together will be

$$z = 0.08x + 0.1y.$$

Since the managers can invest up to \$5000 in the stocks, we must have

$$x + y \leq 5000.$$

Since the investment in stock X can be no more than \$4000 and since the investment in stock Y must be at least \$600, we obtain

$$x \leq 4000$$

and

$$y \geq 600.$$

Since the investment in Y is required to be no greater than one-third of the investment in X, we also have

$$y \leq \tfrac{1}{3}x.$$

Finally, since x and y must be nonnegative, we have

$$x \geq 0 \qquad \text{and} \qquad y \geq 0.$$

Therefore, this problem can be reformulated mathematically as follows:

Find those values of x and y that make

$$z = 0.08x + 0.1y$$

as large as possible, where x and y must satisfy

$$\begin{aligned} x + y &\leq 5000 \\ x &\leq 4000 \\ y &\geq 600 \\ y &\leq \tfrac{1}{3}x \\ x &\geq 0 \\ y &\geq 0. \end{aligned}$$

Allocation of Resources

Example 4 A fuel manufacturer makes two grades of commercial gasoline, type H (high octane) and type R (regular). Each fuel goes through two processes, cracking and refining. Suppose that each liquid unit of type R requires

0.2 hour for cracking
0.5 hour for refining

while each liquid unit of type H requires

0.4 hour for cracking
0.2 hour for refining.

Suppose also that the profits on the two types of gasoline are

$15.00 per liquid unit of R
$30.00 per liquid unit of H.

If the cracking plant can remain open for at most 8 hours a day and the refining plant can also remain open at most 8 hours a day, how many liquid units of each type of gasoline should be manufactured each day to maximize the manufacturer's profit?

To formulate this problem mathematically, let x be the number of liquid units of type R to be manufactured each day, and let y be the number of liquid units of type H. Since the profit on each unit of R is $15.00 and the profit on each unit of H is $30.00, the total daily profit z (in dollars) will be

$$z = 15x + 30y.$$

Since each unit of type R requires 0.2 hours for cracking and each unit of type H requires 0.4 hours for cracking, the two types of fuel will require a total of

$$0.2x + 0.4y$$

hours each day for cracking. Similarly, the two types of fuel will require a total of

$$0.5x + 0.2y$$

hours for refining. Thus, since the cracking and refining plants can be open at most 8 hours a day, we must have

$$0.2x + 0.4y \leq 8$$

$$0.5x + 0.2y \leq 8.$$

Since x and y cannot be negative, we must also have

$$x \geq 0 \qquad \text{and} \qquad y \geq 0.$$

Therefore, this problem can be reformulated mathematically as follows:

Find those values of x and y that make

$$z = 15x + 30y$$

as large as possible, where x and y must satisfy

$$0.2x + 0.4y \leq 8$$

$$0.5x + 0.2y \leq 8$$

$$x \geq 0$$

$$y \geq 0.$$

Observe that the four examples above have much in common; in each problem we want to find nonnegative values of x and y that maximize or minimize an expression of the form

$$z = ax + by \tag{1}$$

called the ***objective function;*** further, x and y must also satisfy certain other inequalities, called ***constraints.*** Mathematically, the expression $ax + by$ on the right-hand side of (1) and the expressions involving x and y in the inequalities are called *linear* functions of x and y—hence the word *linear* in linear programming. The word *programming* is derived from the early applications of the subject to problems in the programming or allocation of supplies.

EXERCISE SET 3.1

Formulate the followng linear programming problems mathematically. Do not attempt to solve the problems.

1. A manufacturer of lawn products wants to prepare a fertilizer by mixing two ingredients, GROW and THRIVE. Each pound of GROW contains 4 ounces of nitrogen and 4 ounces of phosphate, while each pound of THRIVE contains 10 ounces of nitrogen and 2 ounces of phosphate. The final product must contain at least 100 ounces of nitrogen and at least 60 ounces of phosphate. If each pound of GROW costs $1.00 and each pound of THRIVE costs $1.50, how many pounds of each ingredient should be mixed if the manufacturer wants to minimize his cost?

2. John Smith installs and then demonstrates burglar alarms. There are two burglar alarms that he installs, types A and B. Type A requires 1 hour to install and $\frac{1}{4}$ hour to demonstrate. Type B requires 2 hours to install and $\frac{1}{2}$ hour to demonstrate. Union rules require Smith to work a minimum of 20 hours per week as an installer and a maximum of 20 hours as a demonstrator. If he gets paid $3 per hour for installing and $2 per hour for demonstrating, how many alarms of each type should Smith install and demonstrate each week to maximize his earnings?

3. A hospital wants to design a dinner menu containing two items M and N. Each ounce of M provides 1 unit of vitamin A and 2 units of vitamin B. Each ounce of N provides 1 unit of vitamin A and 1 unit of vitamin B. The two dishes must provide at least 7 units of vitamin A and at least 10 units of vitamin B. If each ounce of M costs 8 cents and each ounce of N costs 12 cents how many ounces of each item should the hospital serve to minimize its cost?

4. Repeat Exercise 3 if each ounce of M costs 12 cents and each ounce of N costs 8 cents.

5. A small furniture finishing plant finishes two kinds of tables, A and B. Each table must be sanded, stained, and varnished. Table A requires 10 minutes of sanding, 6 minutes of staining, and 9 minutes of varnishing. Table B requires 5 minutes of sanding, 12 minutes of staining, and 9 minutes of varnishing. The

profit is \$5 on each A table and \$3 on each B table. If the employees who do the sanding and varnishing will each work at most 450 minutes per day and the employees who do the staining will each work at most 480 minutes per day, how many tables of each type should be made each day to maximize the plant's profit?

6. Repeat Exercise 5 if the profit on each A table is \$3 and each B table is \$5.

7. A shipper has trucks that carry cardboard containers. Each container from the Jones Corp. weighs 5 pounds and is 5 cubic feet in volume. Each container from the Jackson Corp. weighs 6 pounds and is 3 cubic feet in volume. For each trip, a contract requires the shipper to charge the Jones Corp. 30 cents for each container and the Jackson Corp. 40 cents for each container. If the truck cannot carry more than 12,000 pounds and cannot hold more than 9000 cubic feet of cargo, how many containers from each customer should the shipper carry to maximize the profit?

8. Repeat Exercise 7 if we are told that the shipper must carry at least 240 containers of the Jones Corp. on the trip.

9. Holiday Airline Service wants to fly 1000 members of a travel club to Rome. The airline owns two types of planes. Type A can carry 100 passengers and type B can carry 200 passengers. Type A will cost the airline \$10,000 for the trip and type B will cost \$12,000 for the trip. If each airplane requires eight stewardesses and there are only 48 stewardesses available, how many planes of each type should be used to minimize the airline's cost for the trip?

10. An investment banker has funds available for investing. He can purchase a type A bond yielding a 5% return on the amount invested and he can purchase a type C bond yielding a 10% return on the amount invested. His client insists that he invest at least twice as much in A as in C, but no more than \$6000 in A, and no more than \$1500 in C. How much should be invested in each kind of bond to maximize the client's return?

11. A farmer has a 120-acre farm on which he plants two crops: alfalfa and soybeans. For each acre of alfalfa planted, his expenses are \$12.00; and for each acre of soybeans planted, they are \$24.00. Each acre of alfalfa requires 32 bushels of storage and yields a profit of \$25; each acre of soybeans requires 8 bushels of storage and yields a profit of \$35. If the total amount of storage available is 160 bushels and the farmer has only \$1200 on hand, how many acres of soybeans should the farmer plant to maximize his profit?

12. A clothes manufacturer has 10 square yards of cotton material, 10 square yards of wool material, and 6 square yards of silk material. A pair of slacks requires 1 square yard of cotton, 2 square yards of wool, and 1 square yard of silk. A skirt requires 2 square yards of cotton, 1 square yard of wool, and 1 square yard of silk. The net profit on a pair of slacks is \$3 and the net profit on a skirt is \$4. How many skirts and how many slacks should be made to maximize profits?

13. A factory uses two kinds of petroleum products in its manufacturing process, regular (R) and low sulfur (L). Each gallon of R used emits 0.03 pound of sulfur dioxide and 0.01 pound of lead pollutants, while each gallon of L used

emits 0.01 pound of sulfur dioxide and 0.01 pound of lead pollutants. Each gallon of R costs $0.50 and each gallon of L costs $0.60. The factory needs to use at least 100 gallons of the petroleum products each day. If federal pollution regulations allow the factory to emit no more than 6 pounds of sulfur dioxide and no more than 4 pounds of lead pollutants daily, how many gallons of each type of product should the factory use to minimize its costs?

14. A handbag manufacturer makes two types of handbags, patent leather and lizard. The sewing machine limits the total daily production to at most 600 handbags. On the other hand, the mechanical clasp fastener limits production to at most 400 patent leather bags and 500 lizard bags. If the net profit on each patent leather bag is 80 cents and on each lizard bag is $1.00, how many bags of each type should the manufacturer produce daily to maximize his profits?

15. A computer user is planning to buy up to 30 minutes of computer time from the Computer Corp. High priority time (H) provides the customer with more memory than does low priority time (L). The user wishes to buy at least 3 minutes of time H. Computer Corp. will not sell more than 10 minutes of time H and they insist on selling at least 18 minutes of L time. Time H sells for $1000 per minute and time L sells for $600 per minute. How much time H and how much time L should the customer buy to minimize his cost?

16. A pharmaceutical firm develops two drugs, Curine I and Curine II. Each gram of Curine I contains 1 milligram of beneficial factor Z and each gram of Curine II contains 2 milligrams of beneficial factor Z. Both drugs contain factors X and Y which produce undesirable side effects. Each gram of Curine I contains 3 milligrams of X and 1 milligram of Y, while each gram of Curine II contains 1 milligram of X and 1 milligram of Y. If the body cannot tolerate more than 6 milligrams of X nor more than 4 milligrams of Y daily, how many grams of Curine I and how many grams of Curine II should be administered daily to provide the patient with the maximum amount of factor Z?

17. (a) Repeat Exercise 16 if each gram of Curine I contains 3 milligrams of Z and each gram of Curine II contains 1 milligram of Z.
(b) Repeat Exercise 16 if each gram of Curine I contains 4 milligrams of Z and each gram of Curine II contains 1 milligram of Z.

3.2 SOLVING LINEAR PROGRAMMING PROBLEMS GEOMETRICALLY

In the previous section we showed how to formulate several linear programming problems mathematically. In this section we illustrate a geometric method for solving these mathematical problems.

Example 1 in the previous section reduced to the following mathematical problem.

Problem Find those values of x and y that make the objective function

$$z = 10x + 12y \tag{1}$$

as large as possible, where x and y must satisy the constraints

$$\begin{aligned} 0.2x + 0.4y &\leq 30 \\ 0.2x + 0.2y &\leq 20 \\ x &\geq 0 \\ y &\geq 0. \end{aligned} \tag{2}$$

Using the techniques studied in Section 2.5, the reader should have no trouble showing that the points (x, y) satisfying the four inequalities in (2) form the shaded region in Figure 3.1. Since each point (x, y) in this region has a chance of being a solution to the problem, we call these points ***feasible solutions.*** To solve the problem, we must look among the feasible solutions for one that makes the objective function (1) as large as possible. Such a solution is called an ***optimal solution.*** This is, however, easier said than done. To see why, consider the two feasible solutions $D(25, 50)$ and $E(60, 20)$ shown in Figure 3.2. At the point $D(25, 50)$ the value of the objective function is

$$z = 10x + 12y = 10(25) + 12(50) = 850$$

while at the point $E(60, 20)$ the value of the objective function is

$$z = 10x + 12y = 10(60) + 12(20) = 840.$$

Since the value of the objective function is larger at the point $D(25, 50)$ than at the point $E(60, 20)$, E cannot be an optimal solution. However, we cannot be sure that D itself is an optimal solution since there may be other feasible solutions at which the objective function has an even larger value than at D. Since there are infinitely many feasible solutions, it is impossible to compute z at each one of them to see where it has the largest value. We

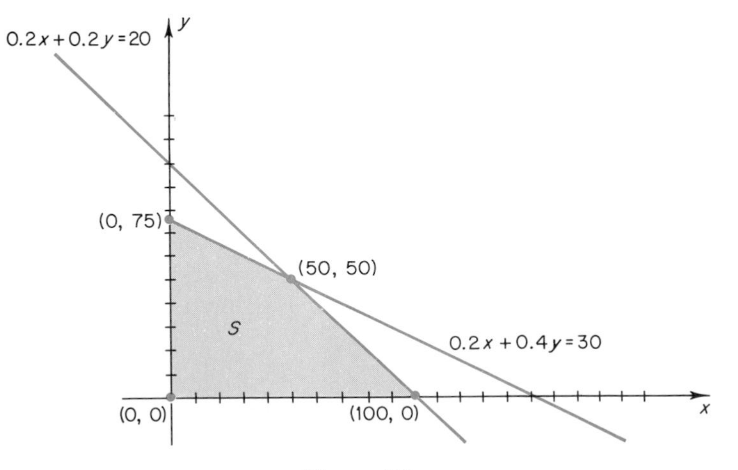

Figure 3.1

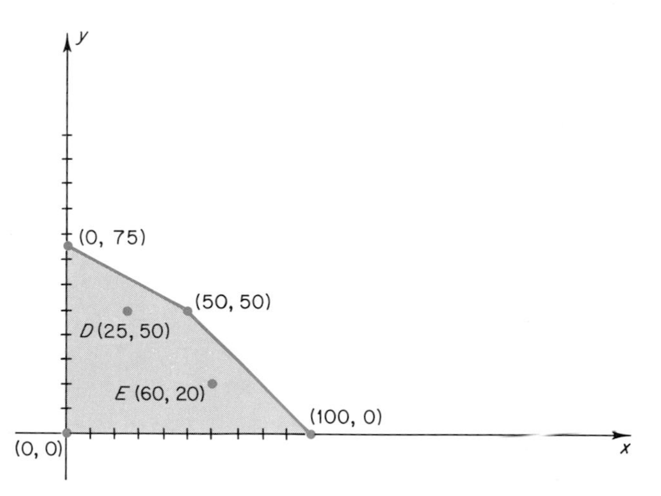

Figure 3.2

shall therefore need some techniques that avoid this difficulty. Before we will be able to describe these techniques we shall need to discuss some preliminary ideas.

The set of feasible solutions shaded in Figure 3.1 is an example of a ***convex set.*** This means that if we connect any two points in the set by a line segment, then the line segment will be completely in the set.

Example 5 The shaded set in Figure 3.3a is *not* convex since the line segment connecting the points P_1 and P_2 does not lie entirely in the set. The shaded set in Figure 3.3b *is* convex since, as illustrated in the figure, a line segment connecting *any* two points of the set will lie in the set.

Convex sets are of two types: ***bounded*** and ***unbounded.*** A bounded set is one that can be enclosed by a circle, and an unbounded set is one that cannot be so enclosed. To illustrate, the convex sets (a) and (b) in Figure 3.4 are bounded, while the convex sets (c) and (d) are unbounded.

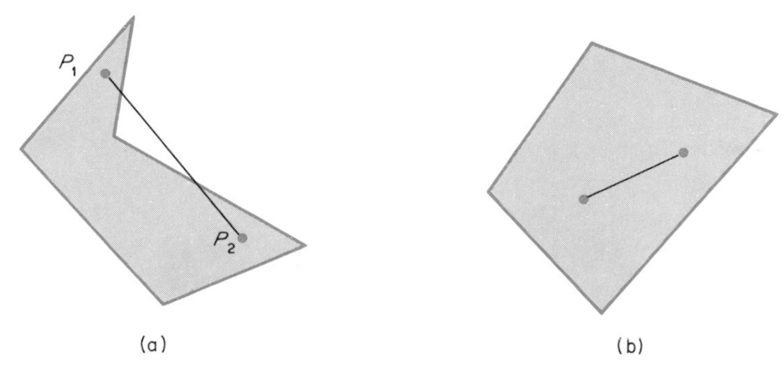

Figure 3.3

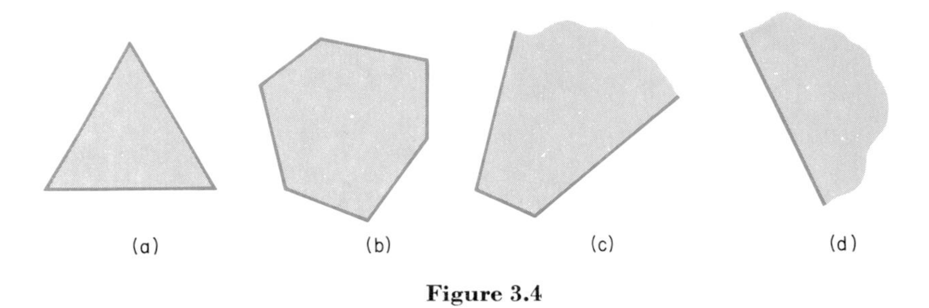

Figure 3.4

We shall also need the following idea.

> A ***corner point*** in a convex set is any point in the set that is the intersection of two boundary lines.

Example 6 In Figure 3.5, we have indicated the corner points of the convex sets in parts (a), (b), and (c) of Figure 3.4. The convex set in part (d) of Figure 3.4 has no corner points.

Example 7 The corner points of the set of feasible solutions shaded in Figure 3.1 are (verify):

$$(0,\ 0) \qquad (0,\ 75) \qquad (50,\ 50) \qquad (100,\ 0).$$

Our interest in convex sets and corner points stems from the following result.

> The set S of feasible solutions to a linear programming problem is convex. Further, if S is bounded, then the objective function
>
> $$z = ax + by$$
>
> has both a largest value and a smallest value on S, and these values occur at corner points of S. If S is unbounded, there may not be a largest or smallest value on S. However, if a largest or smallest value on S exists, it must occur at a corner point.

This result suggests the following procedure for solving linear programming problems when the set of feasible solutions is bounded.

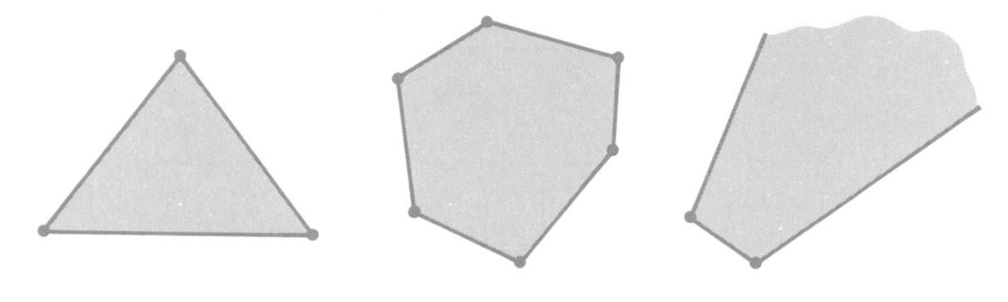

• denotes a corner point

Figure 3.5

Step 1 Determine the set of feasible solutions.

Step 2 Find the corner points of the set.

Step 3 Evaluate the objective function

$$z = ax + by$$

at each corner point to determine where the largest or smallest values of z occur.

Example 8 The problem in Example 1 of the previous section reduced to finding values of x and y that make

$$z = 10x + 12y$$

as large as possible, where x and y must satisfy

$$\begin{aligned} 0.2x + 0.4y &\leq 30 \\ 0.2x + 0.2y &\leq 20 \\ x &\geq 0 \\ y &\geq 0. \end{aligned}$$

The set of feasible solutions was sketched in Figure 3.1 and the corner points

$$(0, 0) \qquad (0, 75) \qquad (50, 50) \qquad (100, 0)$$

were determined in Example 7. Thus, to find values of x and y that satisfy the constraints and make z as large as possible, we must first evaluate z at each corner point. These values are listed in Table 3.1. From this table, the largest value of z is 1100 and this value occurs when

$$x = 50 \qquad \text{and} \qquad y = 50.$$

Table 3.1

Corner point (x, y)	Value of $z = 10x + 12y$
(0, 0)	0
(0, 75)	900
(50, 50)	1100
(100, 0)	1000

We have now solved the problem in Example 1. The meat packer should make $x = 50$ pounds of regular frankfurters and $y = 50$ pounds of deluxe frankfurters, in which case his profit $z = 1100¢ = \$11.00$ will be as large as possible.

Example 9 Solve the problem in Example 3.

Solution In the previous section we reformulated this problem mathematically as follows: Find those values of x and y that make the objective function

$$z = 0.08x + 0.1y \tag{3}$$

as large as possible, where x and y must satisfy

$$\begin{aligned} x + y &\leq 5000 \\ x &\leq 4000 \\ y &\geq 600 \\ y &\leq \tfrac{1}{3}x \\ x &\geq 0 \\ y &\geq 0. \end{aligned} \tag{4}$$

Using the methods studied in Section 2.5, the reader should be able to show that the points (x, y) satisfying the six inequalities in (4) form the shaded region S in Figure 3.6.

The region S is bounded so that the maximum value of the objective function (3) will occur at one of the corner points of S. The values of the

Table 3.2

Corner point (x, y)	Value of $z = 0.08x + 0.1y$
(1800, 600)	204
(4000, 600)	380
(4000, 1000)	420
(3750, 1250)	425

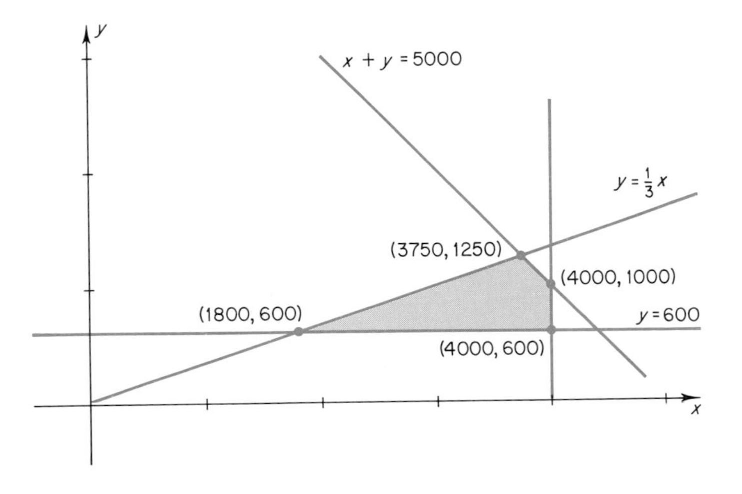

Figure 3.6

objective function at the corner points are listed in Table 3.2. From this table, the largest value z is 425, and this value occurs when

$$x = 3750 \qquad \text{and} \qquad y = 1250.$$

Thus, the managers of the pension plan should invest \$3750 in stock X and \$1250 in stock Y, in which case, their return of \$425 will be as large as possible.

Observe from Table 3.2 that the poorest possible investment would occur if the managers invested \$1800 in stock X and \$600 in stock Y, in which case their return of \$204 would be the smallest possible.

Example 10 We previously stated that if the set S of feasible solutions is bounded, then the objective function in a linear programming problem has both a largest and a smallest value on S, and these values occur at corner points of S. In this example we show that if S is not bounded, then this result does not apply.

Consider the problem: Find those values of x and y that make

$$z = 3x + 4y$$

as large as possible, where x and y satisfy

$$\begin{aligned} x + 2y &\geq 8 \\ 2x + 2y &\geq 10 \\ x &\geq 0 \\ y &\geq 0. \end{aligned}$$

Using the techniques of Section 2.5, the reader should have no trouble obtaining the set S of feasible solutions shaded in Figure 3.7. Observe

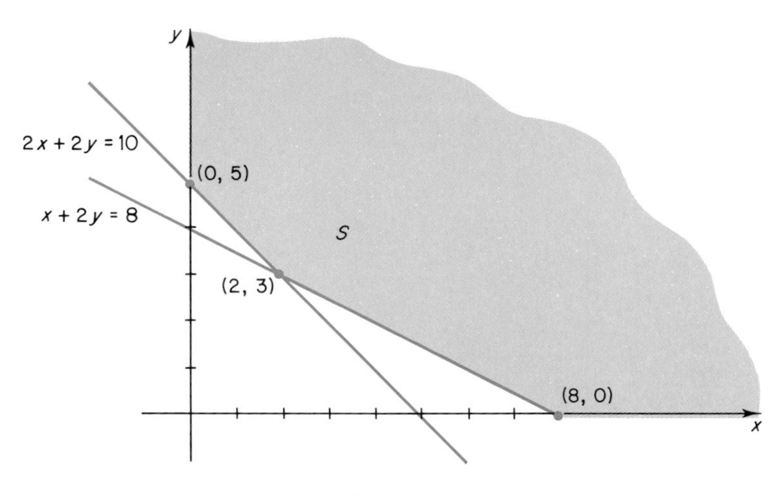

Figure 3.7

that S is an unbounded set. In Table 3.3 we have listed the values of z at the corner points of S. The largest value of z in this table is $z = 24$, which occurs at the point (8, 0). However, this is not the largest possible value of z. For example,

$$z = 3x + 4y$$

has the value $z = 38$ at the point (10, 2). In fact, z does not have a largest value on S. (Can you see why?)

This example illustrates what can happen when the set of feasible solutions is unbounded. In such cases, the objective function can fail to have a largest value, a smallest value, or both. It can be shown, however, that if there is a largest (or smallest) value for the objective function, it must occur at a corner point.

Table 3.3

Corner point (x, y)	Value of $z = 3x + 4y$
(0, 5)	20
(2, 3)	18
(8, 0)	24

Example 11 Solve the problem in Example 4.

Solution In the previous section, we reformulated this problem mathematically as follows. Find those values of x and y that make

$$z = 15x + 30y \tag{5}$$

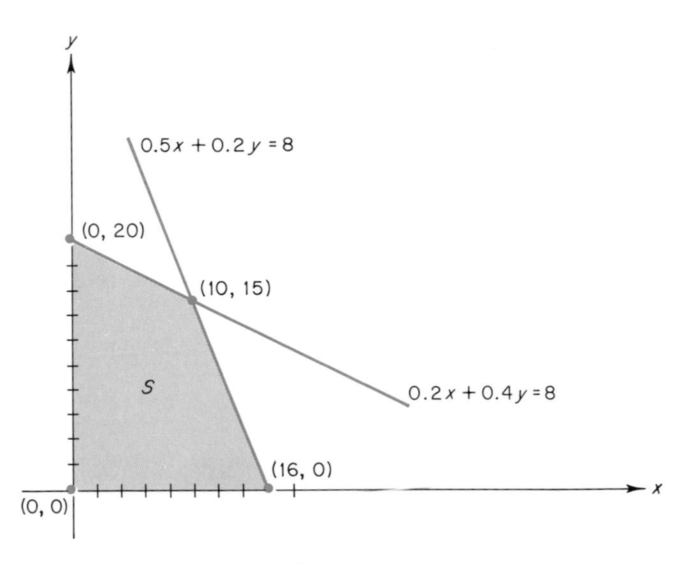

Figure 3.8

as large as possible, where x and y must satisfy

$$\begin{aligned} 0.2x + 0.4y &\leq 8 \\ 0.5x + 0.2y &\leq 8 \\ x &\geq 0 \\ y &\geq 0. \end{aligned} \tag{6}$$

Using the methods of Section 2.5, the reader should be able to show that the points (x, y) satisfying the four inequalities in (6) form the shaded region S in Figure 3.8.

The region S is bounded so that the maximum value of the objective function (5) will occur at one of the corner points of S. The values of the objective function at the corner points are listed in Table 3.4. From this table, the largest value of z is 600, and this value occurs at two different corner points

$$x = 10, \qquad y = 15$$

and

$$x = 0, \qquad y = 20.$$

Table 3.4

Corner point (x, y)	Value of $z = 15x + 30y$
(0, 0)	0
(16, 0)	240
(10, 15)	600
(0, 20)	600

Thus, the fuel manufacturer can maximize his profit in two ways; he can manufacture

$$x = 10 \quad \text{liquid units of type R}$$

and

$$y = 15 \quad \text{liquid units of type H}$$

or he can manufacture

$$x = 0 \quad \text{liquid units of type R}$$

and

$$y = 20 \quad \text{liquid units of type H}$$

each day. In either case, his daily profit of \$600 will be as large as possible.

This example shows that a linear programming problem can have more than one solution. In fact, we shall now show that every point on the line segment connecting the corner points (10, 15) and (0, 20) is also a solution. Using the methods discussed in Section 2.2, the reader should be able to show that the equation of the line passing through the points (0, 20) and (10, 15) is

$$15x + 30y = 600.$$

Comparing this to (5), we see that

$$z = 600$$

at each point on this line. Thus, every point on the line segment connecting (0, 20) and (10, 15) is an optimal solution.

We know that optimal solutions, when they exist, occur at corner points. This example shows that it is also possible to have optimal solutions that do not occur at corner points. However, since the value of the objective function is the same for every optimal solution, it is satisfactory to look for just those optimal solutions that occur at corner points.

The geometric method of solving linear programming problems is quite adequate for problems involving two unknowns. For problems involving three unknowns, a geometric method is applicable, but harder to apply. For most problems involving more than three unknowns, geometric methods are usually not applicable. In Chapter 5 we shall discuss algebraic methods that can be used to solve problems involving any number of unknowns.

EXERCISE SET 3.2

1. For which values of x and y is $z = 18x + 12y$ a maximum subject to

$$x + 3y \leq 12$$
$$x + 2y \leq 9$$
$$x \geq 0, \quad y \geq 0?$$

2. For which values of x and y is $z = 15x - 10y$ a minimum subject to

$$x + 3y \leq 8$$
$$x + 2y \leq 7$$
$$x \geq 0, \quad y \geq 0?$$

3. For which values of x and y is $z = 2x + 3y$ a maximum subject to

$$5x + 2y \geq 9$$
$$6x - 2y \leq 2$$
$$y \leq 6$$
$$x \geq 0$$
$$y \geq 0?$$

4. For which values of x and y is $z = 2x + 3y$ a minimum subject to the constraints of Exercise 3?

5. (a) For which values of x and y is $z = 3x - y$ a maximum subject to the constraints of Exercise 3?
(b) For which values of x and y is it minimum?

For these exercises, solve the indicated problem from Exercises 1–17 of Section 3.1:

6. Exercise 2 **7.** Exercise 5 **8.** Exercise 6
9. Exercise 7 **10.** Exercise 8 **11.** Exercise 9
12. Exercise 10 **13.** Exercise 11 **14.** Exercise 12
15. Exercise 13 **16.** Exercise 14 **17.** Exercise 15
18. Exercise 16 **19.** Exercise 17

4

MATRICES AND LINEAR SYSTEMS

In Section 2.2, we discussed a method for solving systems of two linear equations in two unknowns. In this chapter we shall study systems which may involve more than two equations or more than two unknowns. We shall also develop a new concept (matrices), which has important applications in business, economics, as well as the behavioral and social sciences. We shall study some of these applications both here and in later chapters.

4.1 LINEAR SYSTEMS

Recall from Section 2.2 that an equation of the form

$$ax + by = c$$

represents a straight line in the x, y plane. Thus, such an equation is called a linear equation in the variables x and y. In general, an equation expressible in the form

$$a_1x_1 + a_2x_2 + \cdots + a_nx_n = c,$$

where $a_1, a_2, \ldots, a_n$, and c are constants is called a ***linear equation*** in the variables $x_1, x_2, \ldots, x_n$.

Example 1 The following are linear equations:

$$2x + 3y = 4 \qquad -2x + y + z = 6$$

$$x = 6 \qquad 2x_1 - \tfrac{3}{2}x_2 + 5x_3 + x_4 = 8.$$

Observe that a linear equation does not contain any products or divisions by variables, and all variables occur only to the first power. The following are not linear equations:

$$2xy + 3z = 5 \qquad 2x^2 + 3y = 4$$

$$\sqrt{x_1} + 2x_2 - x_3 = 7 \qquad \frac{1}{x} - 2y + 3y = 9.$$

Application to Production Problems

Example 2 Each day, a pharmaceutical firm produces 100 ounces of a perishable ingredient, called alpha, which is used in the manufacture of drugs A, B, C, and D. These drugs require the following amounts of alpha in their manufacture:

Each ounce of A requires 0.1 ounce of ingredient alpha.
Each ounce of B requires 0.3 ounce of ingredient alpha.
Each ounce of C requires 0.5 ounce of ingredient alpha.
Each ounce of D requires 0.2 ounce of ingredient alpha.

To prevent any waste of the perishable ingredient alpha, the firm wants to produce x_1 ounces of A, x_2 ounces of B, x_3 ounces of C, and x_4 ounces of D each day so that the entire 100 ounces of ingredient alpha is used up. For this to happen, x_1, x_2, x_3, and x_4 must satisfy the condition

$$0.1x_1 + 0.3x_2 + 0.5x_3 + 0.2x_4 = 100, \tag{1}$$

which is a linear equation in x_1, x_2, x_3, and x_4.

A set of linear equations in the variables $x_1, x_2, \ldots, x_n$ is called a ***system of linear equations*** or more briefly a ***linear system*** in the unknowns $x_1, x_2, \ldots, x_n$. A ***solution*** to a linear system in $x_1, x_2, \ldots, x_n$ is a sequence of n numbers $s_1, s_2, \ldots, s_n$ such that each equation is satisfied when we substitute

$$x_1 = s_1, \quad x_2 = s_2, \quad \ldots, \quad x_n = s_n.$$

For a system with $n = 2$ variables, we shall sometimes write the variables as x and y rather than x_1 and x_2; similarly for a system with $n = 3$ variables, we shall sometimes write the variables as x, y, and z, rather than x_1, x_2, and x_3.

Example 3 The linear system

$$\begin{aligned} 2x + 3y + z &= 2 \\ -3x + 2y - 5z &= -7 \\ 5x + 2z &= -4 \end{aligned} \tag{2}$$

is a system of three equations in the unknowns x, y, and z. The sequence

$$x = -2, \qquad y = 1, \qquad z = 3$$

is a solution to (2), as can be verified by substituting these values into every equation in (2). On the other hand, the sequence

$$x = 2, \qquad y = -1, \qquad z = 1$$

is not a solution to (2) since these values do not satisfy the second and third equations.

Example 4 The linear system

$$\begin{aligned} 2x_1 + 3x_2 - x_3 &= 1 \\ 3x_1 + 2x_2 + x_3 &= -1 \end{aligned} \tag{3}$$

is a system of two equations in the unknowns x_1, x_2, and x_3. The sequence

$$x_1 = -2, \qquad x_2 = 2, \qquad x_3 = 1$$

is a solution to the system, and the sequence

$$x_1 = -3, \qquad x_2 = 3, \qquad x_3 = 2$$

is another solution. On the other hand, the sequence

$$x_1 = -2, \qquad x_2 = 3, \qquad x_3 = 4$$

is not a solution to (3) since it does not satisfy the second equation.

Example 5 (Application to Production Problems) Let us expand on the production problem discussed in Example 2. In addition to the ingredient alpha discussed in Example 2, suppose the drugs A, B, C, and D also require the following amounts of a second ingredient, beta, in their manufacture:

Each ounce of A requires 0.4 ounce of ingredient beta.
Each ounce of B requires 0.2 ounce of ingredient beta.
Each ounce of C requires 0.3 ounce of ingredient beta.
Each ounce of D requires 0.8 ounce of ingredient beta.

If the manufacturer makes 300 ounces of ingredient beta each day, and if he wants to produce x_1 ounces of A, x_2 ounces of B, x_3 ounces of C, and x_4 ounces of D each day, so that the entire 300 ounces of ingredient beta is used up, then x_1, x_2, x_3, and x_4 must satisfy the condition

$$0.4x_1 + 0.2x_2 + 0.3x_3 + 0.8x_4 = 300, \tag{4}$$

which is a linear equation in x_1, x_2, x_3, and x_4. Thus, if ingredients alpha and beta are both to be used up, conditions (1) and (4) must both be satisfied. In other words, x_1, x_2, x_3, x_4 must be a solution of the linear system

$$\begin{aligned} 0.1x_1 + 0.3x_2 + 0.5x_3 + 0.2x_4 &= 100 \\ 0.4x_1 + 0.2x_2 + 0.3x_3 + 0.8x_4 &= 300. \end{aligned}$$

There is more than one solution to this system. For example, the firm could produce

$$x_1 = 305, \qquad x_2 = 15, \qquad x_3 = 50, \qquad x_4 = 200$$

ounces of A, B, C, and D respectively, or it could produce

$$x_1 = 104, \qquad x_2 = 32, \qquad x_3 = 40, \qquad x_4 = 300$$

ounces of A, B, C, and D respectively.

Not every linear system has a solution. For example, if we multiply the first equation of the system

$$\begin{aligned} 2x + 3y - z &= 4 \\ 4x + 6y - 2z &= 7 \end{aligned} \tag{5}$$

by 2, system (5) becomes

$$\begin{aligned} 4x + 6y - 2z &= 8 \\ 4x + 6y - 2z &= 7. \end{aligned}$$

It is now evident that there is no solution since these equations contradict each other.

A linear system that has no solutions is said to be ***inconsistent;*** if it has at least one solution, it is called ***consistent.***

In Section 2.2 we showed that a system of two linear equations

$$\begin{aligned} a_1x + b_1y &= c_1 \\ a_2x + b_2y &= c_2 \end{aligned}$$

has either no solutions, exactly one solution, or infinitely many solutions (see Figure 2.19). Although we shall omit the proof, it can be shown that this same result holds for any linear system. In other words, either a linear system is inconsistent (has no solutions) or it is consistent, in which case it has one solution or infinitely many solutions.

Given a system of linear equations, we shall want to decide whether or not the system is consistent; if it is consistent we shall want to *solve* the system, that is, find its solutions. In Section 2.2 we discussed a method for solving systems of two linear equations in two unknowns. We shall now develop a method that can be used to solve any linear system. The method we shall use is based on the idea of replacing the given system by a new system that has the same solutions, but is easier to solve. This new system is obtained in a series of steps, by applying the following operations:

1. Interchange two equations.
2. Multiply an equation by a nonzero constant. (6)
3. Add a multiple of one equation to another equation.

The following example illustrates how these operations can be used to solve a linear system. In reading this example, the reader should not worry about how the steps were selected (we shall discuss this later). Instead he should concentrate on following the computations.

Example 6 Consider the system

$$\begin{aligned} x + y - 2z &= -3 \\ 2x + 3y + z &= 10 \\ 3x - y + 2z &= 11. \end{aligned} \tag{7}$$

To solve this system, first add -2 times the first equation to the second equation (to eliminate x from the second equation). This yields the new system

$$\begin{aligned} x + y - 2z &= -3 \\ y + 5z &= 16 \\ 3x - y + 2z &= 11. \end{aligned}$$

Next, add -3 times the first equation to the third equation (to eliminate x from the third equation). This yields the new system

$$\begin{aligned} x + y - 2z &= -3 \\ y + 5z &= 16 \\ -4y + 8z &= 20. \end{aligned}$$

Next, add -1 times the second equation to the first equation, and add 4 times the second equation to the third equation (to eliminate y from the first and third equations). This yields

$$\begin{aligned} x \quad - 7z &= -19 \\ y + 5z &= 16 \\ 28z &= 84. \end{aligned}$$

Multiply the third equation by $\frac{1}{28}$ to obtain

$$\begin{aligned} x \quad - 7z &= -19 \\ y + 5z &= 16 \\ z &= 3. \end{aligned}$$

Finally, eliminate z from the first and second equations by adding 7 times the third equation to the first equation and -5 times the third equation to the second equation. This gives the simple system

$$\begin{aligned} x \quad &= 2 \\ y \quad &= 1 \\ z &= 3. \end{aligned}$$

Thus

$$x = 2, \qquad y = 1, \qquad z = 3$$

is the solution to system (7).

If we mentally keep track of the location of the +'s, the unknowns, and the ='s, it is possible to abbreviate a system of linear equations by writing only a rectangular array of numbers, called a ***matrix***. For example, the system

$$\begin{aligned} x + y - 2z &= -3 \\ 2x + 3y + z &= 10 \\ 3x - y + 2z &= 11 \end{aligned}$$

can be abbreviated by writing

$$\begin{bmatrix} 1 & 1 & -2 & -3 \\ 2 & 3 & 1 & 10 \\ 3 & -1 & 2 & 11 \end{bmatrix}.$$

This array is called the ***augmented matrix***, for the system. When forming the augmented matrix, the system should be written so that like unknowns are in the same columns and only the constants appear to the right of the equal signs.

Example 7 The augmented matrix

$$\begin{bmatrix} 3 & -1 & 5 & 2 & 3 \\ -2 & 3 & 4 & 1 & -5 \\ 3 & 2 & -1 & 4 & 6 \end{bmatrix}$$

corresponds to the linear system

$$\begin{aligned} 3x_1 - x_2 + 5x_3 + 2x_4 &= 3 \\ -2x_1 + 3x_2 + 4x_3 + x_4 &= -5 \\ 3x_1 + 2x_2 - x_3 + 4x_4 &= 6. \end{aligned}$$

Since the rows (horizontal lines) of an augmented matrix correspond to the equations in the associated linear system, the three operations listed in (6) correspond to the following operations on the rows of the augmented matrix:

1. Interchange two rows.
2. Multiply a row by a nonzero constant. (8)
3. Add a multiple of one row to another row.

The next example illustrates how these row operations on an augmented matrix can be used to solve a linear system.

Example 8 In this example we shall show how to solve the system

$$\begin{aligned} 3x - 4y &= 15 \\ x + 5y &= -14 \end{aligned}$$

by applying the row operations in (8) to the augmented matrix

$$\begin{bmatrix} 3 & -4 & 15 \\ 1 & 5 & -14 \end{bmatrix}.$$

For illustrative purposes, we shall also solve the system by operating on the equations in the system, so the reader can see the two methods side by side.

System	*Augmented Matrix*
$\begin{aligned} 3x - 4y &= 15 \\ x + 5y &= -14 \end{aligned}$	$\begin{bmatrix} 3 & -4 & 15 \\ 1 & 5 & -14 \end{bmatrix}$
Interchange the first and second equations to obtain	Interchange the first and second rows to obtain
$\begin{aligned} x + 5y &= -14 \\ 3x - 4y &= 15 \end{aligned}$	$\begin{bmatrix} 1 & 5 & -14 \\ 3 & -4 & 15 \end{bmatrix}$
Add -3 times the first equation to the second to obtain	Add -3 times the first row to the second to obtain
$\begin{aligned} x + 5y &= -14 \\ -19y &= 57 \end{aligned}$	$\begin{bmatrix} 1 & 5 & -14 \\ 0 & -19 & 57 \end{bmatrix}$
Multiply the second equation by $-\frac{1}{19}$ to obtain	Multiply the second row by $-\frac{1}{19}$ to obtain
$\begin{aligned} x + 5y &= -14 \\ y &= -3 \end{aligned}$	$\begin{bmatrix} 1 & 5 & -14 \\ 0 & 1 & -3 \end{bmatrix}$
Add -5 times the second equation to the first to obtain	Add -5 times the second row to the first to obtain
$\begin{aligned} x \quad &= 1 \\ y &= -3 \end{aligned}$	$\begin{bmatrix} 1 & 0 & 1 \\ 0 & 1 & -3 \end{bmatrix}$

The solution is therefore $x = 1$, $y = -3$.

EXERCISE SET 4.1

1. Which of the following are linear equations?
   (a) $\frac{5}{6}x + 3y - z = 6$ (b) $0.8x^2 + 2y = 5$
   (c) $\sqrt{x_1} + \frac{3}{2}x_2 + 3x_3 - x_4 = 8$ (d) $-3x + 2y + 5z = 16$.
2. Which of the following are linear equations?
   (a) $2x_1 + 3x_2 - \frac{7}{8}x_3 + x_4 + \frac{3}{2}x_5 = 2$ (b) $2x + xy + z = -8$
   (c) $\frac{2}{7}x = 12$ (d) $2x^2 + y - 3z = 7$.

3. Which of the following are linear systems?

(a) $\begin{aligned} 2x + 3y &= 8 \\ \tfrac{3}{2}x - 3y^2 &= -8 \end{aligned}$

(b) $\begin{aligned} 4x - 5y + 4z &= 18 \\ \tfrac{3}{2}x + 7y + 2z &= 6 \end{aligned}$

(c) $\begin{aligned} \tfrac{2}{5}x + \tfrac{4}{7}y + z &= 12 \\ 2x - y + z &= \tfrac{8}{7} \\ x + 3\sqrt[3]{y} + 2z &= 6 \end{aligned}$

(d) $\begin{aligned} 5x_1 + 3x_2 + 2x_3 &= 6 \\ -2x_1 + 2x_2 + \tfrac{7}{5}x_3 &= 6 \\ 3x_1 + 2x_2 &= 4 \\ \tfrac{4}{5}x_1 &= 15. \end{aligned}$

4. Consider the linear system

$$\begin{aligned} 3x + y + z &= -4 \\ 3x + 2y - z &= 4. \end{aligned}$$

Which of the following are solutions?

(a) $x = -5,\ y = 10,\ z = 1$

(b) $x = 1,\ y = 1,\ z = -8$

(c) $x = -2,\ y = 4,\ z = -2.$

5. Consider the linear system

$$\begin{aligned} 3x_1 + 4x_2 + x_3 - 17x_4 &= 15 \\ 3x_1 + 5x_2 + x_3 - 20x_4 &= 20 \\ 2x_1 + 3x_2 + x_3 - 12x_4 &= 13. \end{aligned}$$

Which of the following are solutions?

(a) $x_1 = -2,\ x_2 = 6,\ x_3 = 5,\ x_4 = 1$

(b) $x_1 = -1,\ x_2 = 8,\ x_3 = 3,\ x_4 = 1$

(c) $x_1 = -3,\ x_2 = 5,\ x_3 = 4,\ x_4 = 0.$

6. Write the augmented matrix of each of the following linear systems.

(a) $\begin{aligned} 2x_1 + 3x_2 - x_3 &= 8 \\ 3x_1 + x_2 + x_4 &= 6 \end{aligned}$

(b) $\begin{aligned} x + 2y - 3z &= 6 \\ -2x + z &= -6 \\ 3x + 4y + z &= 8 \end{aligned}$

(c) $\begin{aligned} 4x_1 + 2x_2 + 3x_3 &= 7 \\ 5x_1 + 6x_2 + 7x_3 &= 8 \end{aligned}$

(d) $\begin{aligned} x_1 &= 3 \\ x_2 &= -5. \end{aligned}$

7. Write a linear system corresponding to each augmented matrix.

(a) $\begin{bmatrix} 4 & -2 & 0 & 7 \\ -2 & 3 & 5 & -4 \\ 3 & 2 & -5 & 3 \\ 2 & 1 & 3 & 4 \end{bmatrix}$

(b) $\begin{bmatrix} 2 & 3 & 5 \\ 4 & 2 & 1 \end{bmatrix}$

(c) $\begin{bmatrix} 3 & 2 & 4 & 2 & 5 \\ 2 & 3 & 1 & 3 & 1 \end{bmatrix}$

(d) $\begin{bmatrix} 1 & 0 & 0 & 1 \\ 0 & 1 & 0 & 2 \\ 0 & 0 & 1 & 3 \end{bmatrix}.$

8. For what value(s) of the constant k is the following linear system consistent? Inconsistent?

$$3x - 3y = k$$
$$x - y = 2.$$

9. Carry out the computations in Example 6 using augmented matrices.

10. Consider the linear system

$$ax + by = k$$
$$cx + dy = l$$
$$ex + fy = m.$$

Each of the equations in this system represents a line. Sketch the possible positions of these lines when:

(a) the system has no solutions;

(b) the system has one solution;

(c) the system has infinitely many solutions.

11. A meal containing two foods is proposed for astronauts by a dietician. Each ounce of food I provides 2 units of riboflavin, 3 units of iron, and 2 units of carbohydrates. Each ounce of food II provides 2 units of riboflavin, 1 unit of iron, and 4 units of carbohydrates. Let x_1 and x_2 be the number of ounces of foods I and II contained in the meal. What conditions must x_1 and x_2 satisfy if the meal must provide 12 units of riboflavin, 16 units of iron, and 14 units of carbohydrates?

12. A shipping company has three categories of vessels, A, B and C, which carry containerized cargos of three types I, II, and III. The load capacities of the vessels are given by the matrix

	Types		
Vessel	I	II	III
A	4	3	2
B	5	2	3
C	2	2	3

Let x_1, x_2, and x_3 be the number of vessels in each of the categories A, B, and C. What conditions must x_1, x_2, and x_3 satisfy if the company must ship 42 containers of type I, 27 containers of type II, and 33 containers of type III?

4.2 GAUSS–JORDAN ELIMINATION

In this section we shall develop a systematic procedure that can be used to solve any system of linear equations. In this method, we reduce the augmented matrix of the given system to a simple form, from which the solutions of the system can be obtained by inspection.

Consider the matrix

$$\begin{bmatrix} 1 & 0 & 0 & 2 \\ 0 & 1 & 0 & 1 \\ 0 & 0 & 1 & 3 \\ 0 & 0 & 0 & 0 \end{bmatrix} \tag{1}$$

This is an example of a matrix that is in ***reduced row echelon form.*** This means the matrix has the following four properties:

1. If a row is not made up entirely of zeros, then the leftmost nonzero number in the row is a 1.
2. If there are any rows consisting entirely of zeros, they are all together at the bottom of the matrix.
3. In two successive rows, not consisting entirely of zeros, the first nonzero number in the lower row is to the right of the first nonzero number in the upper row.
4. Each column that contains the first nonzero number of some row has zeros everywhere else.

Example 9 The following matrices are in reduced row echelon form.

$$\begin{bmatrix} 1 & 0 & 0 \\ 0 & 1 & 0 \\ 0 & 0 & 1 \end{bmatrix}, \quad \begin{bmatrix} 1 & 0 & 0 & 1 \\ 0 & 1 & 0 & 5 \\ 0 & 0 & 1 & -2 \\ 0 & 0 & 0 & 0 \end{bmatrix}, \quad \begin{bmatrix} 1 & 7 & 0 & 0 \\ 0 & 0 & 1 & 0 \\ 0 & 0 & 0 & 0 \end{bmatrix}.$$

The reader should check to see that each of the above matrices satisfies all the necessary requirements.

Example 10 The following matrices are not in reduced row echelon form.

$$\begin{bmatrix} 1 & 0 & 0 & 4 \\ 0 & 3 & 0 & 1 \\ 0 & 0 & 1 & 2 \\ 0 & 0 & 0 & 0 \end{bmatrix}, \quad \begin{bmatrix} 0 & 0 & 0 \\ 0 & 1 & 0 \\ 0 & 0 & 1 \end{bmatrix}, \quad \begin{bmatrix} 0 & 0 & 1 \\ 0 & 1 & 0 \\ 1 & 0 & 0 \end{bmatrix},$$

$$\begin{bmatrix} 1 & 2 & 0 \\ 0 & 1 & 0 \\ 0 & 0 & 1 \end{bmatrix}.$$

For each matrix, the reader should determine which of the four conditions is violated.

The strategy for solving a linear system is to perform a sequence of row operations on the augmented matrix for the system until the matrix is in reduced row echelon form. (We shall show how to do this below.) Once the augmented matrix is in reduced row echelon form, the solutions of the system are easy to obtain. The following example illustrates this point.

Example 11 In each part, suppose that the augmented matrix for a linear system has been reduced by row operations to the given matrix (which is in reduced row echelon form). Solve the system.

$$\text{(a)}\quad \begin{bmatrix} 1 & 0 & 0 & -2 \\ 0 & 1 & 0 & 4 \\ 0 & 0 & 1 & 3 \end{bmatrix}, \qquad \text{(b)}\quad \begin{bmatrix} 1 & 0 & 0 & 0 \\ 0 & 1 & 0 & 0 \\ 0 & 0 & 0 & 1 \end{bmatrix},$$

$$\text{(c)}\quad \begin{bmatrix} 1 & 0 & 0 & 2 & 3 \\ 0 & 1 & 0 & 3 & 4 \\ 0 & 0 & 1 & -1 & 1 \\ 0 & 0 & 0 & 0 & 0 \end{bmatrix}, \qquad \text{(d)}\quad \begin{bmatrix} 1 & 2 & 0 & 0 & 3 & 5 \\ 0 & 0 & 1 & 0 & 3 & -2 \\ 0 & 0 & 0 & 1 & -4 & 4 \\ 0 & 0 & 0 & 0 & 0 & 0 \end{bmatrix}.$$

Solution (a) The linear system corresponding to the augmented matrix is

$$\begin{aligned} x \qquad &= -2 \\ y \quad &= 4 \\ z &= 3. \end{aligned}$$

Thus, the solution is $x = -2$, $y = 4$, $z = 3$.

Solution (b) The linear system corresponding to the augmented matrix is

$$\begin{aligned} x \qquad\qquad &= 0 \\ y \qquad &= 0 \\ 0x + 0y + 0z &= 1. \end{aligned}$$

The third equation in this system is not satisfied by any values of x, y, and z, so that the given system is inconsistent (has no solutions).

Solution (c) The linear system corresponding to the augmented matrix is

$$\begin{aligned} x_1 \qquad + 2x_4 &= 3 \\ x_2 \quad + 3x_4 &= 4 \\ x_3 - x_4 &= 1. \end{aligned}$$

Since x_1, x_2, x_3 each "begin" one of these equations, we call them ***beginning variables.*** If we solve for the beginning variables in terms of the remaining variables, we obtain

$$\begin{aligned} x_1 &= 3 - 2x_4 \\ x_2 &= 4 - 3x_4 \\ x_3 &= 1 + x_4. \end{aligned}$$

We can now assign x_4 an arbitrary value t, to obtain infinitely many solutions. These solutions are given by the formulas

$$x_1 = 3 - 2t, \qquad x_2 = 4 - 3t, \qquad x_3 = 1 + t, \qquad x_4 = t.$$

Particular numerical solutions can be obtained from these formulas by substituting values for t. For example, letting $t = 0$ yields the solution

$$x_1 = 3, \qquad x_2 = 4, \qquad x_3 = 1, \qquad x_4 = 0$$

and letting $t = -1$ yields the solution

$$x_1 = 5, \qquad x_2 = 7, \qquad x_3 = 0, \qquad x_4 = -1.$$

Solution (d) The linear system corresponding to the augmented matrix is

$$\begin{aligned} x_1 + 2x_2 \qquad\qquad + 3x_5 &= 5 \\ x_3 \quad + 3x_5 &= -2 \\ x_4 - 4x_5 &= 4. \end{aligned}$$

Here the beginning variables are x_1, x_3, and x_4. If we solve for these beginning variables in terms of the remaining variables, x_2 and x_5, we obtain

$$\begin{aligned} x_1 &= 5 - 2x_2 - 3x_5 \\ x_3 &= -2 \qquad\quad - 3x_5 \\ x_4 &= 4 \qquad\quad + 4x_5. \end{aligned}$$

We can now assign x_2 and x_5 arbitrary values s and t, respectively, to obtain infinitely many solutions. These solutions are given by the formulas

$$x_1 = 5 - 2s - 3t, \qquad x_2 = s, \qquad x_3 = -2 - 3t, \qquad x_4 = 4 + 4t, \qquad x_5 = t.$$

These examples show how to solve a linear system once its augmented matrix is in reduced row echelon form. We now give a procedure, called Gauss–Jordan† elimination, for reducing a matrix to reduced row echelon

† *Carl Friedrich Gauss* (1777–1855)—Sometimes called the "prince of mathematicians," Gauss made profound contributions to number theory, theory of functions, probability, and statistics. He discovered a way to calculate the orbits of asteroids, made basic discoveries in electromagnetic theory, and invented a telegraph.

Camille Jordan (1838–1922)—Jordan was a professor at the École Polytechnique in Paris. He did pioneering work in several branches of mathematics, including matrix theory. He is particularly famous for the Jordan Curve Theorem, which states: A simple closed curve (such as a circle or a square) divides the plane into two nonintersecting connected regions.

form. We shall illustrate the method using the following matrix

$$\begin{bmatrix} 0 & 0 & -2 & 0 & 7 & 6 \\ 2 & 4 & 0 & 6 & 12 & 8 \\ 2 & 4 & 5 & 6 & -5 & -2 \end{bmatrix}.$$

Step 1 Find the leftmost column (vertical line) that does not consist entirely of zeros.

$$\begin{bmatrix} 0 & 0 & -2 & 0 & 7 & 6 \\ 2 & 4 & 0 & 6 & 12 & 8 \\ 2 & 4 & 5 & 6 & -5 & -2 \end{bmatrix}$$

Leftmost nonzero column

Step 2 Interchange the top row with a row below, if necessary, so that the number at the top of the column found in Step 1 is nonzero.

$$\begin{bmatrix} 2 & 4 & 0 & 6 & 12 & 8 \\ 0 & 0 & -2 & 0 & 7 & 6 \\ 2 & 4 & 5 & 6 & -5 & -2 \end{bmatrix}$$

The first and second rows in the preceding augmented matrix were interchanged.

Step 3 If the number that is now at the top of the column found in Step 1 is a, multiply the first row by $1/a$, so that the top entry in the column becomes a 1.

$$\begin{bmatrix} 1 & 2 & 0 & 3 & 6 & 4 \\ 0 & 0 & -2 & 0 & 7 & 6 \\ 2 & 4 & 5 & 6 & -5 & -2 \end{bmatrix}$$

The top row of the preceding augmented matrix was multiplied by $\frac{1}{2}$.

Step 4 Add suitable multiples of the top row to all rows below so that in the column located in Step 1, all numbers below the top row become zeros.

$$\begin{bmatrix} 1 & 2 & 0 & 3 & 6 & 4 \\ 0 & 0 & -2 & 0 & 7 & 6 \\ 0 & 0 & 5 & 0 & -17 & -10 \end{bmatrix}$$

-2 times the first row of the preceding augmented matrix was added to the third row.

Step 5 Cover the top row of the last matrix and begin again with Step 1 applied to the *sub*matrix that remains. Continue in this way until it is impossible to follow the steps any further.

$$\begin{bmatrix} 1 & 2 & 0 & 3 & 6 & 4 \\ 0 & 0 & -2 & 0 & 7 & 6 \\ 0 & 0 & 5 & 0 & -17 & -10 \end{bmatrix}$$

Leftmost nonzero column in the submatrix obtained by covering the first row.

$$\begin{bmatrix} 1 & 2 & 0 & 3 & 6 & 4 \\ 0 & 0 & 1 & 0 & -\frac{7}{2} & -3 \\ 0 & 0 & 5 & 0 & -17 & -10 \end{bmatrix}$$

First row of the preceding submatrix was multiplied by $-\frac{1}{2}$.

$$\begin{bmatrix} 1 & 2 & 0 & 3 & 6 & 4 \\ 0 & 0 & 1 & 0 & -\frac{7}{2} & -3 \\ 0 & 0 & 0 & 0 & \frac{1}{2} & 5 \end{bmatrix}$$

-5 times the first row of the preceding submatrix was added to the second row of the submatrix.

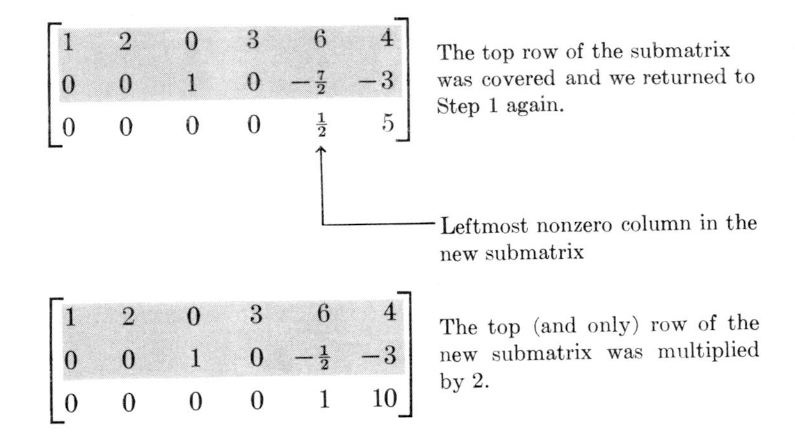

The top row of the submatrix was covered and we returned to Step 1 again.
Leftmost nonzero column in the new submatrix
The top (and only) row of the new submatrix was multiplied by 2.

> ***Step 6*** Uncover all the covered rows. Then beginning with the last nonzero row and working upward, add suitable multiples of each row to the rows above, so that the fourth property in the definition of reduced row echelon form is satisfied.

$$\begin{bmatrix} 1 & 2 & 0 & 3 & 6 & 4 \\ 0 & 0 & 1 & 0 & -\frac{7}{2} & -3 \\ 0 & 0 & 0 & 0 & 1 & 10 \end{bmatrix}$$

All rows were uncovered.

$$\begin{bmatrix} 1 & 2 & 0 & 3 & 0 & -56 \\ 0 & 0 & 1 & 0 & 0 & 32 \\ 0 & 0 & 0 & 0 & 1 & 10 \end{bmatrix}$$

$\frac{7}{2}$ times the third row of the preceding matrix was added to the second row and -6 times the third row was added to the first row.

The final matrix shown is in reduced row echelon form.

We shall now illustrate the Gauss–Jordan elimination method for solving linear systems.

Example 12 Solve the following linear system by Gauss–Jordan elimination:

$$\begin{aligned} 2x_1 + 3x_2 - 3x_3 + 7x_4 + x_5 + 4x_6 &= -2 \\ x_2 - x_3 + 3x_4 - x_5 + 2x_6 &= 2 \\ 2x_1 + 2x_2 - x_3 + 6x_4 + 2x_5 - x_6 &= -7 \\ 2x_1 + x_2 - x_3 + x_4 + 3x_5 \qquad &= -6. \end{aligned} \tag{2}$$

The augmented matrix of this linear system is

$$\begin{bmatrix} 2 & 3 & -3 & 7 & 1 & 4 & -2 \\ 0 & 1 & -1 & 3 & -1 & 2 & 2 \\ 2 & 2 & -1 & 6 & 2 & -1 & -7 \\ 2 & 1 & -1 & 1 & 3 & 0 & -6 \end{bmatrix}.$$

Multiply the first row by $\frac{1}{2}$ to obtain

$$\begin{bmatrix} 1 & \frac{3}{2} & -\frac{3}{2} & \frac{7}{2} & \frac{1}{2} & 2 & -1 \\ 0 & 1 & -1 & 3 & -1 & 2 & 2 \\ 2 & 2 & -1 & 6 & 2 & -1 & -7 \\ 2 & 1 & -1 & 1 & 3 & 0 & -6 \end{bmatrix}.$$

Add -2 times the first row to both the third and fourth rows to obtain

$$\begin{bmatrix} 1 & \frac{3}{2} & -\frac{3}{2} & \frac{7}{2} & \frac{1}{2} & 2 & -1 \\ 0 & 1 & -1 & 3 & -1 & 2 & 2 \\ 0 & -1 & 2 & -1 & 1 & -5 & -5 \\ 0 & -2 & 2 & -6 & 2 & -4 & -4 \end{bmatrix}.$$

Add the second row to the third row and 2 times the second row to the fourth row to obtain

$$\begin{bmatrix} 1 & \frac{3}{2} & -\frac{3}{2} & \frac{7}{2} & \frac{1}{2} & 2 & -1 \\ 0 & 1 & -1 & 3 & -1 & 2 & 2 \\ 0 & 0 & 1 & 2 & 0 & -3 & -3 \\ 0 & 0 & 0 & 0 & 0 & 0 & 0 \end{bmatrix}.$$

Add the third row to the second row to obtain

$$\begin{bmatrix} 1 & \frac{3}{2} & -\frac{3}{2} & \frac{7}{2} & \frac{1}{2} & 2 & -1 \\ 0 & 1 & 0 & 5 & -1 & -1 & -1 \\ 0 & 0 & 1 & 2 & 0 & -3 & -3 \\ 0 & 0 & 0 & 0 & 0 & 0 & 0 \end{bmatrix}.$$

Add $\frac{3}{2}$ times the third row to the first row to obtain

$$\begin{bmatrix} 1 & \frac{3}{2} & 0 & \frac{13}{2} & \frac{1}{2} & -\frac{5}{2} & -\frac{11}{2} \\ 0 & 1 & 0 & 5 & -1 & -1 & -1 \\ 0 & 0 & 1 & 2 & 0 & -3 & -3 \\ 0 & 0 & 0 & 0 & 0 & 0 & 0 \end{bmatrix}.$$

Add $-\frac{3}{2}$ times the second row to the first row to obtain

$$\begin{bmatrix} 1 & 0 & 0 & -1 & 2 & -1 & -4 \\ 0 & 1 & 0 & 5 & -1 & -1 & -1 \\ 0 & 0 & 1 & 2 & 0 & -3 & -3 \\ 0 & 0 & 0 & 0 & 0 & 0 & 0 \end{bmatrix} \tag{3}$$

which is in reduced row echelon form. The linear system corresponding to (3) is

$$\begin{aligned} x_1 \qquad\quad - x_4 + 2x_5 - x_6 &= -4 \\ x_2 \quad + 5x_4 - x_5 - x_6 &= -1 \\ x_3 + 2x_4 \qquad\quad - 3x_6 &= -3. \end{aligned}$$

(We have dropped the last equation $0x_1 + 0x_2 + 0x_3 + 0x_4 + 0x_5 + 0x_6 = 0$ because it will automatically be satisfied.) Solving for the beginning variables, x_1, x_2 and x_3, we obtain

$$\begin{aligned} x_1 &= -4 + x_4 - 2x_5 + x_6 \\ x_2 &= -1 - 5x_4 + x_5 + x_6 \\ x_3 &= -3 - 2x_4 \qquad\quad + 3x_6. \end{aligned}$$

If we assign x_4, x_5, and x_6 the arbitrary values r, s, and t, respectively, we obtain infinitely many solutions. They are given by the formulas

$$x_1 = -4 + r - 2s + t, \quad x_2 = -1 - 5r + s + t, \quad x_3 = -3 - 2r + 3t,$$

$$x_4 = r, \quad x_5 = s, \quad x_6 = t.$$

Example 13 (Application to Production Problems) In Example 5, we showed that if a pharmaceutical manufacturer wants to produce x_1 ounces of A, x_2 of B, x_3 ounces of C, and x_4 ounces of D, each day so that he uses up his daily stock of 100 ounces of alpha and 300 ounces of beta, then x_1, x_2, x_3, and x_4 must satisfy

$$\begin{aligned} 0.1x_1 + 0.3x_2 + 0.5x_3 + 0.2x_4 &= 100 \\ 0.4x_1 + 0.2x_2 + 0.3x_3 + 0.8x_4 &= 300. \end{aligned}$$

We shall solve this system by Gauss–Jordan elimination. The augmented matrix for the system is

$$\begin{bmatrix} 0.1 & 0.3 & 0.5 & 0.2 & 100 \\ 0.4 & 0.2 & 0.3 & 0.8 & 300 \end{bmatrix}.$$

To reduce this matrix to reduced row echelon form, we first multiply the top row by 10 to obtain

$$\begin{bmatrix} 1 & 3 & 5 & 2 & 1000 \\ 0.4 & 0.2 & 0.3 & 0.8 & 300 \end{bmatrix}.$$

Next add -0.4 times the top row to the bottom to obtain

$$\begin{bmatrix} 1 & 3 & 5 & 2 & 1000 \\ 0 & -1 & -1.7 & 0 & -100 \end{bmatrix}.$$

Now multiply the second row by -1. This gives

$$\begin{bmatrix} 1 & 3 & 5 & 2 & 1000 \\ 0 & 1 & 1.7 & 0 & 100 \end{bmatrix}.$$

Finally, add -3 times the bottom row to the top to obtain the reduced row echelon form

$$\begin{bmatrix} 1 & 0 & -0.1 & 2 & 700 \\ 0 & 1 & 1.7 & 0 & 100 \end{bmatrix}.$$

The corresponding system of equations is

$$\begin{aligned} x_1 \quad - 0.1x_3 + 2x_4 &= 700 \\ x_2 + 1.7x_3 \quad &= 100. \end{aligned}$$

Solving for the beginning variables, we obtain

$$\begin{aligned} x_1 &= 700 + 0.1x_3 - 2x_4 \\ x_2 &= 100 - 1.7x_3. \end{aligned}$$

If we assign x_3 and x_4 arbitrary values s and t, respectively, we obtain

$$x_1 = 700 + 0.1s - 2t, \qquad x_2 = 100 - 1.7s, \qquad x_3 = s, \qquad x_4 = t.$$

Thus, the problem has infinitely many solutions. Some of the solutions and the values of s and t that generate them are shown in Table 4.1.

Table 4.1

Ounces of A x_1	Ounces of B x_2	Ounces of C x_3	Ounces of D x_4	Values of s and t
305	15	50	200	($s = 50$, $t = 200$)
104	32	40	300	($s = 40$, $t = 300$)
503	49	30	100	($s = 30$, $t = 100$)
670.5	91.5	5	15	($s = 5$, $t = 15$)
700	100	0	0	($s = 0$, $t = 0$)

In a practical situation, the choice of the values for s and t would be dictated by such considerations as demand for the products, cost of manufacture, available materials and labor, and so on. Also, to obtain a physically sensible solution, the values of s and t would have to be chosen so that $x_1 \geq 0$, $x_2 \geq 0$, $x_3 \geq 0$, and $x_4 \geq 0$.

EXERCISE SET 4.2

1. Which of the following matrices are in reduced row echelon form?

(a) $\begin{bmatrix} 0 & 1 & 0 & 0 & 0 \\ 0 & 0 & 1 & 0 & 1 \\ 0 & 0 & 0 & 0 & 0 \end{bmatrix}$ (b) $\begin{bmatrix} 0 & 1 & 0 & 0 & 2 \\ 0 & 0 & 1 & 0 & -1 \\ 0 & 0 & 0 & 0 & 1 \end{bmatrix}$

(c) $\begin{bmatrix} 0 & 0 & 1 & 0 & 0 & 1 \\ 0 & 0 & 0 & 1 & 0 & -3 \\ 0 & 0 & 0 & 0 & 1 & 4 \end{bmatrix}$ (d) $\begin{bmatrix} 1 & 0 & 2 \\ 0 & 0 & 1 \end{bmatrix}$.

2. Which of the following matrices are in reduced row echelon form?

(a) $\begin{bmatrix} 1 & 0 & 0 & 0 & 0 \\ 0 & 0 & 1 & 0 & 2 \\ 0 & 0 & 0 & 1 & -3 \\ 0 & 0 & 0 & 0 & 0 \end{bmatrix}$ (b) $\begin{bmatrix} 0 & 1 & 0 & 0 \\ 0 & 0 & -1 & 2 \\ 0 & 0 & 0 & 0 \end{bmatrix}$

(c) $\begin{bmatrix} 1 & 0 & 0 & 0 & 1 \\ 0 & 0 & 1 & 2 & 3 \\ 0 & 0 & 1 & 0 & 0 \end{bmatrix}$ (d) $\begin{bmatrix} 0 & 1 & 0 & 0 & -8 \\ 0 & 0 & 1 & 0 & 6 \\ 0 & 0 & 0 & 1 & -2 \end{bmatrix}$.

3. In each of the following, the augmented matrix for a linear system has been transformed by row operations to the given reduced row echelon form. Solve the linear system.

(a) $\begin{bmatrix} 1 & 0 & 0 & 3 \\ 0 & 1 & 0 & -2 \\ 0 & 0 & 1 & 4 \end{bmatrix}$ (b) $\begin{bmatrix} 1 & 0 & 0 & 2 & 3 \\ 0 & 1 & 0 & 3 & 6 \\ 0 & 0 & 1 & 4 & -2 \end{bmatrix}$

(c) $\begin{bmatrix} 1 & 2 & 0 & 0 & 5 & 1 \\ 0 & 0 & 1 & 0 & 4 & -2 \\ 0 & 0 & 0 & 1 & 2 & 3 \\ 0 & 0 & 0 & 0 & 0 & 0 \end{bmatrix}$ (d) $\begin{bmatrix} 1 & 2 & 0 & 0 \\ 0 & 0 & 1 & 0 \\ 0 & 0 & 0 & 1 \end{bmatrix}$.

4. In each of the following, the augmented matrix for a linear system has been transformed by row operations to the given reduced row echelon form. Solve the linear system.

(a) $\begin{bmatrix} 1 & 0 & 5 & 2 \\ 0 & 1 & 3 & 2 \\ 0 & 0 & 0 & 0 \end{bmatrix}$ (b) $\begin{bmatrix} 1 & 4 & 0 & 0 \\ 0 & 0 & 1 & 0 \\ 0 & 0 & 0 & 1 \end{bmatrix}$

(c) $\begin{bmatrix} 1 & 0 & 0 & 0 & 3 & 4 \\ 0 & 1 & 0 & 0 & 2 & 0 \\ 0 & 0 & 0 & 1 & 2 & -1 \end{bmatrix}$ (d) $\begin{bmatrix} 1 & 0 & 3 & 0 & 4 \\ 0 & 1 & 2 & 0 & 2 \\ 0 & 0 & 0 & 1 & -5 \\ 0 & 0 & 0 & 0 & 0 \end{bmatrix}$.

5. Reduce each of the following matrices to reduced row echelon form.

(a) $\begin{bmatrix} 2 & -4 & -2 & 0 & 4 \\ 2 & -2 & -4 & 6 & 10 \\ -2 & 4 & 2 & 0 & 1 \end{bmatrix}$ (b) $\begin{bmatrix} 1 & 1 & 2 & 1 & 5 \\ 1 & 2 & 4 & 2 & 8 \\ 1 & -1 & -2 & -1 & -1 \end{bmatrix}$

(c) $\begin{bmatrix} 1 & 0 & 1 & 5 & -1 \\ 2 & 0 & -1 & 1 & 7 \\ 4 & 0 & 1 & 11 & 5 \end{bmatrix}.$

6. Reduce each of the following matrices to reduced row echelon form.

(a) $\begin{bmatrix} 2 & 4 & 4 & 6 \\ 2 & 2 & -2 & -4 \\ 3 & 3 & 12 & 15 \end{bmatrix}$ (b) $\begin{bmatrix} 1 & 1 & 1 & 6 \\ 1 & 5 & 1 & -2 \\ 1 & 1 & 1 & 3 \\ -1 & 0 & 0 & -3 \end{bmatrix}$ (c) $\begin{bmatrix} 1 & 1 & -1 & 0 & 3 \\ 3 & 5 & 1 & 2 & 11 \\ 1 & 2 & 1 & 1 & 4 \end{bmatrix}.$

7. Solve each of the following linear systems by Gauss–Jordan elimination.

(a)
$$\begin{aligned} 2x - 2y + 4z &= 6 \\ 2x - y + 5z &= 15 \\ -x + y + 3z &= 7 \end{aligned}$$

(b)
$$\begin{aligned} x_1 + 2x_2 + x_3 &= 0 \\ -2x_1 + x_2 + 3x_3 &= 0 \\ 2x_1 + 3x_2 + 5x_3 &= 0 \end{aligned}$$

(c)
$$\begin{aligned} x_1 - x_2 + 3x_3 + 2x_4 - 3x_5 &= 2 \\ 2x_1 + x_2 + x_4 + 3x_5 &= 1 \\ 3x_1 + x_2 - 3x_3 + 2x_4 - x_5 &= 10 \end{aligned}$$

(d)
$$\begin{aligned} -x_1 + 3x_2 - x_3 + x_4 &= 3 \\ x_1 - 2x_2 - x_3 + x_4 &= 1 \\ 3x_1 - 8x_2 + 3x_3 - x_4 &= -5 \end{aligned}$$

8. Solve each of the following linear systems by Gauss–Jordan elimination.

(a)
$$\begin{aligned} 2x + 3y &= 2 \\ -5x + 12y &= -5 \\ -3x + 2y &= -3 \end{aligned}$$

(b)
$$\begin{aligned} x + 2y - z &= 0 \\ 5x + 2y + z &= 0 \\ 3x + 2y + 3z &= 0 \end{aligned}$$

(c)
$$\begin{aligned} 2x_1 + 3x_2 + x_3 + x_4 &= 0 \\ 3x_1 - x_2 + x_3 - x_4 &= 0 \end{aligned}$$

(d)
$$\begin{aligned} x_1 + 2x_2 &= 3 \\ 2x_1 + x_2 + 3x_3 &= 5 \\ x_2 + x_3 &= -3. \end{aligned}$$

9. Solve each of the following linear systems by Gauss–Jordan elimination.

(a)
$$\begin{aligned} x_1 + 2x_2 + 3x_3 + x_4 &= 3 \\ 2x_1 + 3x_2 + 3x_3 + 2x_4 &= -5 \\ -5x_2 - x_3 + 28x_4 &= -13 \\ x_1 + x_2 + 2x_3 - 3x_4 &= -2 \end{aligned}$$

(b)
$$\begin{aligned} x_1 + 2x_2 &= 3 \\ 2x_1 - 3x_2 &= -1 \end{aligned}$$

(c)
$$\begin{aligned} x_1 + 2x_2 - x_3 &= 1 \\ -x_1 + x_2 + x_3 &= 5 \\ -2x_1 - 3x_2 + 4x_3 &= 4 \end{aligned}$$

(d)
$$\begin{aligned} 2x_1 + x_2 - x_3 &= 3 \\ -2x_1 - x_2 + 3x_3 &= 5. \end{aligned}$$

10. For which values of a will the following linear system have no solutions? Exactly one solution? Infinitely many solutions?

$$\begin{aligned} x + \qquad\quad y &= 2 \\ x + (a^2 - 3)y &= a. \end{aligned}$$

11. Show that if $ad \neq 1$, then the linear system

$$\begin{aligned} ax + \ y &= 3 \\ x + dy &= 5 \end{aligned}$$

has exactly one solution.

12. In Exercise 11 of Section 4.1, find the number of ounces of foods I and II that should be included in the astronauts' meal.

13. Solve the problem in Exercise 12 of Section 4.1.

4.3 MATRICES; MATRIX ADDITION AND SCALAR MULTIPLICATION

We have already seen how matrices are used in solving systems of linear equations. Matrices arise in other contexts as well. They occur in problems of business, science, engineering, economics, and the behavioral sciences. In this section we shall study matrices in detail and indicate several applications. Other, more-detailed applications of matrices appear in Chapter 8.

Recall from Section 4.1, that a ***matrix*** is a rectangular array of numbers. The numbers in the array are called the ***entries*** of the matrix.

Example 14 The following are matrices

$$\begin{bmatrix} 1 & 2 \\ 3 & -4 \end{bmatrix}, \qquad \begin{bmatrix} \frac{2}{5} & 3 \end{bmatrix}, \qquad \begin{bmatrix} 0 \\ 1 \\ \sqrt{2} \end{bmatrix}, \qquad \begin{bmatrix} 7 \end{bmatrix}, \qquad \begin{bmatrix} -1 & 1 & 2 \\ 3 & 8 & -5 \end{bmatrix}.$$

A matrix is said to have ***size*** $m \times n$, read "m by n," if it has m rows (horizontal lines) and n columns (vertical lines). If $m = n$, then the matrix is called a ***square matrix.***

Example 15 The matrices in Example 14 are

$$2 \times 2, \qquad 1 \times 2, \qquad 3 \times 1, \qquad 1 \times 1, \qquad \text{and} \qquad 2 \times 3$$

respectively. The first and fourth matrices are square.

We will denote matrices by uppercase letters, such as A, B, and C. In discussions involving matrices, it is common to call numbers ***scalars.*** We shall use lowercase letters, such as a, b, and c to denote scalars. For example, we might write

$$A = \begin{bmatrix} a & b \\ c & d \end{bmatrix} \quad \text{or} \quad B = \begin{bmatrix} 1 & 3 & 5 \\ 7 & 2 & 1 \end{bmatrix}.$$

If A is a matrix, we will often denote the entry in row i and column j of A by a_{ij}. For example, a_{13} would denote the entry in row 1 and column 3 of a matrix A. Thus, a general 2×3 matrix might be written

$$A = \begin{bmatrix} a_{11} & a_{12} & a_{13} \\ a_{21} & a_{22} & a_{23} \end{bmatrix}.$$

Similarly, in a matrix B, we would denote the entry in row i and column j by b_{ij}.

Example 16 If

$$B = \begin{bmatrix} 2 & -4 & 3 & 4 \\ 6 & 0 & 2 & 3 \\ 4 & 2 & -5 & 0 \end{bmatrix},$$

then B is a 3×4 matrix and

$$b_{12} = -4, \quad b_{14} = 4, \quad b_{21} = 6, \quad b_{23} = 2, \quad b_{31} = 4, \quad \text{and} \quad b_{33} = -5.$$

Example 17 A manufacturer has three machines that produce goods. These machines are operated by two shifts, shift 1 and shift 2. If a_{ij} denotes the output on machine j by shift i, then the daily production can be described by a 2×3 matrix A such as

$$\begin{array}{cc} & \text{machines} \\ & \begin{array}{ccc} 1 & 2 & 3 \end{array} \\ A = \begin{array}{c} \text{shift 1} \\ \text{shift 2} \end{array} & \begin{bmatrix} 300 & 217 & 64 \\ 271 & 201 & 83 \end{bmatrix}. \end{array}$$

From this matrix we see that the output on machine 2 by shift 1 is $a_{12} = 217$ units.

Economics

Example 18 Consider a simplified economic society consisting of a farming sector (1), a housing sector (2), and a clothing sector (3). In this economy, just as in our own, each sector uses the products of the other sectors. For example, farmers need houses and clothing, while the people who produce clothes and housing need food. If c_{ij} denotes the fraction of the output of sector j which is consumed by sector i, then the economic relationships between the three sectors can be described by a 3×3 matrix C such as

$$C = \begin{array}{r} \\ \\ \text{farming (1)} \\ \text{housing (2)} \\ \text{clothing (3)} \end{array} \begin{array}{c} \begin{array}{ccc} \text{farming} & \text{housing} & \text{clothing} \\ (1) & (2) & (3) \end{array} \\ \begin{bmatrix} \frac{3}{16} & \frac{4}{7} & \frac{7}{16} \\ \frac{5}{16} & \frac{2}{7} & \frac{5}{16} \\ \frac{1}{2} & \frac{1}{7} & \frac{1}{4} \end{bmatrix} \end{array}.$$

For example, entry $c_{23} = \frac{5}{16}$, so that sector (2) (housing) consumes $\frac{5}{16}$ of the output of sector (3) (clothing).

Two matrices A and B are said to ***equal,*** written $A = B$, if they have the same size and their corresponding entries are equal.

Example 19 In order for the matrices

$$A = \begin{bmatrix} 2 & 3 \\ -1 & 5 \end{bmatrix} \quad \text{and} \quad B = \begin{bmatrix} 2 & a \\ b & 5 \end{bmatrix}$$

to be equal, we must have $a = 3$ and $b = -1$.

Example 20 Consider the matrices

$$A = \begin{bmatrix} 4 & 2 \\ -1 & 2 \end{bmatrix}, \quad B = \begin{bmatrix} 4 & 2 & 0 \\ -1 & 2 & 0 \end{bmatrix}, \quad \text{and} \quad C = \begin{bmatrix} 4 & 2 \\ 3 & 2 \end{bmatrix}.$$

Here, $A \neq B$ since the matrices A and B have different sizes. For the same reason, $B \neq C$. Further, $A \neq C$ since not all the corresponding entries are equal.

If A and B are two matrices with the same size, then the ***sum*** $A + B$ is the matrix obtained by adding the corresponding entries in A and B. We do not define addition for matrices with different sizes.

Example 21 Let

$$A = \begin{bmatrix} 3 & 4 & -1 \\ 2 & 0 & 1 \\ 3 & 1 & 2 \end{bmatrix}, \quad B = \begin{bmatrix} 2 & -1 & 3 \\ -2 & 1 & 2 \\ 6 & 4 & -3 \end{bmatrix},$$

and

$$C = \begin{bmatrix} 1 & 2 & -3 & 4 \\ 2 & 1 & 5 & 0 \\ 6 & 2 & 7 & 1 \end{bmatrix}.$$

Then

$$A + B = \begin{bmatrix} 3+2 & 4+(-1) & -1+3 \\ 2+(-2) & 0+1 & 1+2 \\ 3+6 & 1+4 & 2+(-3) \end{bmatrix} = \begin{bmatrix} 5 & 3 & 2 \\ 0 & 1 & 3 \\ 9 & 5 & -1 \end{bmatrix}.$$

The sum $A + C$ is undefined since A and C have different sizes. Similarly, $B + C$ is undefined.

Obviously, for matrices A and B with the same size, we have

$$A + B = B + A,$$

which is called the ***commutative law for matrix addition.***

Example 22 Let

$$A = \begin{bmatrix} -4 & 3 & -2 \\ 5 & 6 & 3 \end{bmatrix} \quad \text{and} \quad B = \begin{bmatrix} 2 & -1 & 0 \\ 3 & 2 & 4 \end{bmatrix}.$$

Then

$$A + B = \begin{bmatrix} -4+2 & 3+(-1) & -2+0 \\ 5+3 & 6+2 & 3+4 \end{bmatrix} = \begin{bmatrix} -2 & 2 & -2 \\ 8 & 8 & 7 \end{bmatrix},$$

and

$$B + A = \begin{bmatrix} 2+(-4) & -1+3 & 0+(-2) \\ 3+5 & 2+6 & 4+3 \end{bmatrix} = \begin{bmatrix} -2 & 2 & -2 \\ 8 & 8 & 7 \end{bmatrix}.$$

Thus, $A + B = B + A$.

Example 23 An importing firm handles three items imported from Japan: item 1, item 2, and item 3. On each item the firm must pay Japanese export duties, Japanese transportation costs, U.S. import duties, and U.S. transportation costs. These expenses, in dollars per item, are contained in the following matrices:

	Japanese transportation	Japanese export duties	
$A =$	30	12	item 1
	20	7	item 2
	19	8	item 3

	U.S. transportation	U.S. import duties	
$B =$	40	15	item 1
	9	22	item 2
	27	16	item 3.

The matrix representing total transportation costs and total duties is

$$A + B = \begin{array}{c} \\ \\ \\ \end{array}\begin{array}{cc} \text{total transpor-tation} & \text{total duties} \end{array}$$

$$A + B = \begin{bmatrix} 70 & 27 \\ 29 & 29 \\ 46 & 24 \end{bmatrix} \begin{array}{l} \text{item 1} \\ \text{item 2} \\ \text{item 3.} \end{array}$$

If A, B, and C are matrices with the same sizes, then the following equality holds (we omit the proof):

$$A + (B + C) = (A + B) + C$$

This result is called the ***associative law for matrix addition.***

Example 24 Let

$$A = \begin{bmatrix} -4 & 3 & -2 \\ 5 & 6 & 3 \end{bmatrix}, \quad B = \begin{bmatrix} 2 & -1 & 0 \\ 3 & 2 & 4 \end{bmatrix}, \quad \text{and} \quad C = \begin{bmatrix} 3 & -4 & 0 \\ 4 & 2 & 5 \end{bmatrix}.$$

Then

$$A + (B + C) = \begin{bmatrix} -4 & 3 & -2 \\ 5 & 6 & 3 \end{bmatrix} + \left(\begin{bmatrix} 2 & -1 & 0 \\ 3 & 2 & 4 \end{bmatrix} + \begin{bmatrix} 3 & -4 & 0 \\ 4 & 2 & 5 \end{bmatrix} \right)$$

$$= \begin{bmatrix} -4 & 3 & -2 \\ 5 & 6 & 3 \end{bmatrix} + \begin{bmatrix} 5 & -5 & 0 \\ 7 & 4 & 9 \end{bmatrix} = \begin{bmatrix} 1 & -2 & -2 \\ 12 & 10 & 12 \end{bmatrix},$$

$$(A + B) + C = \left(\begin{bmatrix} -4 & 3 & -2 \\ 5 & 6 & 3 \end{bmatrix} + \begin{bmatrix} 2 & -1 & 0 \\ 3 & 2 & 4 \end{bmatrix} \right) + \begin{bmatrix} 3 & -4 & 0 \\ 4 & 2 & 5 \end{bmatrix}$$

$$= \begin{bmatrix} -2 & 2 & -2 \\ 8 & 8 & 7 \end{bmatrix} + \begin{bmatrix} 3 & -4 & 0 \\ 4 & 2 & 5 \end{bmatrix} = \begin{bmatrix} 1 & -2 & -2 \\ 12 & 10 & 12 \end{bmatrix}.$$

Thus $A + (B + C) = (A + B) + C$.

A matrix each of whose entries is zero is called a ***zero matrix.***

Example 25 The following are zero matrices.

$$\begin{bmatrix} 0 & 0 \\ 0 & 0 \end{bmatrix}, \quad \begin{bmatrix} 0 & 0 & 0 & 0 \\ 0 & 0 & 0 & 0 \end{bmatrix}, \quad [0], \quad \begin{bmatrix} 0 & 0 \\ 0 & 0 \\ 0 & 0 \end{bmatrix}.$$

Example 26 Let

$$A = \begin{bmatrix} 4 & -2 \\ 3 & 5 \\ -1 & 2 \end{bmatrix} \quad \text{and} \quad 0 = \begin{bmatrix} 0 & 0 \\ 0 & 0 \\ 0 & 0 \end{bmatrix}.$$

Then

$$A + 0 = \begin{bmatrix} 4 & -2 \\ 3 & 5 \\ -1 & 2 \end{bmatrix} + \begin{bmatrix} 0 & 0 \\ 0 & 0 \\ 0 & 0 \end{bmatrix} = \begin{bmatrix} 4 & -2 \\ 3 & 5 \\ -1 & 2 \end{bmatrix} = A.$$

It is evident from the last example that if A is any matrix and 0 is a zero matrix with the same size, then

$$A + 0 = 0 + A = A.$$

Thus, a zero matrix behaves much like the scalar zero.

If A is any matrix, then the ***negative*** of A, denoted by $-A$, is the matrix whose entries are the negatives of the corresponding entries in A.

Example 27 The negative of the matrix

$$A = \begin{bmatrix} 3 & 4 & -2 \\ -5 & 6 & 3 \end{bmatrix}$$

is

$$-A = \begin{bmatrix} -3 & -4 & 2 \\ 5 & -6 & -3 \end{bmatrix}.$$

If A is any matrix and 0 is the zero matrix with the size as A, then it is evident that

$$A + (-A) = 0.$$

We shall omit the formal proof.

If A and B are matrices with the same size, then the ***difference*** between A and B, denoted $A - B$, is defined by

$$A - B = A + (-B).$$

Example 28 Let

$$A = \begin{bmatrix} -4 & 3 & -2 \\ 5 & 6 & 3 \end{bmatrix} \quad \text{and} \quad B = \begin{bmatrix} 2 & -1 & 0 \\ 3 & 2 & 4 \end{bmatrix}.$$

Then

$$A - B = A + (-B) = \begin{bmatrix} -4 & 3 & -2 \\ 5 & 6 & 3 \end{bmatrix} + \begin{bmatrix} -2 & 1 & 0 \\ -3 & -2 & -4 \end{bmatrix}$$

$$= \begin{bmatrix} -6 & 4 & -2 \\ 2 & 4 & -1 \end{bmatrix}.$$

Observe that the difference $A - B$ can be obtained directly by subtracting each entry in B from the corresponding entry in A.

There are two kinds of multiplications involving matrices, multiplication of matrices by scalars and multiplication of matrices by matrices. To motivate the definition of multiplication by scalars, consider a 2×2 matrix

$$A = \begin{bmatrix} a_{11} & a_{12} \\ a_{21} & a_{22} \end{bmatrix}.$$

From the definition of matrix addition,

$$A + A = \begin{bmatrix} a_{11} & a_{12} \\ a_{21} & a_{22} \end{bmatrix} + \begin{bmatrix} a_{11} & a_{12} \\ a_{21} & a_{22} \end{bmatrix} = \begin{bmatrix} 2a_{11} & 2a_{12} \\ 2a_{21} & 2a_{22} \end{bmatrix}.$$

Thus to add A to itself, we multiply each entry in A by 2. Since it is natural to think of $A + A$ as $2A$, we obtain

$$2A = \begin{bmatrix} 2a_{11} & 2a_{12} \\ 2a_{21} & 2a_{22} \end{bmatrix}.$$

This motivates the following definition.

If A is a matrix and c is a scalar, then the product cA is defined to be the matrix obtained by multiplying each entry of A by c.

Example 29 Let

$$A = \begin{bmatrix} 3 & 4 \\ 6 & 2 \end{bmatrix},$$

then

$$5A = \begin{bmatrix} 5(3) & 5(4) \\ 5(6) & 5(2) \end{bmatrix} = \begin{bmatrix} 15 & 20 \\ 30 & 10 \end{bmatrix}$$

and

$$-\tfrac{1}{2}A = \begin{bmatrix} -\frac{3}{2} & -2 \\ -3 & -1 \end{bmatrix}.$$

If A and B are matrices with the same size and c and d are scalars, then the following rules of matrix arithmetic hold (we omit the proof):

$$(c + d)A = cA + dA$$

$$c(A + B) = cA + cB$$

$$cd(A) = c(dA).$$

Example 30 Let $c = 3$,

$$A = \begin{bmatrix} 3 & 4 & -2 \\ 5 & 2 & 3 \end{bmatrix}, \quad \text{and} \quad B = \begin{bmatrix} -2 & 3 & 4 \\ 5 & 1 & 2 \end{bmatrix}.$$

Then

$$c(A + B) = 3\left(\begin{bmatrix} 3 & 4 & -2 \\ 5 & 2 & 3 \end{bmatrix} + \begin{bmatrix} -2 & 3 & 4 \\ 5 & 1 & 2 \end{bmatrix}\right)$$

$$= 3\begin{bmatrix} 1 & 7 & 2 \\ 10 & 3 & 5 \end{bmatrix} = \begin{bmatrix} 3 & 21 & 6 \\ 30 & 9 & 15 \end{bmatrix}.$$

Also,

$$cA + cB = 3\begin{bmatrix} 3 & 4 & -2 \\ 5 & 2 & 3 \end{bmatrix} + 3\begin{bmatrix} -2 & 3 & 4 \\ 5 & 1 & 2 \end{bmatrix}$$

$$= \begin{bmatrix} 9 & 12 & -6 \\ 15 & 6 & 9 \end{bmatrix} + \begin{bmatrix} -6 & 9 & 12 \\ 15 & 3 & 6 \end{bmatrix} = \begin{bmatrix} 3 & 21 & 6 \\ 30 & 9 & 15 \end{bmatrix}.$$

Thus, $c(A + B) = cA + cB$.

Classification of Information

Example 31 Matrices are useful for recording information that involves a two-way classification. For example, suppose a retail paint store stocks three kinds of exterior paints: standard, deluxe, and commercial. Suppose also, that these paints come in four colors: white, black, yellow, and red. The store's weekly sales can be recorded in matrix form as

$$\begin{array}{cc} & \begin{array}{cccc} \text{white} & \text{black} & \text{yellow} & \text{red} \end{array} \\ S = & \begin{bmatrix} s_{11} & s_{12} & s_{13} & s_{14} \\ s_{21} & s_{22} & s_{23} & s_{24} \\ s_{31} & s_{32} & s_{33} & s_{34} \end{bmatrix} \begin{array}{l} \text{regular} \\ \text{deluxe} \\ \text{commerical} \end{array} \end{array}$$

where s_{32} denotes the number of gallons of commercial black paint that was sold during the week. Similarly, the store's inventory can be recorded in matrix form as

$$A = \begin{bmatrix} a_{11} & a_{12} & a_{13} & a_{14} \\ a_{21} & a_{22} & a_{23} & a_{24} \\ a_{31} & a_{32} & a_{33} & a_{34} \end{bmatrix} \begin{matrix} \text{regular} \\ \text{deluxe} \\ \text{commercial} \end{matrix}$$

(columns: white, black, yellow, red)

If A is the inventory matrix at the beginning of the week and S is the sales matrix for the week, then the matrix

$$A - S$$

represents the inventory at the end of the week. If the store then receives a new shipment of stock represented by a matrix

$$B = \begin{bmatrix} b_{11} & b_{12} & b_{13} & b_{14} \\ b_{21} & b_{22} & b_{23} & b_{24} \\ b_{31} & b_{32} & b_{33} & b_{34} \end{bmatrix} \begin{matrix} \text{regular} \\ \text{deluxe} \\ \text{commercial} \end{matrix}$$

(columns: white, black, yellow, red)

its new inventory would be

$$A - S + B.$$

EXERCISE SET 4.3

1. List the sizes of the following matrices

(a) $\begin{bmatrix} 1 & 2 & -1 \\ 3 & 4 & 2 \end{bmatrix}$ (b) $\begin{bmatrix} 2 & 1 \\ 3 & 2 \\ 4 & -1 \end{bmatrix}$ (c) $[3 \quad 2]$ (d) $[-4]$.

2. If

$$A = \begin{bmatrix} 3 & 2 & -4 \\ -2 & 5 & 0 \\ 2 & 1 & 5 \\ -3 & 4 & 6 \end{bmatrix},$$

give the following elements

(a) a_{12} (b) a_{23} (c) a_{32} (d) a_{41} (e) a_{43}.

3. Let

$$A = \begin{bmatrix} 2 & -3 \\ u & 0 \end{bmatrix} \quad \text{and} \quad B = \begin{bmatrix} v & -3 \\ 5 & w \end{bmatrix}.$$

For what values of u, v, and w are A and B equal?

4. Solve the following equation for a, b, c, and d.

$$\begin{bmatrix} a+b & b+2c \\ 2c+d & 2a-d \end{bmatrix} = \begin{bmatrix} -1 & 4 \\ 8 & 0 \end{bmatrix}.$$

5. Let A and B be 3×4 matrices and let C and D be 3×3 matrices. Which of the following matrix expressions are defined? For those which are defined, give the size of the resulting matrix.

(a) $A + B$ (b) $B + D$ (c) $3A - 2C$ (d) $7C + 2D$.

6. Let

$$A = \begin{bmatrix} -1 & 2 & 3 \\ 4 & 2 & 0 \\ -3 & 2 & 5 \end{bmatrix}, \quad B = \begin{bmatrix} 3 & -1 & 2 \\ -5 & 3 & 4 \\ -3 & -4 & 0 \end{bmatrix}, \quad C = \begin{bmatrix} 2 & -3 & 6 \\ 0 & 4 & -1 \\ -5 & 1 & 3 \end{bmatrix}.$$

Compute, if possible,

(a) $A + 2B$ (b) $3A - 4B$ (c) $(A + B) - A$

(d) $A + (B + C)$ (e) $(A + B) + C$.

7. Let

$$A = \begin{bmatrix} 2 & -3 \\ 4 & 5 \end{bmatrix}, \quad B = \begin{bmatrix} 2 & 5 \\ -1 & 3 \end{bmatrix}, \quad C = \begin{bmatrix} 3 & -1 \\ 0 & 4 \end{bmatrix},$$

$c = 2$, and $d = -4$. Verify that

(a) $A + (B + C) = (A + B) + C$ (b) $(c + d)A = cA + dA$

(c) $c(A + B) = cA + cB$ (d) $cd(A) = c(dA)$.

8. Show that if $kA = 0$, then $k = 0$ or $A = 0$.

9. A number of people were questioned about their income and their political affiliation. The following results were obtained:

205 were Republicans earning under \$10,000 per year
317 were Democrats earning under \$10,000 per year
96 were Independents earning under \$10,000 per year
192 were Republicans earning over \$10,000 per year
128 were Democrats earning over \$10,000 per year
63 were Independents earning over \$10,000 per year.

Display this information in matrix form.

10. In the communication network illustrated in Figure 4.1, station 1 has two transmission channels to receiver A and three transmission channels to receiver B, while station 2 has three transmission channels to A and four transmission channels to B. Show how you might describe this information in matrix form.

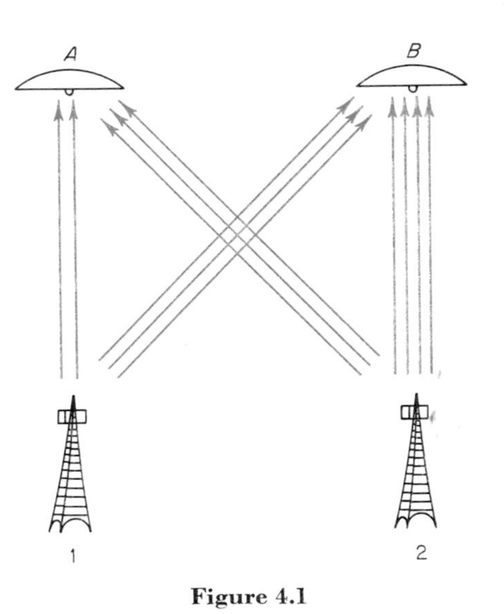

Figure 4.1

4.4 MATRIX MULTIPLICATION

In this section we shall discuss the multiplication of matrices and illustrate some applications.

> If A is an $m \times r$ matrix and B is an $r \times n$ matrix, then the ***product*** AB is the $m \times n$ matrix obtained as follows. To find the entry in row i and column j of AB, multiply the entries in the ith row of A by the corresponding entries in the jth column of B and add up the resulting products.

Example 32 Let

$$A = \begin{bmatrix} 3 & 4 & -1 & 2 \\ 0 & 1 & 2 & 3 \end{bmatrix} \quad \text{and} \quad B = \begin{bmatrix} -1 & 0 & 2 \\ -2 & 4 & 1 \\ 3 & 2 & 3 \\ 5 & -1 & 2 \end{bmatrix}.$$

Since A is 2×4 matrix and B is a 4×3 matrix, the product AB is a 2×3 matrix. Suppose we want the entry in row 2 and column 3 of AB. As illustrated below, we multiply the corresponding entries of row 2 in A and column 3 in B and add the resulting products:

$$\begin{bmatrix} 3 & 4 & -1 & 2 \\ 0 & 1 & 2 & 3 \end{bmatrix} \begin{bmatrix} -1 & 0 & 2 \\ -2 & 4 & 1 \\ 3 & 2 & 3 \\ 5 & -1 & 2 \end{bmatrix} = \begin{bmatrix} \square & \square & \square \\ \square & \square & 13 \end{bmatrix}$$

$$(0)(2) + (1)(1) + (2)(3) + (3)(2) = 13.$$

The entry in row 1 column 2 of AB is computed as follows:

$$\begin{bmatrix} 3 & 4 & -1 & 2 \\ 0 & 1 & 2 & 3 \end{bmatrix} \begin{bmatrix} -1 & 0 & 2 \\ -2 & 4 & 1 \\ 3 & 2 & 3 \\ 5 & -1 & 2 \end{bmatrix} = \begin{bmatrix} \square & 12 & \square \\ \square & \square & \square \end{bmatrix}$$

$$(3)(0) + (4)(4) + (-1)(2) + (2)(-1) = 12.$$

The reader can check that the entire product is

$$\begin{bmatrix} -4 & 12 & 11 \\ 19 & 5 & 13 \end{bmatrix}.$$

Note that the definition of the product AB requires that the number of columns of the first factor, A, must be the same as the number of rows of the second factor, B. If this condition is not satisfied, then the product of A and B cannot be formed.

Example 33 Suppose A is a 3×4 matrix B is a 4×2 matrix, and C is a 4×3 matrix. Then AB is a 3×2 matrix, AC is 3×3, and CA is a 4×4 matrix while BC, CB, and BA are undefined.

Although the definition of matrix multiplication seems complicated, it was devised because it arises naturally in various applications. The following example illustrates one such application.

Ecology (Pollution)

Example 34 A manufacturer makes two types of products, I and II, at two plants, X and Y. In the manufacture of these products, the following pollutants result: sulphur dioxide, carbon monoxide, and particulate matter. At either plant, the daily pollutants resulting from the production of product I are

300 pounds of sulphur dioxide
100 pounds of carbon monoxide
200 pounds of particulate matter

and for product II are

400 pounds of sulphur dioxide
50 pounds of carbon monoxide
300 pounds of particulate matter.

This information can be tabulated in matrix form as

$$A = \begin{bmatrix} 300 & 100 & 200 \\ 400 & 50 & 300 \end{bmatrix} \begin{matrix} \text{product I} \\ \text{product II.} \end{matrix}$$

(columns: sulphur dioxide, carbon monoxide, particulate matter)

To satisfy federal regulations, these pollutants must be removed. Suppose that the daily cost in dollars for removing each pound of pollutant at plant X is

5 dollars per pound of sulphur dioxide
3 dollars per pound of carbon dioxide
2 dollars per pound of particulate matter

and at plant Y is

8 dollars per pound of sulphur dioxide
4 dollars per pound of carbon monoxide
1 dollar per pound of particulate matter.

This cost information can be represented in matrix form as

$$B = \begin{bmatrix} 5 & 8 \\ 3 & 4 \\ 2 & 1 \end{bmatrix} \begin{matrix} \text{sulphur dioxide} \\ \text{carbon monoxide} \\ \text{particulate matter.} \end{matrix}$$

(columns: plant X, plant Y)

From the first row of A and the first column of B, we see that the daily cost of removing all pollutants resulting from the manufacture of product I at plant X is

$$300(5) + 100(3) + 200(2) = 2200 \quad \text{dollars.}$$

But this is the entry in row 1 and column 1 of

$$AB = \begin{bmatrix} 300 & 100 & 200 \\ 400 & 50 & 300 \end{bmatrix} \begin{bmatrix} 5 & 8 \\ 3 & 4 \\ 2 & 1 \end{bmatrix} = \begin{bmatrix} 2200 & 3000 \\ 2750 & 3700 \end{bmatrix}.$$

Similarly, from the second row of A, and the second column of B, the daily cost of removing all pollutants resulting from the manufacture of product

II at plant Y is

$$400(8) + 50(4) + 300(1) = 3700.$$

But this is just the entry in row 2 and column 2 of the product AB. Thus, each entry in AB has a physical interpretation as the daily cost of removing all pollutants from one of the plants in the manufacture of one of the products.

If a and b are numbers, then we know that

$$ab = ba.$$

In matrix arithmetic, however, the product AB and the product BA need not be equal. For example, if A is a 2×3 matrix and B is a 3×4 matrix, then AB is a 2×4 matrix, but BA is undefined (why?). However, as the next example shows, AB and BA may not be equal, even if they are both defined.

Example 35 Let

$$A = \begin{bmatrix} 2 & 3 \\ -4 & 2 \end{bmatrix} \quad \text{and} \quad B = \begin{bmatrix} 3 & -1 \\ 5 & 2 \end{bmatrix}.$$

Then

$$AB = \begin{bmatrix} 21 & 4 \\ -2 & 8 \end{bmatrix} \quad \text{and} \quad BA = \begin{bmatrix} 10 & 7 \\ 2 & 19 \end{bmatrix}.$$

Thus

$$AB \neq BA.$$

In ordinary arithmetic we know:

(i) If $ab = 0$, then $a = 0$ or $b = 0$.
(ii) If $ab = ac$ and $a \neq 0$, then $b = c$ (we can cancel a on both sides of the equation).

These properties do not hold for matrix multiplication, as shown in the next example.

Example 36 Let

$$A = \begin{bmatrix} 0 & 2 \\ 0 & 3 \end{bmatrix}, \quad B = \begin{bmatrix} -1 & 4 \\ 3 & -5 \end{bmatrix}, \quad C = \begin{bmatrix} -2 & 5 \\ 3 & -5 \end{bmatrix}, \quad D = \begin{bmatrix} 2 & 3 \\ 0 & 0 \end{bmatrix}.$$

Then

$$AB = AC = \begin{bmatrix} 6 & -10 \\ 9 & -15 \end{bmatrix}.$$

Thus $AB = AC$, but $B \neq C$.

Although

$$AD = \begin{bmatrix} 0 & 0 \\ 0 & 0 \end{bmatrix},$$

we have $A \neq 0$ and $D \neq 0$.

As the above examples show, some properties of ordinary arithmetic do not carry over to matrix arithmetic. There are many that do, however. We shall now give a partial list of some of the more important of these properties. We shall omit the proofs.

Assuming the sizes of the matrices are such that the indicated operations can be performed, the following rules of matrix arithmetic hold:

$$A(BC) = (AB)C$$ (***Associative law*** for matrix multiplication)

$$A(B + C) = AB + AC$$

$$(A + B)C = AC + BC$$ (***Distributive laws***)

$$A(cB) = c(AB) = (cA)B$$

Example 37 Let

$$A = \begin{bmatrix} 4 & 2 \\ 0 & -3 \\ 3 & -2 \end{bmatrix}, \quad B = \begin{bmatrix} 4 & 2 \\ -2 & 3 \end{bmatrix}, \quad \text{and} \quad C = \begin{bmatrix} 1 & 2 \\ -3 & 4 \end{bmatrix}.$$

Then

$$A(B + C) = \begin{bmatrix} 4 & 2 \\ 0 & -3 \\ 3 & -2 \end{bmatrix} \left(\begin{bmatrix} 4 & 2 \\ -2 & 3 \end{bmatrix} + \begin{bmatrix} 1 & 2 \\ -3 & 4 \end{bmatrix} \right)$$

$$= \begin{bmatrix} 4 & 2 \\ 0 & -3 \\ 3 & -2 \end{bmatrix} \begin{bmatrix} 5 & 4 \\ -5 & 7 \end{bmatrix} = \begin{bmatrix} 10 & 30 \\ 15 & -21 \\ 25 & -2 \end{bmatrix}.$$

On the other hand,

$$AB + AC = \begin{bmatrix} 4 & 2 \\ 0 & -3 \\ 3 & -2 \end{bmatrix} \begin{bmatrix} 4 & 2 \\ -2 & 3 \end{bmatrix} + \begin{bmatrix} 4 & 2 \\ 0 & -3 \\ 3 & -2 \end{bmatrix} \begin{bmatrix} 1 & 2 \\ -3 & 4 \end{bmatrix}$$

$$= \begin{bmatrix} 12 & 14 \\ 6 & -9 \\ 16 & 0 \end{bmatrix} + \begin{bmatrix} -2 & 16 \\ 9 & -12 \\ 9 & -2 \end{bmatrix} = \begin{bmatrix} 10 & 30 \\ 15 & -21 \\ 25 & -2 \end{bmatrix}.$$

Thus

$$A(B + C) = AB + AC$$

as guaranteed by the distributive law.

Example 38 Let $c = 2$,

$$A = \begin{bmatrix} 4 & 2 \\ 0 & -3 \\ 3 & -2 \end{bmatrix} \quad \text{and} \quad B = \begin{bmatrix} 4 & 2 \\ -2 & 3 \end{bmatrix}.$$

Then

$$cB = 2\begin{bmatrix} 4 & 2 \\ -2 & 3 \end{bmatrix} = \begin{bmatrix} 8 & 4 \\ -4 & 6 \end{bmatrix},$$

and

$$A(cB) = \begin{bmatrix} 4 & 2 \\ 0 & -3 \\ 3 & -2 \end{bmatrix} \begin{bmatrix} 8 & 4 \\ -4 & 6 \end{bmatrix} = \begin{bmatrix} 24 & 28 \\ 12 & -18 \\ 32 & 0 \end{bmatrix}.$$

Also

$$c(AB) = 2\left(\begin{bmatrix} 4 & 2 \\ 0 & -3 \\ 3 & -2 \end{bmatrix} \begin{bmatrix} 4 & 2 \\ -2 & 3 \end{bmatrix}\right)$$

$$= 2\begin{bmatrix} 12 & 14 \\ 6 & -9 \\ 16 & 0 \end{bmatrix} = \begin{bmatrix} 24 & 28 \\ 12 & -18 \\ 32 & 0 \end{bmatrix}.$$

so that $A(cB) = c(AB)$. Moreover,

$$cA = 2\begin{bmatrix} 4 & 2 \\ 0 & -3 \\ 3 & -2 \end{bmatrix} = \begin{bmatrix} 8 & 4 \\ 0 & -6 \\ 6 & -4 \end{bmatrix}$$

and

$$(cA)B = \begin{bmatrix} 8 & 4 \\ 0 & -6 \\ 6 & -4 \end{bmatrix} \begin{bmatrix} 4 & 2 \\ -2 & 3 \end{bmatrix} = \begin{bmatrix} 24 & 28 \\ 12 & -18 \\ 32 & 0 \end{bmatrix}.$$

Thus

$$A(cB) = c(AB) = (cA)B.$$

If A is a square matrix, then the entries $a_{11}, a_{22}, \ldots, a_{nn}$ are said to form the ***main diagonal*** of A (see Figure 4.2)

$$A = \begin{bmatrix} a_{11} & a_{12} & \cdots & a_{1n} \\ a_{21} & a_{22} & \cdots & a_{2n} \\ \cdot & \cdot & & \cdot \\ \cdot & \cdot & & \cdot \\ \cdot & \cdot & & \cdot \\ a_{n1} & a_{n2} & \cdots & a_{nn} \end{bmatrix} \qquad \text{(the main diagonal of } A\text{)}.$$

Figure 4.2

Square matrices which have 1's on the main diagonal and 0's elsewhere are of particular importance. They are called ***identity matrices.*** The $n \times n$ identity matrix will be denoted by I_n or sometimes just I.

Example 39 The 2×2 identity matrix is

$$I_2 = \begin{bmatrix} 1 & 0 \\ 0 & 1 \end{bmatrix}$$

and the 3×3 identity matrix is

$$I_3 = \begin{bmatrix} 1 & 0 & 0 \\ 0 & 1 & 0 \\ 0 & 0 & 1 \end{bmatrix}.$$

In ordinary arithmetic, the number 1 has the properties

$$1 \cdot a = a \qquad \text{and} \qquad a \cdot 1 = a.$$

In matrix arithmetic, we have the analogous relationships

$$I_mA = A \qquad \text{and} \qquad AI_n = A \tag{1}$$

for any $m \times n$ matrix A. The following example illustrates the equations in (1).

Example 40 Let

$$A = \begin{bmatrix} a_{11} & a_{12} & a_{13} \\ a_{21} & a_{22} & a_{23} \end{bmatrix}.$$

Then

$$I_2A = \begin{bmatrix} 1 & 0 \\ 0 & 1 \end{bmatrix} \begin{bmatrix} a_{11} & a_{12} & a_{13} \\ a_{21} & a_{22} & a_{23} \end{bmatrix} = \begin{bmatrix} a_{11} & a_{12} & a_{13} \\ a_{21} & a_{22} & a_{23} \end{bmatrix} = A$$

and

$$AI_3 = \begin{bmatrix} a_{11} & a_{12} & a_{13} \\ a_{21} & a_{22} & a_{23} \end{bmatrix} \begin{bmatrix} 1 & 0 & 0 \\ 0 & 1 & 0 \\ 0 & 0 & 1 \end{bmatrix} = \begin{bmatrix} a_{11} & a_{12} & a_{13} \\ a_{21} & a_{22} & a_{23} \end{bmatrix} = A.$$

Matrix multiplication is important in the further study of linear systems. Consider a general linear system of m equations in n unknowns:

$$\begin{aligned} a_{11}x_1 + a_{12}x_2 + \cdots + a_{1n}x_n &= b_1 \\ a_{21}x_1 + a_{22}x_2 + \cdots + a_{2n}x_n &= b_2 \\ \vdots \qquad \vdots \qquad\quad\;\; \vdots \;\;\; & \;\;\; \vdots \\ a_{m1}x_1 + a_{m2}x_2 + \cdots + a_{mn}x_n &= b_m. \end{aligned} \tag{2}$$

Since two matrices are equal if and only if their corresponding entries are equal, we can replace the m equations in (2) by the matrix equation

$$\begin{bmatrix} a_{11}x_1 + a_{12}x_2 + \cdots + a_{1n}x_n \\ a_{21}x_1 + a_{22}x_2 + \cdots + a_{2n}x_n \\ \vdots \\ a_{m1}x_1 + a_{m2}x_2 + \cdots + a_{mn}x_n \end{bmatrix} = \begin{bmatrix} b_1 \\ b_2 \\ \vdots \\ b_m \end{bmatrix}. \tag{3}$$

If we now define

$$A = \begin{bmatrix} a_{11} & a_{12} & \cdots & a_{1n} \\ a_{21} & a_{22} & \cdots & a_{2n} \\ \vdots & \vdots & & \vdots \\ a_{m1} & a_{m2} & \cdots & a_{mn} \end{bmatrix}, \quad X = \begin{bmatrix} x_1 \\ x_2 \\ \vdots \\ x_n \end{bmatrix}, \quad \text{and} \quad B = \begin{bmatrix} b_1 \\ b_2 \\ \vdots \\ b_m \end{bmatrix},$$

then the matrix equation (3) can be written as

$$AX = B. \tag{4}$$

Thus the *entire* system of linear equations in (2) can be represented by the *single* matrix equation in (4). The matrix A in this equation is called the ***coefficient matrix*** of the system. In the next section we shall give some applications of (4).

Example 41 Consider the linear system

$$\begin{aligned} 2x_1 + 3x_2 - 3x_3 - 5x_4 &= 8 \\ -2x_1 \qquad\quad + 5x_3 + 6x_4 &= 12 \\ 3x_1 + 7x_2 - 8x_3 + 2x_4 &= 6. \end{aligned} \tag{5}$$

Letting

$$A = \begin{bmatrix} 2 & 3 & -3 & -5 \\ -2 & 0 & 5 & 6 \\ 3 & 7 & -8 & 2 \end{bmatrix}, \qquad X = \begin{bmatrix} x_1 \\ x_2 \\ x_3 \\ x_4 \end{bmatrix}, \qquad \text{and} \qquad B = \begin{bmatrix} 8 \\ 12 \\ 6 \end{bmatrix}$$

we can write (5) as

$$AX = B.$$

Example 42 The matrix equation

$$\begin{bmatrix} 3 & 4 & -1 & 8 \\ 2 & -3 & 5 & 2 \\ -4 & 2 & 1 & 7 \\ 1 & 1 & -5 & 5 \end{bmatrix} \begin{bmatrix} x_1 \\ x_2 \\ x_3 \\ x_4 \end{bmatrix} = \begin{bmatrix} 8 \\ 2 \\ 7 \\ 5 \end{bmatrix}$$

represents the linear system

$$\begin{aligned} 3x_1 + 4x_2 - x_3 + 8x_4 &= 8 \\ 2x_1 - 3x_2 + 5x_3 + 2x_4 &= 2 \\ -4x_1 + 2x_2 + x_3 + 7x_4 &= 7 \\ x_1 + x_2 - 5x_3 + 5x_4 &= 5. \end{aligned}$$

We conclude this section with a matrix operation that will be needed in later sections. If A is any $m \times n$ matrix, then the ***transpose*** of A, denoted by A^t, is defined to be the $n \times m$ matrix whose first column is the first row of A, whose second column is the second row of A, whose third column is the third row of A, etc. For example, if

$$A = \begin{bmatrix} 1 & -3 & 2 \\ 6 & 1 & 7 \end{bmatrix} \qquad \text{and} \qquad B = \begin{bmatrix} b_{11} & b_{12} \\ b_{21} & b_{22} \end{bmatrix}$$

then

$$A^t = \begin{bmatrix} 1 & 6 \\ -3 & 1 \\ 2 & 7 \end{bmatrix} \qquad \text{and} \qquad B^t = \begin{bmatrix} b_{11} & b_{21} \\ b_{12} & b_{22} \end{bmatrix}.$$

It can be shown that the transpose operation satisfies

(i) $(A^t)^t = A$
(ii) $(A + B)^t = A^t + B^t$
(iii) $(cA)^t = cA^t$
(iv) $(AB)^t = B^tA^t$.

(We omit the proof.)

EXERCISE SET 4.4

1. Let A and B be 3×4 matrices and let C, D, and E be 4×2, 3×2, and 4×3 matrices, respectively. Which of the following matrix expressions are defined? For those that are defined, give the size of the resulting matrix.

   (a) AC (b) AB (c) $BA + C$ (d) $EA + D$

   (e) $E(B + A)$ (f) $AC + D$ (g) $(AE)D$.

2. Let

$$A = \begin{bmatrix} -1 & 2 & 3 \\ 4 & 2 & 0 \\ -3 & 2 & 5 \end{bmatrix}, \quad B = \begin{bmatrix} 3 & -1 & 2 \\ -5 & 3 & 4 \\ -3 & -4 & 0 \end{bmatrix}, \quad C = \begin{bmatrix} 2 & -5 \\ 0 & 4 \\ -5 & 1 \end{bmatrix}$$

$$D = \begin{bmatrix} 1 & 3 & 2 \\ 4 & 5 & -2 \end{bmatrix}, \quad E = \begin{bmatrix} 3 & 5 & -3 \\ 0 & 2 & 4 \end{bmatrix}.$$

   Compute the following, where possible.

   (a) AB (b) BA (c) AC (d) CA

   (e) CD (f) DC (g) EE (h) AA.

3. Using the matrices in Exercise 2, compute the following, where possible.

   (a) $E(A - 2B)$ (b) ABE (c) EBC.

4. Let

$$A = \begin{bmatrix} 2 & -1 \\ 3 & 0 \end{bmatrix}, \quad B = \begin{bmatrix} 4 & 8 \\ 6 & -3 \end{bmatrix}, \quad C = \begin{bmatrix} 1 & 4 \\ 5 & 6 \end{bmatrix}, \quad \text{and} \quad c = -2.$$

   Verify the following.

   (a) $A(BC) = (AB)C$ (b) $A(B + C) = AB + AC$

   (c) $(A + B)C = AC + BC$ (d) $A(cB) = (cA)B$.

5. Let I_3 be the 3×3 identity matrix and let

$$A = \begin{bmatrix} -4 & 3 & 2 \\ 5 & 1 & 2 \\ 3 & -2 & 5 \end{bmatrix}.$$

Verify that

$$AI_3 = I_3A = A.$$

6. Find the matrices A, X, and B that result when the system

$$\begin{aligned} 3x_1 - x_2 + 2x_3 + x_4 + x_5 &= 7 \\ -2x_1 + x_2 + 4x_3 + x_4 - 2x_5 &= -8 \\ 3x_2 + 6x_3 + 3x_4 + x_5 &= 9 \end{aligned}$$

is written in the matrix form $AX = B$.

7. Write out the linear system which is represented in matrix form as

$$\begin{bmatrix} 3 & 2 & 0 \\ 2 & 4 & 5 \\ -3 & 0 & 2 \end{bmatrix} \begin{bmatrix} x_1 \\ x_2 \\ x_3 \end{bmatrix} = \begin{bmatrix} 8 \\ 0 \\ -2 \end{bmatrix}.$$

8. Show that if A has a row of zeros and B is a matrix for which AB is defined, then AB also has a row of zeros.

(b) Discover a similar result involving a column of zeros.

9. Let X_1 be a solution to the linear system

$$AX = B$$

and let Y be a solution to the linear system

$$AX = 0.$$

Show that $X_1 + Y$ is also a solution to the linear system

$$AX = B.$$

10. Suppose the matrix

$$P = \begin{bmatrix} 3 & 4 \\ 2 & 1 \end{bmatrix}$$

describes one hour's output in a small bicycle factory, according to the following table

	Production line 1	Production line 2
3-speed bikes	3	4
10-speed bikes	2	1

Let $M = \begin{bmatrix} 7 \\ 8 \end{bmatrix}$ denote the number of hours in a day that the production lines operate; that is, line 1 operates 7 hours per day and line 2 operates 8 hours per day. Compute PM and give its physical interpretation.

11. Let

$$A = \begin{bmatrix} 1 & 2 \\ -3 & 6 \\ 0 & 1 \end{bmatrix}, \qquad B = \begin{bmatrix} 0 & 3 \\ 5 & 7 \\ 1 & -4 \end{bmatrix}, \qquad \text{and} \qquad c = 3.$$

(a) Compute A^t and B^t.
(b) Verify the transpose properties (i), (ii), and (iii) stated in this section.

12. Let

$$A = \begin{bmatrix} 1 & 0 & -1 \\ 2 & 0 & 6 \end{bmatrix} \qquad \text{and} \qquad B = \begin{bmatrix} 1 & 7 \\ -8 & 4 \\ 0 & 1 \end{bmatrix}.$$

Show that $(AB)^t = B^t A^t$.

4.5 INVERSES OF MATRICES

In Section 4.4 we saw that a system of m linear equations in n unknowns can be written as a single matrix equation

$$AX = B.$$

In this section we shall develop a method that can be used to solve this equation when A is a special kind of square matrix.

If A is a *square* matrix and if we can find a matrix B such that

$$AB = I \qquad \text{and} \qquad BA = I,$$

then A is called an ***invertible*** or ***nonsingular*** matrix. The matrix B is called an ***inverse*** of A.

Example 43 The matrix

$$B = \begin{bmatrix} -3 & 2 \\ 2 & -1 \end{bmatrix}$$

is an inverse of

$$A = \begin{bmatrix} 1 & 2 \\ 2 & 3 \end{bmatrix}$$

since

$$AB = \begin{bmatrix} 1 & 2 \\ 2 & 3 \end{bmatrix} \begin{bmatrix} -3 & 2 \\ 2 & -1 \end{bmatrix} = \begin{bmatrix} 1 & 0 \\ 0 & 1 \end{bmatrix} = I$$

and

$$BA = \begin{bmatrix} -3 & 2 \\ 2 & -1 \end{bmatrix} \begin{bmatrix} 1 & 2 \\ 2 & 3 \end{bmatrix} = \begin{bmatrix} 1 & 0 \\ 0 & 1 \end{bmatrix} = I.$$

The next example shows that not every matrix is invertible.

Example 44 Consider the matrix

$$A = \begin{bmatrix} 2 & 3 \\ 4 & 6 \end{bmatrix}.$$

If A were invertible, we would have a matrix

$$B = \begin{bmatrix} b_{11} & b_{12} \\ b_{21} & b_{22} \end{bmatrix}$$

such that

$$AB = I \qquad \text{and} \qquad BA = I.$$

But the equation

$$AB = I \tag{1}$$

can be written

$$\begin{bmatrix} 2 & 3 \\ 4 & 6 \end{bmatrix} \begin{bmatrix} b_{11} & b_{12} \\ b_{21} & b_{22} \end{bmatrix} = \begin{bmatrix} 1 & 0 \\ 0 & 1 \end{bmatrix}$$

or equivalently

$$\begin{bmatrix} 2b_{11} + 3b_{21} & 2b_{12} + 3b_{22} \\ 4b_{11} + 6b_{21} & 4b_{12} + 6b_{22} \end{bmatrix} = \begin{bmatrix} 1 & 0 \\ 0 & 1 \end{bmatrix}. \tag{2}$$

Since two matrices are equal if and only if their corresponding entries are equal, it follows from (2) that

$$2b_{11} + 3b_{21} = 1$$
$$4b_{11} + 6b_{21} = 0.$$

But this system is inconsistent (why?), so that there can be no matrix B which satisfies (1). Thus, A is not invertible.

By definition, every invertible matrix has an inverse. The following result shows that an invertible matrix cannot have more than one inverse:

Theorem An invertible matrix has exactly one inverse.

It is easy to prove this theorem. Suppose A is an invertible matrix, and B_1 and B_2 are both inverses of A. If we can show that $B_1 = B_2$, then we can conclude that there is only one inverse. This can be done as follows:

Since B_1 is an inverse of A, we know that

$$AB_1 = I \tag{3}$$

and since B_2 is an inverse of A we know that

$$B_2A = I. \tag{4}$$

Multiplying both sides of (3) on the left by B_2 gives

$$B_2(AB_1) = B_2I$$

or

$$B_2(AB_1) = B_2.$$

By the associative law for matrix multiplication, this can be rewritten as

$$(B_2A)B_1 = B_2.$$

From (4), this can be written as

$$IB_1 = B_2$$

or equivalently

$$B_1 = B_2.$$

Thus A has exactly one inverse. ∎

Since an invertible matrix A has exactly one inverse, it is now proper to talk about *the* inverse of A. We shall denote the inverse by A^{-1}. Thus

$$AA^{-1} = I \qquad \text{and} \qquad A^{-1}A = I.$$

We now turn to the problem of finding the inverse of an invertible matrix. Suppose we want to find the inverse of

$$A = \begin{bmatrix} 1 & 2 \\ 2 & 3 \end{bmatrix}.$$

If we let

$$X = \begin{bmatrix} x_1 & x_2 \\ x_3 & x_4 \end{bmatrix}$$

denote the unknown inverse of A, then we must have

$$AX = I \tag{5}$$

and

$$XA = I. \tag{6}$$

The matrix equation (5) can be written

$$\begin{bmatrix} 1 & 2 \\ 2 & 3 \end{bmatrix}\begin{bmatrix} x_1 & x_2 \\ x_3 & x_4 \end{bmatrix} = \begin{bmatrix} 1 & 0 \\ 0 & 1 \end{bmatrix}$$

or

$$\begin{bmatrix} x_1 + 2x_3 & x_2 + 2x_4 \\ 2x_1 + 3x_3 & 2x_2 + 3x_4 \end{bmatrix} = \begin{bmatrix} 1 & 0 \\ 0 & 1 \end{bmatrix}.$$

Since two matrices are equal if and only if their corresponding entries are equal, we must have

$$\begin{aligned} x_1 + 2x_3 &= 1 \\ 2x_1 + 3x_3 &= 0 \end{aligned} \qquad \text{and} \qquad \begin{aligned} x_2 + 2x_4 &= 0 \\ 2x_2 + 3x_4 &= 1. \end{aligned} \tag{7}$$

To solve these systems we must reduce their augmented matrices

$$\begin{bmatrix} 1 & 2 & 1 \\ 2 & 3 & 0 \end{bmatrix} \quad \text{and} \quad \begin{bmatrix} 1 & 2 & 0 \\ 2 & 3 & 1 \end{bmatrix} \tag{8}$$

to reduced row echelon form. The reader can check that these reduced row echelon forms are

$$\begin{bmatrix} 1 & 0 & -3 \\ 0 & 1 & 2 \end{bmatrix} \quad \text{and} \quad \begin{bmatrix} 1 & 0 & 2 \\ 0 & 1 & -1 \end{bmatrix}.$$

Thus

$$\begin{aligned} x_1 &= -3 \\ x_3 &= 2 \end{aligned} \qquad \text{and} \qquad \begin{aligned} x_2 &= 2 \\ x_4 &= -1 \end{aligned}$$

so that the matrix X satisfying (5) is

$$X = \begin{bmatrix} x_1 & x_2 \\ x_3 & x_4 \end{bmatrix} = \begin{bmatrix} -3 & 2 \\ 2 & -1 \end{bmatrix}.$$

The reader can check that this X satisfies (6), so that X is the inverse of A, that is

$$A^{-1} = \begin{bmatrix} -3 & 2 \\ 2 & -1 \end{bmatrix}.$$

We shall now show how to simplify the above computations. To obtain the inverse, we solved the two linear systems in (7) by reducing their augmented matrices (8). Since the two systems have the same coefficient matrix A, we can solve these systems *at the same time* by the following procedure: Write down the coefficient matrix A, and to the right of it, put the constants appearing on the right sides in (7). This gives

$$\left[\begin{array}{cc:cc} 1 & 2 & 1 & 0 \\ 2 & 3 & 0 & 1 \end{array}\right]. \tag{9}$$

(For convenience, we separated A from the constants by a dotted line.) Observe that the matrix to the right of the dotted line is the 2×2 identity matrix I. If we now reduce (9) to reduced row echelon form we will be simultaneously reducing the two augmented matrices in (8), and therefore we will be solving the two systems in (7) at the same time. These computations appear as follows:

$$\begin{array}{c} \quad A \qquad\qquad I \\ \left[\begin{array}{cc:cc} 1 & 2 & 1 & 0 \\ 2 & 3 & 0 & 1 \end{array}\right] \end{array}$$

$$\left[\begin{array}{cc:cc} 1 & 2 & 1 & 0 \\ 0 & -1 & -2 & 1 \end{array}\right]$$ We added -2 times the first row to the last row.

$$\left[\begin{array}{cc:cc} 1 & 2 & 1 & 0 \\ 0 & 1 & 2 & -1 \end{array}\right]$$ We multiplied the last row by -1.

$$\left[\begin{array}{cc:cc} 1 & 0 & -3 & 2 \\ 0 & 1 & 2 & -1 \end{array}\right]$$ We added -2 times the last row to the first.

This 2×4 matrix is now in reduced row echelon form, and the inverse of A appears on the right-hand side of the dotted line. This example suggests the following idea.

> Let A be an invertible $n \times n$ matrix and let $A \,\vdots\, I$ be the matrix obtained by adjoining the $n \times n$ identity matrix on the right side of A. Then the reduced row echelon form of $A \,\vdots\, I$ is $I \,\vdots\, A^{-1}$.

Example 45 Find the inverse of

$$A = \begin{bmatrix} 1 & 2 & 3 \\ 2 & 3 & 4 \\ 1 & 2 & 1 \end{bmatrix}.$$

Solution We adjoin the 3×3 identity matrix on the right side of A and reduce $A \mid I$ to reduced row echelon form.

$$\begin{array}{c} \quad A \qquad\qquad\qquad I \\ \left[\begin{array}{ccc|ccc} 1 & 2 & 3 & 1 & 0 & 0 \\ 2 & 3 & 4 & 0 & 1 & 0 \\ 1 & 2 & 1 & 0 & 0 & 1 \end{array}\right] \end{array}$$

$$\left[\begin{array}{ccc|ccc} 1 & 2 & 3 & 1 & 0 & 0 \\ 0 & -1 & -2 & -2 & 1 & 0 \\ 1 & 2 & 1 & 0 & 0 & 1 \end{array}\right]$$

-2 times the first row was added to the second row.

$$\left[\begin{array}{ccc|ccc} 1 & 2 & 3 & 1 & 0 & 0 \\ 0 & -1 & -2 & -2 & 1 & 0 \\ 0 & 0 & -2 & -1 & 0 & 1 \end{array}\right]$$

-1 times the first row was added to the third row.

$$\left[\begin{array}{ccc|ccc} 1 & 2 & 3 & 1 & 0 & 0 \\ 0 & 1 & 2 & 2 & -1 & 0 \\ 0 & 0 & -2 & -1 & 0 & 1 \end{array}\right]$$

The second row was multiplied by -1.

$$\left[\begin{array}{ccc|ccc} 1 & 2 & 3 & 1 & 0 & 0 \\ 0 & 1 & 2 & 2 & -1 & 0 \\ 0 & 0 & 1 & \frac{1}{2} & 0 & -\frac{1}{2} \end{array}\right]$$

The third row was multiplied by $-\frac{1}{2}$.

$$\left[\begin{array}{ccc|ccc} 1 & 2 & 3 & 1 & 0 & 0 \\ 0 & 1 & 0 & 1 & -1 & 1 \\ 0 & 0 & 1 & \frac{1}{2} & 0 & -\frac{1}{2} \end{array}\right]$$

-2 times the third row was added to the second row.

$$\left[\begin{array}{ccc|ccc} 1 & 2 & 0 & -\frac{1}{2} & 0 & \frac{3}{2} \\ 0 & 1 & 0 & 1 & -1 & 1 \\ 0 & 0 & 1 & \frac{1}{2} & 0 & -\frac{1}{2} \end{array}\right]$$

-3 times the third row was added to the first row.

$$\left[\begin{array}{ccc|ccc} 1 & 0 & 0 & -\frac{5}{2} & 2 & -\frac{1}{2} \\ 0 & 1 & 0 & 1 & -1 & 1 \\ 0 & 0 & 1 & \frac{1}{2} & 0 & -\frac{1}{2} \end{array}\right]$$

-2 times the second row was added to the first row.

This matrix is in reduced row echelon form so that

$$A^{-1} = \begin{bmatrix} -\frac{5}{2} & 2 & -\frac{1}{2} \\ 1 & -1 & 1 \\ \frac{1}{2} & 0 & -\frac{1}{2} \end{bmatrix}.$$

Often, one does not know in advance whether a given matrix A is invertible. If the procedure in the above example is attempted on a matrix that is not invertible, then at some point in the computations a row with all zeros to the left of the dotted line will occur. When this happens, the computations can be stopped and it can be concluded that A is not invertible, because it means that one of the systems of equations we are attempting to solve is inconsistent.

Example 46 Find the inverse of

$$A = \begin{bmatrix} 1 & 3 & 4 \\ 2 & 7 & -1 \\ -1 & -4 & 5 \end{bmatrix}.$$

Solution If we form the matrix $A \mid I_3$ and begin to reduce it to reduced row echelon form, we obtain

$$\left[\begin{array}{rrr|rrr} 1 & 3 & 4 & 1 & 0 & 0 \\ 2 & 7 & -1 & 0 & 1 & 0 \\ -1 & -4 & 5 & 0 & 0 & 1 \end{array}\right]$$

$$\left[\begin{array}{rrr|rrr} 1 & 3 & 4 & 1 & 0 & 0 \\ 0 & 1 & -9 & -2 & 1 & 0 \\ 0 & -1 & 9 & 1 & 0 & 1 \end{array}\right]$$

-2 times the first row was added to the second row and the first row was added to the third row.

$$\left[\begin{array}{rrr|rrr} 1 & 3 & 4 & 1 & 0 & 0 \\ 0 & 1 & -9 & -2 & 1 & 0 \\ 0 & 0 & 0 & -1 & 1 & 1 \end{array}\right]$$

The second row was added to the third row.

Since we have a row with all zeros to the left of the dotted line, the matrix A is not invertible.

The inverse provides a useful alternative way of solving certain systems of n linear equations in n unknowns. To illustrate this method, consider the system

$$\begin{aligned} x + 2y + 3z &= 5 \\ 2x + 3y + 4z &= -2 \\ x + 2y + z &= 3. \end{aligned} \tag{10}$$

We can write this system in matrix form as

$$AX = B \tag{11}$$

where

$$A = \begin{bmatrix} 1 & 2 & 3 \\ 2 & 3 & 4 \\ 1 & 2 & 1 \end{bmatrix}, \qquad X = \begin{bmatrix} x \\ y \\ z \end{bmatrix}, \qquad \text{and} \qquad B = \begin{bmatrix} 5 \\ -2 \\ 3 \end{bmatrix}.$$

In Example 45 we showed that A is an invertible matrix and we computed its inverse

$$A^{-1} = \begin{bmatrix} -\frac{5}{2} & 2 & -\frac{1}{2} \\ 1 & -1 & 1 \\ \frac{1}{2} & 0 & -\frac{1}{2} \end{bmatrix}.$$

If we multiply both sides of (11) on the left by A^{-1} we obtain

$$A^{-1}(AX) = A^{-1}B$$

or

$$(A^{-1}A)X = A^{-1}B$$

or

$$IX = A^{-1}B$$

or

$$X = A^{-1}B. \tag{12}$$

Substituting the matrices X, A^{-1}, and B into (12) gives

$$\begin{bmatrix} x \\ y \\ z \end{bmatrix} = \begin{bmatrix} -\frac{5}{2} & 2 & -\frac{1}{2} \\ 1 & -1 & 1 \\ \frac{1}{2} & 0 & -\frac{1}{2} \end{bmatrix} \begin{bmatrix} 5 \\ -2 \\ 3 \end{bmatrix}$$

or, after multiplying the matrices on the right,

$$\begin{bmatrix} x \\ y \\ z \end{bmatrix} = \begin{bmatrix} -18 \\ 10 \\ 1 \end{bmatrix}.$$

Thus the solution to the given linear system (10) is

$$x = -18, \qquad y = 10, \qquad z = 1.$$

This example illustrates the following general result.

> If $AX = B$ is a linear system of n equations in n unknowns and if the coefficient matrix A is invertible, then the system has one solution, namely $X = A^{-1}B$.

If we are given a linear system, $AX = B$, of n linear equations in n unknowns with an invertible coefficient matrix A, we can solve the system by Gauss–Jordan elimination or by inverting the matrix A and computing $X = A^{-1}B$. The two methods involve about the same amount of computation. In some problems, however, we must solve a number of linear systems

$$AX = B_1, \qquad AX = B_2, \qquad AX = B_3, \qquad \ldots, \qquad \text{and so on.}$$

where the coefficient matrix A is the same for each system, but the constants on the right-hand side change. In this situation, it is best to use the matrix inversion method because once A^{-1} is computed, it can be used to solve each of the systems. The following example illustrates this idea.

Example 47 A plastics firm manufactures two types of commercial Plexiglas rods, flexible and rigid. Each rod undergoes a molding process and a smoothing process. The number of hours per ton required by each process is displayed in the following production matrix:

$$\begin{array}{cc} & \begin{array}{cc}\text{flexible} & \text{rigid}\end{array} \\ A = & \begin{bmatrix} 1 & 2 \\ 1 & 3 \end{bmatrix} \begin{array}{l}\text{molding} \\ \text{smoothing}\end{array} \end{array}$$

If the firm manufactures x_1 tons of flexible rods and x_2 tons of rigid rods, then the matrix product

$$AX = \begin{bmatrix} 1 & 2 \\ 1 & 3 \end{bmatrix} \begin{bmatrix} x_1 \\ x_2 \end{bmatrix} = \begin{bmatrix} x_1 + 2x_2 \\ x_1 + 3x_2 \end{bmatrix}$$

tells us how many total hours will be needed for molding and smoothing; that is, to produce x_1 tons of flexible rods and x_2 tons of rigid rods, the firm will require

$$x_1 + 2x_2$$

hours for molding, and

$$x_1 + 3x_2$$

hours for smoothing.

Let us suppose that the firm's production manager wants answers to the following three questions. How many tons of rods of each type can be produced in a day if

(1) The molding plant operates for a full 16 hours. The smoothing plant operates for a full 20 hours.

(2) The molding plant operates for a full 14 hours. The smoothing plant operates for a full 18 hours.

(3) The molding plant operates for a full 10 hours. The smoothing plant operates for a full 11 hours.

To answer these questions, we form the matrices

$$B_1 = \begin{bmatrix} 16 \\ 20 \end{bmatrix}, \qquad B_2 = \begin{bmatrix} 14 \\ 18 \end{bmatrix}, \qquad \text{and} \qquad B_3 = \begin{bmatrix} 10 \\ 11 \end{bmatrix};$$

these tell us the amount of time available for each process. To answer the production manager's questions, we must solve the three systems

$$AX = B_1, \qquad AX = B_2, \qquad \text{and} \qquad AX = B_3.$$

The reader can verify that the inverse of the matrix A is

$$A^{-1} = \begin{bmatrix} 3 & -2 \\ -1 & 1 \end{bmatrix}.$$

Thus the solution to $AX = B_1$ is

$$X = A^{-1}B_1 = \begin{bmatrix} 3 & -2 \\ -1 & 1 \end{bmatrix} \begin{bmatrix} 16 \\ 20 \end{bmatrix} = \begin{bmatrix} 8 \\ 4 \end{bmatrix}, \tag{13}$$

the solution to $AX = B_2$ is

$$X = A^{-1}B_2 = \begin{bmatrix} 3 & -2 \\ -1 & 1 \end{bmatrix} \begin{bmatrix} 14 \\ 18 \end{bmatrix} = \begin{bmatrix} 6 \\ 4 \end{bmatrix}, \tag{14}$$

and the solution to $AX = B_3$ is

$$X = A^{-1}B_3 = \begin{bmatrix} 3 & -2 \\ -1 & 1 \end{bmatrix} \begin{bmatrix} 10 \\ 11 \end{bmatrix} = \begin{bmatrix} 8 \\ 1 \end{bmatrix}. \tag{15}$$

From (13), (14), and (15) we obtain the following answers to the production manager's questions:

If the molding plant operates for a full 16 hours and the smoothing plant for a full 20 hours, then the firm can produce 8 tons of flexible rods and 4 tons of rigid rods.

If the molding plant operates for a full 14 hours and the smoothing plant for a full 18 hours, then the firm can produce 6 tons of flexible rods and 4 tons of rigid rods.

If the molding plant operates for a full 10 hours and the smoothing plant for a full 11 hours, then the firm can produce 8 tons of flexible rods and 1 ton of rigid rods.

EXERCISE SET 4.5

1. Are either of the following matrices inverses of

$$A = \begin{bmatrix} 1 & 2 \\ 3 & 4 \end{bmatrix}?$$

(a) $\begin{bmatrix} 3 & -2 \\ 1 & 1 \end{bmatrix}$ (b) $\begin{bmatrix} -2 & 1 \\ \frac{3}{2} & -\frac{1}{2} \end{bmatrix}$.

In Exercises 2–5 find the inverse, if possible.

2. (a) $\begin{bmatrix} 2 & -2 \\ 5 & 1 \end{bmatrix}$ (b) $\begin{bmatrix} 2 & 5 \\ 3 & -4 \end{bmatrix}$ (c) $\begin{bmatrix} 1 & -1 \\ 3 & 2 \end{bmatrix}$.

3. (a) $\begin{bmatrix} 1 & 2 & 1 \\ 0 & 1 & 1 \\ 1 & 0 & 1 \end{bmatrix}$ (b) $\begin{bmatrix} 2 & 1 & -1 \\ 1 & 1 & 3 \\ 0 & 1 & 1 \end{bmatrix}$ (c) $\begin{bmatrix} 2 & 1 & -2 \\ 3 & 1 & -5 \\ 3 & 2 & -1 \end{bmatrix}$.

4. (a) $\begin{bmatrix} -1 & 2 & 1 \\ 0 & 1 & 1 \\ 2 & 3 & 1 \end{bmatrix}$ (b) $\begin{bmatrix} 1 & -2 & 2 \\ 0 & -1 & 5 \\ 2 & 1 & 3 \end{bmatrix}$ (c) $\begin{bmatrix} 2 & 0 & 0 & 0 \\ 4 & 3 & 0 & 0 \\ -1 & 3 & -2 & 0 \\ 1 & 1 & 2 & 4 \end{bmatrix}$.

5. (a) $\begin{bmatrix} -1 & -2 & 0 & 0 \\ -4 & -6 & -3 & -3 \\ 2 & 4 & 2 & 3 \\ 2 & 3 & 1 & 1 \end{bmatrix}$ (b) $\begin{bmatrix} 2 & 3 & 1 \\ -2 & 1 & 0 \\ 1 & 1 & 0 \end{bmatrix}$ (c) $\begin{bmatrix} 6 & 5 & -8 \\ 1 & 1 & -2 \\ 3 & 2 & -2 \end{bmatrix}$.

In Exercises 6–9 solve the given linear system by finding the inverse of the coefficient matrix.

6.
$$\begin{aligned} x - 3y &= 5 \\ 4x + 6y &= 7. \end{aligned}$$

7.
$$\begin{aligned} 2x + 3y - z &= 15 \\ x - y + z &= 1 \\ 3x - y + 2z &= 8. \end{aligned}$$

8.
$$\begin{aligned} x_1 - 2x_2 + x_3 &= -2 \\ 2x_1 - x_2 - x_3 &= -10 \\ 3x_1 + x_2 + 2x_3 &= -2. \end{aligned}$$

9.
$$\begin{aligned} 2x_1 + 2x_2 - x_3 &= 2 \\ x_1 + x_2 + x_3 &= 4 \\ x_1 \qquad - x_3 &= -3. \end{aligned}$$

10. By finding the inverse of the coefficient matrix, solve the linear system

$$\begin{aligned} x + 3y &= b_1 \\ 2x - y &= b_2 \end{aligned}$$

when

(a) $b_1 = 1, b_2 = -1$ (b) $b_1 = -2, b_2 = 3$ (c) $b_1 = 3, b_2 = 2$.

11. By finding the inverse of the coefficient matrix, solve the linear system

$$\begin{aligned} 2x_1 - x_2 + x_3 &= b_1 \\ 3x_1 - x_2 - 3x_3 &= b_2 \\ 2x_1 + 2x_2 - x_3 &= b_3 \end{aligned}$$

when

(a) $b_1 = 2,\ b_2 = 1,\ b_3 = 3$ (b) $b_1 = 3,\ b_2 = -4,\ b_3 = 2$

(c) $b_1 = -3,\ b_2 = 2,\ b_3 = 2.$

12. Show that the matrix

$$\begin{bmatrix} a & b & c \\ 0 & 0 & 0 \\ d & e & f \end{bmatrix}$$

is not invertible. Can any square matrix with a row of zeros be invertible?

13. Show that if A is an invertible matrix, then

$$(A^{-1})^{-1} = A.$$

14. Let

$$\begin{bmatrix} 2 & 3 \\ -1 & 2 \end{bmatrix}$$

be the inverse of a matrix A. Find A. (*Hint:* See Exercise 13.)

15. Show that if A and B are invertible matrices, then AB is invertible and $(AB)^{-1} = B^{-1}A^{-1}$.

16. Verify the result of Exercise 15 for the matrices

$$A = \begin{bmatrix} 1 & 1 \\ 2 & -1 \end{bmatrix} \quad \text{and} \quad B = \begin{bmatrix} 1 & 1 \\ 3 & 2 \end{bmatrix}.$$

17. Suppose that a trust fund has \$3000 that must be invested in two different types of bonds. The first bond pays 5% interest per year and the second bond pays 7% interest per year. Determine how to divide the \$3000 among the two types of bonds if the fund must obtain an annual total interest of

(a) \$180? \$200? \$160?

(b) Is it possible to obtain a total annual interest of \$250? Solve this problem using the method of Example 47.

5

LINEAR PROGRAMMING (AN ALGEBRAIC APPROACH)*

In Chapter 3 we showed how linear programming problems involving two variables, x and y, can be solved geometrically. The geometric method we studied there can be extended to linear programming problems involving three variables. However, for problems involving more than three variables geometric techniques must usually be replaced by algebraic ones. In this chapter we shall give an important algebraic technique, called the simplex method, for solving linear programming problems.

5.1 INTRODUCTION; SLACK VARIABLES

The simplex method is applicable only when a linear programming problem satisfies certain conditions. In this section we shall discuss a simplified version of these conditions as well as certain other preliminaries; in the next two sections we shall discuss the simplex method itself.

Consider the following linear programming problems:

(a) Maximize $z = 10x + 12y$

subject to

$$\begin{aligned} 0.2x + 0.4y &\leq 30 \\ 0.2x + 0.2y &\leq 20 \\ x &\geq 0 \\ y &\geq 0 \end{aligned}$$

(b) Minimize $z = 10x - 12y + 2t$

subject to

$$\begin{aligned} 2x - 3y + t &\leq 6 \\ x + y + 9t &\leq 8 \\ x &\geq 0 \\ y &\geq 0 \\ t &\geq 0 \end{aligned}$$

(c) Maximize $z = 0.08x_1 + 0.1x_2$

subject to

$$x_1 + x_2 \leq 5000$$

$$3x_1 - x_2 \leq 17$$

(d) Maximize $z = 3x + 4y$

subject to

$$x + 2y \leq 8$$

$$2x + 2y \geq 10$$

$$x \geq 0$$

$$y \geq 0.$$

Problem (a) is an example of a ***standard linear programming problem.***† This means that:

(i) The objective function is to be maximized.
(ii) The variables are all required to be nonnegative.
(iii) In the other constraints the expressions involving the variables are less than or equal to ($\leq$) a *nonnegative* constant.

Problem (b) is not a standard linear programming problem since the objective function is to be minimized and not maximized as required by condition (i).

Problem (c) is not a standard linear programming problem since the variables x_1 and x_2 are not constrained to be nonnegative, as required by condition (ii).

Problem (d) is not a standard linear programming problem since the constraint

$$2x + 2y \geq 10 \tag{1}$$

violates condition (iii) (the inequality goes the wrong way).

Sometimes linear programming problems that violate condition (iii) can be rewritten in an equivalent form so that this condition is satisfied. For example, consider the problem

(e) Maximize $z = 3x_1 + 2x_2 - x_3$

subject to

$$x_1 - 2x_2 - x_3 \geq -4$$

$$x_1 \geq 0$$

$$x_2 \geq 0$$

$$x_3 \geq 0.$$

† We shall find the term *standard linear programming problem* convenient. This terminology is not used universally however.

As stated, this is not a standard linear programming problem since condition (iii) is violated by the constraint

$$x_1 - 2x_2 - x_3 \geq -4. \tag{2}$$

However, if we multiply both sides of this inequality by -1, we obtain the equivalent inequality†

$$-x_1 + 2x_2 + x_3 \leq 4. \tag{3}$$

If constraint (2) is rewritten in form (3), then problem (e) becomes a standard linear programming problem. To check your understanding of the ideas presented thus far, you may wish to show that this trick cannot be applied to inequality (1) to convert problem (d) into a standard linear programming problem.

In Section 5.4 we shall discuss the solution of certain nonstandard linear programming problems. However, *until that time we shall be working with standard linear programming problems only.*

We now come to the main objective of this section, which is to show that a standard linear programming problem can be reformulated as a problem concerned with systems of equations, rather than inequalities. To illustrate how this can be done, consider the following problem, which we shall refer to as the "original problem."

Original Problem

$$\text{Maximize} \quad z = 10x + 12y$$

subject to

$$0.2x + 0.4y \leq 30 \tag{4a}$$

$$0.2x + 0.2y \leq 20 \tag{4b}$$

$$x \geq 0$$

$$y \geq 0.$$

Since the right-hand sides of (4a) and (4b) are at least as large as the left-hand sides, it is possible to convert these inequalities into equalities by adding *nonnegative* quantities v and w on the left. When this is done (4a) and (4b) become the equations

$$0.2x + 0.4y + v = 30 \tag{4a$'$}$$

$$0.2x + 0.2y + w = 20, \tag{4b$'$}$$

† Recall that if both sides of an inequality are multiplied by a negative number, then the inequality gets reversed. For example, if we multiply both sides of the inequality $-2 \leq 4$ by -1, the result is $2 \geq -4$.

where

$$v \geq 0 \qquad \text{and} \qquad w \geq 0. \tag{5}$$

The quantities v and w are called ***slack variables*** because they take up the "slack" or difference between the two sides.

If we replace inequalities (4a) and (4b) by equalities (4a′) and (4b′) and use the constraints in (5), we obtain a modification of the original problem called the "new problem with slack variables."

New Problem with Slack Variables

$$\text{Maximize} \quad z = 10x + 12y$$

subject to

$$\begin{aligned} 0.2x + 0.4y + v &= 30 \\ 0.2x + 0.2y + w &= 20 \\ x &\geq 0 \\ y &\geq 0 \\ v &\geq 0 \\ w &\geq 0. \end{aligned}$$

Although we shall omit the details, it is not difficult to show that the original problem and the new problem with slack variables are equivalent in the following sense. If

$$x = a, \qquad y = b$$

is any solution of the original problem, then there are values of the slack variables, say $v = c$ and $w = d$, such that

$$x = a, \qquad y = b, \qquad v = c, \qquad w = d$$

is a solution of the new problem with slack variables. Conversely, if

$$x = a, \qquad y = b, \qquad v = c, \qquad w = d$$

is any solution of the new problem with slack variables, then

$$x = a, \qquad y = b$$

is a solution of the original problem.

This result is important since it enables us to concentrate on solving the new problem with slack variables, where it will be possible for us to take advantage of our work on matrices and systems of equations.

Example 1 In Example 11 of Section 3.2 we showed that

$$x = 0, \qquad y = 20$$

is a solution of the problem

$$\text{Maximize} \quad z = 15x + 30y$$

subject to

$$\begin{aligned} 0.2x + 0.4y &\leq 8 \\ 0.5x + 0.2y &\leq 8 \\ x &\geq 0 \\ y &\geq 0. \end{aligned}$$

Use this result to find a solution of the new problem with slack variables.

Solution The new problem with slack variables is

$$\text{Maximize} \quad z = 15x + 30y$$

subject to

$$\begin{aligned} 0.2x + 0.4y + v &= 8 \qquad (6) \\ 0.5x + 0.2y + w &= 8 \qquad (7) \\ x &\geq 0 \\ y &\geq 0 \\ v &\geq 0 \\ w &\geq 0. \end{aligned}$$

Substituting the values $x = 0$, $y = 20$ in (6) and (7) we obtain

$$\begin{aligned} 0.2(0) + 0.4(20) + v &= 8 \\ 0.5(0) + 0.2(20) + w &= 8 \end{aligned}$$

so that $v = 0$ and $w = 4$. Since these values satisfy $v \geq 0$, $w \geq 0$, we have

$$x = 0, \qquad y = 20, \qquad v = 0, \qquad w = 4$$

as a solution of the new problem.

In this example we started with a solution of the original problem and then obtained a solution of the new problem with slack variables. We are now interested in going the other way. The simplex method described in later sections will enable us to solve the new problem with slack variables. Using this solution, we will then obtain a solution of the original problem.

In applied problems, the slack variables have physical interpretations. To illustrate, the problem in the last example arose from the fuel-manufacturing problem in Example 4 of Section 3.1. The slack variable v represents the difference between the total of 8 hours available for cracking and the number of hours $0.2x + 0.4y$ required for cracking. Similarly, w is the difference between the total of 8 hours available for refining and the number of hours $0.5x + 0.2y$ required for refining. Thus, for the optimal solution $x = 0$, $y = 20$ all the available cracking time will be used since $v = 0$. Since $w = 4$, there will be 4 hours of refining time unused.

We conclude this section by showing how linear programming problems can be formulated using matrix terminology. Consider the standard linear programming problem

$$\text{Maximize} \quad z = 10x + 12y \tag{8}$$

subject to

$$0.2x + 0.4y \leq 30 \tag{9a}$$

$$0.2x + 0.2y \leq 20 \tag{9b}$$

$$x \geq 0 \tag{10a}$$

$$y \geq 0 \tag{10b}$$

and let

$$A = \begin{bmatrix} 0.2 & 0.4 \\ 0.2 & 0.2 \end{bmatrix}, \qquad B = \begin{bmatrix} 30 \\ 20 \end{bmatrix}, \qquad X = \begin{bmatrix} x \\ y \end{bmatrix}, \qquad C = \begin{bmatrix} 10 & 12 \end{bmatrix}.$$

If we agree to omit the brackets on 1×1 matrices, then the quantity z in statement (8) can be written as the matrix product

$$z = \begin{bmatrix} 10 & 12 \end{bmatrix} \begin{bmatrix} x \\ y \end{bmatrix}$$

or equivalently

$$z = CX.$$

Further, constraints (9a) and (9b) can be written†

$$\begin{bmatrix} 0.2x + 0.4y \\ 0.2x + 0.2y \end{bmatrix} \leq \begin{bmatrix} 30 \\ 20 \end{bmatrix}$$

or

$$\begin{bmatrix} 0.2 & 0.4 \\ 0.2 & 0.2 \end{bmatrix} \begin{bmatrix} x \\ y \end{bmatrix} \leq \begin{bmatrix} 30 \\ 20 \end{bmatrix}$$

or

$$AX \leq B.$$

† If E and F are matrices of the same size, we shall write $E \leq F$ if each entry in E is $\leq$ the corresponding entry in F.

Finally, constraints (10a) and (10b) can be written

$$\begin{bmatrix} x \\ y \end{bmatrix} \geq \begin{bmatrix} 0 \\ 0 \end{bmatrix}$$

or equivalently

$$X \geq 0.$$

Combining these results, we obtain the following matrix formulation of the problem:

$$\text{Maximize} \quad z = CX$$

subject to

$$AX \leq B$$
$$X \geq 0.$$

EXERCISE SET 5.1

1. Which of the following are standard linear programming problems?

(a) Maximize $z = 2x - y$

subject to

$$x + y \leq 1$$
$$x \geq 0$$
$$y \geq 0.$$

(b) Maximize $z = 3x + y + t$

subject to

$$2x - y + t \leq 3$$
$$x + t \leq 5.$$

(c) Minimize $z = 2x_1 - 3x_2 + 3x_3$

subject to

$$2x_1 + 3x_2 + 3x_3 \leq 20$$
$$3x_1 + 2x_2 - x_3 \leq 30$$
$$x_1 \geq 0, \quad x_2 \geq 0, \quad x_3 \geq 0.$$

(d) Maximize $z = 3x + 2y$

subject to

$$x + 2y \leq 7$$
$$5x - 8y \leq -3$$
$$x \geq 0, \quad y \geq 0.$$

2. Rewrite the following as standard linear programming problems.

(a) Maximize $z = 4x - 2y$

subject to

$$x + 2y \leq 7$$
$$5x - 8y \geq -3$$
$$x \geq 0, \quad y \geq 0.$$

(b) Maximize $z = 2x_1 + 3x_2 + x_3$

subject to

$$2x_1 + x_2 - x_3 \geq -5$$
$$3x_1 + x_2 + 5x_3 \geq -8$$
$$x_1 \geq 0, \quad x_2 \geq 0, \quad x_3 \geq 0.$$

3. Can the following problem be rewritten as a standard linear programming problem? Explain.

$$\text{Maximize} \quad z = 5x - 8y$$

subject to

$$x - 3y \geq -5$$
$$2x + y \geq 4$$
$$x \geq 0, \quad y \geq 0.$$

4. Rewrite the following as equivalent new problems with slack variables.

(a) Maximize $z = 6x - 9y$

subject to

$$\begin{aligned} 6x - 7y &\leq 4 \\ 2x + 5y &\leq 5 \\ x &\geq 0 \\ y &\geq 0. \end{aligned}$$

(b) Maximize $z = 4x_1 + 3x_2 + 2x_3$

subject to

$$\begin{aligned} 2x_1 + x_2 - 3x_3 &\leq 8 \\ x_1 + 2x_2 + 4x_3 &\leq 14 \\ 2x_1 + 3x_2 - x_3 &\leq 12 \\ x_1 \geq 0, \quad x_2 \geq 0, \quad x_3 &\geq 0. \end{aligned}$$

5. Rewrite the following as equivalent new problems with slack variables.

(a) Maximize $z = 5x + 7y$

subject to

$$\begin{aligned} 4x - 6y &\leq 9 \\ 2x + 7y &\leq 3 \\ 5x - 8y &\leq 2 \\ x \geq 0, \quad y &\geq 0. \end{aligned}$$

(b) Maximize $z = -x_1 + x_2 - x_3 + x_4$

subject to

$$\begin{aligned} 2x_1 + 3x_3 + x_4 &\leq 8 \\ x_1 - 2x_2 + x_4 &\leq 6 \\ x_1 \geq 0, \quad x_2 &\geq 0, \\ x_3 \geq 0, \quad x_4 &\geq 0. \end{aligned}$$

6. The following are linear programming problems with slack variables u, v, and w. Find the original problem.

(a) Maximize $z = x - 3y$

subject to

$$\begin{aligned} 2x + 4y + u &= 5 \\ 3x - 2y + v &= 7 \\ x \geq 0, \quad y &\geq 0, \\ u \geq 0, \quad v &\geq 0. \end{aligned}$$

(b) Maximize $z = 3x_1 + 2x_2 - x_3$

subject to

$$\begin{aligned} 2x_1 + 3x_2 - x_3 + u &= 12 \\ 2x_1 + x_2 \qquad + v &= 18 \\ 3x_1 + x_2 + 4x_3 + w &= 13 \\ x_1 \geq 0, \quad x_2 \geq 0, \quad x_3 &\geq 0, \\ u \geq 0, \quad v \geq 0, \quad w &\geq 0. \end{aligned}$$

7. (a) Write the problem in Example 8 of Section 3.2 as a new problem with slack variables.

(b) Use the solution obtained in Example 8 of Section 3.2 to obtain a solution of the new problem with slack variables.

8. The following problems can be written in matrix notation as

Maximize $z = CX$

subject to

$$\begin{aligned} AX &\leq B \\ X &\geq 0. \end{aligned}$$

Find the matrices A, B, C, and X.

(a) Maximize $z = 6x - 9y$

subject to

$$\begin{aligned} 6x - 7y &\leq 4 \\ 2x + 5y &\leq 5 \\ x &\geq 0 \\ y &\geq 0. \end{aligned}$$

(b) Maximize $z = x_1 + x_2 - x_3 + x_4$

subject to

$$\begin{aligned} 2x_1 + 3x_3 + x_4 &\leq 8 \\ x_1 - 2x_2 + x_4 &\leq 6 \\ x_1 \geq 0, \quad x_2 &\geq 0 \\ x_3 \geq 0, \quad x_4 &\geq 0. \end{aligned}$$

9. Repeat the directions of Exercise 8 for the problems

(a) Maximize $z = 4x - 2y + 7t$

subject to

$$\begin{aligned} x - t &\leq 1 \\ y - t &\leq 2 \\ x - y &\leq 3 \end{aligned}$$

$$x \geq 0, \quad y \geq 0, \quad t \geq 0.$$

(b) Maximize $z = x_1 - x_2$

subject to

$$\begin{aligned} x_1 + 3x_2 &\leq 5 \\ x_1 \geq 0, \quad x_2 &\geq 0. \end{aligned}$$

10. Let

$$A = \begin{bmatrix} 3 & 4 & -3 \\ 2 & 5 & -3 \end{bmatrix}, \qquad B = \begin{bmatrix} 2 \\ 5 \end{bmatrix}, \qquad X = \begin{bmatrix} x_1 \\ x_2 \\ x_3 \end{bmatrix}, \qquad C = \begin{bmatrix} 8 & 4 & 2 \end{bmatrix}.$$

Rewrite the problem

Maximize $z = CX$

subject to

$$\begin{aligned} AX &\leq B \\ X &\geq 0 \end{aligned}$$

without using matrix notation.

11. Write the following linear programming problem without using matrix notation:

$$\text{Maximize} \quad z = \begin{bmatrix} 6 & 7 & 9 \end{bmatrix} \begin{bmatrix} x \\ y \\ t \end{bmatrix}$$

subject to

$$\begin{bmatrix} 2 & -4 & 5 \\ 7 & 1 & 3 \end{bmatrix} \begin{bmatrix} x \\ y \\ t \end{bmatrix} \leq \begin{bmatrix} 3 \\ 9 \end{bmatrix}$$

$$\begin{bmatrix} x \\ y \\ t \end{bmatrix} \geq \begin{bmatrix} 0 \\ 0 \\ 0 \end{bmatrix}.$$

12. For the problems in Exercise 8, the new problems with slack variables can be written in matrix notation as

$$\text{Maximize} \quad z = C_s X_s$$

subject to

$$A_s X_s = B$$

$$X_s \geq 0.$$

Find A_s, B, C_s, and X_s for

(a) the problem in Exercise 8a

(b) the problem in Exercise 8b.

5.2 THE KEY IDEAS OF THE SIMPLEX METHOD

In this section we discuss the basic ideas needed to solve linear programming problems by the simplex method. Since the technique is somewhat involved, we shall use this section to motivate and discuss the basic concepts. In the next section we shall formulate a step-by-step procedure for using the ideas we develop here.

In the previous section we showed how to reformulate the standard linear programming problem

$$\text{Maximize} \quad z = CX \tag{1}$$

subject to

$$AX \leq B \tag{2}$$

$$X \geq 0 \tag{3}$$

as a new equivalent problem with slack variables. This was accomplished by adding slack variables to convert each of the inequalities in (2) into equalities. In our discussion of slack variables we emphasized problems with two variables in which (2) had two constraints. However, we shall now want to consider the general case, where we have m variables and $AX \leq B$ has k constraints. Thus to convert each of the k inequalities in $AX \leq B$ into equalities we would have to introduce k slack variables. Therefore, where the original problem contains m variables, the new problem will contain these m variables together with k slack variables, making a total of $m + k$ variables. At first, it might seem that we have complicated affairs by introducing these extra variables. Fortunately, this is not the case. To explain why, we shall need the following result, which we state without proof.

If a standard linear programming problem

$$\begin{aligned} \text{Maximize} \quad & z = CX \\ \text{subject to} \quad & AX \leq B \\ & X \geq 0 \end{aligned}$$

with m unknowns has a solution, then there is a solution of the new problem with slack variables in which at least m variables have value zero. Such a solution is called a ***basic feasible solution*** of the new problem with slack variables.

This result shows that the solutions of a standard linear programming problem can be found by considering the basic feasible solutions of the new problem. The following example illustrates how this result can be used to solve linear programming problems.

Example 2 Consider the problem

$$\text{Maximize} \quad z = -x + y$$

subject to

$$\begin{aligned} 0.2x + 0.4y &\leq 30 \\ 0.2x + 0.2y &\leq 20 \\ x \geq 0, \quad y &\geq 0. \end{aligned}$$

The corresponding new problem with slack variables is

$$\text{Maximize} \quad z = -x + y$$

subject to

$$0.2x + 0.4y + v = 30 \tag{4}$$

$$0.2x + 0.2y + w = 20 \tag{5}$$

$$x \geq 0, \quad y \geq 0, \quad v \geq 0, \quad w \geq 0. \tag{6}$$

According to the result we stated above, this new problem with slack variables has a solution in which at least two of the variables have value zero (since the original problem has $m = 2$ unknowns, x and y). Thus there is a solution of the new problem in which one of the following six situations occurs:

$$x = 0 \quad \text{and} \quad y = 0; \qquad y = 0 \quad \text{and} \quad v = 0;$$
$$x = 0 \quad \text{and} \quad v = 0; \qquad y = 0 \quad \text{and} \quad w = 0;$$
$$x = 0 \quad \text{and} \quad w = 0; \qquad v = 0 \quad \text{and} \quad w = 0.$$

In each of these cases we can use Equations (4) and (5) to find the values of the remaining two variables. For example, if

$$x = 0 \qquad \text{and} \qquad v = 0,$$

we can substitute these values into (4) and (5) to obtain the system of equations

$$\begin{aligned} 0.4y \quad &= 30 \\ 0.2y + w &= 20. \end{aligned}$$

Solving for y and w we obtain

$$y = 75, \qquad w = 5.$$

Thus

$$x = 0, \qquad v = 0, \qquad y = 75, \qquad w = 5$$

is a possible solution of the new problem with slack variables.

If we follow this procedure in each of the six cases, we obtain the following list of possible solutions to the new problem with slack variables:

$$x = 0, \qquad y = 0, \qquad v = 30, \qquad w = 20 \tag{7a}$$
$$x = 0, \qquad v = 0, \qquad y = 75, \qquad w = 5 \tag{7b}$$
$$x = 0, \qquad w = 0, \qquad y = 100, \qquad v = -10 \tag{7c}$$
$$y = 0, \qquad v = 0, \qquad x = 150, \qquad w = -10 \tag{7d}$$
$$y = 0, \qquad w = 0, \qquad x = 100, \qquad v = 10 \tag{7e}$$
$$v = 0, \qquad w = 0, \qquad x = 50, \qquad y = 50. \tag{7f}$$

We know there is a solution of the new problem somewhere among these six possibilities. However, we can eliminate (7c) and (7d) immediately since they are not basic feasible solutions (v and w violate the nonnegativity constraints). This leaves (7a), (7b), (7e), and (7f) as the only candidates. To finish, we need only evaluate the objective function

$$z = -x + y$$

at each basic feasible solution to determine where the maximum value occurs. Table 5.1 summarizes the results of these computations. From this table we see that the maximum value of z is 75, so that

$$x = 0, \qquad v = 0, \qquad y = 75, \qquad w = 5 \tag{8}$$

is a solution of the new problem with slack variables. As discussed in the previous section, a solution of the original problem can be obtained by dis-

Table 5.1

Basic feasible solutions of the new problem with slack variables	Value of $z = -x + y$
$x = 0, \quad y = 0, \quad v = 30, \quad w = 20$	0
$x = 0, \quad v = 0, \quad y = 75, \quad w = 5$	75
$v = 0, \quad w = 0, \quad x = 50, \quad y = 50$	0
$y = 0, \quad w = 0, \quad x = 100, \quad v = 10$	-100

carding the slack variables from a solution of the new problem. Thus, discarding $v = 0$ and $w = 5$ from (8) we obtain the solution

$$x = 0, \qquad y = 75$$

of the original problem. The reader may wish to check this result by solving the problem geometrically, as discussed in Chapter 3. Observe, however, that the solution in this example was completely *algebraic*.

There is a noteworthy relationship between the geometric solution and the algebraic solution. In the geometric solution we look for the maximum value of z at the corner points of the convex set determined by the constraints, while in the algebraic solution we look for the maximum value of z at the basic feasible solutions of the new problem with slack variables. This suggests that there must be a relationship between the basic feasible solutions of the new problem and the corner points in the original problem. This conjecture is correct. To illustrate the relationship, we have sketched in Figure 5.1 the convex set determined by the constraints in the original problem of Example 2; moreover, in Table 5.2 we have listed the corner points of this convex set and next to them the basic feasible solutions of the new problem. (These basic feasible solutions were previously shown in Table 5.1.)

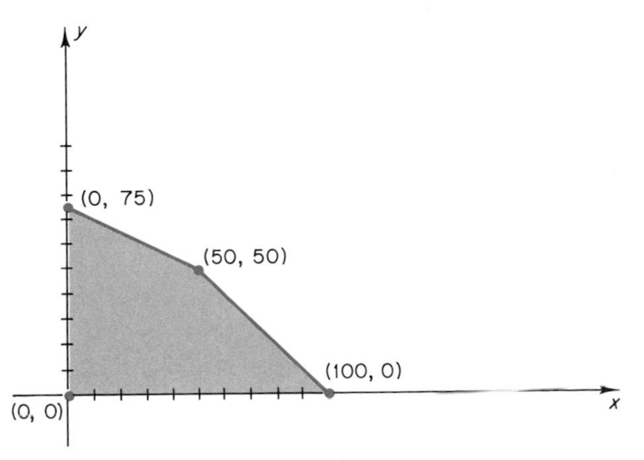

Figure 5.1

Table 5.2

Corner points	Basic feasible solutions
(0, 0)	$x = 0,\quad y = 0,\quad v = 30,\quad w = 20$
(0, 75)	$x = 0,\quad y = 75,\quad v = 0,\quad w = 5$
(50, 50)	$x = 50,\quad y = 50,\quad v = 0,\quad w = 0$
(100, 0)	$x = 100,\quad y = 0,\quad v = 10,\quad w = 0$

The relationship between corner points and basic feasible solutions is now evident from Table 5.2. We need only observe that coordinates of each corner point in the left column are just the x and y values of the corresponding basic feasible solution in the right column. For example, the coordinates of the corner point

$$(0, 75)$$

are just the x and y values in the corresponding basic feasible solution

$$x = 0, \qquad y = 75, \qquad v = 0, \qquad w = 5.$$

Since the objective function

$$z = -x + y$$

does not involve the slack variables v and w, it is clear that evaluating z at a basic feasible solution is equivalent to evaluating z at the corresponding corner point.

The algebraic method we used in Example 2 is fine in theory. However, in practical problems this method breaks down because the amount of computation involved becomes prohibitively large. For example, in a problem with 20 variables and 10 constraints there could conceivably be over thirty million basic feasible solutions to check if we used this method! To remedy this difficulty we shall develop ways of refining the technique in Example 2 to reduce the amount of computation. If we have a linear programming problem with m unknowns, it is evident that the key to solving this problem algebraically is to determine which of the m variables in the final solution of the new problem with slack variables are equal to zero; once this is known, the problem reduces to solving a system of linear equations. Roughly speaking, the simplex method is just an organized way of searching for the m variables that are zero in the solution. Before we can discuss the details we shall need to develop some preliminary ideas.

In Example 2, the new problem with slack variables involved nonnegativity constraints and the equations

$$\begin{aligned} z &= -x + y \\ 0.2x + 0.4y + v &= 30 \\ 0.2x + 0.2y + w &= 20. \end{aligned}$$

If we take the x and y terms in the first equation over to the left, we can rewrite this system with the variables aligned as follows

$$\begin{aligned} x - y \qquad\qquad + z &= 0 \\ 0.2x + 0.4y + v \qquad\qquad &= 30 \\ 0.2x + 0.2y \qquad + w \qquad &= 20. \end{aligned} \tag{9a}$$

Observe that the variables z, v, and w each occur in exactly one equation and where they occur they have a coefficient of $+1$. We call these the ***explicit variables*** or sometimes ***basic variables.*** More precisely, basic variables are those variables that have a coefficient $+1$ in only one equation and coefficient 0 in the remaining equations. We call the remaining variables ***implicit*** or ***nonbasic*** variables. Equations (9a) are particularly convenient for finding the basic feasible solution in which the implicit variables x and y are zero. If we set $x = 0$, $y = 0$ in (9a) we obtain

$$\begin{aligned} z &= 0 \\ v &= 30 \\ w &= 20. \end{aligned}$$

Thus the basic feasible solution in which x and y are zero is

$$x = 0, \qquad y = 0, \qquad v = 30, \qquad w = 20$$

and the value of the objective function at this solution is

$$z = 0.$$

The augmented matrix for (9a) is

$$\begin{bmatrix} 1 & -1 & 0 & 0 & 1 & 0 \\ 0.2 & 0.4 & 1 & 0 & 0 & 30 \\ 0.2 & 0.2 & 0 & 1 & 0 & 20 \end{bmatrix}. \tag{9b}$$

It is called the ***tableau with x and y implicit*** or sometimes the ***tableau corresponding to the basic feasible solution where $x = 0$ and $y = 0$.*** For mnemonic reasons, it is helpful to indicate along the upper margin of the tableau which columns go with which variables, and to indicate along the left-hand side of the tableau which variable is explicit in the corresponding equation. Thus we would write (9b) as

$$\begin{array}{c} \\ z \\ v \\ w \end{array}\begin{array}{c} \begin{array}{cccccc} x & y & v & w & z & \end{array} \\ \begin{bmatrix} 1 & -1 & 0 & 0 & 1 & 0 \\ 0.2 & 0.4 & 1 & 0 & 0 & 30 \\ 0.2 & 0.2 & 0 & 1 & 0 & 20 \end{bmatrix} \end{array}. \tag{9c}$$

Observe that the entries in the last column tell us the values of z, v, and w when the implicit variables are set equal to zero.

If we were interested in the basic feasible solution where y and w have value zero, it would be more convenient to have equations (9a) written so that y and w are implicit (or equivalently so that z, x, and v are explicit). Let us illustrate how to do this. Since z and v are already explicit in (9a), we need only replace the old explicit variable w in the last equation by the new explicit variable x. To do this we must get rid of the x terms in the first two equations and give x a coefficient of $+1$ in the last equation. To accomplish this we first solve the last equation

$$0.2x + 0.2y + w = 20$$

for x by rewriting it as

$$0.2x = -0.2y - w + 20$$

or

$$x = -y - 5w + 100. \tag{10}$$

By substituting (10) into the first two equations of (9a) we can rid ourselves of the unwanted x terms. This gives

$$(-y - 5w + 100) - y + z = 0$$

and

$$0.2(-y - 5w + 100) + 0.4y + v = 30,$$

or upon simplifying

$$-2y - 5w + z = -100 \tag{11}$$

$$0.2y + v - w = 10. \tag{12}$$

Writing (10), (11), and (12) with the variables aligned we obtain the equations

$$\begin{aligned} -2y \quad\quad - 5w + z &= -100 \\ 0.2y + v - w \quad\quad &= 10 \\ x + y \quad\quad + 5w \quad\quad &= 100. \end{aligned} \tag{13}$$

Observe that x, v, and z are now explicit since they each occur in exactly one equation and where they occur they have a coefficient of $+1$. From these equations it is easy to obtain the basic feasible solution in which the implicit variables y and w have value zero; we just set $y = 0$, $w = 0$ in (13) to obtain

$$z = -100$$
$$v = 10$$
$$x = 100.$$

Thus the basic feasible solution in which y and w have value zero is

$$x = 100, \qquad y = 0, \qquad v = 10, \qquad w = 0$$

and the value of the objective function at this solution is

$$z = -100.$$

By writing the augmented matrix for (13) we obtain the tableau with x, v, and z explicitly, namely

$$\begin{array}{c} \\ z \\ v \\ x \end{array} \begin{array}{c} \begin{array}{cccccc} x & y & v & w & z & \end{array} \\ \left[\begin{array}{cccccc} 0 & -2 & 0 & -5 & 1 & -100 \\ 0 & 0.2 & 1 & -1 & 0 & 10 \\ 1 & 1 & 0 & 5 & 0 & 100 \end{array}\right] \end{array}. \tag{14}$$

The reader may wish to check his understanding of these ideas by showing that Equations (13) can be rewritten with z, y, and x explicit as

$$\begin{aligned} 10v - 15w + z &= 0 \\ y + 5v - 5w &= 50 \\ x - 5v + 10w &= 50, \end{aligned} \tag{15}$$

where the corresponding tableau is

$$\begin{array}{c} \\ z \\ y \\ x \end{array} \begin{array}{c} \begin{array}{cccccc} x & y & v & w & z & \end{array} \\ \left[\begin{array}{cccccc} 0 & 0 & 10 & -15 & 1 & 0 \\ 0 & 1 & 5 & -5 & 0 & 50 \\ 1 & 0 & -5 & 10 & 0 & 50 \end{array}\right] \end{array}. \tag{16}$$

The reader can also check that Equations (15) can be rewritten with z, y, and w explicit as

$$\begin{aligned} 1.5x + 2.5v + z &= 75 \\ 0.5x + y + 2.5v &= 75 \\ 0.1x - 0.5v + w &= 5, \end{aligned}$$

where the corresponding tableau is

$$\begin{array}{c} \\ z \\ y \\ w \end{array}\begin{array}{c} \begin{array}{cccccc} x & y & v & w & z & \end{array} \\ \left[\begin{array}{cccccc} 1.5 & 0 & 2.5 & 0 & 1 & 75 \\ 0.5 & 1 & 2.5 & 0 & 0 & 75 \\ 0.1 & 0 & -0.5 & 1 & 0 & 5 \end{array}\right] \end{array}. \tag{17}$$

Example 3 For the problem in Example 2 use a tableau to find the value of z at the basic feasible solution where x and v equal zero.

Solution We look for the tableau in which x and v are implicit. This is Tableau (17). If set $x = 0$ and $v = 0$, then the entries in the last column of this tableau tell us that

$$z = 75$$
$$y = 75$$
$$w = 5.$$

Thus, at the basic feasible solution

$$x = 0, \qquad y = 75, \qquad v = 0, \qquad w = 5$$

the value of z is 75.

In the above discussion, we labeled each row of the tableau with the explicit variable in the corresponding equation. Although convenient, this is really unnecessary since this information can be obtained from the entries in the tableau. To illustrate, consider Tableau (17). The fact that the z column of the tableau is

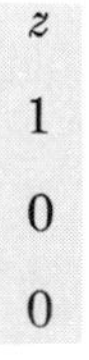

tells us that z appears in the first equation with a coefficient of $+1$ and occurs in no other equation. Thus z is explicit. Similarly, the y and w columns

y	w
0	0
1	0
0	1

tells us that y and w are explicit since y occurs in the second equation with coefficient $+1$ but occurs in no other equation, while w occurs in the third equation with coefficient $+1$ and occurs in no other equation.

These observations will be important in the next section.

Example 4 The explicit variables in the tableau

$$\begin{array}{cccccc} x & y & v & w & z & \end{array}$$
$$\begin{bmatrix} 0 & -2 & 0 & -5 & 1 & 100 \\ 0 & 0.2 & 1 & -1 & 0 & 10 \\ 1 & 1 & 0 & 5 & 0 & -100 \end{bmatrix}$$

are x, v, and z. (Why?)

EXERCISE SET 5.2

1. Solve using the method of Example 2:

$$\text{Maximize} \quad z = 5x + 7y$$
$$\text{subject to}$$
$$x + y \leq 1$$
$$x \geq 0, \quad y \geq 0.$$

2. Solve the problem in Exercise 1 geometrically and match up the corner points with the basic feasible solutions computed in Exercise 1 (as in Table 5.2).

3. Solve using the method of Example 2:

$$\text{Maximize} \quad z = 6x + 2y$$
$$\text{subject to}$$
$$-3x + 2y \leq 6$$
$$5x + 3y \leq 15$$
$$x \geq 0, \quad y \geq 0.$$

4. Solve the problem in Exercise 3 geometrically and match up the corner points with the basic feasible solutions computed in Exercise 3 (as in Table 5.2).

5. Solve using the method of Example 2:

$$\text{Maximize} \quad z = -2x - 5y$$
$$\text{subject to}$$
$$3x + 2y \leq 2$$
$$2x + 5y \leq 8$$
$$x \geq 0, \quad y \geq 0.$$

6. In each of the following, name the explicit variables

(a)
$$\begin{aligned} 3y \qquad - 5w + z &= 105 \\ x + 2y \qquad + 8w \qquad &= 20 \\ 7y + 6v - 4w \qquad &= 13 \end{aligned}$$

(b)
$$\begin{aligned} -10x - 12y \qquad\qquad + z &= 0 \\ x + 2y + v \qquad\qquad &= 150 \\ x + y \qquad + w \qquad &= 40 \end{aligned}$$

(c)
$$\begin{aligned} 6v \qquad + z &= 0 \\ y + 2v + w \qquad &= 75 \\ x \qquad - 2v - 2w \qquad &= 13. \end{aligned}$$

7. Obtain tableau (16) from Equations (13).

8. Obtain tableau (17) from Equations (15).

9. Consider the problem

$$\text{Maximize} \quad z = 2x + 5y$$

subject to

$$\begin{aligned} 3x + 2y &\leq 2 \\ 2x + 5y &\leq 8 \\ x \geq 0, \quad y &\geq 0. \end{aligned}$$

(a) Form the new problem with slack variables v and w.

(b) Find a tableau in which x and y are implicit.

(c) Find a tableau in which x and w are implicit.

(d) Find a tableau in which y and w are implicit.

10. Consider the problem

$$\text{Maximize} \quad z = -3x_1 + 5x_2 + 2x_3$$

subject to

$$\begin{aligned} 2x_1 + 3x_2 + 4x_3 &\leq 18 \\ 3x_1 + 2x_2 + 3x_3 &\leq 12 \\ x_1 \geq 0, \quad x_2 \geq 0, \quad x_3 &\geq 0. \end{aligned}$$

(a) Form the new problem with slack variables v_1 and v_2.

(b) Find a tableau in which x_1, x_2, and x_3 are implicit.

(c) Find a tableau in which x_1, x_2, and v_1 are implicit.

(d) Find a tableau in which x_1, v_1, and v_2 are implicit.

11. Name the explicit variables in the following tableaus.

$$\text{(a)} \quad \begin{array}{cccccc} x & y & v & w & z & \\ \end{array}$$
$$\begin{bmatrix} 0 & 2 & 0 & 4 & 1 & 18 \\ 0 & -3 & 1 & 2 & 0 & 12 \\ 1 & 5 & 0 & 1 & 0 & 22 \end{bmatrix}$$

$$\text{(b)} \quad \begin{array}{cccccc} x & y & t & u & v & z \\ \end{array}$$
$$\begin{bmatrix} 4 & 0 & 4 & 5 & 0 & 1 & -50 \\ -2 & 1 & -2 & 2 & 0 & 0 & 8 \\ 5 & 0 & 3 & 0 & 1 & 0 & 4 \end{bmatrix}$$

(c)
$$\begin{array}{cccccc} x_1 & x_2 & x_3 & v_1 & v_2 & z \end{array}$$
$$\begin{bmatrix} 3 & -5 & 0 & 0 & 2 & 1 & 43 \\ -2 & -1 & 1 & 0 & -3 & 0 & 18 \\ -3 & 2 & 0 & 1 & 4 & 0 & 12 \end{bmatrix}$$

(d)
$$\begin{array}{cccccc} x & y & u & v & w & z \end{array}$$
$$\begin{bmatrix} 0 & 0 & -11 & 0 & 2 & 1 & 12 \\ 1 & 0 & 6 & 0 & 7 & 0 & 13 \\ 0 & 1 & 4 & 0 & 1 & 0 & 14 \\ 0 & 0 & -3 & 1 & 6 & 0 & 2 \end{bmatrix}.$$

12. For each of the tableaus in Exercise 11, find the corresponding basic feasible solution and the value of the objective function at that solution.

5.3 THE SIMPLEX METHOD

In this section we shall give a step-by-step procedure for solving certain linear programming problems. This technique, called the ***simplex method,*** was developed by a contemporary mathematician, George B. Dantzig, during the late 1940's.

In the simplex method, we begin with a certain easily obtainable basic feasible solution; then by a definite procedure we move through a succession of basic feasible solutions in such a way that the objective function increases in value at each step. We continue this process until we hit a basic feasible solution where the objective function is maximum or we find there is no solution. As we discuss the details of each step we shall illustrate the idea using the problem

$$\text{Maximize} \quad z = 10x + 12y$$

subject to

$$\begin{aligned} x + 2y &\leq 150 \\ x + \ \ y &\leq 100 \\ x &\geq 0 \\ y &\geq 0. \end{aligned}$$

The corresponding new problem with slack variables is

$$\text{Maximize} \quad z = 10x + 12y$$

subject to

$$\begin{aligned} x + 2y + v \qquad &= 150 \\ x + \ \ y \qquad + w &= 100 \qquad (1) \end{aligned}$$
$$x \geq 0, \quad y \geq 0, \quad v \geq 0, \quad w \geq 0.$$

Throughout this section we shall assume we have a standard linear programming problem. This being the case, we can always get a basic feasible solution by setting the original (nonslack) variables equal to zero.

We call this the ***initial basic feasible solution.*** To illustrate, if we set $x = 0$ and $y = 0$ in (1), we obtain the initial basic feasible solution

$$x = 0, \qquad y = 0, \qquad v = 150, \qquad w = 100.$$

Moreover, at this solution the objective function $z = 10x + 12y$ has value $z = 0$.

The important information about the initial basic feasible solution can be obtained from the last column of the tableau in which the slack variables are explicit. We call this the ***initial tableau.*** In our illustrative example, the initial tableau is

$$\begin{array}{c} \\ z \\ v \\ w \end{array} \begin{array}{c} \begin{array}{cccccc} x & y & v & w & z & \end{array} \\ \left[\begin{array}{rrrrrr} -10 & -12 & 0 & 0 & 1 & 0 \\ 1 & 2 & 1 & 0 & 0 & 150 \\ 1 & 1 & 0 & 1 & 0 & 100 \end{array}\right] \end{array}. \tag{2}$$

The construction of the initial tableau is the first step in the simplex method. For reference, we formally state this fact.

Step 1 in the Simplex Method (Forming the Initial Tableau)
Construct the tableau in which the slack variables are explicit.

Once the initial tableau is constructed, there are two questions to consider.

(a) Does the tableau correspond to an optimal solution?
(b) If not, how can we obtain a new tableau in which the value of z is closer to the maximum value?

Let us consider question (a) first. In our example, the first row of the initial tableau is

$$\begin{array}{rrrrrr} x & y & v & w & z & \\ -10 & -12 & 0 & 0 & 1 & 0 \end{array} \tag{3}$$

so that the corresponding equation is

$$-10x - 12y + z = 0$$

or

$$z = 10x + 12y. \tag{4}$$

Since x and y occur with positive coefficients $+10$ and $+12$ in this last equation, it is obvious that the value of z will increase whenever x increases in value or y increases in value. With this in mind, we can see that the value $z = 0$ that occurs at the initial basic feasible solution cannot be the maximum possible value of z since larger values of z can be obtained by increasing either x or y. The key to this argument is the fact that the expression for z in terms of the implicit variables contained at least one term with a positive coefficient. Comparing (3) and (4) above, we see that the expression for z will contain terms with positive coefficients whenever the z row of the tableau contains *negative* entries in one of the columns headed by a variable. This suggests the following test for determining from the tableau whether the maximum value of z has been obtained.

> ***Step 2 in the Simplex Method*** (Test for Maximality) If the z row of a tableau contains no negative entries in the columns labeled with variables, we have arrived at an optimal solution; otherwise we have not.

Example 5 In Table 5.1 of the previous section we listed the basic feasible solutions of the new problem with slack variables given in Example 2. From that table it is evident that the maximum value $z = 75$ occurs at the basic feasible solution

$$x = 0, \qquad v = 0, \qquad y = 75, \qquad w = 5.$$

Let us see if we can obtain this same conclusion by examining the tableaus corresponding to the basic feasible solutions. These tableaus were all computed in the last section (see (9c), (14), (16), and (17)), but for convenience we shall list them here.

Basic Feasible Solution — *Corresponding Tableau*

(a) $x = 0,\ y = 0,\ v = 30,\ w = 20$

$$\begin{array}{c} \\ z \\ v \\ w \end{array} \begin{array}{c} \begin{array}{ccccc} x & y & v & w & z \end{array} \\ \left[\begin{array}{cccccc} 1 & -1 & 0 & 0 & 1 & 0 \\ 0.2 & 0.4 & 1 & 0 & 0 & 30 \\ 0.2 & 0.2 & 0 & 1 & 0 & 20 \end{array}\right] \end{array}$$

(b) $y = 0,\ w = 0,\ x = 100,\ v = 10$

$$\begin{array}{c} \\ z \\ v \\ x \end{array} \begin{array}{c} \begin{array}{ccccc} x & y & v & w & z \end{array} \\ \left[\begin{array}{cccccc} 0 & -2 & 0 & -5 & 1 & -100 \\ 0 & 0.2 & 1 & -1 & 0 & 10 \\ 1 & 1 & 0 & 5 & 0 & 100 \end{array}\right] \end{array}$$

(c) $v = 0, \quad w = 0, \; x = 50, \; y = 50$

$$\begin{array}{c} \\ z \\ y \\ x \end{array} \begin{array}{c} \begin{array}{ccccc} x & y & v & w & z \end{array} \\ \begin{bmatrix} 0 & 0 & 10 & -15 & 1 & 0 \\ 0 & 1 & 5 & -5 & 0 & 50 \\ 1 & 0 & -5 & 10 & 0 & 50 \end{bmatrix} \end{array}$$

(d) $x = 0, \; v = 0, \; y = 75, \; w = 5$

$$\begin{array}{c} \\ z \\ y \\ w \end{array} \begin{array}{c} \begin{array}{ccccc} x & y & v & w & z \end{array} \\ \begin{bmatrix} 1.5 & 0 & 2.5 & 0 & 1 & 75 \\ 0.5 & 1 & 2.5 & 0 & 0 & 75 \\ 0.1 & 0 & -0.5 & 1 & 0 & 5 \end{bmatrix} \end{array}.$$

Observe that the z rows of tableaus (a), (b), and (c) all have one or more negative entries in the columns labeled with variables. Thus the corresponding basic feasible solutions are not optimal. However, the z row of tableau (d) contains no negative entries in these columns so that the corresponding basic feasible solution

$$x = 0, \qquad v = 0, \qquad y = 75, \qquad w = 5$$

is optimal. This agrees with our previous observation.

Returning to our illustrative example, we know that the initial basic feasible solution is not optimal since the z row in the initial tableau (2) has negative entries in the x and y columns. The values of x and y in this initial basic feasible solution are $x = 0$ and $y = 0$, and, as we previously pointed out, the value of z can be increased by increasing either x or y. In the simplex method we try to increase only *one* variable at a time. Thus we have two choices:

(1) Look for a new tableau in which x remains at $x = 0$ and y is greater than zero.

or

(2) Look for a new tableau in which y remains at $y = 0$ and x is greater than zero.

To determine which of these alternatives is better, let us reexamine Equation (4). From this equation we see that by increasing y by 1 unit, we increase z by 12 units, while increasing x by 1 unit only increases z by 10 units. Thus, the best strategy is to keep x fixed at $x = 0$ and increase y, since this choice tends to bring us toward the maximum value of z more rapidly. The variable to be increased is usually called the ***entering variable*** (because in going to the next tableau it will leave the set of implicit variables and *enter* the set of explicit variables). For convenience we shall mark the entering variable with an arrow at the top of the tableau as follows.

$$
\begin{array}{c}
\text{entering variable} \\
\downarrow
\end{array}
$$

$$
\begin{array}{c} \\ z \\ v \\ w \end{array}
\begin{array}{c}
\begin{array}{cccccc} x & y & v & w & z & \end{array} \\
\begin{bmatrix}
-10 & -12 & 0 & 0 & 1 & 0 \\
1 & 2 & 1 & 0 & 0 & 150 \\
1 & 1 & 0 & 1 & 0 & 100
\end{bmatrix}
\end{array}. \tag{5}
$$

As tableau (5) illustrates, the entering variable in a tableau is the variable whose entry in the z row is most negative. We can now formally state the next step in the simplex method.

> ***Step 3 in the Simplex Method*** (Selecting the Entering Variable) If the test for maximality in Step 2 shows that an optimal solution has not been reached, then label the variable whose entry in the z row is most negative as the entering variable.

In tableau (5) we labeled y as the entering variable. This means we want to find a new tableau in which the value of y is positive. But if y is to be positive in the new tableau, it will be explicit (the implicit variables in a tableau have value zero). Thus we must look for a new tableau in which y is explicit and some other variable replaces y as an implicit variable. The *new* implicit variable is usually called the ***departing variable*** (since it *departs* from the set of explicit variables and becomes implicit).

In a moment we shall discuss criteria for choosing the departing variable. For illustrative purposes, however, let us suppose that we have somehow settled on making v the departing variable. We mark the departing variable with an arrow on the left side of the tableau as follows

$$
\begin{array}{c}
\text{entering variable} \\
\downarrow
\end{array}
$$

$$
\begin{array}{r} \\ z \\ \text{departing variable} \rightarrow v \\ w \end{array}
\begin{array}{c}
\begin{array}{cccccc} x & y & v & w & z & \end{array} \\
\begin{bmatrix}
-10 & -12 & 0 & 0 & 1 & 0 \\
1 & 2 & 1 & 0 & 0 & 150 \\
1 & 1 & 0 & 1 & 0 & 100
\end{bmatrix}
\end{array}. \tag{6}
$$

Now that we have specified the entering and departing variables, the new tableau is partially determined. The new explicit variables are z, y, and w, so that the new tableau will have the form

$$\begin{array}{c} \\ z \\ y \\ w \end{array} \begin{array}{c} \begin{array}{cccccc} x & y & v & w & z & \end{array} \\ \begin{bmatrix} \cdot & 0 & \cdot & 0 & 1 & \cdot \\ \cdot & 1 & \cdot & 0 & 0 & \cdot \\ \cdot & 0 & \cdot & 1 & 0 & \cdot \end{bmatrix} \end{array} \tag{7}$$

where the dots designate numbers that have yet to be determined. To see how these unknown entries can be obtained, we need only recall that tableaus (6) and (7) are augmented matrices for the *same* system of equations, but written in two different ways. This suggests that tableau (7) can be obtained from tableau (6) using row operations. We now show how this can be done.

Since the w and z columns of (6) and (7) are identical, we need only find row operations that will transform the y column

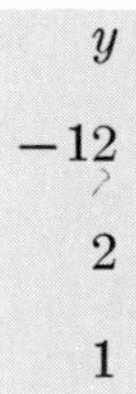

in (6) into the y column

$$\begin{array}{c} y \\ 0 \\ 1 \\ 0 \end{array}$$

in (7) without disturbing the w and z columns. This can be accomplished by a procedure that we shall now describe. If we extend the arrows marking the entering and departing variables in a tableau, they intersect at an entry called the ***pivot entry***. To illustrate, the pivot entry in tableau (6) is the circled entry below

$$
\begin{array}{c} \\ \\ \text{departing variable} \rightarrow \\ \\ \end{array}
\begin{array}{c} \\ z \\ v \\ w \end{array}
\overset{\begin{array}{cccccc} & \text{entering variable} & & & & \\ & \downarrow & & & & \\ x & y & v & w & z & \end{array}}{\begin{bmatrix} -10 & -12 & 0 & 0 & 1 & 0 \\ 1 & \textcircled{2} & 1 & 0 & 0 & 150 \\ 1 & 1 & 0 & 1 & 0 & 100 \end{bmatrix}}. \qquad (6a)
$$

The row and column containing the pivot entry are called, respectively, the ***pivot row*** and ***pivot column.*** To transform tableau (6) into tableau (7) we can use the following method, called ***pivotal elimination.***

Pivotal Elimination

(i) If the pivot entry is k, multiply the pivot row by $1/k$ so that the pivot entry becomes a 1.

(ii) Add suitable multiples of the pivot row to the other rows so that the remaining entries in the pivot column become zeros.

To apply step (i) to tableau (6a), we multiply the pivot row by $\frac{1}{2}$ to obtain

$$
\overset{\begin{array}{cccccc} x & y & v & w & z & \end{array}}{\begin{bmatrix} -10 & -12 & 0 & 0 & 1 & 0 \\ \frac{1}{2} & 1 & \frac{1}{2} & 0 & 0 & 75 \\ 1 & 1 & 0 & 1 & 0 & 100 \end{bmatrix}}.
$$

To apply step (ii) to this matrix, we add $+12$ times the pivot row to the first row and -1 times the pivot row to the third row. This yields the new tableau

$$
\begin{array}{c} z \\ y \\ w \end{array}
\overset{\begin{array}{cccccc} x & y & v & w & z & \end{array}}{\begin{bmatrix} -4 & 0 & 6 & 0 & 1 & 900 \\ \frac{1}{2} & 1 & \frac{1}{2} & 0 & 0 & 75 \\ \frac{1}{2} & 0 & -\frac{1}{2} & 1 & 0 & 25 \end{bmatrix}}. \qquad (8)
$$

Example 6 Use pivotal elimination to find the new tableau if w is taken to be the departing variable in (5).

Solution The pivot entry is the circled entry in the following tableau.

$$\begin{array}{c} \\ \\ z \\ v \\ \text{departing variable} \rightarrow w \end{array} \begin{array}{c} \text{entering variable} \\ \downarrow \\ \begin{array}{cccccc} x & y & v & w & z & \end{array} \\ \left[\begin{array}{cccccc} -10 & -12 & 0 & 0 & 1 & 0 \\ 1 & 2 & 1 & 0 & 0 & 150 \\ 1 & \textcircled{1} & 0 & 1 & 0 & 100 \end{array}\right]. \end{array}$$

Since the pivot entry is a 1, we can go immediately to step (ii) in pivotal elimination. Thus, we add +12 times the pivot row to the first row and −2 times the pivot row to the second row; this yields the new tableau

$$\begin{array}{c} \\ z \\ v \\ y \end{array} \begin{array}{c} \begin{array}{cccccc} x & y & v & w & z & \end{array} \\ \left[\begin{array}{cccccc} 2 & 0 & 0 & 12 & 1 & 1200 \\ -1 & 0 & 1 & -2 & 0 & -50 \\ 1 & 1 & 0 & 1 & 0 & 100 \end{array}\right]. \end{array} \tag{9}$$

So far, we have shown how to select the entering variable in a tableau and how to obtain the new tableau once the entering and departing variables are known. We have not explained how to choose the departing variable.

To illustrate the pitfalls that can occur in choosing the departing variable, let us compare tableaus (8) and (9). Both were obtained from tableau (5), but using different departing variables. In tableau (8), the implicit variables are x and v; if we set these variables equal to zero this tableau tells us that

$$x = 0, \qquad v = 0, \qquad y = 75, \qquad w = 25$$

is a basic feasible solution and the value of the objective function there is

$$z = 900.$$

On the other hand, the implicit variables in tableau (9) are x and w. If we set these variables equal to zero, the tableau tells us that

$$x = 0, \qquad w = 0, \qquad v = -50, \qquad y = 100.$$

But this is not a basic feasible solution since the variable v violates the constraint $v \geq 0$. In other words, by making a wrong choice of the departing variable we can obtain new values of the variables that violate the nonnegativity constraints. Thus, the objectives in selecting the departing variable are twofold:

(a) Get the largest possible increase in z.
(b) Avoid violating the nonnegativity constraints.

For reasons too technical to discuss here, it is difficult to determine which choice of the departing variable will meet objective (a). Nevertheless, we can give a procedure for selecting the departing variable that will meet objective (b) and still give a reasonably good increase in z. Since any increase in z will lead us closer to a solution, this is adequate for our purposes. We shall omit the mathematical theory underlying this procedure.

First we need some terminology. If, as illustrated in Figure 5.2, b is the last element in a row of a tableau and a is the element of that row in the column headed by the entering variable, then we shall call the ratio

$$b/a$$

the ***quotient*** for the row.

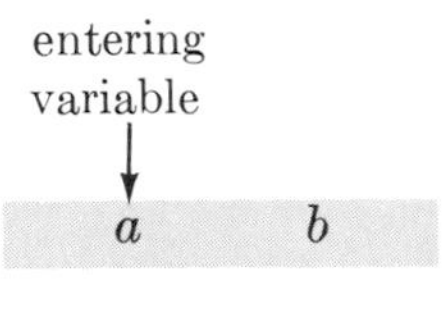

Figure 5.2

Example 7 Since y is the entering variable, the quotients for the v and w rows of tableau (5) are $150/2 = 75$ and $100/1 = 100$ respectively.

Step 4 in the Simplex Method (Selecting the Departing Variable) Compute the quotients for all but the z row of the tableau. Then choose as the departing variable the explicit variable in the row where the quotient is as small as possible, yet nonnegative.

Example 8 As shown in Example 7, the quotients for the v and w rows of tableau (5) are 75 and 100. Since 75 is the smaller of these quotients and is nonnegative, the departing variable is v.

We now have all the necessary ingredients to state the simplex method.

The Simplex Method

Step 1 Compute the initial tableau.

Step 2 Test for maximality. If the test shows we have obtained an optimal solution stop; otherwise go to Step 3.

Step 3 Determine the entering variable.

Step 4 Determine the departing variable.

Step 5 Using pivotal elimination, construct the new tableau and return to Step 2.

Example 9 Solve the following problem by the simplex method.

$$\text{Maximize} \quad z = 10x + 12y$$

subject to

$$\begin{aligned} x + 2y &\leq 150 \\ x + \ \ y &\leq 100 \\ x &\geq 0 \\ y &\geq 0. \end{aligned}$$

(This is the illustrative problem we have been working with in this section.)

Solution The corresponding new problem with slack variables is

$$\text{Maximize} \quad z = 10x + 12y$$

subject to

$$\begin{aligned} x + 2y + v \qquad &= 150 \\ x + \ \ y \qquad + w &= 100 \end{aligned}$$

$$x \geq 0, \quad y \geq 0, \quad v \geq 0, \quad w \geq 0.$$

As shown in (2), the initial tableau is

$$\begin{array}{c} \\ z \\ v \\ w \end{array} \begin{array}{c} \begin{array}{cccccc} x & y & v & w & z & \end{array} \\ \left[\begin{array}{cccccc} -10 & -12 & 0 & 0 & 1 & 0 \\ 1 & 2 & 1 & 0 & 0 & 150 \\ 1 & 1 & 0 & 1 & 0 & 100 \end{array}\right] \end{array}.$$

Since the x and y columns of the z row have negative entries, we have not yet reached an optimal solution. Since -12 is the most negative entry in the z row occurring under a variable, y is the entering variable; that is,

$$\begin{array}{c} \\ \\ \\ z \\ v \\ w \end{array} \begin{array}{c} \begin{array}{cccccc} & \text{entering variable} & & & & \\ & \downarrow & & & & \\ x & y & v & w & z & \end{array} \\ \left[\begin{array}{cccccc} -10 & -12 & 0 & 0 & 1 & 0 \\ 1 & 2 & 1 & 0 & 0 & 150 \\ 1 & 1 & 0 & 1 & 0 & 100 \end{array}\right] \end{array}.$$

The quotient for the v row is $150/2 = 75$ and the quotient for the w row is $100/1 = 100$; thus v is the departing variable since its row has the smallest nonnegative quotient, that is,

$$\begin{array}{r c} & \begin{array}{cccccc} & \text{entering variable} & & & & \\ & \downarrow & & & & \\ x & y & v & w & z & \end{array} \\ \begin{array}{r} z \\ \text{departing} \rightarrow v \\ \text{variable} \quad w \end{array} & \left[\begin{array}{cccccc} -10 & -12 & 0 & 0 & 1 & 0 \\ 1 & \boxed{2} & 1 & 0 & 0 & 150 \\ 1 & 1 & 0 & 1 & 0 & 100 \end{array}\right]. \end{array}$$

As previously shown (see tableau (8)), pivotal elimination yields the new tableau

$$\begin{array}{r c} & \begin{array}{cccccc} x & y & v & w & z & \end{array} \\ \begin{array}{r} z \\ y \\ w \end{array} & \left[\begin{array}{cccccc} -4 & 0 & 6 & 0 & 1 & 900 \\ \frac{1}{2} & 1 & \frac{1}{2} & 0 & 0 & 75 \\ \frac{1}{2} & 0 & -\frac{1}{2} & 1 & 0 & 25 \end{array}\right]. \end{array}$$

Since the z row of this tableau contains a negative entry in the x column, the test for maximality shows we have not yet obtained an optimal solution. In the z row, -4 is the most negative entry under a variable so that x is the new entering variable; that is,

$$\begin{array}{r c} & \begin{array}{cccccc} \text{entering variable} & & & & & \\ \downarrow & & & & & \\ x & y & v & w & z & \end{array} \\ \begin{array}{r} z \\ y \\ w \end{array} & \left[\begin{array}{cccccc} -4 & 0 & 6 & 0 & 1 & 900 \\ \frac{1}{2} & 1 & \frac{1}{2} & 0 & 0 & 75 \\ \frac{1}{2} & 0 & -\frac{1}{2} & 1 & 0 & 25 \end{array}\right]. \end{array}$$

The quotient for the y row is $75/\frac{1}{2} = 150$ and the quotient for the w row is $25/\frac{1}{2} = 50$. Thus, w is the new departing variable since its row has the smallest nonnegative quotient; that is,

$$\begin{array}{r c} & \begin{array}{cccccc} \text{entering variable} & & & & & \\ \downarrow & & & & & \\ x & y & v & w & z & \end{array} \\ \begin{array}{r} z \\ \text{departing} \quad y \\ \text{variable} \rightarrow w \end{array} & \left[\begin{array}{cccccc} -4 & 0 & 6 & 0 & 1 & 900 \\ \frac{1}{2} & 1 & \frac{1}{2} & 0 & 0 & 75 \\ \boxed{\frac{1}{2}} & 0 & -\frac{1}{2} & 1 & 0 & 25 \end{array}\right]. \end{array}$$

Using pivotal elimination, we first multiply the pivot row by 2. This yields

$$\begin{array}{c} \\ z \\ y \\ w \end{array} \begin{array}{c} \begin{array}{cccccc} x & y & v & w & z & \end{array} \\ \left[\begin{array}{cccccc} -4 & 0 & 6 & 0 & 1 & 900 \\ \frac{1}{2} & 1 & \frac{1}{2} & 0 & 0 & 75 \\ 1 & 0 & -1 & 2 & 0 & 50 \end{array}\right] \end{array}.$$

Next we add 4 times the pivot row to the first row and $-\frac{1}{2}$ times the pivot row to the second row; this yields the new tableau

$$\begin{array}{c} \\ z \\ y \\ x \end{array} \begin{array}{c} \begin{array}{cccccc} x & y & v & w & z & \end{array} \\ \left[\begin{array}{cccccc} 0 & 0 & 2 & 8 & 1 & 1100 \\ 0 & 1 & 1 & -1 & 0 & 50 \\ 1 & 0 & -1 & 2 & 0 & 50 \end{array}\right] \end{array}.$$

Since the z row of this tableau has no negative entries in the columns labeled with variables, the test for maximality tells us that we have arrived at an optimal solution. From the tableau, this optimal solution is

$$v = 0, \qquad w = 0, \qquad x = 50, \qquad y = 50$$

and the maximum value of z is

$$z = 1100.$$

To obtain a solution of the original problem, we discard the slack variables. Thus, in the original problem the maximum value $z = 1100$ occurs at $x = 50$, $y = 50$.

Example 10 Use the simplex method to solve the standard linear programming problem

$$\text{Maximize} \quad z = 6x + 9y + 6t$$

subject to

$$\begin{aligned} 2x + 2y + t &\leq 20 \\ 2x + 3y + 3t &\leq 30 \\ 2x + 5y + t &\leq 40 \end{aligned}$$

$$x \geq 0, \quad y \geq 0, \quad t \geq 0.$$

Solution The corresponding problem with slack variables is

$$\text{Maximize} \quad z = 6x + 9y + 6t$$

subject to

$$\begin{aligned} 2x + 2y + t + u \qquad\qquad &= 20 \\ 2x + 3y + 3t \qquad + v \qquad &= 30 \\ 2x + 5y + t \qquad\qquad + w &= 40 \end{aligned}$$

$$x \geq 0, \quad y \geq 0, \quad t \geq 0,$$
$$u \geq 0, \quad v \geq 0, \quad w \geq 0.$$

The initial tableau is

$$\begin{array}{c} \\ z \\ u \\ v \\ w \end{array} \begin{array}{c} \begin{array}{cccccccc} x & y & t & u & v & w & z & \end{array} \\ \begin{bmatrix} -6 & -9 & -6 & 0 & 0 & 0 & 1 & 0 \\ 2 & 2 & 1 & 1 & 0 & 0 & 0 & 20 \\ 2 & 3 & 3 & 0 & 1 & 0 & 0 & 30 \\ 2 & 5 & 1 & 0 & 0 & 1 & 0 & 40 \end{bmatrix} \end{array}$$

Since the z row of this tableau has negative entries in some of the columns labeled with variables, we have not yet obtained an optimal solution. Since -9 is the most negative of these entries, y is the entering variable. Consequently, the quotients for the u, v, and w rows are

$$\frac{20}{2} = 10, \qquad \frac{30}{3} = 10, \qquad \frac{40}{5} = 8$$

respectively. Thus w is the departing variable since its row has the smallest nonnegative quotient; the pivot element is now determined; that is,

entering variable ↓ (over column y); departing variable → w

$$\begin{array}{c} \\ z \\ u \\ v \\ w \end{array} \begin{array}{c} \begin{array}{cccccccc} x & y & t & u & v & w & z & \end{array} \\ \begin{bmatrix} -6 & -9 & -6 & 0 & 0 & 0 & 1 & 0 \\ 2 & 2 & 1 & 1 & 0 & 0 & 0 & 20 \\ 2 & 3 & 3 & 0 & 1 & 0 & 0 & 30 \\ 2 & \boxed{5} & 1 & 0 & 0 & 1 & 0 & 40 \end{bmatrix} \end{array}.$$

Using pivotal elimination we first multiply the pivot row by $\frac{1}{5}$ to obtain

$$\begin{array}{c} \\ z \\ u \\ v \\ w \end{array} \begin{array}{c} \begin{array}{cccccccc} x & y & t & u & v & w & z & \end{array} \\ \begin{bmatrix} -6 & -9 & -6 & 0 & 0 & 0 & 1 & 0 \\ 2 & 2 & 1 & 1 & 0 & 0 & 0 & 20 \\ 2 & 3 & 3 & 0 & 1 & 0 & 0 & 30 \\ \frac{2}{5} & 1 & \frac{1}{5} & 0 & 0 & \frac{1}{5} & 0 & 8 \end{bmatrix} \end{array}.$$

Next we add 9 times the pivot row to the first row, add -2 times the pivot row to the second row, and add -3 times the pivot row to the third row to obtain

$$\begin{array}{c} \\ z \\ u \\ v \\ y \end{array} \begin{array}{c} \begin{array}{cccccccc} x & y & t & u & v & w & z & \end{array} \\ \left[\begin{array}{cccccccc} -\frac{12}{5} & 0 & -\frac{21}{5} & 0 & 0 & \frac{9}{5} & 1 & 72 \\ \frac{6}{5} & 0 & \frac{3}{5} & 1 & 0 & -\frac{2}{5} & 0 & 4 \\ \frac{4}{5} & 0 & \frac{12}{5} & 0 & 1 & -\frac{3}{5} & 0 & 6 \\ \frac{2}{5} & 1 & \frac{1}{5} & 0 & 0 & \frac{1}{5} & 0 & 8 \end{array}\right] \end{array}.$$

This new tableau still contains negative entries in the z row under the variables, so we have not yet reached the optimal solution. We thus repeat the process of choosing entering and departing variables. We leave it to the reader to obtain the following result.

entering variable ↓ (t)

$$\begin{array}{c} \\ z \\ u \\ \text{departing variable} \rightarrow v \\ y \end{array} \begin{array}{c} \begin{array}{cccccccc} x & y & t & u & v & w & z & \end{array} \\ \left[\begin{array}{cccccccc} -\frac{12}{5} & 0 & -\frac{21}{5} & 0 & 0 & \frac{9}{5} & 1 & 72 \\ \frac{6}{5} & 0 & \frac{3}{5} & 1 & 0 & -\frac{2}{5} & 0 & 4 \\ \frac{4}{5} & 0 & \boxed{\frac{12}{5}} & 0 & 1 & -\frac{3}{5} & 0 & 6 \\ \frac{2}{5} & 1 & \frac{1}{5} & 0 & 0 & \frac{1}{5} & 0 & 8 \end{array}\right] \end{array}.$$

After pivotal elimination we obtain the new tableau

$$\begin{array}{c} \\ z \\ u \\ t \\ y \end{array} \begin{array}{c} \begin{array}{cccccccc} x & y & t & u & v & w & z & \end{array} \\ \left[\begin{array}{cccccccc} -1 & 0 & 0 & 0 & \frac{7}{4} & \frac{3}{4} & 1 & \frac{165}{2} \\ 1 & 0 & 0 & 1 & -\frac{1}{4} & -\frac{1}{4} & 0 & \frac{5}{2} \\ \frac{1}{3} & 0 & 1 & 0 & \frac{5}{12} & -\frac{1}{4} & 0 & \frac{5}{2} \\ \frac{1}{3} & 1 & 0 & 0 & -\frac{1}{12} & \frac{1}{4} & 0 & \frac{15}{2} \end{array}\right] \end{array}.$$

Once again we have a negative entry in the z row under a variable, so we do not have an optimal solution. Choosing the new entering and departing variables we obtain

entering variable ↓ (x)

$$\begin{array}{c} \\ z \\ \text{departing variable} \rightarrow u \\ t \\ y \end{array} \begin{array}{c} \begin{array}{cccccccc} x & y & t & u & v & w & z & \end{array} \\ \left[\begin{array}{cccccccc} -1 & 0 & 0 & 0 & \frac{7}{4} & \frac{3}{4} & 1 & \frac{165}{2} \\ \boxed{1} & 0 & 0 & 1 & -\frac{1}{4} & -\frac{1}{4} & 0 & \frac{5}{2} \\ \frac{1}{3} & 0 & 1 & 0 & \frac{5}{12} & -\frac{1}{4} & 0 & \frac{5}{2} \\ \frac{1}{3} & 1 & 0 & 0 & -\frac{1}{12} & \frac{1}{4} & 0 & \frac{15}{2} \end{array}\right] \end{array}.$$

After pivotal elimination we obtain the new tableau

$$\begin{array}{c} \\ z \\ x \\ t \\ y \end{array}\begin{array}{c} \begin{array}{cccccccc} x & y & t & u & v & w & z & \end{array} \\ \left[\begin{array}{cccccccc} 0 & 0 & 0 & 1 & \frac{3}{2} & \frac{1}{2} & 1 & 85 \\ 1 & 0 & 0 & 1 & -\frac{1}{4} & -\frac{1}{4} & 0 & \frac{5}{2} \\ 0 & 0 & 1 & -\frac{1}{3} & \frac{1}{2} & 0 & 0 & \frac{5}{3} \\ 0 & 1 & 0 & -\frac{1}{3} & 0 & \frac{1}{3} & 0 & \frac{20}{3} \end{array}\right] \end{array}.$$

Since the z row in this tableau has no negative entries under the variables, we have obtained an optimal solution. From the tableau, the maximal value $z = 85$ occurs when

$$u = 0, \qquad v = 0, \qquad w = 0, \qquad x = \frac{5}{2}, \qquad t = \frac{5}{3}, \qquad y = \frac{20}{3}.$$

To obtain a solution of the original problem we discard the slack variables. Thus, in the original problem, the maximum value $z = 85$ occurs at

$$x = \frac{5}{2}, \qquad t = \frac{5}{3}, \qquad y = \frac{20}{3}.$$

Not every standard linear programming problem has a solution. If the simplex method is attempted on a standard linear programming problem that has no solution, then at some point in the computations we will obtain a tableau in which the quotients used to select the departing variable are all negative. Thus, the rule for choosing the departing variable will not work. We can then stop the computations and conclude that the problem has no solution (see Exercise 19).

Certain difficulties occur when the minimum nonnegative quotient, computed to determine the departing variable, occurs in two or more rows. This situation is called ***degeneracy.*** Students interested in how degeneracy is handled are referred to more detailed books on linear programming such as

S. I. Gass, *Linear Programming: Methods and Applications* (New York: McGraw-Hill, 1958).

W. W. Garvin, *Introduction to Linear Programming* (New York: McGraw-Hill, 1960).

EXERCISE SET 5.3

In Exercises 1–3 find the initial tableau and the corresponding basic feasible solution.

1. Maximize $z = 4x + 3y$

subject to

$$\begin{aligned} 2x + 3y &\leq 12 \\ -3x + 2y &\leq 6 \\ x \geq 0, \quad y &\geq 0. \end{aligned}$$

2. Maximize $z = -3x_1 + 2x_2 + 3x_3$

subject to

$$\begin{aligned} 2x_1 + x_2 + x_3 &\leq 12 \\ 3x_1 - x_2 + 3x_3 &\leq 8 \\ x_1 \geq 0, \quad x_2 \geq 0, \quad x_3 &\geq 0. \end{aligned}$$

3. Maximize $z = 8x + 6y$

subject to

$$\begin{aligned} x + y &\leq 4 \\ x + 3y &\leq 6 \\ -x + y &\leq 1 \\ x \geq 0, \quad y &\geq 0. \end{aligned}$$

4. (a) Find the equations represented by the tableau

$$\begin{array}{c} \\ z \\ u \\ y \end{array} \begin{array}{c} \begin{array}{cccccc} x & y & u & v & z & \end{array} \\ \left[\begin{array}{rrrrrr} 12 & 0 & 0 & 6 & 1 & 100 \\ -3 & 0 & 1 & 7 & 0 & 2 \\ 5 & 1 & 0 & -5 & 0 & 7 \end{array}\right] \end{array}.$$

(b) Find the basic feasible solution corresponding to this tableau.

(c) Find the value of z at the basic feasible solution in part (b).

5. Repeat the directions of Exercise 4 for the tableau

$$\begin{array}{c} \\ z \\ x_1 \\ x_2 \\ v \end{array} \begin{array}{c} \begin{array}{cccccccc} x_1 & x_2 & x_3 & u & v & w & z & \end{array} \\ \left[\begin{array}{rrrrrrrr} 0 & 0 & 5 & 7 & 0 & 6 & 1 & 40 \\ 1 & 0 & -2 & 1 & 0 & 2 & 0 & 6 \\ 0 & 1 & 4 & 1 & 0 & -1 & 0 & 14 \\ 0 & 0 & 3 & 0 & 1 & 3 & 0 & 12 \end{array}\right] \end{array}.$$

6. Explain why none of the following tableaus correspond to optimal solutions.

(a)
$$\begin{array}{c} \begin{array}{cccccc} x & y & u & v & z & \end{array} \\ \left[\begin{array}{rrrrrr} 4 & 0 & -6 & 0 & 1 & 9 \\ 7 & 1 & 2 & 0 & 0 & 12 \\ 1 & 0 & -3 & 1 & 0 & 4 \end{array}\right] \end{array}$$

(b)
$$\begin{array}{c} \begin{array}{cccccc} x & y & u & v & z & \end{array} \\ \left[\begin{array}{rrrrrr} 0 & 0 & 12 & 6 & 1 & 9 \\ 1 & 0 & -5 & 3 & 0 & -2 \\ 0 & 1 & 2 & 0 & 0 & 1 \end{array}\right] \end{array}$$

(c)
$$\begin{array}{c} \begin{array}{ccccccc} x_1 & x_2 & v_1 & v_2 & v_3 & z & \end{array} \\ \left[\begin{array}{rrrrrrr} 7 & 0 & 0 & 0 & -3 & 1 & 100 \\ 2 & 0 & 0 & 1 & 6 & 0 & 4 \\ 4 & 1 & 0 & 0 & 7 & 0 & 3 \\ 5 & 0 & 1 & 0 & 2 & 0 & 9 \end{array}\right] \end{array}.$$

7. Which of the tableaus in Exercise 6 correspond to basic feasible solutions?

For the tableaus in Exercises 8–10, find the entering variable, the departing variable, and the pivot entry.

8.

$$\begin{array}{c} \\ z \\ u \\ y \end{array} \begin{array}{c} \begin{matrix} x & y & t & u & v & z & \end{matrix} \\ \begin{bmatrix} -8 & 0 & 3 & 0 & 0 & 1 & 140 \\ 2 & 0 & 6 & 1 & 2 & 0 & 100 \\ 3 & 1 & 5 & 0 & 3 & 0 & 180 \end{bmatrix} \end{array}.$$

9.

$$\begin{array}{c} \\ z \\ x \\ w \end{array} \begin{array}{c} \begin{matrix} x & y & t & v & w & z & \end{matrix} \\ \begin{bmatrix} 0 & -3 & -2 & 4 & 0 & 1 & 40 \\ 1 & 4 & 1 & 3 & 0 & 0 & 10 \\ 0 & -3 & 5 & 7 & 1 & 0 & 5 \end{bmatrix} \end{array}.$$

10.

$$\begin{array}{c} \\ z \\ v_1 \\ x_2 \\ x_3 \\ v_2 \end{array} \begin{array}{c} \begin{matrix} x_1 & x_2 & x_3 & v_1 & v_2 & v_3 & v_4 & z & \end{matrix} \\ \begin{bmatrix} -8 & 0 & 0 & 0 & 0 & 5 & -4 & 1 & 120 \\ 4 & 0 & 0 & 1 & 0 & 3 & 0 & 0 & 200 \\ -3 & 1 & 0 & 0 & 0 & 2 & 1 & 0 & 300 \\ 0 & 0 & 1 & 0 & 0 & -1 & 2 & 0 & 150 \\ 6 & 0 & 0 & 0 & 1 & 2 & 3 & 0 & 180 \end{bmatrix} \end{array}.$$

11. Use pivotal elimination on the tableau in Exercise 8 to find the new tableau.

12. Use pivotal elimination on the tableau in Exercise 9 to find the new tableau.

13. Use pivotal elimination on the tableau in Exercise 10 to find the new tableau.

In Exercises 14–18, solve by the simplex method.

14. Maximize $z = 3x_1 - 4x_2 + x_3$

subject to

$$\begin{aligned} 2x_1 + 3x_2 - 2x_3 &\leq 8 \\ 2x_1 - 3x_2 + x_3 &\leq 12 \\ x_1 \geq 0, \quad x_2 \geq 0, \quad x_3 &\geq 0 \end{aligned}$$

15. Maximize $z = 2x + 3y - t$

subject to

$$\begin{aligned} x + 3y + 4t &\leq 12 \\ x + y + 2t &\leq 18 \\ 2x - y - t &\leq 16 \\ x \geq 0, \quad y \geq 0, \quad t &\geq 0. \end{aligned}$$

16. Maximize $z = 3x_1 - 5x_2 + 2x_3 + x_4$

subject to

$$\begin{aligned} 2x_1 + 2x_2 - x_3 \qquad &\leq 14 \\ x_1 - 2x_2 + x_3 + x_4 &\leq 16 \\ 2x_1 + 4x_2 \qquad + 5x_4 &\leq 25 \\ x_1 \geq 0, \quad x_2 \geq 0, \quad x_3 \geq 0, \quad x_4 &\geq 0. \end{aligned}$$

17. Maximize $z = 2x_1 - 4x_2 - x_3$

subject to

$$\begin{aligned} 5x_1 - 4x_2 - 2x_3 &\leq 6 \\ 4x_1 + 3x_2 + x_3 &\leq 5 \\ 3x_1 + 2x_2 - x_3 &\leq 4 \\ x_1 \geq 0, \quad x_2 \geq 0, \quad x_3 &\geq 0. \end{aligned}$$

18. Maximize $z = 2x - 3y$

subject to

$$\begin{aligned} 2x + 6y &\leq 9 \\ 3x - 2y &\leq 8 \\ 5x + y &\leq 5 \\ x \geq 0, \quad y &\geq 0. \end{aligned}$$

19. Use the simplex method to show the following problem has no solution.

Maximize $z = 2x + 5y$

subject to

$$\begin{aligned} -4x + y &\leq 5 \\ -x + y &\leq 4 \\ x \geq 0, \quad y &\geq 0. \end{aligned}$$

20. An investment banker wants to invest $18,000 or less in three types of bonds: type A bond yielding a 5% profit on the amount invested, type B bond yielding a 7% profit on the amount invested, and type C bond yielding a 10% profit on the amount invested. He wants to invest no more than $10,000 in bonds of types A and B together, no more than $8000 in bonds of types B and C together, and no more than $6000 in bonds of type C. How much should be invested in each kind of bond to maximize the profit?

21. An ice cream manufacturer blends ingredients A, B, and C in varying proportions to make three types of ice cream: Low calorie (L), regular (R) and extra-rich (E). Each gallon of L contains 0.2 gallon of A, 0.3 gallon of B, and 0.2 gallon of C. Each gallon of R contains 0.4 gallon of A, 0.2 gallon of B, and 0.2 gallon of C; each gallon of E contains 0.5 gallon of A, 0.2 gallon of B, and 0.1 gallon of C. Suppose that the profits on the ice creams L, R, and E are 20, 40, and 20 cents per gallon respectively. If the manufacturer has 50, 80, and 100 gallons of A, B, and C available, how many gallons of each type of ice cream should be made to maximize the profit?

22. A company has three methods for obtaining high-grade aluminum from recycled cans. With method A, each ton of cans requires 4 hours of smelting, 2 hours of centrifugal separation, and 1 hour of chemical purification; it yields 5 pounds of pure aluminum per ton of cans. With method B, each ton of cans requires 2 hours of smelting, 2 hours of centrifugal separation, and 3 hours of chemical purification; it yields 3 pounds of pure aluminum per ton of cans. With method C, each ton of cans requires 4 hours of smelting, 3 hours of centrifugal separation, and 2 hours of chemical purification; it yields 4 pounds of pure aluminum per ton of cans. Assuming that the smelter is available for 80 hours per week, the centrifugal separator for 50 hours per week, and the chemical purifier for 40 hours per week, how many tons of cans should be processed by each method to maximize the weekly output of pure aluminum?

5.4 NONSTANDARD LINEAR PROGRAMMING PROBLEMS; DUALITY*

Recall that a standard linear programming problem is one that satisfies the following conditions.

(i) The objective function is to be maximized.

(ii) The variables are all constrained to be nonnegative.

(iii) In the remaining constraints, the expressions involving the variables are less than or equal to ($\leq$) a *nonnegative* constant.

In this section we shall discuss methods for solving certain nonstandard problems.

Consider the problem

$$\text{Minimize} \quad z = -10x - 12y$$

subject to

$$\begin{aligned} x + 2y &\leq 150 \\ x + y &\leq 100 \qquad (1) \\ x \geq 0, \quad y &\geq 0. \end{aligned}$$

This problem *fails* to be a standard linear programming problem because it violates condition (i) above. Before we can show how to solve such a problem we will have to develop some preliminary results.

Let S be any set of real numbers and suppose S has a *smallest* number a. If we form the set $\bar{S}$ consisting of the negatives of the numbers in S, then it is evident that $-a$ is the *largest* number in $\bar{S}$. To illustrate, the smallest number in the set

$$S = \{3, 4, 5, 6\}$$

is 3, whereas the largest number in

$$\bar{S} = \{-3, -4, -5, -6\}$$

is -3. In summary, the minimum value in S and the maximum value in S' are related by

$$\text{minimum value in } S = -(\text{maximum value in } \bar{S}).$$

Applying this result to problem (1) above we obtain

$$\text{minimum value of } z = -(\text{maximum value of } z'), \qquad (2)$$

where z' is the negative of z.

The following example shows how this result can be used to solve problem (1) above.

Example 11 Solve the nonstandard linear programming problem

$$\text{Minimize} \quad z = -10x - 12y$$

subject to

$$\begin{aligned} x + 2y &\leq 150 \\ x + y &\leq 100 \\ x \geq 0, \quad y &\geq 0. \end{aligned}$$

Solution From Equation (2) above, the minimum value of z is the negative of the maximum value of

$$z' = -z = 10x + 12y.$$

We thus consider the problem

$$\text{Maximize} \quad z' = 10x + 12y$$

subject to

$$\begin{aligned} x + 2y &\leq 150 \\ x + y &\leq 100 \\ x \geq 0, \quad y &\geq 0. \end{aligned}$$

This problem was solved in Example 9 of Section 5.3, where we showed that the maximum value

$$z' = 1100$$

occurs when $x = 50$, $y = 50$. Thus the minimum value of z is

$$z = -1100$$

and this occurs when $x = 50$, $y = 50$.

We now turn to some other kinds of nonstandard linear programming problems. Consider the problem

$$\text{Minimize} \quad z = 3x_1 + 4x_2$$

subject to

$$\begin{aligned} x_1 + 2x_2 &\geq 8 \\ 2x_1 + 2x_2 &\geq 10 \\ x_1 + 4x_2 &\geq 12 \\ x_1 \geq 0, \quad x_2 &\geq 0. \end{aligned} \tag{3}$$

This fails to be a standard linear programming problem for two reasons. First, the objective function is to be minimized, which violates condition (i) stated at the beginning of this section. Second, the inequalities in the constraints

$$x_1 + 2x_2 \geq 8, \qquad 2x_1 + 2x_2 \geq 10, \qquad x_1 + 4x_2 \geq 12$$

go the wrong way, violating condition (iii). Before explaining how to solve problems of this type we shall develop some preliminary ideas.

In matrix notation problem (3) has the form

$$\text{Minimize} \quad z = CX$$

subject to

$$AX \geq B$$
$$X \geq 0,$$

where

$$C = [3 \quad 4], \qquad X = \begin{bmatrix} x_1 \\ x_2 \end{bmatrix}, \qquad A = \begin{bmatrix} 1 & 2 \\ 2 & 2 \\ 1 & 4 \end{bmatrix}, \qquad B = \begin{bmatrix} 8 \\ 10 \\ 12 \end{bmatrix}.$$

In this problem we have two unknowns, x_1 and x_2, and three constraints in $AX \geq B$. More generally, we shall want to solve

$$\text{Minimize} \quad z = CX$$

subject to

$$AX \geq B \tag{4}$$
$$X \geq 0$$

when there are n unknowns $x_1, x_2, \ldots, x_n$ and k constraints in $AX \geq B$. To solve problems like (4) we begin by forming the following maximization problem involving k new unknowns $y_1, y_2, \ldots, y_k$.

$$\text{Maximize} \quad z' = B^tY$$

subject to

$$A^tY \leq C^t \tag{5}$$
$$Y \geq 0.$$

This problem is called the ***dual problem,*** and the original problem is called the ***primal problem.*** The matrices A^t, B^t, and C^t in the dual problem are the transposes of the matrices A, B, and C in the primal problem. The matrix Y is the $k \times 1$ matrix of new unknowns $y_1, y_2, \ldots, y_k$ and z' is the quantity to be maximized (we cannot use z here since the letter z is already used in the primal problem.)

Example 12 Find the dual problem for (3).

Solution As previously shown, the matrices A, B, and C in problem (3) are

$$A = \begin{bmatrix} 1 & 2 \\ 2 & 2 \\ 1 & 4 \end{bmatrix}, \qquad B = \begin{bmatrix} 8 \\ 10 \\ 12 \end{bmatrix}, \qquad C = [3 \quad 4].$$

Thus

$$A^{t} = \begin{bmatrix} 1 & 2 & 1 \\ 2 & 2 & 4 \end{bmatrix}, \qquad B^{t} = [8 \quad 10 \quad 12], \qquad C^{t} = \begin{bmatrix} 3 \\ 4 \end{bmatrix}.$$

Since the primal problem had $k = 3$ constraints in $AX \geq B$, the dual problem will have three unknowns, y_1, y_2, and y_3. Thus

$$Y = \begin{bmatrix} y_1 \\ y_2 \\ y_3 \end{bmatrix}.$$

Therefore, the dual problem is

$$\text{Maximize} \quad z' = [8 \quad 10 \quad 12] \begin{bmatrix} y_1 \\ y_2 \\ y_3 \end{bmatrix}$$

subject to

$$\begin{bmatrix} 1 & 2 & 1 \\ 2 & 2 & 4 \end{bmatrix} \begin{bmatrix} y_1 \\ y_2 \\ y_3 \end{bmatrix} \leq \begin{bmatrix} 3 \\ 4 \end{bmatrix}$$

$$\begin{bmatrix} y_1 \\ y_2 \\ y_3 \end{bmatrix} \geq \begin{bmatrix} 0 \\ 0 \\ 0 \end{bmatrix}$$

or without matrix notation

$$\text{Maximize} \quad z' = 8y_1 + 10y_2 + 12y_3$$

subject to

$$\begin{aligned} y_1 + 2y_2 + y_3 &\leq 3 \\ 2y_1 + 2y_2 + 4y_3 &\leq 4 \\ y_1 \geq 0, \quad y_2 \geq 0, \quad y_3 &\geq 0. \end{aligned} \tag{6}$$

Observe that this is a standard linear programming problem.

The dual problem is important because of the following result which we state without proof.

> If either the dual problem or the primal problem has a solution, then so does the other. Moreover, the minimum value of z and the maximum value of z' are equal.

The following example illustrates how this result can be used.

Example 13 Find the minimum value of z in (3).

Solution The maximum value of z is the same as the maximum value of z' in the dual problem. We previously showed the dual problem to be (see (6)):

$$\text{Maximize} \quad z' = 8y_1 + 10y_2 + 12y_3$$

subject to

$$\begin{aligned} y_1 + 2y_2 + \ \ y_3 &\leq 3 \\ 2y_1 + 2y_2 + 4y_3 &\leq 4 \\ y_1 \geq 0, \quad y_2 \geq 0, \quad y_3 &\geq 0. \end{aligned}$$

Since this is a standard linear programming problem, we can use the simplex method. The corresponding problem with slack variables is

$$\text{Maximize} \quad z' = 8y_1 + 10y_2 + 12y_3$$

subject to

$$\begin{aligned} y_1 + 2y_2 + \ \ y_3 + v_1 \qquad &= 3 \\ 2y_1 + 2y_2 + 4y_3 \qquad + v_2 &= 4 \\ y_1 \geq 0, \quad y_2 \geq 0, \quad y_3 \geq 0, \quad v_1 \geq 0, \quad v_2 &\geq 0. \end{aligned}$$

The reader can check that the initial tableau and pivot entry are as follows:

$$\begin{array}{r c} & \begin{array}{ccccccc} & & \text{entering variable} & & & & \\ & & \downarrow & & & & \\ y_1 & y_2 & y_3 & v_1 & v_2 & z' & \end{array} \\ \begin{array}{r} z' \\ v_1 \\ \text{departing variable} \rightarrow v_2 \end{array} & \left[\begin{array}{ccccccc} -8 & -10 & -12 & 0 & 0 & 1 & 0 \\ 1 & 2 & 1 & 1 & 0 & 0 & 3 \\ 2 & 2 & \boxed{4} & 0 & 1 & 0 & 4 \end{array}\right]. \end{array}$$

After pivotal elimination we obtain

$$
\begin{array}{c}
\text{entering variable} \\
\downarrow
\end{array}
$$

$$
\begin{array}{rc}
 & \begin{array}{ccccccc} y_1 & y_2 & y_3 & v_1 & v_2 & z' & \end{array} \\
\begin{array}{r} z' \\ \text{departing variable} \rightarrow v_1 \\ y_3 \end{array} &
\left[\begin{array}{ccccccc}
-2 & -4 & 0 & 0 & 3 & 1 & 12 \\
\frac{1}{2} & \frac{3}{2} & 0 & 1 & -\frac{1}{4} & 0 & 2 \\
\frac{1}{2} & \frac{1}{2} & 1 & 0 & \frac{1}{4} & 0 & 1
\end{array}\right].
\end{array}
$$

After pivotal elimination again, we obtain

$$
\begin{array}{c}
\text{entering variable} \\
\downarrow
\end{array}
$$

$$
\begin{array}{rc}
 & \begin{array}{ccccccc} y_1 & y_2 & y_3 & v_1 & v_2 & z' & \end{array} \\
\begin{array}{r} z' \\ y_2 \\ \text{departing variable} \rightarrow y_3 \end{array} &
\left[\begin{array}{ccccccc}
-\frac{2}{3} & 0 & 0 & \frac{8}{3} & \frac{7}{3} & 1 & \frac{52}{3} \\
\frac{1}{3} & 1 & 0 & \frac{2}{3} & -\frac{1}{6} & 0 & \frac{4}{3} \\
\frac{1}{3} & 0 & 1 & -\frac{1}{3} & \frac{1}{3} & 0 & \frac{1}{3}
\end{array}\right].
\end{array}
$$

After one more pivotal elimination, we obtain the final tableau

$$
\begin{array}{rc}
 & \begin{array}{ccccccc} y_1 & y_2 & y_3 & v_1 & v_2 & z' & \end{array} \\
\begin{array}{r} z' \\ y_2 \\ y_1 \end{array} &
\left[\begin{array}{ccccccc}
0 & 0 & 2 & 2 & 3 & 1 & 18 \\
0 & 1 & -1 & 1 & -\frac{1}{2} & 0 & 1 \\
1 & 0 & 3 & -1 & 1 & 0 & 1
\end{array}\right].
\end{array}
$$

From this tableau we see that the maximum value of z' is 18. Therefore the minimum value of z is also 18.

The technique illustrated in this example gave us the minimum value of z; it did not, however, tell us the values of x_1 and x_2 where this minimum occurs. For reasons too technical to pursue here, it can be shown that these values appear in the z' row of the final tableau for the dual problem; the value of x_1 occurs under the slack variable v_1 and the value of x_2 occurs under the slack variable v_2. Thus the minimum value $z = 18$ in problem (3) occurs when

$$x_1 = 2 \qquad \text{and} \qquad x_2 = 3.$$

To summarize, we can state the following result.

The optimal solution to the primal problem occurs under the slack variables in the z' row of the final tableau for the dual problem. The value of x_1 occurs under v_1, the value of x_2 occurs under v_2, the value of x_3 occurs under v_3, and so on.

EXERCISE SET 5.4

1. Solve by the method of Example 11.

$$\text{Minimize} \quad z = -4x + 2y$$

subject to

$$\begin{aligned} 6x + 2y &\leq 18 \\ 3x - 2y &\leq 6 \\ x \geq 0, \quad y &\geq 0. \end{aligned}$$

2. Solve by the method of Example 11.

$$\text{Minimize} \quad z = 2x_1 - 3x_2 + x_3$$

subject to

$$\begin{aligned} x_1 + x_2 + 2x_3 &\leq 5 \\ x_1 - x_2 + 3x_3 &\leq 4 \\ 2x_1 + x_2 \qquad &\leq 6 \\ x_1 \geq 0, \quad x_2 \geq 0, \quad x_3 &\geq 0. \end{aligned}$$

3. Solve by the method of Example 11.

$$\text{Minimize} \quad z = 2x - 2y + t$$

subject to

$$\begin{aligned} 3x + y + t &\geq -6 \\ x - 3y - t &\geq -8 \\ x \geq 0, \quad y \geq 0, \quad t &\geq 0. \end{aligned}$$

4. Find the dual problem for

$$\text{Minimize} \quad z = 3x_1 + 4x_2 + x_3$$

subject to

$$\begin{aligned} 2x_1 + 3x_2 + x_3 &\geq 8 \\ 5x_1 + 2x_2 + 2x_3 &\geq 5 \\ x_1 \geq 0, \quad x_2 \geq 0, \quad x_3 &\geq 0. \end{aligned}$$

5. Solve the problem in Exercise 4.

6. Find the dual problem for

$$\text{Minimize} \quad z = 3x_1 + 5x_2$$

subject to

$$\begin{aligned} 3x_1 + 2x_2 &\geq 6 \\ 4x_1 + x_2 &\geq 4 \\ 14x_1 + 6x_2 &\geq 21 \\ x_1 \geq 0, \quad x_2 &\geq 0. \end{aligned}$$

7. Solve the problem in Exercise 6.

8. Find the dual problem for

$$\text{Minimize} \quad z = 6x_1 + 5x_2$$

subject to

$$\begin{aligned} 2x_1 + 3x_2 &\geq 6 \\ 5x_1 + 2x_2 &\geq 10 \\ x_1 \geq 0, \quad x_2 &\geq 0. \end{aligned}$$

9. Solve the problem in Exercise 8.

10. Solve the problem in Exercise 1 of Section 3.1. (Use the dual problem.)

11. Solve the problem in Exercise 3 of Section 3.1. (Use the dual problem.)

6 PROBABILITY

There is an element of unpredictability in many physical phenomena. Even under apparently identical conditions, many observed quantities vary in an uncertain way. For example, the total number of phone calls arriving at a switchboard varies unpredictably from day to day, the total annual rainfall in Chicago varies unpredictably from year to year, the Dow Jones Industrial Average varies unpredictably from hour to hour, and the number of heads obtained in five tosses of a coin varies unpredictably with each group of tosses. In this chapter we shall study probability theory. This branch of mathematics is concerned with making rational statements about phenomena that are subject to an element of uncertainty.

6.1 INTRODUCTION; SAMPLE SPACE AND EVENTS

The meaning that we shall give to the term *probability* can best be described by considering what happens when we toss a coin. If we toss a coin once, it is impossible to predict (with certainty) in advance whether the outcome will be a head or a tail. On the other hand, if we toss the coin over and over again, something rather interesting occurs. To illustrate, consider Table 6.1. In this table we have recorded what occurred at various stages when an ordinary coin was tossed 20,000 times.

Let us examine the proportions listed in the third column of Table 6.1. In the early stages of tossing, the proportion of heads varied rather considerably. As the coin tossing continued, however, the proportion of heads began to settle down, so that during the last 5000 tosses the proportion of heads changed by only .008. These data suggest that if we continued the coin tossing beyond 20,000 tosses, the proportion of heads would approach some fixed constant value. We think of this value as the *probability of heads*.

Table 6.1 Summary of 20,000 coin tosses

Number of tosses (n)	Number of heads (h)	Proportion of tosses which were heads $\left(\frac{h}{n}\right)$
10	8	.8000
100	62	.6200
1000	473	.4730
5000	2550	.5100
10,000	5098	.5098
15,000	7649	.5099
20,000	10,038	.5019

Intuition suggests that if the coin were perfectly balanced, then the probability of heads would be $\frac{1}{2}$ or .5. However, since no physical coin is perfectly balanced, the probability of heads for the coin used to obtain Table 6.1 would presumably be close to but not exactly equal to $\frac{1}{2}$.

To generalize the above ideas, assume we have an experiment that can be repeated indefinitely under fixed conditions, and suppose that during n repetitions of this experiment, a certain event occurs m times. We shall call the ratio

$$\frac{m}{n}$$

the ***relative frequency*** of the event after n repetitions. If this relative frequency approaches a number p as n gets larger and larger, then p is called the ***probability*** of the event. Thus as n (the number of repetitions) gets larger and larger, the approximation

$$p \simeq \frac{m}{n} \qquad \text{(the symbol } \simeq \text{ means "approximately equal")}$$

gets better and better.

This definition of probability is somewhat unsatisfactory since the meaning of the phrase "this relative frequency *approaches* a number p" has not been precisely explained. Nevertheless, the reader's intuitive feeling for the meaning of this statement should be perfectly adequate for most purposes.

To summarize, it will be satisfactory to interpret the probability of an event intuitively as its "long-term relative frequency."

In some situations, probabilities can be obtained using physical intuition; for example, if we have an ordinary six-sided die that we assume to be perfectly balanced and perfectly symmetrical, then intuition tells us that the probability of tossing a 2 with this die is $\frac{1}{6}$. We obtain this conclusion by arguing that each of the six possible outcomes has an equal chance of occurring, so that over the long term, the number 2 will appear one-sixth of the time. On the other hand, some probabilities cannot be obtained using intuition; they can only be estimated from experimental data. For example, suppose we have a certain production process for manufacturing photographic flashbulbs, and we are interested in the probability that the process will produce a defective bulb. Intuition will not tell us what this probability is. However, if we test 10,000 bulbs and find that three bulbs are defective, we can estimate the probability of a defective bulb to be

$$\frac{3}{10{,}000} = .0003.$$

Throughout this chapter we shall be concerned with outcomes of experiments, where the term *experiment* is used in a broad sense to mean the observation of any physical occurrence. Since we shall be interested in probabilities of events associated with these experiments, it is necessary that the term *event* be given a precise mathematical meaning. This will be our objective for the remainder of this section.

The relative frequency interpretation of probability is appropriate for experiments that can be repeated over and over under fixed conditions. For experiments that cannot be repeated under fixed conditions, the ***subjective*** interpretation of probabilities is more appropriate. With this approach, the probability of an event is viewed as a measure (on a scale of zero to one) of one's "strength of belief" that the event will occur when the experiment is performed:

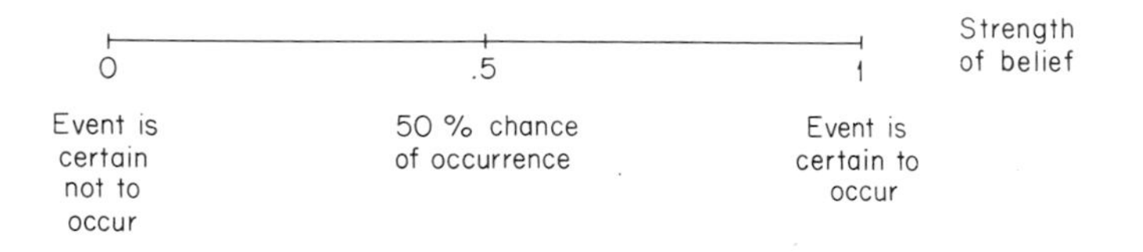

For example, prior to the first soft landing on mars, experts estimated a 40% chance (probability .4) of finding life on mars. This was a subjective assignment of probability based on expert opinion.

In all probability problems, it is important to clearly identify the possible outcomes of the experiment.

> The set of all possible outcomes of an experiment is called the ***sample space*** for the experiment. The outcomes in the sample space are called the ***sample points.***

Example 1 Consider the experiment of tossing a die and recording the number on the top face. There are six possible outcomes or sample points for this experiment. Thus the sample space S is the set

$$S = \{1, 2, 3, 4, 5, 6\}.$$

Example 2 A light-bulb manufacturer tests a bulb by letting it burn until it burns out. He records the total time t, in hours, that the bulb stays lit. Since any nonnegative real number may be recorded for t, the sample space S consists of all nonnegative real numbers; that is,

$$S = \{t \mid t \geq 0\}.$$

Example 3 A traffic engineer records the number of cars entering a tunnel between 9:00 A.M. and 10:00 A.M. The sample space S consists of all nonnegative integers; that is,

$$S = \{0, 1, 2, \ldots\}.$$

Example 4 A stock market analyst observes General Electric common stock for one market day and records whether the stock increases in value, decreases in value, or undergoes no change in value. The sample space S is

$$S = \{i, d, n\},$$

where i, d, and n denote increase, decrease, and no change, respectively.

The sample space of an experiment is partially determined by the type of information we want. To illustrate, in the die-tossing experiment of Example 1, we may be interested in recording only whether the top face is a 3 or not. In this case the sample space S would be

$$S = \{3, \text{non-}3\}.$$

Thus, a given experiment may have more than one sample space, depending on what the experimenter chooses to observe.

We shall be interested in making probability statements about events. Intuitively, we all know examples of events: the event that an odd number is tossed with a die, the event that more than 10 cars enter a certain tunnel between 9:00 A.M. and 10:00 A.M., the event that a bulb burns out in fewer than 10 hours, and so on. We shall now make this intuitive notion of an event mathematically precise.

For motivation, consider the die-tossing experiment in Example 1, and let E denote the event that an odd number comes up when the die is tossed. If, when the experiment is performed, one of the sample points

$$1, \qquad 3, \qquad \text{or} \qquad 5$$

results, then the event E occurs. If, when the experiment is performed, one of the remaining sample points 2, 4, or 6 results, then the event E does not occur. The event E can therefore be described by specifying the set of sample points for which it occurs. Thus, the event E is completely described by the set of sample points

$$\{1, 3, 5\}.$$

This suggests the following definition.

An ***event*** is a subset of the sample space of an experiment.

Intuitively, an event "occurs" when one of its sample points turns up as an outcome of the experiment.

Example 5 Consider the experiment of tossing a die and recording the number on the top face. If E is the event that an odd number occurs, then

$$E = \{1, 3, 5\}.$$

If F is the event that a number larger than 4 occurs, then

$$F = \{5, 6\}.$$

Example 6 In a germination experiment, 10 corn seeds are planted and the number of seeds that germinate within 30 days is recorded. Let E be the event that more than half germinate.

The sample space is

$$S = \{0, 1, 2, 3, 4, 5, 6, 7, 8, 9, 10\},$$

and the event E is

$$E = \{6, 7, 8, 9, 10\}.$$

Example 7 Consider the light-bulb testing experiment in Example 2. Let E be the event that the life of the bulb is between 10 and 20 hours inclusive; then

$$E = \{t \mid 10 \leq t \leq 20\}.$$

If S is the sample space of an experiment, then an event E is called a ***certain event*** if $E = S$ and an ***impossible event*** if $E = \varnothing$ (the empty set). A certain event always occurs and an impossible event never occurs, regardless of the outcome of the experiment. (Why?)

Example 8 Consider the experiment of tossing a coin twice and recording the ordered pair

$$(n_{\mathrm{H}}, n_{\mathrm{T}}),$$

where n_H is the total number of heads tossed and n_T is the total number of tails tossed.

The sample space S is the set of ordered pairs

$$S = \{(2, 0), (1, 1), (0, 2)\}.$$

If E is the event that $n_H + n_T$ is even, then

$$E = \{(2, 0), (1, 1), (0, 2)\} = S,$$

so that E is a certain event.

If F is the event that $n_H + n_T$ is less than 1, then

$$F = \varnothing,$$

so that F is an impossible event.

Since events are sets, we can apply the set operations, union, intersection, and complementation to them. Thus, given events E and F in a sample space S, we can form new events

$$E \cap F, \qquad E \cup F, \qquad \text{and} \qquad E'$$

(see Figure 6.1). Let us try to discover the meaning of these new events.

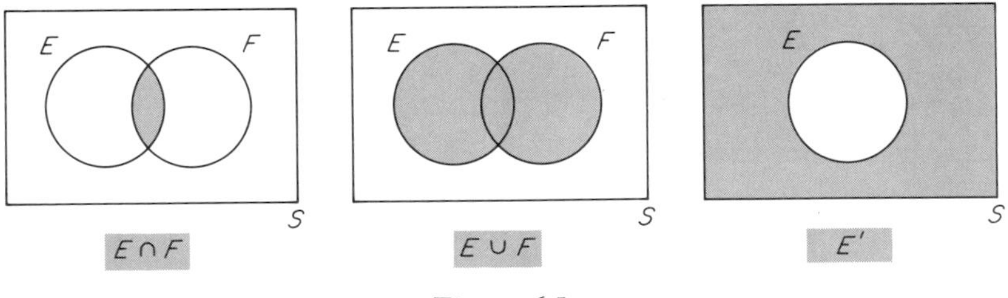

Figure 6.1

The event $E \cap F$ consists of those sample points common to both E *and* F. In other words, $E \cap F$ consists of those sample points for which E and F both occur.

Example 9 An experimenter tosses a die and records the number on the top face. Let E be the event that the number is divisible by 3 and let F be the event that the number is odd. Thus,

$$E = \{3, 6\} \qquad \text{and} \qquad F = \{1, 3, 5\}.$$

The event that the number is divisible by 3 *and* odd is

$$E \cap F = \{3\}.$$

To summarize, the intersection of two events can be interpreted as follows.

> Stating that $E \cap F$ occurs is equivalent to stating that E and F both occur.

If two events E and F have no common sample points, then the two events cannot occur simultaneously; such events are said to be ***mutually exclusive.*** Symbolically, E and F are mutually exclusive if $E \cap F = \varnothing$.

Example 10 Consider the die-tossing experiment in Example 9. Let E be the event that an even number is tossed and let F be the event that an odd number is tossed. Then

$$E = \{2, 4, 6\} \qquad \text{and} \qquad F = \{1, 3, 5\}.$$

Since $E \cap F = \varnothing$, these events are mutually exclusive.

The notion of "mutually exclusive" can be extended to more than two events. We shall call n events $E_1, E_2, \ldots, E_n$ *mutually exclusive* if no two of them have any sample points in common (see Figure 6.2). Observe that to say n events are mutually exclusive means no two of them can occur simultaneously.

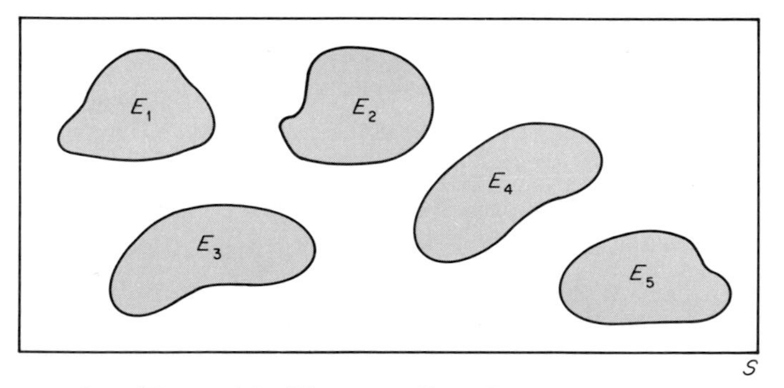

Figure 6.2 Five mutually exclusive events.

The event $E \cup F$ consists of those sample points that belong to E *or* to F (or to both). In other words, $E \cup F$ consists of those sample points for which E occurs or F occurs, or both.

Example 11 Let E and F be the events in Example 9. The event that the number tossed is odd or divisible by 3 is

$$E \cup F = \{1, 3, 5, 6\}.$$

To summarize, the union of two events can be interpreted as follows.

> Stating that $E \cup F$ occurs is equivalent to stating that E or F (or both) occur.

The event E' consists of those sample points that do not belong to E. In other words, E' contains those sample points for which E does not occur.

Example 12 Let E be the event described in Example 9. Then

$$E' = \{1, 2, 4, 5\}.$$

This is just the event that the number tossed is not divisible by 3.

To summarize, the complement of an event can be interpreted as follows.

> Stating that E' occurs is equivalent to stating that E does not occur.

EXERCISE SET 6.1

1. For each of the following experiments, describe the sample space S in set notation.
   (a) Ten people are asked if they smoke regularly, and the number of people answering affirmatively is recorded.
   (b) A traffic engineer records the number of cars crossing the George Washington bridge in a 24-hour period.
   (c) A chemist measures and records the percentage of ethylene glycol in a solution. (Assume he can measure the percentage with perfect accuracy.)
   (d) A transistor is operated until it fails and its life in hours is recorded. (Assume the length of life can be measured with perfect accuracy.)
   (e) A coin is tossed three times and the sequence of h's (heads) and t's (tails) is recorded.
2. A person's weight is measured and recorded. Write the sample space S in set notation assuming:
   (a) The weight is recorded with perfect accuracy.
   (b) The weight is recorded to the nearest half-pound (for example, 161, 161.5, 162, and so on).

3. The sample space of a certain experiment is $S = \{a, b, c\}$. List all possible events.

4. A die is tossed and the number on the top face is recorded. Write the following events in set notation:

(a) $E =$ a number greater than 3 is tossed.

(b) $F =$ a number other than 2 is tossed.

(c) $G =$ the number tossed is either odd or less than 3.

5. A die is tossed twice and the resulting ordered pair of numbers is recorded. Describe the following events *in words*.

(a) $\{(1, 1), (2, 2), (3, 3), (4, 4), (5, 5), (6, 6)\}$

(b) $\{(3, 1), (3, 2), (3, 3), (3, 4), (3, 5), (3, 6)\}$

(c) $\{(4, 4)\}$.

6. Items coming off a production line are checked to see if they are defective (d) or nondefective (n). The checker continues to check until two consecutive defectives are found or four items have been checked, whichever occurs first; he then records the resulting sequence of d's and n's.

(a) Write the sample space S in set notation.

(b) Let E be the event that the checker stops at the third item. Write E in set notation.

(c) Let F be the event that the checker stops at the fourth item. Write F in set notation.

7. A firm that conducts political polls classifies people according to three characteristics:

Sex: male (m), female (f)
Income: high (h), average (a), low (l)
Political registration: Democrat (d), Republican (r), Independent (i)

Thus a person classified (f, h, r) would be a female with high income, who is registered republican.

(a) A person is randomly selected and classified according to the above system. Express the sample space for this experiment in set notation.

(b) Let E be the event that the person selected is a Republican; write E in set notation.

(c) Let F be the event that the person selected is a female registered as an independent; write F in set notation.

(d) Let G be the event that the person selected is either a male or a Democrat. Write G in set notation.

8. A geneticist successively selects three newly hatched fruit flies and observes whether each is long winged (l) or short winged (s). He records the resulting sequence of l's and s's. Write the following events in set notation.

(a) Exactly one long-winged fly is observed.

(b) At least one long-winged fly is observed.

(c) No more than two short-winged flies are observed.

9. A die is tossed and the number on the top face is recorded. Let E, F, and G be the events:

$$E = \text{the number tossed is even}$$
$$F = \text{the number tossed is divisible by 3}$$
$$G = \text{the number tossed is less than 5.}$$

Describe the following events *in words*.

(a) $E \cup G$ (b) $F \cap G$ (c) $F \cup G'$

(d) $F' \cap G$ (e) $E' \cap F' \cap G'$.

10. List the sample points for each event in Exercise 9.

11. Consider the experiment described in Example 2 of the text, and let E and F be the events

$$E = \text{the bulb lasts 10 or more hours}$$
$$F = \text{the bulb lasts fewer than 15 hours.}$$

Express the following events using set-builder notation:

(a) E (b) F (c) E'

(d) F' (e) $E \cap F'$ (f) $E \cup F'$.

12. Are the events E and F in Exercise 11 mutually exclusive? Explain.

13. A medical file is selected from the records of a hospital and information about the person's weight, age, and marital status is noted. Let E, F, G, and H be the events:

$$E = \text{the person is overweight}$$
$$F = \text{the person is underweight}$$
$$G = \text{the person is over 50 years of age}$$
$$H = \text{the person is married.}$$

(a) Are E and F mutually exclusive? Explain.

(b) Are E and H mutually exclusive? Explain.

(c) Is $E \cup F$ a certain event? Explain.

(d) Is $E \cup E'$ a certain event? Explain.

14. Let E, F, and G be three events associated with an experiment. Express the following events in terms of E, F, and G using the operations $\cup$, $\cap$, and $'$.

(a) E occurs or F occurs.

(b) E occurs or G does not occur.

(c) F occurs and G occurs.

(d) E occurs and G does not occur.

(e) At least one of the three events occurs.

(f) Exactly one of the three events occurs.

(g) None of the events occur.

15. A communications transmission line can be open (o) or busy (b). Suppose the line is monitored three times and the resulting sequence of o's and b's is recorded. Determine if the events E, F, and G below are mutually exclusive

$$E = \{ooo, oob, boo\}$$
$$F = \{bob, obo\}$$
$$G = \{bbb, obb\}.$$

Explain the reason for your answer.

6.2 PROBABILITY MODELS FOR FINITE SAMPLE SPACES

Given an experiment, we shall be interested in assigning probabilities to various events. In practice there are usually so many events of importance that it is a tedious task to assign all of them probabilities. One way of overcoming this problem is to assign probabilities only to certain basic events and then use various theorems about probabilities to determine the probabilities of other events as they are needed. Once a procedure is specified for obtaining the probabilities of all the events of interest, we say that a ***probability model*** has been assigned to the experiment. In this section we shall discuss the construction of probability models for an important class of experiments.

The simplest kinds of experiments are those that have a ***finite sample space***; that is, there are a finite number of points in the sample space. If

$$S = \{s_1, s_2, \ldots, s_k\}$$

is the sample space for such an experiment, then the events

$$\{s_1\}, \qquad \{s_2\}, \qquad \ldots, \qquad \{s_k\}$$

that consist of exactly one point are called ***elementary events.***

Example 13 If we toss a coin, then the sample space S is

$$S = \{h, t\}$$

where h = heads and t = tails. The elementary events are

$$\{h\} \qquad \text{and} \qquad \{t\}.$$

Example 14 If we toss a die and observe the number on the top face, then the sample space S is

$$S = \{1, 2, 3, 4, 5, 6\}$$

and the elementary events are

$$\{1\}, \qquad \{2\}, \qquad \{3\}, \qquad \{4\}, \qquad \{5\}, \qquad \text{and} \qquad \{6\}.$$

Example 15 If we pick a name at random from a list of 8000 different names, the sample space has 8000 sample points; consequently there are 8000 different elementary events.

For the remainder of this section *we shall consider only experiments with finite sample spaces.* Once the probabilities of the elementary events are known, the probabilities of all other events can be obtained using the following principle or axiom, called the ***addition principle.***

Addition Principle If an experiment has finitely many points in its sample space, then the probability $P(E)$ of an event

$$E = \{s_1, s_2, \ldots, s_m\}$$

is

$$P(E) = P\{s_1\} + P\{s_2\} + \cdots + P\{s_m\}$$

where $P\{s_1\}, P\{s_2\}, \ldots, P\{s_m\}$ are the probabilities of the elementary events $\{s_1\}, \{s_2\}, \ldots, \{s_m\}$.

To paraphrase, the addition principle states that for experiments with a finite sample space, *the probability of an event is the sum of the probabilities of its sample points.* Since an impossible event has no sample points, the addition principle does not apply. We shall agree to assign an impossible event a probability of zero.

Example 16 A six-sided die is "loaded" so that the numbers from 1 to 6 occur with the following probabilities when the die is tossed,

$$P\{1\} = \frac{2}{12}, \qquad P\{2\} = \frac{1}{12}, \qquad P\{3\} = \frac{4}{12},$$

$$P\{4\} = \frac{1}{12}, \qquad P\{5\} = \frac{2}{12}, \qquad P\{6\} = \frac{2}{12}.$$

Find the probability of

(a) the event E that an even number is tossed;

(b) the event G that a number divisible by 3 is tossed;

(c) the event H that a number greater than 7 is tossed.

Solution The events E, G, and H are

$$E = \{2, 4, 6\}$$

$$G = \{3, 6\}$$

$$H = \varnothing.$$

Thus,

$$P(E) = P\{2\} + P\{4\} + P\{6\} = \frac{1}{12} + \frac{1}{12} + \frac{2}{12} = \frac{1}{3}$$

$$P(G) = P\{3\} + P\{6\} = \frac{4}{12} + \frac{2}{12} = \frac{1}{2}$$

$$P(H) = 0.$$

Of particular importance are experiments that have a finite number of equally likely outcomes in the sample space. We use the phrase *equally likely* in an intuitive sense to mean that all sample points have equal chances of occurring.

Example 17 If we toss a fair coin, that is, a coin that is perfectly balanced, it is reasonable to assume that "heads" and "tails" are equally likely.

Example 18 If we toss a fair die, that is, a die that is perfectly symmetrical and perfectly balanced, it is reasonable to assume that the six possible outcomes are equally likely.

Example 19 If we pick a name at random from a list of 8000 different names, then the 8000 possible outcomes are equally likely.

If an experiment has finitely many *equally likely* outcomes, then it is reasonable to assign each elementary event the same probability.

Example 20 If we toss a *fair* coin then heads (h) and tails (t) should each occur half the time in the long run so that

$$P\{h\} = \frac{1}{2} \quad \text{and} \quad P\{t\} = \frac{1}{2}.$$

Example 21 If we toss a *fair* die, then in the long run each of the six possible outcomes should occur one-sixth of the time so that

$$P\{1\} = P\{2\} = P\{3\} = P\{4\} = P\{5\} = P\{6\} = \frac{1}{6}.$$

The following result is evident from the last two examples.

If the sample space for an experiment has k equally likely outcomes, then each elementary event should be assigned probability $1/k$. This assignment of probabilities is called the ***uniform probability model*** for the experiment.

Example 22 If a card is picked at random from a standard deck of 52 cards, what is the probability that the card selected is the ace of spades?

Solution Since the card is picked at random, it is reasonable to assume that all possible selections are equally likely. Since there are 52 different outcomes for this experiment, each has probability $\frac{1}{52}$. Thus, the probability of picking the ace of spades is $\frac{1}{52}$.

Example 23 The probability model for the die-tossing experiment in Example 16 is not a uniform probability model since the elementary events do not all have the same probability.

Example 24 Consider the experiment of tossing a *fair* die and observing the number on the top face. Find the probability of:

(a) the event E that an even number is tossed;

(b) the event G that a number divisible by 3 is tossed.

Solution The events are

$$E = \{2, 4, 6\}$$

$$G = \{3, 6\}.$$

Since the die is fair we assign the uniform probability model. Therefore

$$P(E) = P\{2\} + P\{4\} + P\{6\} = \frac{1}{6} + \frac{1}{6} + \frac{1}{6} = \frac{1}{2}$$

$$P(G) = P\{3\} + P\{6\} = \frac{1}{6} + \frac{1}{6} = \frac{1}{3}.$$

From this example, the reader should be able to see why the following theorem is true.

Theorem 1 If an experiment can result in any one of k equally likely outcomes and if an event E contains m sample points, then the probability of the event E is

$$P(E) = \frac{m}{k}.$$

Proof Let E be the event

$$E = \{s_1, s_2, \ldots, s_m\}.$$

Since the sample space contains k equally likely outcomes each sample point has probability $1/k$; thus, by the addition principle

$$\begin{aligned} P(E) &= P\{s_1\} + P\{s_2\} + \cdots + P\{s_m\} \\ &= \frac{1}{k} + \frac{1}{k} + \cdots + \frac{1}{k} \quad (m \text{ terms}) = \frac{m}{k}. \end{aligned}$$ ∎

Example 25 A batch of six resistors contains two defectives. If a resistor is selected at random from the batch, what is the probability that it is defective?

Solution Label the two defective resistors d_1 and d_2 and label the nondefective resistors n_1, n_2, n_3, and n_4. The sample space is

$$S = \{d_1, d_2, n_1, n_2, n_3, n_4\}.$$

Since the resistor is selected *at random*, the six points in S are equally likely.

The event F that a defective is picked is $F = \{d_1, d_2\}$ so by Theorem 1,

$$P(F) = \frac{2}{6} = \frac{1}{3}.$$

Example 26 If a pair of fair dice is tossed, what is the probability that the sum of the numbers tossed is seven?

Solution Label the dice a and b. Each die can show one of the integers from 1 to 6, so that the two dice can yield any one of 36 pairs in Table 6.2. Since the dice are fair, we shall assume that each of these 36 possible outcomes is equally likely. The event E that the sum of the numbers is seven contains the six shaded sample points in Table 6.2. Therefore the probability that the sum of the numbers is 7 is $\frac{6}{36} = \frac{1}{6}$.

Table 6.2

$a \backslash b$	1	2	3	4	5	6
1	(1, 1)	(1, 2)	(1, 3)	(1, 4)	(1, 5)	(1, 6)
2	(2, 1)	(2, 2)	(2, 3)	(2, 4)	(2, 5)	(2, 6)
3	(3, 1)	(3, 2)	(3, 3)	(3, 4)	(3, 5)	(3, 6)
4	(4, 1)	(4, 2)	(4, 3)	(4, 4)	(4, 5)	(4, 6)
5	(5, 1)	(5, 2)	(5, 3)	(5, 4)	(5, 5)	(5, 6)
6	(6, 1)	(6, 2)	(6, 3)	(6, 4)	(6, 5)	(6, 6)

When an experimenter assigns a probability model to a physical experiment, he is, in actuality, formulating a model or theory about the physical world. For example, if the experimenter assigns the uniform probability model

$$P\{h\} = P\{t\} = \frac{1}{2}$$

to the experiment of tossing a coin and recording whether the outcome is h = heads or t = tails, he is theorizing that the coin is perfectly balanced and perfectly symmetric. Although no physical coin satisfies these conditions, the coin may come close enough so that there is reasonably good agreement between the relative frequencies predicted by the theory and the relative frequencies observed by physical experimentation. As an illustration, the data in Table 6.1 seem to justify the use of a uniform probability model. On the other hand, a given coin may be so badly weighted that the uniform probability model would not be satisfactory. The only way to determine this, however, is by performing the experiment and comparing the predicted and observed relative frequencies. In Section 7.7 we shall study a test that can be used to determine in a quantitative way whether to accept a model or to reject it in favor of another model.

We conclude this section with an example that shows the importance of experimentally testing the validity of a probability model.

Example 27 In an experiment two fair coins are tossed. Find the probability of tossing exactly one head.

Solution 1 Since the number of heads is zero, one, or two, the sample space is

$$S = \{0, 1, 2\}.$$

The event E that exactly one head is tossed is

$$E = \{1\}.$$

If we argue that the three sample points are equally likely, then

$$P(E) = \frac{1}{3}.$$

Solution 2 Table 6.3 suggests there are four equally likely outcomes s_1, s_2, s_3, and s_4.

Table 6.3

Outcome	First coin	Second coin
s_1	head	head
s_2	head	tail
s_3	tail	head
s_4	tail	tail

If E is the event that one head occurs, then

$$E = \{s_2, s_3\}.$$

If we argue that the four possible outcomes are equally likely, then

$$P(E) = \frac{2}{4} = \frac{1}{2}.$$

Obviously the two solutions in this example are in disagreement. But which is correct? Actually, both are "correct" in the sense that each of them arises from a valid probability model for the experiment, one based on three equally likely outcomes and the other on four equally likely outcomes. A more meaningful question to ask is: Which theory (model) will most accurately describe a physical coin-tossing experiment? The answer to this question cannot be ascertained by logic alone. The answer must be based on *physical* evidence in support of one model over the other. The reader may find it interesting to solve Exercise 13 and show experimentally that the second model most accurately describes the physical situation.

EXERCISE SET 6.2

1. List the elementary events for each of the following experiments.

(a) A coin is tossed two times and the resulting sequence of heads and tails is observed.

(b) A stock on the New York Stock Exchange is observed for one market day and it is recorded whether the stock increases in value (i), decreases in value (d), or undergoes no change in value (n).

(c) The stock in part (b) is observed for two market days.

(d) A letter is picked at random from the set $\{a, b, c, d, e\}$.

(e) A die is tossed and it is observed whether the outcome is even (e) or odd (o).

(f) A card is picked at random from an ordinary deck of 52 cards and the suit (clubs (c), diamonds (d), hearts (h), or spades (s)) is observed.

(g) A lot of 10 items is tested and the number of defectives is recorded.

2. State the number of elementary events for each of the following experiments:

(a) A botanist successively observes four corn plants to decide if the plants have rust (r) or are free from rust (f). He records the resulting sequence of observations.

(b) A card is picked at random from a standard deck of 52 cards and the card picked is recorded.

(c) A set of two letters is picked from the set $\{a, b, c, d, e\}$.

(d) In the investigation of human births, the month of birth and the sex of a child are recorded as an ordered pair, for example (February, male).

3. Use the probability model in Example 16 to compute the probabilities of the following events:

(a) The event A that a number less than 5 is tossed.

(b) The event B that an odd number is tossed.

(c) The event C that a number smaller than 1 is tossed.

4. Compute the probabilities of the events A, B, and C in Exercise 3 assuming that the die is fair.

5. Let $S = \{s_1, s_2, s_3, s_4, s_5\}$ be the sample space for an experiment, and assume the experiment is assigned the probability model

$$P\{s_1\} = \tfrac{4}{11}, \qquad P\{s_2\} = \tfrac{1}{11}, \qquad P\{s_3\} = \tfrac{1}{11}, \qquad P\{s_4\} = \tfrac{2}{11}, \qquad P\{s_5\} = \tfrac{3}{11}.$$

(a) Find the probability of the event $A = \{s_1, s_3\}$.

(b) Find the probability of the event $B = \{s_2, s_3, s_5\}$.

(c) Find the probability of the event S.

6. Using the probability model and the events A and B of Exercise 5 compute

(a) $P(A \cap B)$ (b) $P(A \cup B)$ (c) $P(A \cap B')$

(d) $P(A' \cap B')$ (e) $P(A \cap A')$.

7. In each of the following experiments, list the elementary events and decide if the outcomes are equally likely.

(a) A die is tossed and it is recorded whether the top face is even or odd.

(b) A die is tossed and it is recorded whether the top face is a 3 or not.

(c) A card is picked at random from a standard deck of 52 cards and it is recorded whether the card is an ace or not.

(d) A card is picked at random from a deck of 52 cards and it is recorded whether the card is black or red.

(e) A set of two letters is chosen at random from the set $\{a, b, c, d\}$.

8. Assume a uniform probability model is assigned to each of the following experiments. In each case state the probability assigned to each outcome.

(a) A card is drawn from a standard deck of 52 cards and the card picked is recorded.

(b) A coin is tossed five times and the resulting sequence of outcomes is recorded.

(c) In a three-child family, suppose the sexes of the children are listed from oldest to youngest as an ordered triple (for example, (m, m, f) means the oldest and middle children are males and the youngest is a female).

9. Assume an experiment with sample space $S = \{s_1, s_2, s_3\}$ is assigned the uniform probability model. List all possible events and their probabilities.

10. A survey shows that 20% of all cars entering a certain intersection turn left, 25% of all cars turn right, and 55% of all cars proceed straight ahead. A car entering the intersection is observed and it is recorded whether the car turns left (l), turns right (r), or goes straight (s). Assign a probability model to this experiment and use it to compute the probability that the car makes a turn.

11. Assume that in a three-child family, the eight possible sex distributions of the children *mmm*, *mmf*, ..., *fff* are equally likely and compute the probabilities of the following events:

(a) Exactly one child is a female.

(b) At least one child is a female.

(c) There are more female children than male children.

(d) The family has at most one male child.

12. Compute the probabilities of the following events for the experiment in Example 26.

(a) At least one of the dice shows a 3.

(b) The sum of the numbers on the dice is 6.

(c) Neither a 2 nor a 5 appears.

13. Toss two coins 100 times and record the number of times exactly one head occurs. Use these data to show that the second probability model described in Example 27 is more accurate.

6.3 BASIC THEOREMS OF PROBABILITY

In this section we investigate how the probabilities of various events are related. For example, how is the probability that an event occurs related to the probability it does not occur, and how are the probabilities of $E \cup F$ and $E \cap F$ related to the probabilities of E and F? The results obtained here will be useful in solving complicated probability problems. We shall only consider experiments with a finite sample space and we shall denote the probability of an event E by $P(E)$. As discussed in the previous section, we can assign a probability model to an experiment with a finite sample space by first assigning probabilities to the elementary events and then applying the addition principle to obtain the probabilities of the remaining events. Sometimes the probabilities of the elementary events are determined by intuition and sometimes they can only be estimated by experimentation. In either case, the probabilities that are assigned to the elementary events must satisfy certain conditions. To motivate these conditions, assume that the sample space is

$$S = \{s_1, s_2, \ldots, s_k\}$$

and suppose the experiment is repeated n times under fixed conditions. If

$$\begin{gathered} s_1 \text{ occurs } n_1 \text{ times in the } n \text{ repetitions} \\ s_2 \text{ occurs } n_2 \text{ times in the } n \text{ repetitions} \\ \vdots \\ s_k \text{ occurs } n_k \text{ times in the } n \text{ repetitions,} \end{gathered}$$

then the relative frequencies $f_1, f_2, \ldots, f_k$ of the elementary events

$$\{s_1\}, \qquad \{s_2\}, \qquad \ldots, \qquad \{s_k\}$$

are

$$f_1 = \frac{n_1}{n}, \qquad f_2 = \frac{n_2}{n}, \qquad \ldots, \qquad f_k = \frac{n_k}{n}.$$

Note that each of these relative frequencies is a nonnegative fraction; that is each relative frequency satisfies

$$0 \leq f_i \leq 1.$$

Moreover, since

$$n_1 + n_2 + \cdots + n_k = n \qquad \text{(why?)}$$

the sum of the relative frequencies is 1 because

$$\begin{aligned} f_1 + f_2 + \cdots + f_k &= \frac{n_1}{n} + \frac{n_2}{n} + \cdots + \frac{n_k}{n} \\ &= \frac{n_1 + n_2 + \cdots + n_k}{n} \\ &= \frac{n}{n} \\ &= 1. \end{aligned}$$

Thus the relative frequencies are nonnegative fractions whose sum is 1. But the probabilities $p_1, p_2, \ldots, p_k$ that we assign to the elementary events $\{s_1\}, \{s_2\}, \ldots, \{s_k\}$ are intuitively the "long-term" relative frequencies of these events; this suggests that these probabilities should also be nonnegative fractions whose sum is 1. More precisely,

$$0 \leq p_i \leq 1$$

and (1)

$$p_1 + p_2 + \cdots + p_k = 1.$$

Example 28 A traffic survey shows that 22% of all vehicles turn left at a given intersection, 17% of all vehicles turn right, and 61% of all vehicles make no turn.

An experimenter observes a vehicle approaching the intersection and records whether it turns left, turns right, or makes no turn. The sample space is

$$S = \{s_1, s_2, s_3\}$$

where

$$s_1 = \text{left turn}, \qquad s_2 = \text{right turn}, \qquad s_3 = \text{no turn}.$$

It is reasonable to assign the elementary events the probabilities

$$P\{s_1\} = .22, \qquad P\{s_2\} = .17, \qquad P = \{s_3\} = .61.$$

Observe that each of these values is between 0 and 1 and

$$.22 + .17 + .61 = 1$$

as required.

Recall that an event is a subset of a sample space. In particular, if the sample space is

$$S = \{s_1, s_2, \ldots, s_k\}$$

then S itself is an event since $S \subseteq S$. The event S is certain to occur every time the experiment is performed, so that the probability of S should turn out to be 1, regardless of the probability model assigned to the experiment. This is the content of the following theorem.

Theorem 2 If an experiment has a finite sample space S, then

$$P(S) = 1.$$

Proof Let

$$S = \{s_1, s_2, \ldots, s_k\}$$

be the sample space, and assume that the elementary events are assigned the probabilities

$$P\{s_1\} = p_1, \qquad P\{s_2\} = p_2, \qquad \ldots, \qquad P\{s_k\} = p_k.$$

From the addition principle and the second condition in (1) we obtain

$$\begin{aligned} P(S) &= P\{s_1\} + P\{s_2\} + \cdots + P\{s_k\} \\ &= p_1 + p_2 + \cdots + p_k \\ &= 1. \quad \blacksquare \end{aligned}$$

In the previous section we saw that certain everyday expressions about probabilities of events can be described in terms of the basic operations on sets. In Table 6.4 we have summarized these relationships.

Table 6.4

Everyday expression	Set interpretation
the probability that E does not occur	$P(E')$
the probability that E and F occur	$P(E \cap F)$
the probability that E or F occurs	$P(E \cup F)$

When E and F are mutually exclusive events (that is, $E \cap F = \varnothing$), the probability that E or F occurs is related very simply to the individual probabilities of the events E and F. This is the content of the next theorem.

Theorem 3 If E and F are mutually exclusive events in a finite sample space, then

$$P(E \cup F) = P(E) + P(F). \tag{2}$$

Proof From the addition principle, the probability $P(E \cup F)$ is the sum of the probabilities of the sample points in $E \cup F$. These are the points in the shaded region of Figure 6.3. Since E and F are mutually exclusive, the sum of the probabilities of the sample points in $E \cup F$ can be expressed as the sum of the probabilities of the sample points in E plus the sum of the probabilities of the sample points in F, that is

$$P(E \cup F) = P(E) + P(F). \quad \blacksquare$$

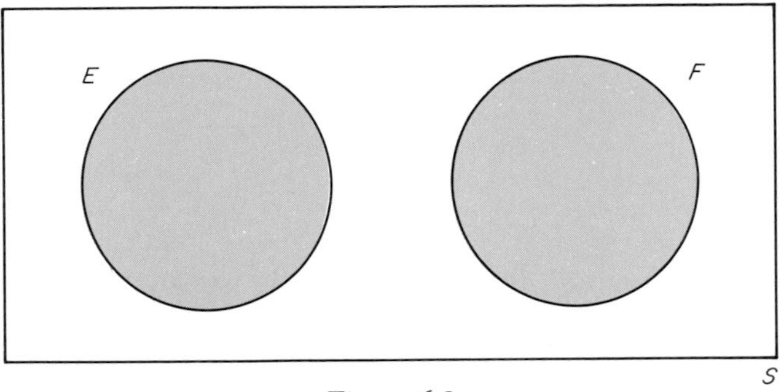

Figure 6.3

This result can be extended to n mutually exclusive events by using a similar argument. More precisely, if

$$E_1, \qquad E_2, \qquad \ldots, \qquad E_n$$

are *mutually exclusive* (Figure 6.2), then

$$P(E_1 \cup E_2 \cup \cdots \cup E_n) = P(E_1) + P(E_2) + \cdots + P(E_n). \quad (3)$$

Example 29 If a fair die is tossed, what is the probability that the number on the top face is even or a 5?

Solution 1 We assign the uniform probability model to this experiment. The event G that the number tossed is either even or a 5 is

$$G = \{2, 4, 5, 6\}$$

Thus

$$P(G) = \frac{4}{6} = \frac{2}{3}.$$

Solution 2 We can write

$$G = E \cup F$$

where E is the event that an even number is tossed and F is the event that a 5 is tossed. The events E and F are

$$E = \{2, 4, 6\} \qquad \text{and} \qquad F = \{5\}$$

Since $E \cap F = \varnothing$, the events E and F are mutually exclusive. Thus,

Theorem 3 tells us that

$$P(E \cup F) = P(E) + P(F) = \frac{3}{6} + \frac{1}{6} = \frac{2}{3}.$$

Therefore

$$P(G) = P(E \cup F) = \frac{2}{3},$$

which agrees with the result obtained in solution (1).

Warning. Formula (2) in Theorem 3 does not apply if E and F are not mutually exclusive. (Can you see where the proof of Theorem 3 breaks down?) The next theorem applies to any two events, mutually exclusive or not.

Theorem 4 If E and F are events in a finite sample space, then

$$P(E \cup F) = P(E) + P(F) - P(E \cap F). \tag{4}$$

Proof As illustrated in Figure 6.4, $E \cup F$ can be expressed as

$$E \cup F = (E \cap F') \cup (E \cap F) \cup (E' \cap F).$$

Since the three sets on the right-hand side of this equation are mutually exclusive, it follows from formula (3) above that

$$P(E \cup F) = P(E \cap F') + P(E \cap F) + P(E' \cap F) \tag{5}$$

But

$$E = (E \cap F') \cup (E \cap F)$$

and

$$F = (E' \cap F) \cup (E \cap F)$$

(see Figure 6.4) so that

$$P(E) = P(E \cap F') + P(E \cap F)$$

and

$$P(F) = P(E' \cap F) + P(E \cap F).$$

Rewriting these expressions we obtain

$$P(E \cap F') = P(E) - P(E \cap F)$$

and

$$P(E' \cap F) = P(F) - P(E \cap F).$$

Substituting these equations into (5) gives

$$P(E \cup F) = P(E) - P(E \cap F) + P(E \cap F) + P(F) - P(E \cap F)$$

or after simplifying

$$P(E \cup F) = P(E) + P(F) - P(E \cap F). \quad \blacksquare$$

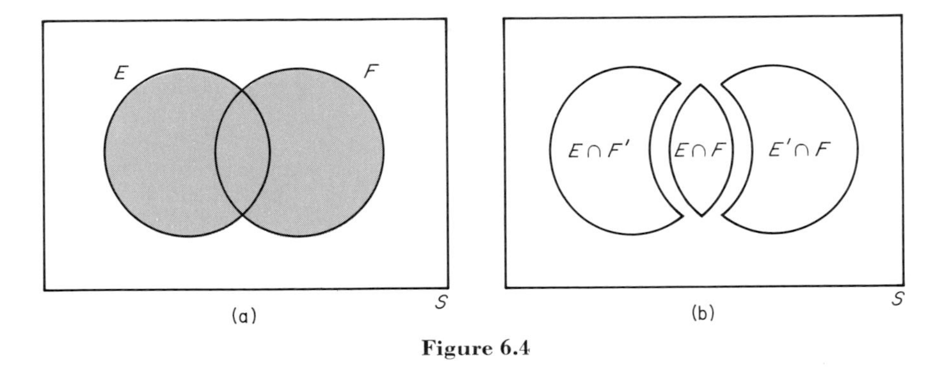

Figure 6.4

Example 30 A card is picked at random from an ordinary deck of 52 cards. Find:

(a) The probability the card is a club.
(b) The probability the card is an ace.
(c) The probability the card is either an ace or a club.

Solution Since the card is picked at random, we shall assign the uniform probability model; thus each sample point will be assigned probability $\frac{1}{52}$.

(a) Let C be the event that the card is a club. Since there are 13 clubs in the deck,

$$P(C) = \frac{13}{52}.$$

(b) Let A be the event that the card is an ace. Since there are four aces in the deck,

$$P(A) = \frac{4}{52}.$$

(c) The probability that the card is either an ace or a club is

$$P(C \cup A).$$

From Theorem 4,

$$P(C \cup A) = P(C) + P(A) - P(C \cap A). \tag{6}$$

To belong to $C \cap A$, a sample point must be both a club and an ace. There is one such sample point, the ace of clubs. Thus

$$P(C \cap A) = \frac{1}{52}.$$

Substituting this value and the values obtained in (a) and (b) into (6) gives

$$P(C \cup A) = \frac{13}{52} + \frac{4}{52} - \frac{1}{52} = \frac{16}{52}.$$

Example 31 Assume that a reel of magnetic tape for a digital computer is classified defective if it has either of the following imperfections: (i) the magnetic coating is improperly applied, or (ii) the tape has a fold or tear; otherwise it is classified nondefective.

A survey of defective reels returned to the manufacturer shows

.5% of all reels produced have improperly applied coatings;
.3% of all reels produced have tape folds or tears;
.1% of all reels produced have both imperfections.

Find the probability that a purchaser will receive a defective reel of tape.

Solution Let E be the event that the purchased reel has an improperly applied coating and let F be the event that the tape has a fold or tear. Thus

$$P(E) = .005, \qquad P(F) = .003, \qquad \text{and} \qquad P(E \cap F) = .001.$$

The event that the purchased reel is defective is $E \cup F$ and the probability of this event is

$$\begin{aligned} P(E \cup F) &= P(E) + P(F) - P(E \cap F) \\ &= .005 + .003 - .001 \\ &= .007. \end{aligned}$$

In other words, .7% of all reels produced are defective.

We conclude this section with a theorem relating the probability that an event occurs to the probability that it does not occur.

Theorem 5 If E is an event in a finite sample space, then

$$P(E') = 1 - P(E).$$

Proof Since $E \cup E' = S$ and $E \cap E' = \varnothing$ (Figure 6.5), we can apply Theorems 2 and 3 to obtain

$$1 = P(S) = P(E \cup E') = P(E) + P(E').$$

Thus

$$P(E') = 1 - P(E). \quad \blacksquare$$

Example 32 In Example 31, we showed that the probability that a purchaser receives a defective reel of tape is .007. Thus, the probability the purchaser receives a nondefective reel is

$$1 - .007 = .993.$$

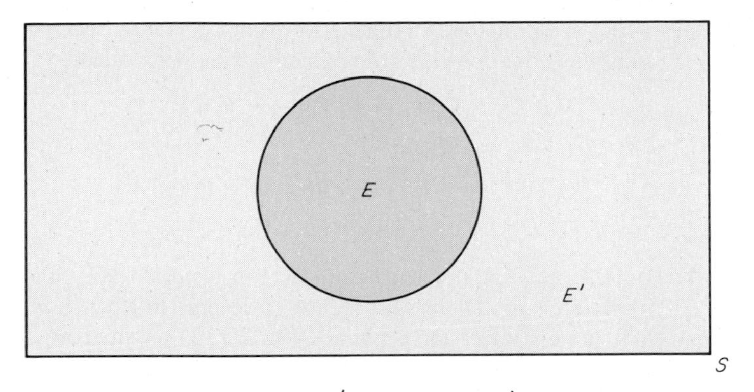

Figure 6.5 $E \cup E' = S$ and $E \cap E' = \varnothing$.

EXERCISE SET 6.3

1. Which of the following are valid probability models for an experiment with sample space $S = \{s_1, s_2, s_3, s_4\}$?

(a) $P(s_1) = .3$, $P(s_2) = .6$, $P(s_3) = .1$, $P(s_4) = .2$

(b) $P(s_1) = -.2$, $P(s_2) = .8$, $P(s_3) = .2$, $P(s_4) = .2$

(c) $P(s_1) = 0$, $P(s_2) = .7$, $P(s_3) = .2$, $P(s_4) = .1$

(d) $P(s_1) = .75$, $P(s_2) = .05$, $P(s_3) = .15$, $P(s_4) = .05$

2. Let $S = \{s_1, s_2, s_3\}$ be the sample space of an experiment, and suppose $P(s_1) = .3$ and $P(s_2) = .2$. Find $P(s_3)$.

3. Let E and F be events such that $P(E) = .3$, $P(F) = .1$, and $P(E \cap F) = .05$. Find

(a) $P(E')$ (b) $P(F')$ (c) $P(E \cup F)$.

4. An item produced by a manufacturing process is either nondefective, has defect a, has defect b, or has both defect a and defect b. In each part, *verbally* describe the complementary event E'.

(a) $E =$ the item has at least one defect.

(b) $E =$ the item has at most one defect.

(c) $E =$ the item has defect a.

(d) $E =$ the item has two defects.

5. Let $S = \{s_1, s_2, s_3, s_4, s_5\}$ be the sample space of an experiment and let

$$P(s_1) = .2, \quad P(s_2) = .1, \quad P(s_3) = .3, \quad P(s_4) = .1, \quad P(s_5) = .3.$$

If

$$E = \{s_1, s_3\}, \quad F = \{s_1, s_4, s_5\}, \quad \text{and} \quad G = \{s_2, s_4, s_5\},$$

find

(a) $P(F')$ (b) $P(E \cup F)$ (c) $P(G \cap F)$

(d) $P(E \cap G)$ (e) $P(E \text{ and } F)$ (f) $P(F \text{ or } G)$.

6. Using the probability model and events given in Exercise 5, verify that the following equations are true by computing both sides

(a) $P(G \cup F) = P(G) + P(F) - P(G \cap F)$

(b) $P(G') = 1 - P(G)$

(c) $P(E \text{ or } F) = P(E) + P(F) - P(E \text{ and } F)$

(d) $P(E \cup G) = P(E) + P(G)$.

7. In order for a bill to come before the President of the United States for signing, it must be passed by the House of Representatives and by the Senate. A lobbyist estimates the probability of his bill passing the House as .5 and the probability of passing the Senate as .7. He also estimates that the probability his bill will be passed either by the House or by the Senate as .8. What is the probability the bill will come before the President?

8. Let E and F be the events described in Exercise 3. Use DeMorgan's laws and Theorems 4 and 5 to find $P(E' \cup F')$.

9. Let E and F be mutually exclusive events such that $P(E) = .4$ and $P(F) = .3$. Find

(a) $P(E \cup F)$ (b) $P(E \cap F)$ (c) $P(E')$ (d) $P(F')$.

10. Let E and F be the events described in Exercise 9. Use DeMorgan's laws and Theorems 3 and 5 to find $P(E' \cap F')$.

11. A study of air traffic patterns at a major metropolitan airport yields the following probabilities for the number of aircraft waiting to land on one of the runways:

Number of aircraft waiting to land	0	1	2	3	more
Probability	.1	.2	.4	.2	.1

Find the probability that

(a) at least three aircraft are waiting to land;

(b) at most two aircraft are waiting to land;

(c) more than two aircraft are waiting to land.

12. River water entering a filtration plant is called polluted if it contains: (i) an intolerable percentage of dangerous organic materials or (ii) an intolerable percentage of inorganic materials; otherwise it is called unpolluted. A survey of the plant's records shows:

27% of all water entering the plant is polluted because of organic materials.
46% of all water entering the plant is polluted because of inorganic materials.
23% of all water entering the plant is polluted because of both organic and inorganic materials.

Find the probability that a quantity of water in the plant is unpolluted.

6.4 COUNTING TECHNIQUES; PERMUTATIONS AND COMBINATIONS

If an experiment with a finite sample space is given the uniform probability model, we know from Theorem 1 in Section 6.2 that the probability of an event is the number of sample points in the event divided by the number of sample points in the sample space. Unfortunately, it is not always an easy task to count the number of points in an event or a sample space. To illustrate, assume a standard deck of 52 cards is well shuffled and a hand of five cards is dealt. What is the probability that the hand contains exactly three aces? Since the deck is well shuffled, it is reasonable to assume that all possible five-card hands are equally likely. Thus, if E is the event that the hand contains exactly three aces, then

$$P(E) = \frac{a}{b},$$

where a is the number of different possible five-card hands with exactly three aces and b is total number of different possible five-card hands. Neither of the numbers a and b is immediately obvious. In this section we shall discuss counting techniques that will be useful in solving probability problems of this type.

There is a general counting rule, called the ***multiplication principle,*** that we shall use repeatedly. It can be stated as follows.

Multiplication Principle If there are k ways to make a decision D_1 and then l ways to make a decision D_2, then there are kl ways to make the two decisions D_1 and D_2.

To see why the multiplication principle holds, let

$$A = \{a_1, a_2, \ldots, a_k\}$$

be the set of possibilities for decision D_1, and let

$$B = \{b_1, b_2, \ldots, b_l\}$$

be the set of possibilities for decision D_2. To make decision D_1 and then decision D_2 amounts to selecting an element from set A and then selecting an element from set B. Some possibilities are

$$(a_1, b_1), \qquad (a_1, b_2), \qquad (a_2, b_1), \qquad \text{and so on.}$$

But these are just ordered pairs from the Cartesian product $A \times B$. As shown in Section 1.3, this Cartesian product has kl members; thus there are kl ways to make the two decisions.

The multiplication principle can be extended to more than two decisions.

Multiplication Principle (Extended) If there are n_1 ways to make a decision D_1, then n_2 ways to make a decision D_2, then n_3 ways to make decision D_3, and so on, then there are

$$n_1 n_2 n_3 \cdots$$

ways to make all the decisions.

Example 33 Assume that an employee identification number consists of two letters of the alphabet followed by a sequence of seven digits selected from the set

$$\{0, 1, 2, 3, 4, 5, 6, 7, 8, 9\}$$

Assuming repetitions are allowed, how many identification numbers are possible?

Solution There are 26 possibilities for each of the two letters and there are 10 possibilities for each of the seven digits. Thus, by the multiplication principle there are

$$26 \cdot 26 \cdot 10 \cdot 10 \cdot 10 \cdot 10 \cdot 10 \cdot 10 \cdot 10 = 26^2 \cdot 10^7 = 676 \cdot 10^7$$

different possible identification numbers.

Example 34 A coin is tossed six times, generating a sequence of h's (heads) and t's (tails). Some of the possibilities are

$$hhhtht, \qquad tttht h, \qquad ththth.$$

How many different sequences are possible?

Solution On each of the six tosses there are two possibilities, h or t. Thus, by the multiplication principle there are

$$2 \cdot 2 \cdot 2 \cdot 2 \cdot 2 \cdot 2 = 2^6 = 64$$

different sequences.

In many counting problems we want to determine the number of different ways of arranging or ordering a set of distinct objects. For example, there are two different ways of ordering the objects in the set $\{a, b\}$. They are

$$ab \qquad \text{and} \qquad ba$$

Given a set of distinct objects, an arrangement of these objects in a definite order without repetitions is called a ***permutation*** of the set.

Example 35 There are six different permutations of the set $\{1, 2, 3\}$. They are

$$\begin{array}{ccc} 123 & 213 & 312 \\ 132 & 231 & 321 \end{array}$$

Example 36 How many different permutations can be formed from the set

$$\{1, 2, 3, 4\}?$$

Solution Each permutation consists of four digits; the first digit can be selected in any one of four ways; then since repetitions are not allowed, the second digit can be selected in any one of three ways, then the third digit in any of two ways, and then the last digit in one way. Thus by the multiplication principle, there are

$$4 \cdot 3 \cdot 2 \cdot 1 = 24$$

different permutations.

Example 37 List the 24 permutations of the set $\{1, 2, 3, 4\}$.

Solution Consider the tree diagram shown in Figure 6.6.

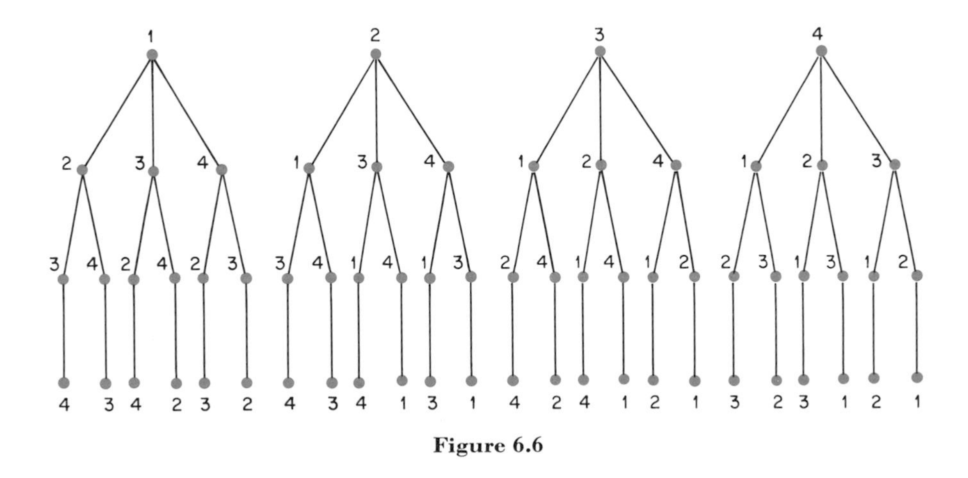

Figure 6.6

The four dots labeled 1, 2, 3, 4 at the top of Figure 6.6 represent the possible choices for the first number in the permutation. The three branches emanating from each of these dots represent the possible choices for the second position in the permutation. Thus, if the permutation begins

$$3 \_\_ \_\_ \_\_$$

the three possibilities for the second position are 1, 2, and 4. The two branches emanating from each dot in the second position represent the possible choices for the third position. Thus, if the permutation begins

$$3\ 2\ \_\ \_$$

the two possible choices for the third position are 1 and 4. Finally, the single branch emanating from each dot in the third position represents the only possible choice for the fourth position. Thus, if the permutation begins

$$3\ 2\ 4\ \_$$

the only possible choice for the fourth position is 1. The different permutations can now be listed by tracing out all possible paths through the tree from the first position to the last position. We obtain the following list by this process:

1 2 3 4	2 1 3 4	3 1 2 4	4 1 2 3
1 2 4 3	2 1 4 3	3 1 4 2	4 1 3 2
1 3 2 4	2 3 1 4	3 2 1 4	4 2 1 3
1 3 4 2	2 3 4 1	3 2 4 1	4 2 3 1
1 4 2 3	2 4 1 3	3 4 1 2	4 3 1 2
1 4 3 2	2 4 3 1	3 4 2 1	4 3 2 1

By imitating the argument given in Example 36, the reader should be able to show that the number of permutations of $\{1, 2, 3, 4, 5\}$ is

$$5 \cdot 4 \cdot 3 \cdot 2 \cdot 1 = 120$$

and the number of permutations of

$$\{a, b, c, d, e, f\}$$

is

$$6 \cdot 5 \cdot 4 \cdot 3 \cdot 2 \cdot 1 = 720.$$

The general principle is now evident.

The number of permutations of a set with n distinct objects is

$$n(n-1)(n-2) \cdots 1.$$

This formula occurs so frequently that we use the symbol $n!$ (read n factorial) to denote the product of the integers from 1 to n; that is

$$n! = n(n-1)(n-2) \cdots 1.$$

Thus,

$$1! = 1$$

$$2! = 2 \cdot 1 = 2$$

$$3! = 3 \cdot 2 \cdot 1 = 6$$

$$4! = 4 \cdot 3 \cdot 2 \cdot 1 = 24$$

$$5! = 5 \cdot 4 \cdot 3 \cdot 2 \cdot 1 = 120$$

$$6! = 6 \cdot 5 \cdot 4 \cdot 3 \cdot 2 \cdot 1 = 720.$$

Once the value of $n!$ is known, this value can be exploited in computing $(n+1)!$. For example, using the value of $6!$ given above, we can compute $7!$ by writing

$$7! = 7 \cdot 6 \cdot 5 \cdot 4 \cdot 3 \cdot 2 \cdot 1 = 7 \cdot (6!) = 7(720) = 5040.$$

Similarly,

$$8! = 8 \cdot (7!) = 8 \cdot (5040) = 40{,}320.$$

In general, $(n+1)!$ and $n!$ are related by

$$(n+1)! = (n+1) \cdot n! \tag{1}$$

Since $n!$ is defined to be the product of the integers from 1 *up to* n, the symbol $0!$ is not yet defined. We shall agree to give $0!$ the value 1. With this definition, Equation (1) will hold when $n = 0$.

Example 38 In a psychological test for extrasensory perception (ESP), three colored cards are used, one red, one yellow, and one blue. The cards are placed in some order and a blindfolded subject is asked to guess the order of the cards. Assuming that the subject has no ESP, what is the probability he will guess the correct order by chance?

Solution There are $3! = 6$ different ways to order the cards, one of which is the correct order. In the absence of ESP, it is reasonable to assume that a person is selecting an order at random. Thus, each of the six possible orders is equally likely to be selected. The probability the correct order will be selected by chance is thus $\frac{1}{6}$.

In forming a permutation of a set, all the elements of the set are used. In some problems, however, we want to consider permutations or arrangements where not all the elements are used. For example, we might want to consider all possible permutations of two letters selected from the set $\{a, b, c\}$; they would be

$$\begin{array}{ccc} ab & ac & bc \\ ba & ca & cb. \end{array}$$

In general, a permutation of r objects selected from a set of n objects is called a ***permutation of the n objects taken r at a time.***

Example 39 List all permutations of

$$\{a, b, c, d\}$$

taken three at a time.

Solution This problem can be solved by constructing the tree shown in Figure 6.7. The dots at the top of the tree correspond to the possible choices for the first letter in the arrangement, the dots in the middle of the tree correspond to the possible choices for the second letter, and the dots at the bottom correspond to the possible choices for the third letter.

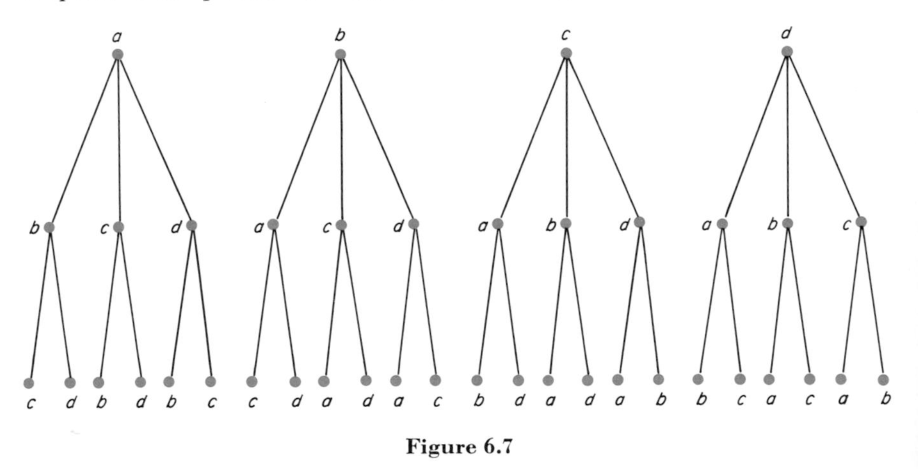

Figure 6.7

By tracing out the different paths from the top of the tree to the bottom, we obtain the following list of possible arrangements.

$$\begin{array}{cccc} abc & bac & cab & dab \\ abd & bad & cad & dac \\ acb & bca & cba & dba \\ acd & bcd & cbd & dbc \\ adb & bda & cda & dca \\ adc & bdc & cdb & dcb \end{array}$$

Thus, there are 24 different permutations of $\{a, b, c, d\}$ taken three at a time.

We shall now derive a formula for the number of permutations of n distinct objects taken r at a time. Since we have n distinct objects, there are n ways to choose the first object in the permutation; then, since repetitions are not allowed, there are $n - 1$ ways to choose the second object in the permutation, then $n - 2$ ways to choose the next object, and so on. Since each permutation contains r objects, it follows from the multiplication principle that there are

$$\underbrace{n(n-1)(n-2)\cdots}_{r\text{-factors}} \tag{2}$$

different permutations.

The expression in (2) occurs so often that it is common to denote it by the symbol $P_{n,r}$; that is, the number of permutations of n distinct objects taken r at a time is

$$P_{n,r} = \underbrace{n(n-1)(n-2)\cdots}_{r\text{-factors}} \tag{3}$$

For example

$$P_{5,3} = 5 \cdot 4 \cdot 3 = 60 \qquad P_{9,1} = 9$$

$$P_{8,2} = 8 \cdot 7 = 56 \qquad P_{4,4} = 4 \cdot 3 \cdot 2 \cdot 1 = 24.$$

Example 40 Find the number of permutations of

$$\{a, b, c, d, e, f, g\}$$

taken four at a time.

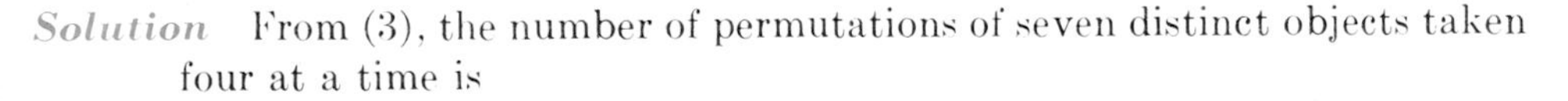

Solution From (3), the number of permutations of seven distinct objects taken four at a time is

$$P_{7,4} = 7 \cdot 6 \cdot 5 \cdot 4 = 840.$$

Formula (3) can be rewritten in another way which is sometimes useful. In (3),

the first factor is n

the second factor is $n - 1$

the third factor is $n - 2$

$\vdots$

From the pattern emerging here, the reader should be able to see that the rth or last factor will be

$$n - (r - 1) = n - r + 1.$$

Therefore (3) can be written in the form

$$P_{n,r} = n(n-1)(n-2) \cdots (n-r+1).$$

If we multiply this expression by

$$\frac{(n-r)!}{(n-r)!},$$

we obtain

$$P_{n,r} = \frac{n!}{(n-r)!}. \tag{4}$$

(See Exercise 11.)

Example 41 Compute $P_{7,4}$ in two ways, first by formula (3) and then by formula (4).

Solution From formula (3),

$$P_{7,4} = 7 \cdot 6 \cdot 5 \cdot 4 = 840.$$

From formula (4) with $n = 7$ and $r = 4$

$$P_{7,4} = \frac{7!}{3!} = \frac{7 \cdot 6 \cdot 5 \cdot 4 \cdot 3 \cdot 2 \cdot 1}{3 \cdot 2 \cdot 1} = 7 \cdot 6 \cdot 5 \cdot 4 = 840.$$

In many counting problems we want to determine the number of different subsets with r objects than can be selected from a set with n distinct objects. For example, there are four different subsets with three objects that can be selected from the set

$$\{a, b, c, d\}.$$

They are

$$\{a, b, c\}, \qquad \{a, b, d\}, \qquad \{a, c, d\}, \qquad \{b, c, d\}.$$

Note, however, that there are 24 different *permutations* of $\{a, b, c, d\}$ taken three at a time (see Example 39). There are many more permutations with three objects than subsets with three objects because permutations distinguish among different orderings of the objects, while subsets do not. For example,

$$acd \qquad \text{and} \qquad cad$$

are different permutations of $\{a, b, c, d\}$ taken three at a time, while the sets

$$\{a, c, d\} \qquad \text{and} \qquad \{c, a, d\}$$

are not different (they have the same members).

To illustrate why one might be interested in counting subsets, suppose we are interested in finding the number of different five-card hands that can be dealt from a standard deck of 52 cards. A hand of five cards can be viewed as a set of five objects selected from a set of 52 distinct objects. Thus the number of different five-card hands that can be dealt is the same as the number of different subsets with five objects that can be selected from a set with 52 distinct objects.

Observe that it would be *incorrect* to argue that the number of five-card hands is the number of permutations of 52 objects taken five at a time. The reason is that permutations distinguish between different orders so that

2H 3H 4H 5H 6H

(2H = 2 of hearts, 3H = 3 of hearts, and so on) and

3H 4H 6H 2H 5H

would *incorrectly* be counted as different hands, even though they involve the same cards.

For historical reasons, a subset of r objects selected from a set of n objects is usually called a ***combination*** of the n objects taken r at a time. We shall now try to develop a formula for the number of different combinations of n distinct objects taken r at a time; we shall denote this quantity by the symbol $C_{n,r}$.† To obtain a formula for $C_{n,r}$ we need only observe that a list of all *permutations* of n objects taken r at a time can be obtained by the following two-step procedure:

1. List all combinations of the n objects taken r at a time.
2. Take each combination and list all the permutations of its elements.

When these two steps are completed we will have a complete list of all permutations of the n objects taken r at a time. For example, the combinations of

$$\{a, b, c, d\}$$

taken three at a time are

$$\{a, b, c\}, \qquad \{a, b, d\}, \qquad \{a, c, d\}, \qquad \{b, c, d\}.$$

If we now take each of these combinations and form all possible permutations of the elements we obtain the list shown in the tinted block of Figure 6.8. The reader may wish to check that this list is the same as that obtained in Example 39 using a tree.

† The symbol $\binom{n}{r}$ is often used rather than $C_{n,r}$; we shall use both notations.

$\{a, b, c\}$	*abc*	*acb*	*bac*	*bca*	*cab*	*cba*
$\{a, b, d\}$	*abd*	*adb*	*bad*	*bda*	*dab*	*dba*
$\{a, c, d\}$	*acd*	*adc*	*cad*	*cda*	*dac*	*dca*
$\{b, c, d\}$	*bcd*	*bdc*	*cbd*	*cdb*	*dbc*	*dcb*

Figure 6.8

If we are given a set with n distinct objects, there are $C_{n,r}$ different combinations taken r at a time. Since each of these combinations has r elements, the elements can be permuted in $r!$ different ways. Thus, by the multiplication principle, there are

$$C_{n,r} \cdot r!$$

different ways of forming a combination and permuting its elements. Since this process yields all permutations of the n objects taken r at a time, we obtain

$$P_{n,r} = C_{n,r} \cdot r!$$

or

$$C_{n,r} = \frac{P_{n,r}}{r!}$$

or using formula (4),

$$C_{n,r} = \frac{n!}{r!(n-r)!}.$$

To summarize:

The number of combinations of n distinct objects taken r at a time is

$$C_{n,r} = \frac{n!}{r!(n-r)!}. \qquad (5)$$

Example 42 How many different five-card hands can be dealt from a deck of 52 cards?

Solution As previously discussed, the number of five-card hands is the number of combinations of 52 objects taken five at a time. From formula (5) we obtain

$$C_{52,5} = \frac{52!}{5!47!} = \frac{52 \cdot 51 \cdot 50 \cdot 49 \cdot 48}{5 \cdot 4 \cdot 3 \cdot 2 \cdot 1} = 2{,}598{,}960.$$

Sometimes it takes a little thought to decide whether to count permutations or combinations. The following rule should always be kept in mind.

> If order matters, use permutations.
> If order does not matter, use combinations.

Example 43 The board of directors authorizes a bank to promote three of its top 10 executives to the rank of Vice President. In how many different ways can the men for promotion be selected?

Solution We must first decide if order matters. To decide, label the executives 1, 2, 3, . . ., 10. Should we, for example, regard

$$1, 5, 7 \qquad \text{and} \qquad 7, 1, 5$$

as different selections, in which case order counts, or should we regard them as the same selection, in which case order does not matter?

Clearly, the two selections should be regarded as the same since the same men receive the promotions in both cases. Thus, order does not matter and the number of ways the men can be selected is the number of *combinations* of 10 objects taken three at a time. Thus, there are

$$C_{10,3} = \frac{10!}{3!7!} = \frac{10 \cdot 9 \cdot 8}{3 \cdot 2 \cdot 1} = 120$$

different possible selections.

Example 44 The board of directors authorizes a bank to promote three of its top 10 executives, one to the rank of First Vice President, one to the rank of Second Vice President and one to the rank of Third Vice President. In how many ways can the candidates for promotion be selected?

Solution As in Example 43, label the executives 1, 2, 3, . . ., 10 and suppose that we first choose the First Vice President, then the Second Vice President, then the Third Vice President. In this case

$$1, 5, 7 \qquad \text{and} \qquad 7, 1, 5$$

are different selections, so that order matters. Thus, there are

$$P_{10,3} = 10 \cdot 9 \cdot 8 = 720$$

possible selections.

Example 45 A fair coin is tossed six times. What is the probability that exactly two of the tosses are heads?

Solution If we denote heads by h and tails by t, then six tosses of a coin results in a sequence of h's and t's. As shown in Example 34, there are $2^6 = 64$ different sequences possible. Since the coin is fair, it is reasonable to assume that these 64 sequences are equally likely. Thus, by Theorem 1 in Section 6.2,

$$P(\text{exactly two heads}) = \frac{m}{64} \tag{6}$$

where m is the number of sequences with exactly two h's. The value of m can be obtained by arguing as follows. A sequence with two h's will necessarily have four t's. Some possibilities are

$$hhtttt \qquad \text{and} \qquad tththt \tag{7}$$

It is evident that any sequence with two h's and four t's is completely determined once we specify the two tosses on which the heads occur. For example, if we state that the heads occur on tosses 1 and 2, we obtain the first sequence in (7), and if we state that the heads occur on tosses 3 and 5 we obtain the second sequence in (7). Thus, the number of different sequences with two h's and four t's is just the number of ways of selecting a set of two toss numbers for the h's from

$$\{1, 2, 3, 4, 5, 6\}.$$

Thus, the value of m in Equation (6) is

$$m = C_{6,2} = \frac{6!}{2!4!} = \frac{6 \cdot 5}{2 \cdot 1} = 15.$$

Therefore

$$P(\text{exactly two heads}) = \frac{15}{64}.$$

Example 46 If a hand of seven cards is dealt from a deck of 52 cards, what is the probability that the hand contains two spades and five red cards?

Solution The number of different seven card hands is

$$C_{52,7}$$

To find the number of seven-card hands with two spades and five red cards we can argue as follows. Since the deck has 13 spades; there are,

$$C_{13,2}$$

possible ways to choose a set of two spades from the 13. Moreover, the deck has 26 red cards so that there are

$$C_{26,5}$$

to choose five red cards from the 26. Thus, by the multiplication principle, there are

$$C_{13,2} \cdot C_{26,5}$$

different ways to choose two spades and five red cards to form a hand. Therefore

$$P(\text{two spades and five red cards}) = \frac{C_{13,2} \cdot C_{26,5}}{C_{52,7}} = \frac{(78)(65{,}780)}{133{,}784{,}560} \approx .038.$$

Example 47 A lot of 100 items from a manufacturing process is known to contain 10 defectives and 90 nondefectives. If a sample of eight items is selected at random, what is the probability that:

(a) the sample has three defectives and five nondefectives?

(b) the sample has at least one defective?

(c) the sample has more than six defectives?

Solution (a) This problem has a close relationship with the card-dealing problem in the previous example. We can imagine the 100 items to be a deck of 100 cards. In this deck, 10 cards are marked $D_1, D_2, \ldots, D_{10}$ (corresponding to the 10 defectives) and 90 of the cards are marked $N_1, N_2, \ldots, N_{90}$ (corresponding to the 90 nondefectives). The sample of eight items corresponds to a hand of eight cards dealt from this deck. Thus the probability that the sample has three defectives and five nondefectives is just the probability that the hand of eight cards contains three D's and five N's. By imitating the argument in Example 46 the reader should be able to show that this probability is

$$\frac{C_{10,3} \cdot C_{90,5}}{C_{100,8}}.$$

Solution (b) By Theorem 5 in Section 6.3, the probability that the sample has at least one defective is one minus the probability the sample does *not* have at least one defective, or equivalently,

$$P(\text{at least one defective}) = 1 - P(\text{no defectives}). \tag{8}$$

If, as in part (a), we view this as a card-dealing problem then the probability of no defectives is just the probability of dealing a hand with eight N's from the deck with 90 N's and 10 D's described in part (a). Since there are

$$C_{90,8}$$

different ways to select eight N's from the 90 N's and since there are

$$C_{100,8}$$

different eight-card hands possible, the probability of no defectives is

$$\frac{C_{90,8}}{C_{100,8}}.$$

Substituting this value in (8) we obtain

$$P(\text{at least one defective}) = 1 - \frac{C_{90,8}}{C_{100,8}}.$$

Solution (c) The probability that the sample has more than six defectives is just the probability that the sample has exactly seven defectives or exactly eight defectives. Thus, if we let E and F denote the events:

$$E = \text{the sample has exactly seven defectives}$$

$$F = \text{the sample has exactly eight defectives}$$

then

$$P(\text{more than six defectives}) = P(E \cup F).$$

But E and F are mutually exclusive. (Why?) Thus, by Theorem 3 in Section 6.3

$$P(\text{more than six defectives}) = P(E) + P(F). \tag{9}$$

By viewing this as a card-dealing problem the reader should now be able to show that

$$P(E) = P(\text{exactly seven defectives}) = \frac{C_{10,7}C_{90,1}}{C_{100,8}}$$

and

$$P(F) = P(\text{exactly eight defectives}) = \frac{C_{10,8}}{C_{100,8}}.$$

Substituting these values in (9) yields

$$P(\text{more than six defectives}) = \frac{C_{10,7}C_{90,1}}{C_{100,8}} + \frac{C_{10,8}}{C_{100,8}}.$$

EXERCISE SET 6.4

1. A communications signal can be transmitted from point a to one of seven possible satellites, after which it is relayed to one of 50 possible ground stations and sent to point b. Use the multiplication principle to compute the number of different ways of sending a signal from a to b.

2. A license plate consists of two letters of the alphabet followed by three of the integers from 0 to 9. Use the multiplication principle to determine how many license plates are possible.

3. A medical researcher classifies humans according to one of three skin colors, one of eight blood types, and one of two sexes. Use the multiplication principle to determine how many classes are possible.

4. A coin is tossed ten times and the sequences of heads and tails is recorded. Use the multiplication principle to determine how many sequences are possible. How many sequences are possible if the coin is tossed n times?

5. In an IBM 360 digital computer a *bit* is one of the integers 0 or 1 and a *word* is a sequence of 32 bits. Use the multiplication principle to determine how many words are possible.

6. A psychological test consists of ten equations, each of which can be answered, "always," "never," or, "sometimes." Use the multiplication principle to determine the number of ways the ten-question test can be answered.

7. A die is tossed five times and the sequence of numbers is recorded. Use the multiplication principle to determine how many sequences are possible. How many sequences are possible if the die is tossed n times?

8. How many different permutations can be formed from the following sets?

   (a) $\{1, 2\}$ (b) $\{a, b, c, d, e\}$ (c) $\{1, 2, \ldots, 10\}$.

9. Use a tree to help list all permutations of $\{a, b, c, d\}$.

10. Compute

    (a) $9!$ (b) $10!$ (c) $11!$ (d) $\dfrac{22!}{19!}$.

11. Show that

$$n(n-1)(n-2)\cdots(n-r+1) = \frac{n!}{(n-r)!}.$$

12. Compute

    (a) $\dfrac{(n+1)!}{n!}$ (b) $\dfrac{n!}{(n-2)!}$ (c) $\dfrac{(n!)(n+1)}{(n+1)!}$.

13. Use formula (3) to evaluate

    (a) $P_{7,2}$ (b) $P_{12,4}$ (c) $P_{5,5}$

    (d) $P_{3,1}$ (e) $P_{8,5}$ (f) $\dfrac{P_{6,3}}{3!}$.

14. Use formula (4) to evaluate the quantities in Exercise 13.

15. Use a tree to help list all permutations of $\{a, b, c, d, e\}$ taken two at a time.

16. (a) How many permutations can be formed from $\{a, b, c, d, e\}$ taken three at a time?

    (b) Use a tree to help list them.

17. In the runoff of a state lottery, there are ten ticket holders in a drawing for three prizes, a $50,000 prize, a $25,000 prize, and a $10,000 prize. If no person can win more than one prize, how many outcomes are possible?

18. How many three-letter words (real or fictitious) can be formed from the first ten letters of the alphabet if:

(a) repetitions are not allowed;

(b) repetitions are allowed?

19. A combination lock is opened by making three turns to the left, stopping at a digit a, then two turns to the right, stopping at digit b, and finally one turn to the left stopping at digit c. How many lock combinations are possible if a, b, and c are selected from the digits from 0 to 9 inclusive?

20. A testing program to determine the pollutants present in river water involves selecting one of three filtering processes, then one of five precipitation procedures, then one of two evaporation methods. How many testing programs are possible?

21. In a customer preference survey, a person is asked to test three out of ten breakfast cereals and rate them 1, 2, and 3 in order of preference. How many ratings are possible?

22. Compute

(a) $C_{7,2}$ (b) $C_{12,4}$ (c) $C_{5,5}$

(d) $C_{3,1}$ (e) $C_{8,5}$ (f) $3!C_{5,3}$

23. Show

(a) $C_{10,2} = C_{10,8}$ and $C_{7,5} = C_{7,2}$

(b) $C_{n,r} = C_{n,n-r}$

24. List all combinations of $\{a, b, c, d, e\}$ taken three at a time.

25. Use the results of Exercise 24 and the two-step procedure illustrated in Figure 6.8 to obtain a list of all permutations of $\{a, b, c, d, e\}$ taken three at a time. Compare your results to those obtained in Exercise 16b.

26. (a) How many different 11-card hands can be dealt from a deck of 52 cards?

(b) How many different 11-card hands with four aces can be dealt from a deck of 52 cards?

(c) How many different 11-card hands with seven black cards and four red cards can be dealt from a deck of 52 cards?

27. In a state election, there are nine candidates running for three positions entitled "state judge." How many different election outcomes are possible?

28. At a national presidential convention, party rules require that the Credentials Committee contain three men and two women. If there are nine men and six women to choose from, in how many different ways can the committee be formed?

29. A soil engineer divides a proposed building site into 25 equal square plots and randomly chooses four plots for testing. In how many different ways can the four plots be selected?

30. A die is tossed seven times in succession and the resulting sequence of numbers is observed.

(a) How many sequences are possible?

(b) How many sequences with exactly one 6 are possible?

(c) How many sequences with exactly two 6's are possible?

(d) How many sequences are possible which contain at most two 6's?

31. A fair coin is tossed seven times. What is the probability that exactly five of the tosses are heads?

32. (a) In Exercise 31, what is the probability that at most two of the tosses are heads?

(b) In Exercise 31, what is the probability of at least three heads? (*Hint:* Use the result of part (a).)

33. A lot of 50 items from a manufacturing process is known to contain 20 defectives and 30 nondefectives. If a sample of ten items is selected at random, what is the probability that:

(a) the sample has six defectives and four nondefectives?

(b) the sample has no defectives?

(c) the sample has at least one defective?

34. (Birthday Problem) A group of n people are gathered together. What is the probability that at least two of them have the same birthday (that is, the same day and month of birth, although not necessarily the same year)? (*Hint:* Let E be the event that at least two of the people have the same birthday. Instead of finding $P(E)$ directly, it is easier to use the formula $P(E) = 1 - P(E')$, where E' is the event that no two people in the group have the same birthday. To simplify the problem, we neglect leap years and assume that each year has 365 days. To finish the problem, the reader should show that

$$P(E') = \frac{P_{365,n}}{365^n} = \frac{365 \cdot 364 \cdot 363 \cdots (365 - n + 1)}{365^n}.$$

A Comment: It is a messy job to evaluate $P(E')$ for a specific value of n. However, with the aid of a digital computer we obtained the following rather surprising table of probabilities.

n = number of people	5	10	20	22	23	30	50	60
$P(E)$ = probability that at least two have the same birthday	.027	.117	.411	.476	.507	.706	.970	.994

This table shows that when 23 people are gathered together, the probability is greater than $\frac{1}{2}$ that at least two will have the same birthday. Further, when 60 people are gathered together, it is almost certain that at least two will have the same birthday!

35. The street system in a certain city can be represented schematically by the diagram in Figure 6.9, in which A, B, C, D, E, and F denote intersections. Use a tree diagram to help determine the number of ways in which a car can travel from A to F without passing through the same intersection twice.

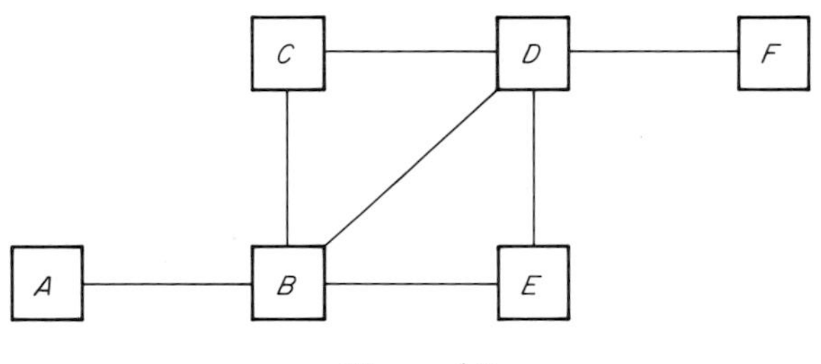

Figure 6.9

36. A department store receives eight dresses, four green and four yellow. In how many different ways can these be distributed between two show windows if each window must display at least three dresses, and at least one dress of each color. (Assume that dresses of the same color are of different types, and thus distinguishable.)

37. A modeling agency agrees to provide a fashion house three models each week for 52 weeks. The contract requires that in no two weeks will exactly the same three models be sent. What is the minimum number of models that will be needed?

38. A psychologist has a maze with five compartments. For a certain experiment, each compartment is to be painted with one color, red black, or orange.

(a) In how many different ways can the maze be painted?

(b) In how many different ways can the maze be painted if no more than two colors can be used?

6.5 CONDITIONAL PROBABILITY; INDEPENDENCE

From our previous work, we know that the probability of tossing a one with a fair die is $\frac{1}{6}$. Suppose, however, that a fair die is tossed and we are told that the number on the top face is odd. What is the probability that the number tossed is a one? Since the number tossed must be odd, we have three equally likely outcomes for the experiment,

$$1, \quad 3, \quad \text{or} \quad 5.$$

Thus, the probability that the number tossed is a one, given that the number is odd is $\frac{1}{3}$. This simple illustration shows that probabilities can be affected by information about the outcome of the experiment. In this section we shall develop systematic methods for computing probabilities when additional information about the outcome of an experiment is known.

In the die-tossing experiment discussed above, let E and F be the events

$$E = \text{a one is tossed}$$

$$F = \text{an odd number is tossed.}$$

We can summarize our above discussion succinctly by stating that

$$P(E) = \frac{1}{6},$$

while

$$P(E \text{ occurs given that } F \text{ occurs}) = \frac{1}{3}. \tag{1}$$

In probability problems, it is usual to abbreviate the phrase,

"E occurs given that F occurs"

by writing

$$E|F.$$

With this notation (1) can be written

$$P(E|F) = \frac{1}{3}.$$

The probability $P(E|F)$ is called the ***conditional probability of E given F***.

Suppose now that A and B are two events associated with an experiment having a finite sample space. Out next objective is to formulate a relationship between the conditional probability $P(A|B)$ and the probabilities $P(A)$ and $P(B)$. To motivate the ideas, we shall consider the simplest situation, an experiment with n equally likely outcomes. Suppose, as illustrated in Figure 6.10, that A and B are two events in the sample space S

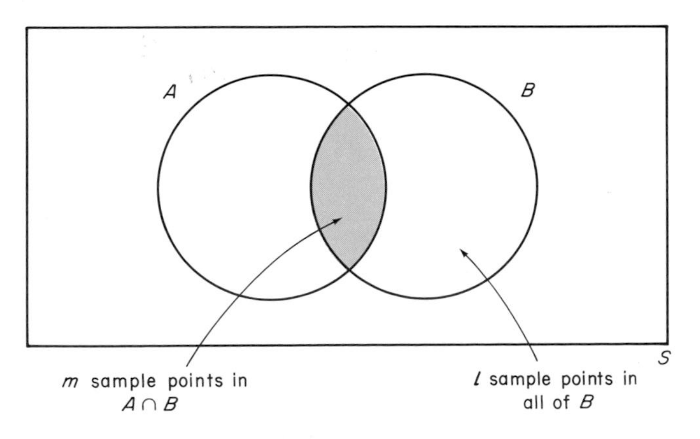

Figure 6.10

for this experiment. If we assume that the event B contains l sample points and the event $A \cap B$ contains m sample points, then

$$P(B) = \frac{l}{n} \qquad \text{and} \qquad P(A \cap B) = \frac{m}{n}. \tag{2}$$

We are interested in finding $P(A|B)$, that is, the probability that A occurs given that B occurs.

Since we are given that the event B occurs, the outcome of the experiment must be one of the l sample points in B. Of these l sample points, there are m sample points for which A occurs; these are the m sample points in $A \cap B$. Thus, since all outcomes are equally likely,

$$P(A|B) = \frac{m}{l}. \tag{3}$$

If we divide the numerator and denominator of (3) by n, we obtain

$$P(A|B) = \frac{m/n}{l/n},$$

which can be rewritten, using (2), as

$$P(A|B) = \frac{P(A \cap B)}{P(B)}. \tag{4}$$

Although formula (4) was derived under the assumption that the outcomes in S are equally likely, it serves to motivate the following definition which applies to any sample space.

If A and B are events and $P(B) \neq 0$, then we define the ***conditional probability of A given B*** by

$$P(A|B) = \frac{P(A \cap B)}{P(B)}. \tag{5}$$

Example 48 If two fair dice are tossed, what is the probability that the sum is seven if we know that at least one face shows a 4?

Solution Let A and B be the events

A = sum is seven

B = at least one face shows a 4.

We want to compute $P(A|B)$. From Table 6.2 in Section 6.2 we obtain

$$A = \{(6, 1), (5, 2), (4, 3), (3, 4), (2, 5), (1, 6)\}$$

$$B = \{(4, 1), (4, 2), (4, 3), (4, 4), (4, 5), (4, 6), (1, 4), (2, 4), (3, 4), (5, 4), (6, 4)\},$$

so that

$$A \cap B = \{(4, 3), (3, 4)\}.$$

Thus

$$P(A \cap B) = \frac{2}{36} \quad \text{and} \quad P(B) = \frac{11}{36},$$

so that, from formula (5),

$$P(A|B) = \frac{2/36}{11/36} = \frac{2}{11}.$$

Example 49 A survey of 1100 traffic accidents resulted in the data shown in Table 6.5. The table shows the cause of the accident and the sex of the driver at fault

Table 6.5

	Mechanical failure	Intoxication	Poor judgment
Male	310	102	208
Female	280	45	155

(a) Given that an accident was due to mechanical failure, what is the probability the driver was a male?

(b) Given that the driver was a male, what is the probability that the accident was due to mechanical failure?

Solution (a) Let A and B be the events

A = the driver was a male

B = the accident was due to mechanical failure.

We want to find $P(A|B)$. Of the 1100 traffic accidents, it follows from Table 6.5 that 310 + 280 were due to mechanical failure and 310 were due to mechanical failure with a male driver. Thus,

$$P(B) = \frac{310 + 280}{1100} = \frac{590}{1100}$$

and

$$P(A \cap B) = \frac{310}{1100}.$$

Thus, from formula (5),

$$P(A|B) = \frac{P(A \cap B)}{P(B)} = \frac{310}{590} = \frac{31}{59}.$$

Solution (b) In this part we want to find $P(B|A)$. From formula (5) with the roles of A and B interchanged we have

$$P(B|A) = \frac{P(B \cap A)}{P(A)}.$$

From Table 6.5,

$$P(A) = \frac{310 + 102 + 208}{1100} = \frac{620}{1100}$$

and

$$P(B \cap A) = \frac{310}{1100},$$

so that

$$P(B|A) = \frac{P(B \cap A)}{P(A)} = \frac{310}{620} = \frac{1}{2}.$$

If we multiply both sides of (5) by $P(B)$ we obtain

The Product Principle for Probabilities

$$P(A \cap B) = P(A|B)P(B).$$

In problems where the value of $P(A|B)$ is obvious, this formula can be used to compute $P(A \cap B)$. The next example illustrates this idea.

Example 50 In 95% of all manned lunar flights, a midcourse trajectory correction is required. This is done by sending a "fire signal" from ground control to ignite small correction thrusters. For technical reasons, this fire signal is sometimes not executed by the thrusters. Tests show that when a fire signal is required, the probability is .0001 that it will not be executed. If the correction is required and not executed, the rocket will plunge into the sun's gravitational field. What is the probability that this will happen?

Solution Let A and B be the events

$$A = \text{correction is not executed}$$

$$B = \text{correction is required.}$$

We want to find $P(A \cap B)$. From the data in the problem we have

$$P(B) = .95, \qquad P(A|B) = .0001.$$

Thus

$$P(A \cap B) = P(A|B)P(B) = (.0001)(.95) = .000095.$$

Many practical problems involve a sequence of experiments in which the possible outcomes and probabilities associated with any one experiment depend on the outcomes of the preceding experiments. Such a sequence of experiments is called a ***stochastic process.*** As a simple example, suppose we draw a card from a standard deck of 52 cards (experiment 1) and then draw another card from the remaining 51 cards (experiment 2). A typical question one might ask about this stochastic process is:

What is the probability that the second card drawn is a spade?

The analysis can be simplified by considering the tree diagram in Figure 6.11. There are two stages to this diagram, the first representing the outcomes for experiment 1 and the second for experiment 2. Since 13 of the 52 original cards are spades, the chances of obtaining a spade on draw one are $\frac{13}{52}$ and the chances of a nonspade are $\frac{39}{52}$. These probabilities are shown at the appropriate branches of the first stage of the tree. The chances of obtaining a spade on the second draw are influenced by the

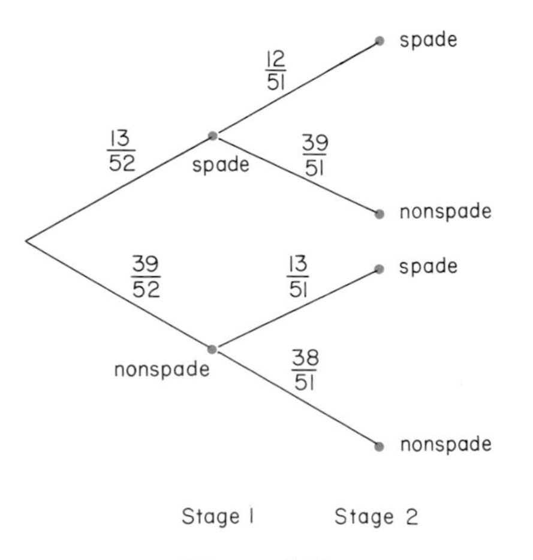

Figure 6.11

outcome of the first draw. For example, if a spade is obtained on the first draw, only 12 of the remaining 51 cards are spades, so in this case, the probability of obtaining a spade on the second draw is $\frac{12}{51}$ and the probability of a nonspade is $\frac{39}{51}$. These probabilities are noted on the appropriate branches at the second stage of the tree. Similarly, if a nonspade turns up on the first draw, then 13 of the remaining 51 cards are spades so that in this case the probability of obtaining a spade on the second draw is $\frac{13}{51}$ and the probability of a nonspade is $\frac{38}{51}$.

To find the probability of drawing a spade as the second card, we must take all the possibilities into account: we can get a spade on draw one and a spade on draw two (which corresponds to the colored path in Figure 6.12a) or we can get a nonspade on draw one and a spade on draw two (which corresponds to the colored path in Figure 6.12b). By

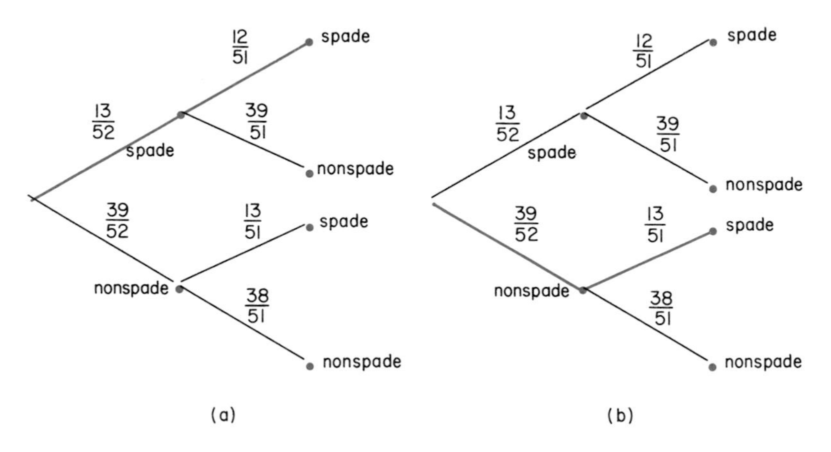

Figure 6.12

the probability product principle, the probability of following the path in Figure 6.12a is

$$\frac{13}{52} \cdot \frac{12}{51}$$

and the probability of following the path in Figure 6.12b is

$$\frac{39}{52} \cdot \frac{13}{51}.$$

Thus the probability of following the path in (a) *or* (b) is

$$\frac{13}{52} \cdot \frac{12}{51} + \frac{39}{52} \cdot \frac{13}{51}$$

(why?) so that

$$P(\text{second card is a spade}) = \frac{13}{52}\cdot\frac{12}{51} + \frac{39}{52}\cdot\frac{13}{51}$$
$$= \frac{1}{4}\cdot\frac{12}{51} + \frac{3}{4}\cdot\frac{13}{51}$$
$$= \frac{51}{204}$$
$$= \frac{1}{4}. \tag{6}$$

Example 51 Three cards are drawn in succession from a standard deck without replacing those already drawn. What is the probability that the third card drawn is a spade?

Solution Using an analysis similar to that above, we obtain the three-stage tree diagram (Figure 6.13). There are four paths through the tree that result in a spade on the third draw. If we multiply the probabilities along each path and then add the resulting products, we obtain

$$P\begin{pmatrix}\text{spade}\\ \text{on third}\\ \text{draw}\end{pmatrix} = \frac{13}{52}\cdot\frac{12}{51}\cdot\frac{11}{50} + \frac{13}{52}\cdot\frac{39}{51}\cdot\frac{12}{50} + \frac{39}{52}\cdot\frac{13}{51}\cdot\frac{12}{50} + \frac{39}{52}\cdot\frac{38}{51}\cdot\frac{13}{50}$$
$$= \frac{1716}{132600} + \frac{6084}{132600} + \frac{6084}{132600} + \frac{19266}{132600}$$
$$= \frac{33150}{132600}$$
$$= \frac{1}{4}.$$

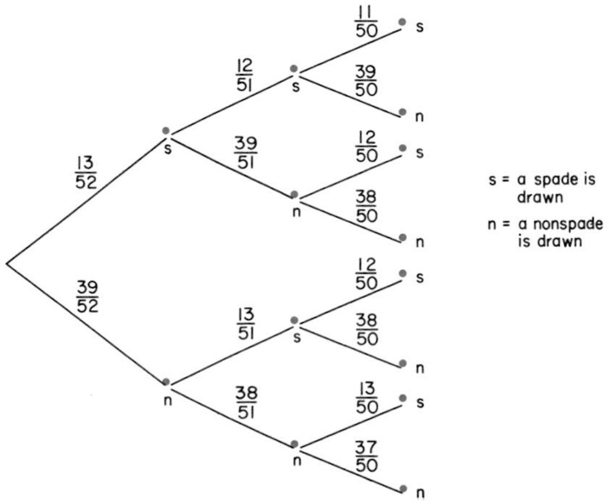

Figure 6.13

Example 52 Three manufacturing plants A, B, and C supply, respectively, 20%, 30%, and 50% of all shock absorbers used by a certain automobile manufacturer. Records show that the percentage of defective items produced by A, B, and C is 3%, 2%, and 1% respectively. What is the probability that a randomly chosen shock absorber installed by the manufacturer will be defective?

Solution The possibilities are shown in the two-stage tree diagram of Figure 6.14. At the first stage of the tree, the probabilities .2, .3, and .5

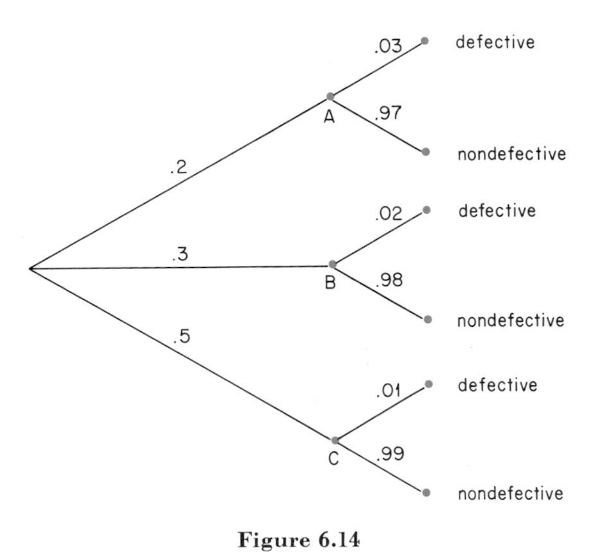

Figure 6.14

represent the chances of the selected item being produced by A, B, or C. The probabilities at the second stage represent the chances of the item being defective or not.

There are three paths through the tree that result in a defective shock absorber. Multiplying the probabilities along each such path and adding the resulting products yields:

$$\begin{aligned} P(\text{defective is selected}) &= (.2)(.03) + (.3)(.02) + (.5)(.01) \\ &= .006 + .006 + .005 \\ &= .017 \end{aligned}$$

In other words, 1.7% of all shock absorbers installed will be defective.

Example 53 A fair coin is tossed twice. Find the probability that:

(a) the second toss is a head;

(b) the second toss is a head given that the first toss is a head.

Solution If A and B are the events

$$A = \text{the second toss is a head}$$
$$B = \text{the first toss is a head},$$

then the sample space S for this experiment is

$$S = \{hh, ht, th, tt\}.$$

The events A and B are

$$A = \{hh, th\} \qquad B = \{hh, ht\},$$

and the event $A \cap B$ is

$$A \cap B = \{hh\}.$$

Since the coin is fair, we shall assume the four sample points in S are equally likely so that

$$P(A) = \frac{2}{4} = \frac{1}{2} \qquad \text{and} \qquad P(A \cap B) = \frac{1}{4}.$$

Thus, the probability that the second toss is a head is

$$P(A) = \frac{1}{2} \tag{8}$$

and the probability that the second toss is a head given that the first toss is a head is

$$P(A|B) = \frac{P(A \cap B)}{P(B)} = \frac{\frac{1}{4}}{\frac{1}{2}} = \frac{1}{2}. \tag{9}$$

This example illustrates an important phenomenon that occurs in many probability problems. Comparing (8) and (9) we see that

$$P(A|B) = P(A).$$

In other words, the probability of A is unaffected by the additional knowledge that B occurs. Intuitively, this tells us that the two events are "independent" or do not influence the probabilities of one another. This result is not unexpected. It states that the coin has no memory. When the second toss is made, the coin does not "remember" what happened on the first toss. Thus, the chances of a head on the second toss will be $\frac{1}{2}$ regardless of the outcome of the first toss.

Let us pursue these ideas in a little more detail. Suppose A and B are two events such that

$$P(A|B) = P(A)$$

or equivalently

$$\frac{P(A \cap B)}{P(B)} = P(A).$$

If we multiply both sides of this equation by $P(B)$, we obtain

$$P(A \cap B) = P(A)P(B).$$

This suggests the following definition

Two events A and B are called ***independent*** if

$$P(A \cap B) = P(A)P(B). \tag{10}$$

We leave it for the exercises to show the following result.

Let A and B be independent events with nonzero probabilities, then

$$P(A|B) = P(A) \quad \text{and} \quad P(B|A) = P(B) \tag{11}$$

Conversely, if either of the equations in (11) hold, then A and B are independent.

Thus when A and B are independent events, knowledge that B occurs does not affect the probability that A occurs and knowledge that A occurs does not affect the probability that B occurs.

Example 54 The events A and B in Example 53 are independent since

$$P(A) = \frac{1}{2}, \quad P(B) = \frac{1}{2}, \quad \text{and} \quad P(A \cap B) = \frac{1}{4}$$

so that

$$P(A \cap B) = P(A)P(B).$$

Example 55 A card is drawn from a standard deck of 52 cards and then a second card is drawn without replacing the first card. Let A be the event that the second card is a spade and let B be the event that the first card is a spade. Determine whether A and B are independent.

Solution Intuitively we should suspect that the events are *not* independent since the chances of getting a spade on the second draw are better if a spade is not removed from the deck on the first draw than if one is removed. Let us try to confirm our suspicions mathematically. To show that A and B are not independent we can proceed in one of three ways. We can show

$$P(A \cap B) \neq P(A)P(B)$$

or we can show

$$P(B|A) \neq P(B)$$

or we can show

$$P(A|B) \neq P(A).$$

We shall use the last approach.

The conditional probability $P(A|B)$ is the probability of getting a spade on the second draw given that a spade was obtained on the first draw. In this case a spade is removed from the deck on the first draw, so that 12 of the 51 cards available for the second draw are spades. Thus

$$P(A|B) = \frac{12}{51} = \frac{4}{17}.$$

On the other hand, it was shown in the discussion preceding Example 51 that

$$P(A) = \frac{1}{4}.$$

Thus

$$P(A|B) \neq P(A)$$

so that A and B are not independent events.

In many practical problems, it is difficult or even impossible to determine with certainty whether two events are independent. In such situations, the experimenter often relies on intuition to decide whether he should assume the events to be independent or not. For example, suppose we know from past records that a certain sharpshooter will hit the bull's-eye on a target with probability .9. Suppose also we want to find the probability that this sharpshooter will fire two bull's-eyes on two successive shots. We might begin by introducing the events

A = the sharpshooter gets a bull's-eye on the first shot

B = the sharpshooter gets a bull's-eye on the second shot.

Thus $A \cap B$ corresponds to getting bull's-eyes on both shots. If we *assume* that A and B are independent events, then

$$P(A \cap B) = P(A)P(B) = (.9)(.9) = .81.$$

Since this computation is erroneous if A and B are not independent, we should carefully examine this independence assumption. In order for A and B to be independent, the chances of getting a bull's-eye on the second shot should not be affected by the result of the first shot. It is possible, however, that a miss on the first shot might disturb the sharpshooter's concentration and reduce his chances of a bull's-eye on the second shot, or it might increase his desire for a bull's-eye and thus improve his concentration, thereby bettering his chances. In this situation the independence assumption would not be warranted. On the other hand, if the experimenter does not believe this psychological factor is significant, he may choose to accept the independence assumption. Although there are statistical tests that can sometimes be applied to test for independence experimentally, they are beyond the scope of this text. In any event, these tests cannot always be applied so that the independence assumption often depends on the judgment of the experimenter.

Example 56 A certain manufacturing process produces defective items 1% of the time. What is the probability that two items selected at random will both be defective? Assume independence.

Solution Let A and B be the events

$$A = \text{the first item is defective}$$

$$B = \text{the second item is defective.}$$

We want to find $P(A \cap B)$. Assuming A and B are independent, we obtain

$$P(A \cap B) = P(A)P(B) = (.01)(.01) = .0001.$$

The notion of independence can be extended to more than two events as follows.

A set of events is called ***independent*** if for each finite subset $\{A_1, A_2, \ldots, A_n\}$ we have

$$P(A_1 \cap A_2 \cap \cdots \cap A_n) = P(A_1)P(A_2)\cdots P(A_n)$$

Example 57 For a set of three events $\{A, B, C\}$ to be independent we must have

$$P(A \cap B) = P(A)P(B)$$

$$P(A \cap C) = P(A)P(C)$$

$$P(B \cap C) = P(B)P(C)$$

and

$$P(A \cap B \cap C) = P(A)P(B)P(C).$$

Example 58 In the manufacture of light bulbs, filaments, glass casings, and bases are manufactured separately and then assembled into the final product.

From past records we know

2% of all filaments are defective

3% of all glass casings are defective

1% of all bases are defective.

What is the probability that one of these bulbs will have no defects?

Solution Let A, B, and C be the events

A = the bulb has no defect in the filament

B = the bulb has no defect in the glass casing

C = the bulb has no defect in the base.

From the given data,

$$P(A) = .98, \qquad P(B) = .97, \qquad \text{and} \qquad P(C) = .99.$$

We are interested in finding

$$P(A \cap B \cap C).$$

Since filaments, glass casings, and bases are manufactured separately, it seems reasonable to assume A, B, and C are independent. Thus

$$P(A \cap B \cap C) = P(A)P(B)P(C) = (.98)(.97)(.99) \simeq .94.$$

EXERCISE SET 6.5

1. Let A and B be events such that $P(A) = .7$, $P(B) = .4$, and $P(A \cap B) = .2$. Find

   (a) $P(A|B)$ (b) $P(B|A)$.

2. Let A and B be the events in Exercise 1. Find

   (a) $P(A|B')$ (b) $P(B|A')$.

3. Let A and B be events such that $P(A|B) = .5$ and $P(B) = .2$. Find $P(A \cap B)$.

4. A certain missile guidance device contains 100 components. Some of the components are resistors (r) and some of the components are transistors (t). Some of the components are made in the United States (u), while others are made in Japan (j). The number of components in each category is listed in the table below. Assume each component is equally likely to malfunction during a lunar flight.

   (a) If a malfunction occurs in a transistor, what is the probability it was made in the United States?

(b) If a malfunction occurs in a component made in Japan, what is the probability that it is a resistor?

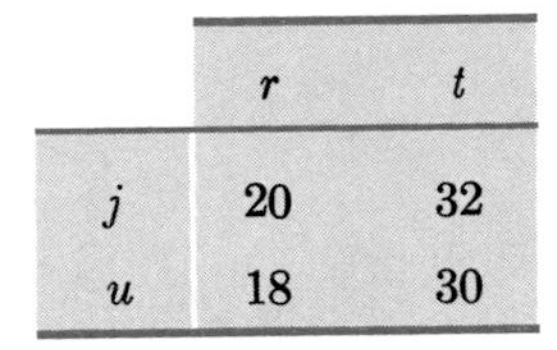

	r	t
j	20	32
u	18	30

5. A lot contains 25 defective items and 75 nondefective items. If we choose two items in succession without replacement, what is the probability that both items are defective?

6. A telephone repairman knows that two circuits are working properly and two circuits are working improperly. He tests the circuits, one by one, until both defective circuits are located.

(a) What is the probability that the two defective circuits will be located on the first two tests?

(b) What is the probability that the last defective circuit is located on the third test?

7. In a certain state, 45% of all major manufacturers violate some federal pollution standard and 30% violate both a state and federal pollution standard. Given that a major manufacturer violates a federal standard, what is the probability he violates a state pollution standard?

8. A card is drawn from a standard deck of 52 cards. What is the probability that it is an ace given that it is not a king?

9. A student must answer a multiple choice question with one of five possible answers.

(a) What is the probability he will get the correct answer if he guesses?

(b) What is the probability he will get the correct answer if he can rule out the first two possibilities?

10. Identical twins come from a single egg and must therefore be of the same sex. Fraternal twins, on the other hand, come from different eggs and can be of opposite sex. Assuming the probability that two twins are fraternal is $\frac{2}{3}$ and the probability that a fraternal twin is female is $\frac{1}{2}$, find the probability that two twins have the same sex. (*Hint:* Let $S =$ two twins have the same sex, $I =$ two twins are identical, and $F =$ two twins are fraternal. Then $S = (S \cap I) \cup (S \cap F)$.)

11. In a medical study of the common cold, 1000 cold sufferers exhibited the following symptoms.

18 people had fevers

32 people had coughs

50 people had stuffy noses

4 people had coughs and stuffy noses

5 people had fevers and stuffy noses

7 people had fevers and coughs

2 people had fevers, coughs, and stuffy noses.

If a person, selected at random from the group, has a stuffy nose, what is the probability the person also has a cough?

12. Let A and B be independent events such that $P(A) = \frac{1}{5}$ and $P(B) = \frac{1}{6}$. Find $P(A \cap B)$.

13. Let $S = \{s_1, s_2, s_3, s_4, s_5\}$ be the sample space for an experiment and assume $P(s_1) = P(s_2) = P(s_3) = \frac{1}{4}$ and $P(s_4) = P(s_5) = \frac{1}{8}$.

(a) Show that $A = \{s_1, s_2\}$ and $B = \{s_2, s_3\}$ are independent events.

(b) Show that $C = \{s_3, s_4\}$ and $D = \{s_4, s_5\}$ are not independent events.

14. Let E and F be independent events such that $P(E) = \frac{1}{2}$ and $P(E \cup F) = \frac{3}{4}$. Find $P(F)$

15. In a federal study of unemployment, 1000 people were classified as employed, unemployed, having a high school diploma, or not having a high school diploma; the following data were obtained:

	Has high school diploma (H)	Does not have high school diploma (H')
Employed (E)	917	34
Unemployed (E')	9	40

(a) Find $P(E|H)$ and $P(E)$.

(b) Are E and H independent events?

16. A submarine detection system consists of three units, a sonar device, a magnetic detector, and a visual spotter. To enter a certain area without being detected, a submarine must escape detection by all three units. The probability of escaping the sonar is .5, the probability of escaping the magnetic detector is .4, and the probability of escaping the visual spotter is .8. If we assume that the units act independently, what is the probability that a submarine can enter the area undetected?

17. Prove statement (11).

18. In each part, decide if E and F can reasonably be assumed to be independent events.

(a) A person is picked at random;

E = the person is over 6-feet tall
F = the person weighs more than 150 pounds.

(b) The weather is observed for two consecutive days;

E = rain occurs on the first day
F = rain occurs on the second day.

(c) A person is picked at random;

$$E = \text{the person has blue eyes}$$
$$F = \text{the person is wearing a ring.}$$

19. Let A and B be two events associated with an experiment, and suppose $P(A) = .5$, while $P(A \cup B) = .8$. Let $P(B) = p$. For which values of p are

(a) A and B mutually exclusive?

(b) A and B independent?

20. Assume we have two identical urns. Urn A contains three black balls and one white ball, while urn B contains two black balls and four white balls. An urn is chosen at random and then a ball is selected at random from this urn. What is the probability that the ball is white?

21. A box contains ten light bulbs, two of which are defective. Suppose three bulbs are selected in succession without replacing those already selected.

(a) What is the probability that all three bulbs selected will be good?

(b) What is the probability that the third bulb selected will be good?

(c) What is the probability that two are good and one is defective?

(d) What is the probability that at least two are good?

22. A fair coin is tossed four times in succession. Use a tree diagram to find:

(a) the probability of tossing four heads;

(b) the probability of tossing three heads and one tail.

23. In a manufacturing plant a certain part is made by the day shift and the night shift. Of the parts made by the day shift, 80% are good and 20% defective. The night shift produces good parts only 75% of the time. Suppose one of the shifts is chosen at random and two pieces of its output are selected independently for inspection.

(a) What is the probability that both pieces are good?

(b) What is the probability that the pieces are good, given that they came from the day shift?

(c) What is the probability that the pieces came from the night shift, given that both are defective?

6.6 BAYES' FORMULA*

In many practical problems an experimenter observes the outcome of an experiment and then asks for the probability that the outcome was caused by one of several possible factors. For example, a skin specialist who observes a certain type of facial rash in one of his patients may be interested in the probability that it is caused by German measles, an astronaut who observes a certain configuration of error lights in a display panel may be interested in the probability that it is caused by a fault in the cooling system of his rocket, or a political analyst who observes a change in traditional voting patterns may be interested in the probability that it was caused by the winning candidate's position on foreign policy. In this section we shall discuss techniques for solving certain problems of this type.

In 1763 the Reverend Thomas Bayes† published a paper entitled, "An Essay Toward Solving a Problem in the Doctrine of Chances." In that paper, Bayes derived the following important result relating the conditional probabilities $P(A|B)$ and $P(B|A)$

$$P(B|A) = \frac{P(A|B)P(B)}{P(A)}. \tag{1}$$

With the use of set theory, the proof of this result is much easier for us than it was for Bayes. We can establish formula (1) as follows:

$$\begin{aligned} P(B|A) &= \frac{P(B \cap A)}{P(A)} && \text{(definition of } P(B|A)\text{)} \\ &= \frac{P(A \cap B)}{P(A)} && (A \cap B = B \cap A) \\ &= \frac{P(A|B)P(B)}{P(A)} && \text{(from formula (5) in Section 6.5).} \end{aligned}$$

In some problems $P(B|A)$ is hard to obtain, while $P(A|B)$ is easy. In such problems formula (1) enables us to use the value of $P(A|B)$ to obtain the value $P(B|A)$. The following example illustrates such a situation.

Example 59 A card is drawn from a standard deck of 52 cards and then a second card is drawn without replacing the first card. If we know the second card to be a spade, what is the probability that the first card is a spade?

Solution Let A and B be the events

$$A = \text{the second card is a spade}$$

$$B = \text{the first card is a spade.}$$

We want to find $P(B|A)$. In this problem, $P(A|B)$ is easy to obtain. The conditional probability $P(A|B)$ is the probability of getting a spade on the second draw given that a spade was obtained on the first draw. In this case a spade is removed from the deck on the first draw, so that 12 of the 51

† *Thomas Bayes* (1702–1763)—Bayes was the son of a Presbyterian minister. He began his own ministry by assisting his father, and he continued in the ministry until his death. Bayes published several theological papers, but he is most famous for his paper on probability, which was published after his death by a friend who found it in his effects. This work is noteworthy because it is the first discussion of inductive inference in precise quantitative form.

cards available for the second draw are spades, thus

$$P(A|B) = \frac{12}{51} = \frac{4}{17}.$$

From Example 51 we know that

$$P(A) = \frac{1}{4} \quad \text{and} \quad P(B) = \frac{1}{4},$$

so that from formula (1)

$$P(B|A) = \frac{P(A|B)P(B)}{P(A)} = \frac{\frac{4}{17} \cdot \frac{1}{4}}{\frac{1}{4}} = \frac{4}{17}.$$

We shall show next how formula (1) can be used to derive a result, called Bayes' theorem, which is useful in solving problems like those discussed in the introduction of this section. The following example will motivate the basic ideas.

Example 60 A certain item is manufactured by three factories, I, II, and III. Suppose we have a stockpile of these items in which

30% of the items were made in factory I

20% of the items were made in factory II

50% of the items were made in factory III.

Moreover, suppose

2% of all items produced by I are defective

3% of all items produced by II are defective

4% of all items produced by III are defective.

Assume that an item selected at random from the stockpile is observed to be defective.

(a) What is the probability that the item came from factory I?
(b) From factory II?
(c) From factory III?

Solution Let us introduce the events

$$A = \text{the item is defective}$$

$$B_1 = \text{the item came from factory I}$$

$$B_2 = \text{the item came from factory II}$$

$$B_3 = \text{the item came from factory III.}$$

In parts (a), (b), and (c) we want to find

$$P(B_1|A), \qquad P(B_2|A), \qquad \text{and} \qquad P(B_3|A)$$

respectively. Before attempting to find these values, let us see what information is immediately available from the data. Since 30% of all items in the stockpile come from factory I, we have

$$P(B_1) = .3.$$

Similarly,

$$P(B_2) = .2 \qquad \text{and} \qquad P(B_3) = .5.$$

Since 2% of all items from factory I are defective we have

$$P(A|B_1) = .02.$$

Similarly,

$$P(A|B_2) = .03 \qquad \text{and} \qquad P(A|B_3) = .04.$$

To solve problems (a), (b), and (c) in this example, we shall try to express the unknown quantities

$$P(B_1|A), \qquad P(B_2|A), \qquad \text{and} \qquad P(B_3|A)$$

in terms of the known quantities

$$P(B_1), \quad P(B_2), \quad P(B_3), \quad P(A|B_1), \quad P(A|B_2), \quad \text{and} \quad P(A|B_3). \qquad (2)$$

We shall do this first for $P(B_1|A)$. From formula (1) we can write

$$P(B_1|A) = \frac{P(A|B_1)P(B_1)}{P(A)}. \qquad (3)$$

The numerator in (3) contains only known quantities. Thus to finish we must express the denominator, $P(A)$, in terms of the quantities in (2). To do this, let the sample space S consist of all items in the stockpile. Since every item in S comes from factory I, factory II, or factory III and since no item can come from two different factories, the events B_1, B_2, and B_3 divide the sample space into three mutually exclusive parts, as illustrated in Figure 6.15a. Therefore, as shown in Figure 6.15b, A can be written as a union of three mutually exclusive events

$$A = (A \cap B_1) \cup (A \cap B_2) \cup (A \cap B_3).$$

Thus

$$P(A) = P(A \cap B_1) + P(A \cap B_2) + P(A \cap B_3). \qquad (4)$$

But the terms in (4) can be rewritten as

$$P(A \cap B_1) = P(A|B_1)P(B_1)$$

$$P(A \cap B_2) = P(A|B_2)P(B_2)$$

$$P(A \cap B_3) = P(A|B_3)P(B_3).$$

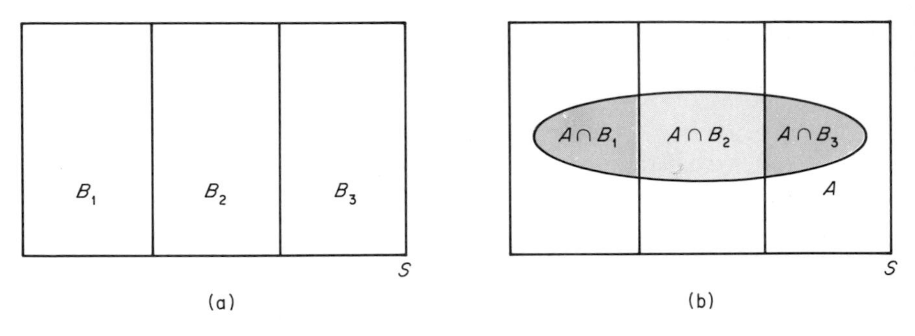

Figure 6.15

Substituting these expressions into (4) we obtain

$$P(A) = P(A|B_1)P(B_1) + P(A|B_2)P(B_2) + P(A|B_3)P(B_3) \tag{5}$$

Substituting (5) into (3) yields

$$P(B_1|A) = \frac{P(A|B_1)P(B_1)}{P(A|B_1)P(B_1) + P(A|B_2)P(B_2) + P(A|B_3)P(B_3)} \tag{6}$$

Formula (6) expresses $P(B_1|A)$ in terms of the known quantities in (2). Substituting these values yields

$$P(B_1|A) = \frac{(.02)(.3)}{(.02)(.3) + (.03)(.2) + (.04)(.5)} = \frac{3}{16}.$$

This solves the problem in part (a). To solve (b) and (c) we can imitate the derivation of formula (6) to obtain

$$P(B_2|A) = \frac{P(A|B_2)P(B_2)}{P(A|B_1)P(B_1) + P(A|B_2)P(B_2) + P(A|B_3)P(B_3)} \tag{7}$$

and

$$P(B_3|A) = \frac{P(A|B_3)P(B_3)}{P(A|B_1)P(B_1) + P(A|B_2)P(B_2) + P(A|B_3)P(B_3)}. \tag{8}$$

Substituting the known values into the right-hand sides of (7) and (8) we obtain

$$P(B_2|A) = \frac{(.03)(.2)}{(.02)(.3) + (.03)(.2) + (.04)(.5)} = \frac{3}{16}$$

$$P(B_3|A) = \frac{(.04)(.5)}{(.02)(.3) + (.03)(.2) + (.04)(.5)} = \frac{10}{16}.$$

Formulas (6), (7), and (8) are special cases of the following result, called *Bayes' theorem.*

Theorem 6 (Bayes' theorem) If a sample space S can be divided into finitely many mutually exclusive events $B_1, B_2, B_3, \ldots$ and if A is an event with nonzero probability, then for each B_i we have

$$P(B_i|A) = \frac{P(A|B_i)P(B_i)}{P(A|B_1)P(B_1) + P(A|B_2)P(B_2) + P(A|B_3)P(B_3) + \cdots} \tag{9}$$

Formula (9) is often called ***Bayes' formula.***

It is not essential to memorize Bayes' formula since all problems to which it applies can be solved graphically using tree diagrams. The technique is as follows (Figure 6.16):

Step 1 From a common initial point draw a separate branch to represent each of the events $B_1, B_2, B_3, \ldots$ and label these branches with the probabilities $P(B_1), P(B_2), P(B_3), \ldots$.

Step 2 From the end of each of these branches, draw a single branch to represent the event A and label these new branches with the conditional probabilities $P(A|B_1), P(A|B_2), P(A|B_3), \ldots$.

Step 3 Each possible path through the tree ends at a different point. Label each endpoint with the product of the probabilities on the path leading to it. We call these the ***path probabilities.***

A tree diagram constructed in this way will be called a ***Bayes tree.***

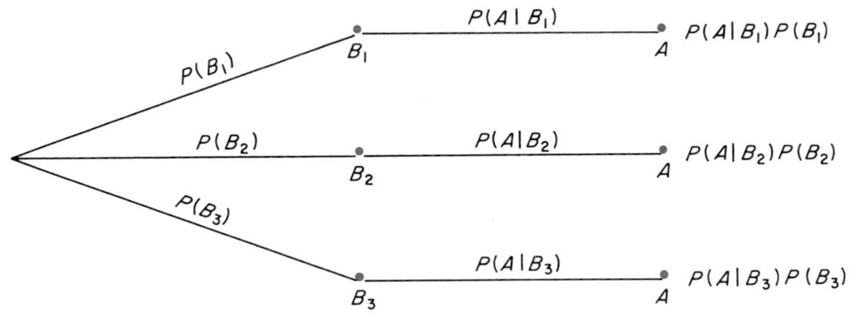

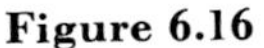

Figure 6.16

Observe that the sum of all the path probabilities is precisely the denominator in Bayes' formula and that the numerator

$$P(A|B_i)P(B_i)$$

in Bayes' formula is precisely the path probability for the path through events A and B_i. Thus, the probability

$$P(B_i|A)$$

can be obtained from the tree diagram as follows:

Obtaining Conditional Probabilities from a Bayes Tree

$$P(B_i|A) = \frac{\text{path probability for the path through } A \text{ and } B_i}{\text{sum of all path probabilities}}$$

Example 61 As an illustration, let us solve the problem in Example 60 using a Bayes tree. From the data in the example, the tree is as shown in Figure 6.17. From the tree, the sum of all the path probabilities is

$$.006 + .006 + .02 = .032.$$

Thus from the tree

$$P(\text{item is from I} \mid \text{item is defective}) = \frac{.006}{.032} = \frac{3}{16}$$

$$P(\text{item is from II} \mid \text{item is defective}) = \frac{.006}{.032} = \frac{3}{16}$$

$$P(\text{item is from III} \mid \text{item is defective}) = \frac{.02}{.032} = \frac{10}{16}.$$

These results agree with those obtained in Example 60 using Bayes' formula.

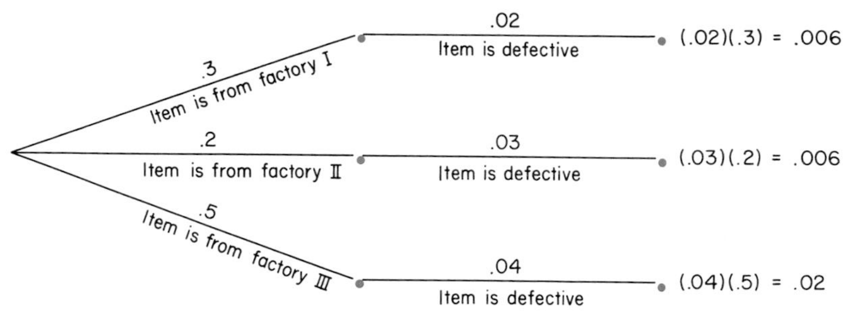

Figure 6.17

Example 62 A box contains four coins. One of the coins has two heads and the other three coins are fair. One of the coins is picked at random and tossed. If we know the outcome is heads, what is the probability that the coin is fair?

Solution We shall solve this problem using a Bayes tree. Consider the events

$$B_1 = \text{the coin is fair}$$

$$B_2 = \text{the coin has two heads}$$

$$A = \text{the outcome is heads when the coin is tossed.}$$

We want to find

$$P(B_1|A) = P(\text{coin is fair}|\text{heads is tossed})$$

Since three of the four coins are fair and one of the four has two heads, we have

$$P(B_1) = \frac{3}{4} \quad \text{and} \quad P(B_2) = \frac{1}{4}.$$

If the coin is fair, then the probability of tossing heads is $\frac{1}{2}$ so that

$$P(A|B_1) = \frac{1}{2}.$$

If the coin has two heads, then the probability of tossing heads is 1 so that

$$P(A|B_2) = 1.$$

Thus the Bayes tree for the problem is as shown in Figure 6.18. Thus the sum of the path probabilities is

$$\frac{1}{4} + \frac{3}{8} = \frac{5}{8}$$

and

$$P(\text{coin is fair}|\text{heads is tossed}) = \frac{\frac{3}{8}}{\frac{5}{8}} = \frac{3}{5}.$$

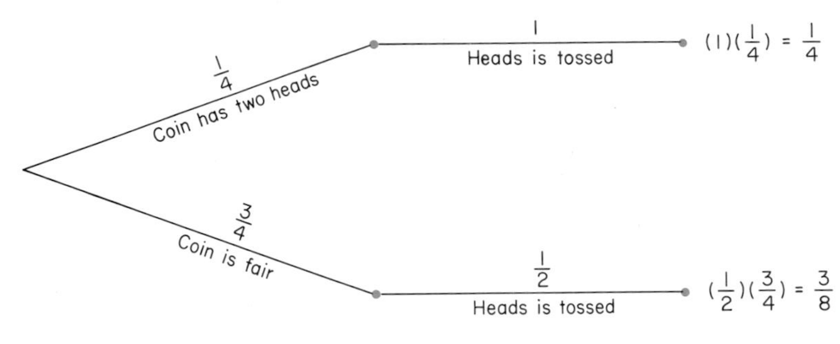

Figure 6.18

Oil Exploration

Example 63 Records show that when drilling for oil, the probability of a successful strike is .1. However, it has been observed that if there is oil, the probability is .6 that there is permeable, porous, sedimentary rock present. Records also show that when no oil is below, the probability is .3 that such rock formations are present. What is the probability of oil beneath permeable, porous, sedimentary rock?

Solution We will solve this problem using Bayes' theorem. The reader may want to try solving it using a Bayes tree. Consider the events

A = there is oil present

B = there is permeable, porous, sedimentary rock present

We want to find

$$P(A|B)$$

From the given data we have

$$P(A) = .1 \qquad P(A') = .9$$
$$P(B|A) = .6 \qquad P(B|A') = .3$$

Thus, from Bayes' theorem

$$P(A|B) = \frac{P(B|A)P(A)}{P(B|A)P(A) + P(B|A')P(A')}$$

$$= \frac{(.6)(.1)}{(.6)(.1) + (.3)(.9)}$$

$$= \frac{2}{11}.$$

Applications of Bayes' Theorem

Bayes' theorem has many varied and interesting applications. One of the most fascinating was given by F. Mosteller and D. L. Wallace in a book called, *Inference and Disputed Authorship: The Federalist* (Reading, Massachusetts: Addison Wesley, 1964). In that book Mosteller and Wallace used Bayes' theorem to resolve a longstanding historical question: Between 1787 and 1788 a series of 77 essays, called the *Federalist Papers,* appeared in New York newspapers under the pen name Publius. It is known that James Madison wrote 14 of the essays, John Jay wrote 5, Alexander Hamilton wrote 43, and three were jointly authored. The authorship of the remaining 12 essays has been in dispute for years. By analyzing the writing styles and applying Bayes' theorem, Mosteller and Wallace gave a convincing argument to show that Madison was the most probable author of the disputed papers. The reader will find a more extensive discussion of the Mosteller–Wallace method in an excellent elementary text called *Sets, Functions and Probability* by Johnston, Price, and Van Vleck (Reading, Massachusetts: Addison Wesley, 1968).

At this point the reader may wish to go to Section 8.1 where we give an application of Bayes' theorem to the medical diagnosis of disease.

EXERCISE SET 6.6

1. In a tire factory, assembly lines A, B, and C account for 60%, 30%, and 10% of the total production. If .3% of the tires from line A are defective, .6% of the tires from B are defective, and .8% of the tires from line C are defective, what is the probability that:

   (a) a defective tire comes from line A?

   (b) a defective tire comes from line B?

   (c) a defective tire comes from line C?

2. Company insurance records show that a new driver who has finished a driver training program has a probability of .9 of completing the first year of driving without an accident, while a new driver who has not finished a driver training program, has a probability of only .7. If 60% of all new drivers have completed a driver training program, what is the probability that a person who is involved in an accident during his first year of driving finished a driver training program?

3. Records show that 20% of all students taking calculus at a certain university fail the course. To reduce the failure rate, the university designs a screening test to determine beforehand which students are likely to fail. These students are to be given remedial coursework. An experiment shows that the probability is .9 that a student who passes calculus can pass the screening test and the probability is .3 that a student who fails calculus can pass the screening test. A student passes the screening test. What is the probability he will pass calculus?

4. A pediatrician knows from past experience that when a mother calls and states that her child has a low grade fever, the probabilities for various causes are:

	Cause	Probability
(A)	Flu	.3
(B)	Strep infection	.2
(C)	Inaccurate thermometer	.1
(D)	Other causes	.4

He also knows that when he advises the mother to administer aspirin and give lots of liquid, the probability that the fever will be gone (G) within 24 hours is

$$P(G|A) = .6 \qquad P(G|B) = .5$$
$$P(G|C) = 0 \qquad P(G|D) = .7$$

(a) Given that a mother has called and has been advised to administer aspirin and lots of liquid, what is the probability that the child's fever will be gone within 24 hours?

(b) Given that the child's fever is not gone within 24 hours after being advised to administer aspirin and lots of liquid, what is the probability that the child has the flu?

5. Teleview and Radion Corporation are competing for a contract to build television equipment for a Mars probe; they both have equal chances of getting the contract. If Teleview gets the contract, the probability of a total success is .6, the probability of a partial success is .1 and the probability of failure is .3. If Radion gets the contract, the probability of a total success is .3, the probability of a partial success is .5 and the probability of failure is .2. If the contract is awarded and the result is a total success, what is the probability that Teleview got the contract?

6. A shipper must send fresh produce to a distribution center. The produce can be sent by one of four different routes, and the probabilities of the produce reaching the center within 24 hours on the four routes are .3, .2, .5, and .1 respectively. The shipper picks a route at random.

(a) If the produce arrives within 24 hours, what is the probability that the first route was picked?

(b) If the produce does not arrive within 24 hours, what is the probability that the third route was picked?

7. Theoretical genetic considerations suggest that three-fourths of all marigold seeds produced by a certain crop of marigolds should yield pure yellow offspring; the rest should yield multicolored offspring. A seed packager has a test to separate the seeds that will produce pure yellow marigolds from the seeds that will produce multicolored marigolds. However, the test is not perfect, and 5% of the pure yellow will be classified as multicolored and 10% of the multicolored will be classified as pure yellow. Out of 1000 seeds packaged as pure yellow, how many will actually be pure yellow?

8. A test is designed to detect cancer. If a person has cancer, the probability that the test will detect it is .95, and if the person does not have cancer, the probability that the test will erroneously indicate he does is .1. If 3% of the population who take the test have cancer, what is the probability that a person described by the test as having cancer does not really have it?

9. A patient has one of two diseases, D_1 or D_2. Each disease can result in one of the following sets of symptoms.

$$S_1 = \{\text{fever, aching}\} \qquad \text{and} \qquad S_2 = \{\text{vomiting, tiredness}\}$$

It is known that

$$P(S_1|D_1) = .4 \qquad P(S_2|D_1) = .6$$
$$P(S_1|D_2) = .8 \qquad P(S_2|D_2) = .2$$

Moreover D_1 and D_2 occur equally often and never occur together. If a patient has fever and aching, what is the probability that he has disease D_1?

10. A box contains two coins, an ordinary fair dime and a dime with two heads. One of the coins is picked at random and tossed twice. If both tosses are heads, what is the probability the coin is two-headed?

11. In the sheik's palace are three vaults; each vault contains two chests and each chest contains one gem. In one vault each chest contains a ruby; in the second vault each chest contains a diamond; and in the last vault one chest contains a ruby and the other contains a diamond. A thief randomly chooses a vault and steals one of the chests. When the chest is opened, the thief finds a ruby. What is the probability that the other chest of the same vault also contains a ruby?

7 STATISTICS

Historically, statisticians were people who collected masses of numerical data and organized this information into charts and graphs. Although some statisticians still do this kind of work, the majority are concerned with the problem of *statistical inference,* that is using numerical data to make logical decisions and inferences. Our primary concern will be this aspect of the statistics field. In Chapter 6 we saw that some probability models can be determined by intuition while others can be obtained only from experimental data. One broad area of statistical inference, called *estimation theory,* is concerned with devising precise ways to construct probability models from numerical data. As we discussed in the previous chapter, a probability model is a theory or hypothesis about the physical world. Like all theories, probability models are useful only if they accurately predict or describe physical phenomena. If the results predicted by the model do not agree with physical observations, then the model must be abandoned in favor of a more-accurate model. This leads to a second main area of statistical inference, *hypothesis testing.* Hypothesis testing is concerned with procedures for determining if results predicted by a probability model are in reasonable agreement with physical observations. In this chapter we shall touch on both estimation theory and hypothesis testing.

7.1 INTRODUCTION; RANDOM VARIABLES

For many purposes it is convenient to have a numerical way of describing the outcomes of an experiment. By so doing it is often possible to bring to bear the tools of arithmetic, algebra, and other branches of mathematics to help solve probability and statistics problems. In this section we shall study this idea.

Suppose we toss a coin three times and record the resulting sequence of heads and tails; the sample space is

$$S = \{hhh, hht, hth, thh, htt, tht, tth, ttt\}.$$

For this experiment we have outcomes that are sequences of three letters. Suppose, however, we are interested in making probability statements about the number of heads that occur in the three tosses. For example, what is the probability of obtaining two heads in the three tosses? In this case we are not primarily concerned with the particular sequence of h's and t's in the outcome, but rather we are interested in the number of h's that occur. Thus, associated with each outcome of this experiment, we have a number that is of importance to us, the number of h's in the sequence. This idea of associating a number with an outcome of an experiment occurs so frequently that statisticians have found it convenient to introduce the following terminology.

A ***random variable*** is a rule that assigns a numerical value to each outcome of an experiment.

We shall denote random variables by uppercase letters such as X, Y, and Z.

Example 1 Suppose we toss a coin three times and introduce the random variable X to denote the number of heads that occur in the three tosses. In Table 7.1 we have listed the sample points for the experiment and next to each sample point we have given the associated value of the random variable X.

Table 7.1

Sample point	Value of X
hhh	3
hht	2
hth	2
thh	2
htt	1
tht	1
tth	1
ttt	0

Example 2 Suppose we toss a coin repeatedly until a head occurs and then stop. Let the random variable Y denote the number of tosses performed in the experiment. In Table 7.2 we have listed some of the sample points and the associated values of the random variable.

Table 7.2

Sample point	Value of Y
h	1
th	2
tth	3
$ttth$	4
$tttth$	5
$ttttth$	6
$\vdots$	$\vdots$

Example 3 A traffic engineer records the times at which two consecutive buses arrive at a checkpoint. Let the random variable Z denote the time elapsed in minutes between the arrivals of the buses. Thus, if the arrival times are

$$9{:}07 \text{ A.M.} \qquad \text{and} \qquad 9{:}36 \text{ A.M.},$$

then the value of Z associated with this experimental outcome is

$$Z = 29 \quad \text{minutes.}$$

Example 4 A geneticist records the length of life in hours of a fruitfly. Let the random variable W denote the number recorded. If we assume for simplicity that the time can be recorded with perfect accuracy, then the value of W can be any nonnegative real number.

A random variable is often described according to the number of values it can take on. A random variable is called ***finite discrete*** if it can take on only finitely many possible values. For example, the random variable X In Example 1 is finite discrete since it can take on only the values

$$X = 0, \quad 1, \quad 2, \quad \text{or} \quad 3.$$

A random variable is called ***infinite discrete*** if it can take on infinitely many values which can be arranged in a sequence. For example, the random variable Y in Example 2 is infinite discrete since its possible values can be arranged in the sequence

$$Y = 1, 2, 3, 4, \ldots.$$

Finally, a random variable is called ***continuous*** if its possible values form an entire interval of numbers. For example, the random variable W in Example 4 can take on any nonnegative value. The variable W is thus continuous since its possible values form the interval shown in Figure 7.1.

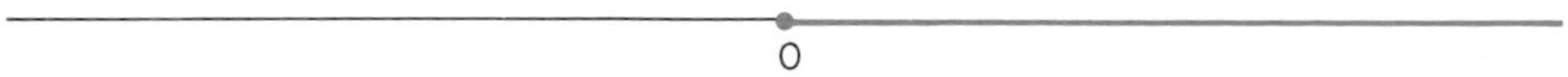

0

Figure 7.1

For the remainder of this section *we shall restrict our discussion to finite discrete random variables.*

Since the value that a random variable takes on is determined by the outcome of an experiment, and since the outcomes of an experiment have various probabilities of occurring, it follows that the values of the random variable also occur with varying probabilities. To illustrate, consider the coin-tossing experiment discussed in Example 1. In the left column of Table 7.1 we listed the eight possible outcomes for this experiment. If we assume the coin is fair, then each of these eight outcomes will be equally likely so that each outcome will have a probability $\frac{1}{8}$ of occurring. From the right column of Table 7.1 we see that X takes on the value 2 for three of these sample points; thus the probability that $X = 2$ will be $\frac{3}{8}$. We denote this by writing

$$P(X = 2) = \frac{3}{8},$$

read "the probability that X equals 2 is $\frac{3}{8}$." Similarly, Table 7.1 shows that

$$P(X = 0) = \frac{1}{8}$$

$$P(X = 1) = \frac{3}{8}$$

and

$$P(X = 3) = \frac{1}{8}.$$

Example 5 Let X denote the sum of the numbers tossed with two fair dice. Find the probabilities of the various possible X values.

Solution This experiment was discussed in Example 26 of Section 6.2 and the possible outcomes were shown in Table 6.2. It was shown in that example that $P(X = 7) = \frac{6}{36}$. Similarly,

$$P(X = 2) = \frac{1}{36}, \qquad P(X = 3) = \frac{2}{36}, \qquad P(X = 4) = \frac{3}{36}$$

$$P(X = 5) = \frac{4}{36}, \qquad P(X = 6) = \frac{5}{36}, \qquad P(X = 7) = \frac{6}{36}$$

$$P(X = 8) = \frac{5}{36}, \qquad P(X = 9) = \frac{4}{36}, \qquad P(X = 10) = \frac{3}{36}$$

$$P(X = 11) = \frac{2}{36}, \qquad P(X = 12) = \frac{1}{36}.$$

Example 6 Company records show that 5% of all resistors produced at a certain electronics plant are defective. Suppose two resistors are picked independently from the plant's production line and assume the random variable X denotes the number of defectives in the sample. Find

$$P(X = 0), \qquad P(X = 1), \qquad P(X = 2)$$

assuming that resistors are picked independently.

Solution If we let n denote a nondefective resistor and d a defective, then the outcomes can be viewed as

$$(n, d), \qquad (d, n), \qquad (n, n), \qquad (d, d),$$

where, for example, (n, d) means that the first resistor picked was nondefective and the second was defective. Since 5% of all resistors produced are defective, the probability of picking a nondefective resistor is .05 and the probability of picking a nondefective resistor is .95. Moreover, since we are assuming that the resistors are selected independently, the probability of selecting a nondefective and then a defective will be the product (.95)(.05); that is, the probability of the outcome (n, d) is

$$(.95)(.05) = .0475.$$

The probabilities of all the outcomes and the associated values of X are listed in Table 7.3.

Table 7.3

Outcome	Probability	Value of X associated with the outcome
(n, d)	$(.95)(.05) = .0475$	$X = 1$
(d, n)	$(.05)(.95) = .0475$	$X = 1$
(n, n)	$(.95)(.95) = .9025$	$X = 0$
(d, d)	$(.05)(.05) = .0025$	$X = 2$

From Table 7.3 we obtain

$$\begin{aligned} P(X = 0) &= .9025 \\ P(X = 1) &= .0475 + .0475 = .0950 \\ P(X = 2) &= .0025. \end{aligned} \tag{1}$$

If X is a finite discrete random variable, there is a simple way of describing the various probabilities of X geometrically. To illustrate, consider the experiment of tossing three fair coins with X denoting the total number of heads tossed. Table 7.4 summarizes the results we previously obtained (preceding Example 5). Using this table, we can pair up each value of X with its probability to obtain the four ordered pairs

$$\left(0, \frac{1}{8}\right) \quad \left(1, \frac{3}{8}\right) \quad \left(2, \frac{3}{8}\right) \quad \left(3, \frac{1}{8}\right).$$

Table 7.4

Value of X	0	1	2	3
Probability that the value occurs	$\frac{1}{8}$	$\frac{3}{8}$	$\frac{3}{8}$	$\frac{1}{8}$

These ordered pairs can now be plotted in a Cartesian coordinate system, as shown in Figure 7.2a. Since the first coordinate of each point is a value of X and the second coordinate is a probability, we have labeled the coordinate axes x and p.

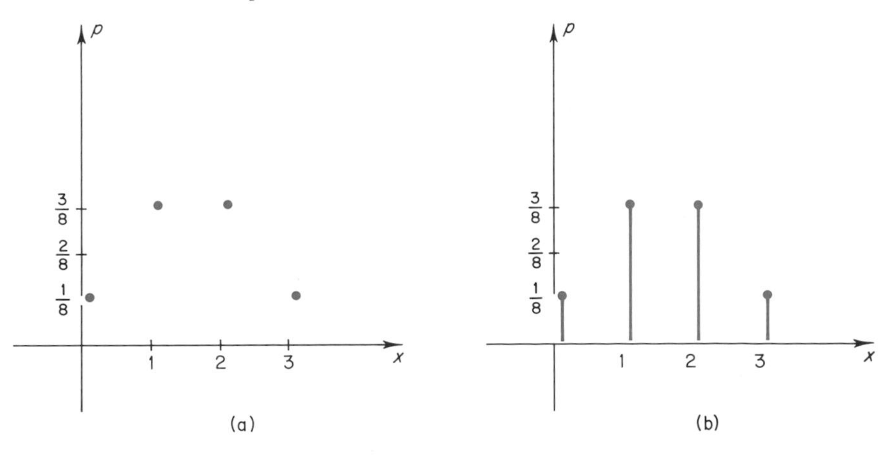

Figure 7.2

The set of ordered pairs relating the values of a discrete random variable with their probabilities is called the ***probability function*** for the random variable; the configuration of points obtained by plotting these ordered pairs is called the ***graph*** of the probability function. For visual emphasis, it is usual to join each point in the graph of the probability function to the x axis by a vertical line (see Figure 7.2b). The resulting *bar graph* is usually easier to read.

Example 7 Graph the probability function of the random variable X in Example 6.

Solution From the calculations in Example 6 the values of X and their probabilities are summarized by Table 7.5. The probability function thus consists of the ordered pairs

Table 7.5

Value of X	0	1	2
Probability that the value occurs	.9025	.0950	.0025

$$(0, .9025) \qquad (1, .0950) \qquad (2, .0025)$$

and the graph of the probability function is shown in Figure 7.3.

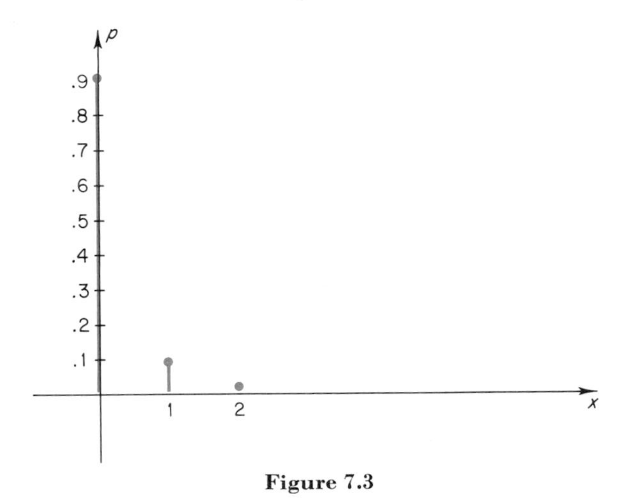

Figure 7.3

EXERCISE SET 7.1

In Exercises 1–5 list the values of X together with their probabilities of occurrence.

1. A fair die is tossed once. The random variable X is the number showing on the top face.
2. A fair coin is tossed four times. The random variable X is the number of heads obtained on the four tosses.

3. Let X be the number of male children in a family with two children. (Assume the probability of a male birth is $\frac{1}{2}$.)

4. A coin is weighted so that on each toss the probability of a head is $\frac{2}{3}$ and the probability of a tail is $\frac{1}{3}$. The coin is tossed three times and X is the total number of tails obtained.

5. A student takes a three-question true–false examination, and guesses at every answer. Let X be the number right minus the number wrong.

6–10. Graph the probability functions for the random variables in Exercises 1–5.

11. Classify the following random variables as finite discrete, infinite discrete, or continuous.

(a) X is the number of defectives in a lot of 1000 items.

(b) X is the number of calls entering a switchboard in a 24-hour period.

(c) X is the length of time a person must wait in line to check out at a supermarket.

(d) A fair die is tossed until a 2 occurs, and X is the number of tosses.

12. Let X be a random variable whose probability function is graphed in Figure 7.4.

(a) Find $P(X = 3)$.

(b) Find $P(X \leq 3)$ (that is, the probability that X takes on a value ≤ 3).

(c) Find $P(X = \frac{1}{2})$.

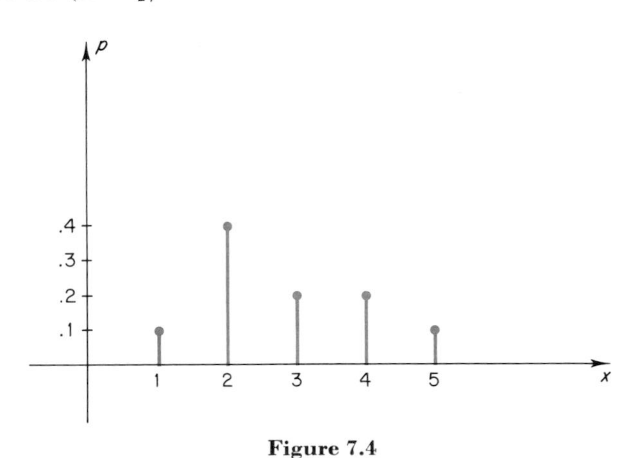

Figure 7.4

7.2 EXPECTED VALUE OF A RANDOM VARIABLE

We are all familiar with the idea of using an average value to summarize a collection of numerical data. The use of averages has been ingrained in most of us by our technological environment so that we are constantly comparing aspects of ourselves against the average. Are we overweight or underweight compared to the average? Is our temperature too high or too low compared to the average? Is our I.Q. high or low compared to the average?

In the process of melting down data into an average, we lose certain information. For example, knowing that the average weight of males inducted into the Army since 1950 is 161.5 pounds does not enable us to determine the highest and lowest weights recorded. Since these numbers could have been obtained from the original data, we have lost this information in the averaging process. However, since the original data would take volumes to list individually, the conciseness of the average as a description of the data may more than compensate for this loss of information.

In this section we shall develop a notion called the *expected value* of a random variable. The expected value is akin to the idea of an average in the sense that it is a single number that is meant to summarize certain data, namely the values of a random variable and their probabilities of occurrence. We shall also establish an important mathematical relationship between expected values and averages.

Recall that the *arithmetic average* of a finite set of numbers is computed by adding the numbers in the set and then dividing by the number of terms. For example, the average of the numbers

$$1.2, \qquad 3.3, \qquad 2.6, \qquad 1.7$$

is

$$\frac{1.2 + 3.3 + 2.6 + 1.7}{4} = \frac{8.8}{4} = 2.2.$$

We can formalize this idea as follows.

The ***arithmetic average*** or ***mean*** of the numbers $x_1, x_2, \ldots, x_n$ is denoted by the symbol $\bar{x}$ and is defined by the formula

$$\bar{x} = \frac{x_1 + x_2 + \cdots + x_n}{n}. \tag{1}$$

The arithmetic average of a collection of numbers is also sometimes called simply the ***average*** of the numbers.

Example 8 Find the arithmetic average of the numbers

$$1.1, \quad 1.1, \quad 1.1, \quad 4.7, \quad 4.7, \quad 5.3, \quad 5.3, \quad 5.3, \quad 5.3, \quad 5.3.$$

Solution

$$\bar{x} = \frac{1.1 + 1.1 + 1.1 + 4.7 + 4.7 + 5.3 + 5.3 + 5.3 + 5.3 + 5.3}{10} \tag{2}$$

$$= \frac{39.2}{10}$$

$$= 3.92.$$

In the last example, the data to be averaged had repetitions. When this occurs, it is possible to simplify the computation of the arithmetic average. To see how, observe that the numerator in (2) contained

3 terms of 1.1

2 terms of 4.7

5 terms of 5.3.

Thus, we could have written (2) in the alternative form

$$\bar{x} = \frac{3(1.1) + 2(4.7) + 5(5.3)}{10} \tag{3}$$

$$= \frac{3.3 + 9.4 + 26.5}{10}$$

$$= \frac{39.2}{10}$$

$$= 3.92.$$

To formalize this idea, suppose we want to compute the arithmetic average of data with repetitions. Specifically, suppose that the *distinct* numbers occurring in the data are $x_1, x_2, \ldots, x_t$ and that

x_1 occurs f_1 times

x_2 occurs f_2 times

x_3 occurs f_3 times

$\vdots$

x_t occurs f_t times.

The following result is evident from Equation (3).

The arithmetic average of the above data with repetitions is given by the formula

$$\bar{x} = \frac{f_1x_1 + f_2x_2 + f_3x_3 + \cdots + f_tx_t}{n}, \tag{4}$$

where n is the total number of values being averaged.

Warning: In this formula, $x_1, x_2, \ldots, x_t$ denote the *distinct* values that occur in the data. Since some of these distinct values may be repeated in the original data, the total number of values being averaged, denoted by n in formula (4), is generally greater than t. In other words, unlike formula (1), the denominator in formula (4) is *not* simply the number of terms in the numerator. It is true, however, that

$$n = f_1 + f_2 + \cdots + f_t. \tag{5}$$

Can you see why? The numbers $f_1, f_2, \ldots, f_t$ are called the ***frequencies*** of the numbers $x_1, x_2, \ldots, x_t$.

Example 9 Suppose that the Bureau of the Census surveys a group of married couples and records the data shown in Table 7.6 concerning the number of children born to them. Find the average number of children born to the couples surveyed.

Table 7.6

Number of couples	Number of children born to the couples
50	0
107	1
201	2
102	3
25	4
9	5
3	6
2	7
1	9
Total = 500	

Solution The numbers denoted by $x_1, x_2, \ldots, x_t$ in formula (4) are the numbers in the right-hand column of Table 7.6 and their frequencies $f_1, f_2, \ldots, f_t$ are the numbers in the left-hand column of the table. As indicated in (5), the total number n of values being averaged is obtained by adding the frequencies; thus

$$n = 500.$$

On substituting the data into formula (4) we obtain

$$\begin{aligned}\bar{x} &= \frac{50(0) + 107(1) + 201(2) + 102(3) + 25(4) + 9(5) + 3(6) + 2(7) + 1(9)}{500} \\ &= \frac{0 + 107 + 402 + 306 + 100 + 45 + 18 + 14 + 9}{500} \\ &= \frac{1001}{500} \\ &= 2.002.\end{aligned}$$

Suppose now that X is a discrete random variable associated with a certain experiment. If we perform the experiment repeatedly n times, we will observe n values of X. We might ask if it is possible to predict the arithmetic average of these n values. Obviously, it is impossible to predict the average *exactly* since the values of X obtained in one group of n repetitions need not be the same as the values obtained in another group of n repetitions. For example, if we toss a die three times and obtain the numbers

$$6, \quad 6, \quad 3$$

whose arithmetic mean is

$$\frac{6+6+3}{3} = 5,$$

it is conceivable that on the next three tosses, we might obtain the numbers

$$2, \quad 1, \quad 3$$

whose arithmetic mean is

$$\frac{2+1+3}{3} = 2.$$

Nevertheless, it is possible to make a reasonable guess about the average of n observed values of X when the number n is large. To see why, consider a finite discrete random variable whose only possible values are

$$x_1, \quad x_2, \quad \text{and} \quad x_3.$$

Assume also that these values occur with known probabilities

$$p_1, \quad p_2, \quad \text{and} \quad p_3.$$

Suppose that after n repetitions of the experiment

$$x_1 \text{ occurs } f_1 \text{ times}$$

$$x_2 \text{ occurs } f_2 \text{ times}$$

$$x_3 \text{ occurs } f_3 \text{ times.}$$

Applying formula (4) we obtain as the arithmetic mean of these values

$$\bar{x} = \frac{f_1x_1 + f_2x_2 + f_3x_3}{n}$$

which can be rewritten as

$$\bar{x} = \frac{f_1}{n}x_1 + \frac{f_2}{n}x_2 + \frac{f_3}{n}x_3. \tag{6}$$

Since

$$\frac{f_1}{n}, \quad \frac{f_2}{n}, \quad \text{and} \quad \frac{f_3}{n}$$

represent the proportion of x_1's, x_2's, and x_3's in the n repetitions, it follows that as n gets larger and larger, these proportions approach the probabilities

$$p_1, \qquad p_2, \qquad \text{and} \qquad p_3.$$

That is, when n is large the approximations

$$\frac{f_1}{n} \simeq p_1, \qquad \frac{f_2}{n} \simeq p_2, \qquad \frac{f_3}{n} \simeq p_3$$

should be good. Thus, from (6) we obtain the following approximation to $\bar{x}$ which is likely to be good when n is large:

$$\bar{x} \simeq p_1x_1 + p_2x_2 + p_3x_3.$$

More generally, if X is a finite discrete random variable whose possible values are

$$x_1, \qquad x_2, \qquad \ldots, \qquad x_k$$

and if these values occur with probabilities

$$p_1, \qquad p_2, \qquad \ldots, \qquad p_k$$

then when n is large, the arithmetic average of n observed values of X will be approximately

$$\bar{x} \simeq p_1x_1 + p_2x_2 + \cdots + p_kx_k. \tag{7}$$

Example 10 An experiment consists of tossing a fair die and observing the number X showing on the top face. Use the approximation in (7) to estimate the arithmetic average of the observed values if the experiment is repeated many times.

Solution The possible values of X are

$$x_1 = 1, \qquad x_2 = 2, \qquad x_3 = 3, \qquad x_4 = 4, \qquad x_5 = 5, \qquad x_6 = 6$$

and since the die is fair, each of these values has a probability $\frac{1}{6}$ of occurring; that is

$$p_1 = \frac{1}{6}, \qquad p_2 = \frac{1}{6}, \qquad p_3 = \frac{1}{6}, \qquad p_4 = \frac{1}{6}, \qquad p_5 = \frac{1}{6}, \qquad p_6 = \frac{1}{6}.$$

Substituting these values in (7) gives

$$\bar{x} \simeq \left(\frac{1}{6}\right)(1) + \left(\frac{1}{6}\right)(2) + \left(\frac{1}{6}\right)(3) + \left(\frac{1}{6}\right)(4) + \left(\frac{1}{6}\right)(5) + \left(\frac{1}{6}\right)(6)$$

or

$$\bar{x} \simeq \frac{21}{6} = 3.5.$$

The quantity on the right-hand side of (7) is of such importance that it has its own notation and name.

If the values

$$x_1, \quad x_2, \quad \ldots, \quad x_k$$

of a finite discrete random variable X occur with probabilities

$$p_1, \quad p_2, \quad \ldots, \quad p_k$$

then the ***expected value of*** X, denoted by $E(X)$, is defined by

$$E(X) = p_1x_1 + p_2x_2 + \cdots + p_kx_k. \tag{8}$$

In other words, *the expected value of X is the sum of the possible values of X times their probabilities of occurrence.* The expected value of X is also called the ***mean of X*** or the ***expectation of X***. It is also denoted by μ_X or possibly just μ when the random variable involved is evident.

Example 11 If X is the number showing on the top face when a fair die is tossed, then, as shown in Example 10,

$$E(X) = \mu_X = 3.5.$$

As shown in this example, the expected value $E(X)$ need not be a possible value of X; the expected value represents the long-term average value of X.

Example 12 In Example 6 of Section 7.1, we considered a random variable X denoting the number of defective resistors in a sample of two resistors from a production process in which 5% of all resistors produced are defective. Find the expected value of X.

Solution Since the expected value of a finite discrete random variable is the sum of the values of the random variable times their probabilities of occurrence we obtain

$$\begin{aligned} E(X) &= (.9025)(0) + (.0950)(1) + (.0025)(2) \\ &= .1 \end{aligned}$$

(see Equations (1) in Example 6). This result can also be written

$$\mu_X = .1 \qquad \text{or} \qquad \mu = .1$$

The expected value of a finite discrete random variable has an interesting physical interpretation. Suppose the values

$$x_1, \quad x_2, \quad \ldots, \quad x_k$$

of a random variable X occur with probabilities

$$p_1, \quad p_2, \quad \ldots, \quad p_k$$

and imagine the x axis to be a solid, uniform rigid bar. If we place weights of

$$p_1 \text{ pounds}, \quad p_2 \text{ pounds}, \quad \ldots, \quad p_k \text{ pounds}$$

on the x axis at the points

$$x_1, \quad x_2, \quad \ldots, \quad x_k$$

as illustrated in Figure 7.5, it can be shown that the bar would balance on a knife-edge located at the point $E(X)$.

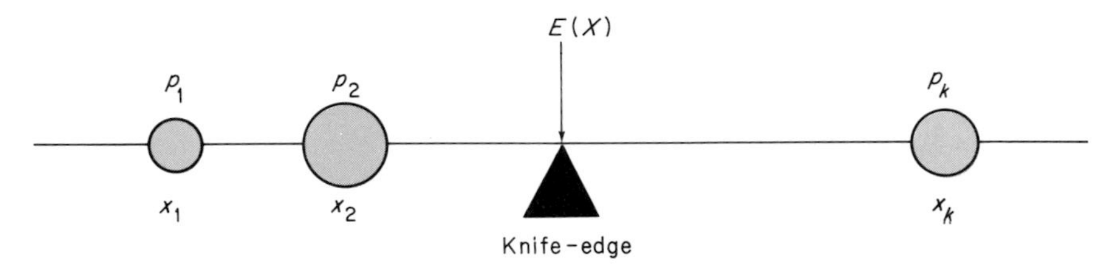

Figure 7.5

Thus, the expected value of a random variable is intuitively the "center of probability" for the random variable. Just as the balance point for a seesaw will be closer to the heavier child, so the expected value or center of probability tends to be close to the portion of the x axis which contains the largest concentration of probability.

To summarize, the expectation $E(X)$ of a finite discrete random variable has two useful interpretations:

(1) It is approximately the average of the observed values of X when the experiment is repeated many times.

(2) It is the center of probability for the random variable.

EXERCISE SET 7.2

1. Find the arithmetic average of

$$3.2, \quad 1.7, \quad 6.3, \quad .2, \quad 5.6.$$

2. Find the mean of the following numbers in two ways, using formulas (1) and (2).

$$15, \quad 14, \quad 14, \quad 16, \quad 13, \quad 15, \quad 15, \quad 14, \quad 17, \quad 16.$$

3. The table below is a record of visits for dental care made by 250 clinic patients during a one-year period. Find the average number of visits per patient.

Number of visits	0	1	2	3	4	5	6	7	8	9	10
Frequency	41	25	33	49	60	24	11	0	6	0	1

4. Table 7.7 is a record of the yearly income for a group of people eligible for Medicare benefits. Find the mean yearly income for the group.

Table 7.7

Yearly income	Frequency	Yearly income	Frequency
\$1,000	225	\$6,000	315
2,000	342	7,000	263
3,000	516	8,000	101
4,000	822	9,000	94
5,000	491	10,000	16

5. Find the expected value of the random variable X in Exercise 2 of Section 7.1.

6. Find the expected value of the random variable X in Exercise 3 of Section 7.1.

7. Find the expected value of the random variable X in Exercise 4 of Section 7.1.

8. A random variable Y has the probability distribution below. Find the expectation of Y.

y	-2	-1	0	1	7	12	18
$P(Y = y)$	.01	.09	.2	.1	.4	.005	.195

9. (a) Find the expectation of the random variable X whose probability function is graphed in Figure 7.6.

(b) Find the expectation of the random variable X whose probability function is graphed in Figure 7.7.

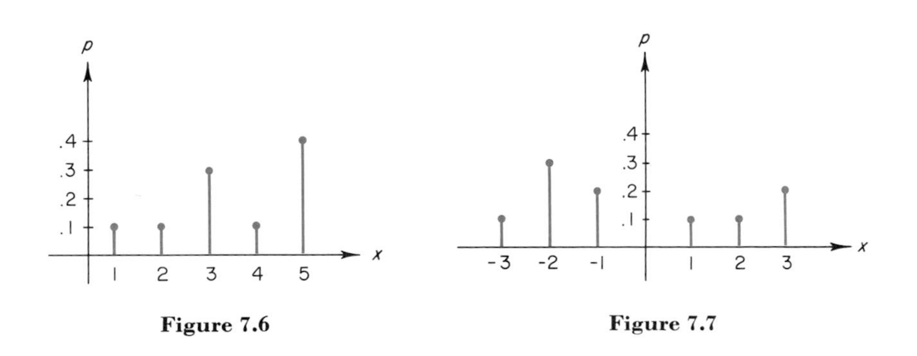

Figure 7.6 **Figure 7.7**

10. The number of accidents on the New Jersey Turnpike during the Friday rush hour is 0, 1, 2, or 3 with probabilities .93, .02, .03, and .02 respectively.

(a) Find the expected number of accidents during the Friday rush hour.

(b) How many accidents can be expected to occur during Friday rush hours over a one year period (52 weeks)?

11. A retailer purchases men's shirts for \$3.00 each and sells them for \$5.00 each. Based on past sales records he estimates the following probabilities for his weekly sales.

Number of shirts sold	0	1	2	3	4
Probability	.1	.1	.3	.4	.1

Compute the expected value of the retailer's profit X.

12. A U.S. lumber firm is allowed to bid for rights to Soviet forest reserves in one of two regions of Siberia, region A or region B. If the firm gets rights to region A, its estimated profit is \$3,000,000 and if the firm gets rights to region B its estimated profit is \$5,000,000. The cost of submitting bids for the two regions is \$100,000 for A and \$300,000 for B. Due to a difference in the number of competitors, the firm estimates its probability of getting rights to A as .7 and its probability of getting rights to B as .5. Should the firm bid for region A or region B? (*Hint:* Compare the expected profits for the bids.)

13. For each part of Figure 7.8, determine $E(X)$ *by inspection* (no calculations) from the given graph.

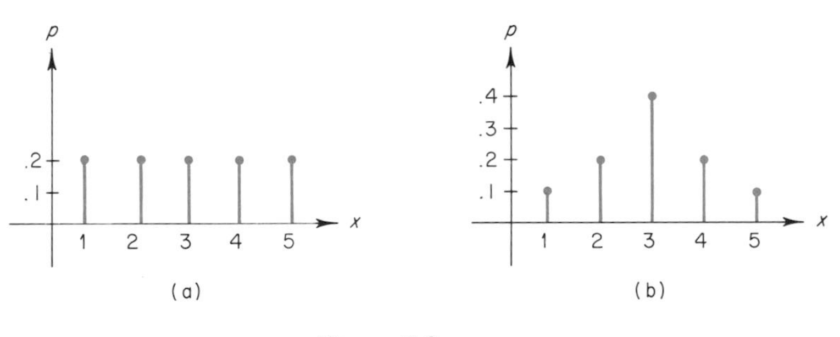

Figure 7.8

14. A certain insurance company charges a 45-year-old man \$150 for a one-year term life insurance policy that will pay the beneficiary \$10,000 if the man dies within the year. Assuming the probability to be .995 that a 45-year-old man survives another year, determine the expected profit per policy if the insurance company sells many such policies.

15. Pilferage is a serious problem in business. Suppose that the manager of a store finds from the analysis of the records that the store loses, on the average, \$500 of material each week from pilfering. Further, suppose that for \$300 a week a security program can be implemented. The table below

shows the results found by other managers of comparable stores where such a program has been used.

Pilferage total per week in dollars	Probability
500	.05
400	.10
300	.15
200	.60
100	.10

(a) If the program is implemented, how much of a loss per week from pilferage should be expected over the long term?

(b) Should the security program be undertaken?

7.3 VARIANCE OF A RANDOM VARIABLE

The expected value of a finite discrete random variable X gives us information about the "long-term" average value of X when the experiment is repeated over and over. For many purposes, however, an average does not supply sufficient information. For example, suppose a pharmaceutical manufacturer wants to buy automatic machinery designed for counting and then packaging 10,000 aspirin tablets in bulk containers. Since no production process is perfectly accurate, sometimes the device may package too many aspirin and sometimes not enough. What criteria might the pharmaceutical firm use to judge the quality of the automatic machinery it intends to buy? One criterion is immediately obvious. Since the machine is designed to package 10,000 aspirin, it seems reasonable to require that over a long period of time the average number of aspirin packaged should be 10,000. More formally, if X denotes the number of aspirin packaged in a container, then we would want the expected value of X to be 10,000. This criterion alone, however, does not adequately describe the quality of the machinery. To see why, suppose machine A packages aspirin according to the pattern

$$10{,}001, \quad 9{,}999, \quad 10{,}001, \quad 9{,}999, \quad \ldots,$$

while machine B packages aspirin according to the pattern

$$0, \quad 20{,}000, \quad 0, \quad 20{,}000, \quad \ldots.$$

In both cases, the long-term average number of aspirin packaged will be 10,000. However, machine A is obviously preferable to B since it is less variable in its performance. Thus, not only should the pharmaceutical firm look for a machine that packages a long-term average of 10,000 aspirin, but it should also seek a machine for which most of the output is close to the average. In terms of X, the machine should have a high probability of producing values of X close to the expected value. In this section we shall introduce a way of measuring the variability of a finite discrete random variable.

Let X and Y be the random variables whose probability functions are graphed in Figure 7.9. Since the origin is a center of symmetry for both probability functions, the two random variables have 0 as their expected values (why?); that is,

$$\mu_X = 0 \qquad \text{and} \qquad \mu_Y = 0.$$

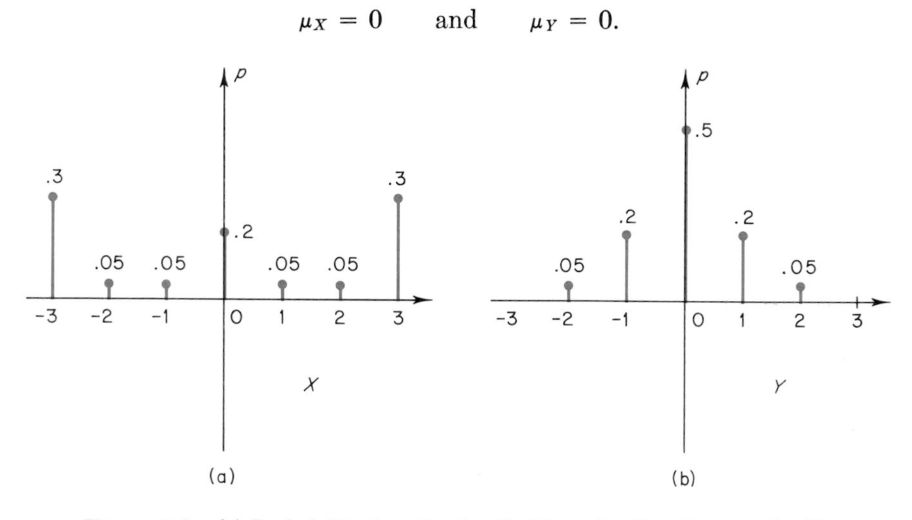

Figure 7.9 (a) Probability function for X; (b) probability function for Y.

Observe, however, that it is more likely for the random variable Y to take on values close to the mean than it is for the random variable X. For example, the random variable Y takes on one of the values -1, 0, or 1 with probability

$$.2 + .5 + .2 = .9$$

while the random variable X takes on one of the values -1, 0, or 1 with probability

$$.05 + .2 + .05 = .3$$

Thus, Y takes on values within 1 unit (on either side) of the mean 90% of the time, whereas X takes on values within 1 unit of the mean only 30% of the time. We can summarize this idea loosely by stating that X is more variable about its mean than is Y. Our next objective is to give a numerical way of describing the variability of a finite discrete random variable about its mean.

Let X be a finite discrete random variable with expected value μ. If, when the experiment is performed, X takes on the value x, then the difference

$$x - \mu,$$

called the ***deviation of x from μ***, tells us how far the observed value x is from μ. For example, a deviation of $+3$ means x is three units to the right of μ and a deviation of -3 means x is three units to the left of μ. For most purposes, it is unnecessary and bothersome to worry about the side of μ on which x falls. For this reason as well as others, statisticians prefer to work with the square of the deviation

$$(x - \mu)^2.$$

For example, a squared deviation of 4 tells us that x is 2 units from μ, but does not tell whether x is 2 units to the left of μ or 2 units to the right of μ.

Suppose now that X is a finite discrete random variable and assume that we perform the associated experiment repeatedly n times. We will observe n values of X and for each of these n values of X we can compute the squared deviation from the mean μ. If we now average these squared deviations, we will produce a single number that gives us some indication of how the data were spread out. The following example illustrates this idea.

Example 13 An experiment consists of tossing a fair die and observing the number X showing on the top face. Suppose, after four repetitions of this experiment, we obtain the values

$$4, \quad 3, \quad 2, \quad 3. \tag{1}$$

Since $\mu_X = 3.5$ (see Example 11), the squared deviations from the mean are

$$(4 - 3.5)^2, \qquad (3 - 3.5)^2, \qquad (2 - 3.5)^2, \qquad (3 - 3.5)^2$$

and the average of these squared deviations is

$$\begin{aligned}\frac{(4 - 3.5)^2 + (3 - 3.5)^2 + (2 - 3.5)^2 + (3 - 3.5)^2}{4}\\ &= \frac{0.25 + 0.25 + 2.25 + 0.25}{4}\\ &= .75.\end{aligned}$$

On the other hand, if we had obtained the values

$$1, \quad 2, \quad 6, \quad 5, \tag{2}$$

then the average of the squared deviations would be

$$\begin{aligned}\frac{(1 - 3.5)^2 + (2 - 3.5)^2 + (6 - 3.5)^2 + (5 - 3.5)^2}{4}\\ &= \frac{6.25 + 2.25 + 6.25 + 2.25}{4}\\ &= 4.25.\end{aligned}$$

Since data set (2) has a larger average squared deviation than data set (1), we describe data set (2) as more spread out or more variable about the mean than data set (1).

Suppose now that X is a finite discrete random variable associated with a certain experiment. We might ask if it is possible to predict what the average of the squared deviations will be after n repetitions of the experiment. Obviously, we cannot predict this value exactly since the values of X obtained in one group of n repetitions need not be the same as the values obtained in another group of n repetitions. Nevertheless, when n is large it is possible to make a reasonable guess about the average of the squared deviations. To see why, consider a finite discrete random variable whose only possible values are

$$x_1, \qquad x_2, \qquad \text{and} \qquad x_3.$$

Assume also that these values occur with known probabilities

$$p_1, \qquad p_2, \qquad \text{and} \qquad p_3.$$

Suppose that in n repetitions of the experiment,

$$x_1 \text{ occurs } f_1 \text{ times}$$

$$x_2 \text{ occurs } f_2 \text{ times}$$

$$x_3 \text{ occurs } f_3 \text{ times.}$$

Upon computing the squared deviations of these values we see that

$$(x_1 - \mu)^2 \text{ occurs } f_1 \text{ times}$$

$$(x_2 - \mu)^2 \text{ occurs } f_2 \text{ times}$$

$$(x_3 - \mu)^2 \text{ occurs } f_3 \text{ times.}$$

Thus, the average of the squared deviations will be

$$\text{average squared deviation} = \frac{f_1(x_1 - \mu)^2 + f_2(x_2 - \mu)^2 + f_3(x_3 - \mu)^2}{n}$$

which can be rewritten

$$\text{aver. squared deviation} = \frac{f_1}{n}(x_1 - \mu)^2 + \frac{f_2}{n}(x_2 - \mu)^2 + \frac{f_3}{n}(x_3 - \mu)^2. \quad (3)$$

Since

$$\frac{f_1}{n}, \qquad \frac{f_2}{n}, \qquad \text{and} \qquad \frac{f_3}{n}$$

represent the proportion of x_1's, x_2's, and x_3's in the n repetitions, it follows that when n is large, these proportions are likely to be close to the probabilities

$$p_1, \qquad p_2, \qquad \text{and} \qquad p_3;$$

that is, when n is large the approximations

$$\frac{f_1}{n} \simeq p_1, \qquad \frac{f_2}{n} \simeq p_2, \qquad \frac{f_3}{n} \simeq p_3$$

should be good. Thus from (3) we obtain the following approximation to the average squared deviation, good when n is large:

$$\text{average squared deviation} \simeq p_1(x_1 - \mu)^2 + p_2(x_2 - \mu)^2 + p_3(x_3 - \mu)^2.$$

More generally, if X takes on values

$$x_1, \qquad x_2, \qquad \ldots, \qquad x_k$$

with probabilities

$$p_1, \qquad p_2, \qquad \ldots, \qquad p_k,$$

then for large n, the average squared deviation will be approximately

$$\text{average squared deviation} \simeq p_1(x_1 - \mu)^2 + p_2(x_2 - \mu)^2 + \cdots + p_k(x_k - \mu)^2. \tag{4}$$

Example 14 An experiment consists of tossing a fair die and observing the number X showing on the top face. Use the approximation in (4) to estimate the average squared deviation of the observed values if the experiment is repeated many times.

Solution The possible values of X are

$$x_1 = 1, \qquad x_2 = 2, \qquad x_3 = 3, \qquad x_4 = 4, \qquad x_5 = 5, \qquad x_6 = 6$$

and since the die is fair, each of these values has probability $\frac{1}{6}$ of occurring; that is

$$p_1 = \frac{1}{6}, \qquad p_2 = \frac{1}{6}, \qquad p_3 = \frac{1}{6}, \qquad p_4 = \frac{1}{6}, \qquad p_5 = \frac{1}{6}, \qquad p_6 = \frac{1}{6}.$$

Moreover, from Example 11, we have

$$\mu = 3.5.$$

Substituting these values in (4) we obtain the approximation

$$\begin{aligned}\text{average squared deviation} &\simeq \frac{1}{6}(1 - 3.5)^2 + \frac{1}{6}(2 - 3.5)^2 + \frac{1}{6}(3 - 3.5)^2 \\ &\quad + \frac{1}{6}(4 - 3.5)^2 + \frac{1}{6}(5 - 3.5)^2 \\ &\quad + \frac{1}{6}(6 - 3.5)^2 \\ &= \frac{17.5}{6} \simeq 2.92.\end{aligned}$$

The quantity on the right-hand side of (4) is of such importance that it has its own notation and name.

> If the values
>
> $$x_1, \quad x_2, \quad \ldots, \quad x_k$$
>
> of a finite discrete random variable X occur with probabilities
>
> $$p_1, \quad p_2, \quad \ldots, \quad p_k$$
>
> then the ***variance of X***, denoted by $\mathrm{Var}(X)$, is defined by
>
> $$\mathrm{Var}(X) = p_1(x_1 - \mu)^2 + p_2(x_2 - \mu)^2 + \cdots + p_k(x_k - \mu)^2 \tag{5}$$
>
> where μ is the expected value of X.

In other words, *the variance of X is the sum of the squared deviations of the possible values of X times their probabilities of their occurrence.*

Example 15 If X is the random variable in the die-tossing experiment of Example 14, then the computations in that example show that

$$\mathrm{Var}(X) = \frac{17.5}{6} \simeq 2.92.$$

Since the variance formula involves the squares of the deviations, the units of $\mathrm{Var}(X)$ are the squares of the units of X. For example, if the values of X represent feet, then the units of $\mathrm{Var}(X)$ would be square feet. Many people prefer to have the variability of X about its mean described by a number having the same units as X. This can be accomplished by taking the positive square root of $\mathrm{Var}(X)$. The resulting number, called the ***standard deviation*** of X, is denoted either by σ_X or just σ when the random variable is evident. Symbolically we have

$$\sigma_X = \sqrt{\mathrm{Var}(X)}$$

and

$$\sigma_X{}^2 = \mathrm{Var}(X).$$

Example 16 The standard deviation of the random variable X in Example 15 is

$$\sigma_X = \sqrt{\mathrm{Var}(X)} = \sqrt{\frac{17.5}{6}} \simeq 1.71.$$

The computation of Var(X) can be simplified by rewriting Equation (5) in an alternative form. To see how, consider a random variable X that takes on values

$$x_1, \quad x_2, \quad x_3$$

with probabilities

$$p_1, \quad p_2, \quad p_3.$$

Equation (5) states

$$\text{Var}(X) = p_1(x_1 - \mu)^2 + p_2(x_2 - \mu)^2 + p_3(x_3 - \mu)^2. \tag{6}$$

If we expand the squares, the terms on the right-hand side of this equation become

$$p_1(x_1 - \mu)^2 = p_1x_1^2 - 2p_1x_1\mu + p_1\mu^2$$

$$p_2(x_2 - \mu)^2 = p_2x_2^2 - 2p_2x_2\mu + p_2\mu^2$$

$$p_3(x_3 - \mu)^2 = p_3x_3^2 - 2p_3x_3\mu + p_3\mu^2.$$

Summing up both sides of these equations and using (6) we obtain

$$\begin{aligned}\text{Var}(X) = {} & (p_1x_1^2 + p_2x_2^2 + p_3x_3^2) - 2\mu(p_1x_1 + p_2x_2 + p_3x_3) \\ & + \mu^2(p_1 + p_2 + p_3)\end{aligned} \tag{7}$$

But

$$p_1x_1 + p_2x_2 + p_3x_3 = \mu$$

and

$$p_1 + p_2 + p_3 = 1 \qquad \text{(why?)}$$

Substituting these relations in (7) yields

$$\text{Var}(X) = (p_1x_1^2 + p_2x_2^2 + p_3x_3^2) - 2\mu^2 + \mu^2$$

or

$$\text{Var}(X) = (p_1x_1^2 + p_2x_2^2 + p_3x_3^2) - \mu^2.$$

More generally, if X takes on values

$$x_1, \quad x_2, \quad \ldots, \quad x_k$$

with probabilities

$$p_1, \quad p_2, \quad \ldots, \quad p_k,$$

then the formula for variance can be written in the alternative form

$$\text{Var}(X) = (p_1x_1^2 + p_2x_2^2 + \cdots + p_kx_k^2) - \mu^2 \tag{8}$$

Observe that this formula involves only one subtraction, while formula (5) involves k subtractions. This is one reason why formula (8) is preferable for numerical computations.

Example 17 Use formula (8) to compute the variance of the random variable X in the die-tossing experiment of Example 14.

Solution In solving this problem, we shall arrange our computations in a format the reader may wish to follow whenever a variance must be computed without the help of a computer. See Table 7.8. To construct this table, we

Table 7.8

x_i	p_i	$p_i x_i$	$p_i x_i^2$
1	$\frac{1}{6}$	$\frac{1}{6}$	$\frac{1}{6}$
2	$\frac{1}{6}$	$\frac{2}{6}$	$\frac{4}{6}$
3	$\frac{1}{6}$	$\frac{3}{6}$	$\frac{9}{6}$
4	$\frac{1}{6}$	$\frac{4}{6}$	$\frac{16}{6}$
5	$\frac{1}{6}$	$\frac{5}{6}$	$\frac{25}{6}$
6	$\frac{1}{6}$	$\frac{6}{6}$	$\frac{36}{6}$
		Sum $= \frac{21}{6}$	Sum $= \frac{91}{6}$

listed the values of X in the first column and the probabilities of these values in the second column. We multiplied each x_i in column 1 by the corresponding p_i in column 2 to obtain the third column. Finally, we multiplied each $p_i x_i$ in column 3 by the corresponding x_i from column 1 to obtain $p_i x_i^2$ in column 4. Observe that the sum of the entries in column 3 is

$$\mu = \frac{21}{6}.$$

(See formula (8) in the previous section. Since the sum of the entries in column 4 is the first term on the right-hand side of formula (8) of this section, we obtain

$$\begin{aligned}\text{Var}(X) &= \frac{91}{6} - \left(\frac{21}{6}\right)^2 \\ &= \frac{105}{36} \\ &\simeq 2.92\end{aligned}$$

which agrees with the result obtained in Example 15.)

Example 18 Let X be the random variable in the resistor sampling problem of Example 6. Find the variance and standard deviation of X.

Solution As shown in Example 6, X takes on values

$$0, \quad 1, \quad 2$$

with probabilities

$$.9025, \quad .0950, \quad .0025,$$

so that its variance can be computed using Table 7.9 and formula (8).

Table 7.9

x_i	p_i	$p_i x_i$	$p_i x_i^2$
0	.9025	0.0000	0.0000
1	.0950	0.0950	0.0950
2	.0025	0.0050	0.0100
		Sum = 0.1000	Sum = 0.1050

The variance is

$$\begin{aligned} \sigma_X^2 = \text{Var}(X) &= (0.1050) - (0.1000)^2 \\ &= 0.1050 - 0.0100 \\ &= 0.0950 \end{aligned}$$

and the standard deviation is

$$\sigma_X = \sqrt{0.0950} \simeq 0.3082.$$

EXERCISE SET 7.3

1. Let X be the random variable whose probability function is described in the following table. Use formulas (5) and (8) to compute the variance of X in two different ways

Value of X	1	2	3	4
Probability that the value occurs	.2	.4	.3	.1

In Exercises 2–4 compute the variance and standard deviation of the random variable whose probability function is described in the given table. (A table of square roots appears in the Appendix.)

2.

Value of X	-2	-3	2	3
Probability that the value occurs	.3	.2	.3	.2

3.

Value of X	0.01	0.25	0.20	0.9	0.4
Probability that the value occurs	.4	.2	.2	.1	.1

4.

Value of X	4000	2500	3000	9100	2600
Probability that the value occurs	.8	.01	.01	.02	.16

5. Find the variance of the random variable X in Exercise 2 of Section 7.1.

6. Find the variance of the random variable X in Exercise 3 of Section 7.1.

7. Find the variance of the random variable X in Exercise 4 of Section 7.1.

8. Compute the variance of the random variable whose probability function is graphed in Figure 7.10.

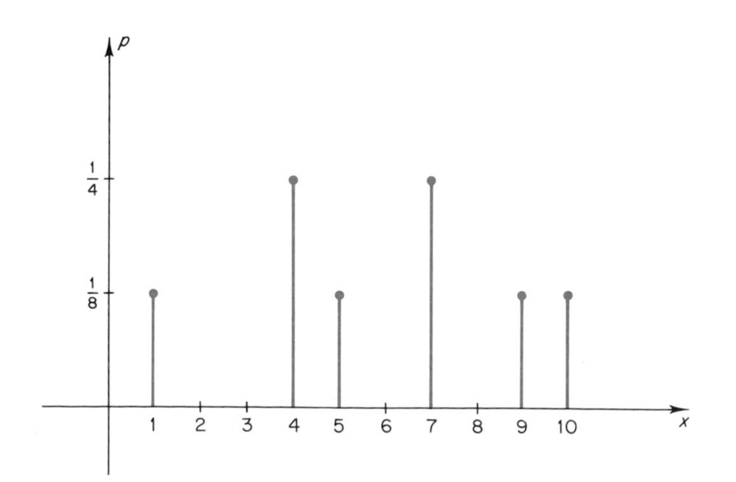

Figure 7.10

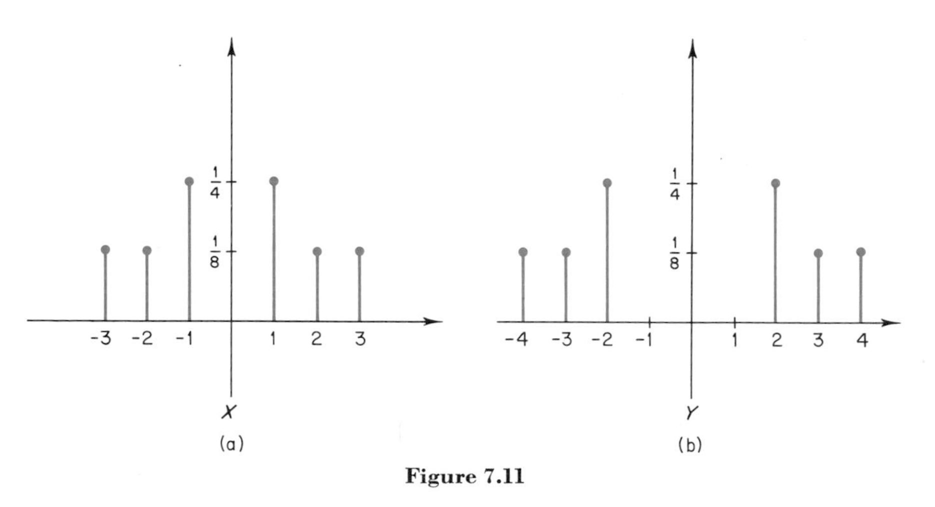

Figure 7.11

9. In Figure 7.11 we have graphed probability functions for random variables X and Y. By *inspection* (no computations), determine which variable has the larger variance.

10. In Figure 7.12 we have graphed probability functions for random variables X and Y. By *inspection* (no computations), determine which variable has the larger variance.

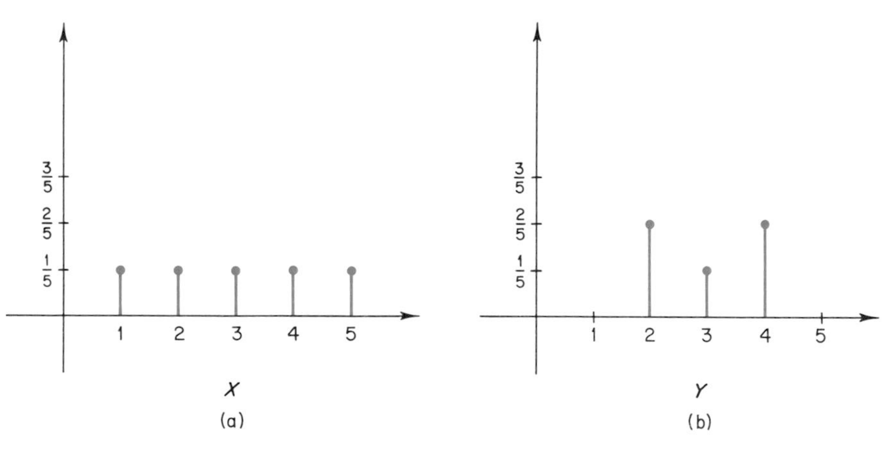

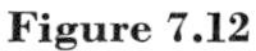

Figure 7.12

7.4 CHEBYSHEV'S INEQUALITY; APPLICATIONS OF MEAN AND VARIANCE*

In the previous section we discussed criteria that a pharmaceutical firm might consider when planning to purchase automatic machinery for counting and packaging 10,000 aspirin tablets in bulk containers. If X denotes the number of tablets packaged in a container, we observed that the firm would want

(1) the mean of X to be 10,000,

and

(2) the probability of obtaining values of X close to the mean to be high.

In this section we shall develop some mathematical techniques for investigating the second of these criteria.

Every time the automatic machinery produces a value of X that is greater than the mean of 10,000 the pharmaceutical firm loses money since it is giving the customer more aspirin than he is paying for. By the same token, every time the machinery produces a value of X that is less than 10,000, the firm risks losing the customer since it has not given the customer all the aspirin he is expecting. Thus a value of X too far from the mean, on either side, is undesirable. Recognizing that no machinery is perfect, the pharmaceutical firm may settle on a certain error that it feels is tolerable. For example, the firm's marketing staff might decide that an error of 10 aspirins in either direction would not be too detrimental; that is, overpackaging by 10 aspirin would not be too costly and underpackaging by 10 aspirin would still be good enough to satisfy the customer. Thus, the pharmaceutical firm will be content if the machinery it buys always produces values of X that lie within 10 units of the mean; that is values of X satisfying

$$9990 \leq X \leq 10{,}010. \tag{1}$$

Unfortunately, even this relaxed requirement may be too stringent. Occasionally, for example, aspirin may chip and disturb the counting mechanism or still other unforeseen problems may occur. Recognizing this fact of life, the firm, after studying the problem, may relax its requirements still further and require only that (1) hold most of the time, say at least 95% of the time. This requirement can be formulated mathematically as follows. If X denotes the number of aspirin packed in a container and if

$$P(9990 \leq X \leq 10{,}010)$$

denotes the probability that X is between 9990 and 10,010, inclusive, then this probability should be at least .95, that is,

$$P(9990 \leq X \leq 10{,}010) \geq .95. \tag{2}$$

We shall show next that once the mean μ and variance σ^2 of the random variable X are known, it is possible to devise a mathematical procedure for determining if a criterion like (2) is satisfied. But first we shall need to develop some preliminary ideas.

Let X be a random variable with mean μ and standard deviation σ. Statisticians often find it convenient to measure distance from the mean in terms of ***σ-units***. For example, the point A in Figure 7.13 can be described as 2 σ-units to the right of the mean μ and the point B as 3.5 σ-units to the left of the mean. In other words, to say that a point is h σ-units from μ means that its distance from μ is

$$h\sigma.$$

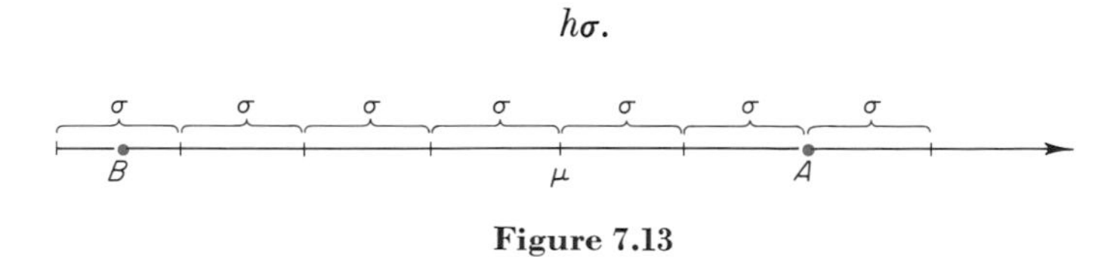

Figure 7.13

Example 19 Let X be a random variable with mean $\mu = 3$ and variance $\sigma^2 = 25$. In terms of σ-units, how far to the right of μ is the point $x = 9$?

Solution The distance between x and μ is

$$x - \mu = 9 - 3 = 6.$$

Since $\sigma = 5$, this distance in terms of σ-units is

$$\frac{x - \mu}{\sigma} = \frac{6}{5} = 1.2 \quad \sigma\text{-units.}$$

The following result, named in honor of P. L. Chebyshev,† shows that once the mean and variance of a random variable X are known, it is possible to estimate the chances that a value of X will fall within a specified distance of the mean.

Chebyshev's Theorem The probability that a random variable will take on a value within h σ-units of the mean is at least $1 - \dfrac{1}{h^2}$.

Symbolically, Chebyshev's theorem states that for any random variable X, we have

$$P(\mu - h\sigma \leq X \leq \mu + h\sigma) \geq 1 - \frac{1}{h^2}.$$

We shall omit the proof of this important result and concentrate, instead, on some of its applications.

† *Pafnuti Liwowich Chebyshev* (1821–1894)—Chebyshev was one of the most distinguished Russian mathematicians. He served most of his life as a professor of mathematics at the University of St. Petersburg. In addition to his work in probability and statistics, he made important contributions to the field of number theory, he solved basic problems in the construction of geographical maps, and he devised a way for mechanically linking bars to obtain rectilinear motion.

From Chebyshev's theorem, the probability that a random variable takes on a value within 2 σ-units of the mean (that is, $h = 2$) is at least

$$1 - \frac{1}{h^2} = 1 - \frac{1}{4} = .75.$$

This result has the following geometric interpretation. Let X be a random variable with mean μ and standard deviation σ. Suppose, as shown in Figure 7.14, we construct an interval extending a distance 2σ on each side of μ.

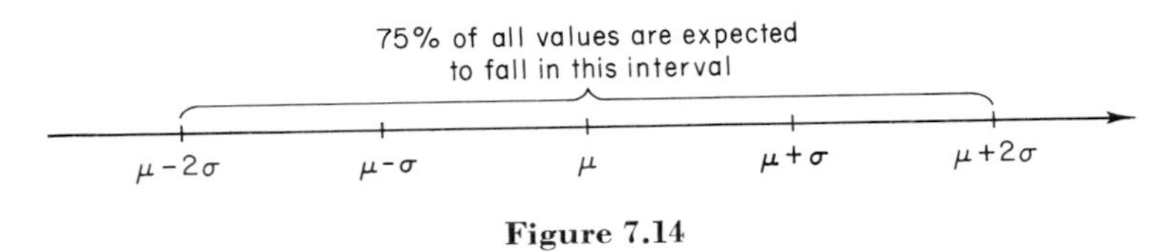

Figure 7.14

If we observe many values of X, then we expect at least 75% of these values to fall in the constructed interval and consequently no more than 25% of the values to fall outside the interval. For example, if X is a random variable with mean 10 and standard deviation .1, then the probability is at least .75 that X takes on a value between 9.8 and 10.2. Thus, if we observe a large number of X values, 75% of these values can be expected to fall between 9.8 and 10.2.

Example 21 Let X be a random variable with mean $\mu = 100$ and variance $\sigma^2 = 4$. If we are going to observe many X values and if we want to state that at least 96% of the observed values are likely to fall between $100 - k$ and $100 + k$, what value should we take for k? Use Chebyshev's theorem to obtain the answer.

Solution To ask that at least 96% of the observed values fall between $100 - k$ and $100 + k$ is equivalent to asking that each observed value fall between $100 - k$ and $100 + k$ with probability at least .96.

If we choose h in Chebyshev's theorem so that

$$1 - \frac{1}{h^2} = .96 \tag{3}$$

then, since $\mu = 100$, X will fall between $100 - h\sigma$ and $100 + h\sigma$ with probability at least .96. To solve (3) for h^2 we multiply both sides by h^2 to obtain

$$h^2 - 1 = .96h^2$$

or

$$0.04h^2 = 1$$

or

$$h^2 = 25.$$

Taking positive square roots gives $h = 5$. Thus, the probability is at least .96 that X falls between

$$100 - 5\sigma \quad \text{and} \quad 100 + 5\sigma.$$

Since $\sigma = 2$, we obtain

$$k = 5\sigma = 10.$$

Observe that Chebyshev's theorem applies to *all* random variables. Since the theorem has such broad applicability, it is natural to expect that for some random variables the estimate $1 - (1/h^2)$ may be rather poor. It is, in fact, the case that for many random variables the probability is much greater than $1 - (1/h^2)$ of obtaining a value within h σ-units of the mean. For this reason, statisticians usually regard the estimate $1 - (1/h^2)$ as very conservative.

Example 22 (Light Bulb Life) A certain high-quality commercial light bulb has an expected lifetime of $\mu = 750$ hours with a variance $\sigma^2 = 100$.

(a) Estimate the probability that one of these light bulbs will live between 735 and 765 hours.

(b) If 20,000 of these bulbs are installed in city street lights, how many bulbs do you estimate would need replacement between 735 and 765 hours of use?

Solution (a) As shown in Figure 7.15, the time interval from 735 hours to 765 hours extends 15 units to the left of the mean $\mu = 750$ and 15 units to the right of the mean. Thus, we are interested in finding the probability that the lifetime is within 15 units of the mean. To apply Chebyshev's theorem, we must express 15 in terms of σ-units. Since $\sigma^2 = 100$, we have $\sigma = \sqrt{100} = 10$ so that 15 units is

$$h = \frac{15}{\sigma} = \frac{15}{10} = \frac{3}{2} \quad \sigma\text{-units.}$$

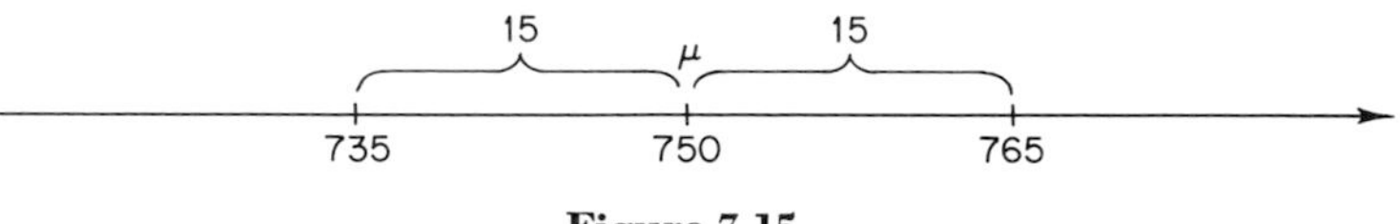

Figure 7.15

By Chebyshev's theorem the probability that the lifetime is within $h = \frac{3}{2}$ σ-units of the mean is at least

$$1 - \frac{1}{h^2} = 1 - \frac{1}{(\frac{3}{2})^2} = \frac{5}{9}.$$

Solution (b) From (a) the probability is at least $\frac{5}{9}$ that one of the bulbs will burn out between 735 and 765 hours, so that we would expect at least

$$\left(\frac{5}{9}\right)(20{,}000) \simeq 11{,}111 \quad \text{bulbs}$$

to need replacement during the given time interval.

Example 23 (Reliability) A manufacturer of automatic machinery for counting and then packaging aspirin tablets in bulk containers claims that his machine will package a mean of 10,000 tablets per container with a variance of $\sigma^2 = 25$. If a pharmaceutical firm wants to purchase a machine that will package between 9990 tablets and 10,010 tablets per container with probability at least .95, can the firm be certain that this machine will meet its needs? Assume that the manufacturer's claims are correct and base your conclusion on Chebyshev's inequality.

Solution The interval from 9990 tablets to 10,010 tablets extends 10 units to the left of the mean $\mu = 10{,}000$ and 10 units to the right of the mean. To apply Chebyshev's inequality, we must express 10 units in terms of σ-units. Since $\sigma^2 = 25$, we have

$$\sigma = \sqrt{25} = 5,$$

so that 10 units is

$$h = \frac{10}{5} = 2 \quad \sigma\text{-units.}$$

By Chebyshev's theorem, the probability that the number of tablets packaged is within $h = 2$ σ-units of the mean is at least

$$1 - \frac{1}{h^2} = 1 - \frac{1}{2^2} = .75.$$

Since the pharmaceutical firm wants a probability of at least .95, the firm cannot be certain that this machine will meet its needs.

Example 24 (Reliability) Consider the automatic machinery in the previous example. For what value of k can the manufacturer claim that his machine will package between

$$10{,}000 - k \qquad \text{and} \qquad 10{,}000 + k$$

tablets with probability at least .95?

Solution We shall choose h in Chebyshev's theorem so that

$$1 - \frac{1}{h^2} = .95. \tag{4}$$

Then since $\mu = 10{,}000$, the probability will be at least .95 that the number of tablets packaged is between $10{,}000 - h\sigma$ and $10{,}000 + h\sigma$.

Solving (4) for h, as in Example 21, we obtain

$$h^2 - 1 = .95h^2$$

or

$$0.05h^2 = 1$$

or

$$h^2 = 20.$$

Taking positive square roots we obtain

$$h = \sqrt{20} \simeq 4.47.$$

Thus the probability is at least .95 that the number of tablets falls between

$$10{,}000 - 4.47\sigma \qquad \text{and} \qquad 10{,}000 + 4.47\sigma.$$

Since $\sigma = 5$, we obtain

$$k \simeq 4.47\sigma = 4.47(5) = 22.35.$$

Since only integer numbers are sensible in this problem, we round k upward to the next whole integer and take

$$k \simeq 23.$$

Thus with probability at least .95 the machine will package between 9977 tablets and 10,023 tablets. (Can you see why it would be incorrect to round off 22.35 to 22 even though it is closer to 22 than 23?)

EXERCISE SET 7.4

1. Let X be a random variable with mean $\mu_X = 0$ and variance $\sigma_X{}^2 = 16$. In each part, express the distance between a and μ_X in terms of σ-units.

   (a) $a = 4$ (b) $a = 8$ (c) $a = 6$ (d) $a = -11$.

2. Let X be a random variable with mean $\mu_X = 5$ and variance $\sigma_X{}^2 = 4$. In each part, express the distance between a and μ_X in terms of σ-units.

   (a) $a = 7$ (b) $a = 11$ (c) $a = 2$ (d) $a = 0$.

3. Let X be a random variable with mean $\mu_X = 10$ and variance $\sigma_X{}^2 = 9$. In each part, the distance between a point a and the mean μ_X is expressed in terms of σ-units. Find the value of a.

   (a) a is 2 σ-units to the right of μ_X.

   (b) a is 3.5 σ-units to the left of μ_X.

   (c) a is 4.7 σ-units to the right of μ_X.

   (d) a is 0.3 σ-units to the left of μ_X.

4. Let X be a random variable with mean $\mu_X = 0$ and variance $\sigma_X{}^2 = 4$. Use Chebyshev's theorem to estimate the probability that X takes on a value

within

(a) 1 unit of the mean

(b) 2 units of the mean

(c) 3 units of the mean.

5. Let X be a random variable with mean $\mu_X = 12$ and variance $\sigma_X{}^2 = 9$. Use Chebyshev's theorem to estimate the probability that

(a) X takes on a value between 9 and 15

(b) X takes on a value between 11 and 13

(c) X takes on a value between 0 and 24.

6. Let X be a random variable with mean $\mu_X = 10$ and variance $\sigma_X{}^2 = 4$. If we are going to observe many X values and if we want to state that at least 90% of the observed values are likely to fall between $10 - k$ and $10 + k$, what value should we take for k? Use Chebyshev's theorem to obtain the answer.

7. A camera used in a weather satellite has an expected lifetime of $\mu = 900$ days with a variance of $\sigma^2 = 81$. Use Chebyshev's theorem to estimate the probability that this camera will live between 870 and 930 days.

8. Heavy-duty resistors have an expected lifetime of 5 years with a variance of 0.16.

(a) If 3000 such resistors are installed, how many do you estimate will live between 3 and 7 years?

(b) If 3000 such resistors are installed, how many do you estimate will live between 2 and 8 years?

9. When a raw 10-carat diamond is cut and polished, the finished diamonds have an expected weight of 6.3 carats with a variance of .04. Estimate the probability that a raw 10-carat diamond will yield finished diamonds with a weight between 5.5 and 7.1 carats.

10. A manufacturer of an automatic device for filling bottles with 16 ounces of soda claims that his machine will fill bottles to a mean of 16 ounces of soda with a variance of 0.04. If a bottling plant wants a device that will fill bottles to between 15.5 and 16.5 ounces at least 70% of the time, will this device meet its needs? Assume the manufacturer's claims are correct and base your conclusion on Chebyshev's inequality.

11. Consider the automatic device described in Exercise 10. For what value of k can the manufacturer claim that his machine will fill bottles to between $16 - k$ and $16 + k$ ounces at least 90% of the time?

12. Let X be a random variable whose probability function is described by

Value of X	1	2	3	4	5
Probability that the value occurs	$\frac{1}{5}$	$\frac{1}{5}$	$\frac{1}{5}$	$\frac{1}{5}$	$\frac{1}{5}$

(a) Find μ_X.

(b) Find σ_X.

(c) Use Chebyshev's inequality to estimate the probability $P(2 \leq X \leq 4)$.

(d) Find the exact value of $P(2 \leq X \leq 4)$.

(e) Are the results obtained in (c) and (d) contradictory?

7.5 BINOMIAL RANDOM VARIABLES

In the previous section we saw how Chebyshev's theorem could be used to *estimate* certain probabilities when the mean and variance of a random variable were known. In cases where it is possible to find the probability function for the random variable, probabilities can, in theory, be computed exactly so that approximations often become unnecessary. In this section we shall discuss an important class of problems where probability functions are readily obtainable.

In many experiments we are primarily interested in whether a certain result does or does not occur. For example, does a tossed coin show a head or not, does an individual innoculated with a flu vaccine contract flu or not, is an item selected from the output of a production process defective or not, is a newborn child a girl or not, and so on.

Experiments for which there are only two possible outcomes are called ***Bernoulli*† *experiments*** or sometimes ***Bernoulli trials.*** It is traditional in statistical work to arbitrarily label one of the two outcomes *success* and the other *failure*. For example, if we toss a coin we might call a head a success and a tail a failure. The selection of the outcome to be labeled success is completely arbitrary and need not connote a successful outcome in everyday life. For example, an investigator studying the effect of a new drug for the prevention of malignant tumors in rats, may possibly choose to call the appearance of a tumor a "success" even though one would not ordinarily think of this outcome as successful.

Experiments with more than two outcomes can, if desired, be viewed as Bernoulli experiments by grouping the outcomes into two categories. For example, if we toss a die, there are six possible outcomes

$$1, \quad 2, \quad 3, \quad 4, \quad 5, \quad 6.$$

However, if we call the outcome 3 a success and call the outcomes 1, 2, 4, 5, and 6 failures, then we can view this as a Bernoulli experiment with only two outcomes, success and failure.

† *Jacob Bernoulli* (1654–1705)—Bernoulli was one of a family of distinguished Swiss mathematicians. He was born in Basel, Switzerland and studied theology at the insistence of his father. Later, he refused a church appointment and began lecturing on experimental physics at the University of Basel. His work included studies in astronomy, the motion of comets, and applications of (the newly discovered) calculus.

In a Bernoulli experiment, we shall use the letter p for the probability of a success and the letter q for the probability of a failure. Since success and failure are the only two outcomes, we must have

$$p + q = 1$$

or equivalently

$$q = 1 - p.$$

Example 25 Toss a fair die and let

$$\text{success} = 3$$

$$\text{failure} = 1, 2, 4, 5, \text{ or } 6.$$

Then

$$p = \frac{1}{6} \quad \text{and} \quad q = \frac{5}{6}.$$

Many experiments consist of a sequence of Bernoulli trials. For example, a single toss of a coin is a Bernoulli trial and an experiment consisting of five tosses of a coin is a sequence of five Bernoulli trials. We shall be concerned with the following problem:

Problem If a Bernoulli experiment is repeated n times, what is the probability of obtaining exactly x successes in the n repetitions?

This problem is most easily solved when we assume that the trials are *independent*. Physically, this means we are assuming that the occurrence of a success or failure in one trial has no effect on the chances of success in any other trial. Mathematically, independence allows us to multiply probabilities. For example, if we have three independent repetitions of a Bernoulli experiment, where the probability of success on each trial is p, the probability of the outcome

$$ssf,$$

denoted $P(ssf)$, would be

$$P(ssf) = P(s)P(s)P(f) = ppq = p^2q$$

and the probability of the outcome fff would be

$$P(fff) = P(f)P(f)P(f) = qqq = q^3.$$

The following examples further illustrate this idea.

Example 26 A fair die is tossed three times. What is the probability of obtaining exactly two 4's in the three tosses?

Solution On each toss of the die, let success indicate that a 4 is tossed and failure that a 4 is not tossed. Thus

$$p = \text{probability of success} = \frac{1}{6}$$

$$q = \text{probability of failure} = \frac{5}{6}.$$

We are interested in the probability of obtaining exactly two successes in the three trials.

In the experiment of tossing a die three times, the event of obtaining exactly two successes contains three sample points. They are

$$ssf \qquad sfs \qquad fss.$$

Thus,

$$P(\text{exactly two successes}) = P(ssf) + P(sfs) + P(fss). \tag{1}$$

If we assume that the tosses are independent, then the terms in (1) can be computed as follows:

$$P(ssf) = P(s)P(s)P(f) = \frac{1}{6} \cdot \frac{1}{6} \cdot \frac{5}{6} = \frac{5}{216}$$

$$P(sfs) = P(s)P(f)P(s) = \frac{1}{6} \cdot \frac{5}{6} \cdot \frac{1}{6} = \frac{5}{216}$$

$$P(fss) = P(f)P(s)P(s) = \frac{5}{6} \cdot \frac{1}{6} \cdot \frac{1}{6} = \frac{5}{216}.$$

Substituting these values in (1), we find that the probability of exactly two successes is

$$\frac{15}{216} = \frac{5}{72};$$

this is the probability of obtaining exactly two 4's in three tosses.

Example 27 In the die-tossing experiment of the previous example, find the probability of obtaining: no 4's, exactly one 4, and three 4's.

Solution

$$P(\text{no 4's}) = P(0 \text{ successes}) = P(fff)$$

$$= P(f)P(f)P(f) = \frac{5}{6}\frac{5}{6}\frac{5}{6} = \frac{125}{216}.$$

$$\begin{aligned} P(\text{exactly one 4}) &= P(\text{exactly one success}) \\ &= P(sff) + P(fsf) + P(ffs) \\ &= P(s)P(f)P(f) + P(f)P(s)P(f) + P(f)P(f)P(s) \\ &= \frac{1}{6}\frac{5}{6}\frac{5}{6} + \frac{5}{6}\frac{1}{6}\frac{5}{6} + \frac{5}{6}\frac{5}{6}\frac{1}{6} \\ &= \frac{25}{72}. \end{aligned}$$

$$\begin{aligned} P(\text{three 4's}) &= P(\text{three successes}) \\ &= P(sss) = P(s)P(s)P(s) \\ &= \frac{1}{216}. \end{aligned}$$

We now generalize the ideas in the previous two examples.

Example 28 In a certain Bernoulli experiment, the probability of success is p and the probability of failure is $q = 1 - p$. What is the probability of exactly x successes in three independent repetitions of this experiment, when $x = 0, 1, 2, 3$?

Solution The situation in this problem is similar to that in the previous two examples with the exception that the value of p is unspecified. Following the same procedures as before, we obtain

$$\begin{aligned} P(x = 0) &= P(fff) = qqq = q^3 \\ P(x = 1) &= P(sff) + P(fsf) + P(ffs) \\ &= pqq + qpq + qqp \\ &= 3pq^2. \\ P(x = 2) &= P(ssf) + P(sfs) + P(fss) \\ &= ppq + pqp + qpp \\ &= 3p^2q. \\ P(x = 3) &= P(sss) = ppp = p^3. \end{aligned}$$

The probabilities in this example were easy to obtain since the number of repetitions was small. The next sample shows how to handle cases where the number of repetitions is larger.

Example 29 Assuming the probability of success in a single trial of a Bernoulli experiment is p, find the probability of exactly four successes in 10 independent repetitions of this experiment.

Solution One way of obtaining exactly four successes is

$$ssssffffff.$$

Another way is

$$ssffsfsfff$$

and still another way is

$$fffssfffss.$$

Observe that each of these results is an arrangement of 4 s's and 6 f's. Obviously, the number of different ways of obtaining four successes is just the number of different arrangements of 4 s's and 6 f's. It is evident that any such arrangement is completely determined once we specify the four positions where the s's occur. Thus the number of arrangements with 4 s's and 6 f's is just the number of ways of selecting four positions for the s's from 10 possibilities. Therefore, the number of such arrangements is

$$C_{10,4} = \frac{10!}{4!6!} = 210.$$

Consequently

$$P(\text{exactly four successes}) = \underbrace{P(ssssffffff) + P(ssffsfsfff) + \cdots}_{210 \text{ terms}}$$

Since each probability on the right side has value p^4q^6 (why?), we obtain

$$P(\text{exactly four successes}) = 210p^4q^6.$$

We are now in a position to solve the following general problem:

Problem If the probability of success in a certain Bernoulli experiment is p, what is the probability of obtaining exactly x successes in n independent repetitions of the experiment?

Solution One way of obtaining exactly x successes in the n repetitions is to have the x successes immediately, followed by $n - x$ failures; that is

$$\underbrace{sss \cdots s}_{x \text{ successes}} \quad \underbrace{fff \cdots f}_{n - x \text{ failures}} \tag{2}$$

The other ways of obtaining x successes can be obtained by rearranging these x s's and $n - x$ f's. As illustrated in Example 29 there are

$$C_{n,x}$$

different arrangements of these letters. Since each of these arrangements has x s's and $n - x$ f's, the probability of each arrangement is

$$p^x q^{n-x};$$

for example,

$$P(\underbrace{sss \cdots s}_{x \text{ successes}} \quad \underbrace{fff \cdots f}_{n-x \text{ failures}})$$

$$= \underbrace{P(s)P(s)P(s) \cdots P(s)}_{x \text{ factors}} \quad \underbrace{P(f)P(f)P(f) \cdots P(f)}_{n-x \text{ factors}}$$

$$= \underbrace{ppp \cdots p}_{x \text{ factors}} \quad \underbrace{qq \cdots q}_{n-x \text{ factors}}$$

$$= p^x q^{n-x}.$$

Thus:

The probability of exactly x successes in n independent repetitions of a Bernoulli experiment with success probability p is given by

$$P(\text{exactly } x \text{ successes}) = C_{n,x} p^x q^{n-x}. \tag{3}$$

Example 30 Find the probability of obtaining exactly four successes in 10 independent repetitions of a Bernoulli experiment if the probability of success on each trial is .6.

Solution Applying formula (3) with

$$n = 10, \qquad p = .6, \qquad q = .4, \qquad x = 4,$$

we obtain

$$P(\text{exactly four successes}) = C_{10,4}(.6)^4(.4)^6$$

$$= \frac{10!}{4!6!}(.6)^4(.4)^6$$

$$\simeq .111.$$

An experiment that consists of n independent repetitions of a Bernoulli experiment is called a ***binomial experiment.*** If the random variable X denotes the number of successes in a binomial experiment, then X is called a ***binomial random variable.***

As evidenced by Example 30, it can be a tedious job to compute the probabilities of a binomial random variable. In applications where many binomial probabilities must be computed, digital computers are usually used. However, in problems where only a few binomial probabilities are needed, the cost of using a computer is usually unwarranted so that statisticians often resort to ready-made tables.

In Table II in the Appendix we give the binomial probabilities, accurate to three decimal places, for certain values of n and p. These tables are somewhat incomplete so that in practical problems the reader may have to consult more extensive tables.† To illustrate how these tables are used, we shall use them to solve the problem in Example 30.

Example 31 Use Table II at the back of the text to find the probability of obtaining four successes in 10 independent repetitions of a Bernoulli experiment if the probability of success on each trial is .6.

Solution The answer can be found as follows:

(a) Find the table for $n = 10$.

(b) Locate the column labeled $p = .60$.

(c) Go down the column to the block in the row labeled $x = 4$. We obtain the number .111 which is the probability we want.

Example 32 Find the probability of obtaining *at least* 9 successes in 12 independent repetitions of a Bernoulli experiment, where the probability of success on each trial is .8.

Solution If X denotes the number of successes in the 12 repetitions, then we are interested in the probability that $X \geq 9$ or equivalently the probability that X has one of the values 9, 10, 11, or 12. From Table II we obtain

$$\begin{aligned} P(X \geq 9) &= P(X = 9) + P(X = 10) + P(X = 11) + P(X = 12) \\ &= \quad .236 \quad + \quad .283 \quad + \quad .206 \quad + \quad .069 \\ &= .794. \end{aligned}$$

We conclude this section with a number of examples that illustrate how binomial random variables may arise in practical problems. In each of these examples, we assume that the trials are independent.

Propagation of Disease

Example 33 Medical evidence shows that one out of every five cows coming in contact with alpha-virus will contract a disease called bovina syndrome.

† National Bureau of Standards, *Tables of the Binomial Probability Distribution*, Applied Mathematics Series 6 (1950). ($n = 1$ to $n = 49$.) Harry G. Romig, *Binomial Tables* (New York: Wiley, 1953). ($n = 50$ to $n = 100$.)

If a small herd of 20 cows is exposed to alpha-virus, then the number X of cows contracting bovina syndrome is a binomial random variable.

To see this, imagine that each cow in turn is subjected to the virus and let success denote that the cow contracts bovina syndrome. Thus, X represents the number of successes in 20 repetitions of a Bernoulli experiment where the probability of success on each trial is $\frac{1}{5}$.

Radioactive Decay

Example 34 Physical evidence shows that the probability a given type of atom will split in a 24-hour period is $1/10^{24}$. In a sample containing 10^{30} of these atoms, the number X of atoms that will split in a 24-hour period is a binomial random variable.

To see this, imagine that each atom in turn is observed for a 24-hour period and let success indicate that the atom splits. Thus, X represents the number of successes in 10^{30} repetitions of a Bernoulli experiment, where the probability of success on each trial is $1/10^{24}$.

Genetic Mutations

Example 35 A generation of 2500 fruit fly larvae is subjected to radiation to induce mutations. Genetic evidence suggests that the probability of a mutation under the given conditions is .001. The number X of mutations is a binomial random variable.

To see this, imagine that each larva in turn is observed for evidence of mutation and let success denote that a mutation occurs. Thus, X is the number of successes in 2500 repetitions of a Bernoulli experiment, where the probability of success on each trial is .001.

Divorce Proceedings

Example 36 In a certain state "mental cruelty" is the legal reason given in four out of five divorce proceedings. If 25 divorce cases are selected at random, then the number X of cases based on mental cruelty is a binomial random variable.

To see this, imagine that each case in turn is observed and let success indicate that the legal reason for the divorce is mental cruelty. Thus, X is the number of successes in 25 repetitions of a Bernoulli experiment, where the probability of success on each trial is $\frac{4}{5} = .8$.

These examples, diverse as they are, all follow the same basic pattern. In each situation we can interpret the problem as a repetition of a simple Bernoulli experiment. It is important to keep in mind, however, that in order to have a binomial experiment the trials must be *independent*. In each of the examples above, we assumed we had independence. In many physical problems, however, the assumption of independence is unwarranted. For example, suppose medical records show that 40% of all U.S. citizens will contract German measles by age 25. What is the probability that in a group of six people over 25, exactly two have contracted German measles at some time before their twenty-fifth birthday? If we let success indicate that an individual has contracted German measles and let X be the number of successes in the group of six, then we might argue that we have six repetitions of a Bernoulli experiment where the probability of success on each trial is .4. If we assume that the trials are independent, then X is a binomial random variable and from Table II, the probability of exactly two people out of the six having contracted German measles is

$$P(X = 2) \simeq .311$$

However, let us examine the independence assumption. Suppose that the six people selected are brothers. Since German measles is contagious, it is evident that if one brother contracts the disease, then the chances that some of the other brothers will also contract it are greatly increased over the national probability of .4. Thus success or failure on one trial affects the chances for success on the other trials so that the assumption of independence is unwarranted. If, on the other hand, the six people selected were not from the same family, then the independence assumption would be warranted if we knew that the six people had no direct contact with one another and had no indirect contact through mutual acquaintances who might have carried the disease from one individual to the other.

EXERCISE SET 7.5

1. Assume there are four independent repetitions of a Bernoulli experiment with success probability $p = \frac{1}{5}$. Use formula (3) to find the probability of

   (a) no successes in the four repetitions;

   (b) exactly one success;

   (c) exactly two successes;

   (d) exactly three successes;

   (e) exactly four successes;

   (f) at most three successes;

   (g) at least two successes.

2. Use Table II in the Appendix to check your work in Exercise 1.

3. Let X be the number of successes in 12 independent repetitions of a Bernoulli experiment with success probability $p = .6$. Find

   (a) $P(X = 4)$ (b) $P(X = 11)$ (c) $P(X = 0)$

   (d) $P(X \geq 10)$ (e) $P(X \leq 5)$ (f) $P(X > 11)$.

4. In each part, assume X is a binomial random variable with the given values of n and p. Use Table II in the Appendix to find the requested probability.

   (a) $n = 3$, $p = .5$; find $P(X = 2)$

   (b) $n = 8$, $p = .8$; find $P(X = 6)$

   (c) $n = 11$, $p = .2$; find $P(X \leq 5)$

   (d) $n = 9$, $p = .4$; find $P(X > 6)$.

5. Let X be a binomial random variable with $n = 7$ and $p = .4$. Graph the probability function for X.

6. A certain system for betting on horses produces winners 40% of the time. What is the probability the system produces exactly three winners out of eight on a race day?

7. On a large boulevard, ten traffic lights operate independently. Each light is red for one minute and green for four minutes.

   (a) What is the probability a car traveling on the boulevard will not meet any red lights?

   (b) What is the probability that the car will meet at least one red light?

   (c) What is the probability the car will meet exactly three red lights?

8. Assuming the probability of a male birth is $\frac{1}{2}$, find the probability that among 12 births at Columbia Presbyterian Hospital, more than half will be females.

9. A package of crimson giant radish seeds states that the probability of germination for each seed is .9. If ten such seeds are planted, what is the probability that

   (a) all ten seeds germinate?

   (b) eight or more seeds germinate?

   (c) five or fewer germinate?

10. Assume that the probability is .7 that a medical technician can detect the presence of lung cancer from an X ray. An X ray from an individual with lung cancer is independently examined by each of eight medical technicians.

    (a) What is the probability that at least one technician will detect the cancer?

    (b) What is the probability that none will detect the cancer?

    (c) What is the probability that more than half of the technicians will fail to detect the cancer?

11. A fair die is tossed four times; what is the probability of obtaining exactly two 5's in the four tosses?

12. Suppose that in a lunar rocket flight, each midcourse-correction thruster has a probability of .7 of working, and that the rocket has four thrusters that work independently. If the rocket can carry out its correction when at least two of the thrusters are working, what is the probability that the correction can be carried out? What is this probability if the rocket has six thrusters?

13. Assume that 70% of the voting population supports the President's foreign policy. If ten people are selected at random and asked for an opinion, what is the probability that the majority will indicate that they oppose the President's

foreign policy? (This problem shows how errors due to chance can arise in opinion polls.)

14. Which has a better chance of occurring: tossing five heads in a row with a fair coin or obtaining eight heads out of ten tosses with a fair coin?

15. Suppose that 5% of all items coming off a production line are defective. Assume the manufacturer packages his items in boxes of six and guarantees "double your money back" if more than two items in a box are defective. On what percentage of the boxes will the manufacturer have to pay double money back?

16. A manufacturer of polygraphs (lie detectors) claims that his machine can correctly distinguish between the truth and a lie 90% of the time. The machine is to be tested on a series of ten questions. If it correctly identifies at least eight out of the ten questions as a truth or a lie, the machine will be purchased by the Los Angeles Police Department, otherwise it will be rejected:

(a) What is the probability that the machine will be purchased if it is absolutely worthless (that is, its probability of producing a correct result is $\frac{1}{2}$)?

(b) What is the probability that the machine will be rejected even if it can perform as the manufacturer claims?

17. Let X be a binomial random variable with $n = 2$ and success probability p. Show that $\mu_X = 2p$ and $\sigma_X{}^2 = 2pq$. (This is a special case of a general result which states that the mean of a binomial random variable is np and the variance is npq.)

7.6 THE NORMAL APPROXIMATION TO THE BINOMIAL

In Example 34 of the previous section we considered a binomial experiment concerned with radioactive decay. In that problem, X (the number of atoms that split in a 24-hour period) was a binomial random variable representing the number of successes in $n = 10^{30}$ repetitions, where the probability of success on each trial was $p = 1/10^{24}$. If, for example, we were interested in the probability that exactly three atoms split in the 24-hour period, we would have had to compute (see formula (3) in Section 7.5)†

$$P(X = 3) = \binom{10^{30}}{3}\left(\frac{1}{10^{24}}\right)^3\left(1 - \frac{1}{10^{24}}\right)^{10^{30}-3} \tag{1}$$

No binomial tables are available for $n = 10^{30}$ and $p = 1/10^{24}$, and even with a digital computer it would be difficult to find the value of (1) directly with any degree of accuracy. To handle such problems, statisticians have devised methods for approximating binomial probabilities. In this section we shall investigate one such method.

† Recall that $\binom{n}{x}$ is an alternative notation for $C_{n,x}$.

Consider an experiment that consists of three independent repetitions of a Bernoulli experiment with a success probability $p = \frac{2}{5}$. Substituting $n = 3$ and $p = \frac{2}{5}$ into formula (3) of Section 7.5, we see that the probability of getting x successes in the three repetitions is given by

$$P(X = x) = C_{3,x}\left(\frac{2}{5}\right)^{x}\left(\frac{3}{5}\right)^{3-x}.$$

From this formula or from Table II in the Appendix, we obtain the probabilities

$$P(X = 0) = \frac{27}{125} \simeq .216$$

$$P(X = 1) = \frac{54}{125} \simeq .432$$

$$P(X = 2) = \frac{36}{125} \simeq .288$$

$$P(X = 3) = \frac{8}{125} \simeq .064.$$

As discussed in Section 7.1, these results can be described geometrically by graphing the probability function of X; this graph is shown in Figure 7.16a. For our purposes, however, it will be more convenient to describe the random variable by means of a ***histogram.*** To draw the histogram of a discrete random variable X, we locate the possible values of X on the x axis; then, centered on each of these values we draw a rectangle whose base has length 1 and whose height is the probability of the x value on which the rectangle is centered. The histogram of the binomial random variable with $n = 3$, $p = \frac{2}{5}$ is shown in Figure 7.16b.

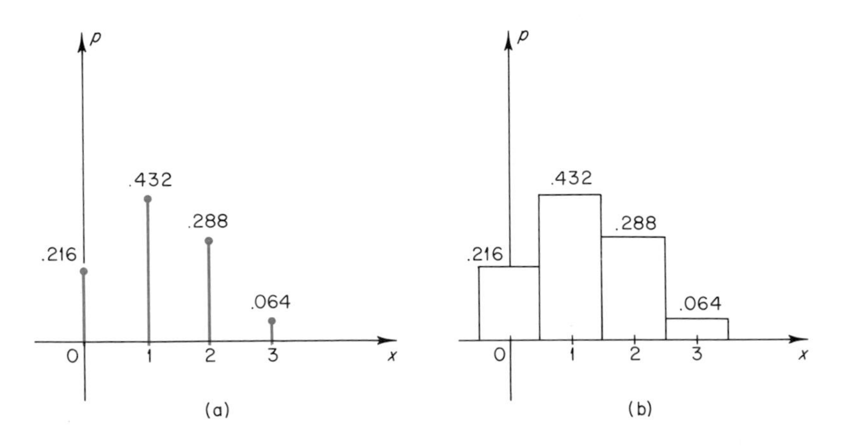

Figure 7.16

Histograms are important since they can be used to describe probabilities geometrically as *areas*. For example, in Figure 7.16b the area of the rectangle centered at $x = 2$ is

$$\begin{aligned} \text{area} &= \text{base} \times \text{height} \\ &= 1 \times .288 \\ &= .288. \end{aligned}$$

Thus, the probability that $X = 2$ is the same as the area of the rectangle centered above the point $x = 2$. Since the rectangles in the histogram have bases of length 1, the area of each rectangle is the probability of the point on the x axis above which the rectangle is centered.

Let us now see what happens to the histogram of a binomial random variable as n, the number of repetitions, gets larger and larger. To illustrate, we have constructed in Figure 7.17 three histograms for binomial random variables. In these histograms the number of repetitions increases from $n = 5$ to $n = 25$ with the probability of success in each case fixed at $p = .2$.

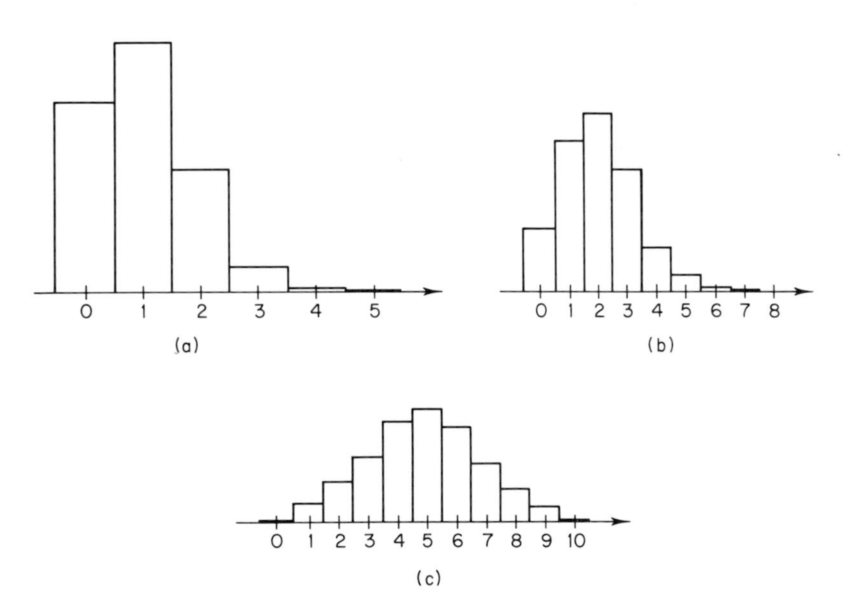

Figure 7.17 (a) $n = 5$, $p = .2$; (b) $n = 10$, $p = .2$; (c) $n = 25$, $p = .2$.

The histograms in Figure 7.17 suggest that as n increases, the area occupied by the rectangles in the histogram begins to look more and more like the area under a bell-shaped curve (see Figure 7.18). These curves are called ***normal curves*** or sometimes ***Gaussian curves.***

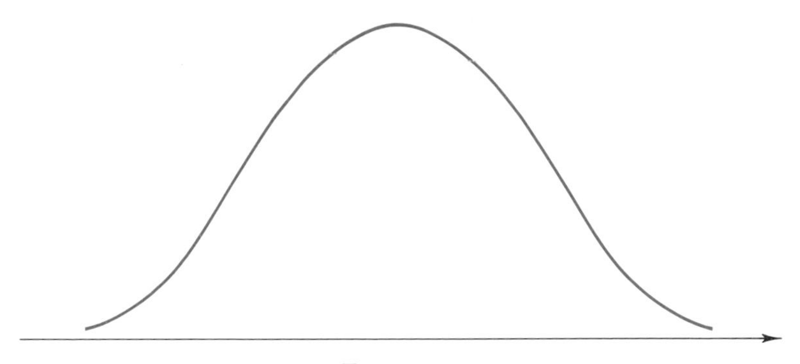

Figure 7.18

Statisticians have studied normal curves in great detail. Equations for normal curves are known and extensive tables of areas between points under normal curves have been constructed using calculus. We shall be concerned with using these tables of areas to approximate probabilities of binomial random variables. First, we shall need some preliminary ideas.

If X is a binomial random variable, then X can be interpreted as the number of successes in n independent repetitions of a Bernoulli experiment with success probability p. Since X is a random variable, we can ask for its expected value μ and its variance σ^2. Although we shall omit the mathematical details, the following result can be proved.

If a binomial random variable X represents the number of successes in n independent repetitions of a Bernoulli experiment with success probability p, then the expected value and variance of X are given by

$$\mu = E(X) = np$$

$$\sigma^2 = \mathrm{Var}(X) = npq$$

where $q = 1 - p$.

Example 37 If X is a binomial random variable with $n = 48$ and $p = \frac{1}{4}$, then

$$\mu = E(X) = np = 48\left(\frac{1}{4}\right) = 12$$

and

$$\sigma^2 = \mathrm{Var}(X) = npq = 48\left(\frac{1}{4}\right)\left(\frac{3}{4}\right) = 9.$$

A normal curve never touches the x axis, the total area under a normal curve is 1, and every normal curve is symmetric about a vertical line through its highest point. To prove that the area under a normal curve is 1 requires calculus. However, if we remember that areas under normal curves approximate areas under binomial histograms, this result seems reasonable since the total area encompassed by a binomial histogram is always 1 (why?). If we travel along a normal curve away from the center, we pass through points where the curvature changes from the downward direction to the upward direction; these are called ***inflection points*** for the curve (see Figure 7.19a). Note that between the inflection points, the curve looks like an upside-down bowl; to the left or right of the inflection points, the curve looks like part of a right-side-up bowl. If X is a binomial random variable, and if n is large enough so that the area occupied by the histogram of X can be approximated by the area under a normal curve, it can be shown that the approximating normal curve will have its center of symmetry at the point

$$np$$

and its inflection points at

$$np + \sqrt{npq} \qquad \text{and} \qquad np - \sqrt{npq}.$$

Equivalently, since $np = \mu = E(X)$ and since $npq = \sigma^2 = \text{Var}(X)$, the center of symmetry is at

$$\mu$$

and the inflection points are at

$$\mu + \sigma \qquad \text{and} \qquad \mu - \sigma$$

(see Figure 7.19b).

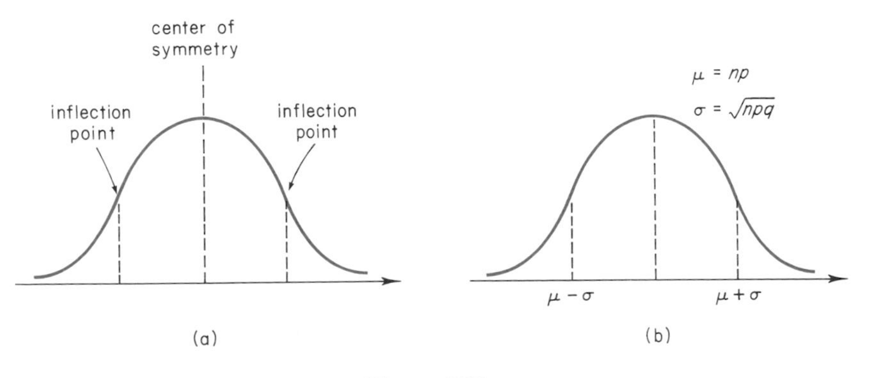

Figure 7.19

Example 38 If X is a binomial random variable with $n = 6400$ and $p = \frac{1}{2}$, then the normal curve that approximates the histogram for X has its center of symmetry at

$$\mu = np = (6400)\left(\frac{1}{2}\right) = 3200$$

and since

$$\sigma = \sqrt{npq} = \sqrt{(6400)\left(\frac{1}{2}\right)\left(\frac{1}{2}\right)} = 40,$$

the inflection points occur at

$$\mu - \sigma = np - \sqrt{npq} = 3160 \qquad \text{and} \qquad \mu + \sigma = np + \sqrt{npq} = 3240.$$

(see Figure 7.20).

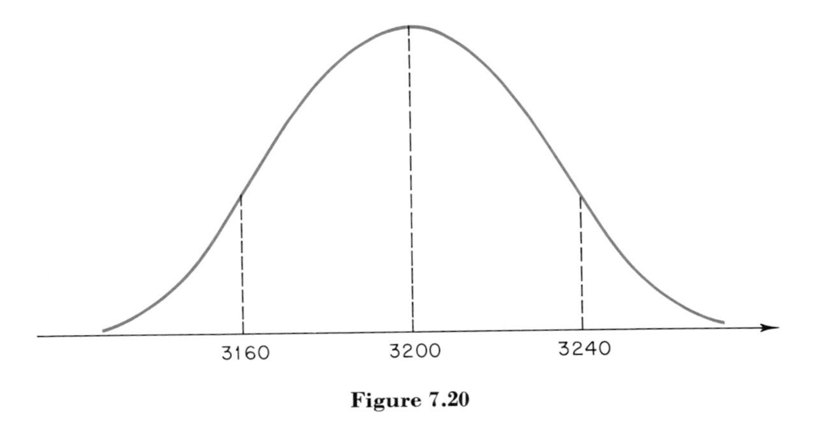

Figure 7.20

The shape of a normal curve is determined by the values of μ and σ. The value of μ determines the center and the value of σ determines how flat the curve is. To illustrate, we superimposed three normal curves in Figure 7.21, each with $\mu = 0$. As shown in this figure, the larger the value of σ, the flatter the curve. The normal curve with $\mu = 0$ and $\sigma = 1$ is of particular importance; it is called the ***standard normal curve.***

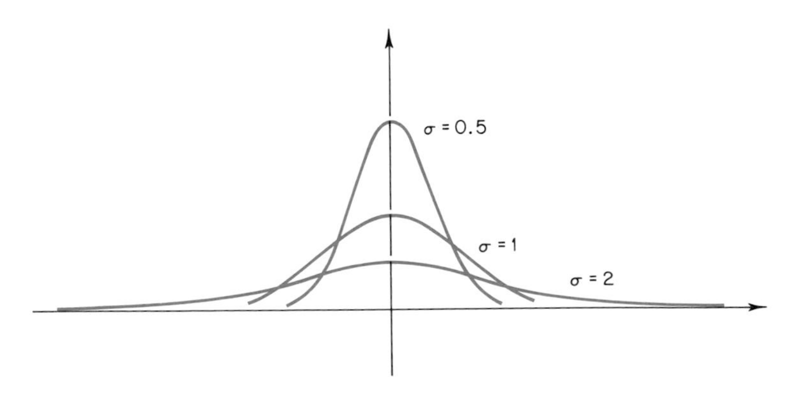

Figure 7.21

Presently we shall show how to obtain areas under normal curves. For the moment, however, assume we know how to do this. Suppose we are interested in the probability that a binomial variable X takes on the value x, that is

$$P(X = x).$$

In terms of the histogram, this probability is just the area of the rectangle centered at x. Since the base of this rectangle extends over the interval from $x - \frac{1}{2}$ to $x + \frac{1}{2}$, its area is approximately the area between $x - \frac{1}{2}$ to $x + \frac{1}{2}$ under the approximating normal curve. (See Figure 7.22.) We shall show how to obtain this area from tables.

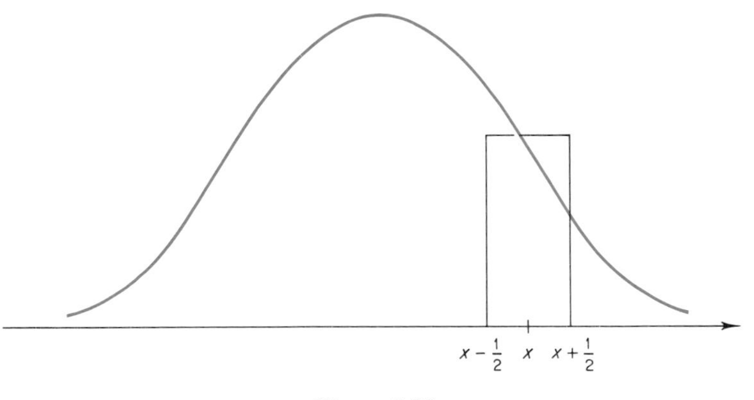
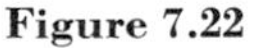

Figure 7.22

Since the approximating normal curve depends on the values of n and p, it seems we need different tables of areas for different n and p. Fortunately, this is not the case. By using a technique that we shall now describe, it will always be possible to work with tables of areas under the standard normal curve. This technique is based on the following result, the proof of which we omit.

> For a normal curve with center of symmetry μ and inflection points at $\mu + \sigma$ and $\mu - \sigma$, the area under the curve to the left of a point x is the same as the area to the left of
>
> $$z = \frac{x - \mu}{\sigma}$$
>
> under the standard normal curve (see Figure 7.23).

Example 39 For the normal curve with $\mu = 3200$ and $\sigma = 40$, find the area under the curve to the left of the point $x = 3286$.

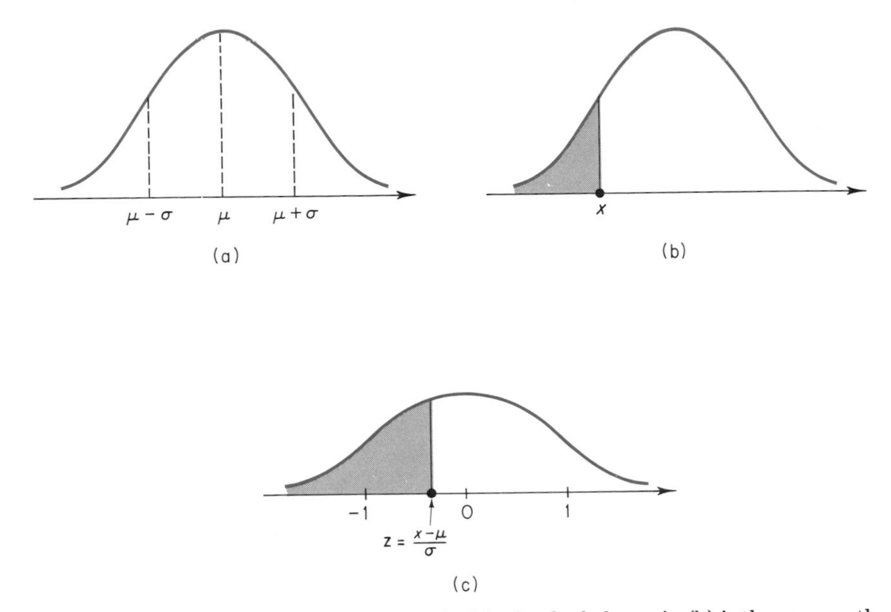

Figure 7.23 For the normal curve shown in (a), the shaded area in (b) is the same as the shaded area under the standard normal curve in (c).

Solution The area under this curve to the left of $x = 3286$ is the same as the area to the left of

$$z = \frac{x - \mu}{\sigma} = \frac{3286 - 3200}{40} = 2.15$$

under the standard normal curve. We can find this area from Table I in the Appendix as follows:

(i) Since the first two digits of 2.15 are 2.1, we locate the row in the table labeled 2.1 in the left margin.

(ii) Since the next digit in 2.15 is a 5, we go across the row found in (i) until we reach the column headed by a 5.

(iii) The number in the block reached in (ii) is the area to the left of $z = 2.15$. This area (to four decimal places) is .9842.

Example 40 For the normal curve with $\mu = 3200$ and $\sigma = 40$, find the area between the points 3260 and 3270.

Solution The area between $x_1 = 3260$ and $x_2 = 3270$ can be obtained by subtracting the area to the left of x_1 from the area to the left of x_2 (see Figure 7.24). The area to the left of x_1 is the area under the standard normal curve to the left of

$$z_1 = \frac{x_1 - \mu}{\sigma} = \frac{3260 - 3200}{40} = 1.5$$

which from Table I is .9332. Similarly, the area to the left of x_2 is the area under the standard normal curve to the left of

$$z_2 = \frac{x_2 - \mu}{\sigma} = \frac{3270 - 3200}{40} = 1.75$$

which from Table I is .9599.

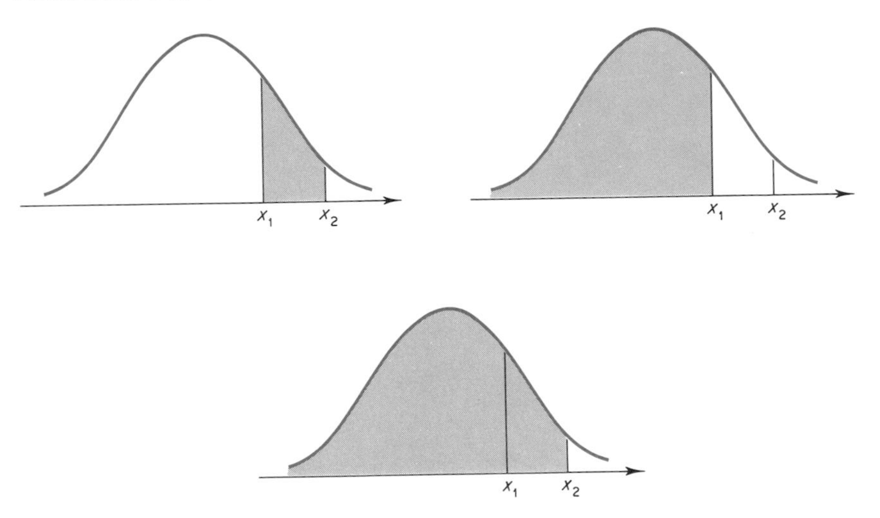

Figure 7.24 The area between x_1 and x_2 is the area to the left of x_2 minus the area to the left of x_1.

Thus, the area between x_1 and x_2 is

$$.9599 - .9332 = .0267.$$

Example 41 For the normal curve with $\mu = 3200$ and $\sigma = 40$, find the area to the left of the point 3120.

Solution The area to the left of $x = 3120$ is the area to the left of

$$z = \frac{x - \mu}{\sigma} = \frac{3120 - 3200}{40} = -2$$

under the standard normal curve.

Since Table I only gives areas to the left of *positive* values of z, we shall need some tricks here. The following facts about normal curves will be helpful:

(i) The standard normal curve is symmetric about $z = 0$.

(ii) The total area under a normal curve is 1.

Armed with these two facts, we can now find the area to the left of -2 under a standard normal curve. From the symmetry it follows that the area to the left of -2 is the same as the area to the right of $+2$ (see Figure 7.25). Moreover, since the total area is 1 we have

$$\begin{aligned}\text{area to the right of } 2 &= (\text{total area under the curve}) \\ &\quad - (\text{area to the left of } 2) \\ &= 1 - (\text{area to the left of } 2) \\ &= 1 - .9772 \\ &= .0228.\end{aligned}$$

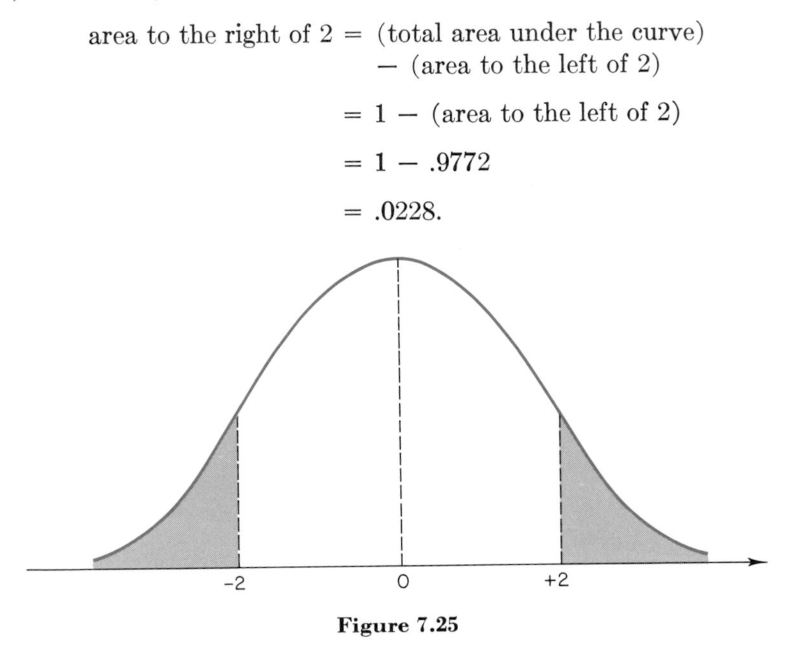

Figure 7.25

Thus, the area to left of $x = 3120$ is .0228.

The next three examples illustrate the basic computational techniques for approximating binomial probabilities by areas under normal curves.

Example 42 Use a normal curve to estimate the probability that in 324 tosses of a fair coin, the number of heads is between 171 and 180 inclusive.

Solution If X denotes the number of heads in 324 tosses, then X is a binomial random variable with $n = 324$ and $p = \frac{1}{2}$. We want to estimate

$$P(171 \leq X \leq 180)$$

This probability is the area between 170.5 and 180.5 in the histogram of X (see Figure 7.26). This is approximately the area between

$$x_1 = 170.5 \qquad \text{and} \qquad x_2 = 180.5$$

under the normal curve with

$$\mu = np = (324)\left(\frac{1}{2}\right) = 162$$

and

$$\sigma = \sqrt{npq} = \sqrt{(324)\left(\frac{1}{2}\right)\left(\frac{1}{2}\right)} = 9.$$

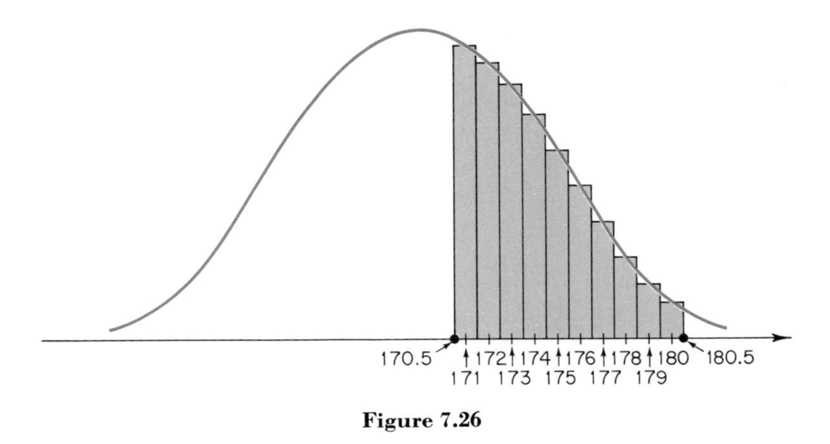

Figure 7.26

Proceeding as in Example 40, the area between x_1 and x_2 is the area between

$$z_1 = \frac{x_1 - np}{\sqrt{npq}} = \frac{170.5 - 162}{9} \simeq .94$$

and

$$z_2 = \frac{x_2 - np}{\sqrt{npq}} = \frac{180.5 - 162}{9} \simeq 2.05.$$

From Table I, the area between z_1 and z_2 is

$$.9798 - .8264 = .1534.$$

Thus,

$$P(171 \leq X \leq 180) \simeq .1534.$$

Example 43 Use a normal curve to estimate the probability that in 150 independent Bernoulli trials with success probability .4 we observe 70 or fewer successes.

Solution If X is the number of successes, then X is a binomial random variable with $n = 150$ and $p = .4$. We want to estimate

$$P(X \leq 70).$$

This probability is the area between -0.5 and 70.5 in the histogram of X, which is approximately the area between

$$x_1 = -0.5 \qquad \text{and} \qquad x_2 = 70.5$$

under the normal curve with

$$\mu = np = 150(.4) = 60$$

and

$$\sigma = \sqrt{npq} = \sqrt{150(.4)(.6)} = 6.$$

In terms of the standard normal curve, this is the area between

$$z_1 = \frac{x_1 - np}{\sqrt{npq}} = \frac{-.5 - 60}{6} = -10.08$$

$$z_2 = \frac{x_2 - np}{\sqrt{npq}} = \frac{70.5 - 60}{6} = 1.75.$$

As in Example 38, the area between z_1 and z_2 is the area to the left of z_2 minus the area to the left of z_1. However, in problems like this, where we are interested in a probability of the form

$$P(X \leq k)$$

it turns out that the area to the left of z_1 is usually so close to zero that it can be neglected. Taking advantage of this fact, we can use Table I to obtain the estimate

$$P(X \leq 70) \simeq \text{area to the left of } z_2 = .9599.$$

The idea of using a normal curve to estimate binomial probabilities is based on the assumption that n is large enough so that areas under the histogram are approximately the same as the corresponding areas under the normal curve. But how large should n be in order to use this method of approximation? The answer to this question depends to some degree on the nature of the problem. However, as a rule of thumb, most statisticians only use normal curve approximations when $npq \geq 3$. Thus, if $p = \frac{1}{2}$, then n can be as small as 12. However, if p is near 0 or 1 larger values of n are required; for example, if $p = \frac{9}{10}$, we would need

$$n\left(\frac{9}{10}\right)\left(\frac{1}{10}\right) \geq 3$$

or

$$n(.09) \geq 3.$$

Thus n would have to be at least as large as 34.

EXERCISE SET 7.6

1. In each part, draw the histogram for the binomial random variable with the given values of n and p.

   (a) $n = 3, p = \frac{1}{2}$ (b) $n = 5, p = .2$

   (c) $n = 5, p = .8$ (d) $n = 10, p = .6$.

2. Let X be the random variable whose histogram is shown in Figure 7.27.

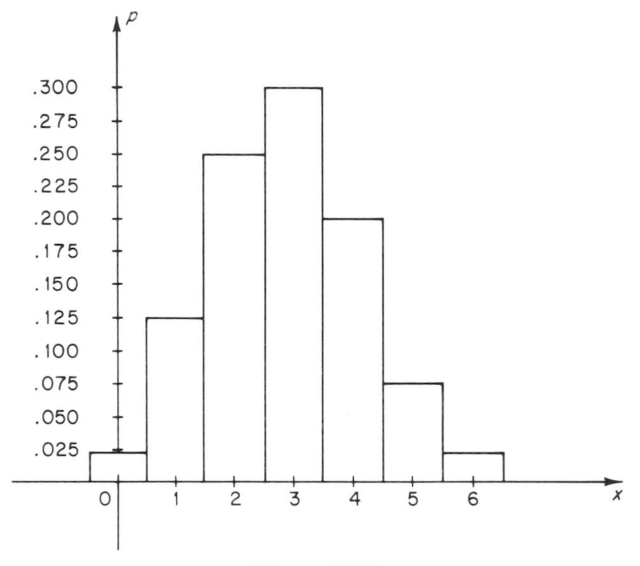

Figure 7.27

Determine the following:

(a) $P(X = 2)$ (b) $P(1 \leq X \leq 4)$

(c) $P(X \geq 3)$ (d) $P(X < 1)$.

3. In each part, find the mean and variance of the binomial random variable with the given values of n and p.

(a) $n = 3,\ p = \frac{1}{2}$ (b) $n = 5,\ p = .2$

(c) $n = 1000,\ p = \frac{1}{4}$ (d) $n = 2500,\ p = .9$

4. In each part, let X be the binomial random variable with the given values of n and p. Find the center of symmetry and the inflection points of the normal curve that approximates the histogram for X.

(a) $n = 6400,\ p = \frac{1}{2}$ (b) $n = 2500,\ p = \frac{4}{5}$

(c) $n = 1000,\ p = \frac{1}{4}$ (d) $n = 500,\ p = .6$

5. Find the following areas under the standard normal curve.

(a) The area to the left of 1.25.

(b) The area to the right of .84.

(c) The area between .37 and 1.84.

6. Find the following areas under the standard normal curve.

(a) The area to the left of -0.72.

(b) The area to the right of -1.80.

(c) The area between -1.31 and -0.38.

(d) The area between -2.81 and 1.47.

7. In each part, find the value of c that satisfies the given conditions.

(a) The area to the left of c under the standard normal curve is .8749.

(b) The area to the right of c under the standard normal curve is .1423.

(c) The area to the right of c under the standard normal curve is .9726.

(d) The area to the left of c under the standard normal curve is .0078.

8. Find the following areas under the normal curve with $\mu = 10$ and $\sigma = 2$.

(a) The area to the left of 13.

(b) The area to the right of 8.

(c) The area between 11.6 and 13.8.

(d) The area between 5.2 and 15.4.

9. Find the following areas under the normal curve with $\mu = 1500$ and $\sigma^2 = 3600$.

(a) The area to the right of 1400.

(b) The area to the left of 1375.

(c) The area between 1420 and 1580.

10. Use a normal approximation to find the following binomial probabilities:

(a) $P(X \leq 10)$ when X is a binomial random variable with $n = 25$ and $p = \frac{1}{2}$.

(b) $P(25 \leq X \leq 35)$ when X is a binomial random variable with $n = 80$ and $p = .4$.

(c) $P(X \geq 155)$ when X is a binomial random variable with $n = 200$ and $p = .75$.

Solve Exercises 11–15 using a normal approximation.

11. Assuming that a penicillin injection causes an allergic reaction in one out of ten people, estimate the probability that fewer than 470 reactions will occur among 5000 people injected with penicillin.

12. A bag of 400 dimes is dumped onto a table. Estimate the probability that the number of heads is between 175 and 225 inclusive.

13. Assume 10% of the items coming off a production line are defective; estimate the probability that more than 14% of the items in a sample of size n will be defective if

(a) $n = 100$ (b) $n = 500$ (c) $n = 1000$.

14. The death rate for yellow fever is five deaths per 1000 cases. If 700 individuals contract yellow fever, what is the probability of

(a) exactly three deaths? (b) more than three deaths?

15. Genetic theory predicts that when the right two varieties of peas are crossed, the probability of obtaining a smooth seed is .5. How many seeds must be produced to assure with probability .9 that at least 100 smooth seeds are obtained?

7.7 HYPOTHESIS TESTING; THE CHI-SQUARE TEST

A probability model, like any theory, is useful only if it accurately describes the physical world. If results predicted by the model do not agree with physical observations, then the model must be abandoned in favor of one that is more exact. In this section we discuss a statistical procedure, called the chi-square test, which can be used to test the accuracy of some probability models.

To motivate some basic ideas, consider the following situation. A good friend, Mr. X, produces a coin which he insists is fair. We watch him flip this coin three times, and each time the coin turns up heads. If the coin really is fair, as Mr. X claims, the probability of tossing three heads in a row is

$$\frac{1}{2^3} = \frac{1}{8}.$$

Although this probability is small, Mr. X insists that the coin is fair so we have no strong reason to doubt his claim. Mr. X continues to flip the coin and still after 10 tosses every single flip has turned up heads. If the coin is fair, as Mr. X claims, the probability of tossing 10 heads in a row is

$$\frac{1}{2^{10}} = \frac{1}{1024}.$$

Since this probability is so small, we are beginning to have doubts that the coin is fair. But Mr. X insists so vigorously that the coin is fair that we are willing to give him the benefit of the doubt. We view his feat of tossing 10 heads in a row as a freak occurrence. Mr. X continues to flip the coin and still after 30 tosses every flip has turned up heads. If the coin is fair, as Mr. X claims, the probability of tossing 30 heads in a row is

$$\frac{1}{2^{30}} \simeq .000000001.$$

Thus, the chances of flipping 30 heads in a row with a fair coin are less than 1 in a billion. The odds against tossing 30 heads in a row with a fair coin are now so great that we are forced to reject X's claim and conclude that the coin is unfair.

This simple example illustrates an important principle about probability models. In any probabilistic experiment we expect some discrepancy between observed frequencies and frequencies suggested by the model. If the discrepancy is reasonably small, we are inclined to accept the validity of the model and attribute the discrepancy to chance. If, as in the case of tossing 30 heads in a row, there is a big difference between observed frequencies and the frequencies suggested by the model, we are inclined to reject the validity of the model.

Before we try to make these ideas mathematically more precise, let us consider some examples that illustrate the types of problems we want to consider.

Example 44 A die is tossed 60 times. If the die is fair, we expect each sample point in the sample space

$$S = \{1, 2, 3, 4, 5, 6\}$$

to occur 10 times in the 60 tosses. These theoretical frequencies together with the frequencies actually observed are shown in Table 7.10. We shall want to decide if the observed data in this problem are consistent with the assumption that the die is fair. In other words, is the difference between the observed and theoretical frequencies due to chance or is the die loaded?

Table 7.10

	1	2	3	4	5	6
Observed frequencies	7	12	8	11	9	13
Theoretical frequencies	10	10	10	10	10	10

Example 45 A coin is tossed 1000 times. If the coin is fair we expect each point in the sample space

$$S = \{h, t\}$$

to occur 500 times in the 1000 tosses. These theoretical frequencies together with the frequencies actually observed are shown in Table 7.11. We shall want to decide if the difference between the observed and theoretical frequencies is due to chance or whether the coin is unfair. In other words, should we accept or reject the uniform probability model for this experiment?

Table 7.11

	h	t
Observed frequencies	450	550
Theoretical frequencies	500	500

Genetics

Example 46 The Mendelian† inheritance theory in genetics states that in crossing two kinds of peas, four types of seeds will result

$$A, \quad B, \quad C, \quad \text{and} \quad D.$$

According to the theory, the different types occur with probabilities

$$P(A) = \frac{9}{16}, \qquad P(B) = \frac{3}{16}, \qquad P(C) = \frac{3}{16}, \qquad P(D) = \frac{1}{16}.$$

If 320 seeds are obtained from crossing the two kinds of peas, then Mendelian theory predicts that the number of type A seeds will be

$$\frac{9}{16}(320) = 180.$$

Similarly, the number of seeds of the remaining types should theoretically be

$$\frac{3}{16}(320) = 60 \qquad \text{type } B \text{ seeds}$$

$$\frac{3}{16}(320) = 60 \qquad \text{type } C \text{ seeds}$$

and

$$\frac{1}{16}(320) = 20 \qquad \text{type } D \text{ seeds.}$$

These theoretical frequencies together with the number of seeds of each type actually observed are shown in Table 7.12. We shall want to decide if the difference between the observed and theoretical values is due to chance or whether the data in this experiment conflict with the Mendelian theory.

Table 7.12

	A	B	C	D
Observed frequencies	168	65	68	19
Theoretical frequencies	180	60	60	20

† *Gregor Johann Mendel* (1822–1884)—Mendel was an Austrian, Augustinian monk. Working in a small monestary garden, he discovered the first laws of heredity and thereby laid the foundation for the science of genetics. For many years Mendel taught science in the technical high school at Brünn, Austria (later called Brno, Czechoslovakia) without a teacher's license. The reason—he failed the *biology* portion of the license examination!

Each of the above examples is a special case of the general problem we shall now describe. Suppose an experiment with a finite sample space

$$S = \{s_1, s_2, \ldots, s_k\}$$

has been assigned a probability model. Specifically, suppose that the elementary events have probabilities

$$P\{s_1\} = p_1, \qquad P\{s_2\} = p_2, \qquad \ldots, \qquad P\{s_k\} = p_k.$$

If the experiment is repeated n times, then the sample points should theoretically occur with frequencies

$$f_1 = np_1, \qquad f_2 = np_2, \qquad \ldots, \qquad f_k = np_k \tag{1}$$

in the n repetitions. (For example, if $p_1 = P\{s_1\} = .3$, then in $n = 50$ repetitions of the experiment, we would expect sample point s_1 to occur $np_1 = 50(.3) = 15$ times.) We shall call the frequencies $f_1, f_2, \ldots, f_k$ in (1) the ***expected frequencies*** or ***theoretical frequencies.*** When the experiment is performed physically n times, we observe certain frequencies

$$o_1, \qquad o_2, \qquad \ldots, \qquad o_k \tag{2}$$

for the sample points $s_1, s_2, \ldots, s_k$. We call the numbers in (2) the ***observed frequencies.***

Usually the observed frequencies and the theoretical frequencies will be different, and the problem is to decide whether the difference can be attributed to chance or whether the difference results from an incorrect probability model. To make this decision, we need a way of measuring how much the observed frequencies

$$o_1, \qquad o_2, \qquad \ldots, \qquad o_k$$

differ from the theoretical frequencies

$$f_1, \qquad f_2, \qquad \ldots, \qquad f_k.$$

Although there are many possible ways of measuring this difference, statisticians have settled on the following quantity called ***chi-square*** (written χ^2)†:

$$\chi^2 = \frac{(o_1 - f_1)^2}{f_1} + \frac{(o_2 - f_2)^2}{f_2} + \cdots + \frac{(o_k - f_k)^2}{f_k}. \tag{3}$$

Example 47 Find the value of χ^2 for the data in Example 46.

† *Chi* (χ) is a letter in the greek alphabet.

Solution From Table 7.12, the observed frequencies are

$$o_1 = 168, \qquad o_2 = 65, \qquad o_3 = 68, \qquad o_4 = 19$$

and the theoretical frequencies are

$$f_1 = 180, \qquad f_2 = 60, \qquad f_3 = 60, \qquad f_4 = 20.$$

Substituting these values in formula (3) yields

$$\begin{aligned}\chi^2 &= \frac{(168 - 180)^2}{180} + \frac{(65 - 60)^2}{60} + \frac{(68 - 60)^2}{60} + \frac{(19 - 20)^2}{20} \\ &= \frac{420}{180} \\ &\simeq 2.33.\end{aligned}$$

If the observed frequencies are the same as the theoretical frequencies then

$$o_1 - f_1 = 0, \qquad o_2 - f_2 = 0, \qquad \ldots, \qquad o_k - f_k = 0,$$

so that from formula (3), $\chi^2 = 0$. As the differences

$$o_1 - f_1, \qquad o_2 - f_2, \qquad \ldots, \qquad o_k - f_k$$

get larger, the value of χ^2 increases. Thus, χ^2 is a measure of how closely the observed frequencies agree with the theoretical frequencies predicted by the probability model; a small value of χ^2 means good agreement and a large value of χ^2 means poor agreement.

This suggests the following procedure for using experimental data to decide whether to accept or reject the correctness of a probability model:

(1) Pick a cutoff point that divides "large values" of χ^2 from "small values" (we shall show how to do this in a moment).
(2) If the value of χ^2 obtained from the data is "small," then accept the correctness of the model, and if the value of χ^2 is "large" then reject the correctness of the model. (See Figure 7.28.)

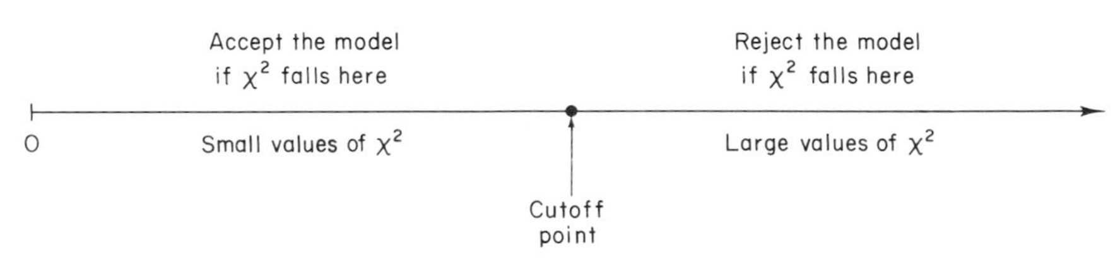

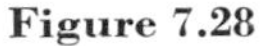

Figure 7.28

Before we can show how to select the cutoff point, we shall need some preliminary ideas. If a probability model is assigned to an experiment with a finite sample space, the quantity χ^2 becomes a random variable whose value is determined by the frequencies observed in n repetitions of the experiment. Moreover, since the original experiment has a finite sample space, χ^2 will be a finite discrete random variable. (We shall omit the details.)

Thus the values of χ^2 have various probabilities of occurring and these probabilites can be represented as areas in a histogram. Just as we were able to approximate areas in a binomial histogram by areas under a normal curve, so it is possible, in certain cases, to approximate areas in a chi-square histogram by areas under a curve called a ***chi-square curve.*** The exact shape of the chi-square curve depends on the value of k in formula (3). Some typical chi-square curves are shown in Figure 7.29. As was the case for normal curves, the area under every chi-square curve is 1.

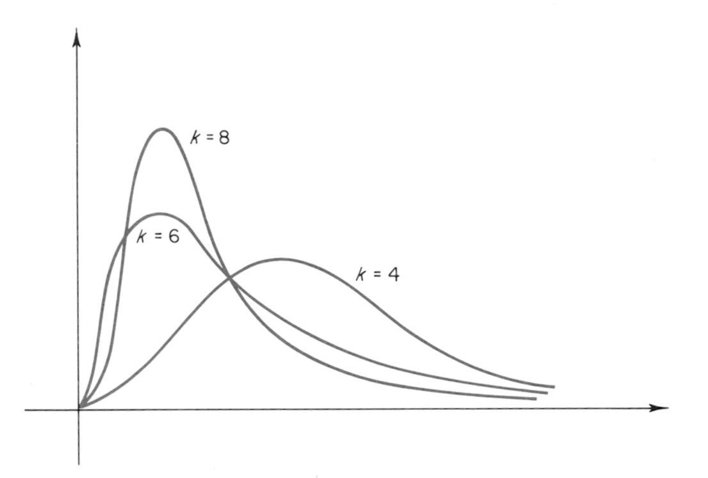

Figure 7.29 Typical chi-square curves.

In selecting the cutoff point c that divides "large" values of χ^2 from "small" values, we must recognize that no matter how the value of c is chosen we will make an error a certain percentage of the time. Even if the probability model is correct, there is an outside chance that the value of χ^2 will be greater than c, in which case we would *incorrectly* reject the model. It is common in statistical work to select the cutoff point c so that this kind of error occurs only 5% of the time. In this case c is called the 5% ***critical level.*** To illustrate how the 5% critical level can be obtained, let us assume that the areas of the rectangles in the chi-square histogram can be approximated by the corresponding areas under a chi-square curve. Since the total area under a chi-square curve is 1, there is, on the x axis, a point c that divides the area under the curve into two parts: an area of .95 to the left of c and an area of .05 to the right of c (see Figure 7.30). Since areas under the chi-square curve correspond to probabilities for the χ^2 random variable, the probability of obtaining a value of χ^2 less than c is .95 and the probability of obtaining a value of χ^2 greater than c is .05 when the probability model being tested is correct. Thus, if we use c as the cutoff point between large and small values of χ^2, the probability of rejecting a correct model will be .05.

Table III in the Appendix lists the 5% critical levels for various chi-square curves. Observe that the first row in Table III is labeled ν

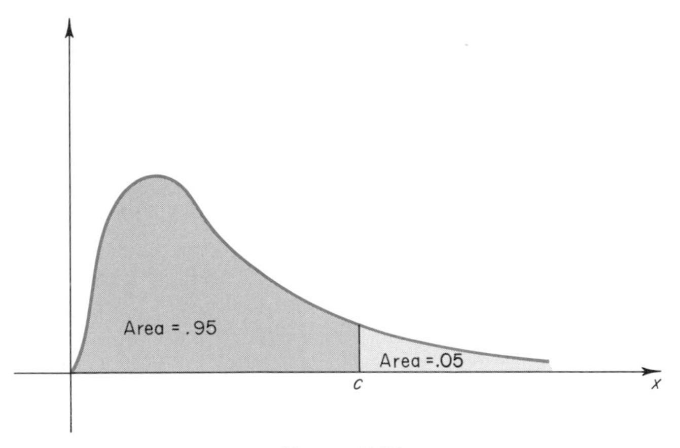

Figure 7.30

(greek letter *nu*). The quantity ν is called the ***number of degrees of freedom*** for the chi-square curve and its value is

$$\nu = k - 1, \tag{4}$$

where k is the number of terms in the formula for χ^2 (formula (3)). To use the table, determine the value of ν for the problem and read off the critical level c next to the value of ν just determined. To illustrate, consider the die-tossing problem in Example 44. Since the formula for χ^2 contains $k = 6$ terms, the approximating chi-square curve has $\nu = k - 1 = 5$ degrees of freedom, so that from Table III the 5% critical level is $c = 11.1$.

To summarize, we have the following procedure for using experimental data to decide whether to accept or reject the validity of a probability model for a finite-sample-space experiment:

(1) Compute the value of χ^2 from the experimental data.

(2) Compute the number of degrees of freedom from formula (4).

(3) Determine the 5% critical level c from Table III in the Appendix.

(4) If the value of χ^2 is more than c, reject the validity of the probability model, and if the value of χ^2 is less than c accept the validity of the model.

This four-step procedure is called the ***chi-square test at the 5% significance level.*** The phrase "5% significance level" reminds us that this test is designed with the understanding that there is a probability of .05 or 5% that a correct model will be rejected.

Example 48 Use the chi-square test at the 5% significance level to determine if the data in Table 7.10 are consistent with the assumption that the die is fair.

Solution From the data in Table 7.10 and from formula (3) we obtain

$$\chi^2 = \frac{(7-10)^2}{10} + \frac{(12-10)^2}{10} + \frac{(8-10)^2}{10} + \frac{(11-10)^2}{10} + \frac{(9-10)^2}{10} + \frac{(13-10)^2}{10} = 2.8.$$

Since the above formula for χ^2 contains $k = 6$ terms, the number of degrees of freedom is $\nu = k - 1 = 5$, so that from Table 3 in the Appendix the cutoff point is $c = 11.1$. Since the value of χ^2 is less than 12.6, we accept the probability model and conclude that the observed data are consistent with assumption that the die is fair.

Example 49 Use the chi-square test at the 5% significance level to determine if the data in Table 7.11 support the assumption that the coin is fair.

Solution From the data in Table 7.11 and from formula (3) we obtain

$$\chi^2 = \frac{(450-500)^2}{500} + \frac{(550-500)^2}{500} = 10.$$

Since this formula for χ^2 contains $k = 2$ terms, the number of degrees of freedom is $\nu = k - 1 = 1$, so that from Table III in the Appendix the 5% critical level is $c = 3.84$. The value of χ^2 is greater than c so we *reject* the probability model and conclude that the coin is not fair.

Births of Presidents

Example 50 Studies have been made to determine if there is any relationship between month or season of birth and traits like artistic talent, I.Q., and leadership ability. In Table 7.13 we have listed the season of birth for the first 36 men to serve as U.S. Presidents. Use the chi-square test at the 5% significance level to determine if the chance of becoming president is affected by season of birth.

Table 7.13 Season of birth for the first 36 U.S. presidents[a]

Season of birth	Frequency
Summer	6
Fall	11
Winter	10
Spring	9

[a] *The World Almanac* (New York: Newspaper Enterprise Association, 1973).

Solution Let us hypothesize that season of birth has *no* effect on one's chances to become president. With this hypothesis, the probability is $\frac{1}{4}$ that a U.S. president will be born in any one of the four seasons. Thus, among the 36 U.S. presidents, we expect

$$\frac{1}{4}(36) = 9$$

presidents to be born in each season. In Table 7.14 we have listed these theoretical frequencies and the observed frequencies in Table 7.13. From Table 7.14 and formula (3) we obtain

$$\chi^2 = \frac{(6-9)^2}{9} + \frac{(11-9)^2}{9} + \frac{(10-9)^2}{9} + \frac{(9-9)^2}{9} \simeq 1.556.$$

Table 7.14

	Summer	Fall	Winter	Spring
Observed frequencies	6	11	10	9
Theoretical frequencies	9	9	9	9

Since there are $k = 4$ terms in the expression for χ^2, we have $\nu = k - 1 = 3$ degrees of freedom. Thus, from Table III in the appendix, the critical 5% level is $c = 7.81$. Since the value of χ^2 is below the critical level we *accept* the hypothesis that season of birth has no effect on one's chances to become president.

In this section we have restricted our discussion to chi-square tests at the 5% significance level, which means that a correct probability model will be erroneously rejected 5% of the time. In problems where severe consequences result from rejecting a correct model, it may be desirable to reduce this error below 5%. In still other kinds of problems an error of more than 5% may be tolerable. For such problems, there are chi-square tests at other significance levels. These tests are carried out exactly as those at the 5% level with the exception that different tables are used to obtain the cutoff points.

EXERCISE SET 7.7

1. Find the value of χ^2 for the data in the following tables:

(a)

Observed frequencies	41	59
Theoretical frequencies	50	50

(b)

Observed frequencies	147	316	537
Theoretical frequencies	200	300	500

(c)

Observed frequencies	27	33	31	49
Theoretical frequencies	20	30	40	50

2. Find the 5% critical level for the chi-square curve with

(a) $\nu = 4$ degrees of freedom (b) $\nu = 10$ degrees of freedom.

3. In New York City, 210 automobile accidents involving an intoxicated driver were classified according to the day of the week on which the accident occurred, and the following table was obtained:

Day	Mon	Tues	Wed	Thurs	Fri	Sat	Sun
Number of accidents	27	18	28	20	31	46	40

Use a chi-square test at the 5% significance level to determine if these data support the assumption that an accident involving an intoxicated driver is equally likely to occur on any day of the week.

4. In an experiment designed to control the sex of baby rabbits by chemical injections of the mother, a sample of 50 baby rabbits yielded 30 males and 20 females. Use a chi-square test at 5% significance level to determine if these data support the hypothesis that male births and female births are equally likely.

5. In a survey of consumer opinion, 500 people were asked to use toothpastes A, B, and C, each for 1 week. At the end of three weeks, the individuals were asked to name the product which they felt was most effective. The results were:

Toothpaste	A	B	C
Number of people naming it most effective	176	159	165

Use a chi-square test at the 5% significance level to determine if these data support the conclusion that the consumers regard the three products to be equally effective.

6. According to genetic theory, the offspring of a certain cross of bearded iris plants should be colored pink, blue, or red with probabilities $\frac{9}{16}$, $\frac{3}{16}$, and $\frac{4}{16}$ respectively. If an experiment yields 68, 34, and 42 in each category, does the experiment support the theory? Use a chi-square test at the 5% significance level.

7. In a certain manufacturing plant, 800 of the employees are men and 200 are women. Company records show that during the previous year 33 job-related accidents involved men and seven job-related accidents involved women. Use a chi-square test at the 5% significance level to decide if these data support the contention that men and women are equally likely to be involved in job-related accidents.

8 APPLICATIONS

8.1 AN APPLICATION OF CONDITIONAL PROBABILITY TO MEDICAL DIAGNOSIS

Most diseases produce one or more observable symptoms that make the presence of the disease apparent. A doctor, by observing these symptoms, tries to deduce the kind of disease present; this is medical diagnosis. A difficulty often arises, however, because different diseases, in their early stages, can produce some of the same observable symptoms. Thus, the doctor is often faced with deciding which of several possible diseases is the *most probable cause* of the symptoms. Although this kind of detective work is usually based on the doctor's judgment and intuition, some researchers have suggested the use of probability theory to aid in medical diagnosis (see references [1–4] at the end of this section). In this section we shall discuss some of these ideas.

Let

$$\{d_1, d_2, \ldots, d_k\}$$

be a finite set of diseases and let

$$\{s_1, s_2, \ldots, s_k\}$$

be the set of observable symptoms produced by the diseases. For example, if one of the diseases is measles, then fever and rash would be among the observable symptoms; and if tuberculosis is one of the diseases, then certain chest X-ray patterns would appear among the observable symptoms.

Let

$S =$ the set of all people in the world who have exactly one of the diseases and let

A = the set of people in S who exhibit one or more of the observable symptoms.

We are allowing for the possibility that some of the people in S may not exhibit any of the observable symptoms. We shall need the events

$$\begin{aligned} D_1 &= \text{the set of people in } S \text{ with disease } d_1 \\ D_2 &= \text{the set of people in } S \text{ with disease } d_2 \\ &\vdots \\ D_k &= \text{the set of people in } S \text{ with disease } d_k. \end{aligned}$$

We shall assume that a person coming to a doctor for diagnosis is a sample point from S. Moreover, he is coming to the doctor because he has a symptom of an illness, so we can take it as given that the person belongs to A; that is, the event A occurs. Since the doctor is concerned with finding the most probable cause of the illness, he would like to know each of the probabilities

$$P(D_1|A), \qquad P(D_2|A), \qquad \ldots, \qquad P(D_k|A),$$

where $P(D_i|A)$ is the probability that the person has disease d_i given that he has one of the observable symptoms. We can calculate these probabilities from Bayes' formula

$$P(D_i|A) = \frac{P(A|D_i)P(D_i)}{P(A|D_1)P(D_1) + P(A|D_2)P(D_2) + \cdots + P(A|D_k)P(D_k)}$$

provided we can obtain the probabilities

$$P(A|D_1), \qquad P(A|D_2), \qquad \ldots, \qquad P(A|D_k) \tag{1}$$

and

$$P(D_1), \qquad P(D_2), \qquad \ldots, \qquad P(D_k). \tag{2}$$

The probabilities in (2) can be estimated from available medical records as follows. Randomly select a large number of medical records of persons who had exactly one of the diseases $d_1, d_2, d_3, \ldots$. Assume that there are N such medical records. For each disease d_i, determine from these records the number of people who had the disease. Assume there are N_i such people. Since $P(D_i)$ is approximately the relative frequency of people who have disease d_i among all people who have exactly one of the diseases in

$$\{d_1, d_2, \ldots, d_k\},$$

we can estimate $P(D_i)$ by

$$P(D_i) \simeq \frac{N_i}{N}. \tag{3}$$

To estimate the probabilities $P(A|D_i)$ in (1), we use the definition of conditional probability

$$P(A|D_i) = \frac{P(A \cap D_i)}{P(D_i)}. \tag{4}$$

Since the denominator in the right-hand side of (4) has already been estimated in (3), we need only find the probabilities $P(A \cap D_i)$. This can be done as follows. For each disease d_i, determine, from the records, the number of people who exhibited one or more of the observable symptoms and had disease d_i. Assume there are N_i' such people. Among all people who have exactly one of the diseases in

$$\{d_1, d_2, \ldots, d_k\},$$

$P(A \cap D_i)$ is approximately the proportion of people who exhibit one or more of the observable symptoms and also have disease d_i. Thus we can estimate $P(A \cap D_i)$ by

$$P(A \cap D_i) \simeq \frac{N_i'}{N}. \tag{5}$$

Now, by substituting the approximations (3) and (5) in (4), we can estimate the conditional probabilities $P(A|D_i)$.

The following example illustrates these ideas using simulated medical data.

Example 1 Suppose that each of the diseases d_1, d_2, and d_3 can produce one or more of the following observable symptoms:

s_1 = loss of appetite

s_2 = chest pains

s_3 = shortness of breath

s_4 = dilation of the pupils.

Assume also that a nationwide survey of hospital records produced the data in Table 8.1 on 10,000 people who had *exactly one* of the diseases, d_1, d_2, or d_3. From Table 8.1 and formula (3) we obtain

$$P(D_1) \simeq \frac{3750}{10000} = .375$$

$$P(D_2) \simeq \frac{2250}{10000} = .225$$

$$P(D_3) \simeq \frac{4000}{10000} = .400$$

Table 8.1

	Number of people with disease d_i	Number of people with disease d_i who also had one or more of the symptoms s_1, s_2, s_3, and s_4
Disease d_1	3750	3000
Disease d_2	2250	2050
Disease d_3	4000	3500

and from Table 8.1 and formula (5) we obtain

$$P(A \cap D_1) \simeq \frac{3000}{10000} = .300$$

$$P(A \cap D_2) \simeq \frac{2050}{10000} = .205$$

$$P(A \cap D_3) \simeq \frac{3500}{10000} = .350.$$

Thus, from formula (4)

$$P(A|D_1) = \frac{P(A \cap D_1)}{P(D_1)} \simeq \frac{.300}{.375} \simeq .800$$

$$P(A|D_2) = \frac{P(A \cap D_2)}{P(D_2)} \simeq \frac{.205}{.225} \simeq .911$$

$$P(A|D_3) = \frac{P(A \cap D_3)}{P(D_3)} \simeq \frac{.350}{.400} \simeq .875.$$

Therefore the Bayes tree for this problem is

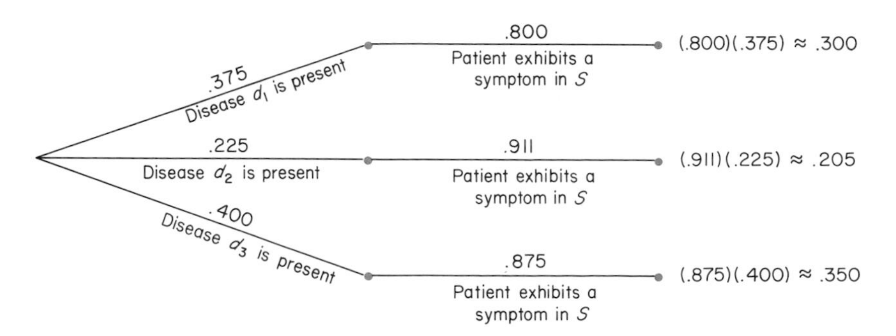

From the tree we find the sum of the path probabilities to be approximately

$$.300 + .205 + .350 = .855.$$

Therefore

$$P(D_1|A) = P\left(\text{disease } d_1 \text{ is present} \,\middle|\, \begin{array}{l}\text{patient exhibits a}\\ \text{symptom in } S\end{array}\right) \approx \frac{.300}{.855} \approx .351$$

$$P(D_2|A) = P\left(\text{disease } d_2 \text{ is present} \,\middle|\, \begin{array}{l}\text{patient exhibits a}\\ \text{symptom in } S\end{array}\right) \approx \frac{.205}{.855} \approx .240$$

$$P(D_3|A) = P\left(\text{disease } d_3 \text{ is present} \,\middle|\, \begin{array}{l}\text{patient exhibits a}\\ \text{symptom in } S\end{array}\right) \approx \frac{.350}{.855} \approx .409.$$

Thus, if a patient with one of the diseases d_1, d_2, or d_3 shows one or more of the symptoms s_1, s_2, s_3, or s_4, then the probability he has disease d_1 is .351, the probability he has disease d_2 is .240, and the probability he has disease d_3 is .409. Therefore, in absence of additional information, these data suggest that the physician should treat the patient for disease d_3.

Readers interested in pursuing these ideas in more detail can consult the following references:

[1] H. R. Warner, A. F. Toronto, L. G. Veasey, and R. Stephenson, A Mathematical Approach to Medical Diagnosis, *Journal of the American Medical Association* **177** (1961) 177–183.

[2] J. E. Overall and C. M. Williams, Conditional Probability Program for Diagnosis of Thyroid Function, *Journal of the American Medical Association* **183** (1963) 307–313.

[3] J. E. Overall and D. R. Gorham, A Pattern Probability Model for the Classification of Psychiatric Patients, *Behavioral Science* **8** (1963) 108–116.

[4] J. B. Johnston, G. B. Price, and Fred S. Van Vleck, *Sets, Functions, and Probability* (Reading, Massachusetts: Addison Wesley, 1968).

EXERCISE SET 8.1

1. Suppose that each of the diseases d_1 and d_2 can produce one or more of the observable symptoms

$$s_1 = \text{fever}, \qquad s_2 = \text{aching}, \qquad s_3 = \text{coughing}.$$

Assume also that a nationwide survey of hospital records produced the following data on 100,000 people who had exactly one of the diseases d_1 and d_2:

	Number of people with disease d_i	Number of people with disease d_i who also had one or more of the symptoms s_1, s_2, and s_3
Disease d_1	40,000	35,000
Disease d_2	60,000	45,000

(a) Find the probabilities of diseases d_1 and d_2, given that a patient has one or more of the observable symptoms.

(b) If a patient has one or more of the observable symptoms, for which disease should the doctor treat him?

2. Suppose that each of the diseases d_1, d_2, and d_3 can produce one or more of the observable symptoms

$$s_1 = \text{high blood pressure}, \qquad s_2 = \text{fainting spells}.$$

Assume also that a nationwide survey of hospital records produced the following data on 50,000 people who had exactly one of the diseases d_1, d_2, and d_3:

	Number of people with disease d_i	Number of people with disease d_i who also had one or more of the symptoms s_1, s_2, and s_3
Disease d_1	15,000	9,000
Disease d_2	25,000	20,000
Disease d_3	10,000	4000

(a) Find the probabilities of diseases d_1, d_2, and d_3, given that a patient nas one or more of the observable symptoms.

(b) If a patient has one or more of the observable symptoms, for which disease should the doctor treat him?

8.2 AN APPLICATION OF PROBABILITY TO GENETICS; THE HARDY-WEINBERG STABILITY PRINCIPLE

Probability theory plays a major part in the study of heredity. In this section we shall briefly discuss some of the genetic principles formulated by Gregor Mendel (see p. 329), and then we shall use some probability theory to derive one of the fundamental results in genetics.

One aspect of genetics is concerned with ***heredity,*** that is, how traits such as eye color and hair texture in humans, or petal color and disease resistance in plants are passed on from parents to offspring. In the following discussion we shall describe the theory of heredity as originally formulated by Mendel. Mendel's theory has undergone considerable refinement in the evolution of present-day genetics; however, our simplified discussion should serve to illustrate the role probability plays in the study of heredity.

We can illustrate Mendel's observations by considering a certain plant that can have red flowers (R), pink flowers (P), or white flowers (W). Mendel observed that when two of the red-flowered plants are crossed, the offspring always have red flowers. Also, when two of the white-flowered plants are crossed, the offspring always have white flowers, and when a red-flowered plant is crossed with a white-flowered plant, the offspring always have pink flowers; schematically, we describe this by writing

$$\mathrm{R} \times \mathrm{R} \to \mathrm{R} \tag{1a}$$

$$\mathrm{W} \times \mathrm{W} \to \mathrm{W} \tag{1b}$$

and

$$\mathrm{R} \times \mathrm{W} \to \mathrm{P} \tag{1c}$$

Mendel observed that the remaining crosses, R × P, W × P, and P × P, yield variable results. For example, R × P yields about 50% red offspring and 50% pink offspring. Schematically, the results of the above three crosses are

$$\mathrm{P} \times \mathrm{P} \to 25\%\ \mathrm{R}, \quad 50\%\ \mathrm{P}, \quad 25\%\ \mathrm{W} \tag{1d}$$

$$\mathrm{R} \times \mathrm{P} \to 50\%\ \mathrm{R}, \quad 50\%\ \mathrm{P} \tag{1e}$$

$$\mathrm{W} \times \mathrm{P} \to 50\%\ \mathrm{P}, \quad 50\%\ \mathrm{W} \tag{1f}$$

To explain these results, Mendel postulated the existence of certain structures, called ***genes,*** which are passed on from parents to offspring. In the simplest case, a single physical characteristic, such as flower color in the above example, is determined by one pair of genes in the offspring. One member of the pair is inherited from the male parent and one from the female parent. Each gene in the pair can assume one of two forms, called ***alleles;*** it is common to denote the two possible alleles by A and a. Thus, there are three possible forms for a pair of genes

$$AA, \qquad Aa, \qquad aa. \tag{2}$$

(Biologically, there is no difference between Aa and aA.) Each of the combinations in (2) is called a ***genotype.*** In the flower example, each genotype corresponds to a different flower color: a plant of genotype AA will have red flowers, a plant of genotype Aa will have pink flowers, and a plant of genotype aa will have white flowers.

According to Mendelian theory, a parent of genotype AA can only transmit an A gene to its offspring, a parent of genotype aa can only transmit an a gene to its offspring, and a parent of genotype Aa can transmit either an A gene or an a gene, each with probability $\frac{1}{2}$. It follows that if both parents have genotype AA, then every offspring will also be of genotype AA since each parent must transmit an A gene. Schematically,

$$AA \times AA \to AA. \tag{3a}$$

Similarly,

$$aa \times aa \to aa. \tag{3b}$$

In the flower example, genotype AA corresponds to red flowers (R) and genotype aa corresponds to white flowers (W), so that (3a) and (3b) explain observations (1a) and (1b). The following examples will help us explain some of the other results.

Example 2 Assume an individual of genotype Aa mates with another individual of genotype Aa. Find the probability that the offspring is:

(i) of genotype AA

(ii) of genotype aa

(iii) of genotype Aa.

Solution (i) To be of genotype AA, the offspring must inherit an A gene from each parent. The probability of getting an A gene from the first parent is $\frac{1}{2}$ and the probability of getting an A gene from the second parent is $\frac{1}{2}$; thus, assuming that the genes are inherited *independently* from each parent, the probability of genotype AA is

$$\frac{1}{2} \cdot \frac{1}{2} = \frac{1}{4}.$$

Solution (ii) The same argument used in (i) will show that the probability of genotype aa is $\frac{1}{4}$.

Solution (iii) To be of genotype Aa, the offspring can inherit an A gene from the male parent and an a gene from the female parent (call this event E) *or* it can inherit an a gene from the male parent and an A gene from the female parent (call this event F). Thus

$$P(\text{genotype } Aa) = P(E \cup F) = P(E) + P(F). \qquad (4)$$

The probability of receiving an A gene from the father is $\frac{1}{2}$ and the probability of receiving an a gene from the mother is $\frac{1}{2}$; thus, assuming independence,

$$P(E) = \frac{1}{2} \cdot \frac{1}{2} = \frac{1}{4}.$$

Similarly,

$$P(F) = \frac{1}{2} \cdot \frac{1}{2} = \frac{1}{4},$$

so that from (4)

$$P(\text{genotype } Aa) = \frac{1}{4} + \frac{1}{4} = \frac{1}{2}.$$

Schematically, the results in this example can be written

$$Aa \times Aa \rightarrow 25\%\ AA, \quad 50\%\ Aa, \quad 25\%\ aa. \qquad (5)$$

In the flower example, genotype AA corresponds to red flowers (R), aa to white flowers (W), and Aa to pink flowers (P), so that (5) explains the result (1d) where two pink flowered plants were crossed.

Example 3 Assume an individual of genotype AA mates with an individual of genotype Aa (that is, $AA \times Aa$). Find the probability that the offspring is:

(i) of genotype AA
(ii) of genotype aa
(iii) of genotype Aa.

Solution (i) To be of genotype AA, the offspring must inherit an A gene from the first parent and an A gene from the second parent. Since the first parent has genotype AA, the probability of inheriting an A gene is 1. Since the second parent has genotype Aa, the probability of inheriting an a gene is $\frac{1}{2}$. Thus, by independence,

$$P(\text{genotype } AA) = 1 \cdot \frac{1}{2} = \frac{1}{2}.$$

Solution (ii) To be of genotype aa, the offspring must inherit an a gene from each parent. Since the AA parent cannot transmit an a gene, we obtain

$$P(\text{genotype } aa) = 0.$$

Solution (iii) The only way the offspring can be of genotype Aa is to receive an A gene from the AA parent and an a gene from the Aa parent. The probability of getting an A gene from the AA parent is 1 and the probability of getting an a gene from the Aa parent is $\frac{1}{2}$, so that

$$P(\text{genotype } Aa) = 1 \cdot \frac{1}{2} = \frac{1}{2}.$$

Schematically, the results in this example can be written

$$AA \times Aa \rightarrow 50\%\ AA, \quad 50\%\ Aa$$

which explains observation (1e).

In Table 8.2 we have given a complete list of genotype probabilities for the offspring that result from various crosses. The reader may wish to check his understanding of the previous discussion by verifying the results listed in this table.

Some physical traits, such as right- or left-handedness in humans, exhibit only two possible physical forms. In such cases the genotypes

$$AA \quad \text{and} \quad Aa$$

produce one of the forms and the genotype

$$aa$$

produces the second form. The trait produced by genotypes AA and Aa is called the ***dominant trait*** and the trait produced by genotype aa is called

Table 8.2

Mating (male × female)	Genotype probabilities for the offspring		
	AA	Aa	aa
$AA \times AA$	1	0	0
$Aa \times AA$	$\frac{1}{2}$	$\frac{1}{2}$	0
$AA \times Aa$	$\frac{1}{2}$	$\frac{1}{2}$	0
$Aa \times Aa$	$\frac{1}{4}$	$\frac{1}{2}$	$\frac{1}{4}$
$aa \times Aa$	0	$\frac{1}{2}$	$\frac{1}{2}$
$Aa \times aa$	0	$\frac{1}{2}$	$\frac{1}{2}$
$aa \times AA$	0	1	0
$AA \times aa$	0	1	0
$aa \times aa$	0	0	1

the ***recessive trait.*** In humans, for example, right-handedness is dominant and left-handedness is recessive.

Table 8.2 suggests that "overall" there is a much better chance for the offspring to have the dominant trait than the recessive one. Based on this observation, many scientists in the late 1800's and early 1900's argued that for large populations, Mendel's theory implied that over successive generations, the number of offspring with the recessive trait should diminish and eventually disappear. Since this conclusion was at odds with physical evidence (lefthandedness in humans has not died out), they rejected Mendel's theory. In 1908 G. H. Hardy† and W. Weinberg‡ independently proved that Mendel's theory did not imply the disappearance of recessive traits. As a result, the work of Mendel eventually became the cornerstone of modern genetic theory. We shall conclude this section with a discussion of the arguments given by Hardy and Weinberg.

Let us consider a large population of individuals (plants, people, or animals) and let us fix our attention on a characteristic determined by a single pair of genes. We shall investigate the percentage of offspring of each genotype in successive generations under the assumption that matings occur randomly in the population. Suppose that in the initial parental population,

† *G. H. Hardy* (1877–1947)—Hardy was a world renowned English mathematician. He taught at Cambridge and Oxford and was a prolific researcher who produced over three hundred research papers. He received numerous medals and honorary degrees for his accomplishments.

‡ *Wilhelm Weinberg*—Weinberg was a German physician who did research in blood types.

the fraction of the population of each genotype is:

$$\text{fraction with genotype } AA = r$$
$$\text{fraction with genotype } Aa = s$$
$$\text{fraction with genotype } aa = t.$$

Further, assume that these fractions are the same among the males and females. It follows that if an individual is picked at random, the probability that it provides its offspring an A gene is

$$p = r + \frac{s}{2}. \tag{6}$$

(Why?) Similarly, the probability that a random individual provides its offspring an a gene is

$$q = t + \frac{s}{2}. \tag{7}$$

(Why?) Thus, the probability that the offspring is of genotype AA is

$$p \cdot p = p^2$$

and the probability that the offspring is of genotype aa is

$$q \cdot q = q^2.$$

Finally, there are two different ways that the offspring can be of genotype Aa; the male parent can provide the A gene and the female parent the a gene, or vice versa. Each of these possibilities occurs with probability pq, so that the probability that the offspring is of genotype Aa is

$$pq + pq = 2pq.$$

Table 8.3 summarizes the results we have obtained so far.

Table 8.3 Genotype probabilities for parents and offspring

	AA	Aa	aa	
Parents	r	s	t	
Offspring (first generation)	p^2	$2pq$	q^2	where $p = r + \frac{1}{2}s$ and $q = t + \frac{1}{2}s$

Let us now investigate the genotype probabilities for the offspring of the offspring, that is, the second generation. Again, we shall assume that mating is random and the genotype probabilities are the same for both males and females. First, we shall need a preliminary observation. Since r, s, and t denote the fraction of parents of each genotype in the original population, we have

$$r + s + t = 1.$$

Thus,

$$\begin{aligned} p + q &= \left(r + \frac{s}{2}\right) + \left(t + \frac{s}{2}\right) \\ &= r + s + t \\ &= 1. \end{aligned}$$

Now we are in a position to compute the genotype probabilities for the second generation. If an individual is picked at random from the first generation of offspring, then from Table 8.3, the probability that it provides its offspring an A gene is

$$p^2 + \frac{1}{2}(2pq).$$

(Why?) But this expression can be simplified by writing

$$\begin{aligned} p^2 + \frac{1}{2}(2pq) &= p^2 + pq \\ &= p(q + p) \\ &= p \cdot 1 \\ &= p. \end{aligned} \tag{8}$$

Similarly, the probability that it provides its offspring with an a gene is

$$\begin{aligned} q^2 + \frac{1}{2}(2pq) &= q^2 + pq \\ &= q(q + p) \\ &= q \cdot 1 \\ &= q. \end{aligned} \tag{9}$$

From (8) and (9) and computations like those used for the first generation we see that for the second generation of offspring, the probability of genotype AA is

$$p \cdot p = p^2,$$

the probability of genotype aa is

$$q \cdot q = q^2,$$

and the probability of genotype Aa is

$$pq + pq = 2pq.$$

Table 8.4 summarizes the results we have obtained so far. From this table, we see that the genotype probabilities are the same for the first and second generations of offspring. Moreover, since the genotype probabilities of any generation depend only on the genotype probabilities of the previous generation, it follows that the third generation and in fact all future genera-

Table 8.4

	AA	Aa	aa	
Parents	r	s	t	
Offspring (first generation)	p^2	$2pq$	q^2	
Offspring (second generation)	p^2	$2pq$	q^2	where $p = r + \frac{1}{2}s$ and $q = t + \frac{1}{2}s$

tions of offspring will have the same genotype probabilities as the first and second generations. This result is the ***Hardy–Weinberg stability principle.*** It is often stated in the following form.

> **Hardy–Weinberg Stability Principle** In a population with random mating, the genotype probabilities stabilize after one generation.

The Hardy–Weinberg principle shows that, according to Mendelian theory, recessive traits should not die out. Rather, the proportion of the population exhibiting the recessive trait should stabilize over a period of time. We note, however, that the Hardy–Weinberg principle involves certain assumptions, such as random mating, which are not always satisfied. In such situations, the genetic composition of the population can slowly change from generation to generation.

Example 4 Assume that in an initial population of plants, 50% of the plants have red flowers (R), 40% have pink flowers (P), and 10% have white flowers (W). Assuming that the conditions of the Hardy–Weinberg principle are satisfied, find the percentage of each kind of flower that will occur in the first generation of offspring.

Solution Since red flowers correspond to genotype AA, pink flowers to genotype Aa, and white flowers to genotype aa, the fraction of each genotype in the original population is

$$r = .5 \qquad \text{for genotype } AA$$

$$s = .4 \qquad \text{for genotype } Aa$$

and

$$t = .1 \qquad \text{for genotype } aa.$$

From formulas (6) and (7) we obtain

$$p = r + \frac{s}{2} = .7$$

and

$$q = t + \frac{s}{2} = .3.$$

Thus, from Table 8.4 the probabilities for each genotype in the first generation are

$$p^2 = (.7)^2 = .49 \qquad \text{for genotype } AA$$

$$2pq = 2(.7)(.3) = .42 \qquad \text{for genotype } Aa$$

and

$$q^2 = (.3)^2 = .09 \qquad \text{for genotype } aa.$$

Therefore 49% of the first generation will be red flowered, 42% pink flowered, and 9% white flowered.

EXERCISE SET 8.2

1. Compute the genotype probabilities that result from the mating $aa \times Aa$, and check your results against those given in Table 8.2.

2. Use your results from Exercise 1 to explain observation (1f) in the text.

In Exercises 3–5, assume a certain dominant trait is produced by genotypes AA and Aa and the recessive trait by genotype aa.

3. For the mating $Aa \times Aa$, what percentage of the offspring will have the dominant trait?

4. For the mating $Aa \times aa$, what percentage of the offspring will have the dominant trait?

5. If a female of genotype Aa is mated with a male whose genotype is unknown, what is the probability that an offspring will have the dominant trait?

6. Suppose that in an initial population, the genotypes AA, Aa, and aa occur in equal numbers. Assuming that the conditions of the Hardy–Weinberg principle are satisfied, find the percentage of each genotype that will occur in the first generation of offspring.

7. For the problem in Exercise 6, find the percentage of each genotype that will occur in the second generation. Explain your answer.

8. Suppose that in an initial population of plants, 50% have red flowers, 20% have pink flowers, and 30% have white flowers. Assuming that the conditions of the Hardy–Weinberg principle are satisfied, find the percentage of each color that will occur in the first generation of offspring.

9. For the problem in Exercise 8, find the percentage of each color that will occur in the second generation. Explain your answer.

10. Consider families with two children in which the male parent has genotype AA and the female parent has genotype Aa. Show that $\frac{1}{4}$ of the families have two AA children, $\frac{1}{2}$ of the families have one AA child and one Aa child, and $\frac{1}{4}$ of the families have two Aa children.

11. Consider families with three children in which the male parent has genotype AA and the female parent has genotype Aa. Show that $\frac{1}{8}$ of the families will have three AA children, $\frac{3}{8}$ will have two AA and one Aa, $\frac{3}{8}$ will have one AA and two Aa, and $\frac{1}{8}$ will have three Aa children.

12. A number of human diseases, such as amaurotic idiocy, are inherited recessive traits that are fatal to infants. Some of these diseases are apparently determined by a single pair of genes. Individuals of genotype AA and Aa are healthy, while individuals of genotype aa are affected and die in infancy. Because of the fatal nature of the a gene, it is commonly called a ***lethal gene*** and a parent of genotype Aa is called a ***carrier*** of the lethal gene.

(a) If an adult had a brother or sister affected by the disease, then both parents of the adult must be carriers. Explain why.

(b) Show that if an adult had a brother or sister affected by the disease, then his genotype probabilities are

Genotype of adult	AA	Aa
Probability	$\frac{1}{3}$	$\frac{2}{3}$

(*Hint:* This is a conditional probability problem. We are given that the individual has survived to adulthood, so that he is not of genotype aa.)

8.3 APPLICATIONS OF PROBABILITY AND EXPECTED VALUE TO LIFE INSURANCE AND MORTALITY

Probability and statistics play an important role in the insurance industry. Companies that want to insure individuals against fire, death, or automobile accidents use probability and statistics to determine their ***premiums,*** that is, what they will charge for the insurance. In order to determine the premiums, the company must be able to estimate the probability that it will have to pay off on each policy it sells. In this section we shall illustrate how this is done in the field of life insurance.

Life insurance companies have compiled statistics, called ***mortality tables,*** which they use to determine the probability that a policyholder will die while his insurance policy is in effect. Mortality tables are constructed by selecting a large sample of individuals and recording how many deaths occur among individuals in the sample at each age. Mortality tables are constantly revised to reflect changes in medical technology, health standards, and environmental conditions. There are many different mortality tables available, and each insurance company tries to pick a mortality table based on individuals whose characteristics closely match those of its policyholders. Table 8.5 is a typical mortality table, called the *Commissioners Standard Ordinary Mortality Table.*

Table 8.5 Commissioners Standard Ordinary Mortality Table

Age	Number living	Age	Number living	Age	Number living
0	10,000,000	34	9,396,358	67	6,355,865
1	9,929,200	35	9,373,807	68	6,114,088
2	9,911,725	36	9,350,279	69	5,859,253
3	9,896,659	37	9,325,594	70	5,592,012
4	9,882,210	38	9,299,482	71	5,313,586
5	9,868,375	39	9,271,491	72	5,025,855
6	9,855,053	40	9,241,359	73	4,731,089
7	9,842,241	41	9,208,737	74	4,431,800
8	9,829,840	42	9,173,375	75	4,129,906
9	9,817,749	43	9,135,122	76	3,826,895
10	9,805,870	44	9,093,740	77	3,523,881
11	9,794,005	45	9,048,999	78	3,221,884
12	9,781,958	46	9,000,587	79	2,922,055
13	9,769,633	47	8,948,114	80	2,626,372
14	9,756,737	48	8,891,204	81	2,337,524
15	9,743,175	49	8,829,410	82	2,058,541
16	9,728,950	50	8,762,306	83	1,792,639
17	9,713,967	51	8,689,404	84	1,542,781
18	9,698,230	52	8,610,244	85	1,311,348
19	9,681,840	53	8,524,486	86	1,100,037
20	9,664,994	54	8,431,654	87	909,929
21	9,647,694	55	8,331,317	88	741,474
22	9,630,039	56	8,223,010	89	594,477
23	9,612,127	57	8,106,161	90	468,174
24	9,593,960	58	7,980,191	91	361,365
25	9,575,636	59	7,844,528	92	272,552
26	9,557,155	60	7,698,698	93	200,072
27	9,538,423	61	7,542,106	94	142,191
28	9,519,442	62	7,374,370	95	97,165
29	9,500,118	63	7,195,099	96	63,037
30	9,480,358	64	7,003,925	97	37,787
31	9,460,165	65	6,800,531	98	19,331
32	9,439,447	66	6,584,614	99	6,415
33	9,418,208				

The following examples illustrate how this table can be used to compute probabilities important to an insurance company. In these examples, we will use L_x to denote the event that an individual is alive on his xth birthday and we will use n_x to denote the number of people living at their xth birthday according to Table 8.5. For example, $n_{22} = 9{,}630{,}039$ and L_{22} denotes the event that an individual is alive on his 22nd birthday.

Example 5 Use Table 8.5 to estimate the probability that an individual will be alive at age 50.

Solution From the Table 8.5, 8,762,306 of the original 10,000,000 people were alive at age 50, so that the probability that an individual is alive at age 50 is

$$P(L_{50}) = \frac{8{,}762{,}306}{10{,}000{,}000} \simeq .876. \tag{1}$$

Example 6 Use Table 8.5 to estimate the probability that an individual who is alive at age 50 will die between the ages of 50 and 51.

Solution In this problem we are given that the individual is alive at age 50, and we want the probability of the event L'_{51}, that he is not alive at age 51. Thus, we must find the conditional probability

$$P(L'_{51}|L_{50}). \tag{2}$$

From the definition of conditional probability, (2) can be written

$$P(L'_{51}|L_{50}) = \frac{P(L_{50} \cap L'_{51})}{P(L_{50})}. \tag{3}$$

Of the original 10,000,000 individuals, there were 8,762,306 alive at age 50 and only 8,689,404 still alive at age 51. Thus the probability an individual is alive at age 50 and not alive at age 51 is

$$P(L_{50} \cap L'_{51}) = \frac{8{,}762{,}306 - 8{,}689{,}404}{10{,}000{,}000} = \frac{72{,}902}{10{,}000{,}000}. \tag{4}$$

(Why?) We can now use results (1) and (4) to compute the probability in (3); we obtain

$$P(L'_{51}|L_{50}) = \frac{72{,}902}{10{,}000{,}000} \div \frac{8{,}762{,}306}{10{,}000{,}000} = \frac{72{,}902}{8{,}762{,}306} \simeq .00832.$$

Using the method illustrated in the last example, we can derive a simple formula for computing the probability that an individual who is alive at age x will not be alive at age y. Mathematically, this probability is

$$P(L'_y|L_x) = \frac{P(L_x \cap L'_y)}{P(L_x)}. \tag{5}$$

As illustrated in Equation (4), the numerator of (5) is

$$P(L_x \cap L'_y) = \frac{\text{number living at age } x - \text{number living at age } y}{10{,}000{,}000}$$

$$= \frac{n_x - n_y}{10{,}000{,}000} \tag{6}$$

and as illustrated in Equation (1), the denominator in (5) is

$$P(L_x) = \frac{\text{number living at age } x}{10{,}000{,}000} = \frac{n_x}{10{,}000{,}000}. \tag{7}$$

From (6) and (7), the probability in (5) is

$$P(L'_y|L_x) = \frac{n_x - n_y}{10{,}000{,}000} \div \frac{n_x}{10{,}000{,}000} = \frac{n_x - n_y}{n_x}. \tag{8}$$

Thus, Equation (8) gives the following useful result.

The probability that an individual who is alive at age x will not be alive at age y is given by

$$\frac{n_x - n_y}{n_x}, \tag{9}$$

where n_x is the number of individuals alive at age x and n_y is the number of individuals alive at age y.

Example 7 Use Formula (9) and Table 8.5 to compute the probability that an individual who is alive at age 30 will die before age 40.

Solution From formula (9) and Table 8.5 this probability is

$$\frac{n_{30} - n_{40}}{n_{30}} = \frac{9{,}480{,}358 - 9{,}241{,}359}{9{,}480{,}358}$$

$$= \frac{238{,}999}{9{,}480{,}358} \simeq .0252.$$

From formula (9),

$$\frac{n_x - n_y}{n_x}$$

is the probability that an individual who is alive at age x will *not* be alive at age y. Therefore,

$$1 - \frac{n_x - n_y}{n_x}$$

is the probability that an individual who is alive at age x will *still* be alive at age y. This formula can be simplified by writing

$$1 - \frac{n_x - n_y}{n_x} = \frac{n_x}{n_x} - \frac{n_x - n_y}{n_x} = \frac{n_x - n_x + n_y}{n_x} = \frac{n_y}{n_x}.$$

Thus, we obtain the following useful result.

The probability that an individual who is alive at age x will still be alive at age y is given by

$$\frac{n_y}{n_x} \tag{10}$$

where n_x is the number of individuals alive at age x and n_y is the number of individuals alive at age y.

Example 8 Use formula (10) and Table 8.5 to compute the probability that an individual who is alive at age 60 will still be alive at age 70.

Solution From formula (10) and Table 8.5 this probability is

$$\frac{n_{70}}{n_{60}} = \frac{5{,}592{,}012}{7{,}698{,}698} \simeq .726.$$

The following two examples illustrate how the notion of expected value of a random variable enters into the computation of life insurance premiums.

Example 9 An individual, on his 45th birthday, wants to buy a life insurance policy that will pay his beneficiary \$2000 if he dies within one year. How much should the company charge for such a policy to make a profit?

Solution For simplicity, we shall ignore the company's overhead costs and the fact that it can obtain income by investing the money it receives for the policy. If we denote the company's profit by X, then X can be viewed as a random variable defined on the sample space of an experiment. The experiment consists of observing the policyholder for one year and recording whether he lives or dies. Thus, the sample space S for the experiment can be written

$$S = \{s_1, s_2\},$$

where

$$s_1 = \text{the policyholder lives through the year}$$

$$s_2 = \text{the policyholder dies during the year.}$$

If we assume that the company charges m dollars for the policy, then its profit X will be m dollars if the policyholder lives and $m - 2000$ dollars if he dies.

Since the company does not know whether the policyholder will live or die it cannot determine its profit with certainty. The best it can do is compute the expected value of the profit

$$E(X) = mP(s_1) + (m - 2000)P(s_2). \tag{11}$$

Since the policyholder is celebrating his 45th birthday, it follows from formulas (9) and (10) and Table 8.5 that

$$P(s_1) = \frac{n_{46}}{n_{45}} = \frac{9{,}000{,}587}{9{,}048{,}999} \simeq .99465$$

and

$$P(s_2) = \frac{n_{45} - n_{46}}{n_{45}} = \frac{48{,}412}{9{,}048{,}999} \simeq .00535.$$

Thus, (11) becomes

$$E(X) = m(.99465) + (m - 2000)(.00535). \tag{12}$$

If the company expects to break even, then the expected profit $E(X)$ should be zero, so that from (12) the value of m for which the company breaks even must satisfy

$$0 = m(.99465) + (m - 2000)(.00535)$$

or

$$0 = m(.99465) + m(.00535) - (2000)(.00535)$$

or

$$0 = m - 10.7$$

or

$$m = 10.7 \quad \text{dollars.}$$

Thus, the company must charge more than \$10.70 to make a profit.

It is important to interpret the results in this example properly. By charging \$10.70 for the policy, the insurance company does not *guarantee* that it will break even on the policy. Obviously, if the individual dies, the company will lose money when it pays off the \$2000 to the beneficiary. However, if the company sells many such policies, it will break even in the long run if it charges \$10.70 for each policy.

In this section we have only scratched the surface of probability and its applications to business and insurance. Readers interested in pursuing this subject in more detail will find the following references helpful:

[1] J. E. Freund, *College Mathematics with Business Applications* (Englewood Cliffs, New Jersey: Prentice Hall, 1969).

[2] J. Kemeny, A. Schleifer, J. Snell, and G. Thompson, *Finite Mathematics with Business Applications* (Englewood Cliffs, New Jersey: Prentice Hall, 1962).

[3] E. Monzon and P. Rees, *Mathematics of Finance,* Ginn and Company, 1952.

In addition, the reader may be interested in writing for a pamphlet entitled *Sets, Probability, and Statistics* published by the Institute of Life Insurance, Educational Division, 277 Park Avenue, New York, New York 10017.

EXERCISE SET 8.3

The following exercises are to be solved using the *Commissioners Standard Ordinary Mortality Table* (Table 8.5).

1. Estimate the probability that an individual will be alive at age

(a) 16 (b) 37 (c) 92.

2. Estimate the probability that an individual will be alive at age

(a) 42 (b) 51 (c) 78.

3. Estimate the probability that an individual will not be alive at age

(a) 21 (b) 46 (c) 63.

4. Estimate the probability that an individual will not be alive at age

(a) 51 (b) 29 (c) 88.

5. Estimate the probability that

(a) an individual who is alive at age 20 will still be alive at age 40;

(b) an individual who is alive at age 51 will still be alive at age 60;

(c) an individual who is alive at age 48 will still be alive at age 49.

6. Estimate the probability that

(a) an individual who is alive at age 10 will still be alive at age 80;

(b) an individual who is alive at age 36 will still be alive at age 45;

(c) an individual who is alive at age 76 will still be alive at age 77.

7. Estimate the probability that an individual will die between the ages of 49 and 50.

8. Estimate the probability that an individual will die between the ages of 62 and 63.

9. An individual, on his 35th birthday, wants to buy a life insurance policy that will pay his beneficiary $10,000 if he dies within 10 years. How much should the company charge for such a policy to make a profit? (As in Example 9, ignore the overhead costs and any income the company can obtain by investing the money it receives for the policy.)

10. An individual, on his 60th birthday, wants to buy a life insurance policy that will pay his beneficiary $20,000 if he dies within 5 years. How much should the company charge for such a policy to make a profit? (Make the assumptions stated at the end of Exercise 9.)

8.4 INTRODUCTION TO GAME THEORY AND APPLICATIONS

Many activities in business, economics, politics, and the sciences call for decisions in competitive situations. As a result of the pioneering efforts by John von Neumann,† Emil Borel,‡ and Oskar Morgenstern,§ beginning in the 1920s, it is now recognized that many of these competitive situations are analogous to games, played according to formal rules. The players are groups or individuals, each pursuing their own objectives in direct conflict with the other players. In this section and the next we shall consider the simplest kinds of games; we shall show how matrices and probability theory can be used to help determine optimal strategies for the players, and we shall discuss several applications.

We shall begin with some examples that illustrate the kinds of problems to which game theory is applicable.

Franchise Location: A Business Game

Example 10 A new regional shopping center, Oxford Mall, plans to have only two restaurants, one to be operated by MacDonald's and the other by Gino's.

† *John von Neumann* (1903–1957)—Von Neumann was born in Budapest, Hungary and was a child prodigy in mathematics and general science. In 1923 he received a doctorate in mathematics from the University of Budapest and a degree in chemical engineering from the Federal Institute of Technology at Zurich, Switzerland. After lecturing in Europe he came to Princeton University, where he stayed until his death. Von Neumann was probably the greatest mathematical genius of this century. He contributed to quantum mechanics, economics, computers, many areas of pure mathematics, and he developed a technique that accelerated the production of the first atomic bomb. His genius was recognized with many awards and honors.

‡ *Emil Borel* (1871–1956)—Borel was a French mathematician who made important contributions to several branches of mathematics. His pioneering work helped launch the field of measure theory on which the modern advanced notions of length, area, volume, and probability rest. In addition to his work in mathematics, he sat in the Chamber of Deputies for a period of time and was Minister of the Navy.

§ *Oskar Morgenstern* (1902–   )—Morgenstern is a German-born American economist at Princeton University. His book, *Theory of Games and Economic Behavior*, coauthored with John von Neumann is a landmark work in economics and game theory.

As illustrated in Figure 8.1, MacDonald's can locate at position ① (an end) or at position ② (near the center), and Gino's can locate at position ③ (near the center) or at position ④ (an end). If both franchises choose central locations or both choose end locations, they will each get 50% of the business. If MacDonald's chooses the center and Gino's the end, then MacDonald's will capture 75% of the business (and Gino's 25%), while if MacDonald's chooses an end and Gino's the center, then MacDonald's will get 30% of the business (and Gino's 70%).

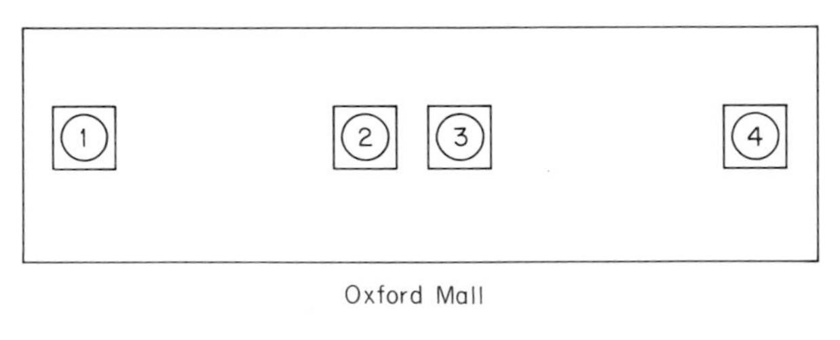

Oxford Mall

Figure 8.1

This situation is analogous to a game; each player (franchise) can control where it locates, but it cannot control where its opponent locates. The objective of each player is to develop an "optimal strategy" that will ensure his franchise the maximum possible business.

Example 11 (Penny Matching) Consider a game involving two players, I and II, in which each player selects one side of a penny without knowing his opponent's choice. If the choices match, then player I pays player II \$1; otherwise player II pays player I \$1. As in the previous example, each player can control his own move, but cannot control his opponent's move. The objective of each player is to develop an optimal strategy that will ensure him the best possible result.

Example 12 (A Matrix Game) Consider the 3×3 matrix

$$\begin{bmatrix} 1 & 6 & -1 \\ 3 & -2 & -3 \\ 4 & 5 & 3 \end{bmatrix}.$$

We can associate a game with this matrix as follows. Consider two players, R and C. Player R picks a row without telling C his choice, and player C picks a column without telling R. After the players make their choices, money or some other item of value changes hands as specified by the entry in the chosen row and column of the matrix. Positive entries in the matrix represent amounts player C pays to player R and negative signs precede amounts player R pays to player C. For example, if player R picks row 2 and player C picks column 1, then player C pays 3 units to player R; and if player R picks row 1 and player C picks column 3, then player R pays 1 unit to player C.

Games of the type described in this example are called ***matrix games*** and the matrix is called the ***payoff*** matrix.

The games in Examples 10, 11, and 12 are ***two-person*** games, which means that they are played by two opposing players. These games are also examples of ***constant-sum*** games, which means that the sum of the payoffs realized by the players is a constant that does not depend on which moves are made. To illustrate, in Example 10, MacDonald's percentage of the business plus Gino's percentage of the business is always 100% regardless of their moves. In Example 11, player I's payoff plus player II's payoff is always zero, regardless of their moves, since the amount lost by one player is won by the other player. Constant-sum games in which the constant is zero are called ***zero-sum*** games. Examples 11 and 12 are zero-sum games. Example 12 is of particular importance because any two-person constant-sum game in which each player has only a finite number of possible moves can be viewed as a matrix game. The following example illustrates this point.

Example 13 Consider the business game described in Example 10. The players, MacDonald's and Gino's, each have two possible moves: locate in the center or locate at an end. The possible outcomes resulting from the moves can be tabulated in matrix form as

$$\begin{array}{ccc} & & \text{Gino's moves} \\ & & \begin{array}{cc} \text{Center} & \text{End} \end{array} \\ \text{MacDonald's moves} & \begin{array}{c} \text{Center} \\ \text{End} \end{array} & \begin{bmatrix} 50 & 75 \\ 30 & 50 \end{bmatrix} \end{array} \qquad (1)$$

where the numbers in this matrix indicate the percentage of shoppers that MacDonald's will service as a result of the player's moves. Thus, if MacDonald's locates at an end and Gino's locates in the center, then MacDonald's will get 30% of the business.

Using (1) as the payoff matrix, Example 10 can be viewed as a matrix game, where MacDonald's picks a row and Gino's picks a column; the entry in the chosen row and column then indicates what percentage of the shoppers Gino's "pays" or concedes to MacDonald's.

Example 14 Consider the penny-matching game described in Example 11. The players, I and II, each have two possible moves: show a head or show a tail. The possible outcomes resulting from the moves can be tabulated in matrix form as

$$\begin{array}{c c} & \text{Player II} \\ & \begin{array}{cc} \text{Heads} & \text{Tails} \end{array} \\ \text{Player I} \begin{array}{c} \text{Heads} \\ \text{Tails} \end{array} & \begin{bmatrix} -1 & 1 \\ 1 & -1 \end{bmatrix} \end{array}$$

Using this as the payoff matrix, Example 11 can be viewed as a matrix game, where player I picks a row and player II picks a column; the entry in the chosen row and column then indicates the amount that player II pays player I. For example, if player I picks row 2 (tails) and player II picks column 1 (heads), then II pays I \$1. Similarly, if player I picks row 2 and player II picks column 2, then I pays II \$1.

We shall now describe how the players of a matrix game can determine their optimal moves. In our analysis, we shall make the following assumptions:

1. Each player must choose his move without knowing what move his opponent has made or is planning to make.

2. Each player chooses his move with the assumption that he is facing an intelligent opponent who will make the best possible move.

The next example illustrates the reasoning process the players of a matrix game might use to determine their optimal moves.

Example 15 Consider a game with payoff matrix

$$\begin{bmatrix} -2 & 5 & -1 \\ 4 & -4 & 0 \\ 3 & 4 & 1 \\ 3 & 3 & 2 \end{bmatrix}.$$

Recall that positive entries in the matrix are amounts won by the row player R and negative signs precede amounts won by the column player C.

In this game, player R would like to win the 5 units appearing in row 1. If he plays row 1, however, then he must assume that his opponent will make the best possible move and play column 1. Thus, instead of winning

5 units he would *lose* 2 units. With this in mind, it is clear that row 3, for example, would be a better move than row 1 since player R is assured of winning at least 1 unit, even against player C's best countermove (column 3). By the same token, row 4 would be an even better move than row 3 since player R could then assure himself of winning at least 2 units against C's best countermove.

The analysis in the last example suggests the following method for player R to determine his best move:

Strategy for Player R

Step (a) For each row, player R should find the smallest element in the row (called the ***row minimum***); this number represents the payoff to R when R chooses that row and C makes the best possible countermove.

Step (b) Then, to obtain his best possible move, player R should choose a row that yields the largest value among the row minima. This assures that player R will get the largest possible payoff against player C's best countermoves.

In Figure 8.2, we have used these two steps to determine player R's best move in the game from Example 15.

$$\begin{bmatrix} -2 & 5 & -1 \\ 4 & -4 & 0 \\ 3 & 4 & 1 \\ 3 & 3 & 2 \end{bmatrix} \quad \begin{matrix} \text{Row minima} \\ -2 \\ -4 \\ 1 \\ 2 \end{matrix} \qquad \begin{bmatrix} -2 & 5 & -1 \\ 4 & -4 & 0 \\ 3 & 4 & 1 \\ 3 & 3 & 2 \end{bmatrix} \quad \begin{matrix} \text{Row minima} \\ -2 \\ -4 \\ 1 \\ 2 \leftarrow \text{largest row minimum} \end{matrix}$$

Step (a) Compute the row minima.

Step (b) Row 4 is player R's best move.

Figure 8.2

Player C will use a similar analysis to determine his best move (see next page). In Figure 8.3 we have used these steps to determine player C's best move in the game from Example 15.

Strategy for Player C

Step (a) For each column, player C should find the largest element in the column (called the ***column maximum***); this number represents the payoff to R when C plays that column and R makes the best possible countermove.

Step (b) To obtain his best possible move, player C should choose a column that yields the smallest value among the column maxima. This assures that player C will give the smallest possible payoff against player R's best countermoves.

$$\begin{bmatrix} -2 & 5 & -1 \\ 4 & -4 & 0 \\ 3 & 4 & 1 \\ 3 & 3 & 2 \end{bmatrix}$$

Column maxima 4 5 2

Step (a) Compute the column maxima.

$$\begin{bmatrix} -2 & 5 & -1 \\ 4 & -4 & 0 \\ 3 & 4 & 1 \\ 3 & 3 & 2 \end{bmatrix}$$

Column maxima 4 5 2 ↑ Smallest column maximum

Step (b) Column 3 is player C's best move.

Figure 8.3

Example 16 Find the best moves for Gino's and MacDonald's in the business game of Example 10.

Solution As shown in Example 13, the payoff matrix for this game is

		Gino's Center	Gino's End
MacDonald's	Center	50	75
	End	30	50

Computing the row minima and column maxima we obtain

			Row minima
	50	75	50
	30	50	30
Column maxima	50	75	

Since the largest row minimum is 50, MacDonald's should play row 1, and since the smallest column maximum is 50, Gino's should play column 1. Thus, both MacDonald's and Gino's should choose central locations, in which case they will each get 50% of the business.

The following notions are useful in the analysis of matrix games.

Definition A matrix game is called ***strictly determined*** if there is an entry in the payoff matrix that is both the smallest element in its row and the largest element in its column. Such an entry is called a ***saddle point*** for the game.

To illustrate, the game in Example 15 is strictly determined, since the entry 2, which appears in row 4 and column 3, is a saddle point.

As the next example shows, there are matrix games that are not strictly determined, that is, have no saddle points.

Example 17 Consider the game with payoff matrix

$$\begin{array}{r} \\ \\ \\ \text{Column maxima} \end{array} \begin{array}{c} \\ \left[\begin{array}{rrr} 2 & -1 & 3 \\ 0 & 1 & -2 \\ -3 & 2 & 1 \end{array}\right] \\ \begin{array}{rrr} 2 & 2 & 3 \end{array} \end{array} \begin{array}{r} \text{Row minima} \\ -1 \\ -2 \\ -3 \\ \\ \end{array}$$

It is clear from the indicated row minima and column maxima that no element is simultaneously a minimum for its row and a maximum for its column; thus the game has no saddle points.

The next example shows that a game can have more than one saddle point.

Example 18 Consider the game with payoff matrix

$$\begin{bmatrix} 2 & 3 & -2 & -1 \\ 6 & 7 & 8 & 6 \\ 0 & 1 & -2 & 3 \\ 6 & 8 & 9 & 6 \end{bmatrix}$$

Each of the entries marked in color is a minimum for its row and maximum for its column, so that this game has four saddle points.

When a game has more than one saddle point, player R or player C or both have more than one optimal move. Thus, in Example 18, player R has two best moves, row 2 and row 4; player C also has two best moves, column 1 and column 4.

It is not accidental that the same number appears at each saddle point in Example 18. It can be proved, although we shall not do it, that the same number *must* occur at every saddle point. This number is called the ***value*** of the game. Thus, in a strictly determined game, the value of the game is the payoff that results when both players make their best moves.

It is interesting to note that in strictly determined games, neither player can benefit from discovering which move his opponent will make. In other words, *it does not do any good to spy on one's opponent when playing a strictly determined game.* To see why, consider the game from Example 15. Even if player C announced in advance that he would play column 3 (see Figures 8.2 and 8.3), it would still be best for player R to play row 4, which was the choice he made with no knowledge of player C's intentions.

On the other hand, consider the penny-matching game from Example 11, where the payoff matrix was

$$\begin{array}{ccc} & & \text{Player II} \\ & & \begin{array}{cc}\text{Heads} & \text{Tails}\end{array} \\ \text{Player I} & \begin{array}{c}\text{Heads}\\ \text{Tails}\end{array} & \begin{bmatrix} -1 & 1 \\ 1 & -1 \end{bmatrix} \end{array}$$

This game has no saddle point and consequently is not strictly determined. Since the minimum in each row is -1, both rows are equally good moves for player I. Similarly, since both column maxima are 1, both columns are equally good moves for player II. Suppose, however, that player I discovers in advance that player II will play the first column. Player I would then definitely play row 2 and win \$1. Thus, for games that are not strictly determined, discovery of the opponent's intended move affects the play of the game.

We conclude this section with a game theory analysis of an important battle that occurred in the Pacific Theatre during World War II.†

A War Game: The Battle of Rabaul-Lae in World War II

Example 19 In 1943 General Kenney was commander of the Allied Air Forces in the Southwest Pacific Area. In the critical stages of the struggle for New Guinea, intelligence reports indicated that a Japanese troop and supply

† This example was presented by O. G. Haywood, Jr. in the *Journal of the Operations Research Society of America* **2** (1954) 365–385. It is also discussed in reference [2] at the end of Section 8.5.

convoy was going to be moved from the port of Rabaul in New Britain to Lae in New Guinea (Figure 8.4). The Japanese commander had two possible courses of action: he could either travel north of New Britain or south of New Britain. Either trip would take three days. Rain and poor visibility were predicted for the northern route; clear weather was predicted for the southern route. General Kenney could concentrate most of his reconnaissance aircraft on either the northern or southern route. Once the convoy was sighted, it would be bombed until its arrival in Lae.

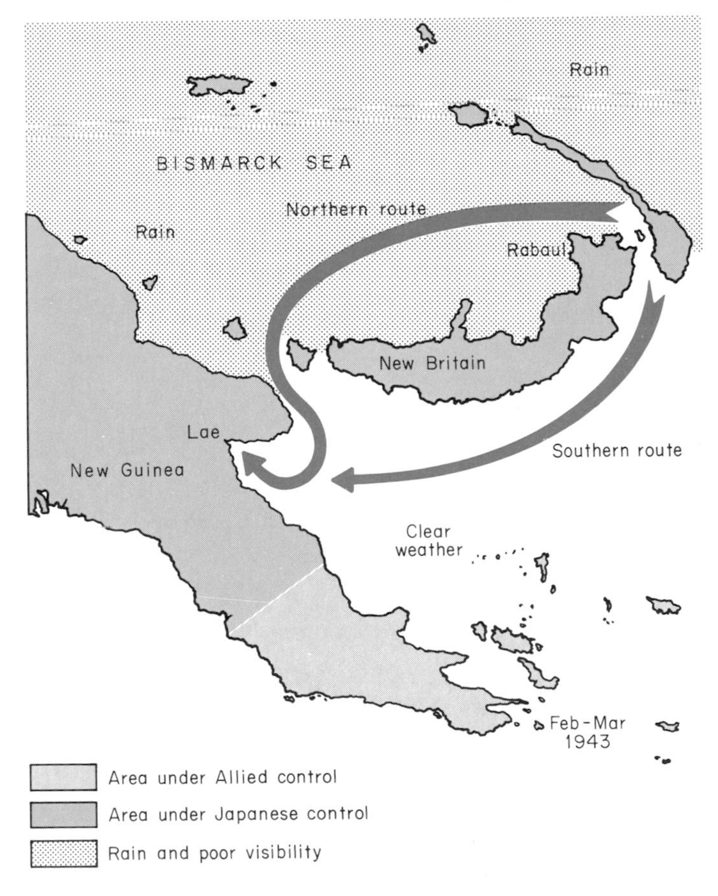

Figure 8.4

General Kenney's staff analyzed the possible alternatives as follows:

Situation 1 If Kenney concentrated on the northern route and the Japanese also went the northern route then, due to bad weather, the convoy would be discovered on the second day, allowing two days of bombing (Figure 8.5a).

If Kenney concentrated on the northern route and the Japanese went the southern way, the convoy would be missed on the first day (most of the aircraft would be on the northern route), allowing two days of bombing (Figure 8.5b).

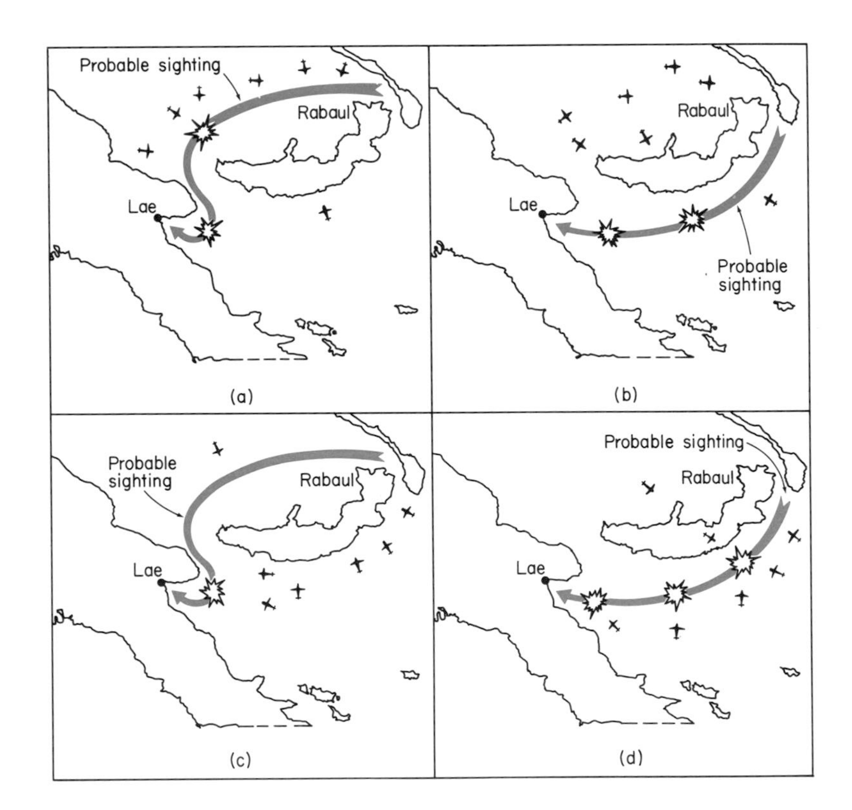

Figure 8.5

Situation 3 If Kenney concentrated on the southern route and the Japanese went the northern way, then, due to poor visibility and limited aircraft on the northern route, the convoy would not be spotted for the first two days, allowing one day of bombing (Figure 8.5c).

Situation 4 If Kenney concentrated on the southern route and the Japanese went the southern way, then the convoy would be spotted at once, giving three days of bombing (Figure 8.5d).

This military problem can be viewed as a two-person game; the Japanese commander wanted to minimize the number of days he would be bombed, while General Kenney wanted to maximize the number of days he could bomb the Japanese convoy. The number of bombing days in each of the four possible situations described above is given by the following payoff matrix:

		Japanese	
		Northern route	Southern route
Kenney	Northern route	2 days	2 days
	Southern route	1 day	3 days

Computing the row minima and column maxima for this game we obtain

$$\begin{array}{rcc|c} & & & \text{Row minima} \\ & 2 & 2 & 2 \\ & 1 & 3 & 1 \\ \text{Column maxima} & 2 & 3 & \end{array}$$

Since the largest row minimum is 2, Kenney's best move is row 1; and since the smallest column maximum is 2, the best move for the Japanese commander is column 1. Thus, the best course of action for each side is to pick the northern route.

As a historical note, these were precisely the choices made by the opposing sides. The convoy sailed along the northern route; it was sighted one day after it sailed and was bombed for two days, resulting in a disastrous defeat for the Japanese. As our game-theory analysis shows, however, the Japanese commander made the best possible decision in a hopeless situation.

It is interesting to observe that the Rabaul-Lae game is strictly determined since the entry in row 1 and column 1 is a saddle point. (It is the minimum for its row and a maximum for its column.) Thus, neither commander would have benefited if an intelligence operation had uncovered his opponent's decision in advance.

EXERCISE SET 8.4

1. Which of the following matrix games are strictly determined?

(a) $\begin{bmatrix} -1 & 2 \\ 3 & -2 \end{bmatrix}$ (b) $\begin{bmatrix} -1 & -2 \\ 3 & 2 \end{bmatrix}$ (c) $\begin{bmatrix} -1 & 1 & 0 \\ 4 & 5 & 3 \\ -1 & 2 & 0 \end{bmatrix}$

(d) $\begin{bmatrix} 1 & 0 & 2 \\ 3 & 2 & -1 \\ -1 & 2 & 4 \end{bmatrix}$ (e) $\begin{bmatrix} 1 & 0 & 0 & -1 \\ 2 & 3 & 1 & 4 \\ 2 & -1 & 3 & 2 \\ 3 & -2 & -2 & 3 \end{bmatrix}$ (f) $\begin{bmatrix} 1 & -2 & 0 & 1 \\ 1 & 1 & 1 & 3 \\ 1 & 1 & 1 & 3 \\ 0 & 0 & 0 & 3 \end{bmatrix}$.

2. Find all saddle points for the following two-person constant sum matrix games

(a) $\begin{bmatrix} 4 & 6 \\ 2 & -7 \end{bmatrix}$ (b) $\begin{bmatrix} 7 & 3 \\ 3 & -4 \end{bmatrix}$

(c) $\begin{bmatrix} -4 & 1 & 1 \\ -3 & 4 & 7 \\ -10 & 3 & 5 \end{bmatrix}$ (d) $\begin{bmatrix} 0 & 5 & 2 \\ 3 & 3 & 7 \\ 5 & 7 & -1 \end{bmatrix}$ (e) $\begin{bmatrix} 1 & -2 & 0 & 4 \\ 1 & 1 & 1 & 3 \\ 1 & 1 & 1 & 3 \\ 0 & 0 & 0 & 3 \end{bmatrix}$.

3. Find the best moves for players R and C in the following strictly determined games, and give the payoff when both players use their best moves.

(a) $\begin{bmatrix} 3 & -5 \\ 2 & -6 \end{bmatrix}$ (b) $\begin{bmatrix} 1 & 1 \\ 2 & 0 \end{bmatrix}$ (c) $\begin{bmatrix} 1 & -4 & 7 \\ 6 & -5 & 9 \end{bmatrix}$

(d) $\begin{bmatrix} 7 & 4 & 6 \\ -2 & 3 & 1 \\ 1 & 2 & 3 \end{bmatrix}$.

4. Follow the directions of Exercise 3 for the following games:

(a) $\begin{bmatrix} 0 & 2 \\ -1 & 3 \end{bmatrix}$ (b) $\begin{bmatrix} 4 & 2 & 8 & 3 \\ -6 & 0 & 7 & -1 \\ 3 & 2 & 4 & 5 \end{bmatrix}$ (c) $\begin{bmatrix} 1 & 2 & 1 \\ 1 & 3 & 1 \\ 1 & -8 & 1 \end{bmatrix}$.

In Exercises 5–8 write the payoff matrix for each game.

5. Each of two players has two cards, a 3 and a 4. Each player selects one of his cards and they simultaneously show their cards. If the sum of the numbers is even, player II pays player I an amount equal to the sum; if the sum is odd, player I pays player II an amount equal to the sum.

6. Each of two players shows one or two fingers. Player R pays player C a number equal to the total number of fingers shown.

7. (The children's game "stone, scissors, paper") Each of two players writes down one of the words, *stone, scissors,* or *paper*. Stone beats scissors, scissors beats paper, and paper beats stone. The winner receives $1 from the loser and the payoff is $0 in the case of a tie.

8. (Morra) Each of two players shows one or two fingers and at the same time each guesses at the total number of fingers that will be shown. If one player guesses the correct number, he wins that amount from his opponent. If both or neither guesses correctly, no money changes hands. (*Hint:* A move is one of the ordered pairs (1, 2), (1, 3), (2, 3), or (2, 4), where, for example, (2, 4) means 2 fingers shown, number 4 called.)

9. Show that the game in Exercise 6 is strictly determined and find the best move for each player.

10. A region is served by two shopping centers, center A and center B. Center A serves 60% of the region and center B serves 40% of the region. Valley Bank and Colonial Bank are each planning to open a branch in one of the centers. If both open at the same center, they will split the business in that center and if they open in different centers, each will get all the business in its center.

(a) Show that the percentage of *regional* shoppers that will be serviced by Valley Bank, in each case, can be expressed in matrix form as

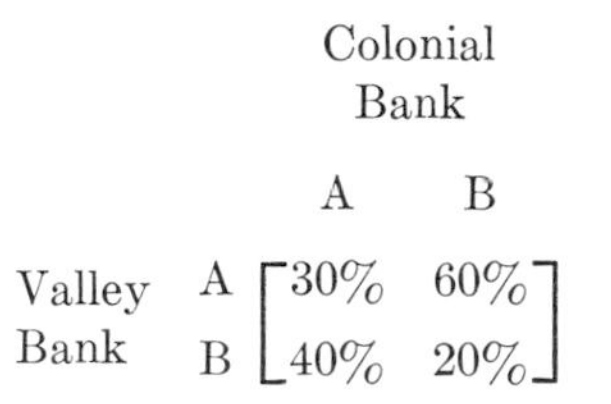

(b) Show that this game is not a constant-sum game.

11. Show that the matrix game

$$\begin{bmatrix} a & -2 \\ 7 & -1 \end{bmatrix}$$

is strictly determined, regardless of the value of a.

12. (The Bootlegger's game)† A bootlegger can cross a border by two different routes: over a highway or through the mountains. The border patrol has three possible plans: guard only the highway heavily, guard only the mountain road heavily, or guard both the highway and mountain routes lightly. If the bootlegger can travel the highway undetected, he will be able to smuggle in a fully loaded truck and make a profit of $500. If he travels the highway and there is a light patrol on the highway, he will avoid arrest, but will have to abandon his load, losing $200. If he travels the highway and there is a heavy patrol, he will be arrested, losing his load ($200), and fined $300. The mountain road is much narrower and can only allow a smaller truck. If the road is unguarded, he will have no problem in getting through and will make a profit of $200. If it is lightly or heavily guarded, he will still be able to get through, but will have to bribe the locals to get him past the border patrol. In either of these cases his profit will be $100. What is the best course of action for the border patrol and for the bootlegger?

13. (Media Selection in Advertising)† Firms A and B, competing for the same customers, each have one million dollars for an advertising campaign. Each firm decides to use no advertising or to use exactly one of the media: radio, television, or printed matter. A neutral consulting firm has developed the

† Exercises 12 and 13 are variations of problems discussed in an article, by Martin Shubik, Game Theory and Management Science, *Management Science* **2** (1955) 40–54.

following payoff matrix, where each entry is the amount of extra revenue (in millions of dollars) that will be earned by firm A, in each case.

		Firm B			
		Radio	Television	Printed matter	No advertising
Firm A	Radio	0	−.5	0	2.5
	Television	2	0	1.5	5
	Printed matter	1	−.5	0	3.5
	No advertising	−2	−4	−3	0

How should each firm best spend its advertising money?

8.5 GAMES WITH MIXED STRATEGIES

In the previous section we observed that the players of a strictly determined game cannot improve on their best moves, even if they discover their opponent's move in advance. We pointed out, however, that for games that are not strictly determined, a player can benefit by uncovering his opponent's intended move. In this section we shall study games that are *not* strictly determined and which the players intend to play over and over again. In this situation a player must avoid making the same move in every game; otherwise his opponent will detect the pattern and use the information to his own advantage. Our objective in this section is to explain how the players should mix their moves from game to game to obtain the most favorable results.

Consider a two-person matrix game with payoff matrix

$$\begin{bmatrix} 3 & -4 \\ 0 & 5 \end{bmatrix}.$$

As usual, positive entries in the matrix represent the amounts that the column player (player C) pays to the row player (player R) after each move, and negative signs precede amounts that player R pays to player C. Since no entry is simultaneously the smallest element in its row and the largest element in its column, this game has no saddle point and consequently is not strictly determined.

The row minima and column maxima are

$$\begin{array}{rcc|c} & & & \text{Row minima} \\ & \left[\begin{matrix} 3 \\ 0 \end{matrix}\right. & \left.\begin{matrix} -4 \\ 5 \end{matrix}\right] & \begin{matrix} -4 \\ 0 \end{matrix} \\ \text{Column maxima} & 3 & 5 & \end{array}$$

Since row 2 contains the largest row minimum and column 1 contains the smallest column maximum, player R might initially play row 2 and player C might initially play column 1, resulting in a payoff of zero. If however, player C continues playing the first column game after game, player R will eventually detect the pattern and switch to row 1, thereby winning 3 units. By the same token, if player R continues to play row 1 game after game, then player C will detect the pattern and choose column 2, resulting in a loss of 4 units for player R.

It is evident from this analysis that each player should avoid using the same move in every game; rather the players should mix their moves in an unpredictable fashion. Such mixtures of moves are called ***mixed strategies*** to distinguish them from ***pure strategies,*** where the player uses the same move in every game. Each player can mix his moves by using a chance device such as cards, dice, or tables of random numbers to determine which move to make in each game. The question that remains, however, is how often each move should be used. For example, should player R in the game above choose each row 50% of the time, or should he mix his moves in some other proportion, say the first row 80% of the time and the second row 20% of the time? To answer this question, we need some way to evaluate the effectiveness of various mixtures of moves. One way of doing this is to compare the ***expected winnings*** that result from different mixtures of moves. The following example illustrates this idea.

Example 20 Suppose that the matrix game whose payoff matrix is

$$\begin{bmatrix} 3 & -4 \\ 0 & 5 \end{bmatrix} \tag{1}$$

will be played over and over, and assume that player C will play each column 50% of the time. Which would be the better strategy for player R:

Strategy a: to play the first row 80% of the time and the second row 20% of the time?

or

Strategy b: to play each row 50% of the time?

Solution We can regard each play of the game as a probability experiment in which the four combinations of moves

row 1 and column 1	row 1 and column 2
row 2 and column 1	row 2 and column 2

are the possible outcomes. To illustrate how the probabilities of these outcomes can be obtained, consider strategy (a) where player R plays the first row 80% of the time and the second row 20% of the time. Since neither player knows in advance what the other's choice will be, it is reasonable to assume that the row and column are chosen *independently*. Thus we obtain the probabilities

$$\begin{aligned} P(\text{row 1 and column 1}) &= P(\text{row 1})\ P(\text{column 1}) \\ &= (.8)(.5) \\ &= .4 \\ P(\text{row 2 and column 1}) &= P(\text{row 2})\ P(\text{column 1}) \\ &= (.2)(.5) \\ &= .1 \end{aligned}$$

and so forth. In Table 8.6 we have listed the probabilities of the outcomes for strategy (a) and strategy (b). Also, we have listed the payoff that results to player R in each case.

Table 8.6

Outcome	Strategy (a) Payoff to R	Probability of outcome
row 1 — column 1	3	$(.8)(.5) = .4$
row 2 — column 1	0	$(.2)(.5) = .1$
row 1 — column 2	−4	$(.8)(.5) = .4$
row 2 — column 2	5	$(.2)(.5) = .1$

Outcome	Strategy (b) Payoff to R	Probability of outcome
row 1 — column 1	3	$(.5)(.5) = .25$
row 2 — column 1	0	$(.5)(.5) = .25$
row 1 — column 2	−4	$(.5)(.5) = .25$
row 2 — column 2	5	$(.5)(.5) = .25$

If we denote the payoff to R by X, then X is a random variable whose expected value represents the long-term average payoff to R when the game is played over and over. Thus, to compare the relative merits of strategy (a) and strategy (b), player R can examine the expected payoff $E(X)$ in each case to determine which strategy will produce the larger payoff.

Recall from Section 7.2 that the expected value of a random variable X is the sum of the possible values of X times their probabilities of occurrence.

Thus from Table 8.6, the expected payoff to R for strategy (a) is

$$\begin{aligned} E(X) &= 3(.4) + 0(.1) + (-4)(.4) + 5(.1) \\ &= 1.2 - 1.6 + .5 \\ &= .1 \end{aligned}$$

and the expected payoff to R for strategy (b) is

$$\begin{aligned} E(X) &= 3(.25) + 0(.25) + (-4)(.25) + 5(.25) \\ &= .75 - 1.00 + 1.25 \\ &= 1.00. \end{aligned}$$

Thus strategy (b) is better than strategy (a) from player R's viewpoint *if C plays each column 50% of the time.* Of course, if C uses some other mixed strategy, then it is possible that strategy (a) might turn out to be better than strategy (b) for player R.

Let us examine each player's objectives in a nonstrictly determined game.

Objectives of Player R

Player R wants to find a mixed strategy that will give him the largest expected payoff. Whatever mixed strategy he uses, however, player R must assume that player C will oppose him with his best counterstrategy. Thus, from among all possible mixed strategies, player R wants to pick one that will give him the largest possible expected payoff against player C's best counterstrategy. This is called an ***optimal strategy for player R.***

Objectives of Player C

Player C wants to find a mixed strategy that will give player R the smallest possible expected payoff. Whatever mixed strategy he uses, however, player C must assume that player R will oppose him with his best counterstrategy. Thus, from among all possible mixed strategies, player C wants to pick one which will give player R the smallest expected payoff against player R's best counterstrategy. This is called an ***optimal strategy for player C.***

Presently, we shall discuss methods for finding optimal strategies in certain kinds of games, but first we need some preliminary ideas. Consider a two-person constant-sum game with payoff matrix

$$A = \begin{bmatrix} a_{11} & a_{12} \\ a_{21} & a_{22} \end{bmatrix}.$$

Suppose that player R chooses row 1 with probability p_1 and row 2 with probability p_2 (where $p_1 + p_2 = 1$), and suppose, independently, player C chooses column 1 with probability q_1 and column 2 with probability q_2 (where $q_1 + q_2 = 1$). Each play of the game results in one of four possible outcomes

row 1 and column 1   row 1 and column 2

row 2 and column 1   row 2 and column 2

The probabilities of these outcomes can be obtained using an argument like that in Example 20. These probabilities together with the payoffs resulting from the outcomes are shown in Table 8.7.

Table 8.7

Outcome	Payoff	Probability of outcome
row 1 — column 1	a_{11}	p_1q_1
row 2 — column 1	a_{21}	p_2q_1
row 1 — column 2	a_{12}	p_1q_2
row 2 — column 2	a_{22}	p_2q_2

From Table 8.7, we obtain

$$\text{expected payoff to player R} = p_1q_1a_{11} + p_2q_1a_{21} + p_1q_2a_{12} + p_2q_2a_{22}. \quad (2)$$

The result in (2) can also be obtained using matrices as follows. Use the probabilities in player R's mixed strategy to form a 1×2 matrix

$$P = [p_1 \quad p_2]$$

and use the probabilities in player C's mixed strategy to form a 2×1 matrix

$$Q = \begin{bmatrix} q_1 \\ q_2 \end{bmatrix}.$$

If we form the product PAQ, we obtain

$$PAQ = [p_1 \quad p_2] \begin{bmatrix} a_{11} & a_{12} \\ a_{21} & a_{22} \end{bmatrix} \begin{bmatrix} q_1 \\ q_2 \end{bmatrix}$$

$$= p_1q_1a_{11} + p_2q_1a_{21} + p_1q_2a_{12} + p_2q_2a_{22}$$

which as shown above in Equation (2) is the expected payoff to player R.† This is a special case of the following general result.

If a matrix game has an $m \times n$ payoff matrix A, and if the matrices

$$P = [p_1 \quad p_2 \quad \cdots \quad p_m] \qquad \text{and} \qquad Q = \begin{bmatrix} q_1 \\ q_2 \\ \vdots \\ q_n \end{bmatrix}$$

list the probabilities of the moves for players R and C, respectively, then the expected payoff E to player R is

$$E = PAQ. \tag{3}$$

Example 21 Consider the matrix game whose payoff matrix is

$$A = \begin{bmatrix} 5 & -9 \\ -4 & 8 \end{bmatrix}.$$

Use matrices to find the expected payoff to player R if player R plays row 1 with probability $\frac{5}{6}$ and row 2 with probability $\frac{1}{6}$, while player C plays column 1 with probability $\frac{1}{3}$ and column 2 with probability $\frac{2}{3}$.

Solution The strategies for the players can be written in matrix form as

$$P = [\tfrac{5}{6} \quad \tfrac{1}{6}] \qquad \text{and} \qquad Q = \begin{bmatrix} \frac{1}{3} \\ \frac{2}{3} \end{bmatrix}.$$

Thus, from (3), the expected payoff to player R is

$$\begin{aligned} E = PAQ &= [\tfrac{5}{6} \quad \tfrac{1}{6}] \begin{bmatrix} 5 & -9 \\ -4 & 8 \end{bmatrix} \begin{bmatrix} \frac{1}{3} \\ \frac{2}{3} \end{bmatrix} \\ &= [\tfrac{21}{6} \quad -\tfrac{37}{6}] \begin{bmatrix} \frac{1}{3} \\ \frac{2}{3} \end{bmatrix} \\ &= -\tfrac{53}{18}. \end{aligned}$$

Therefore, the long-term average payoff will be $\frac{53}{18}$ units to player C.

Example 22 Consider a two-person zero-sum matrix game with payoff matrix

$$A = \begin{bmatrix} 3 & -2 & 3 \\ -1 & 2 & 3 \end{bmatrix}.$$

† Strictly speaking PAQ is the 1×1 matrix $[p_1q_1a_{11} + p_2q_1a_{21} + p_1q_2a_{12} + p_2q_2a_{22}]$; however, we have followed the standard practice of omitting the brackets on 1×1 matrices.

If player R plays row 1 with probability $\frac{1}{4}$ and row 2 with probability $\frac{3}{4}$, while player C plays column 1 with probability $\frac{1}{6}$, column 2 with probability $\frac{1}{6}$, and column 3 with probability $\frac{2}{3}$, then the expected payoff is

$$E = PAQ = \begin{bmatrix} \frac{1}{4} & \frac{3}{4} \end{bmatrix} \begin{bmatrix} 3 & -2 & 3 \\ -1 & 2 & 3 \end{bmatrix} \begin{bmatrix} \frac{1}{6} \\ \frac{1}{6} \\ \frac{2}{3} \end{bmatrix}$$

$$= \begin{bmatrix} 0 & 1 & 3 \end{bmatrix} \begin{bmatrix} \frac{1}{6} \\ \frac{1}{6} \\ \frac{2}{3} \end{bmatrix}$$

$$= \tfrac{13}{6}.$$

Thus, the long-term average payoff will be $\frac{13}{6}$ units to player R.

Let us now consider a two-person matrix game with a 2×2 payoff matrix

$$A = \begin{bmatrix} a & b \\ c & d \end{bmatrix}. \tag{4}$$

It can be shown that if the game is not strictly determined then

$$a + d - b - c \neq 0$$

and the optimal strategies for the players are as follows

The optimal strategy for player R is $P = \begin{bmatrix} p_1 & p_2 \end{bmatrix}$, where

$$p_1 = \frac{d - c}{a + d - b - c} \quad \text{and} \quad p_2 = \frac{a - b}{a + d - b - c} \tag{5}$$

The optimal strategy for player C is

$$Q = \begin{bmatrix} q_1 \\ q_2 \end{bmatrix},$$

where

$$q_1 = \frac{d - b}{a + d - b - c} \quad \text{and} \quad q_2 = \frac{a - c}{a + d - b - c} \tag{6}$$

Moreover:

If both players use their optimal strategies, then the expected payoff is

$$E = PAQ = \frac{ad - bc}{a + d - b - c} \tag{7}$$

Example 23 In the previous section we observed that the penny-matching game with payoff matrix

$$A = \begin{bmatrix} -1 & 1 \\ 1 & -1 \end{bmatrix}$$

is not strictly determined. Let us now try to find the optimal strategies for the players.

Solution Comparing A to matrix (4) above we obtain

$$a = -1, \qquad b = 1, \qquad c = 1, \qquad d = -1.$$

Substituting these values in (5) we obtain

$$p_1 = \frac{d - c}{a + d - b - c} = \frac{-1 - 1}{-1 - 1 - 1 - 1} = \frac{1}{2}$$

$$p_2 = \frac{a - b}{a + d - b - c} = \frac{-1 - 1}{-1 - 1 - 1 - 1} = \frac{1}{2};$$

similarly, from (6) we obtain

$$q_1 = \frac{d - b}{a + d - b - c} = \frac{-1 - 1}{-1 - 1 - 1 - 1} = \frac{1}{2}$$

$$q_2 = \frac{a - c}{a + d - b - c} = \frac{-1 - 1}{-1 - 1 - 1 - 1} = \frac{1}{2}.$$

Thus player I should choose each row 50% of the time and player II should choose each column 50% of the time. If each player uses his optimal strategy, then from (7) the expected payoff will be

$$E = \frac{ad - bc}{a + d - b - c} = \frac{(-1)(-1) - (1)(1)}{-1 - 1 - 1 - 1} = 0.$$

Example 24 Consider the game, discussed in Example 20; the payoff matrix is

$$A = \begin{bmatrix} 3 & -4 \\ 0 & 5 \end{bmatrix}$$

Comparing A to matrix (4) above we obtain

$$a = 3, \qquad b = -4, \qquad c = 0, \qquad d = 5.$$

Substituting these values in (5) we obtain the optimal strategy for player R, which is

$$p_1 = \frac{d - c}{a + d - b - c} = \frac{5 - 0}{3 + 5 - (-4) - 0} = \frac{5}{12}$$

$$p_2 = \frac{a - b}{a + d - b - c} = \frac{3 - (-4)}{3 + 5 - (-4) - 0} = \frac{7}{12};$$

similarly, from (6) we obtain the optimal strategy for player C, which is

$$q_1 = \frac{d - b}{a + d - b - c} = \frac{5 - (-4)}{3 + 5 - (-4) - 0} = \frac{9}{12}$$

$$q_2 = \frac{a - c}{a + d - b - c} = \frac{3 - 0}{3 + 5 - (-4) - 0} = \frac{3}{12}.$$

If each player uses his optimal strategy, then from (7) the expected payoff will be

$$E = \frac{ad - bc}{a + d - b - c} = \frac{3(5) - (-4)(0)}{3 + 5 - (-4) - 0} = \frac{15}{12} = \frac{5}{4}.$$

Thus, in the long run, the game is favorable to R since he will obtain an expected payoff of $\frac{5}{4}$ units.

Although the optimal strategies given by formulas (5) and (6) apply only to games with a 2×2 payoff matrix, it is sometimes possible, with a little ingenuity, to use these formulas in other cases. The following example illustrates this point.

Example 25 Consider the two-person constant-sum game with payoff matrix

$$\begin{bmatrix} 3 & -4 & -2 \\ 0 & 5 & 5 \\ -2 & 1 & 3 \end{bmatrix}.$$

The reader can check that this game has no saddle point and consequently is not strictly determined; thus the players should use mixed strategies. Since the payoff matrix is not 2×2, the optimal strategies cannot be determined from formulas (5) and (6). If, however, we compare the second and third rows of the payoff matrix, we see that each element in the third row is less than the corresponding element in the second row. Thus, a third-row move is always worse than a second-row move for player R, regardless of what player C decides to do. It follows that player R will *never* use the third row as a move. Thus, we can cross out the third row from the payoff matrix, to obtain

$$\begin{bmatrix} 3 & -4 & -2 \\ 0 & 5 & 5 \\ -2 & 1 & 3 \end{bmatrix}$$

In the remaining 2×3 matrix we see that each element in the third column is at least as large as the corresponding element in the second column. Thus, a second column move for player C is always at least as good as a third column move, regardless of what player R decides to do. It follows that player C will *never* use the third column as a move. Thus, we can cross out the third column from the payoff matrix to obtain

$$\begin{bmatrix} 3 & -4 & -2 \\ 0 & 5 & 5 \\ -2 & 1 & 3 \end{bmatrix}$$

The remaining 2×2 matrix is the one we considered in Example 24. In that example we used formulas (5) and (6) to obtain the optimal strategy

$$p_1 = \frac{5}{12}, \qquad p_2 = \frac{7}{12}$$

and

$$q_1 = \frac{9}{12}, \qquad q_2 = \frac{3}{12}.$$

Thus, in this example, the optimal strategies are

$$p_1 = \frac{5}{12}, \qquad p_2 = \frac{7}{12}, \qquad p_3 = 0$$

and

$$q_1 = \frac{9}{12}, \qquad q_2 = \frac{3}{12}, \qquad q_3 = 0.$$

Competitive Pricing of Products†

Example 26 Assume that a certain region has only two gasoline stations, R and C, and suppose the stations are faced with the following pricing problem. Each station can sell its gasoline for either 35 or 40 cents a gallon, and each

† This example is a variation of one discussed in an excellent book by Andrei Rogers, *Matrix Methods in Urban and Regional Analysis* (San Francisco: Holden-Day, 1971).

dealer's total percentage of the regional business will depend on the prices charged. Assume also that a market survey produces the payoff matrix

		Station C	
		35 cents	40 cents
Station R	35 cents	60%	50%
	40 cents	40%	70%

where the entries represent the percentage of the regional business that will be captured by station R.

How should each station price its gasoline?

Solution No entry in the matrix is both the minimum in its row and the maximum in its column, so that this game is not strictly determined. Thus, the stations should use mixed strategies. Comparing the payoff matrix in Figure 8.6 to the matrix in (4) we obtain

$$a = 60, \qquad b = 50, \qquad c = 40, \qquad d = 70.$$

Substituting these values in (5) we obtain the optimal strategy for station R, which is

$$p_1 = \frac{d - c}{a + d - b - c} = \frac{70 - 40}{60 + 70 - 50 - 40} = \frac{3}{4}$$

$$p_2 = \frac{a - b}{a + d - b - c} = \frac{60 - 50}{60 + 70 - 50 - 40} = \frac{1}{4};$$

similarly, from (6) we obtain the optimal strategy for station C, which is

$$q_1 = \frac{d - b}{a + d - b - c} = \frac{70 - 50}{60 + 70 - 50 - 40} = \frac{1}{2}$$

$$q_2 = \frac{a - c}{a + d - b - c} = \frac{60 - 40}{60 + 70 - 50 - 40} = \frac{1}{2}.$$

Thus, the optimal strategy for station R is to charge 35 cents a gallon three-fourths of the time and 40 cents a gallon the rest of the time, while the optimal strategy for station C is to charge 35 cents half the time and 40 cents half the time.

If both stations use their optimal strategies, then from (7), the expected payoff to station R is

$$E = \frac{ad - bc}{a + d - b - c} = \frac{(60)(70) - (50)(40)}{60 + 70 - 50 - 40} = 55.$$

Thus, over the long term, station R will have a slight competitive advantage, capturing 55% of the regional business.

Columbus' Discovery of America as a Game Theory Problem†

Example 27 History tells us that Christopher Columbus, faced with a mutiny by his crew, had to decide between two alternatives: to turn back to Spain as his crew demanded or to continue his voyage until land was sighted. Undoubtedly, Columbus must have analyzed the possible consequences of his decision in the way we have indicated in Table 8.8.

Table 8.8

		Unknown location of land	
		Land near	No land near
Columbus' decision	Turn back	Bitter disappointment in later life	Satisfaction for saving lives
	Keep going	Personal glory	Death for himself and his men

To analyze Columbus' plight as a game theory problem, it is necessary to assign numerical values to each of the outcomes, indicating their relative importance. It is clear that personal glory and satisfaction for saving lives are favorable outcomes for Columbus and should be assigned positive values, while disappointment and death are unfavorable outcomes which should be assigned negative values. Beyond this, it is difficult to pick appropriate numerical values. For example, which is more favorable, satisfaction for saving lives or personal glory? And even if we could agree that personal glory is more favorable, we must still decide how much more favorable it is. Is glory five times as favorable as satisfaction? Ten times? A hundred times? Obviously these are difficult *psychological* questions which everyone would answer differently. This difficulty is often the stumbling block in applying game theory to behavioral problems. Nevertheless, once the values are agreed on and the payoff matrix constructed, we have a matrix game which can be analyzed mathematically. For example, suppose Columbus had appraised the relative values of the outcomes in Table 8.8 as in Table 8.9. Since this game has no saddle point, Columbus should have used a mixed strategy. Comparing the payoff matrix in Table 8.9 to (4), we obtain

$$a = -10, \qquad b = 1, \qquad c = 5, \qquad d = -50.$$

Substituting these values in (5) gives Columbus' optimal strategy, which is

$$p_1 = \frac{d-c}{a+d-b-c} = \frac{-50-5}{-10-50-1-5} = \frac{-55}{-66} \simeq .83$$

$$p_2 = \frac{a-b}{a+d-b-c} = \frac{-10-1}{-10-50-1-5} = \frac{-11}{-66} \simeq .17.$$

† This problem is based on an article by Leonid Hurwicz in the book, *Mathematical Thinking in Behavioral Sciences* (San Francisco: Freeman, 1968).

Table 8.9

		Unknown location of land	
		Land near	No land near
Columbus' decision	Turn back	−10	1
	Keep going	5	−50

Thus with probability .83 Columbus should have turned around and never discovered America!

It is important to keep in mind that the conclusion reached in this last example depends on accepting the validity of the values selected for the payoff matrix.

There are many topics in game theory we have not had time to discuss; readers interested in studying game theory in more detail can consult the following references:

[1] Ira Buchler and Hugo Nutini, *Game Theory in the Behavioral Sciences* (Pittsburgh, Pennsylvania: Univ. of Pittsburgh Press, 1969).
[2] R. Duncan Luce and Howard Raiffa, *Games and Decisions* (New York: Wiley, 1957).
[3] Daniel P. Maki and Maynard Thompson, *Mathematical Models and Applications* (Englewood Cliffs, New Jersey: Prentice Hall, 1973).
[4] J. C. C. McKinsey, *Introduction to the Theory of Games* (New York: McGraw Hill, 1952).
[5] G. Owen, *Game Theory* (Philadelphia: Saunders, 1968).

EXERCISE SET 8.5

1. Consider a game with payoff matrix

$$\begin{bmatrix} -2 & 3 \\ 3 & 1 \end{bmatrix}.$$

In each part, use formula (3) to find the expected payoff that results from the given strategies

(a) $p_1 = \frac{1}{3}$, $p_2 = \frac{2}{3}$ and $q_1 = \frac{3}{4}$, $q_2 = \frac{1}{4}$

(b) $p_1 = \frac{2}{5}$, $p_2 = \frac{3}{5}$ and $q_1 = \frac{3}{5}$, $q_2 = \frac{2}{5}$

(c) $p_1 = 0$, $p_2 = 1$ and $q_1 = \frac{1}{6}$, $q_2 = \frac{5}{6}$.

2. Consider a game with payoff matrix

$$\begin{bmatrix} -3 & 2 & 1 \\ -2 & 1 & 2 \end{bmatrix}.$$

In each part, use formula (3) to find the expected payoff that results from the given strategies.

(a) $p_1 = 0$, $p_2 = 1$ and $q_1 = \frac{1}{3}$, $q_2 = \frac{1}{3}$, $q_3 = \frac{1}{3}$

(b) $p_1 = \frac{1}{2}$, $p_2 = \frac{1}{2}$ and $q_1 = \frac{5}{12}$, $q_2 = \frac{1}{12}$, $q_3 = \frac{1}{2}$

(c) $p_1 = \frac{1}{4}$, $p_2 = \frac{3}{4}$ and $q_1 = 0$, $q_2 = 0$, $q_3 = 1$.

In Exercises 3–8, find the optimal strategies and the expected payoff.

3. $\begin{bmatrix} 2 & 4 \\ 3 & 1 \end{bmatrix}$. **4.** $\begin{bmatrix} -1 & 3 \\ 5 & 0 \end{bmatrix}$.

5. $\begin{bmatrix} 7 & 1 \\ -6 & 2 \end{bmatrix}$. **6.** $\begin{bmatrix} 4 & 2 \\ -8 & 3 \end{bmatrix}$.

7. $\begin{bmatrix} 2 & -1 \\ -1 & 2 \end{bmatrix}$. **8.** $\begin{bmatrix} 0 & -2 \\ 1 & 1 \end{bmatrix}$.

In Exercises 9–12, use the method of Example 25 to find the optimal strategies and the expected payoff.

9. $\begin{bmatrix} 0 & 2 & 4 \\ 1 & -1 & 3 \\ -3 & -4 & 0 \end{bmatrix}$. **10.** $\begin{bmatrix} 2 & 6 & 7 \\ 8 & 3 & 9 \\ 0 & 2 & 5 \end{bmatrix}$.

11. $\begin{bmatrix} 1 & -9 & -2 \\ 6 & 5 & -1 \\ 3 & -8 & 0 \end{bmatrix}$. **12.** $\begin{bmatrix} -1 & -2 & 0 \\ 6 & 3 & 9 \\ 3 & 6 & 7 \end{bmatrix}$.

13. In the problem of Example 27, find Christopher Columbus' optimal strategy if the payoff matrix is

$$\begin{bmatrix} -100 & 70 \\ 1 & -100 \end{bmatrix}.$$

14. **(A Marketing Problem)** An automobile manufacturer can stress either physical appearance or sound engineering in its magazine advertising. Market studies show that ads based on physical appearance are effective on 70% of the female readers and effective on 20% of the male readers, while ads based on sound engineering are effective on 40% of the female readers and 80% of the male readers. If the ratio of male to female readers is unknown, in what proportions should the two kinds of ads be mixed to be most effective? (*Hint:* This problem can be viewed as a game between the manufacturer and the market place with payoff matrix

		Marketplace	
		Male	Female
Manufacturer	Appearance	.2	.7
	Engineering	.8	.4

15. In Exercise 14, what proportion of male and female readers would be least favorable to the advertiser?

8.6 MARKOV CHAINS AND APPLICATIONS

In Section 7.5 we studied binomial random variables by examining independent sequences of Bernoulli experiments. In this section we shall study other kinds of sequences in which the experiments need not be independent. Broadly speaking, we shall be interested in studying physical ***systems*** and their ***states.*** To explain what this means, we can imagine a pond, dotted with lily pads, and a lazy frog who suns himself by randomly leaping about from pad to pad whenever he pleases. The pond, the frog, and the lily pads constitute a *physical system*. When the frog leaps from one pad to another, the physical system looks different, and we say that the system has changed its *state*.

To pursue the frog discussion a little further, let us assume that the pond contains only two lily pads, pad 1 and pad 2, and let us also assume that we record the state of the system at regular time intervals. A typical record of 11 observations might look like

$$1,\ 1,\ 2,\ 1,\ 2,\ 2,\ 2,\ 2,\ 1,\ 2,\ 2, \tag{1}$$

where a 1 indicates that the frog was observed on pad 1 and a 2 indicates that the frog was observed on pad 2.

In many problems, one knows the present state of a physical system and is interested in predicting the state of the system at the next observation or even at some more remote future observation. In general, it is impossible to predict a future state with *certainty*; however, it is often possible to predict the *chances* that the system will be in a particular state at a later time if its present state is known. To illustrate, let us consider the following questions about the jumping frog problem:

(a) If the system is presently in state 1, what is the probability that it will still be in state 1 at the next observation?
(b) If the system is presently in state 1, what is the probability that it will be in state 2 at the next observation?
(c) If the system is presently in state 2, what is the probability that it will be in state 1 at the next observation?
(d) If the system is presently in state 2, what is the probability that it will still be in state 2 at the next observation?

From the data in (1), we see that the system was in state 1 four times. In three of these cases, the system changed to state 2 at the next observation and in one case the system was still in state 1 at the next observation. Thus, based on the data, the probabilities requested in (a) and (b) are:

$$P(\text{system remains in state 1 given that it is presently in state 1}) = \tfrac{1}{4}, \tag{2a}$$

$$P(\text{system changes to state 2 given that it is presently in state 1}) = \tfrac{3}{4}. \tag{2b}$$

Similarly, using the *first ten* observations in (1), the system was in state 2 six times. In two of these cases, the system changed to state 1 at the next observation; and in four cases, the system was still in state 2 at the next observation. Thus, based on the data, the probabilities requested in (c) and (d) are:

$$P(\text{system changes to state 1 given that it is presently in state 2}) = \tfrac{2}{6} = \tfrac{1}{3}, \tag{2c}$$

$$P(\text{system remains in state 2 given that it is presently in state 2}) = \tfrac{4}{6} = \tfrac{2}{3}. \tag{2d}$$

The information in (2a), (2b), (2c), and (2d) can be organized conveniently in matrix form as follows:

$$\begin{array}{c} \\ \text{State 1} \\ \text{State 2} \end{array} \begin{array}{c} \begin{array}{cc} \text{State 1} & \text{State 2} \end{array} \\ \begin{bmatrix} \frac{1}{4} & \frac{3}{4} \\ \frac{1}{3} & \frac{2}{3} \end{bmatrix} \end{array}. \tag{3}$$

Matrix (3) is called the ***transition matrix*** for the system; each entry is the probability of moving from a given state to another state at the next observation. For example, in matrix (3) the probability $\frac{1}{3}$ of moving from state 2 to state 1 appears in row 2 and column 1 and the probability $\frac{1}{4}$ of moving from state 1 to state 1, appears in row 1 and column 1. We note that the transition matrix (3) is based on the rather small data sample (1). More extensive data would yield a more accurate transition matrix.

Although the illustration of the frog in the pond is more descriptive than practical, it contains all the ingredients of the following more realistic examples:

Stock Market Fluctuations

Example 28 In a stable stock market, many stocks show a tendency to cancel out one day's price change with a change in the opposite direction on the following day. Thus an increase one day tends to be followed by a decrease the next day, and similarly a decrease tends to be followed by an increase.

We can view a stock as a physical system with two possible states, increase or decrease. Suppose that the probability of a price increase, given that the previous change was a decrease, is .7. (Thus, the probability of a price decrease, given that the previous price change was a decrease, is $1 - .7 = .3$.) Suppose also that the probability of a price decrease, given that the previous change was an increase, is .8. (Thus, the probability of a price increase, given that the previous change was an increase, is .2.) Then the transition matrix for the system is

$$\begin{array}{cc} & \\ \text{Today's} & \text{increase} \\ \text{change} & \text{decrease} \end{array} \begin{array}{c} \text{Tomorrow's change} \\ \begin{array}{cc} \text{increase} & \text{decrease} \end{array} \\ \begin{bmatrix} .2 & .8 \\ .7 & .3 \end{bmatrix} \end{array}. \tag{4}$$

Adjustment of Equipment in Space Exploration

Example 29 Much of the instrumentation used in lunar exploration or deep-space probes is "self-adjusting" in the sense that the equipment is designed to correct alignment errors automatically when they occur. Often the adjustment takes place over a period of time, so that when a monitoring device finds the instrumentation out of alignment at a certain time, there is a fairly good chance it will still be out of alignment a short time later. Likewise, if the equipment is correctly aligned, it is likely to be aligned a short time later.

We can view a self-adjusting instrument as a physical system with two possible states, aligned or unaligned. A typical transition matrix for such a system might be

		State of the instrument one second later	
		aligned	unaligned
Present state of the instrument	aligned	.9	.1
	unaligned	.2	.8

$$\begin{bmatrix} .9 & .1 \\ .2 & .8 \end{bmatrix}. \qquad (5)$$

From this matrix we see that the probability that the instrument is unaligned one second from now, given that it is presently unaligned, is .8; and the probability that the instrument is unaligned one second from now, given that it is presently aligned, is .1.

A Transportation Problem

Example 30 The Borough of Manhattan in New York City can be divided into three sectors: uptown (sector 1), midtown (sector 2), and lower Manhattan (sector 3). Taxicabs that operate only in Manhattan can pick up a passenger in any sector and drop him off in any sector. We can view a taxicab and the Borough of Manhattan as a physical system. If we observe the system when the cab is picking up or discharging a passenger, the system can be in one of three states: the cab can be in sector 1, in sector 2, or in sector 3. A typical transition matrix for such a system is

		Discharge sector		
		1	2	3
Pickup sector	1	.5	.4	.1
	2	.3	.6	.1
	3	.2	.1	.7

$$\begin{bmatrix} .5 & .4 & .1 \\ .3 & .6 & .1 \\ .2 & .1 & .7 \end{bmatrix}.$$

Thus, given that a passenger is picked up in sector 3, the probability is .2 that he will be discharged in sector 1, and given that a passenger is picked up in sector 1, the probability is .5 that he will be discharged in sector 1.

A Genetics Problem

Example 31 (For readers who have read Section 8.2.) Suppose a parent of unknown genotype is crossed with an individual of genotype Aa; then an offspring of unknown genotype is selected from this mating and crossed with another individual of genotype Aa; then an offspring of unknown genotype is selected from this mating and crossed with still another individual of genotype Aa. If we continue this process, we obtain a succession of matings

$$\begin{array}{ll} ⓘ \times Aa & \text{(first mating)} \\ ? \times Aa & \text{(second mating)} \\ ? \times Aa & \text{(third mating)} \\ \vdots & \vdots \end{array}$$

in which one parent always has unknown genotype and the other has genotype Aa. We can think of the two parents as a physical system. The system can be in one of three states, AA, Aa, or aa corresponding to the genotype of parent ⓘ.

To obtain the transition matrix for the system, let us suppose, first, that the system is in state AA. In other words, we suppose that parent ⓘ has genotype AA. In this case the mating

$$? \times Aa$$

is

$$AA \times Aa$$

and from Table 8.2, the genotype probabilities for the offspring are

AA	Aa	aa
$\frac{1}{2}$	$\frac{1}{2}$	0

Since the new parent of unknown genotype is selected from these offspring, these probabilities form the first row of the transition matrix

$$\begin{array}{ll} & \text{Genotype of next parent} \\ & \quad AA \quad\; Aa \quad\; aa \\ \begin{array}{l}\text{Unknown} \\ \text{genotype} \\ \text{of parent}\end{array} \begin{array}{c} AA \\ Aa \\ aa \end{array} & \begin{bmatrix} \frac{1}{2} & \frac{1}{2} & 0 \\ \frac{1}{4} & \frac{1}{2} & \frac{1}{4} \\ 0 & \frac{1}{2} & \frac{1}{2} \end{bmatrix}. \end{array} \tag{6}$$

The second and third rows in this transition matrix were obtained using lines four and five of Table 8.2.

If we observe a physical system for a period of time, the system will progress from its ***initial state*** (its state at the first observation), through a succession of possibly different states, thereby generating a sequence or chain of observations. (See, for example, the chain of observations in (1).) If a future state of the system cannot be predicted with certainty, that is, it is only possible to give the probability of the occurrence of the future state, then the evolution of the system is an example of a ***stochastic process.*** To illustrate, consider the space instrumentation problem in Example 29. If we know that the system is currently aligned, we cannot tell with certainty whether it will be aligned at the next observation. At best, we can assert that it will be aligned with probability .9 and unaligned with probability .1. Thus, Example 29 illustrates a stochastic process. Similarly, the other examples in this section are stochastic processes. The stochastic processes in all of our examples have an additional important property. In each case, once the present state of the system is known, the state probabilities for the next observation can be determined without any other information. The following examples illustrate this point.

Example 32 Consider the stock fluctuation problem of Example 28. If the present state of the system (today's change) is known to be an increase, then from the transition matrix (4), the state probabilities for tomorrow are

Increase	Decrease
.2	.8

No information, beside the present state, was needed to obtain these probabilities. In particular, it was unnecessary to know the past history of the stock that led up to today's state.

Example 33 (For readers who have read Section 8.3.) Consider the genetics problem of Example 31. If the present state of the system is Aa (that is, the parent ? has genotype Aa), then from transition matrix (6), the state probabilities for the next observation are

AA	Aa	aa
$\frac{1}{4}$	$\frac{1}{2}$	$\frac{1}{4}$

No information, beside the present state, was needed to obtain these probabilities.

Stochastic processes in which the probabilities for the next state are completely determined by the present state are called ***Markov† processes*** or sometimes ***Markov chains.***

† *Andrei Andreevich Markov* (1856–1922)—Markov was an outstanding Russian mathematician. His work was the starting point for the modern theory of stochastic processes. Markov studied and taught at the University of St. Petersburg and was a member of the Soviet Academy of Sciences.

In many applications of Markov processes, we know the initial state of the system and are interested in finding the state probabilities not only for the next observation, but also for more remote future observations. The next example illustrates one way of obtaining such probabilities.

Example 34 Consider the space equipment problem of Example 29. Let us assume that at our initial observation the instrument is in the aligned state and let us try to find the state probabilities two seconds later. The analysis is easy to picture using the tree diagram shown in Figure 8.6.

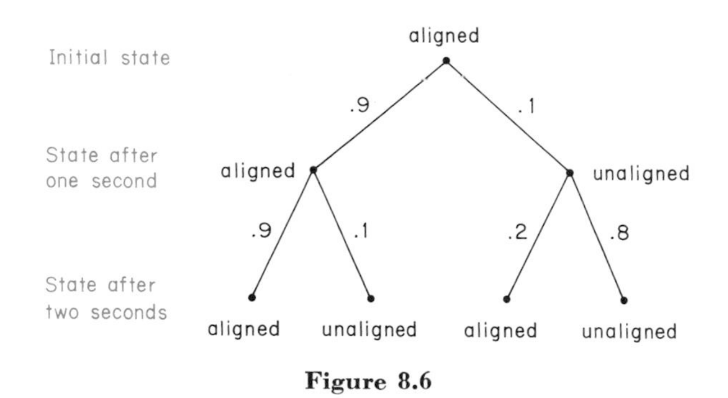

Figure 8.6

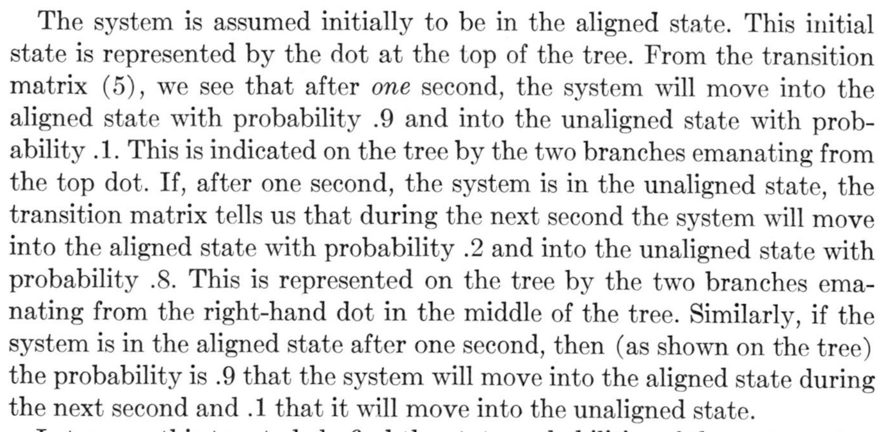
The system is assumed initially to be in the aligned state. This initial state is represented by the dot at the top of the tree. From the transition matrix (5), we see that after *one* second, the system will move into the aligned state with probability .9 and into the unaligned state with probability .1. This is indicated on the tree by the two branches emanating from the top dot. If, after one second, the system is in the unaligned state, the transition matrix tells us that during the next second the system will move into the aligned state with probability .2 and into the unaligned state with probability .8. This is represented on the tree by the two branches emanating from the right-hand dot in the middle of the tree. Similarly, if the system is in the aligned state after one second, then (as shown on the tree) the probability is .9 that the system will move into the aligned state during the next second and .1 that it will move into the unaligned state.

Let us use this tree to help find the state probabilities of the system after two seconds. By tracing the paths from the top of the tree to the bottom, we see that there are two ways the system can move from its initial aligned state to an aligned state two seconds later. We can have

$$\text{aligned} \rightarrow \text{aligned} \rightarrow \text{aligned} \tag{7}$$

or

$$\text{aligned} \rightarrow \text{unaligned} \rightarrow \text{aligned}. \tag{8}$$

Using the probabilities marked on the branches of the tree, we see that the probability that (7) occurs is

$$(.9)(.9) = .81$$

and the probability that (8) occurs is

$$(.1)(.2) = .02.$$

Thus the probability that the system is aligned after two seconds is

$$.81 + .02 = .83.$$

Similarly, there are two ways the system can move from its initial aligned state to an unaligned state two seconds later. We can have

$$\text{aligned} \rightarrow \text{aligned} \rightarrow \text{unaligned} \tag{9}$$

or

$$\text{aligned} \rightarrow \text{unaligned} \rightarrow \text{unaligned}. \tag{10}$$

Using the probabilities marked on the branches of the tree, we see that the probability of (9) is

$$(.9)(.1) = .09$$

and the probability of (10) is

$$(.1)(.8) = .08.$$

Thus the probability that the system is unaligned after two seconds is

$$.09 + .08 = .17.$$

Example 35 For the equipment problem of Example 29, find the state probabilities after two seconds, assuming that the system is initially unaligned.

Solution Using the transition matrix (5), we obtain the tree diagram in Figure 8.7. The system can move from its initial unaligned state to an aligned state two seconds later in two ways

$$\text{unaligned} \rightarrow \text{aligned} \rightarrow \text{aligned} \tag{11}$$

$$\text{unaligned} \rightarrow \text{unaligned} \rightarrow \text{aligned}. \tag{12}$$

From the tree diagram, (11) occurs with probability

$$(.2)(.9) = .18$$

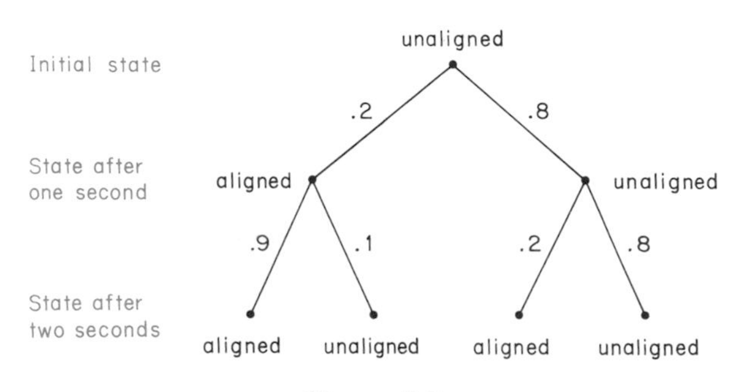

Figure 8.7

and (12) occurs with probability

$$(.8)(.2) = .16.$$

Thus, the probability that the system is aligned after two seconds is

$$.18 + .16 = .34.$$

Similarly, the probability that the system is unaligned after two seconds is

$$(.2)(.1) + (.8)(.8) = .66.$$

If, in the last two examples, we had been interested in the state probabilities after five seconds rather than two seconds, the tree diagram would have contained so many branches that it would have been too cumbersome to use. To remedy this problem, we shall show how matrices can be used to compute state probabilities.

Consider a physical system that can be in any one of k states. If, at a certain observation, the system is in

$$\begin{gathered}\text{state 1 with probability } p_1\\ \text{state 2 with probability } p_2\\ \vdots\\ \text{state } k \text{ with probability } p_k\end{gathered}$$

then the $k \times 1$ matrix of probabilities

$$[p_1 \quad p_2 \quad \cdots \quad p_k]$$

is called the ***state matrix*** or sometimes the ***state vector*** for the system.

Example 36 Consider the space equipment problem discussed in Example 34. At each observation, the system is in one of two states:

$$\begin{array}{ll}\text{state 1} & \text{(aligned)}\\ \text{state 2} & \text{(unaligned)}\end{array}$$

In Example 34, the system was assumed to be aligned initially, so that at the initial observation, the probability is $p_1 = 1$ that the system is aligned and $p_2 = 0$ that the system is unaligned. Thus, the initial state matrix for the system is

$$[p_1 \quad p_2] = [1 \quad 0]. \tag{13}$$

As shown in Figure 8.6, the probability that the system is aligned after one second is $p_1 = .9$ and the probability that it is unaligned after one second is $p_2 = .1$. Thus the state matrix for the system after one second is

$$[p_1 \quad p_2] = [.9 \quad .1]. \tag{14}$$

In Example 34 we showed that after two seconds the probability is $p_1 = .83$ that the system is aligned and $p_2 = .17$ that the system is unaligned; thus the state matrix after two seconds is

$$[p_1 \quad p_2] = [.83 \quad .17]. \tag{15}$$

The following result provides a quick way for finding the state matrix for any observation from the state matrix for the preceding observation.

If P is the transition matrix for a Markov process and if X is the state matrix for any observation, then

$$XP \tag{16}$$

is the state matrix for the next observation.

Example 37 For the problem in Example 36, the initial state matrix is

$$X = [1 \quad 0]$$

(see (13)), and the transition matrix is

$$P = \begin{bmatrix} .9 & .1 \\ .2 & .8 \end{bmatrix}.$$

Thus, from result (16) the state matrix after one second is

$$XP = [1 \quad 0]\begin{bmatrix} .9 & .1 \\ .2 & .8 \end{bmatrix} = [.9 \quad .1],$$

which agrees with (14). Similarly, the state matrix after two seconds is

$$[.9 \quad .1]\begin{bmatrix} .9 & .1 \\ .2 & .8 \end{bmatrix} = [.83 \quad .17],$$

which agrees with (15).

Application of Markov Processes in Marketing

Example 38† A shopper's weekly purchases of a laundry cleaner (soap or detergent) can be assumed to constitute a two-state Markov process with transition

† The data in this problem is based on a 26-week study by G. P. Styon and H. Smith which appeared in the paper, Markov chains applied to marketing, *Journal of Marketing Research* **1,** 50–55 (1964).

matrix

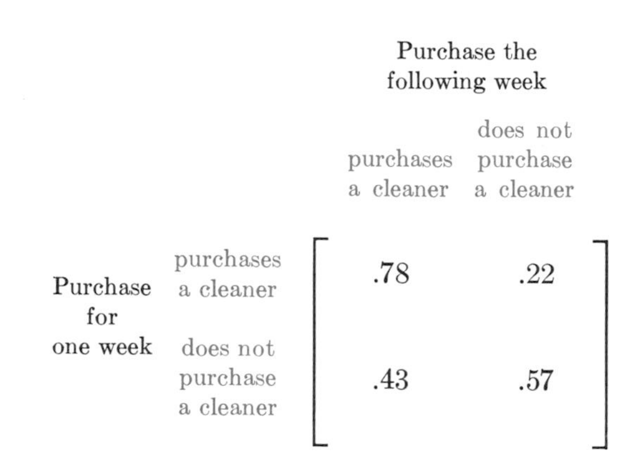

Purchase the following week
does not purchase a cleaner
purchases a cleaner
Purchase for one week
purchases a cleaner
does not purchase a cleaner
.78
.22
.43
.57

Assuming that a shopper initially purchases a cleaner, find the probability that he will purchase a cleaner: one week later, two weeks later, and three weeks later.

Solution Since we are assuming that the shopper initially purchases a cleaner, the initial probability is $p_1 = 1$ that a cleaner is purchased and $p_2 = 0$ that a cleaner is not purchased. Thus, the initial state matrix is

$$\begin{bmatrix} 1 & 0 \end{bmatrix}.$$

From (16) the state matrix after one week will be

$$\begin{bmatrix} 1 & 0 \end{bmatrix} \begin{bmatrix} .78 & .22 \\ .43 & .57 \end{bmatrix} = \begin{bmatrix} .78 & .22 \end{bmatrix},$$

the state matrix after two weeks will be

$$\begin{bmatrix} .78 & .22 \end{bmatrix} \begin{bmatrix} .78 & .22 \\ .43 & .57 \end{bmatrix} = \begin{bmatrix} .703 & .297 \end{bmatrix}$$

and the state matrix after three weeks will be

$$\begin{bmatrix} .703 & .297 \end{bmatrix} \begin{bmatrix} .78 & .22 \\ .43 & .57 \end{bmatrix} = \begin{bmatrix} .676 & .324 \end{bmatrix}.$$

Thus, the probability the shopper purchases a cleaner one week later is .78, two weeks later is .703, and three weeks later is .676.

In many applications of Markov processes, one is interested in the "long-run" behavior of the physical system. The following example illustrates what we have in mind.

Example 39 Consider the stock market fluctuation problem of Example 28. To illustrate how this system behaves over a period of time we computed the state matrices for the system for a period of ten market days, assuming that the initial state of the system was an increase. We obtained (to two decimal places):

Initial state matrix $[1 \quad 0]\begin{bmatrix} .2 & .8 \\ .7 & .3 \end{bmatrix} = [.20 \quad .80]$ (after one day)

$$[.20 \quad .80]\begin{bmatrix} .2 & .8 \\ .7 & .3 \end{bmatrix} = [.60 \quad .40] \qquad \text{(after two days)}$$

$$[.60 \quad .40]\begin{bmatrix} .2 & .8 \\ .7 & .3 \end{bmatrix} = [.40 \quad .60] \qquad \text{(after three days)}$$

$$[.40 \quad .60]\begin{bmatrix} .2 & .8 \\ .7 & .3 \end{bmatrix} = [.50 \quad .50] \qquad \text{(after four days)}$$

$$[.50 \quad .50]\begin{bmatrix} .2 & .8 \\ .7 & .3 \end{bmatrix} = [.45 \quad .55] \qquad \text{(after five days)}$$

$$[.45 \quad .55]\begin{bmatrix} .2 & .8 \\ .7 & .3 \end{bmatrix} = [.48 \quad .53] \qquad \text{(after six days)}$$

$$[.48 \quad .53]\begin{bmatrix} .2 & .8 \\ .7 & .3 \end{bmatrix} = [.46 \quad .54] \qquad \text{(after seven days)}$$

$$[.46 \quad .54]\begin{bmatrix} .2 & .8 \\ .7 & .3 \end{bmatrix} = [.47 \quad .53] \qquad \text{(after eight days)}$$

$$[.47 \quad .53]\begin{bmatrix} .2 & .8 \\ .7 & .3 \end{bmatrix} = [.47 \quad .53] \qquad \text{(after nine days)}$$

$$[.47 \quad .53]\begin{bmatrix} .2 & .8 \\ .7 & .3 \end{bmatrix} = [.47 \quad .53] \qquad \text{(after ten days).}$$

Thus, after eight market days, the probability of an increase stabilizes at .47 and the probability of decrease stabilizes at .53. Similarly, if the initial state of the system had been a decrease, we would have obtained

$$[0 \quad 1]\begin{bmatrix} .2 & .8 \\ .7 & .3 \end{bmatrix} = [.70 \quad .30] \qquad \text{(after one day)}$$

$$[.70 \quad .30]\begin{bmatrix} .2 & .8 \\ .7 & .3 \end{bmatrix} = [.35 \quad .65] \qquad \text{(after two days)}$$

$$[.35 \quad .65]\begin{bmatrix} .2 & .8 \\ .7 & .3 \end{bmatrix} = [.53 \quad .48] \qquad \text{(after three days)}$$

$$[.53 \quad .48]\begin{bmatrix} .2 & .8 \\ .7 & .3 \end{bmatrix} = [.44 \quad .56] \qquad \text{(after four days)}$$

$$[.44 \quad .56]\begin{bmatrix} .2 & .8 \\ .7 & .3 \end{bmatrix} = [.48 \quad .52] \qquad \text{(after five days)}$$

$$[.48 \quad .52]\begin{bmatrix} .2 & .8 \\ .7 & .3 \end{bmatrix} = [.46 \quad .54] \qquad \text{(after six days)}$$

$$[.46 \quad .54]\begin{bmatrix} .2 & .8 \\ .7 & .3 \end{bmatrix} = [.47 \quad .53] \qquad \text{(after seven days)}$$

$$[.47 \quad .53]\begin{bmatrix} .2 & .8 \\ .7 & .3 \end{bmatrix} = [.47 \quad .53] \qquad \text{(after eight days)}$$

$$[.47 \quad .53]\begin{bmatrix} .2 & .8 \\ .7 & .3 \end{bmatrix} = [.47 \quad .53] \qquad \text{(after nine days)}$$

$$[.47 \quad .53]\begin{bmatrix} .2 & .8 \\ .7 & .3 \end{bmatrix} = [.47 \quad .53] \qquad \text{(after ten days).}$$

Thus, once again the probability of an increase stabilizes at .47 and the probability of a decrease stabilizes at .53 after a period of time.

The results observed in the previous example are not accidental. As the following theorem shows, this stabilization phenomenon occurs in many Markov processes.

Let P be the transition matrix for a Markov process. If some power of P has all positive entries, then:

(1) regardless of the initial state of the system, the successive state matrices will approach the same fixed state matrix Q;

(2) the matrix Q satisfies the equation

$$QP = Q \tag{17}$$

and is called the ***steady-state matrix*** for the system.

We omit the proof. A transition matrix P is called ***regular*** if some power of P has all positive entries. Thus (17) applies only when the transition matrix P is regular.

Example 40 The calculations in Example 39 show that the steady-state matrix for the stock-market system is (to two decimal places):

$$Q \simeq [.47 \quad .53].$$

This same result can be obtained as follows by using Equation (17). Let

$$Q = [q_1 \quad q_2]$$

be the unknown steady-state matrix, where q_1 is the steady-state probability of an increase and q_2 is the steady-state probability of a decrease. (Note that $q_1 + q_2 = 1$.) Substituting Q and the transition matrix (4) into (17) we obtain

$$[q_1 \quad q_2]\begin{bmatrix} .2 & .8 \\ .7 & .3 \end{bmatrix} = [q_1 \quad q_2]$$

or

$$[.2q_1 + .7q_2 \quad .8q_1 + .3q_2] = [q_1 \quad q_2]. \tag{18}$$

In order for the matrices in (18) to be equal, their corresponding entries must be equal; that is,

$$\begin{aligned} .2q_1 + .7q_2 &= q_1 \\ .8q_1 + .3q_2 &= q_2 \end{aligned}$$

or equivalently,

$$\begin{aligned} .8q_1 - .7q_2 &= 0 \\ -.8q_1 + .7q_2 &= 0. \end{aligned}$$

Upon solving this system, the reader will see that there are infinitely many solutions, and they are given by the formulas

$$q_1 = \tfrac{7}{8}t, \qquad q_2 = t \tag{19}$$

where t is arbitrary. However $Q = [q_1 \quad q_2]$ is a state matrix so that, as observed above, $q_1 + q_2 = 1$. Thus from (19) we obtain

$$1 = q_1 + q_2 = \tfrac{7}{8}t + t = \tfrac{15}{8}t,$$

or $t = \frac{8}{15}$. Substituting this value into (19) we obtain the steady state matrix;

$$Q = [q_1 \quad q_2] = [\tfrac{7}{15} \quad \tfrac{8}{15}],$$

which agrees with the results obtained to two decimal places in Example 39.

Application of Markov Processes to Problems in Mass Transit

Example 41 The following example illustrates how Markov processes can enter into the long-range planning of urban transit facilities.

Commuters who work in the Delaware Valley can commute to and from their jobs either by public transportation or automobile. After the installation of a new public transit system, the Planning Commission predicts that each year 30% of those using public transportation will change to automobile, while 70% will continue to use public transportation. The Commission also predicts that each year 60% of those using automobiles will change to public transportation, while 40% will continue to use their automobiles. This information is contained in the following transition matrix

		Mode of transportation the next year	
		public transportation	automobile
Mode of transportation one year	public transportation	.7	.3
	automobile	.6	.4

$$\begin{bmatrix} .7 & .3 \\ .6 & .4 \end{bmatrix} \qquad (20)$$

(In this matrix we have recorded the percentage changes as probabilities.)

Initially, assume that 20% of the commuters use public transportation and 80% use their automobile. Assuming that the population of the Delaware Valley remains constant, find:

(a) the percentage of commuters using each mode of transportation after two years;

(b) the percentage of commuters using each mode of transportation after a "long period of time."

Solution (a) Since 20% initially use public transportation and 80% initially use their automobiles, the initial state matrix is

public transportation	automobile
.2	.8

$$[.2 \quad .8].$$

Thus, after one year the new state matrix will be

$$[.2 \quad .8]\begin{bmatrix} .7 & .3 \\ .6 & .4 \end{bmatrix} = [.62 \quad .38],$$

and after two years the new state matrix will be

$$[.62 \quad .38]\begin{bmatrix} .7 & .3 \\ .6 & .4 \end{bmatrix} = [.662 \quad .338].$$

Thus, after two years, 66.2% of the commuters will be using public transportation and 33.8% will be using automobiles.

Solution (b) To find the percentage of commuters using each mode after a long period of time, we must find the steady-state matrix for the system. Using (17) and (20) we see that the steady-state matrix

$$Q = [q_1 \quad q_2]$$

must satisfy

$$[q_1 \quad q_2]\begin{bmatrix} .7 & .3 \\ .6 & .4 \end{bmatrix} = [q_1 \quad q_2],$$

or

$$[.7q_1 + .6q_2 \quad .3q_1 + .4q_2] = [q_1 \quad q_2],$$

or

$$\begin{aligned} .7q_1 + .6q_2 &= q_1 \\ .3q_1 + .4q_2 &= q_2, \end{aligned}$$

or

$$\begin{aligned} .3q_1 - .6q_2 &= 0 \\ -.3q_1 + .6q_2 &= 0. \end{aligned}$$

Upon solving this system, the reader will see that there are infinitely many solutions, and they are given by the formulas

$$q_1 = 2t, \qquad q_2 = t \tag{21}$$

where t is arbitrary. Since $Q = [q_1 \quad q_2]$ is a state matrix, we must have $q_1 + q_2 = 1$. Thus from (21) we obtain

$$1 = q_1 + q_2 = 2t + t = 3t,$$

so that $t = \frac{1}{3}$. Substituting this value in (21) yields the steady state matrix

$$Q = [q_1 \quad q_2] = [\tfrac{2}{3} \quad \tfrac{1}{3}].$$

Thus, after a long period of time, the proportion of riders using public transportation will stabilize at $\frac{2}{3}$ and the proportion of riders using their automobiles will stabilize at $\frac{1}{3}$.

Application of Markov Processes to Genetics

Example 42 (For readers who have read Section 8.3.) In this example we shall determine the long-term genotype probabilities for the offspring that result from the mating pattern described in Example 31. Let P denote the transition matrix (6). To apply result (17), some power of P must have all positive entries. Although P itself does not have all positive entries, the matrix

$$P^2 = \begin{bmatrix} \frac{3}{8} & \frac{1}{2} & \frac{1}{8} \\ \frac{1}{4} & \frac{1}{2} & \frac{1}{4} \\ \frac{1}{8} & \frac{1}{2} & \frac{3}{8} \end{bmatrix}$$

does; thus (17) can be used to compute the steady-state matrix

$$Q = [q_1 \quad q_2 \quad q_3],$$

where q_1, q_2, and q_3 are the long-term probabilities of genotypes AA, Aa, and aa respectively. Substituting Q and P into (17) gives

$$[q_1 \quad q_2 \quad q_3]\begin{bmatrix} \frac{1}{2} & \frac{1}{2} & 0 \\ \frac{1}{4} & \frac{1}{2} & \frac{1}{4} \\ 0 & \frac{1}{2} & \frac{1}{2} \end{bmatrix} = [q_1 \quad q_2 \quad q_3],$$

or equivalently

$$\begin{aligned} \tfrac{1}{2}q_1 + \tfrac{1}{4}q_2 \qquad\quad &= q_1 \\ \tfrac{1}{2}q_1 + \tfrac{1}{2}q_2 + \tfrac{1}{2}q_3 &= q_2 \\ \tfrac{1}{4}q_2 + \tfrac{1}{2}q_3 &= q_3, \end{aligned}$$

or equivalently

$$\begin{aligned} \tfrac{1}{2}q_1 - \tfrac{1}{4}q_2 \qquad\quad &= 0 \\ -\tfrac{1}{2}q_1 + \tfrac{1}{2}q_2 - \tfrac{1}{2}q_3 &= 0 \\ -\tfrac{1}{4}q_2 + \tfrac{1}{2}q_3 &= 0. \end{aligned}$$

Upon solving this system, the reader will see that there are infinitely many solutions, and they are given by the formulas

$$q_1 = t, \qquad q_2 = 2t, \qquad q_3 = t. \tag{22}$$

Since $Q = [q_1 \quad q_2 \quad q_3]$ is a state matrix we must have $q_1 + q_2 + q_3 = 1$. Thus from (22) we obtain

$$1 = q_1 + q_2 + q_3 = t + 2t + t = 4t,$$

so that $t = \frac{1}{4}$. Substituting this value in (22) yields the steady state matrix

$$Q = [q_1 \quad q_2 \quad q_3] = [\tfrac{1}{4} \quad \tfrac{1}{2} \quad \tfrac{1}{4}].$$

Thus, if the mating pattern is carried out repeatedly, the genotype probabilities of the offspring will stabilize at

$$\tfrac{1}{4} \text{ for } AA, \qquad \tfrac{1}{2} \text{ for } Aa, \qquad \tfrac{1}{4} \text{ for } aa,$$

regardless of the original unknown genotype.

In this section we have just touched the surface of Markov processes and their applications. Readers interested in studying this subject in more detail can consult the following references:

[1] J. Kemeny, H. Mirkil, J. Snell, and G. Thompson, *Finite Mathematical Structures* (Englewood Cliffs, New Jersey: Prentice Hall, 1959).

[2] J. Kemeny, A. Schleifer, J. Snell, and G. Thompson, *Finite Mathematics with Business Applications* (Englewood Cliffs, New Jersey: Prentice Hall, 1962).

[3] D. P. Maki and M. Thompson, *Mathematical Models and Applications* (Englewood Cliffs, New Jersey: Prentice Hall, 1973).

EXERCISE SET 8.6

1. For the frog-jumping problem described in this section, replace the data in (1) by:

$$1,\ 2,\ 1,\ 2,\ 2,\ 2,\ 1,\ 1,\ 1,\ 2,\ 1$$

Based on these data, find the transition matrix for the system.

2. Explain why

$$\begin{bmatrix} \frac{1}{2} & \frac{1}{2} & \frac{1}{2} \\ 0 & 1 & 0 \\ \frac{1}{4} & \frac{3}{4} & \frac{1}{8} \end{bmatrix}$$

cannot be a transition matrix for any physical system.

3. Consider a Markov process with transition matrix

$$\begin{array}{c} \\ \text{State 1} \\ \text{State 2} \end{array} \begin{array}{c} \begin{array}{cc} \text{State 1} & \text{State 2} \end{array} \\ \begin{bmatrix} \frac{1}{4} & \frac{3}{4} \\ \frac{2}{5} & \frac{3}{5} \end{bmatrix} \end{array}$$

(a) What does the entry $\frac{3}{4}$ in this matrix represent?

(b) Assuming that the system is initially in State 1, find the state matrix one observation later.

(c) Assuming that the system is initially in State 2, find the state matrix one observation later.

4. Consider a Markov process with transition matrix

$$\begin{array}{c} \\ \text{State 1} \\ \text{State 2} \end{array} \begin{array}{c} \begin{array}{cc} \text{State 1} & \text{State 2} \end{array} \\ \begin{bmatrix} .6 & .4 \\ .9 & .1 \end{bmatrix} \end{array}$$

If the state matrix at a certain observation is $[.75 \quad .25]$, what is the state matrix at the next observation?

5. Consider the Markov process in Exercise 3.

(a) Assume the system is initially in State 1 and use a tree diagram to find the state matrix two observations later.

(b) Solve the problem in part (a) using result (16) in the text.

6. Consider a Markov process with transition matrix

$$\begin{array}{c} \\ \text{State 1} \\ \text{State 2} \\ \text{State 3} \end{array} \begin{array}{c} \begin{array}{ccc} \text{State 1} & \text{State 2} & \text{State 3} \end{array} \\ \begin{bmatrix} .4 & .1 & .5 \\ .3 & .3 & .4 \\ .6 & .2 & .2 \end{bmatrix} \end{array}$$

(a) Assume the system is initially in State 2 and use a tree diagram to find the state matrix three observations later.

(b) Solve the problem in part (a) using result (16) in the text.

In Exercises 7–10, find the steady-state matrix for the Markov process whose transition matrix is given.

7. $\begin{bmatrix} .75 & .25 \\ .65 & .35 \end{bmatrix}$.

8. $\begin{bmatrix} .5 & .5 \\ .8 & .2 \end{bmatrix}$.

9. $\begin{bmatrix} .1 & .2 & .7 \\ .3 & .1 & .6 \\ .5 & .4 & .1 \end{bmatrix}$.

10. $\begin{bmatrix} .4 & .4 & .2 \\ .7 & .1 & .2 \\ .5 & .3 & .2 \end{bmatrix}$.

11. (a) Consider a Markov process with transition matrix

$$P = \begin{bmatrix} .7 & .3 & 0 \\ .5 & .5 & 0 \\ .6 & .4 & 0 \end{bmatrix}.$$

Explain why result (17) does not apply.

(b) Does result (17) apply to a Markov process with transition matrix

$$P = \begin{bmatrix} .6 & .2 & .2 \\ .7 & .3 & 0 \\ .5 & 0 & .5 \end{bmatrix}?$$

Explain.

12. Recently, some resorts have been offering insurance to pay vacationers' hotel bills when it rains. Suppose that an insurer of a Florida resort has found that, after a clear day, there is a probability of .2 for rain the next day, while after a rainy day, there is a probability of .4 for rain the next day.

(a) Write down an appropriate transition matrix.

(b) Over the long term, what fraction of the time will the company have to pay off?

13. Suppose that children whose parents are in a middle-to-high-income bracket will, as adults, be in the middle-to-high-income bracket with probability .9 and in a low-income bracket with probability .1. Suppose also that children whose parents are in a low-income bracket will, as adults, be in a middle-to-high income bracket with probability .5 and in a low-income bracket with probability .5.

(a) What is the probability that a grandchild of a low-income family will be in the low-income bracket as an adult?

(b) Over the long-term, what fraction of the adult population will be in the low-income bracket?

14. A regional planning commission estimates that each year 10% of the people in the eastern part of the region will move to the western part and 3% of the people in the western part will move to the eastern part. Assuming that the population in the region remains constant and, initially, 40% of the population lives in the eastern region and 60% in the western region, find the percentage of population in each region:

(a) after one year;

(b) after three years;

(c) over the long term.

15. The New York metropolitan area is serviced by three major airports: J. F. Kennedy, LaGuardia, and Newark. Suppose that a car rental agency with a fleet of 500 cars has rental and storage facilities at each of the three airports. Customers rent and then return their cars to the various airports with the probabilities in the transition matrix

		Returned to		
		Kennedy	LaGuardia	Newark
Rented from	Kennedy	.8	.1	.1
	LaGuardia	.3	.2	.5
	Newark	.2	.6	.2

How many parking spaces should the company allocate at each airport? (*Hint*: Use the steady-state matrix.)

9 MATHEMATICS OF FINANCE

9.1 EXPONENTS AND LOGARITHMS—A REVIEW*

In this section we briefly review some results from algebra about exponents and logarithms. We shall need this material for our study of growth and decay problems in the next section.

The ***integers*** are the numbers $\ldots, -3, -2, -1, 0, 1, 2, 3, \ldots$ In Table 9.1 we recall some basic definitions involving integer exponents.

Warning. There is some common confusion about the symbol $\sqrt{\ }$. Every positive number has two square roots, one positive and one negative; for example, 16 has -4 and 4 as square roots. The symbol $\sqrt{a}$ always denotes the positive square root of a; thus $\sqrt{16} = 4$.

In Table 9.2 we recall some of the properties of exponents.

Example 1

$$3^{-4} = \frac{1}{3^4} = \frac{1}{81}$$

$$2^5 \cdot 2^3 = 2^8 = 256$$

$$\frac{4^5}{4^2} = 4^{5-2} = 4^3 = 64$$

$$\frac{4^2}{4^5} = 4^{2-5} = 4^{-3} = \frac{1}{4^3} = \frac{1}{64}$$

$$(2^3)^2 = 2^{3 \cdot 2} = 2^6 = 64$$

$$25^{3/2} = (25^{1/2})^3 = (\sqrt{25})^3 = 5^3 = 125$$

or

$$25^{3/2} = (25^3)^{1/2} = \sqrt{15{,}625} = 125.$$

Table 9.1

Definition	Examples
Positive integer exponents	$4^2 = 4 \cdot 4 = 16$
$a^n = \underbrace{a \cdot a \cdot \cdots \cdot a}_{n \text{ factors}}$	$(-2)^3 = (-2)(-2)(-2) = -8$
	$\left(\frac{1}{5}\right)^4 = \frac{1}{5} \cdot \frac{1}{5} \cdot \frac{1}{5} \cdot \frac{1}{5} = \frac{1}{625}$
$a^1 = a$	$(1.6)^1 = 1.6$
Zero exponent	$6^0 = 1$
$a^0 = 1 \qquad (a \neq 0)$	$(-1)^0 = 1$
	$(\frac{1}{2})^0 = 1$
Negative integer exponents	$2^{-4} = \frac{1}{2^4} = \frac{1}{2 \cdot 2 \cdot 2 \cdot 2} = \frac{1}{16}$
$a^{-n} = \frac{1}{a^n} \qquad (a \neq 0)$	$\left(\frac{1}{3}\right)^{-2} = \frac{1}{\frac{1}{3} \cdot \frac{1}{3}} = \frac{1}{\frac{1}{9}} = 1 \cdot \frac{9}{1} = 9$
Fractional exponents	$9^{1/2} = \sqrt{9} = 3$
$a^{1/n} = \sqrt[n]{a}$	$8^{1/3} = \sqrt[3]{8} = 2$ (since $2^3 = 8$)
(recall: $\sqrt[n]{a}$ denotes a number	$(-8)^{1/3} = \sqrt[3]{-8} = -2$ (since $(-2)^3 = -8$
whose nth power is a)	$(16)^{.5} = 16^{1/2} = \sqrt{16} = 4$

Example 2 Use Table IV in the Appendix to find approximate numerical values of

(a) $4^{-1/3}$ (b) $2^{2/3}$ (c) $64^{1/4}$ (c) $56^{1/2}$.

Solution

(a) $4^{-1/3} = \dfrac{1}{4^{1/3}} = \dfrac{1}{\sqrt[3]{4}} \approx \dfrac{1}{1.587} \approx .6301$

(b) $2^{2/3} = (2^2)^{1/3} = \sqrt[3]{4} \approx 1.587$

(c) $64^{1/4} = (64^{1/2})^{1/2} = 8^{1/2} = \sqrt{8} \approx 2.828$

(d) $56^{1/2} = (8 \cdot 7)^{1/2} = 8^{1/2} \cdot 7^{1/2} = \sqrt{8}\,\sqrt{7} \approx (2.828)(2.646) \approx 7.483.$

Those numbers that are expressible as ratios of integers are called ***rational*** numbers. Some examples are

$$\frac{1}{2}, \quad \frac{12}{5}, \quad -\frac{100}{25}, \quad 8\left(=\frac{8}{1}\right), \quad 0.$$

Table 9.2

Property	Examples	Numerical examples
$a^r \cdot a^s = a^{r+s}$	$a^3 \cdot a^5 = a^{3+5} = a^8$ $a^6 \cdot a^{-2} = a^{6+(-2)} = a^4$ $a^3 \cdot a^{-7} = a^{3+(-7)} = a^{-4} = \frac{1}{a^4}$ $a \cdot a^8 = a^1 \cdot a^8 = a^{1+8} = a^9$	$2^3 \cdot 2^5 = 2^8 = 256$ $2^6 \cdot 2^{-2} = 2^4 = 16$ $3^3 \cdot 3^{-7} = 3^{-4} = \frac{1}{3^4} = \frac{1}{81}$ $2 \cdot 2^8 = 2^9 = 512$
$\frac{a^r}{a^s} = a^{r-s}$	$\frac{a^8}{a^3} = a^{8-3} = a^5$ $\frac{a^2}{a^7} = a^{2-7} = a^{-5} = \frac{1}{a^5}$ $\frac{a^4}{a^4} = a^{4-4} = a^0 = 1$ $\frac{a^{-2}}{a^5} = a^{-2-5} = a^{-7} = \frac{1}{a^7}$	$\frac{4^8}{4^3} = 4^5 = 1024$ $\frac{(-5)^2}{(-5)^7} = \frac{1}{(-5)^5} = -\frac{1}{3125}$ $\frac{9^4}{9^4} = 9^0 = 1$ $\frac{2^{-2}}{2^5} = 2^{-7} = \frac{1}{2^7} = \frac{1}{128}$
$(a^r)^s = a^{rs}$	$(a^3)^4 = a^{3\cdot 4} = a^{12}$ $(a^{-2})^3 = a^{-2\cdot 3} = a^{-6} = \frac{1}{a^6}$ $(a^4)^{-5} = a^{4\cdot(-5)} = a^{-20} = \frac{1}{a^{20}}$ $(a^{1/3})^{1/2} = a^{1/3\cdot 1/2} = a^{1/6} = \sqrt[6]{a}$	$(2^3)^4 = 2^{12} = 4096$ $(3^{-2})^3 = 3^{-6} = \frac{1}{729}$ $((\sqrt{2})^4)^{-5} = \frac{1}{(\sqrt{2})^{20}} = \frac{1}{2^{10}} = \frac{1}{1024}$ $((64)^{1/3})^{1/2} = \sqrt[6]{64} = 2$
$a^{r/s} = (a^{1/s})^r = (a^r)^{1/s}$	$a^{3/4} = (a^{1/4})^3 = (\sqrt[4]{a})^3$ or $a^{3/4} = (a^3)^{1/4} = \sqrt[4]{a^3}$	$16^{3/4} = (\sqrt[4]{16})^3 = 2^3 = 8$ or $16^{3/4} = (16^3)^{1/4} = \sqrt[4]{4096} = 8$
$\left(\frac{a}{b}\right)^r = \frac{a^r}{b^r}$	$\left(\frac{a}{b}\right)^3 = \frac{a^3}{b^3}$ $\left(\frac{a}{b}\right)^{-4} = \frac{a^{-4}}{b^{-4}}$	$\left(\frac{2}{3}\right)^3 = \frac{2^3}{3^3} = \frac{8}{27}$ $\left(\frac{2}{3}\right)^{-4} = \frac{2^{-4}}{3^{-4}} = \frac{1/16}{1/81} = \frac{81}{16}$
$(ab)^r = a^r b^r$	$(ab)^3 = a^3 b^3$ $(ab)^{-3} = a^{-3}b^{-3} = \frac{1}{a^3 b^3}$	$(7 \cdot 8)^3 = 7^3 \cdot 8^3$ $(2 \cdot 4)^{-3} = 2^{-3} \cdot 4^{-3} = \frac{1}{2^3 \cdot 4^3}$

Rational numbers can be characterized by the fact that from some point on the digits in their decimal expansions repeat in a regular pattern; for example

$$\frac{1}{3} = .\underbrace{3}\underbrace{3}\underbrace{3}\underbrace{3}\ldots \qquad \frac{1}{8} = .125\underbrace{0}\underbrace{0}\underbrace{0}\ldots$$

$$\frac{12}{99} = .\underbrace{12}\underbrace{12}\ldots \qquad \frac{8}{13} = .\underbrace{615384}\underbrace{615384}\ldots$$

Numbers not expressible as ratios of integers are called ***irrational numbers***; their decimal representations do not repeat. For example,

$$.101001000100001\ldots$$

represents an irrational number. It can be shown that $\sqrt{2}$ is irrational and with great difficulty that π is irrational. Thus we are assured that the decimals

$$\pi = 3.141592654\ldots$$

$$\sqrt{2} = 1.4142135\ldots$$

do not repeat in a regular pattern.

One of the most important irrational numbers in mathematics is denoted by the letter e. To eleven decimal places its value is

$$e = 2.71828182845\ldots$$

As we shall see, this number plays a fundamental role in problems ranging from population growth to interest on monetary investments.

It can be shown using calculus that the quantity

$$\left(1 + \frac{1}{m}\right)^m$$

gets closer and closer to the number e as the value of m increases. This is illustrated by Table 9.3, constructed using a hand-held calculator.

Table 9.3

m	1	10	100	1000	10,000
$\left(1 + \frac{1}{m}\right)^m$	2.000	2.593	2.705	2.717	2.718

So far, we have reviewed the definition for expressions of the form a^k, where k is a rational number. For expressions like

$$a^{\pi}, \quad a^{\sqrt{2}}, \quad \text{and} \quad a^{-\sqrt{5}},$$

where k is irrational, the precise definition requires calculus. However, for practical purposes an irrational exponent can always be approximated to any degree of accuracy by a rational exponent. For example, in an expression like

$$1024^{\sqrt{2}}$$

we might approximate $\sqrt{2}$ as

$$\sqrt{2} \approx 1.4 = \frac{14}{10}$$

to obtain

$$1024^{\sqrt{2}} \approx (1024)^{14/10} = (\sqrt[10]{1024})^{14} = 2^{14} = 16{,}384.$$

The accuracy of this approximation can be improved by using more decimal places in the approximation of $\sqrt{2}$.

It can be shown that the properties of rational exponents listed in Table 9.2 are also valid for irrational exponents.

We now review some ideas about logarithms; these notions will be needed in the next section. Recall from algebra that *a logarithm is an exponent.* More precisely, if a is a positive number

$$\log_a x$$

(read, "the logarithm to the base a of x") represents that power to which a must be raised to obtain x. Thus

$$\log_{10} 1000 = 3$$

since 10 must be raised to the third power to obtain 1000, and

$$\log_2 16 = 4$$

since $2^4 = 16$. In short,

$$y = \log_a x \qquad \text{and} \qquad x = a^y$$

are equivalent statements.

Of special importance are logarithms using base 10 (***common logarithms***) and logarithms using the irritational base e (***natural logarithms***). It is usual to write:

$$\log x \qquad \text{rather than} \qquad \log_{10} x$$

and

$$\ln x \qquad \text{rather than} \qquad \log_e x.$$

Example 3

$$\log 100 = 2 \qquad \text{since} \qquad 10^2 = 100$$

$$\log \frac{1}{100} = -2 \qquad \text{since} \qquad 10^{-2} = \frac{1}{100}$$

$$\log 1 = 0 \qquad \text{since} \qquad 10^0 = 1$$

$$\ln 1 = 0 \qquad \text{since} \qquad e^0 = 1$$

$$\ln \frac{1}{e^5} = -5 \qquad \text{since} \qquad e^{-5} = \frac{1}{e^5}$$

$$\ln e = 1 \qquad \text{since} \qquad e^1 = e.$$

In Table 9.4 we review some properties of logarithms.

Table 9.4

Property	Examples
(a) $\log_a 1 = 0$	$\log 1 = 0, \quad \ln 1 = 0$
(b) $\log_a a = 1$	$\log 10 = 1, \quad \ln e = 1$
(c) $\log_a xy = \log_a x + \log_a y$	$\log(5 \cdot 6) = \log 5 + \log 6$ $\ln 6 = \ln(3 \cdot 2) = \ln 3 + \ln 2$
(d) $\log_a \frac{x}{y} = \log_a x - \log_a y$	$\log \frac{3}{4} = \log 3 - \log 4$ $\ln \frac{1}{5} = \ln 1 - \ln 5 = 0 - \ln 5 = -\ln 5$
(e) $\log_a x^k = k \log_a x$	$\log 4^3 = 3 \log 4$ $\log 4^{-3} = -3 \log 4$ $\ln \sqrt{3} = \ln 3^{1/2} = \frac{1}{2} \ln 3$

Example 4 Properties (b) and (e) yield a useful fact about logarithms:

$$\log_a a^c = c \log_a a \qquad \text{(Property (e))}$$

or

$$\log_a a^c = c \qquad \text{(Property (b))}$$

In particular, we obtain the following useful results:

$$\log 10^c = c \qquad \text{and} \qquad \ln e^c = c.$$

In this text we shall be concerned primarily with natural logarithms. We shall use them to solve equations like those in the following examples. For reference we have included a brief table of values for natural logarithms in Table V of the Appendix.

Example 5 Find the value for t if $600 = 200e^t$.

Solution

$$\begin{aligned} 600 &= 200e^t \\ 3 &= e^t && \text{(divide through by 200)} \\ \ln 3 &= \ln e^t && \text{(take ln of both sides)} \\ \ln 3 &= t && \text{(Example 4)} \\ t &= 1.0986 && \text{(Table V, Appendix).} \end{aligned}$$

Example 6 Find the value for t if $90 = 100e^{-.5t}$.

Solution

$$\begin{aligned} 90 &= 100e^{-.5t} & & \\ .9 &= e^{-.5t} & &\text{(divide by 100)} \\ \ln(.9) &= \ln(e^{-.5t}) & &\text{(take ln of both sides)} \\ \ln(.9) &= -.5t & &\text{(Example 4)} \\ -.1054 &= -.5t & &\text{(Table V, Appendix)} \\ t &= \frac{.1054}{.5} = .2108. & & \end{aligned}$$

EXERCISE SET 9.1

1. Evaluate each of the following:
(a) 3^5 (b) 2^{-8} (c) 6^0
(d) $(-4)^3$ (e) $(-4)^{-3}$ (f) $\left(\frac{2}{5}\right)^4$
(g) $(64)^{2/3}$ (h) $(625)^{-1/4}$ (i) $\left(\frac{1}{27}\right)^{1/3}$.

2. Evaluate the following, using Appendix Table IV, where needed:
(a) $5^{-1/3}$ (b) $7^{2/3}$ (c) $81^{1/6}$
(d) $48^{1/2}$ (e) $3^{-2} \cdot 3^{9/2}$ (f) $\left(\frac{7^{-3} \cdot 7^5}{7^{-2}}\right)^{1/8}$.

3. Rewrite the following using only positive exponents:
(a) 5^{-9} (b) $\left(\frac{1}{2}\right)^{-8}$ (c) a^{-2}
(d) $\frac{a^{-7}}{a^{-2}}$ (e) $t^5 \cdot t^{-3}$ (f) $\frac{x^{-3}}{y^{-3}}$.

4. Evaluate the following without using any tables:
(a) $\log 1000$ (b) $\log \frac{1}{1000}$ (c) $\log .00001$
(d) $\log(100^{20})$ (e) $\log \sqrt{1000}$ (f) $\log_3 81$.

5. Express the following without using logarithms:
(a) $\ln e^7$ (b) $\ln \sqrt[3]{e}$
(c) $\ln e^{-.6t}$ (d) $\ln \frac{e^{.8}}{e^{.2}}$.

6. In each part find the numerical value of x. Use Appendix Table VI, where needed.

(a) $\log x = 4$ (b) $\log x = -6$ (c) $\log x^2 = 4$

(d) $\ln x = 2$ (e) $\ln e^x = 5$ (f) $\ln x = 1$

(g) $\ln(x + 1) = 3.2$ (h) $2 \ln(x - 2) = 7$ (i) $\ln x = -5$.

7. Given that $\log 2 = .3010$ and $\log 3 = .4771$, find:

(a) $\log 6$ (b) $\log\left(\frac{2}{3}\right)$ (c) $\log 9$

(d) $\log \sqrt{3}$ (e) $\log \frac{1}{2000}$ (f) $\log 54$.

8. In each part obtain a numerical value for t. Use the tables in the Appendix, where necessary.

(a) $e^t = 3$ (b) $500 = 200e^t$

(c) $80 = 100e^{8t}$ (d) $40 = 10e^{-2t}$

(e) $40 - 5e^{-6t} = 20$ (f) $3^t = 8$.

9. Follow the directions of Exercise 8:

(a) $105 - 35e^{-.2t} = 35$ (b) $e^{-0.04t} = 0.03$

(c) $3 = 2 \ln t$.

10. Use Appendix Tables V and VI and properties of exponents and logarithms to evaluate the following:

(a) $\ln 1000$ (b) $\ln 54$

(c) $\ln .001$ (d) $e^{15.01}$.

11. Use Appendix Tables V and VI to help graph the following:

(a) $y = \ln t$ (use $t = 0.1, 0.4, 0.7, 1.0, 2.0, \ldots, 6.0$)

(b) $y = e^t$ (use $t = -3.0, -2.0, -1.0, -0.5, 0, 0.5, 1.0, 1.5, 2.0$)

(c) $y = e^{-t}$ (use $t = -2.0, -1.5, -1.0, -0.5, 0, 0.5, 1.0, 2.0, 3.0$).

12. Use Appendix Tables V and VI to help graph the following. (In each part choose 10 appropriate values of t.)

(a) $y = 2e^{0.5t}$

(b) $y = 2 \ln 2t$

(c) $y = 4e^{-0.25t}$.

13. The repeating decimal .125125125... can be expressed as a ratio of integers as follows: Let $N = .125125125\ldots$ so that

$$1000N = 125.125125\ldots$$

$$1000N - N = 125.000\ldots$$

$$999N = 125$$

$$N = \frac{125}{999}.$$

Use this idea, where needed, to express the following repeating decimals as ratios of integers:

(a) .172172172... (b) .565656...

(c) .1293000... (d) .8888....

14. Is the sum of two rational numbers always rational? The product? The quotient?

15. Is $\sqrt{2} + 3$ rational? Explain.

16. Use the approximations $\sqrt{2} \approx 1.4$, $\sqrt{3} \approx 1.7$, $\sqrt{5} \approx 2.2$ to help estimate the values of

(a) $(1024)^{\sqrt{3}}$ (b) $(1024)^{-\sqrt{2}}$ (c) $(32)^{\sqrt{5}}$ (d) $(1024)^{\sqrt{3}-\sqrt{2}}$

17. (**Advertising Models**) Natural logarithms arise in certain models for advertising response. Suppose such a model is given by the equation

$$N = 3000 + 200 \ln(x + 1),$$

where

$$N = \text{number of units sold}$$

$$x = \text{amount spent for advertising (in thousands of dollars).}$$

(a) How many units would be sold with no advertising expenditure?

(b) Find the number of units sold with an expenditure of 39,000 dollars. (Round off to the nearest integer.)

(c) According to the model, how much advertising money must be spent to sell 5000 units?

18. (**Earthquake Intensity**) Magnitudes of earthquakes are measured using the Richter scale. On this scale the magnitude R of an earthquake is given by

$$R = \log \frac{I}{I_0}$$

where I_0 is a standard intensity used for comparison, and I is the intensity of the earthquake being measured. Thus a magnitude of $R = 3$ on the Richter scale means

$$3 = \log \frac{I}{I_0}$$

or

$$\frac{I}{I_0} = 1000$$

or

$$I = 1000 I_0$$

which states that the intensity of the earthquake measured is 1000 times as great as the standard.

(a) The San Francisco earthquake of 1906 was estimated to have registered 8.2 on the Richter scale. Compare its intensity to the standard.

(b) The devastating Chile earthquake of 1960 had an intensity of $10^{8.5}$ times the standard. What did it register on the Richter scale?

(c) How many times more intense is an earthquake registering $R = 6$ than an earthquake measuring $R = 5$?

9.2 MATHEMATICS OF FINANCE

Simple Interest

In Section 2.3 we discussed simple interest as an application of linear equations. Let us review the basic ideas. Simple interest involves three considerations:

P = ***principal*** (amount borrowed)

r = ***interest rate*** (a percentage per year)

t = ***time*** (in years that the principal is held).

By definition, ***simple interest*** is computed by the formula

$$Prt = \text{principal} \times \text{rate} \times \text{time}.$$

Thus if money is borrowed at simple interest, the amount S that must be repaid after t years consists of the principal P plus the simple interest, Prt. Therefore,

$$S = P + Prt$$

or

Simple Interest Formula

$$S = P(1 + rt).$$

In this formula, S is called the ***amount*** or the ***sum.*** Moreover, it is common to call the principal P the ***present value of*** S, and to call S the ***future value*** of P.

Example 7 A principal of \$1000 is borrowed at 5% per year simple interest. Find the future value after

(a) two years,

(b) three months,

(c) 180 days.

Solution (a) From the information given, $P = 1000$, $r = .05$, and $t = 2$, so that from the simple interest formula

$$S = 1000[1 + (.05)(2)] = 1000(1.1) = \$1100.$$

(b) Since three months is three-twelfths of a year we have $t = \frac{3}{12} = .25$, so that from the simple interest formula

$$S = 1000[1 + (.05)(.25)] = 1000(1.0125) = \$1012.50.$$

(c) For monetary transactions where the time is specified in days, it is common to assume a year of 360 days (sometimes called an ***ordinary interest year***†). With this convention, 180 days corresponds to $t = \frac{180}{360} = .5$ of a year. Thus from the simple interest formula

$$S = 1000[1 + (.05)(.5)] = 1000[1.025] = \$1025.$$

In the language of the financial world, the result in part (a) states that the future value of \$1000 two years hence at 5% simple interest is \$1100; or equivalently, the present value of \$1100 two years hence at 5% simple interest is \$1000.

Compound Interest

For many transactions interest is added to the principal at regular time intervals so that the interest itself earns interest. This is called ***compounding*** of interest. The time interval between successive additions of interest is called the ***conversion period.*** Typical conversion periods are given in Table 9.5.

Table 9.5

Conversion period	Common description
1 day	compounded daily
1 month	compounded monthly
3 months	compounded quarterly
6 months	compounded semiannually
12 months	compounded annually (simple interest)

Example 8 Find the interest on \$1000 for one year at

(a) 8 percent per year simple interest

(b) 8 percent per year compounded semiannually

(c) 8 percent per year compounded quarterly.

† In this text, we assume an ordinary interest year for all problems in which time is specified in days. Another common convention uses a year of 365 days (called an ***exact interest year.***)

Solution (a) The simple interest is

$$\begin{aligned} Prt &= \text{principal} \times \text{rate} \times \text{time} \\ &= (1000)(.08)(1) \\ &= \$80. \end{aligned}$$

(b) Since the yearly interest rate is 8%, the semiannual interest rate is $\frac{1}{2}(8\%) = 4\%$. Thus, at the end of the first six months the interest earned is

$$(\$1000)(.04) = \$40$$

which, when added to the principal, yields \$1040. For the second six months this new principal earns interest of

$$(\$1040)(.04) = \$41.60.$$

Combining the interest for the two six-month periods we obtain a total interest of

$$\$40.00 + \$41.60 = \$81.60.$$

(c) Since the annual interest rate is 8%, the quarterly interest rate is $\frac{1}{4}(8\%) = 2\%$. Thus

$$\begin{aligned} \text{first quarter interest} &= (\$1000)(.02) = \$20.00 \\ \text{second quarter interest} &= (\$1020)(.02) = \$20.40 \\ \text{third quarter interest} &= (\$1040.40)(.02) = \$20.81 \\ \text{fourth quarter interest} &= (\$1061.21)(.02) = \underline{\$21.22} \\ &\qquad\qquad\text{total} \quad \$82.43 \end{aligned}$$

This example illustrates an important point: *the more frequent the compounding the greater the interest.*

Our next objective is to obtain the following result:

Compound Interest Formula If P dollars are invested at an annual interest rate r, compounded k times a year, then after n conversion periods the investment will have grown to an amount S given by

$$S = P\left(1 + \frac{r}{k}\right)^n \tag{1}$$

To see this, let V_{beg} denote the value of the investment at the beginning of any conversion period. The conversion period will have a duration of

$1/k$ years (because there are k conversion periods per year). Since the annual interest rate is r, it follows that the interest earned during the conversion period will be

$$\text{principal} \times \text{rate} \times \text{time} = V_{\text{beg}} \cdot r \cdot \frac{1}{k} = V_{\text{beg}} \cdot \frac{r}{k}$$

Thus, if we let V_{end} be the value of the investment at the end of the conversion period we shall have

$$V_{\text{end}} = V_{\text{beg}} + \left(V_{\text{beg}} \cdot \frac{r}{k}\right)$$

or

$$V_{\text{end}} = V_{\text{beg}}\left(1 + \frac{r}{k}\right).$$

In short, the value of the investment at the end of a period is the value at the beginning of the period times the factor $1 + (r/k)$. Thus with an initial investment of P, the value grows as shown in Table 9.6. It is

Table 9.6

Period	Value at period beginning	Value at period end
1	P	$P\left(1 + \frac{r}{k}\right)$
2	$P\left(1 + \frac{r}{k}\right)$	$P\left(1 + \frac{r}{k}\right)^2$
3	$P\left(1 + \frac{r}{k}\right)^2$	$P\left(1 + \frac{r}{k}\right)^3$
4	$P\left(1 + \frac{r}{k}\right)^3$	$P\left(1 + \frac{r}{k}\right)^4$
⋮	⋮	⋮

clear from the pattern in this table that the value of the investment at the end of n periods is $P(1 + (r/k))^n$, which is precisely the result stated in (1).

The quantity r/k in formula (1) represents the interest rate per conversion period. It is common to denote this quantity by the letter i so that formula (1) becomes

$$S = P(1 + i)^n. \tag{2}$$

Appendix Table VII lists values of

$$(1 + i)^n \qquad \text{and} \qquad \frac{1}{(1 + i)^n} = (1 + i)^{-n}$$

for various common values of i and n.

The terms ***principal, amount, future value,*** and ***present value*** have the same meaning in compound interest problems that they have in simple interest problems.

Example 9 If \$10,000 is invested at an annual interest rate of 8% per year compounded quarterly, what is the value of the investment after two years? (Equivalently, what is the future value of \$10,000 two years hence at 8% per year compounded quarterly?)

Solution From the data

$$P = \$10{,}000, \qquad r = .08, \qquad k = 4 \quad \text{(quarterly = 4 yearly conversion periods)},$$

and

$$n = 8 \qquad \text{(2 years = 8 quarterly conversion periods)}.$$

Thus,

$$i = \frac{r}{k} = \frac{.08}{4} = .02 = 2\%,$$

so that from (2) the value S of the investment will be

$$\begin{aligned} S &= P(1 + i)^n \\ &= 10{,}000(1 + .02)^8 \\ &= 10{,}000(1.17165938) \qquad \text{(Appendix Table VII, } i = 2\%, \; n = 8) \\ &= \$11{,}716.59. \end{aligned}$$

Because compound interest is affected by both the annual interest rate (often called the ***nominal interest rate***) and the frequency of compounding, it is sometimes hard to tell offhand which of two compound interest procedures is the better. For example, would it be better to invest at 4% compounded monthly or $4\frac{1}{4}$% compounded semiannually? To make such comparisons, it is common to utilize the notion of ***effective rate of interest.*** By definition this is the simple interest rate that produces the same yearly return as the compound interest procedure.

A formula for effective rate of interest can be derived as follows. Suppose a principal P is invested at a nominal or annual interest rate of r, compounded k times a year, and let r_{eff} be the effective rate of interest.

It follows from formula (1) that in *one* year the compound interest procedure yields an amount

$$A = P\left(1 + \frac{r}{k}\right)^k$$

($n = k =$ number of conversion periods in one year), while the amount resulting in one year from the simple interest rate r_{eff} is

$$A = P(1 + r_{\text{eff}}) \qquad (t = 1 \text{ for one year}).$$

Since the two amounts are the same we have

$$P(1 + r_{\text{eff}}) = P\left(1 + \frac{r}{k}\right)^k.$$

Dividing both sides by P yields

$$1 + r_{\text{eff}} = \left(1 + \frac{r}{k}\right)^k,$$

from which we obtain

Effective Rate of Interest Formula

$$r_{\text{eff}} = \left(1 + \frac{r}{k}\right)^k - 1$$

or

$$r_{\text{eff}} = (1 + i)^k - 1.$$

The "Truth in Lending Law" requires the effective interest rate to appear on all contracts.

Example 10 Find the effective rate of interest for 8% per year compounded quarterly.

Solution We have $r = .08$ and $k = 4$ (conversion periods per year) so that

$$\begin{aligned} r_{\text{eff}} &= \left(1 + \frac{.08}{4}\right)^4 - 1 \\ &= (1 + .02)^4 - 1 \\ &= 1.08243216 - 1 \qquad (\text{Appendix Table VII}, \quad i = 2\%, \quad n = 4) \\ &= .08243216, \end{aligned}$$

which is an effective interest rate of 8.243216%.

Example 11 Which is a better investment, 3% per year compounded monthly or 3.1% per year simple interest?

Solution We shall compute the effective rate of interest for the first investment. We have

$$r = .03, \qquad k = 12 \quad \text{(conversion periods per year)},$$

$$i = \frac{r}{k} = \frac{.03}{12} = .0025.$$

Thus

$$\begin{aligned} r_{\text{eff}} &= (1 + .0025)^{12} - 1 \\ &= 1.03041596 - 1 \qquad \text{(Appendix Table VII,} \quad i = \tfrac{1}{4}\%, \quad n = 12) \\ &= .03041596, \end{aligned}$$

which is an effective interest rate of 3.041596%. Thus, the simple interest of 3.1% per year is the better investment.

Sometimes one is interested in finding the principal P that must be invested now to obtain a certain desired amount S at some future time. For example, how much should be invested now at 6% per year compounded semiannually to yield $20,000 eighteen years from now for a child's education? In other words, what is the present value of $20,000 eighteen years hence at 6% per year compounded semiannually? Problems like this can be solved by rewriting formula (2) in the form:

Present Value Formula

$$P = S(1 + i)^{-n}. \tag{3}$$

Example 12 A principal P is to be invested at 6% per year compounded semiannually to yield $20,000 for a child's college education eighteen years later. How much should the principal be?

Solution From the data

$$S = \$20{,}000, \qquad r = .06, \qquad k = 2 \quad \text{(two semiannual conversion periods per year)}$$

$$i = \frac{r}{k} = \frac{.06}{2} = .03, \qquad n = 36 \quad \text{(18 years = 36 conversion periods).}$$

Thus, from formula (3)

$$\begin{aligned} P &= S(1+i)^{-n} \\ &= 20{,}000(1+.03)^{-36} \\ &= 20{,}000(.34503243) \qquad \text{(Appendix Table VII, } i = 3\%, \quad n = 36) \\ &= \$6{,}900.65. \end{aligned}$$

Continuous Compounding of Interest

For a fixed principal, time period, and annual interest rate, the more frequent the compounding, the greater will be the return on the investment. One might ask if the return can be increased indefinitely by increasing the frequency of compounding. The answer is no; there is a theoretical upper limit on the return that can be achieved in this way. If we imagine the number of annual conversions to increase indefinitely, we approach a situation in which interest is compounded "continuously." That is, at each instant of time the investment grows in proportion to its current value. This is called ***continuous compounding.*** In Exercises 35 and 36 we help the reader to show that if a principal P is invested at an annual interest rate of r, then with continuous compounding the amount of the investment after t years will be:

Continuous Compound Interest Formula

$$S = Pe^{rt} \tag{4}$$

and the resulting effective rate of interest will be:

Effective Interest Rate for Continuous Compounding

$$r_{\text{eff}} = e^{r} - 1. \tag{4a}$$

Example 13 Suppose $10,000 is invested at 8% per year compounded continuously.

(a) What is the value of the investment after two years?

(b) What is the effective rate of interest?

Solution (a) Substituting the principal

$$P = \$10{,}000$$

and the annual interest rate

$$r = .08$$

in (4) yields

$$S = 10{,}000e^{.08t}.$$

After $t = 2$ years have elapsed the value S of the investment will be

$$\begin{aligned} S &= 10{,}000e^{.08(2)} \\ &= 10{,}000e^{.16} \\ &= 10{,}000(1.1735) \qquad \text{(Appendix Table VI).} \end{aligned}$$

Thus, the value will be \$11,735.

(b) Substituting the interest rate $r = .08$ in (4a) yields

$$\begin{aligned} r_{\text{eff}} &= e^{.08} - 1 \\ &= 1.0833 - 1 \qquad \text{(Appendix Table VI)} \\ &= .0833. \end{aligned}$$

Thus the effective rate of interest is 8.33%.

Simple Discount

It is a common business procedure for a lender to deduct the interest due in advance. For example, if one makes a "bank loan of \$300" it is not unusual for the bank to compute the interest due and deduct it in advance, turning over \$300 minus the interest to the borrower. At the end of the lending period the borrower then pays the bank \$300. This procedure is called ***discounting.*** The money deducted in advance is called the ***discount*** and the money received by the borrower is called the ***proceeds.***

Simple discount is computed in much the same way as simple interest, with the exception that it is based on the amount rather than the principal. More precisely, let

$$\begin{aligned} P &= \text{proceeds (amount received by borrower)} \\ d &= \text{discount rate (a percentage per year)} \\ t &= \text{time (in years that the proceeds will be held)} \\ S &= \text{amount (to be paid back by borrower).} \end{aligned}$$

By definition, ***simple discount*** is computed as

$$Sdt = \text{amount} \times \text{discount rate} \times \text{time}$$

so that the proceeds received by the borrower is the amount S less the discount. Thus

$$P = S - Sdt$$

or

Simple Discount Formula

$$P = S(1 - dt).$$

Example 14 Jones borrows \$600 from his bank for two years at 8% simple discount. Thus the proceeds are

$$P = S(1 - dt) = 600[1 - (.08)(2)] = 600(.84) = \$504.$$

Therefore Jones will receive \$504 and pay \$600 at the end of two years.

Example 15 From Jones' point of view in the above example, Jones paid interest of

$$\$600 - \$504 = \$96$$

for the use of \$504 for two years. His effective rate of interest can be obtained from the formula

$$\text{simple interest} = \text{principal} \times \text{rate} \times \text{time} = Prt.$$

This formula yields

$$96 = (504)(r)(2) = 1008r.$$

Thus the effective rate of interest is

$$r = \frac{96}{1008} \approx .095$$

or an effective interest rate of $9\frac{1}{2}\%$.

Annuities

An ***annuity*** is a sequence of equal payments paid or received at equal time intervals. Some possibilities are:

(1) A sequence of equal monthly investments (the annuity) is made with the objective of accumulating a certain lump sum at a specified future time (for example, monthly deposits in a Christmas Club savings account).

(2) A sequence of equal monthly payments (the annuity) to pay off an interest-bearing debt (for example, a home mortgage).
(3) A lump sum is placed with an insurance company to be invested and paid back with interest in ten equal annual installments (the annuity) on retirement.

The time period between successive annuity payments is called the ***payment period*** or ***payment interval*** for the annuity. Usually the payments are due at the *end* of the conversion periods, in which case the annuity is called an ***ordinary annuity***. The time from the beginning of the first payment period to the end of the last period is called the ***term*** of the annuity (see Figure 9.1 for a five-term annuity).

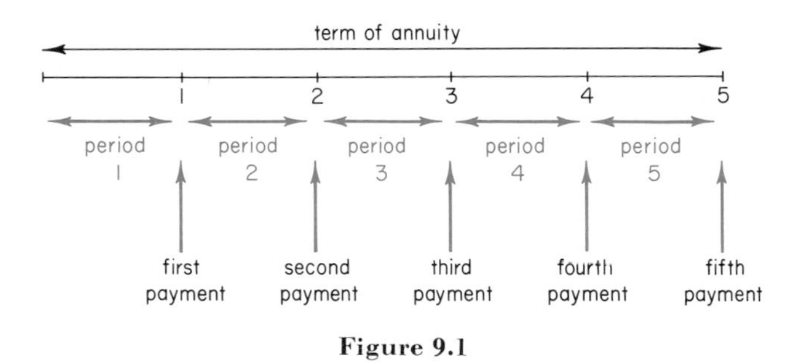

Figure 9.1

Example 16 At the end of each month \$100 is invested in a savings certificate paying 6% per year compounded monthly. What is the value S of this ordinary annuity after the fifth payment?

Solution The payment program is shown in Figure 9.2. Observe that the first \$100 payment draws interest for four months, the second for three

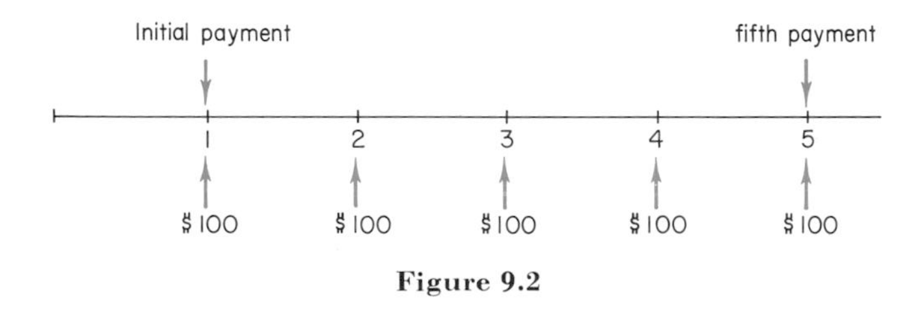

Figure 9.2

months, the third for two months, the fourth for one month, and the fifth payment draws no interest. To find the value S of the annuity after the fifth payment we shall apply formula (2) separately to each \$100 payment and add these amounts. For each payment we have

$$P = \$100, \qquad i = \frac{r}{k} = \frac{.06}{12} = .005.$$

Thus, from formula (2) and Appendix Table VII with $i = \frac{1}{2}\%$, we obtain the results shown in Figure 9.3. Thus, $S = \$505.03$.

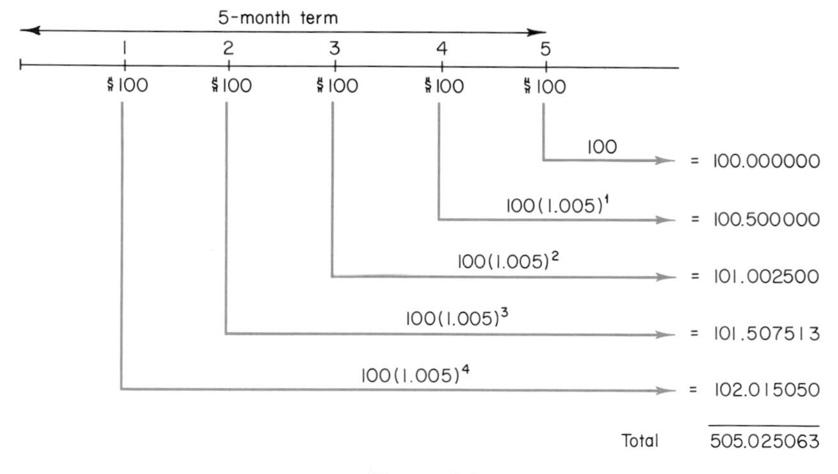

Figure 9.3

To consider a more general problem, suppose we are interested in the value S of an ordinary annuity at the end of n payment periods, where

each payment $= R$ dollars

annual interest rate $= r$

compounding is k times per year

$$i = \frac{r}{k}.$$

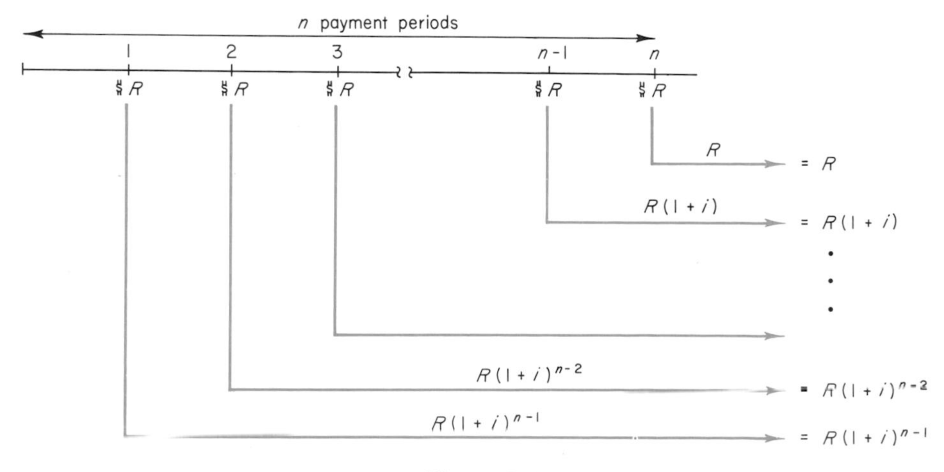

Figure 9.4

Then, as suggested by Figure 9.4, the value S of the annuity will be

$$S = R(1+i)^{n-1} + R(1+i)^{n-2} + \cdots + R(1+i) + R$$

or

$$S = R[(1+i)^{n-1} + (1+i)^{n-2} + \cdots + (1+i) + 1]. \tag{5}$$

In formula (5), the quantity S is called the ***amount, sum,*** or ***future value*** of the annuity and the quantity R is called the ***payment size*** or ***rent.*** The expression in the square brackets of (5) is commonly denoted by the symbol

$$s_{\overline{n}|i}$$

read, "s sub n bracket i." With this notation, formula (5) becomes:

Future Value of an Ordinary Annuity

$$S = R \cdot s_{\overline{n}|i}. \tag{6}$$

For convenience, values of $s_{\overline{n}|i}$ and $1/s_{\overline{n}|i}$ are given in Table VII of the Appendix.

Example 17 Solve the problem of Example 16 using formula (6).

Solution

$$\begin{aligned} S &= R \cdot s_{\overline{n}|i} \\ &= 100 \cdot s_{\overline{5}|.005} \\ &= 100 \cdot (5.05025063) \qquad \text{(Appendix Table VII,} \quad i = \tfrac{1}{2}\%, \quad n = 5) \\ &= 505.025063 \\ &\approx \$505.03, \end{aligned}$$

which agrees with the result in Example 16.

Example 18 To save for a child's education, a family decides to invest \$300 at the end of each 6-month period in a fund paying 8% per year compounded semiannually. Find the amount of the investment at the end of 18 years.

Solution From the given information

$$r = \text{annual interest rate} = .08$$

$$k = \text{number of annual conversion periods} = 2$$

$$i = \frac{r}{k} = \frac{.08}{2} = .04$$

$$R = \text{dollars per payment} = 300.$$

Thus, at the end of $n = 36$ payment periods (18 years), the amount S of the investment will be

$$\begin{aligned} S &= R \cdot s_{\overline{n}|i} \\ &= 300 s_{\overline{36}|.04} \\ &= 300(77.59831385) \qquad \text{(Appendix Table VII,} \quad i = 4\%, \quad n = 36) \\ &= 23{,}279.49416 \\ &\approx \$23{,}279.49. \end{aligned}$$

Sinking Funds

A ***sinking fund*** is a fund that is accumulated for the purpose of paying off a financial obligation at some future date.

Example 19 A firm anticipates a capital expenditure of $10,000 for new equipment in five years. How much should be deposited quarterly in a sinking fund earning 10% per year compounded quarterly to provide for the purchase?

Solution From the given information

$$r = \text{annual interest rate} = .10$$

$$k = \text{number of annual conversion periods} = 4$$

$$i = \frac{r}{k} = \frac{.10}{4} = .025$$

$$n = 5 \text{ years} \times 4 \text{ quarters per year} = 20 \text{ quarters or periods.}$$

The future value of the investment after five years must be $S = \$10{,}000$ and our problem is to determine the value R of each payment. From formula (6),

$$S = R \cdot s_{\overline{n}|i}$$

or

$$10{,}000 = R \cdot s_{\overline{20}|.025}$$

or

$$R = 10{,}000 \cdot \frac{1}{s_{\overline{20}|.025}}$$

$$= 10{,}000(0.03914713) \quad (\text{Appendix Table VII,} \quad i = 2\tfrac{1}{2}\%, \quad n = 20)$$

$$\approx \$391.47$$

Thus, $391.47 should be invested quarterly in the sinking fund.

Present Value of an Annuity

So far we have considered annuity problems in which payments are made with the objective of obtaining a certain lump sum at a future date. We now consider the reverse problem. A lump sum is invested at compound interest with the objective of obtaining a series of payments (an annuity) over some future period of time. The lump sum investment is often called the *present value* of the annuity.

Example 20 On retirement, a couple wants to make a lump sum investment paying 8% per year compounded annually in order to receive annuity payments of $10,000 per year for the following five years. How much must they invest?

Solution The investment objectives are pictured in Figure 9.5. We can imagine the lump sum deposit to consist of five parts, each part to finance one of

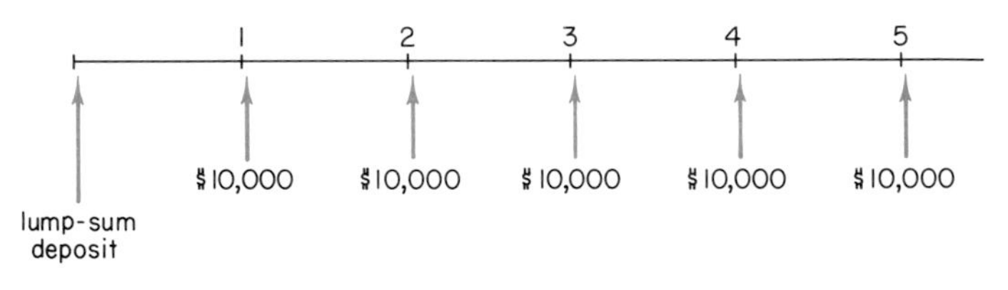

Figure 9.5

the $10,000 payments. As illustrated in Figure 9.6, the part to finance the first payment is the present value of $10,000 for one year, the part to finance the second payment is the present value of $10,000 for two years, and so forth. Thus, applying formula (3) to each of the parts with

$$S = \$10{,}000$$

$$r = \text{annual interest rate} = .08$$

$$k = \text{number of annual conversion periods} = 1$$

$$i = \frac{r}{k} = \frac{.08}{1} = .08,$$

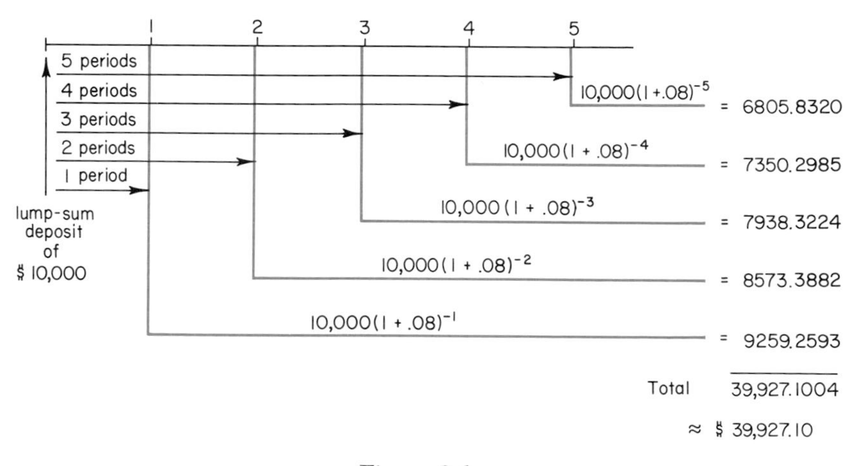

Figure 9.6

we find (Figure 9.6 and Appendix Table VII with $i = 8\%$) that a lump sum deposit of \$39,927.10 is needed for the annuity.

To consider a more general problem, suppose we are interested in finding the lump sum A that must be invested at an interest rate of i per conversion period to obtain a simple annuity of n payments of R dollars with payment beginning one conversion period after the lump sum deposit.

Proceeding as in Example 20, we regard the lump sum A as consisting of n parts, each of which must finance one of the n payments of R dollars. Thus, each of the n parts represents the present value for one of the payments. It follows from formula (3) with $S = R$ and Figure 9.7 that the

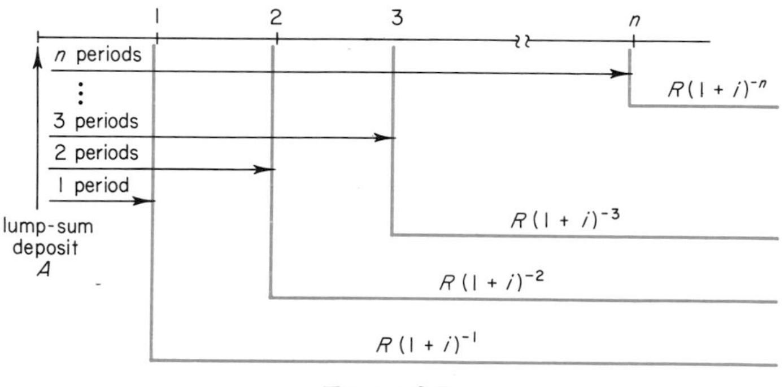

Figure 9.7

lump sum deposit must be

$$A = R(1 + i)^{-n} + \cdots + R(1 + i)^{-2} + R(1 + i)^{-1}$$

or

$$A = R[(1 + i)^{-n} + \cdots + (1 + i)^{-2} + (1 + i)^{-1}]. \tag{7}$$

The quantity A is the ***present value*** of the annuity. The expression in the square brackets of (7) is commonly denoted by the symbol

$$a_{\overline{n}|i}$$

read, "a sub n bracket i." With this notation, formula (7) becomes

Present Value of an Ordinary Annuity

$$A = R \cdot a_{\overline{n}|i} \tag{8}$$

Values of $a_{\overline{n}|i}$ and $1/a_{\overline{n}|i}$ are given in Table VII of the Appendix.

Example 21 Solve the problem in Example 20 using formula (8).

Solution

$$\begin{aligned} A &= R \cdot a_{\overline{n}|i} \\ &= 10{,}000 \cdot a_{\overline{5}|.08} \\ &= 10{,}000(3.99271004) \quad \text{(Appendix Table VII,} \quad i = 8\%, \quad n = 5) \\ &= \$39{,}927.10, \end{aligned}$$

which agrees with the result in Example 20.

Example 22 An automobile advertisement reads: "No money down, \$100 per month for 36 months; interest rate equals 12% per year compounded monthly on the unpaid balance." How much of the \$3600 to be paid will go for interest and how much for the car itself?

Solution Let A be the unknown cost of the car. If we imagine this amount to be invested at 12% per year compounded monthly to yield an annuity of \$100 per month for 36 months, the purchaser would meet his payments exactly. Thus, the cost A of the car is the present value of such an annuity. Therefore, from formula (8) with

$$R = 100, \qquad n = 36, \qquad i = \frac{r}{k} = \frac{.12}{12} = .01$$

we obtain as the cost of the car

$$\begin{aligned} A &= 100 \cdot a_{\overline{36}|.01} \\ &= 100(30.10750504) \qquad \text{(Appendix Table VII,} \quad i = 1\%, \quad n = 36) \\ &= 3010.750504 \\ &\approx \$3010.75 \end{aligned}$$

It follows that the interest paid is

$$\$3600 - \$3010.75 = \$589.25.$$

Amortization; Mortgages

An interest-bearing debt is said to be ***amortized*** if the principal and interest are paid by a sequence of equal payments made over equal time periods. For example, most home mortgages are paid in this way.

Example 23 A developer purchases a parcel of land for \$10,000 to be amortized over two years at 9% per year compounded monthly. How much will each monthly payment be?

Solution Let R be the unknown amount of each monthly payment and consider the problem from the lender's viewpoint. The lender is investing a lump sum of \$10,000 (present value) at 9% per year compounded monthly to receive a monthly annuity of R dollars for two years. Thus, from formula (8) for the present value of an annuity

$$A = R \cdot a_{\overline{n}|i}$$

with

$$A = \$10{,}000, \qquad n - 24, \qquad i = \frac{r}{k} = \frac{.09}{12} = .0075,$$

we obtain

$$10{,}000 = R \cdot a_{\overline{24}|.0075}$$

or

$$\begin{aligned} R &= (10{,}000)\frac{1}{a_{\overline{24}|.0075}} \\ &= (10{,}000)(.04568474) \quad (\text{Appendix Table VII}, \quad i = \tfrac{3}{4}\%, \quad n = 24) \\ &= 456.8474. \end{aligned}$$

Thus, each monthly payment will be \$456.85.

EXERCISE SET 9.2

1. Find the interest on \$5000 for one year at
   (a) 6% per year simple interest
   (b) 6% per year compounded monthly
   (c) 6% per year compounded quarterly
   (d) 6% per year compounded semiannually.

2. Find the interest on \$6000 for
   (a) 2 years at 7% per year compounded quarterly
   (b) 4 years at 7% per year compounded annually
   (c) 6 years at 7% per year compounded annually
   (d) 9 months at 7% per year compounded quarterly.

3. Find the effective rate of interest for each investment in Exercise 1.

4. Find the effective rate of interest for each investment in Exercise 2.

5. Which is better for the investor, an investment paying 9% per year compounded quarterly or 9.1% per year simple interest?

6. Which is better for the investor, an investment paying 9% per year compounded semiannually or 8% per year compounded continuously?

7. A principal P is to be invested to yield \$5000 in six years. Find the principal if the interest is
   (a) 8% per year compounded annually
   (b) 8% per year compounded quarterly
   (c) 8% per year compounded semiannually.

8. (a) Find the present value of \$10,000 invested for four years at 9% per year compounded monthly.
   (b) Find the future value of \$10,000 invested for four years at 9% per year compounded monthly.

9. How much should a 35-year-old man invest now at 7% per year compounded semiannually to obtain a lump sum of \$50,000 for retirement at age 65?

10. With semiannual compounding, what nominal interest rate is equivalent to $3\frac{1}{2}$% per year simple interest?

11. Suppose \$5000 is invested at 6% per year, compounded continuously.
    (a) What is the value of the investment after one year?
    (b) After 18 months?
    (c) How long does it take for the investment to double in value?
    (d) What is the effective rate of interest?

12. A principal P is to be invested at continuous compound interest of 8% per year to yield \$5000 in five years. How much should be invested?

13. From a practical viewpoint there is not much difference between daily compounding and continuous compounding. For example, with an annual interest rate of 10% the effective rate of interest with daily compounding is

$$r_{\text{eff}} = \left(1 + \frac{.1}{360}\right)^{360} \approx .10516 \qquad \text{or} \qquad 10.516\%.$$

Calculate the effective interest rate with continuous compounding.

14. You borrow $400 for 6 months from a bank that charges 8% simple discount.

(a) How much will you receive from the bank?

(b) What is the discount?

(c) How much will you repay at the end of six months?

(d) From your point of view, what is the simple interest rate you are paying for the loan?

15. (a) Find the simple discount on $5000 for two years at 9%.

(b) Find the proceeds.

16. How much should be borrowed at 12% simple discount for 5 months to obtain proceeds of $1000?

17. The proceeds of a $2000 loan for one year at simple discount were $1800. What was the simple discount rate?

18. At the end of each month $200 is invested in bonds paying 6% per year compounded monthly. What is the value of this ordinary annuity after

(a) the third payment?

(b) the tenth payment?

19. Find the future value of the following ordinary annuities:

(a) $1000 a year for 5 years at 7% per year compounded annually

(b) $500 per quarter for 10 years at 8% per year compounded quarterly

(c) $4000 each six months for 15 years at 5% per year compounded semiannually

(d) $40 each month for five years at 6% per year compounded monthly.

20. A firm anticipates a capital expenditure of $50,000 for new equipment in 10 years. How much should be deposited annually in a sinking fund earning 8% per year compounded annually to provide for the purchase?

21. A municipality issues $25,000,000 worth of revenue bonds that are due in 30 years. To pay off the debt a sinking fund earning 4% per year compounded semiannually will be established. What semiannual deposit is needed?

22. To finance his son's college education, a father wants to make a lump sum investment paying 8% per year compounded annually in order to receive annuity payments of $6000 per year for the following four years. The first payment is to be made one year after the investment. How much must he invest?

23. Find the present value of the following ordinary annuities:

(a) $1000 a year for five years at 7% per year compounded annually

(b) $500 per quarter for 10 years at 8% per year compounded quarterly

(c) $4000 every six months for 15 years at 5% per year compounded semiannually

(d) $40 each month for five years at 6% per year compounded monthly.

24. A boat is financed for \$200 per month for 24 months at an interest rate of 15% per year compounded monthly on the unpaid balance. How much of the total amount paid goes for interest and how much for the boat itself?

25. A real estate speculator purchases a lot for \$25,000 to be amortized over three years at 8% per year compounded semiannually. How much will each semiannual payment be?

26. For the problem in Exercise 25, fill in the following amortization schedule:

Payment	Outstanding principal before payment	Interest due	Payment	Amount of principal repaid	Outstanding principal after payment
1	\$25,000				
2					
3					
4					
5					
6					

27. If a woman wants to borrow money for six months at 8% per year simple interest and if the interest is payable in advance, how much should she borrow if she needs \$500 immediately?

28. What annual rate of interest compounded quarterly is needed to double the principal in three years? (*Hint*: The approximation $\sqrt[12]{2} \approx 1.06$ will be helpful.)

29. If David charges \$500 on his credit card at $1\frac{1}{2}\%$ simple interest per month, how much will he owe at the end of three months if he makes no payments?

30. What is the effective interest rate in Exercise 29?

31. If Pat saved \$200 every month at 8% interest, compounded quarterly, how much would she have at the end of 20 years?

32. An automobile costs \$6000 and can be financed at 9% compounded monthly. If a down payment of 10% is made, what is the size of the monthly payment if

(a) the automobile is financed over three years?

(b) the automobile is financed over four years?

33. For Exercise 32:

(a) What is the amount of each annuity?

(b) How much interest is charged in each case?

34. The XYZ Corporation must retire (buy back) \$200,000 worth of bonds in 20 years. If they will receive 6% interest compounded semiannually, how much must they deposit every six months to retire the bonds?

35. In Section 9.1 we observed that the value of the quantity $(1 + (1/m))^m$ gets closer and closer to the number e as m gets larger and larger. In this exercise we shall help the reader to derive the continuous compound interest formula from this result.

(a) Suppose a principal P is invested at a nominal interest rate of r, compounded k times a year. Show that the amount of the investment after t years is

$$S = P\left(1 + \frac{r}{k}\right)^{tk}.$$

(b) Let $m = k/r$, and show that the result in (a) can be written

$$S = P\left(1 + \frac{1}{m}\right)^{trm}.$$

(c) Explain why the result in (b) can be written

$$S = P\left[\left(1 + \frac{1}{m}\right)^{m}\right]^{rt}.$$

(d) Since $m = k/r$, it follows that m gets larger and larger as k (the number of annual conversion periods) increases. Use this observation to deduce the continuous compound interest formula from the result in part (c).

36. Derive formula (4a) for the effective rate of interest with continuous compounding. (*Hint*: Review the derivation of the effective interest rate for compound interest.)

10 COMPUTERS*

Few inventions have had such a profound influence in such a short time as the computer. Computers are involved in virtually every facet of our lives. They address labels, handle financial records, control air traffic, predict weather, assist in studies of ecology and population growth, and help make important decisions in matters of national defense. We could easily continue this list with thousands of additional applications. In this chapter we give a brief introduction to computers and computer programming. It is not the purpose of this chapter to make the reader into a computer expert; rather, we are concerned with providing an intelligent understanding of what a computer is and how it works.

10.1 WHAT IS A COMPUTER?

Simply speaking, a computer is a device that accepts information, called ***data***, processes it according to some scheme or program determined in advance, and provides results. Data may consist of numbers, names, or combination of letters, symbols, and numbers that mean something to the computer user. Examples of data are:

JOHN JONES
12.362
SS 169-30-6506
J2+
ACCT-NUMBER K2

Data can be processed in a computer in various ways. The computer can perform arithmetic operations on the data, it can scan large lists of data and pick out pieces of data of importance to the user, or it can perform more complicated operations of data handling. Once the computer has performed its work on the data, it can provide the results to the user in a wide variety of forms, ranging from a simple listing of answers to elaborate reports, complete with headings. Although a computer is often called an "electronic brain," it is not a thinking machine. Every operation the computer performs is mapped out in advance by a human being, from the steps in the data processing, to the form in which the answers appear. In the next few sections, we shall pursue these ideas in more detail.

In the mid-nineteenth century a mathematician, Charles Babbage,† designed the first computer. This computer was completely mechanical but was never built because its mechanical intricacy exceeded the technology of the times. The work of Babbage was hardly noticed by his contemporaries. It was not until 1944 that the first mechanical computer, called the MARK I, was built by Dr. Howard Aiken of Harvard University. In 1948, Dr. John Mauchly and J. P. Eckert built ENIAC, the first electronic computer. It used an enormous number of vacuum tubes, and its size and weight are legendary. In 1948, they built a simpler and better version of ENIAC, which they called BINAC. Eckert and Mauchly formed a company which eventually became the UNIVAC Division of Sperry Rand Corporation. In 1951, they delivered UNIVAC I, the first electronic computer designed for business applications, to the United States Bureau of the Census. After this, IBM and other companies entered the computer field and developments in the industry proceeded at an astounding rate.

Computers can be described according to the type of job they can perform. A ***special purpose computer*** is designed to do just one specific job, such as guide a space ship or control production in a petroleum refinery. A ***general purpose computer*** is more flexible; it can do many different jobs, such as payroll problems, engineering problems, inventory control, etc. We confine our attention to general purpose computers. Computers can also be described according to the way they operate. A ***digital computer*** does its work by operating with information representing digits and using the rules of arithmetic. An ***analog computer*** does its work by operating with information represented by quantities such as electrical voltages. We shall confine our attention to digital computers.

10.2 COMMUNICATING WITH THE COMPUTER

As we have already noted, a computer takes certain data, processes or operates on it, and then gives the results. To carry out these functions we

† *Charles Babbage* (1791–1871)—Babbage was an English mathematician. He studied at Trinity College, but never took the tripos examinations. Instead he helped found the Analytical Society, a group interested in introducing new ideas in English mathematics. He devoted much of his life to developing computing machines and he helped found a number of astronomical and statistics societies. He wrote several papers in mathematics, physics, machine design, and geology.

need:

(a) someone to put together a list of instructions telling the computer what steps to use in processing the data;
(b) a way of getting the instructions and the data into the computer;
(c) a way of getting the answers out of the computer.

We shall now discuss each of these steps in detail.

A detailed list of instructions telling the computer what operations to perform is called a ***computer program.*** The job of preparing this list of instructions is called ***computer programming;*** the person doing this job is called a ***computer programmer.*** Once a program has been written to do a specific job, it can be used over and over again. This is one of the very appealing features of computers. After the program has been prepared, the program and the data must be entered into a part of the computer, called the ***memory.*** For many problems the data are coded on a card using a ***key punch*** (Figure 10.2). This is a machine resembling a typewriter which is used to punch holes representing the data and the instructions on ***punch cards*** (see Figure 10.1). Then an input device, called a ***card reader***

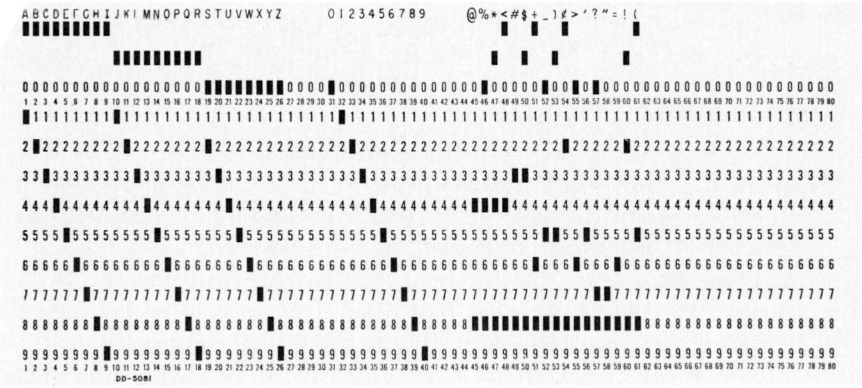

Figure 10.1 Punch card.

(Figure 10.2), scans the card to sense the positions of the holes and electronically sends the data or instructions represented by the holes to the memory.

The standard punch card contains 80 columns which are labeled with numbers from 1 to 80 near the top and bottom of the card. It also contains 12 rows, ten of which are marked from 0 to 9, and two of which are near the top of the card and unmarked.

Example 1 A major publisher publishes five different magazines: A, B, C, D, and E. For each subscriber to one of the magazines, there is a card giving the subscriber's last name, first name, middle initial, the account number, the magazine to which he subscribes, and the month and year when

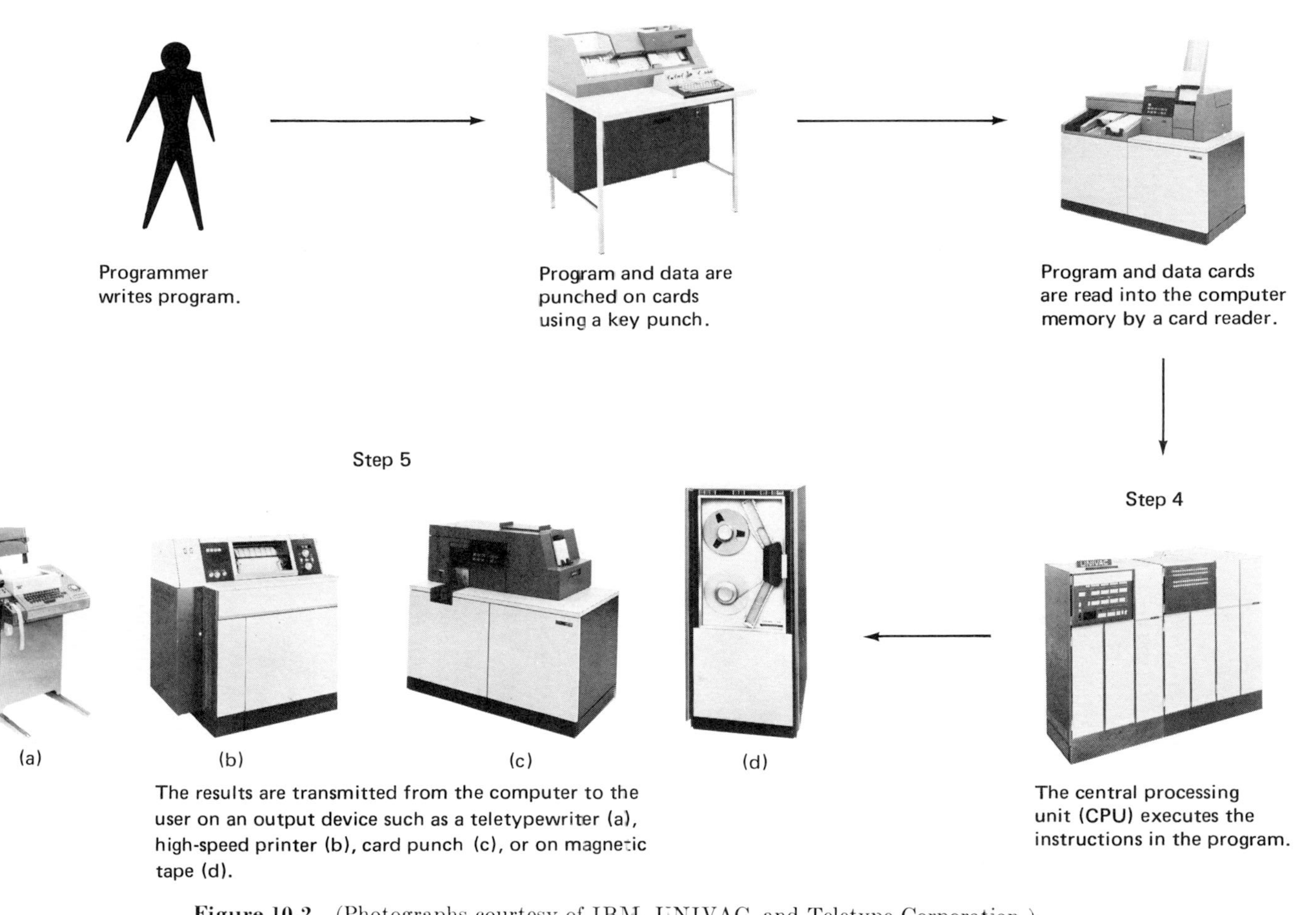

Step 1
Programmer
writes program.
Step 2
Program and data are
punched on cards
using a key punch.
Step 3
Program and data cards
are read into the computer
memory by a card reader.
Step 4
The central processing
unit (CPU) executes the
instructions in the program.
Step 5
(a)
(b)
(c)
(d)
The results are transmitted from the computer to the
user on an output device such as a teletypewriter (a),
high-speed printer (b), card punch (c), or on magnetic
tape (d).

Figure 10.2 (Photographs courtesy of IBM, UNIVAC, and Teletype Corporation.)

the subscription expires. A way of organizing the data on a punch card is shown in Figure 10.3.

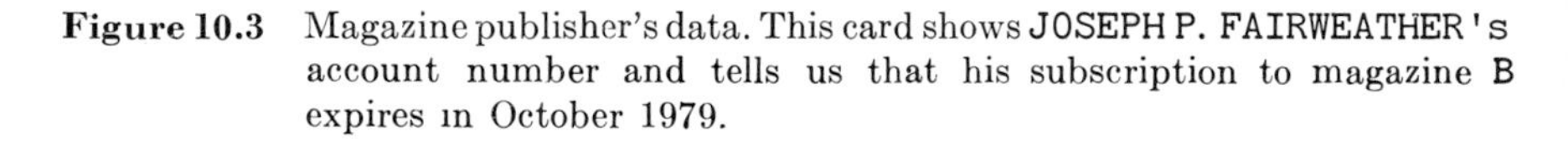

Figure 10.3 Magazine publisher's data. This card shows JOSEPH P. FAIRWEATHER's account number and tells us that his subscription to magazine B expires in October 1979.

The punch card is not the only medium for entering information into the computer; other input media are ***magnetic tape,*** which resembles recording tape, ***punched paper tape,*** where characters are represented by holes punched on a roll of paper tape, ***paper tape*** with ink characters which are scanned optically, and magnetic cards.

In another common system for using the computer, information is entered by a ***remote terminal.*** This is a typewriterlike device on which the user types the information, after which it can be transmitted, often by telephone lines, to the computer. By linking many remote terminals to one computer, many users in widely scattered locations have simultaneous access to the same machine. This is called ***time sharing.***

Once the instructions and data have been loaded into the memory, the ***Central Processing Unit*** (CPU) of the computer (Figure 10.2) takes over and begins the process of carrying out the instructions specified in the program. In a sense the CPU is the "brain" of the machine; it selects an instruction from the program (which has been stored in the memory) and then electronically "orders" either the ***arithmetic*** or ***logic unit*** to carry out the instruction. The arithmetic unit carries out such operations as addition, subtraction, multiplication, and division, while the logic unit carries out instructions of a nonarithmetic nature (some of which we shall study later).

Once the computer has completed its work, the results, called ***output,*** must be transmitted to the outside world. This can be done in various ways, depending on what the user finds most useful. For example, the output can be printed on paper using a high speed printer which prints about ten lines per second (120 characters per line) (Figure 10.2). The output can also be placed on magnetic tapes or punched cards

(Figure 10.2); these forms are useful when the output will be used as data for future computer programs. When the output appears on punch cards, the cards are often taken to a machine, called an ***interpreter,*** whose job is to print on the card the information punched on the card. The reader has probably seen electric and phone bills printed on punch cards by an interpreter. Output can also be typed out by a printing unit or displayed on a video screen.

Above, we talked about storing information in the memory of a computer. We shall now briefly discuss how this is done. It was observed, when electronic computers were first conceived, that the decimal system of numbers is not well suited for digital computers. The problem is that there are ten different decimal digits, 0, 1, 2, 3, 4, 5, 6, 7, 8, and 9, so that a computer would have to be able to distinguish between these ten digits electronically. This difficulty was solved by exploiting the ***binary number system,*** which requires only two digits, 0 and 1. To see how this number system works, let us first reexamine the structure of the decimal system. When we write a number such as

$$7435$$

in the decimal system, we mean

$$7(1000) + 4(100) + 3(10) + 5(1)$$

or equivalently

$$7(10^3) + 4(10^2) + 3(10^1) + 5(10^0).$$

Thus, reading from right to left, the successive digits represent increasing powers of 10. In the binary number system, the only digits required are 0 and 1, and numbers are represented by increasing powers of 2, reading from right to left. For example, in the binary system

$$11010$$

represents

$$\begin{aligned} 1(2^4) + 1(2^3) + 0(2^2) + 1(2^1) + 0(2^0) &= 16 + 8 + 0 + 2 + 0 \\ &= 26. \end{aligned}$$

Example 2 In the binary number system, the decimal number 62 would be written

$$111110$$

since

$$\begin{aligned} 111110 &= 1(2^5) + 1(2^4) + 1(2^3) + 1(2^2) + 1(2^1) + 0(2^0) \\ &= 32 + 16 + 8 + 4 + 2 \\ &= 62. \end{aligned}$$

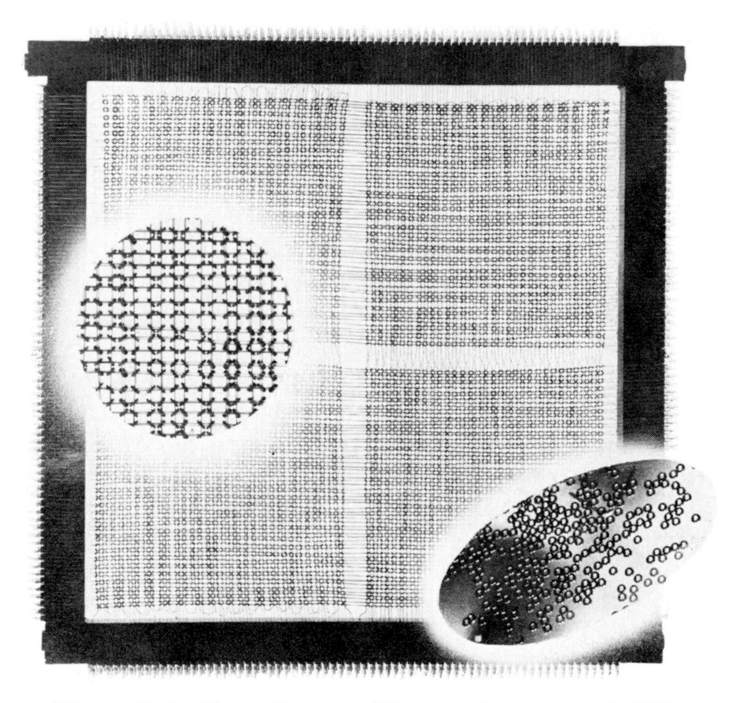

Figure 10.4 Magnetic cores. (Photograph courtesy of IBM.)

The binary system is well suited for use in computers because the two digits, 1 and 0, can be represented by two possible states of an electrical device. For example, the memory units in many computers consist of elements, called magnetic cores (see Figure 10.4). Each core can be magnetized in two different ways, which we can think of as being clockwise

or counterclockwise

Each core represents one binary digit; a binary digit is commonly called a ***bit.*** With the convention that

 represents 0

and

represents 1

binary numbers can be stored in the computer by properly magnetizing a string of cores. For example, a string of cores magnetized as follows:

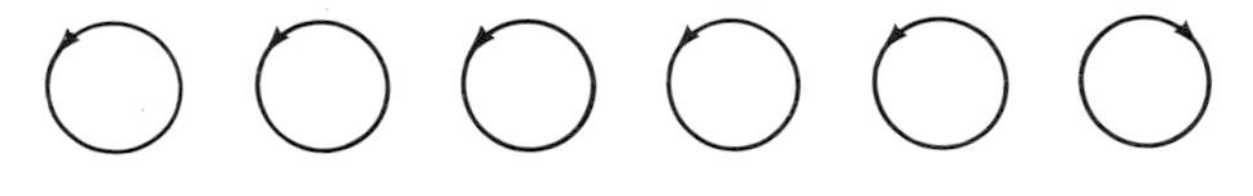

represents the binary number

111110

which as shown in Example 2 has value 62 in the decimal system.

Example 3 The cores

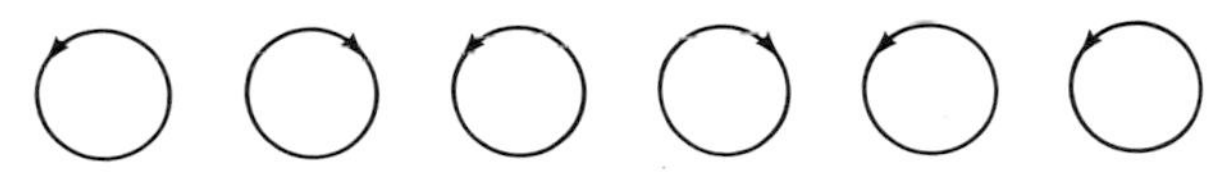

represent the binary number

101011

which has value 43 in the decimal system.

As shown in this example, each core stores one bit (binary digit) so that a number of bits or cores must be strung together to represent a binary number. A string of cores or bits is called a ***word.*** Thus, the six bits in Example 3 form the word

101011

Some computers work with a fixed word size, while others use a variable word size.

The sections of memory where words are stored have ***addresses*** that can be used to locate stored information. For example, in a computer having a 6-bit word length, we can visualize a portion of the memory to be structured as in Figure 10.5. Thus, in the location with the address 709, the word

010001

is stored; and, in the location with the address 711, the word

001001

is stored.

The memory of a computer can be compared to the mailboxes in a post office, where each box has its own address for receiving and storing information.

In summary, a computer reads in instructions and data, carries out the operations specified in the instructions, and transmits the results to the outside world. Moreover, the computer can do all of these operations at

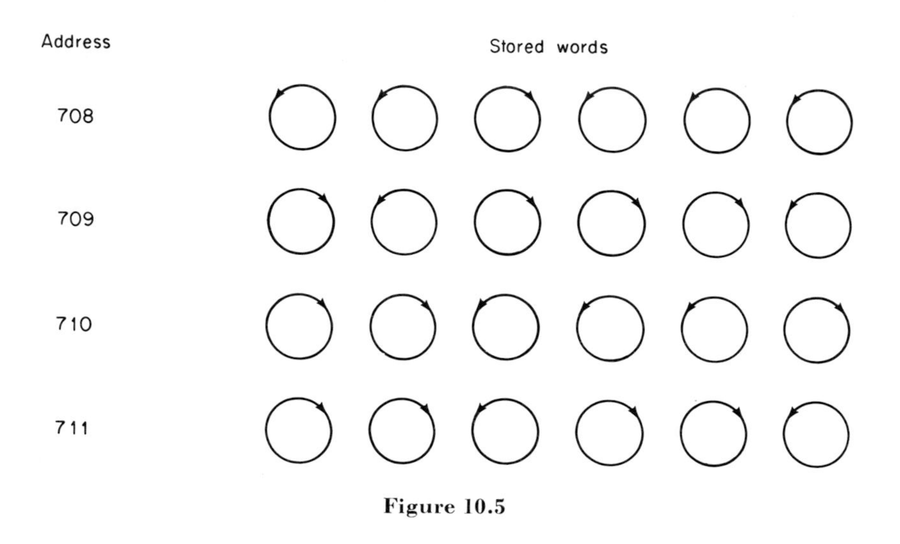

Figure 10.5

fantastic speeds. For example, a fast computer can perform several million additions per second. However, a computer can give incorrect answers. Errors can arise through incorrect problem analysis, programming errors, or a computer malfunction. Fortunately, this last kind of error occurs infrequently.

EXERCISE SET 10.2

1. In the card in Figure 10.6, what four-digit number is punched in
   (a) columns 1–4?
   (b) columns 10–13?
   (c) columns 20–23?

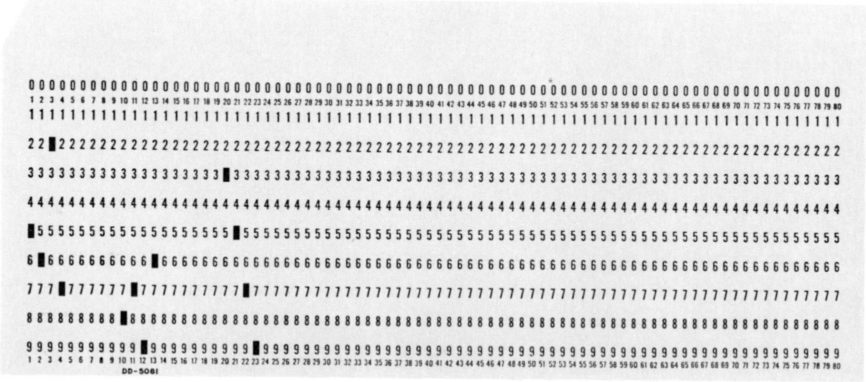

Figure 10.6

2. Write the following binary numbers in decimal form.

(a) 10101 (b) 101 (c) 10111 (d) 111101.

3. Write the following decimal numbers in binary form.

(a) 1 (b) 2 (c) 3 (d) 4 (e) 5 (f) 6 (g) 7.

4. Using the convention that the magnetized core ◯ denotes 0 and the magnetized core ◯ denotes 1, find the decimal value of the number stored by the given string of cores

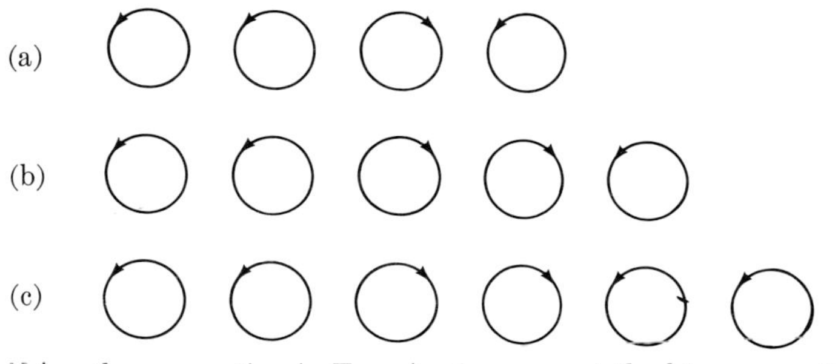

5. Using the convention in Exercise 4, represent the binary number 1100 by a string of four magnetized cores.

6. Referring to Figure 10.5, give the decimal value of the number stored in

(a) location 708

(b) location 710.

10.3 BASIC, A PROGRAMMING LANGUAGE

The job of the programmer is to formulate a list of instructions that the computer can follow to solve a given problem. The programmer and the computer speak different languages, however; the programmer (in our case) uses the English language, while the computer is designed to receive coded instructions that are suited for its electrical circuits. We can think of this as the ***machine language.*** In order to avoid burdening the programmer with the problem of formulating his instructions in machine language, various compromise languages, called ***programming languages,*** have been developed. They are close enough to English to make the programmer comfortable, yet they are structured so that they can be translated readily into machine language by a program called a ***compiler.*** (Compilers are supplied by the computer manufacturer.) In this section we shall introduce a language which is widely used in time-sharing systems. It is called BASIC, which stands for Beginner's All-purpose Symbolic Instruction Code, and was originally developed at Dartmouth College by Professors J. Kemeny and T. Kurtz in the 1960s.

BASIC is well suited for problems that do not use large volumes of data. In addition, BASIC has the added feature that it can be used in an ***interactive mode.*** This means that the programmer and the computer can communicate while the program is being typed or run. For example, the computer can often tell the programmer that a typing or programming error has been made before the program is executed. This allows the programmer to correct the error immediately. In addition, the programmer can arrange to have the computer type out intermediate results and stop until the programmer types in additional instructions or data. This allows the programmer to make decisions based on preliminary output. In short, BASIC allows for a constant two-way communication between programmer and computer. In Section 10.7 we shall discuss briefly some other programming languages.

There are a number of minor variations of BASIC, so that before running any programs, the reader is advised to check on the variations appropriate to the computer being used.

The remote terminal or teletypewriter is used as an input as well as an output device. The keyboard of the console is shown in Figure 10.7 and

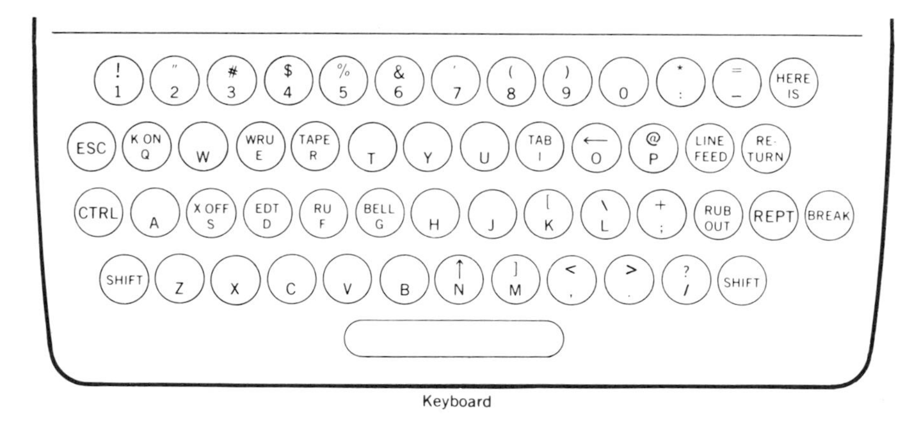

Figure 10.7 (Courtesy of Teletype Corporation.)

resembles an ordinary typewriter. However, the only letters that can be typed are uppercase letters; the shift key is used to type other symbols, such as ↑,],\, <, >, and so on.

One of the most important instructions in BASIC is the command **LET**, which means *store*.

Example 4 The BASIC statement

```
LET A = 7.2                                                    (1)
```

directs the computer to store the number 7.2 in a location that will be designated by A throughout the program.

The letter A appearing on the left side of statement (1) is called a ***variable.*** For each variable that is introduced in a BASIC program, the compiler sets aside a memory location which can be referenced throughout the program by using that variable. BASIC, like all languages, has certain guidelines or rules of grammar that must be followed. Here are the restrictions for selecting names of variables

A variable name in BASIC may consist of:

(1) exactly one letter of the alphabet

or

(2) exactly one letter of the alphabet followed by exactly one of the digits from 0 to 9.

Examples of valid BASIC variables are

```
A4    B2    Y    X2    D1    R
```

The following variable names are improper in BASIC:

AR	(the second character cannot be a letter)
A22	(too many characters)
#A	(the first character is not a letter)
3C	(the first character is not a letter)
A*	(the second character is not a digit).

Example 5 Suppose the tax due on a certain income is $705.36. The BASIC statement

```
LET  T = 705.36
```

directs the computer to store the number 705.36 in a location that will be called T throughout the program. The programmer can just as easily store the number 705.36 in the computer memory by writing

```
LET  T1 = 705.36
```

or

```
LET  C  = 705.36
```

or

```
LET  A  = 705.36
```

or by using any other allowable variable. The choices `T` or `T1` are desirable, however, since they help the programmer remember that the number 705.36 means "tax."

Numbers used in BASIC programs are ***constants;*** they can be written in the usual way. For example,

`8        3.72        8672        +1.93        -9.56        .0008`

are all correct ways of writing constants in BASIC. The number of decimal digits that can be used in a constant depends on the computer involved. In this chapter we shall allow up to nine decimal digits in each constant. However, before writing an actual program, the reader should check on the limitations of the computer to be used.

A constant can be preceded by a + or − sign (which does not count as a digit). As usual, the + sign can be omitted preceding positive constants.

Example 6 The constants

`14213789,        21321.5,` and `-9673.21842`

are all valid because they involve nine or fewer decimal digits. The constants

`7821635698` and `.0003257134`

are invalid because they involve more than nine decimal digits.

Certain numbers with more than nine decimal digits can be used as BASIC constants by writing them in ***scientific*** or ***exponential notation.*** For example,

1,936,000,000

has 10 digits. However, it can be written in scientific notation as

$$1936 \times 10^{6}$$

and written in BASIC as

`1936E6`

Another possibility is to write the number as

$$1.936 \times 10^{9}$$

in which case it would be written in BASIC as

`1.936E9`

Example 7

BASIC constant	Value
3.2164E7	$3.2164 \times 10^7 = 32{,}164{,}000$
4.1E-8	$4.1 \times 10^{-8} = .000000041$
-7.32E-4	-7.32×10^{-4} } = $-.000732$
-.0732E-2	$-.0732 \times 10^{-2}$ } = $-.000732$
-.00732E-1	$-.00732 \times 10^{-1}$ } = $-.000732$

On many computers a BASIC constant written in exponential notation cannot have more than nine digits before the E nor more than two digits after the E. However, there are variations between different computers.

Example 8 The following constants are not valid in BASIC:

.0023456789E2 (too many digits before E)

.002345678E-320 (too many digits after E).

In addition to instructions for naming variables, there are instructions for carrying out arithmetic operations. In BASIC, the standard arithmetic operations are denoted as follows:

+ means addition. Thus, A + B means that we add A and B.
− means subtraction. Thus, A − B means that we subtract B from A.
* means multiplication. Thus, A * B means that we multiply A and B.
/ means division. Thus, A/B means that we divide A by B.
↑ means exponentiation. Thus, A ↑ B means that we raise A to the B power.

Example 9 The BASIC statement

```
LET  A = 6.2 * 3.1
```

tells the computer to multiply 6.2 and 3.1 together and store the result in the location denoted by the variable A.

Example 10 The BASIC statement

```
LET  C = A/6.2
```

tells the computer to divide the contents of location A by 6.2 and store the result in location C.

Example 11 The BASIC statement

```
LET X = A ↑ 2
```

tells the computer to raise the contents of location A to the second power (square A) and store the result in location X.

Example 12 The following sequence of instructions, if performed in succession, would store the numbers 18.3 and 14.5 in locations A and B and store their sum in location C.

```
LET A = 18.3
LET B = 14.5
LET C = A + B
```

Example 13 Parentheses are used in BASIC arithmetic statements just as in algebra. To illustrate, we have indicated below some algebraic expressions and some possible ways these expressions can be written in BASIC.

Algebraic expression	Expression in BASIC
$\frac{AC}{B}$	(A * C)/B
X^2Y	(X ↑ 2) * Y
$X^2 + Y^2$	(X ↑ 2) + (Y ↑ 2)
$\frac{X + Y}{Z}$	(X + Y)/Z
$X + \frac{Y}{Z}$	X + (Y/Z)

Example 14 The following sequence of instructions, if performed in succession, would store the numbers 11.2, 12.3, and 18.5 in locations X, Y, and Z and store their arithmetic average in location A.

```
LET X = 11.2
LET Y = 12.3
LET Z = 18.5
LET A = (X + Y + Z)/3
```

Example 15 In the next section we shall need the instruction

```
LET  I = I + 1
```

Although this statement is nonsensical if = is interpreted as equality, it makes perfectly good sense in BASIC because = does not mean equality; it means store. This statement instructs the computer to add 1 to the number in the location denoted by I and then store the result back in I. Thus, if I initially contains the value 3, it will contain the value $3 + 1 = 4$ after the command

```
I = I + 1
```

is executed.

The BASIC instruction that tells the computer to print out results stored in the memory is called a PRINT instruction. The programmer must tell the computer where to locate the information to be printed. For example, by writing

```
PRINT  A
```

the programmer tells the computer to print out the contents of the memory location denoted by the variable A. Similarly, the instruction

```
PRINT  T1
```

tells the computer to print out the contents of the memory location denoted by the variable T1.

A PRINT instruction does not alter information in the memory. Thus, if location A contains the number 7.3, it will still contain this number after the instruction

```
PRINT  A
```

is executed. Unlike a PRINT statement, a LET statement *does* alter information in the memory. Thus, if location A contains the number 7.3, execution of the instruction

```
LET  A = 12.5
```

will replace the value 7.3 by the value 12.5.

Example 16 Consider the following sequence of instructions

```
LET  A = 7.3
PRINT  A
LET  A = 12.5
PRINT  A
```

If these instructions are executed in order, the computer will first enter the constant 7.3 into memory location A. Next the computer will print the value in A, namely 7.3. This PRINT instruction does not alter the value in A so that location A still contains the constant 7.3. However, the next instruction LET A = 12.5 changes the value in A from 7.3 to 12.5 and the final instruction prints out the constant in A, namely 12.5.

Example 17 The following sequence of instructions, if executed in order, would compute the arithmetic average of the numbers 11.2, 12.3, and 18.5 and then print out the three numbers and their average in the format shown in Figure 10.8.

```
LET  X = 11.2
LET  Y = 12.3
LET  Z = 18.5
LET  A = (X + Y + Z)/3
PRINT  X
PRINT  Y
PRINT  Z
PRINT  A
```

```
11.2
12.3
18.5
14
```

Figure 10.8

Each PRINT statement produces a separate line of results; thus the four PRINT statements in Example 17 produced data typed on four separate lines. Sometimes it is desirable to have two or more pieces of data typed on the same line. If, for example, we want the data in locations R and S typed on the *same* line we would use the single instruction

```
PRINT  R, S
```

rather than two separate instructions

```
PRINT  R
PRINT  S
```

which would cause the data to be typed on separate lines.

Example 18 If we modify the sequence of instructions in Example 17 as follows

```
LET  X = 11.2
LET  Y = 12.3
LET  Z = 18.5
LET  A = (X + Y + Z)/3
PRINT  X, Y, Z
PRINT  A
```

then the output would appear as in Fig. 10.9 rather than as in Figure 10.8.

```
11.2                    12.3                    18.5
14
```

Figure 10.9

As we mentioned earlier, a computer is not a "brain" capable of thinking for itself. Every action it takes must be specified by an instruction, including when to stop its work. For this reason, every BASIC program must contain an instruction called

```
END
```

which tells the computer that the program is over. The computer will then know that it can move on to the next job.

The computer must also be told the order in which instructions are to be executed. To do this we type each instruction on a separate line and assign each line a different ***line number*** (an integer between 1 and 9999 on most computers). The line number is typed to the left of the instruction. Some examples are

```
  11  LET  A = 12.3
  17  PRINT  X, Z
1342  END
```

Unless it is instructed otherwise, the computer executes the instructions in ascending order of line number (regardless of the order in which they are typed). Thus, in absence of instructions to the contrary, the instruction with smallest line number is executed first and the instruction with largest line number last. It follows that the instruction END must always have the largest line number in any program.

Line numbers need not begin with 1 and need not be consecutive integers. In fact it would be preferable to use 10, 20, 30, 40, . . . for consecutive line numbers rather than 1, 2, 3, 4, . . . since this allows us easily to insert additional instructions if we wish to modify the program at a later time.

Example 19 The program in Example 18 with line numbers and END instruction included can be written

```
10  LET  X = 11.2
20  LET  Y = 12.3
30  LET  Z = 18.5
40  LET  A = (X + Y + Z)/3
50  PRINT  X, Y, Z
60  PRINT  A
70  END
```

Assume now that we have written a BASIC program which we would like to have executed on a computer. The first step is to have the teletypewriter linked to the computer. The exact procedure varies from installation to installation. Once this is done the computer on most systems types out

READY

which means that the programmer can type in his program. The next step is to type each program instruction on the teletypewriter. This constitutes what is called the ***source program.*** After the source program has been typed the programmer types, on a new line, the instruction

RUN

The computer then goes through two steps, ***compilation*** and ***execution.*** In the compilation stage, the compiler translates the BASIC language statements into machine language statements, and stores them in the computer memory. When the END statement is reached, the computer knows that it is has come to the end of the source program. In the course of compiling, the computer checks for certain errors in the program. If it finds them, it prints out a description of the errors for the programmer. If no errors are found,† the computer executes the program; that is, it carries out the instructions specified in the program. In the course of executing the program, any output generated by the computer is printed on the console teletypewriter (or displayed on a video screen).

After the program has been executed, the word READY again appears, meaning that the computer is ready for the next job. If the programmer has no more work, the terminal is disconnected from the computer by typing in the command BYE (or in some systems OFF).

† The computer can only detect certain kinds of errors. There is no guarantee that the program is correct just because the compiler fails to locate any errors.

EXERCISE SET 10.3

1. Write a BASIC statement which stores the constant 48.7 in location M.

2. Which of the following statements violate the rules for naming BASIC variables?

   (a) `LET AA = 73.2` (b) `LET K = 43.7`
   (c) `LET MAX = 123` (d) `LET W1 = -7.2`
   (e) `LET 1A = 274.8` (f) `LET 3% = .5`

3. Explain what the following BASIC instructions do:

   (a) `LET X = X + 2` (b) `LET X = X ↑ 2` (c) `LET I = K - I`

4. Find three errors in the following BASIC program:

```
10 LET K = 7.5
20 LET M = K² + 5
   PRINT K, M
```

5. Which of the following are valid BASIC constants?

   (a) -34752.21 (b) .0003258142
   (c) -.000000001 (d) +47254298.3

6. Which of the following are valid BASIC constants? For those which are not, explain why.

   (a) 1347E-2 (b) 213479254.3E2
   (c) .0023451E130 (d) -7349E3

7. Let A, B, and C have the values

   A = 2.5, B = .5, C = 4.5

   Find the number stored in X by the BASIC instruction

   (a) `LET X = A/B` (b) `LET X = A ↑ 3`
   (c) `LET X = (A + B) + C` (d) `LET X = A * (B + C)`

8. Find numerical values for the following BASIC constants:

   (a) 4.218E5 (b) 3.2E-6 (c) -7.43E-3
   (d) -.0432E-4 (e) -.000531E2 (f) 8E8

9. Write BASIC expressions for the following algebraic expressions:

   (a) $\frac{A+B}{C}$ (b) $\frac{C}{D+E}$ (c) $\frac{C}{D^2}$

   (d) $\frac{C+E}{D^2}$ (e) $\frac{A+B}{2}$ (f) $\frac{4}{3}\pi r^3$ (take π as 3.14)

10. Let A, B, and C have the values given in Exercise 8. Find the number stored in X by the BASIC instruction

(a) `X = A * (B ↑ 2)`

(b) `X = (A - 2) * (B/A)`

11. Consider the BASIC program

```
10  LET  K = 5
20  PRINT  K
30  LET  K = 10
40  PRINT  K
50  END
```

(a) What will the computer print out?

(b) At the end of the program, what will be stored in the location represented by the variable K?

12. What number will be printed out by the following BASIC program?

```
10  LET  X = 2
20  LET  Y = 8
30  LET  Z = (X + Y) ↑ 2
40  PRINT  Z
50  END
```

13. Write a BASIC program that computes the arithmetic average of the numbers 7.2, 3.8, 1.6, 2.7 and prints out the answer.

14. Consider the BASIC program

```
10  LET  A = 3
20  LET  B = 4
30  PRINT A
40  PRINT B
50  PRINT  A, B
60  END
```

What will the computer print out?

15. Write a program that computes the sum of the squares of the first five positive integers and prints out the answer.

10.4 BASIC CONTINUED (PRINT AND DATA INSTRUCTIONS)

In this section we discuss more BASIC instructions: instructions that print words, instructions that facilitate the storage of input data, and instructions that stop the execution of a program so the programmer can observe intermediate results or enter additional data.

It is often useful to have the computer print words and sentences. These can be used as labels and headings for output data and can provide helpful information about the program when it is used by someone unfamiliar with its contents. Words and sentences are printed by using a BASIC instruction of the form

```
PRINT "          "
```

When this instruction is executed, the words and spaces inside the quotation marks will be reproduced. Thus the instruction

```
PRINT "INCOME"
```

causes the computer to print

```
INCOME
```

and the instruction

```
PRINT "10% MARKUP"
```

causes the computer to print

```
10% MARKUP
```

The computer reproduces spacing exactly as it appears inside the quotation marks. Thus, if we want to print the words GROSS and NET with seven spaces between them we can type the instruction

7 spaces

```
PRINT "GROSS       NET"
```

The printing area on the output of the teletypewriter is divided into four or five parts (depending on the computer) called ***fields.*** Each time a comma is inserted into a PRINT statement, the teletypewriter moves over to the next field. The following example illustrates how to print column headings for output data.

Example 20 Assume that the data 10.2, 74.2, 14.2, 12.2, 78.7, 15.1, 11.1, 76.3, 14.8 have been stored in locations A1, B1, C1, A2, B2, C2, A3, B3, and C3. The sequence of statements

```
PRINT "VOLUME", "TEMPERATURE", "PRESSURE"
PRINT A1, B1, C1
PRINT A2, B2, B3
PRINT A3, B3, C3
```

will cause the computer to print out the headings and data in three fields as follows:

```
VOLUME            TEMPERATURE       PRESSURE
 10.2              74.2              14.2
 12.2              78.7              15.1
 11.1              76.3              14.8
```

Sometimes it is desirable to have words and variables printed on the same line to label output. For example, the instruction

```
PRINT  "INCOME TAX IS ";T
```

causes the computer to print the words INCOME TAX IS followed by the contents of location T. Thus if the program has computed a tax T (in dollars) of 942.25 this instruction would cause the computer to print

```
INCOME TAX IS  942.25
```

The semicolon in the above instruction caused the computer to print the contents of location T immediately next to the material enclosed in the quotation marks. Had we used a comma instead, the contents of location T would have been printed in the next field. Thus

```
PRINT  "INCOME TAX IS ",T
```

would cause the computer to print:

```
INCOME TAX IS             942.25
```

As further examples, the instruction

```
PRINT  "INCOME TAX IS $ ";T
```

would cause the computer to print

```
INCOME TAX IS $942.25
```

Finally, the instruction

```
PRINT T; " = INCOME TAX"
```

would cause the computer to print

```
942.25 = INCOME TAX
```

Entering Data

One way to enter data into the computer memory is by a series of LET instructions. For example

```
LET  A = 1.2
LET  B = 2.7
LET  C = -3.8
LET  D = .04
```

stores the data 1.2, 2.7, −3.8, and .04 in locations A, B, C, and D. The same result can be achieved more efficiently by using the two instructions

```
READ  A, B, C, D
DATA  1.2, 2.7, -3.8, .04
```

The **READ** and **DATA** instructions are used together. The **READ** instruction lists the variables that are to receive the values listed in the **DATA** statement; the first variable in the **READ** is assigned the first value in the **DATA**, the second variable is assigned the second value, and so forth.

Example 21 The program in Example 19 can be rewritten using **READ** and **DATA** as follows:

```
10  READ  X, Y, Z
20  DATA  11.2, 12.3, 18.5
30  LET  A = (X + Y + Z)/3
40  PRINT  X, Y, Z
50  PRINT  A
60  END
```

Remark It is not essential that **READ** and **DATA** statements appear on successive lines in a program; moreover, it is allowable to have the **DATA** statement precede the **READ** statement. However, for proofreading purposes, it is desirable to arrange **READ** and **DATA** statements in an orderly fashion.

One of the most important features of BASIC is that it allows the user to interact with the computer and make decisions during the execution of the program. This is done by means of an **INPUT** command. This command is used like a **READ** statement with the exception that the user types in the data from the console instead of specifying it in a **DATA** statement. This allows the user to decide on the data to be entered after observing some partial results.

To illustrate, suppose the computer encounters the instruction

```
INPUT  G, N
```

The computer stops (saving everything it has done to that point) and types

?

The user must then respond by typing in two pieces of data and depressing the typewriter key marked

RETURN

The computer will then store the typed data in locations G and N and continue with the execution of the program.

Because the **INPUT** statement produces only a question mark to signal for data, it is a good idea to have a **PRINT** statement just before each **INPUT** statement to make the computer print out a description of the data to be supplied from the console. These directions are essential for an operator who is unfamiliar with the program. For example, if we want the user to type in two values, the first representing a gross income and the second representing a net income, we might write

```
PRINT  "TYPE GROSS, NET"
INPUT  G, N
```

When the program is executed the computer will print

```
TYPE GROSS,NET
?
```

These instructions tell the user to type in values for gross income and net income separated by a comma. The user might then respond by typing

```
28,24
```

and then depressing the RETURN key.

Example 22 The following program computes the average of three test scores that are provided at the console by the operator and then prints out the answer:

```
 5  PRINT "TYPE IN: SCORE 1, SCORE 2, SCORE 3"
10  INPUT X, Y, Z
15  LET  A = (X + Y + Z)/3
20  PRINT  "AVERAGE IS ";A
25  END
```

When this program is executed the computer will type

```
TYPE IN: SCORE 1, SCORE 2, SCORE 3
?
```

The user will type in the data in the format stated; for example

```
87,92,64
```

The user will then depress the RETURN key and the computer will complete the program, finally typing

```
AVERAGE IS 81
```

It is generally advisable to type along with a program a description of the program's purpose and perhaps other information such as the name of the programmer, dates, code numbers, and so forth. In order to assure that the computer does not interpret such descriptions as program commands, we must precede the descriptive material by the letters

REM

(standing for remark). The computer ignores the material following REM, but a person reading the program would be aided by its inclusion.

Example 23 The program in Example 22 with an identifying description might be written as

```
5   REM   PROGRAM TO COMPUTE TEST SCORE AVERAGES
10  PRINT   "TYPE IN: SCORE 1, SCORE 2, SCORE 3"
15  INPUT  X, Y, Z
20  LET  A = (X + Y + Z)/3
25  PRINT   "AVERAGE IS ";A
30  END
```

EXERCISE SET 10.4

1. Write a BASIC instruction that causes the computer to print the word INVENTORY.

2. Assume that the variable M represents the market value of a certain stock. Write a BASIC instruction that causes the computer to print the value of M preceded by

   MARKET VALUE = $

3. Assume K = 600,000. What do the following BASIC instructions cause the computer to print?

   (a) PRINT "BACTERIA COUNT = K"

   (b) PRINT "BACTERIA COUNT = ";K

   (c) PRINT "BACTERIA COUNT K = ";K

   (d) PRINT K;" = BACTERIA COUNT"

4. Assume A = 12.8, B = −4.2, C = 7.8 are stored in memory. What will the following BASIC programs cause the computer to print?

   (a)
   ```
   10  PRINT  A
   20  PRINT  B
   30  PRINT  C
   40  END
   ```
   (b)
   ```
   10  PRINT  A, B, C
   20  END
   ```
   (c)
   ```
   10  PRINT  A, B
   20  PRINT  C
   30  END
   ```

5. Assume T1 = 98.4, T2 = 98.6, T3 = 99.1 are stored in memory. What will the following BASIC programs cause the computer to print?

(a)
```
100  PRINT  "TEMP 1, TEMP 2, TEMP 3"
200  PRINT  T1, T2, T3
300  END
```

(b)
```
100  PRINT  "TEMP T1";T1
200  PRINT  "TEMP T2"
300  PRINT  T2
400  PRINT  "T3"
500  END
```

6. Rewrite the following BASIC program using READ and DATA instructions to replace the first two LET instructions:

```
10  LET  C = 1.6
20  LET  D = 2.4
30  LET  E = C + D
40  PRINT  E
50  END
```

7. For the following BASIC program what does the computer print out?

```
10  READ  K, L
20  DATA  2, 4
30  LET  M = (K + L)/2
40  PRINT  K, L, M
50  END
```

8. Using READ, DATA, and PRINT statements, write a program that causes the computer to print out:

(a)
```
PRESSURE = 29.7     TEMP = 76     HUMIDITY = 50
```

(b)
```
PRESSURE            TEMP          HUMIDITY
  29.7               76             50
```

9. Determine what the following BASIC program does:

```
10  PRINT  "TYPE X"
20  INPUT  X
30  LET  Y = X ↑ 2
40  PRINT  X, Y
50  END
```

10. What does the following BASIC program print out?

```
100  LET  X = 3
200  REM  LET  X = 4
300  PRINT X
400  END
```

11. The formula for converting Celsius temperature to Fahrenheit is

$$F = \frac{9}{5}C + 32$$

Write a BASIC program that converts the Celsius temperatures 3°, 20°, 37° to Fahrenheit and prints out the results on three separate lines.

12. Write a BASIC program that computes the area of a rectangle using values for length and width typed by the operator. Have the computer print out the length, width, and area with headings.

13. Write a BASIC program to compute the square and cube of three given numbers. Print out the given numbers and the answers in three columns with headings

```
NUMBER      SQUARE      CUBE
```

Have the data 2, 7, 9 supplied by a READ--DATA statement.

14. Repeat Exercise 13, but have the operator supply the data.

10.5 BASIC CONTINUED (LOOPS AND DECISIONS)

In this section we shall discuss some additional BASIC instructions. These instructions enable the computer to make logical decisions and to modify the program instructions. By properly using these new instructions, programs that might otherwise require thousands of instructions can be written using just a few instructions.

Example 24 Most savings banks offer their customers various interest rates, depending on the type of account. In this problem, we shall assume that a bank teller enters the following information for one of the customers:

the account number	N
the current balance	B
the interest rate	R

The bank wants a program that will compute the dividend from the formula

$$\text{dividend} = \text{balance} \times \text{interest rate}$$

and print out the account number and the dividend. The following is such a program:

```
10  REM  PROGRAM TO COMPUTE DIVIDEND
20  REM  N = ACCT NO, B = CURRENT BAL, R = INTEREST
    RATE, D = DIVIDEND
30  PRINT  "ENTER N, B, R"
40  INPUT  N, B, R
```

```
50  LET  D = B * R
60  PRINT  N, "DIVIDEND = ";D
70  END
```

Example 25 To take the problem in the previous example a little further, suppose the bank wants a program that will print out the account number and dividend for five accounts. The following is such a program.

```
10  REM  PROGRAM TO COMPUTE DIVIDEND FOR FIVE
    ACCOUNTS
20  REM  N = ACCT NO, B = CURRENT BAL, R = INTEREST
    RATE, D = DIVIDEND
30  PRINT  "ENTER N, B, R"
40  INPUT  N, B, R
50  LET  D = B * R
60  PRINT  N, "DIVIDEND = ";D
70  PRINT  "ENTER N, B, R"
80  INPUT  N, B, R
90  LET  D = B * R
100 PRINT  N, "DIVIDEND = ";D
110 PRINT  "ENTER N, B, R"
120 INPUT  N, B, R
130 LET  D = B * R
140 PRINT  N, "DIVIDEND = ";D
150 PRINT  "ENTER N, B, R"
160 INPUT  N, B, R
170 LET  D = B * R
180 PRINT  N, "DIVIDEND = ";D
190 PRINT  "ENTER N, B, R"
200 INPUT  N, B, R
210 LET  D = B * R
220 PRINT  N, "DIVIDEND = ";D
230 END
```

From this example we can see a problem. If the bank wanted to perform these operations on 100 accounts instead of just 5, a program written like the one above would require more than 400 instructions, so that the job of preparing the source program would be both tedious and expensive. Fortunately, there are BASIC instructions that enable us to handle problems of this type in a more efficient manner. We shall now show how this can be done.

The instructions in BASIC are executed in order of increasing statement number unless the computer is directed otherwise. One way of changing the order of instructions is to use a GO TO command. For example, suppose the programmer writes the instruction

```
GO TO 10
```

When this instruction is reached the computer will execute the instruction labeled 10 next and then continue executing the commands in increasing order of line number from statement 10 on. To illustrate the effect of a GO TO command, consider the following program:

```
10  REM  PROGRAM TO COMPUTE DIVIDEND
20  REM  N = ACCT NO, B = CURRENT BAL, R = INTEREST
         RATE, D = DIVIDEND
30  PRINT  "ENTER  N, B, R"
40  INPUT  N, B, R
50  LET  D = B * R
60  PRINT  N, "DIVIDEND =";D
70  GO TO 30
80  END
```

As the program begins, it asks for the values of N, B, and R for the first account; then it computes the dividend and prints out the account number N and dividend D. At this point the computer reaches the instruction

```
GO TO 30
```

which causes it to return to line number 30. Thus, new values of N, B and R are entered by the teller for the next account, the dividend computed and the account number and dividend printed out. Again, the computer reaches the instruction

```
GO TO 30
```

which causes it to return to line number 30, and the process begins again.

In this program we have constructed what is called a ***loop***, a series of instructions that are executed over and over again.

There is a problem with this program, however. The computer will never reach the END instruction; it will try to execute the loop forever. To avoid this kind of error, the programmer must provide a way of breaking out of a loop at the right time. This can be done by using another kind of computer instruction, called an IF–THEN statement. In a sense, IF–THEN statements enable the computer to make certain decisions.

A typical IF–THEN statement in BASIC tells the computer to go to a specified instruction if a certain condition is satisfied, and to proceed to the next instruction in line if the condition is not satisfied. To illustrate, consider the instruction

```
IF I > 5  THEN 30
```

This instruction is read, "if I is greater than 5, then go to instruction 30." To execute this instruction the computer examines the number in the location denoted by the variable I. If that number is greater than 5, then the computer will execute instruction 30 next; if the number in I is less than or equal to 5, then the computer will simply continue to execute the next instruction in line. The following example illustrates how an IF–THEN statement can be used to break out of a loop.

This program does the same job as the program in Example 25. The bank teller types in the account number, balance, and interest for five accounts and the computer prints out the account number and dividend for the five accounts. In Example 25 the statements

```
PRINT "ENTER N, B, R"
INPUT N, B, R
LET  D = B * R
PRINT N, "DIVIDEND = ";D
```

were written down five times, once for each account. In the program below, we shall write this group of instructions only once and use a loop to have them executed over and over, five times. We shall use the variable I as a ***counter*** to keep track of the number times the instructions in the loop are executed; and as soon as the counter tells us that we have gone through the loop five times, we will use an IF–THEN statement to break out of the loop. The program in the following example will do all of this.

Example 26

```
10  REM  PROGRAM TO COMPUTE DIVIDEND
20  REM  N = ACCT NO, B = CURRENT BAL, R = INTEREST
      RATE, D = DIVIDEND
30  LET  I = 1
```

```
 40  PRINT  "ENTER N, B, R"
 50  INPUT  N, B, R
 60  LET  D = B * R
 70  PRINT  N, "DIVIDEND = ";D
 80  LET  I = I + 1
 90  IF I > 5  THEN 110
100  GO TO 40
110  END
```

Let us analyze this program. The statement

```
30  LET  I = 1
```

sets the counter I equal to 1, indicating that we are beginning the first pass through the loop. The instructions

```
40  PRINT  "ENTER N, B, R"
50  INPUT  N, B, R
60  LET  D = B * R
70  PRINT  N, "DIVIDEND = ";D
```

ask that the data for the first account be entered, calculate the dividend, and print out the account number and dividend for the first account.

We are now ready for the second pass through the loop, so that we must increase the number in the counter I by 1. This is accomplished by the instruction

```
80  LET  I = I + 1
```

At this point, the number in the counter will be 2. The next statement

```
90  IF I > 5  THEN 110
```

checks the number in the counter to see it if is greater than 5. Since it is not, the computer proceeds to the next statement in line, which is

```
100  GO TO 40
```

This causes the computer to go to instruction 40 and begin the second pass through the loop. Thus, the data for the second account are entered, the dividend is computed, and the account number and dividend are printed out. The instruction

```
80  LET  I = I + 1
```

then increases the counter from 2 to 3. The next statement

```
90  IF I > 5  THEN 110
```

checks the number in the counter again to see if it is greater than 5. Since it is not, the computer proceeds to the next statement in line, which is

```
100  GO TO 40
```

The computer then asks that the data for the third account be entered and makes the third pass through the loop. This process continues until we have made five passes through the loop. After the fifth account has been processed, we reach the instruction

```
80  LET I = I + 1
```

which changes the value in the counter from 5 to 6. The next instruction

```
90  IF I > 5  THEN 110
```

checks the number in the counter to see if it is greater than 5. Since it is, the computer goes to instruction 110, which is

```
110  END
```

so that the program is finished.

The above program can be easily modified to process any number of accounts. For example, to process 100 accounts, we need only change the **IF–THEN** statement (statement 90) in the program to read

```
90  IF I > 100  THEN 110
```

There are a number of variations of the **IF–THEN** statement available in BASIC. In each of these statements the computer is instructed to transfer to a stated instruction when a certain condition is satisfied. If the condition is not satisfied, the computer ignores the **IF–THEN** statement and proceeds to the next instruction in line. The reader should find these **IF–THEN** statements self-explanatory.

```
IF K < 50  THEN 150          (< means "less than")
IF K = 0  THEN 140           (= means "equal")
IF K < = 2.5 THEN 200        (< = means "less than or equal")
IF K > 25  THEN 170          (> means "greater than")
If K > = 2.8  THEN 110       (> = means "greater than or equal")
If K < > 4.6  THEN 180       (< > means "not equal")
```

The program in Example 25 was designed to handle five accounts. The following variation will process any number of accounts. The operator supplies the number of accounts to be processed from the console.

Example 27

```
 10  REM  PROGRAM TO COMPUTE DIVIDEND
 20  REM N = ACCT NO, B = CURRENT BAL, R = INTEREST
     RATE, D = DIVIDEND, Q = NO OF ACCTS
 30  PRINT  "ENTER Q"
 40  INPUT  Q
 50  LET  I = 1
 60  PRINT "ENTER N, B, R"
 70  INPUT  N, B, R
 80  LET  D = B * R
 90  PRINT  N, "DIVIDEND = ";D
100  LET  I = I + 1
110  IF I > Q  THEN 130
120  GO TO 60
130  END
```

We consider another variation of the dividend computation problem. We assume this time that the operator does not know in advance how many accounts are to be processed. The idea is to keep executing the loop until the operator signals from the console that all accounts have been processed. To do this we invent a fictitious account number (which we are sure will never be used). The program will be designed to stop when the operator enters this fictitious account number. This stopping signal is called a ***sentinel*** or ***trailer.***

Example 28 The following dividend computation program is designed to stop when the operator enters the fictitious account number N = −99 and the fictitious data B = 0, R = 0.

```
10  REM  PROGRAM TO COMPUTE DIVIDEND
20  REM  N = ACCT NO, B = CURRENT BAL, R = INTEREST
    RATE, D = DIVIDEND
30  REM  OPERATOR SHOULD TYPE N = -99, B = 0, R = 0
    TO STOP PROCESSING
40  PRINT  "ENTER N, B, R"
50  INPUT  N, B, R
60  IF N = -99  THEN 100
70  LET  D = B * R
```

```
 80  PRINT  N, "DIVIDEND = ";D
 90  GO TO 40
100  END
```

In this program, each time the data for an account are entered, it is checked to see if the account number is −99. If it is, we stop, since we have reached the trailer, which tells us that all of our accounts have been processed. If the account number is not −99, we process another account.

BASIC provides another simple method for constructing loops. It uses two instructions, a **FOR** instruction to mark the top of the loop and indicate how many times the loop is to be executed and a **NEXT** instruction to mark the bottom of the loop and increase the counter on each pass through the loop. To illustrate its use we shall rewrite the program in Example 26 using **FOR** and **NEXT** instructions.

Example 29

```
10  REM  PROGRAM TO COMPUTE DIVIDEND
20  REM  N = ACCT NO, B = CURRENT BAL, R = INTEREST
    RATE, D = DIVIDEND
30  FOR I = 1 TO 5
40  PRINT  "ENTER N, B, R"
50  INPUT  N, B, R
60  LET  D = B * R
70  PRINT  N, D
80  NEXT I
90  END
```

The loop consists of statements 30 to 80, inclusive. Statement 30, the **FOR** instruction, causes the computer to execute instructions 40, 50, 60, and 70 (all the instructions between the **FOR** statement and the **NEXT** statement) for the values `I = 1`, `I = 2`, `I = 3`, `I = 4`, and `I = 5`, successively. That is, statement 30 first sets `I` to 1, and then executes statements 40, 50, 60, and 70, which process the first account. As soon as statement 80 is reached, the value of `I` is increased by 1 to 2, and statements 40, 50, 60, and 70 are again executed. Finally after executing statements 40, 50, 60, and 70 with `I = 5`, the computer leaves the loop and goes to the next statement in line, in this case the **END** statement.

It is important to note that the variable in the **NEXT** statement must be the same as the variable in the **FOR** statement; it is called the ***index*** of the loop.

EXERCISE SET 10.5

1. What does the following BASIC program do?

```
10  INPUT  I, J
20  IF I > J THEN 50
30  PRINT  J
40  GO TO 60
50  PRINT  I
60  END
```

2. What does the following BASIC program do?

```
10  LET  I = 1
20  PRINT  I
30  LET  I = I + 1
40  IF  I > 10  THEN 60
50  GO TO 20
60  END
```

3. What does the following BASIC program do?

```
10  FOR I = 1 TO 7
20  INPUT  A, B
30  LET  C = (A + B)/2
40  PRINT  C
50  NEXT I
60  END
```

4. Rewrite the program in Exercise 2 using FOR–NEXT instructions.

5. Write a BASIC program that prints the first 200 even integers.

6. Write a BASIC program that prints the first n positive integers, where n is provided by the operator.

7. If P dollars are borrowed at an annual interest rate of i compounded annually, then the amount S owed after n years is given by

$$S = P(1 + i)^n.$$

Write a program to print out a list showing the amount owed if \$1000 is borrowed at a rate of 5% ($i = .05$) for 1, 2,. . ., 25 years.

8. Given the formula

$$X = .0728T + .534,$$

write a BASIC program that will print out the values of X and T in two columns with headings, for $T = 1, 2, \ldots, 100$.

9. For a series of items, a computer operator is to enter the following data provided by the Accounting Department:

Stock number   Number of items sold   Sale price per item

Write a BASIC program to print out in two columns the stock number and total receipts for each item. Stop the computer using the fictitious stock number 0 as a trailer.

10. A computer operator is to enter a list of 500 numbers. Write a program that will print out only those numbers that are greater than 10.

10.6 FLOW CHARTS

The job of preparing a program for a computer requires a careful, logical analysis of the problem being solved. Since the computer carries out the programmer's instructions precisely, the programmer must be sure that the sequence of instructions constructed correctly solves the problem. Sometimes programs are so complicated that it is helpful to have an organized way of outlining the logical steps in the program before writing the instructions. In this section we shall discuss a method, called ***flow charting,*** for graphically representing the logical steps in a programming problem. ***Flow charts*** are also important because they make it possible to understand what a program does without reading through the list of instructions. The following symbols are commonly used in flow charting:

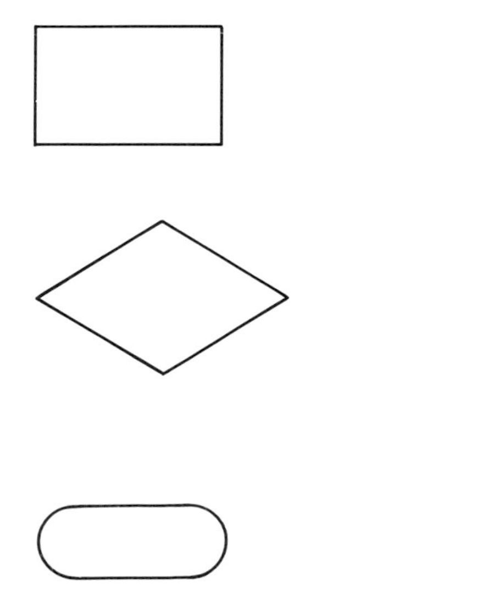

Indicates an operation is to be performed.

Indicates a decision to be made.

Indicates the beginning or end of a program.

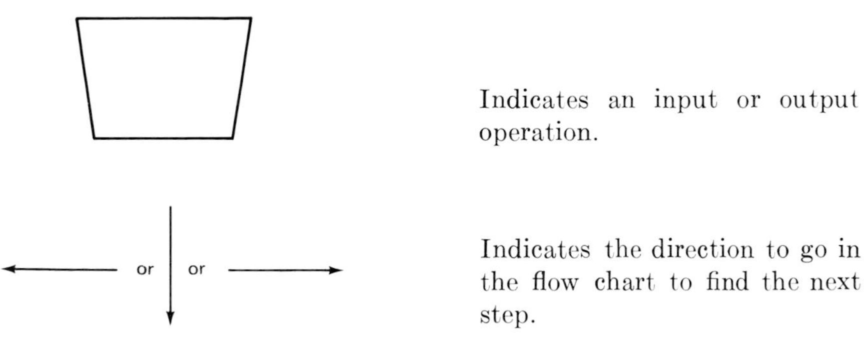

Indicates an input or output operation.
or
or
Indicates the direction to go in the flow chart to find the next step.

Although these symbols are fairly standard, some programmers use different ones. It really does not matter, however, since the essence of flow charting is the logical analysis of a programming problem.

To illustrate how flow charts are constructed, we give in Figure 10.10 a flow chart for the program in Example 26 of Section 10.5. The reader should trace through this flow chart and verify that it describes that program.

Example 30 The flow chart in Figure 10.11 describes the program in Example 28 of Section 10.5.

For complicated problems, most programmers construct a flow chart for the program first and then follow the flow chart to make the list of program instructions. The following example illustrates how this might be done.

Example 31 Suppose a book club wants to send a letter to its customers at the end of the year. Each customer has agreed to buy four books during the year. The club records for each customer: N, the customer's account number, and B, the number of books he bought during the year. If he bought less than four books during the year, the club will send him letter number 1, mildly reprimanding him for not fulfilling his contract. If he bought four books, he will receive letter number 2, congratulating him on meeting his quota and urging him to do better next year. If he bought more than four books, he will receive an enthusiastic letter, letter number 3, from the president of the company, along with a bonus book. The operator is to enter the data for each customer and the computer is to print the customer's account number, and then print a 1, 2, or 3 according to which letter he is to receive. To stop the program the operator is to enter the fictitious account number N = 0 and the value B = 0. The flow chart in Figure 10.12 suggests one possible way of solving this programming problem. Working from this flow chart, we

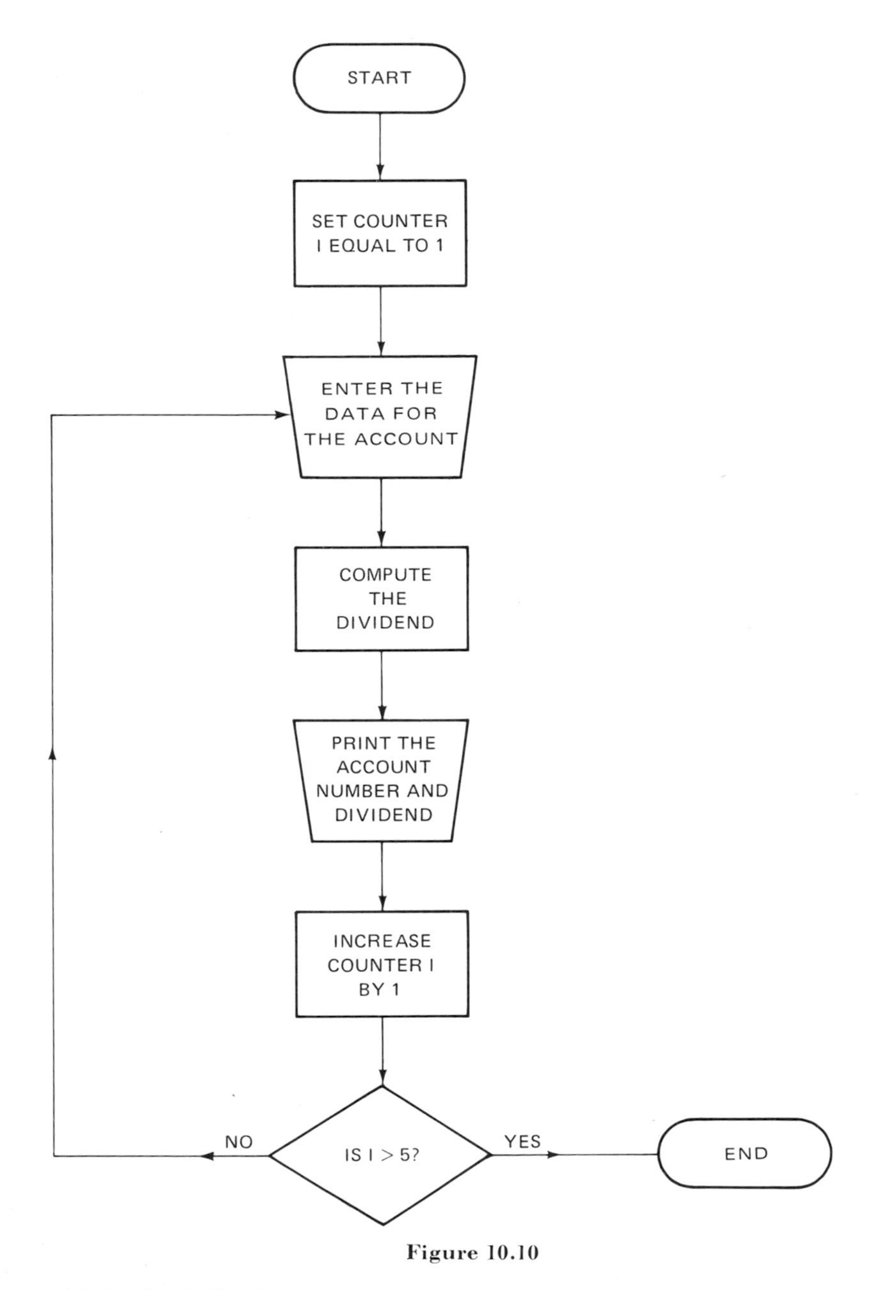

Figure 10.10

obtain the following program:

```
10  REM  BOOK CLUB PROGRAM
20  REM  N = ACCOUNT NO, B = NO BOOKS PURCHASED
30  REM  TO STOP OPERATOR TYPES N = 0, B = 0
40  PRINT  "ENTER N, B"
50  INPUT  N, B
```

```
 60  IF N = 0  THEN 150
 70  IF B < 4  THEN 110
 80  IF B = 4  THEN 130
 90  PRINT  N, "3"
100  GO TO 40
110  PRINT  N, "1"
120  GO TO 40
130  PRINT N, "2"
140  GO TO 40
150  END
```

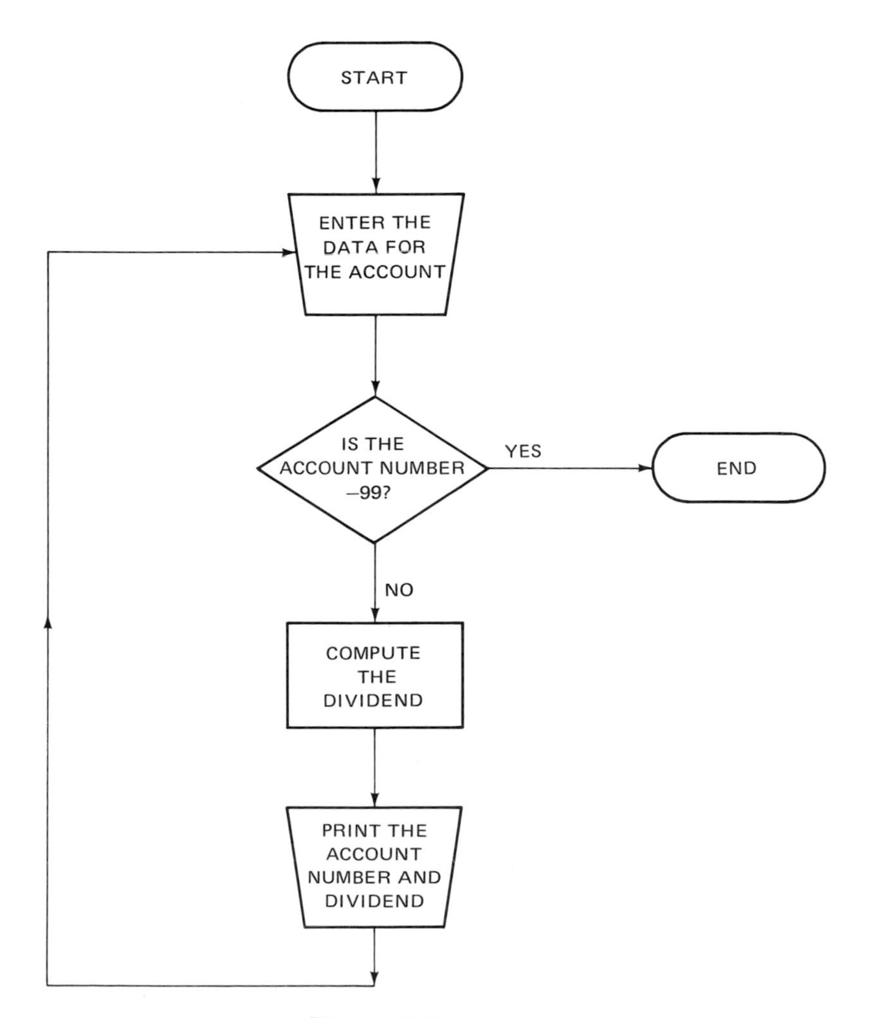

Figure 10.11

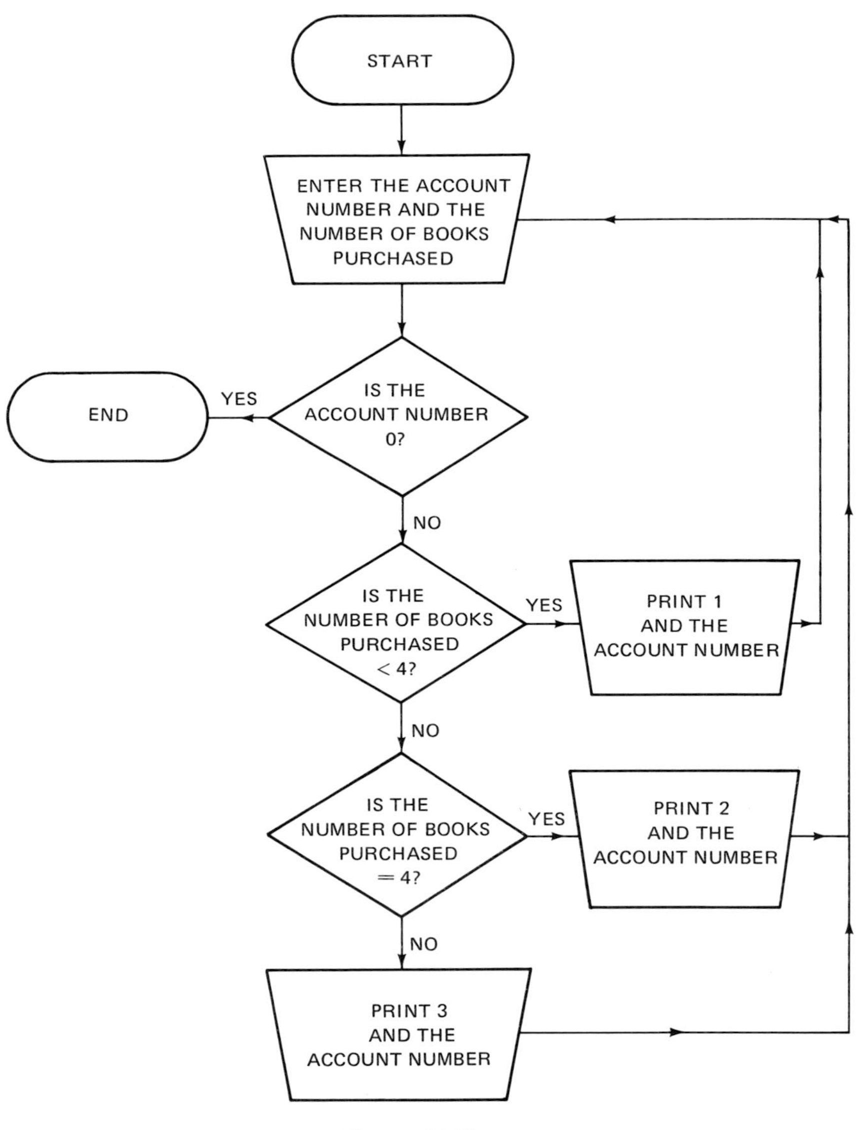

Figure 10.12

Although flow charts can be a great help in avoiding logical errors in the writing of programs, there are countless ways in which errors can be introduced: the programmer may have analyzed the problem incorrectly, used an incorrect instruction, inadvertently put instructions in the wrong order, or made a key-punching or typing error, and so forth. Careful planning and attention to detail is the key to successful programming. Moreover, a good programmer always checks the correctness of the final results in some way. For example, when writing a program to compute the inverse of a matrix A, one might compute AA^{-1} or $A^{-1}A$ to see if the result is the identity matrix; and, when writing a program to compute dividends, one might check the work by taking a few sample accounts and verifying that the computer has printed the correct data for each account.

In the last two sections we have been able to provide only a brief introduction to the BASIC language. There are many instructions and details we have omitted. Readers interested in pursuing BASIC in more depth might consult the following books:

[1] Samuel L. Marateck, *BASIC* (New York: Academic Press, 1975).

[2] John G. Kemeny and Thomas E. Kurtz, *BASIC Programming* (New York: Wiley, 1967).

[3] C. Joseph Sass, *BASIC Programming and Applications* (Boston: Allyn and Bacon, 1976).

Readers interested in the interaction between social issues and the widespread use of the computer may consult

[4] C. C. Gotlieb and A. Borodin, *Social Issues in Computing* (New York: Academic Press, 1973).

EXERCISE SET 10.6

1. Explain what the procedure described by the flow chart in Figure 10.13 accomplishes.

2. Write a BASIC program for the flow chart in Exercise 1.

3. Assume that `K` is a positive integer. Explain what the procedure described by the flow chart in Figure 10.14 accomplishes.

4. Write a program for the flow chart in Exercise 3.

5. Assume that a taxpayer's yearly salary is available. Explain what the flow chart in Figure 10.15 accomplishes.

6. Write a program for the flow chart in Exercise 5.

7. A union determines its dues according to the following schedule:

Annual salary		Annual dues
Less than	\$10,000	\$150
Between \$10,000 and	\$14,999.99	\$150 + 6% of the excess over \$10,000
At least	\$15,000	\$150 + 7% of the excess over \$10,000

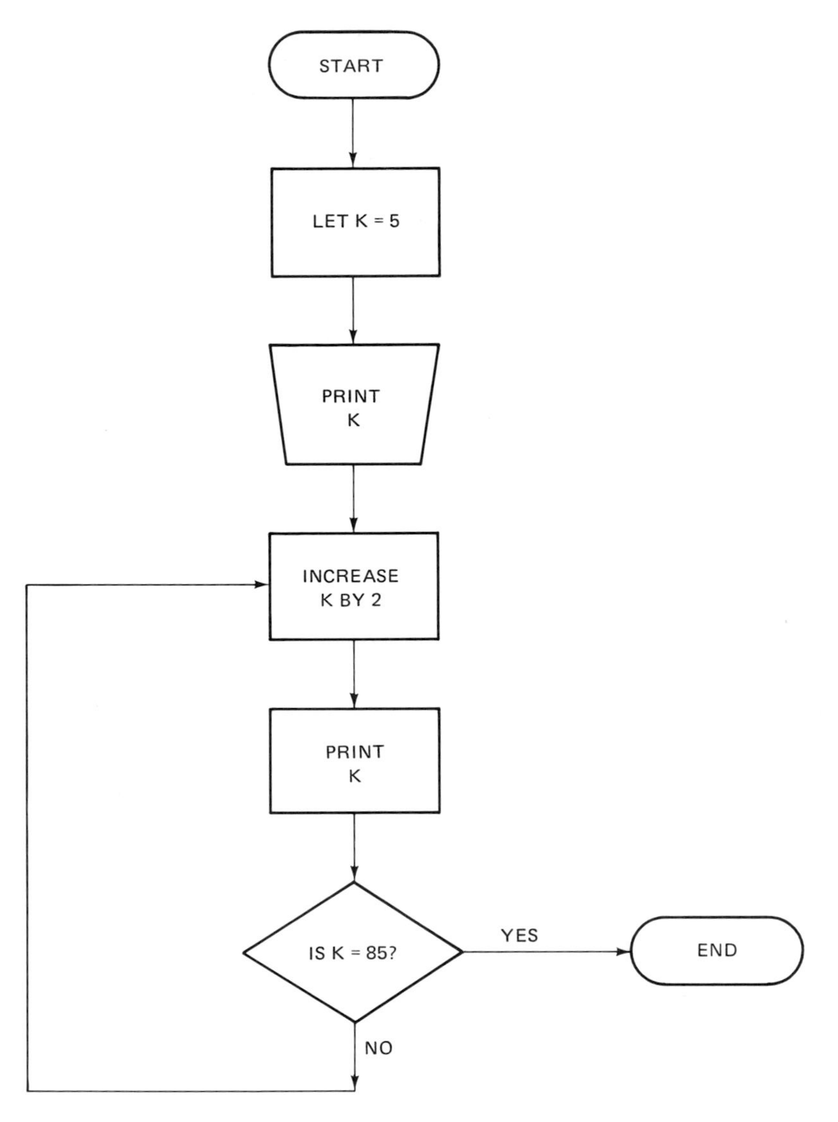

Figure 10.13

The union keeps for each employee a five-digit employee number along with the annual salary. Prepare a flow chart to print out each employee's employee number and annual dues.

8. Write a program for the flow chart in Exercise 7.

9. A brokerage firm uses the following method to determine its brokerage commission:

		Price per share	
		Less than \$3.00	At least \$3.00
Number of shares bought or sold	Less than 100 shares	3.25%	2.75%
	At least 100 shares	2.75%	2.50%

Each transaction is described by giving the customer's account number, the number of shares bought or sold, and the price per share. Prepare a

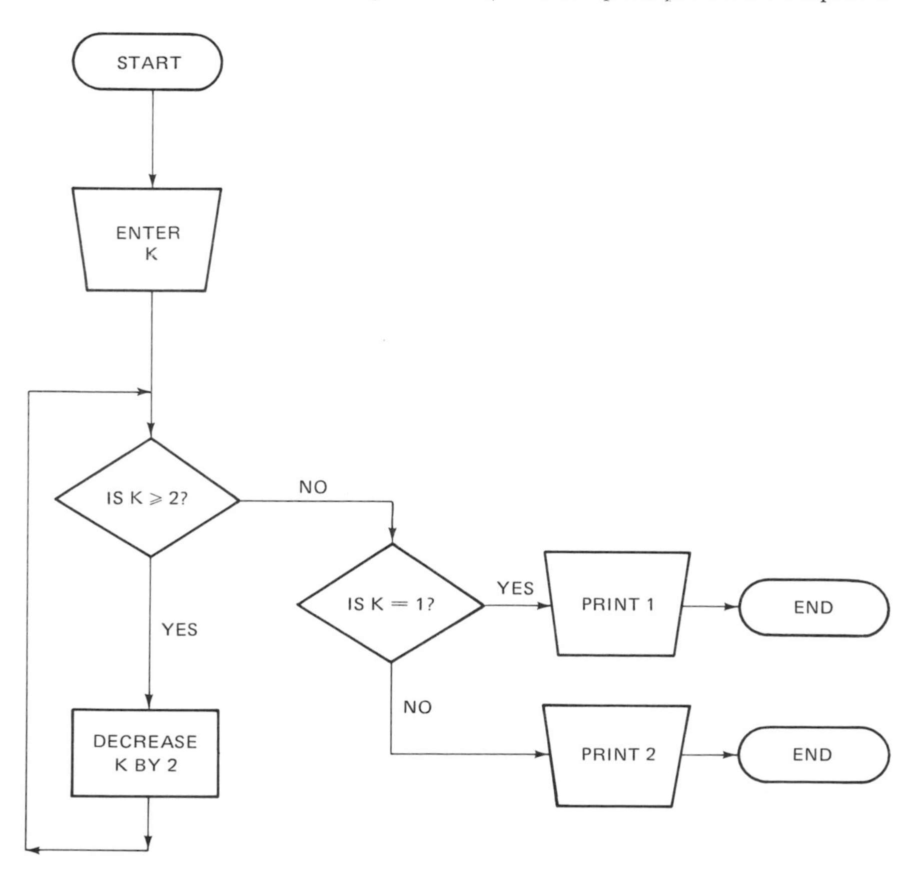

Figure 10.14

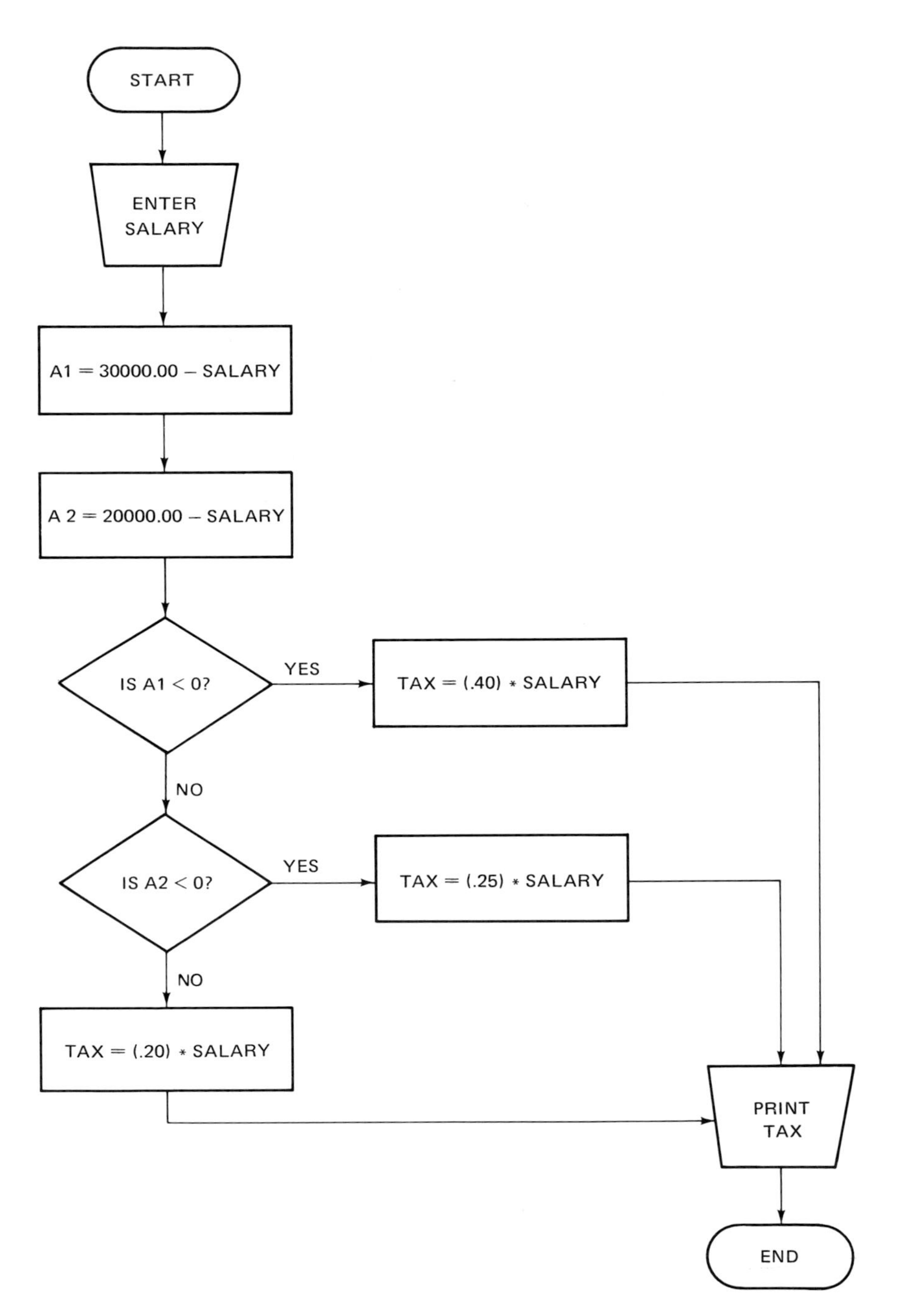
START
ENTER
SALARY
A1 = 30000.00 – SALARY
A 2 = 20000.00 – SALARY
IS A1 < 0?
YES
TAX = (.40) * SALARY
NO
IS A2 < 0?
YES
TAX = (.25) * SALARY
NO
TAX = (.20) * SALARY
PRINT
TAX
END

Figure 10.15

flow chart for a program that has the data for the day's transactions entered by the operator and prints out each account number along with the commission due. To stop the program the operator is to enter the fictitious account number zero.

10. Write a program for the flow chart in Exercise 9.

10.7 OTHER PROGRAMMING LANGUAGES

In the previous two sections we discussed BASIC, an easy-to-learn programming language. BASIC compilers are available for many computers. However, there are other widely used programming languages, each having its own special features. The following is a list of the most commonly used languages:

FORTRAN one of the most widely used programming languages for scientific applications and data handling

COBOL a programming language developed for business applications

ALGOL a science-oriented language widely used in Europe, but less commonly used in the United States than FORTRAN

PL/I a general purpose language that is suitable for both business and science applications

APL a programming language well suited for mathematical work, particularly matrix manipulations.

11 FUNCTIONS, LIMITS, AND RATES OF CHANGE

11.1 INTRODUCTION

The development of calculus in the seventeenth century by Isaac Newton† and Gottfried Wilhelm Leibniz‡ is one of the greatest achievements in the entire history of science and mathematics. Calculus is concerned with the mathematical study of change. Since we live in a universe of constant change, it is not surprising that calculus has been used as a fundamental tool in the physical sciences, engineering, economics, business, and the social sciences.

There are two basic geometric problems that call for the use of calculus:

(a) finding a tangent to a curve;
(b) finding the area under a curve.

† *Isaac Newton* (1642–1727)—Newton was born in Wools Thorpe, Lincolnshire, England. He is considered one of the greatest names in the history of human thought because of his fundamental contributions in mathematics, physics, and astronomy. In a period of 18 months (1665–1667) he developed the calculus, discovered the theory of gravity, and made contributions to the theory of light and color (for example, showing that white light is made up of many colors). His work in astronomy provided an explanation of the motion of planets in their orbits around the sun. He constructed a new type of telescope, using a reflecting mirror.

‡ *Gottfried Wilhelm von Leibniz* (1646–1716)—Leibniz was a German philosopher, mathematician, physicist, and historian. He was a universal genius who made major contributions to mathematics, logic, philosophy, mechanics, geology, law, and theology, all while pursuing a career as a civil servant. He was named a Baron in 1700. Leibniz invented differential and integral calculus later than, but independently of, Isaac Newton, and in his later years was embroiled in a bitter dispute with friends of Newton over who invented calculus.

As we shall see, there is a close relationship between the tangent problem and the problem of determining the rate at which a variable quantity is changing in value. The portion of calculus concerned with this problem is commonly called *differential calculus*. We shall also see that the area problem is related to the problem of finding a variable quantity whose rate of change is known. The portion of calculus concerned with these ideas is commonly called *integral calculus*.

It is not our objective to make the reader an expert in the theoretical aspects of calculus. Rather, our aims are to give an intuitive presentation of the basic ideas and to illustrate the kinds of applied problems that can be solved by using calculus.

11.2 FUNCTIONS

In this section we consider one of the most fundamental concepts in mathematics, the notion of a ***function.*** Loosely stated, a function is a rule for describing the way in which one quantity depends on another. For example, the area A of a circle depends on its radius r by the formula

$$A = \pi r^2;$$

thus we say, this formula describes "A as function of r." The quantities A and r in this equation are called ***variables.*** If we specify a value for the variable r, then this equation determines a unique value for the variable A. Accordingly, we call r the ***independent variable*** and A the ***dependent variable.*** The quantity $\pi \approx 3.14159\ldots$ has a fixed value which does not depend on A or r; thus we call π a ***constant.***

Sometimes we shall want to indicate that a dependent variable y is a function of an independent variable x without actually specifying a formula relating x and y. We do this by writing

$$y = f(x) \tag{1}$$

(read, "y equals f of x"). In this equation $f(x)$ does not mean f times x. Rather, f denotes the abstract concept of a function, that is, a rule for associating a *unique* value of y with a value of x. In the context of (1), $f(x)$ denotes the value for y that the function f associates with x.

Example 1 If the function f is given by $f(x) = x^2$, then f associates the number x^2 with the number x. Thus

$$\begin{aligned} f(2) &= 2^2 = 4 \\ f(-1) &= (-1)^2 = 1 \\ f(\sqrt{3}) &= (\sqrt{3})^2 = 3. \end{aligned}$$

In the context of the equation $y = f(x)$ these computations show that f associates a value of $y = 4$ with $x = 2$, a value of $y = 1$ with $x = -1$, and a value of $y = 3$ with $x = \sqrt{3}$.

There is nothing special about the letter f; any letter can be used to name a function. Thus

$$y = f(x), \qquad y = g(x), \qquad y = F(x), \qquad \text{and} \qquad y = \phi_1(x)$$

are all possible ways to indicate that y is a function of x.

Example 2 Let g be the function given by

$$g(x) = \frac{1}{x^2 - 4}.$$

Find $g(3)$ and $g(2)$.

Solution To find $g(3)$ we substitute $x = 3$ in the formula for g. This yields

$$g(3) = \frac{1}{3^2 - 4} = \frac{1}{9 - 4} = \frac{1}{5}.$$

To find $g(2)$ we substitute $x = 2$ in the formula for g. This yields

$$g(2) = \frac{1}{2^2 - 4} = \frac{1}{0}.$$

Since we cannot divide by zero, $g(2)$ is undefined.

As this example shows, there may exist values of the independent variable where a function is undefined. Those values of the independent variable at which a function is defined and yields a real value form a set of real numbers called the ***domain*** of the function.

Example 3 The function

$$g(x) = \frac{1}{x^2 - 4}$$

has a value for every x except $x = 2$ and $x = -2$ since these values result in a division by zero. Thus the domain of g consists of all real numbers except 2 and -2.

Example 4 The function

$$f(x) = \sqrt{x}$$

does not yield a real value if $x < 0$ since negative numbers do not have real square roots. The domain of f consists of all $x \geq 0$.

The following definition will enable us to study functions geometrically.

> **The Graph of a Function** The graph of a function f is defined to be the graph of the equation $y = f(x)$.

Example 5 Sketch the graphs of the functions

$$\text{(a) } f(x) = x + 2 \qquad \text{(b) } f(x) = x^2 + 1 \qquad \text{(c) } f(x) = 2.$$

Solution (a) By definition, the graph of f is the graph of the equation $y = x + 2$. As discussed in Chapter 2 this is the equation of a line, so the graph can be obtained by plotting any two points on the graph and drawing a line through them. Setting $x = 0$ yields $y = 2$, so that (0, 2) is on the graph. Setting $y = 0$ yields $x = -2$, so that $(-2, 0)$ is on the graph. Thus the graph of f is the line in Figure 11.1a.

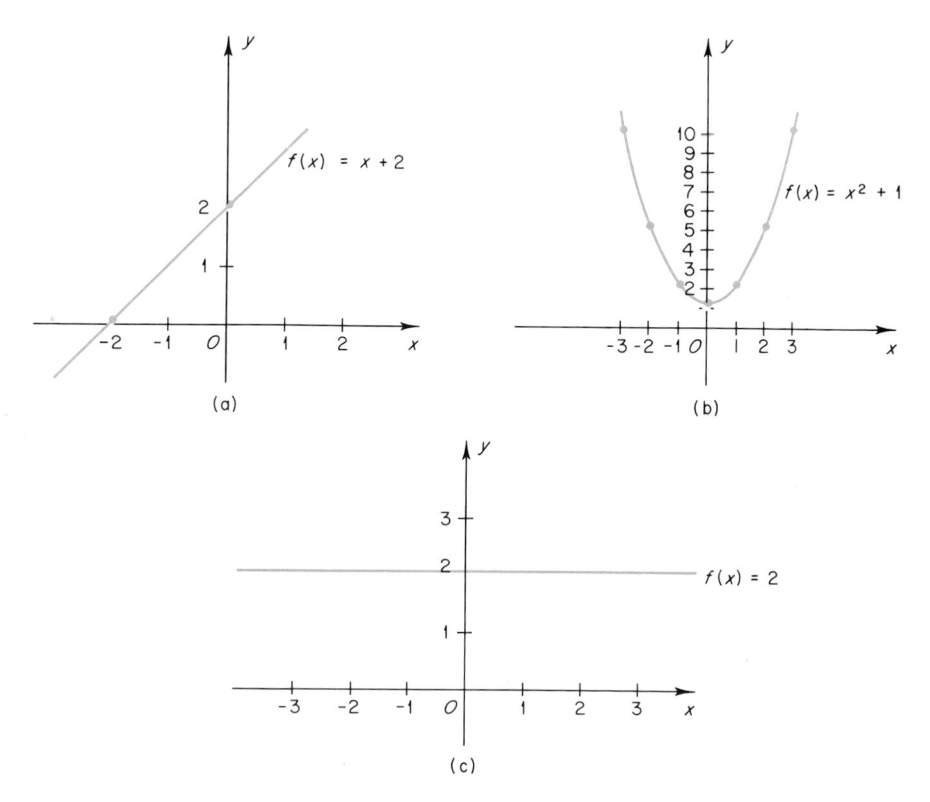

Figure 11.1

(b) By definition, the graph of f is the graph of the equation $y = x^2 + 1$. This equation was graphed in Example 4 of Chapter 2 by point plotting. The graph is shown in Figure 11.1b.

(c) By definition, the graph of f is the graph of the equation $y = 2$. As discussed in Chapter 2, this graph is the horizontal line shown in Figure 11.1c.

The function in part (c) of this example is called a ***constant function*** because it assigns the same value (namely 2) to every x.

Not every curve in the xy plane can be the graph of a function. For example, suppose the curve in Figure 11.2 is the graph of $y = f(x)$ for

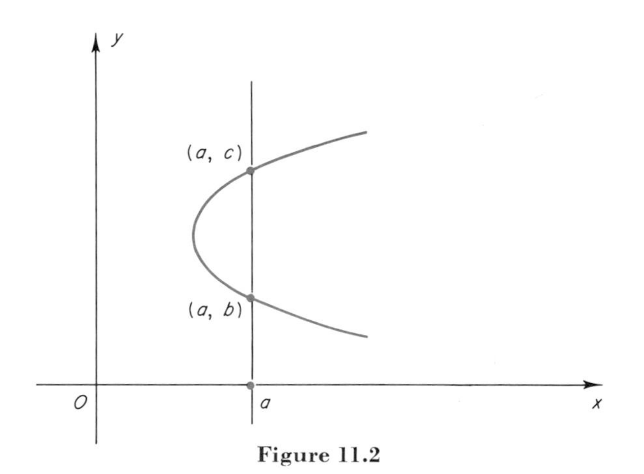

Figure 11.2

some function f. The vertical line $x = a$ shown in the figure cuts the curve at two points, say (a, b) and (a, c). Because (a, b) and (a, c) lie on the curve $y = f(x)$, the coordinates of these points satisfy this equation. Thus

$$b = f(a) \qquad \text{and} \qquad c = f(a).$$

But this is impossible since f cannot associate two different values with a. Thus the curve shown in Figure 11.2 is not the graph of any function f of x. In light of this discussion we can state the following useful result:

The Vertical Line Test If some vertical line cuts a curve in more than one point, then the curve is not the graph of any function of x.

We conclude with a brief introduction to some functions that will be of importance in later sections.

The ***absolute value*** of a number is its distance from the origin when the number is plotted on a coordinate line. The absolute value of a is written $|a|$.

Example 6 As shown in Figure 11.3,

$$|3| = 3, \qquad |-2| = 2, \qquad |0| = 0.$$

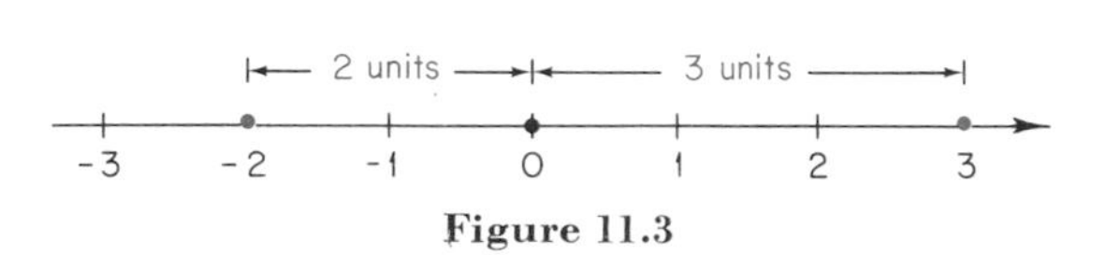

Figure 11.3

In general,

$$|a| = \begin{cases} a & \text{if} \quad a \geq 0 \\ -a & \text{if} \quad a < 0. \end{cases}$$

Example 7 Sketch the graph of the function $f(x) = |x|$.

Solution By definition, the graph of f is the graph of the equation $y = |x|$, or equivalently

$$y = \begin{cases} x & \text{if} \quad x \geq 0 \\ -x & \text{if} \quad x < 0. \end{cases}$$

We shall sketch the graph in two steps. If $x \geq 0$, the equation is $y = x$, which gives a line of slope 1 and y intercept 0 (Figure 11.4). if $x < 0$, the equation is $y = -x$, which gives a line of slope -1 and y intercept 0 (Figure 11.4).

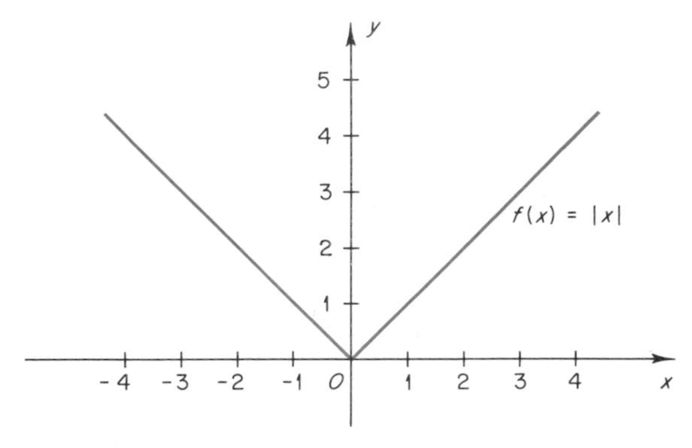

Figure 11.4

In high school it is learned that every nonnegative real number a has exactly one nonnegative square root; this root is denoted by $\sqrt{a}$.

For example, 16 has two square roots, $+4$ and -4. The symbol $\sqrt{16}$ denotes the nonnegative root $+4$; that is, $\sqrt{16} = 4$.

Perhaps the most common error when working with the $\sqrt{\ }$ symbol is to write

$$\sqrt{a^2} = a.$$

Although this is correct if $a \geq 0$, it is not correct if $a < 0$. For example, if $a = -4$, then $a^2 = 16$, so that

$$\sqrt{a^2} = \sqrt{16} = 4 \neq a.$$

To obtain a statement that is true for all values of a, positive or negative, one should write

$$\sqrt{a^2} = |a|$$

EXERCISE SET 11.2

1. Find the value of

(a) $|-3|$ (b) $|7|$ (c) $\sqrt{9}$ (d) $\sqrt{(-8)^2}$.

2. Find all values of x that satisfy the condition

(a) $|x| = 8$ (b) $|x - 2| = 3$ (c) $\sqrt{x^2} = 4$.

3. Consider the function f defined by

$$f(x) = x^2 - 4.$$

Find: (a) $f(3)$ (b) $f(2)$ (c) $f(-4)$
(d) $f(a)$ (e) $f(x - 2)$ (f) $f(x + h)$.

4. Consider the function f defined by

$$f(x) = 3x^2 + 2x - 5.$$

Find: (a) $f(2)$ (b) $f(-3)$ (c) $f(5)$
(d) $f(s)$ (e) $f(a - 3)$ (f) $f(2 + h)$.

5. Consider the function g defined by

$$g(x) = \frac{3}{x - 2}.$$

Find: (a) $g(5)$ (b) $g(-3)$ (c) $g(0)$
(d) $g(b)$ (e) $g(b + 7)$ (f) $g(b + h)$.

6. Consider the function F defined by

$$F(x) = \frac{3x + 2}{2x^2 - 5x + 4}.$$

Find: (a) $F(4)$ (b) $F(1)$ (c) $F(-2)$
(d) $F(r)$ (e) $F(r - 1)$ (f) $F(r + h)$.

In Exercises 7–12 specify the domain of the given function.

7. $f(x) = x + 2$

8. $f(x) = \sqrt{x - 1}$

9. $f(x) = \dfrac{1}{x - 3}$

10. $f(x) = \dfrac{1}{x^2 - 3x + 2}$

11. $f(x) = \dfrac{x - 2}{x^2 - 2x + 1}$

12. $f(x) = \sqrt{2 - x}$.

In Exercises 13–18 sketch the graph of the given function.

13. $f(x) = 2x - 3$

14. $f(x) = 2x^2 - 8$

15. $f(x) = x^2 - 2x$

16. $f(x) = |x - 1|$

17. $f(x) = \dfrac{1}{x^2 + 1}$

18. $f(x) = \sqrt{x}$.

In Exercises 19–24 determine which of the indicated curves are graphs of functions.

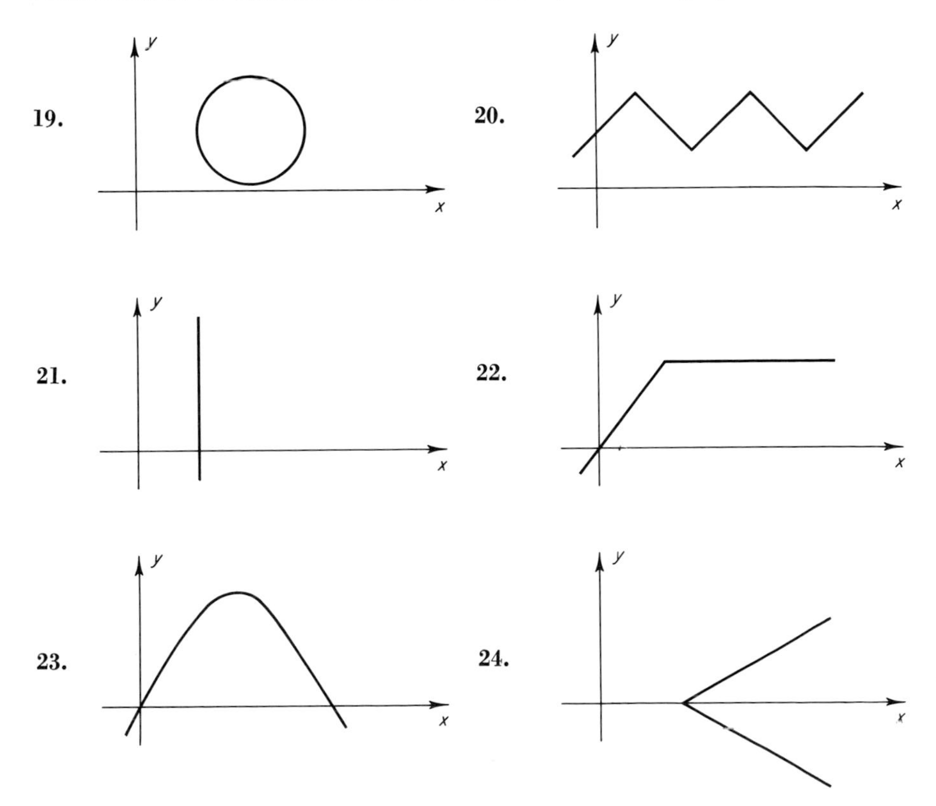

25. The Jones Corporation finds that its profits depend on the amount of money it spends on advertising. If x dollars are spent on advertising, the profit is given by

$$P(x) = 2x^2 - x + 1000.$$

Determine the profit if the amount spent on advertising is:

(a) \$0 (b) \$100 (c) \$50 (d) \$1000.

11.3 LIMITS

In this section we introduce a concept that forms the cornerstone of calculus, the notion of a *limit*. Loosely speaking, limits are used to indicate the behavior of a function as the independent variable gets closer and closer to a fixed value a. The results we develop here will be used repeatedly in later sections.

In many problems one is concerned with describing the behavior of a function $f(x)$ in the vicinity of a point $x = a$. There are usually two questions of importance:

(a) What is the value of f when $x = a$?

(b) What do the values of f look like when x is near, but different from a?

For example, if x represents the distance between a lunar lander and the moon's surface, and $f(x)$ represents the speed of the lander at distance x, then it is important to know $f(0)$ (the speed of the lander at impact); but it is just as important to know how the speed $f(x)$ varies when x is near, but different from 0. For it is this information that enables the control center to make corrections to bring about a soft landing.

The distinction between the value of a function f at a point $x = a$ and the value of f for x near, but different from, a will become clear in the following examples.

Example 8 Let f be the function defined by $f(x) = 3x + 1$, and let us investigate the behavior of f at the point $x = 2$ and near the point $x = 2$.

The first part is easy; at $x = 2$ the value of the function is

$$f(2) = 3(2) + 1 = 7.$$

To investigate the behavior near $x = 2$, we have evaluated the function f at a succession of x values closer and closer to (but different from) $x = 2$. The results are shown in Figure 11.5. It seems intuitively clear from these computations that the value of $f(x) = 3x + 1$ gets closer and closer to 7 as x gets closer and closer to 2 from either side. The number 7 is called the ***limit*** of $f(x) = 3x + 1$ as x approaches 2 and we write

$$\lim_{x \to 2} (3x + 1) = 7. \tag{1}$$

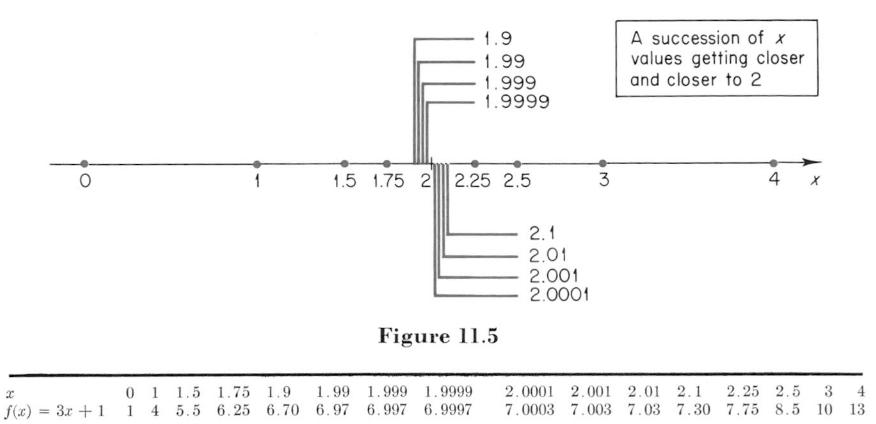

Figure 11.5

x	0	1	1.5	1.75	1.9	1.99	1.999	1.9999	2.0001	2.001	2.01	2.1	2.25	2.5	3	4
$f(x) = 3x + 1$	1	4	5.5	6.25	6.70	6.97	6.997	6.9997	7.0003	7.003	7.03	7.30	7.75	8.5	10	13

In this expression $x \to 2$ indicates that x is approaching (but different from) 2. The entire expression is read, "the limit as x approaches 2 of $3x + 1$ equals 7."

It is important to keep in mind that the limit in (1) is only a guess based on our computations in Figure 11.5. It is theoretically possible that if we enlarged our table of computations to include values of x still closer to 2, the values for $f(x)$ might change their pattern and approach some number different from 7 or perhaps approach no limit at all. To be absolutely certain (1) is correct, we need mathematical proof; and to give a mathematical proof we need a precise mathematical definition of a limit. In this textbook we shall not be concerned with such proofs and formal definitions. However, the reader will have a satisfactory grasp of the limit concept if it is viewed as follows.

The Intuitive Notion of a Limit Given a function f, we interpret the statement

$$\lim_{x \to a} f(x) = L$$

to mean that the values of $f(x)$ get closer and closer to the number L as x gets closer and closer to (but remains different from) a. Stated another way, we can make the value of $f(x)$ as close as we like to L by making x sufficiently close to (but different from) a.

For the function $f(x) = 3x + 1$ in Example 8, it turned out that

$$f(2) = 7 \qquad \text{and} \qquad \lim_{x \to 2} f(x) = 7,$$

so that the limit as x *approaches* 2 and the value *at* 2 were the same. As we shall see in subsequent examples, this is not always the case. However, by graphing the function $f(x) = 3x + 1$, it is easy to visualize why $f(2)$ and $\lim_{x \to 2} f(x)$ have the same value in this case (Figure 11.6).

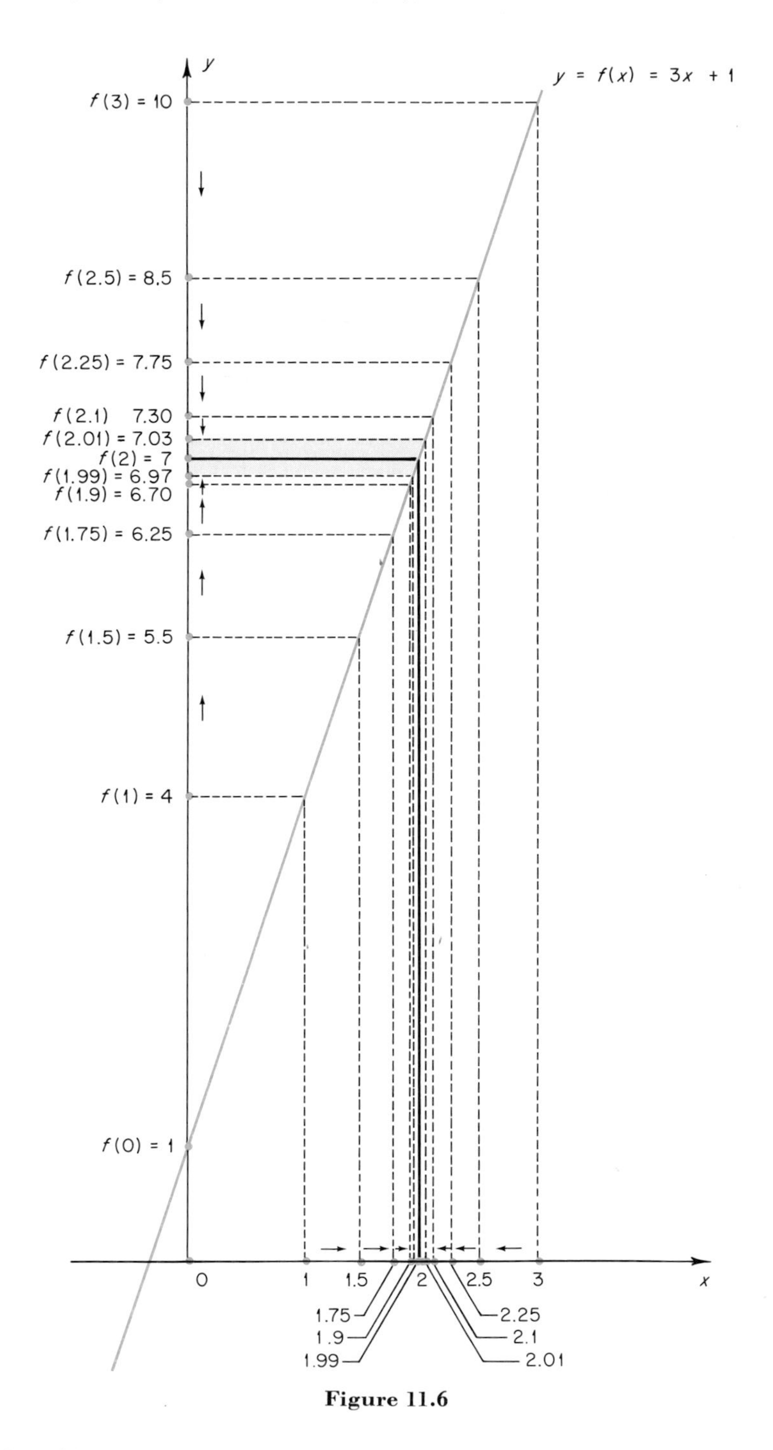

Figure 11.6

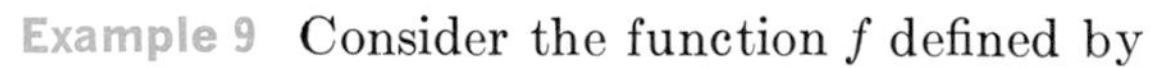
Example 9 Consider the function f defined by

$$f(x) = \frac{x^2 - 9}{x - 3}.$$

First, observe that $f(3)$ is undefined since substituting $x = 3$ in the formula for f leads to a division by zero. However, as Table 11.1 suggests, there is a value for the limit of $f(x)$ as x approaches 3, namely,

$$\lim_{x \to 3} \frac{x^2 - 9}{x - 3} = 6. \tag{2}$$

Table 11.1

$x \to 3$ from the right side

x	5	4	3.5	3.1	3.01	3.001	3.0001
$f(x) = \frac{x^2 - 9}{x - 3}$	8	7	6.5	6.1	6.01	6.001	6.0001

$x \to 3$ from the left side

x	1	2	2.5	2.9	2.99	2.999	2.9999
$f(x) = \frac{x^2 - 9}{x - 3}$	4	5	5.5	5.9	5.99	5.999	5.9999

To gain some geometric insight into this limit, let us graph the function f. If $x \neq 3$, then we can write

$$f(x) = \frac{x^2 - 9}{x - 3} = \frac{(x + 3)(x - 3)}{x - 3} = x + 3.$$

Thus for $x \neq 3$, the graph of $y = f(x)$ coincides with the graph of the line $y = x + 3$; and for $x = 3$, there is no point on the graph since $f(3)$ is undefined. As shown in Figure 11.7, the graph of f is a straight line

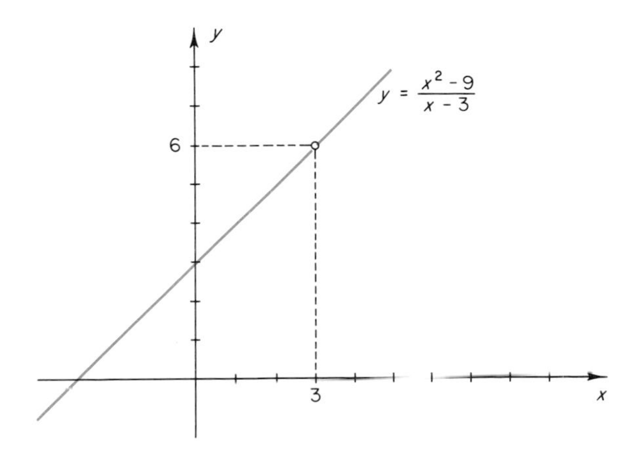

Figure 11.7

with a hole in it at $x = 3$. The limit in (2) should be intuitively evident from this graph; when x gets closer and closer to 3, $f(x)$ gets closer and closer to 6.

Example 10 Consider the function f defined by

$$f(x) = \frac{1}{x - 2}.$$

Observe first that $f(2)$ is undefined since substituting $x = 2$ results in a division by zero. Next let us investigate

$$\lim_{x \to 2} f(x) = \lim_{x \to 2} \frac{1}{x - 2}.$$

As Table 11.2 shows, the values of $f(x)$ get larger and larger as $x \to 2$ from the right side and get smaller and smaller as $x \to 2$ from the left side. Thus, $f(x)$ does not approach any fixed number as x approaches 2.

Table 11.2

$x \to 2$ from the right side

x	3	2.5	2.2	2.1	2.05	2.01	2.001	2.0001
$f(x) = \frac{1}{x-2}$	1	2	5	10	20	100	1000	10,000

$x \to 2$ from the left side

x	1	1.5	1.8	1.9	1.95	1.99	1.999	1.9999
$f(x) = \frac{1}{x-2}$	−1	−2	−5	−10	−20	−100	−1000	−10,000

The graph of f is shown in Figure 11.8.

If, as in the last example, the quantity $f(x)$ approaches no single finite value as x approaches a point a, then we say

$$\lim_{x \to a} f(x)$$

does not exist.

Example 11 Let f be the function given by the formula

$$f(x) = \begin{cases} 1 & \text{if } x > 0 \\ 0 & \text{if } x = 0 \\ -1 & \text{if } x < 0. \end{cases}$$

The graph of f is shown in Figure 11.9.

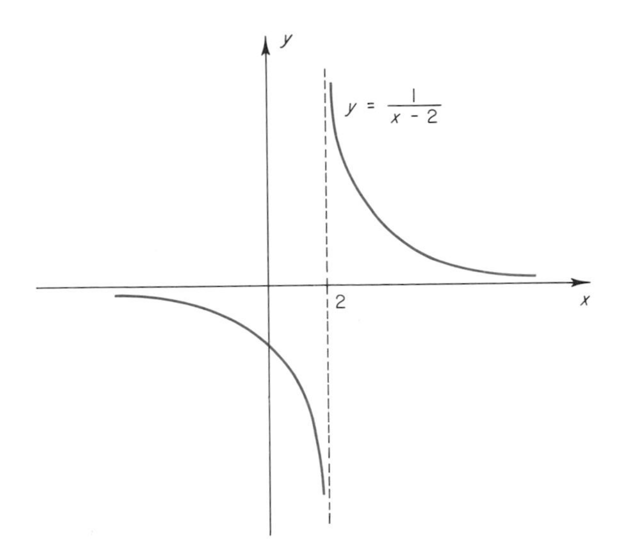

Figure 11.8

For this function, $f(0)$ has a value, namely $f(0) = 0$. However, $\lim_{x \to 0} f(x)$ does not exist. To see this we need only observe that as $x \to 0$ from the right side the value of $f(x)$ is constantly $+1$, while as $x \to 0$ from the left side the value of $f(x)$ is constantly -1. Thus $f(x)$ does not approach a *single* finite value as $x \to 0$.

Example 12 Consider the function f defined by

$$f(x) = \begin{cases} -\frac{2}{3}x + 6 & \text{if} \quad x \geq 3 \\ x + 1 & \text{if} \quad x < 3. \end{cases}$$

The graph of f is shown in Figure 11.10. To the left of 3 the formula for f is $f(x) = x + 1$, so that $f(x) = x + 1$ approaches 4 as x approaches 3

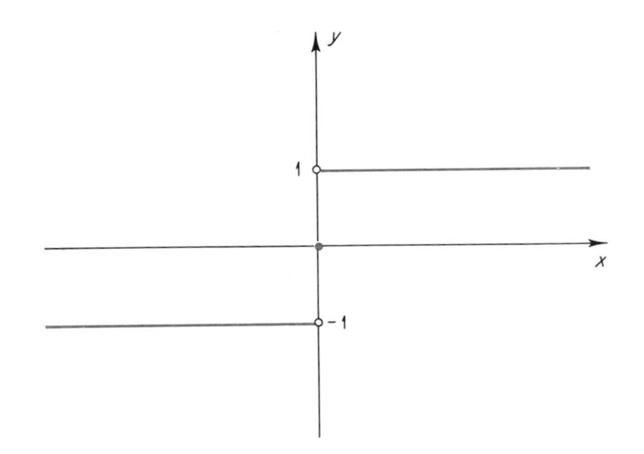

Figure 11.9

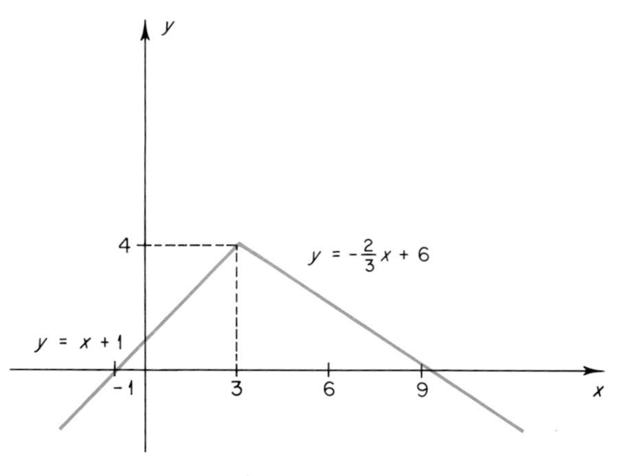

Figure 11.10

from the left. To the right of 3 the formula for f is $f(x) = -\frac{2}{3}x + 6$, so that $f(x) = -\frac{2}{3}x + 6$ approaches 4 as x approaches 3 from the right. Since the values of $f(x)$ approach 4 as $x \to 3$ from either side, we have

$$\lim_{x \to 3} f(x) = 4.$$

The following results facilitate the calculation of limits. We omit the proofs.

Properties of Limits

Property 1 The limit of a constant function $f(x) = k$ is the constant k; that is

$$\lim_{x \to a} k = k.$$

Example 13

$$\lim_{x \to 3} 5 = 5, \qquad \lim_{x \to 0} 5 = 5, \qquad \lim_{x \to -8} 5 = 5.$$

Property 2

$$\lim_{x \to a} x = a.$$

Example 14

$$\lim_{x\to 0} x = 0, \qquad \lim_{x\to -2} x = -2, \qquad \lim_{x\to 6} x = 6.$$

Property 3 The limit of a sum is the sum of the limits; the limit of a difference is the difference of the limits; and the limit of a product is the product of the limits; that is:

$$\lim_{x\to a} [f(x) + g(x)] = \lim_{x\to a} f(x) + \lim_{x\to a} g(x)$$

$$\lim_{x\to a} [f(x) - g(x)] = \lim_{x\to a} f(x) - \lim_{x\to a} g(x)$$

$$\lim_{x\to a} [f(x)g(x)] = \lim_{x\to a} f(x) \cdot \lim_{x\to a} g(x).$$

Example 15

$$\lim_{x\to 4} (x + 5) = \lim_{x\to 4} x + \lim_{x\to 4} 5 = 4 + 5 = 9$$

$$\lim_{x\to 4} (x - 5) = \lim_{x\to 4} x - \lim_{x\to 4} 5 = 4 - 5 = -1$$

$$\lim_{x\to 4} x(x + 5) = \lim_{x\to 4} x \cdot \lim_{x\to 4} (x + 5) = 4 \cdot 9 = 36$$

$$\lim_{x\to 2} x^2 = \lim_{x\to 2} x \cdot \lim_{x\to 2} x = 2 \cdot 2 = 4$$

$$\lim_{x\to 3} 7x = \lim_{x\to 3} 7 \cdot \lim_{x\to 3} x = 7 \cdot 3 = 21.$$

Example 16 If k is any constant, then using Properties 1 and 3 we can write

$$\lim_{x\to a} kf(x) = \lim_{x\to a} k \cdot \lim_{x\to a} f(x)$$

or

$$\lim_{x\to a} kf(x) = k \lim_{x\to a} f(x).$$

Stated verbally, *a constant k can be moved past a limit sign.*

Example 17

$$\begin{aligned}\lim_{x\to 3} (2x^2 - 4x + 5) &= \lim_{x\to 3} 2x^2 - \lim_{x\to 3} 4x + \lim_{x\to 3} 5 \\ &= 2 \lim_{x\to 3} x^2 - 4 \lim_{x\to 3} x + \lim_{x\to 3} 5 \\ &= 2(\lim_{x\to 3} x \cdot \lim_{x\to 3} x) - 4 \lim_{x\to 3} x + \lim_{x\to 3} 5\end{aligned}$$

$$= 2(9) - 4(3) + 5$$
$$= 11.$$

Property 4 The limit of a ratio is the ratio of the limits, provided the limit of the denominator is not zero; that is

$$\lim_{x \to a} \frac{f(x)}{g(x)} = \frac{\lim_{x \to a} f(x)}{\lim_{x \to a} g(x)} \quad \text{if} \quad \lim_{x \to a} g(x) \neq 0.$$

Example 18

$$\lim_{x \to 4} \frac{3x^2 - 5}{x - 2} = \frac{\lim_{x \to 4} (3x^2 - 5)}{\lim_{x \to 4} (x - 2)} = \frac{43}{2}.$$

Example 19 In Example 9 we showed

$$\lim_{x \to 3} \frac{x^2 - 9}{x - 3} = 6.$$

This result cannot be obtained using Property 4 since $\lim_{x \to 3} (x - 3) = 0$. However, we can write

$$\lim_{x \to 3} \frac{x^2 - 9}{x - 3} = \lim_{x \to 3} \frac{(x - 3)(x + 3)}{x - 3} = \lim_{x \to 3} (x + 3) = 6.$$

EXERCISE SET 11.3

1. (a) Complete the following table of values for $f(x) = 2x - 3$.

x	2.6	2.7	2.8	2.9	2.99	2.999
$f(x) = 2x - 3$						

x	3.4	3.3	3.2	3.1	3.01	3.001
$f(x) = 2x - 3$						

(b) Use your table to help find $\lim_{x \to 3} (2x - 3)$.

2. (a) Complete the following table of values for $g(x) = x^2 + 2$. (A hand calculator may help.)

x	-1.3	-1.2	-1.1	-1.01	-1.001	-1.0001
$g(x) = x^2 + 2$						

x	-0.7	-0.8	-0.9	-0.99	-0.999	-0.9999
$g(x) = x^2 + 2$						

(b) Use your table to help find $\lim_{x \to -1} (x^2 + 2)$.

3. (a) Complete the following table of values for $f(x) = |x|/x$.

x	-1	-0.1	-0.01	-0.001	-0.0001	0.0001	0.001	0.01	0.1	1
$f(x) = \frac{\lvert x\rvert}{x}$										

(b) From your table, what can you say about $\lim_{x \to 0} (|x|/x)$?

In Exercises 4–13 find the indicated limit or state that the limit does not exist.

4. $\lim_{x \to 3} 2x$

5. $\lim_{x \to 3} (2x + 4)$

6. $\lim_{x \to 4} (3x^2 + 2x - 5)$

7. $\lim_{x \to 3} \dfrac{2x + 5}{x^2 - 4}$

8. $\lim_{x \to 0} \dfrac{2}{x}$

9. $\lim_{x \to 0} \dfrac{x^2 - 2x}{x}$

10. $\lim_{x \to -1} (1 - x^2)$

11. $\lim_{x \to 1} \dfrac{x^2 - 36}{x - 6}$

12. $\lim_{x \to -1} \dfrac{x^2 + 2x + 1}{x + 1}$

13. $\lim_{x \to 4} \dfrac{x + 1}{x^2 - 3x - 4}$.

14. (a) Find $\lim_{x \to 1} f(x)$, if it exits, for

$$f(x) = \begin{cases} x + 3 & \text{if } x \geq 1 \\ 7x - 3 & \text{if } x < 1 \end{cases}$$

and sketch the graph of f.

(b) Find $\lim_{x \to 0} g(x)$, if it exists, for

$$g(x) = \begin{cases} x + 1 & \text{if } x \geq 0 \\ 3x + 2 & \text{if } x < 0 \end{cases}$$

and sketch the graph of f.

In Exercises 15–19 find the indicated limit, if it exists; and sketch the graph of $f(x)$.

15. $\lim\limits_{x \to 4} \dfrac{x^2 - 16}{x - 4}$; $f(x) = \dfrac{x^2 - 16}{x - 4}$ **16.** $\lim\limits_{x \to -5} \dfrac{x^2 - 25}{x + 5}$; $f(x) = \dfrac{x^2 - 25}{x + 5}$

17. $\lim\limits_{x \to 1} \dfrac{x^3 + x^2 - 2x}{x(x + 2)}$; $f(x) = \dfrac{x^3 + x^2 - 2x}{x(x + 2)}$ **18.** $\lim\limits_{x \to 5} \dfrac{1}{x - 5}$; $f(x) = \dfrac{1}{x - 5}$

19. $\lim\limits_{x \to 3} \dfrac{|x - 3|}{x - 3}$; $f(x) = \dfrac{|x - 3|}{x - 3}$.

In Exercises 20–23 use the graph to find the indicated limit if it exists.

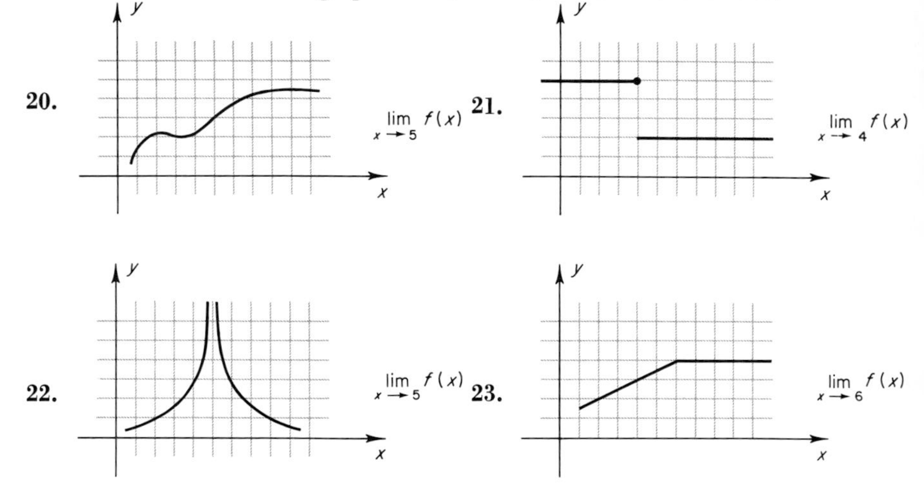

24. Use the Appendix Table VI to help find

$$\lim_{x \to 0} \frac{1 - e^x}{x}$$

11.4 CONTINUITY

In this section we investigate conditions under which the graph of a function f is assured of forming a continuous, unbroken curve.

Loosely speaking, we use the term *continuous* to mean: "without gaps or jumps." Thus we might conceive of a continuous curve as one that can be drawn without having to lift the pencil from the paper (Figure 11.11).

In order to make this intuitive notion of continuity precise, let us consider some of the ways in which a curve can *fail* to be continuous. In Figure 11.12 we have sketched some curves that, because of their behavior at $x = a$, are not continuous. The curve in Figure 11.12a has a hole in it at the point $x = a$ indicating that the function f is undefined

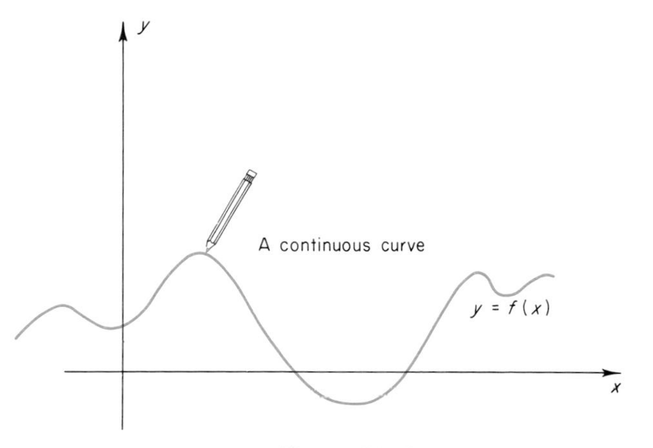

Figure 11.11

there. For the curves in Figures 11.12b and 11.12c, the function f is defined at $x = a$, but

$$\lim_{x \to a} f(x)$$

does not exist, thereby causing a break in the graph. For the curve in (d), f is defined at $x = a$ and $\lim_{x \to a} f(x)$ exists, yet the graph has a break

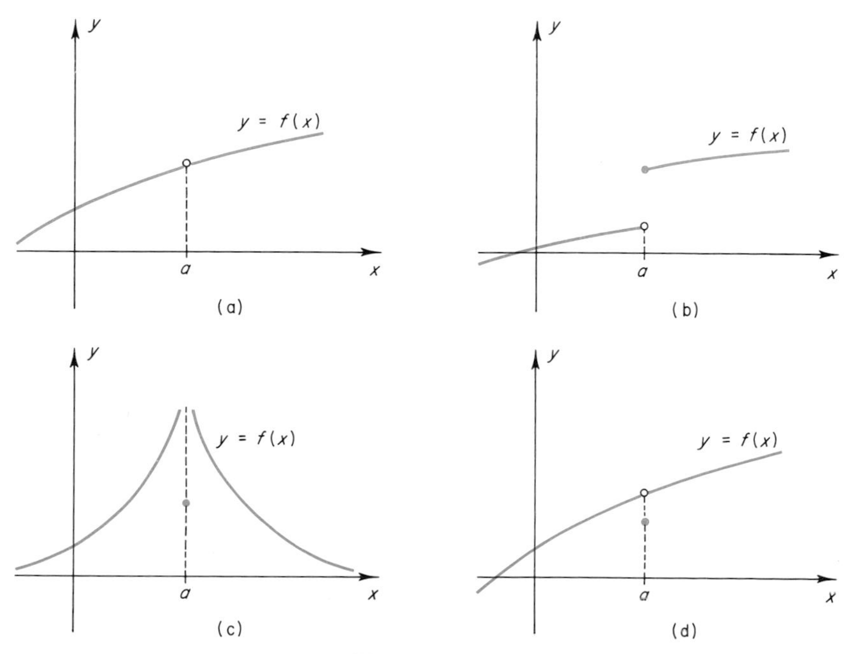

Figure 11.12

at the point $x = a$ because

$$\lim_{x \to a} f(x) \neq f(a).$$

As a result of this discussion, it should be evident that a function must satisfy the following three conditions in order that its graph not have a hole or break at the point $x = a$:

(1) $f(a)$ must be defined;

(2) $\lim_{x \to a} f(x)$ must exist;

(3) $\lim_{x \to a} f(x) = f(a)$.

A function that satisfies these conditions is said to be ***continuous at $x = a$.*** If f is not continuous at $x = a$, we say that f is ***discontinuous at $x = a$*** or that a is a ***point of discontinuity*** for f. A function that is continuous at every point is called ***continuous;*** if there is at least one point of discontinuity, then the function is called ***discontinuous.***

Example 20 The function $f(x) = x$ is continuous because at any point a,

(1) f is defined; the value of f is $f(a) = a$.

(2) $\lim_{x \to a} f(x)$ exists since $\lim_{x \to a} x = a$.

(3) $\lim_{x \to a} f(x) = f(a) = a$.

Example 21 Every constant function is continuous; for if $f(x) = c$ is constant, then at any point a

$$f(a) = c \qquad \text{and} \qquad \lim_{x \to a} f(x) = \lim_{x \to a} c = c.$$

The reader may recall that a ***polynomial*** is a function of the form

$$f(x) = c_0 + c_1x + c_2x^2 + \cdots + c_nx^n$$

where the c's are constants. A polynomial is said to have ***degree n*** if x^n is the highest power of x in the polynomial. A nonzero constant is regarded as a polynomial of degree zero.

Example 22 The following are polynomials:

$f(x) = -3$	degree 0	(constant)
$f(x) = 2x + 5$	degree 1	(linear)

$$f(x) = x^2 - 3x + 7 \qquad \text{degree 2} \qquad \text{(quadratic)}$$

$$f(x) = 7x^3 - 5x^2 + x + 1 \qquad \text{degree 3} \qquad \text{(cubic)}$$

Functions that are ratios of polynomials are called ***rational*** functions. Some examples are:

$$\frac{x+1}{x^2 - x + 2}, \quad \frac{1}{x}, \quad \frac{x^2 - \sqrt{2}\,x + 8}{x+3}, \quad x^3 + 2 \left(= \frac{x^3 + 2}{1}\right).$$

The following results, which can be proved using the limit properties in Section 11.3, will help us to study the continuity properties of polynomials and rational functions.

Continuity Properties

(1) If $f(x)$ and $g(x)$ are continuous at a point, then $f(x) + g(x)$, $f(x) - g(x)$, and $f(x) \cdot g(x)$ are also continuous at that point.

(2) If $f(x)$ and $g(x)$ are continuous at a point a, then $f(x)/g(x)$ is also continuous at a unless $g(a) = 0$, in which case a is a point of discontinuity for $f(x)/g(x)$.

In view of Property 1 of limits and Examples 20 and 21, any function of the form

$$cx^n = c \cdot \underbrace{x \cdot x \cdot \cdots \cdot x}_{n \text{ factors}} \tag{1}$$

is continuous since it is a product of continuous functions. Consequently:

all polynomials are continuous

since they are sums of functions of the form in (1). Thus from Property 4 of limits:

rational functions are continuous except at the points where the denominator has value zero.

Example 23 The functions

$$f(x) = 2 + x$$

$$f(x) = x^2 - 2x + 3$$

$$f(x) = 4 + 5x - 6x^2 + 7x^3$$

are continuous since they are polynomials.

Example 24 The rational function

$$f(x) = \frac{1}{x - 3}$$

is continuous everywhere except at $x = 3$ since the denominator is zero at this point. Note the break in the graph at $x = 3$ (Figure 11.13).

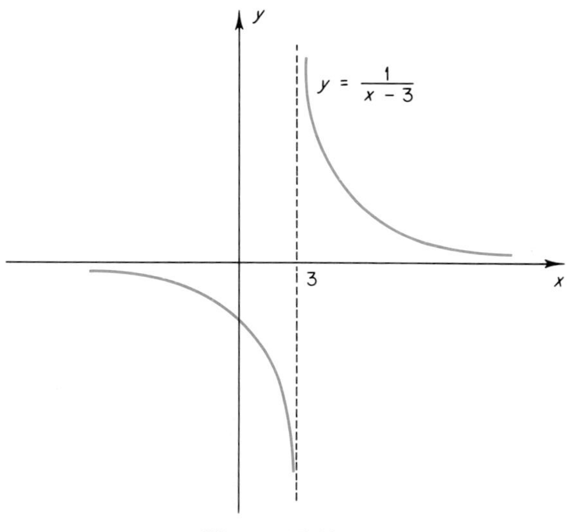

Figure 11.13

EXERCISE SET 11.4

1. In each part determine whether the function whose graph is shown is continuous.

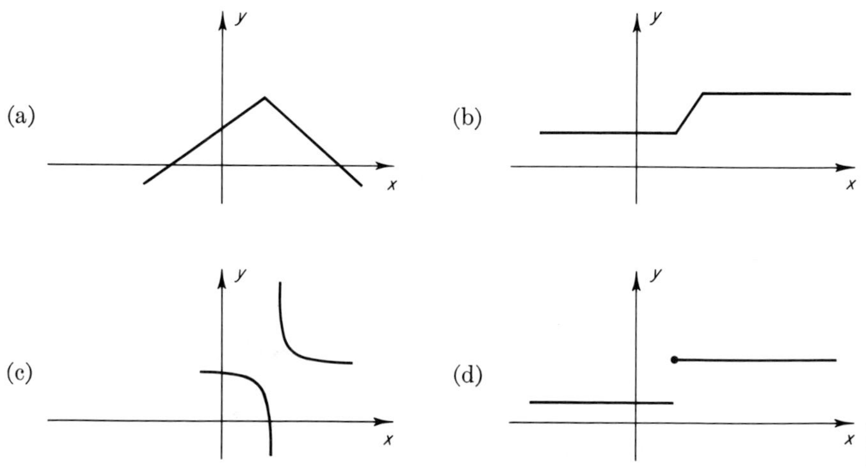

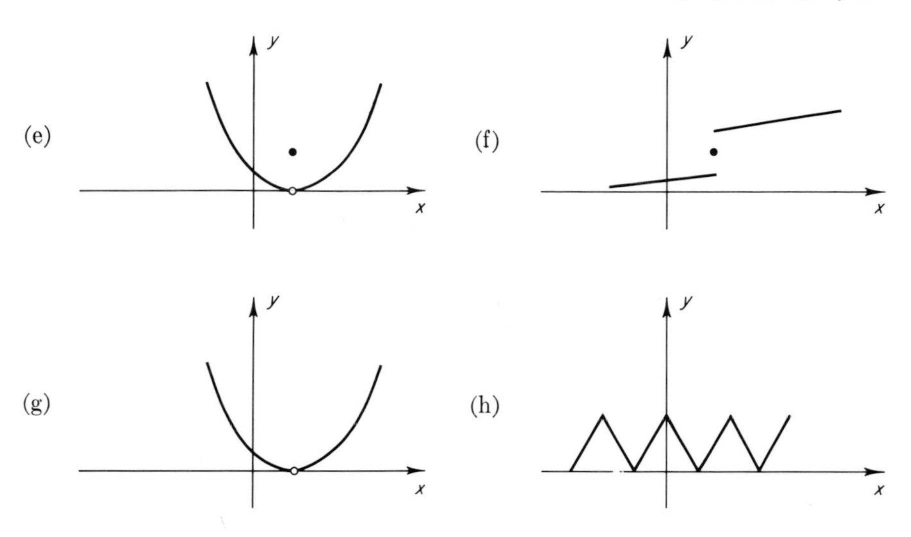

2. In each part, the given function is discontinuous at the point $x = 6$. In each case, state which of the three requirements for continuity is (are) violated at $x = 6$.

(a) $f(x) = \dfrac{x^2 - 36}{x - 6}$

(b) $f(x) = \begin{cases} x + 6 & \text{if} \quad x \neq 6 \\ 0 & \text{if} \quad x = 6, \end{cases}$

(c) $f(x) = \begin{cases} x + 1 & \text{if} \quad x \geq 6 \\ x - 1 & \text{if} \quad x < 6. \end{cases}$

3. Consider the function f defined by

$$f(x) = \begin{cases} \dfrac{x^2 - 36}{x - 6} & \text{if} \quad x \neq 6 \\ 12 & \text{if} \quad x = 6. \end{cases}$$

Is f continuous at $x = 6$? Justify your answer.

4. Consider the function f defined by

$$f(x) = \begin{cases} \dfrac{x^2 - 4}{x - 2} & \text{if} \quad x \neq 2 \\ 8 & \text{if} \quad x = 2. \end{cases}$$

Is f continuous at $x = 2$? Justify.

In Exercises 5–11 determine whether the function is continuous at the indicated point.

5. $f(x) = 2x^2 - 5x + 1; \quad x = 4$

6. $f(x) = \dfrac{1}{2x - 1}; \quad x = \frac{1}{2}$

7. $f(x) = \dfrac{(x - 1)(x - 2)}{(x - 2)}; \quad x = 2$

8. $f(x) = \begin{cases} 2x & \text{if} \quad x \leq 2 \\ 4 & \text{if} \quad x > 2 \end{cases}; \quad x = 2$

9. $f(x) = \begin{cases} 2x & \text{if} \quad x \neq 2 \\ 5 & \text{if} \quad x = 2 \end{cases}; \quad x = 2$

10. $f(x) = \begin{cases} 2x & \text{if} \quad x \leq 2 \\ 3x + 1 & \text{if} \quad x > 2 \end{cases}; \quad x = 2$

11. $f(x) = |x - 3|; \quad x = 3.$

In Exercises 12–20 determine all points of discontinuity, if any, of the given function.

12. $f(x) = 4x^2 - 2x + 5$

13. $f(x) = \dfrac{x^2 - 9}{x - 3}$

14. $f(x) = \dfrac{3}{x - 3}$

15. $f(x) = |x + 1|$

16. $f(x) = \dfrac{3}{(x - 2)(x + 3)}$

17. $f(x) = \begin{cases} 3x & \text{if} \quad x \leq 1 \\ 2x + 1 & \text{if} \quad x > 1 \end{cases}$

18. $f(x) = \dfrac{x}{x^2 - 5x + 6}$

19. $f(x) = \dfrac{2x}{x}$

20. $f(x) = \dfrac{(x - 1)(x + 3)}{4}.$

21. A coffee merchant reorders as soon as his stock falls below 100 pounds. When this occurs, he reorders enough coffee to bring his stock up to 500 pounds. Let $f(x)$ denote the stock on hand at time x. Is f a continuous function? Explain.

11.5 AVERAGE AND INSTANTANEOUS RATE OF CHANGE

Most physical quantities are in a state of constant change—the speed of a rocket changes with time, the cost of an object changes with the available supply, the profits of a manufacturer change with sales, the size of a tumor varies with the quantity of radiation to which it is subjected, and so on. In this section we shall see that there is a close relationship between two apparently unrelated problems, drawing a tangent to a curve and finding the *rate* at which one quantity changes relative to another. The two problems are tied together by the concept of a "derivative" which we begin to develop in this section.

The rate at which one quantity changes relative to another can be illustrated with a familiar example, the *speed* of a moving object. Speed is the rate at which distance traveled changes with time. There are two very different ways to describe the speed of a moving object: *average speed* and *instantaneous speed*. For example, suppose a car moves along a straight road on which we have introduced a coordinate line (Figure

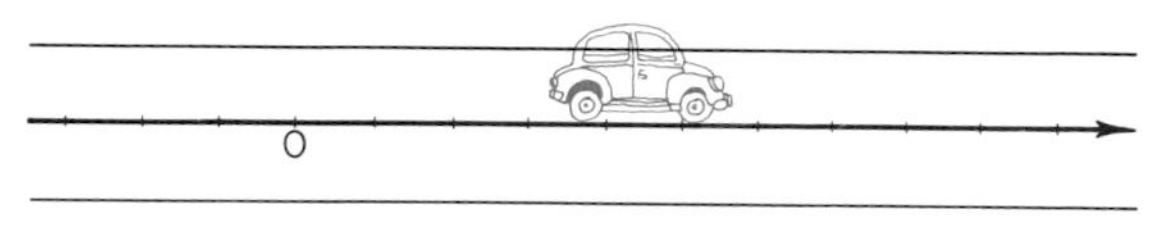

Figure 11.14

11.14) and suppose the data in Table 11.3 are recorded at three check-

Table 11.3

	Distance d from the origin to the checkpoint (miles)	Time t elapsed from start of trip (hours)
A	4	$\frac{1}{4}$
B	10	$\frac{3}{4}$
C	70	$1\frac{3}{4}$

points A, B, and C. To find the ***average speed*** of the car between two checkpoints we would divide the distance traveled between the checkpoints by the time elapsed, that is,

$$\text{average speed} = \frac{\text{distance traveled}}{\text{time elapsed}}.$$

For example, between checkpoints B and C the average speed is

$$\text{average speed} = \frac{70 - 10}{1\frac{3}{4} - \frac{3}{4}} = \frac{60}{1} = 60 \quad \text{mi/hr},$$

while the average speed between checkpoints A and B is

$$\text{average speed} = \frac{10 - 4}{\frac{3}{4} - \frac{1}{4}} = \frac{6}{\frac{1}{2}} = 12 \quad \text{mi/hr}.$$

More generally, if the car is d_1 miles from the origin after t_1 hours and is d_2 miles from the origin after t_2 hours, then its average speed over this period is

$$\text{average speed} = \frac{d_2 - d_1}{t_2 - t_1} \quad \text{mi/hr}.$$

Average speed can be interpreted geometrically as follows. If we make a graph of the distance traveled versus time elapsed, then the average speed between times t_1 and t_2 will be the slope of the line joining the points (t_1, d_1) and (t_2, d_2) on this graph (Figure 11.15).

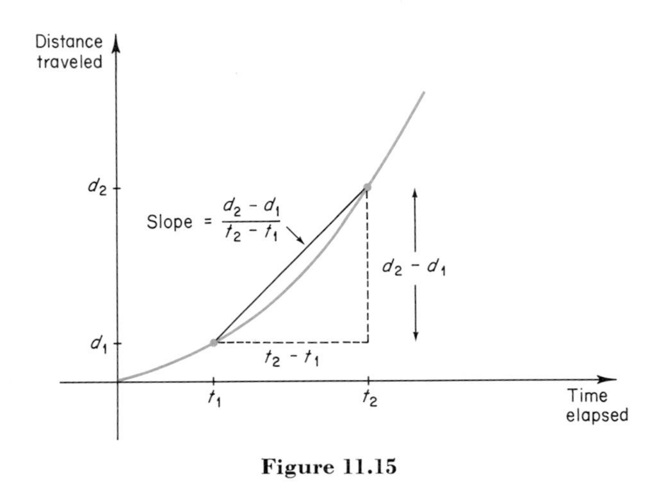

Figure 11.15

The notion of an average rate of change is applicable to problems other than speed. We make the following definition.

Definition If a quantity y is a function of a quantity x and if y changes from y_1 to y_2 as x changes from x_1 to x_2, then the ***average rate of change*** of y with respect to x between x_1 and x_2 is

$$\frac{y_2 - y_1}{x_2 - x_1}. \tag{1}$$

In this definition, if y and x are related by the equation $y = f(x)$, then

$$y_1 = f(x_1) \qquad \text{and} \qquad y_2 = f(x_2);$$

so (1) can be expressed in the alternative form

$$\text{average rate of change between } x_1 \text{ and } x_2 = \frac{f(x_2) - f(x_1)}{x_2 - x_1}. \tag{2}$$

In terms of the graph of $y = f(x)$ this average rate of change can be interpreted as the slope of the line joining the points $A(x_1, y_1)$ and $B(x_2, y_2)$ (Figure 11.16). The line joining A and B is called the ***secant line*** joining (x_1, y_1) and (x_2, y_2).

Example 25 A manufacturer of an industrial liquid determines that the cost in dollars for manufacturing x gallons of liquid is given by the formula

$$C(x) = x^2 - 2x + 5.$$

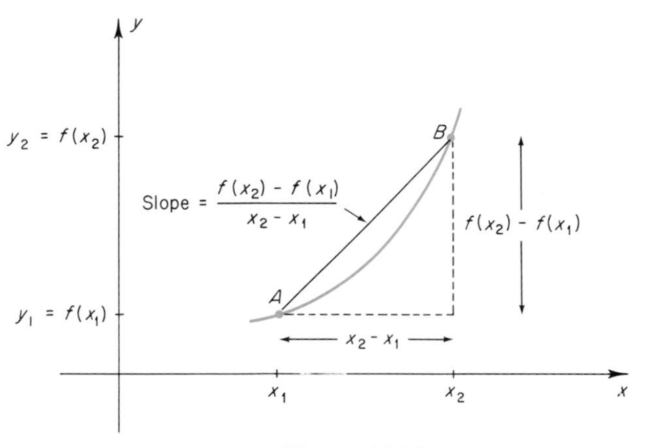

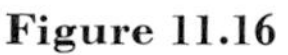

Figure 11.16

The cost for manufacturing 5 gallons is $C(5) = 20$ dollars and the cost for manufacturing 10 gallons is $C(10) = 85$ dollars, so that by increasing production from 5 to 10 gallons the unit cost increases an average of

$$\text{average increase in cost} = \frac{C(10) - C(5)}{10 - 5} = \frac{85 - 20}{5} = \frac{65}{5}$$
$$= 13 \text{ dollars per gallon.}$$

Geometrically, this average increase in cost is the slope of the secant line joining the points (5, 20) and (10, 85) on the graph of $C(x)$ (Figure 11.17).

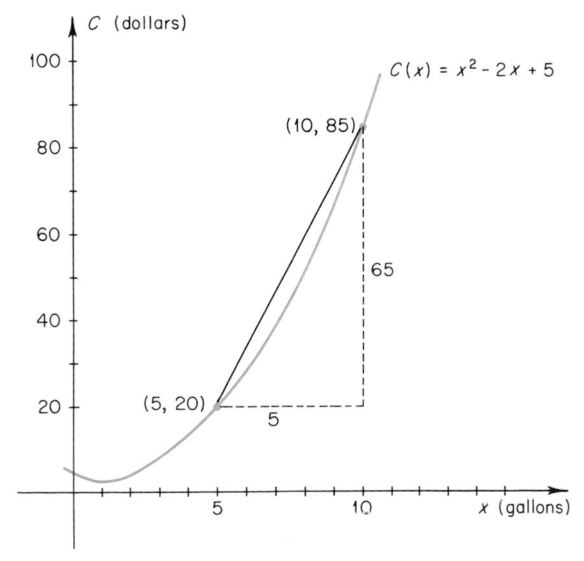

Figure 11.17

Example 26 Following the administration of an experimental drug, a cancerous tumor undergoes a weight change that seems to follow the formula

$$W(t) = -\tfrac{1}{3}t^2 + 3$$

where $W(t)$ is the weight in grams and t is the time elapsed in months. At the end of one month ($t = 1$) the weight of the tumor is $W(1) = \frac{8}{3}$ gm; and at the end of two months ($t = 2$), the weight is $W(2) = \frac{5}{3}$ gm. Thus the average rate of change in weight over this time period is

$$\frac{W(2) - W(1)}{2 - 1} = \frac{\frac{5}{3} - \frac{8}{3}}{1} = -1 \quad \text{gm/month.}$$

The negative sign tells us that the weight has *decreased* over the time period. Geometrically, this average decrease is represented by the slope of the secant line joining the points $(1, \frac{8}{3})$ and $(2, \frac{5}{3})$ on the graph of $W(t)$ (Figure 11.18).

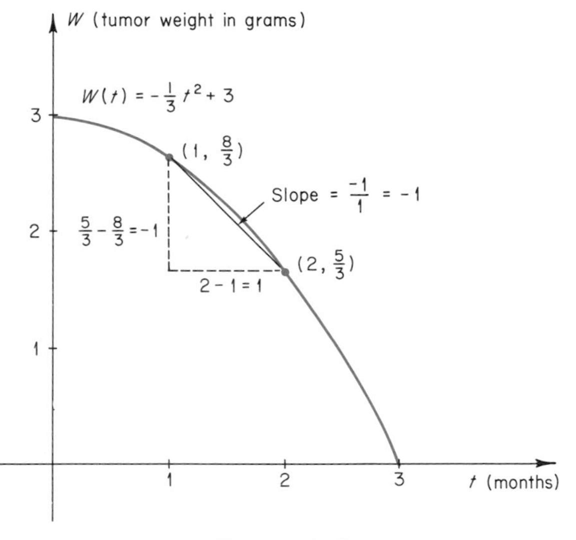

Figure 11.18

While average rate of change is useful in many problems, there are times when we must use another concept, called ***instantaneous rate of change.*** For example, if a moving car strikes a tree, the damage sustained is not determined by the average speed during the trip but rather by the instantaneous speed at the precise moment of impact. This suggests that we try to define the rate at which a quantity is changing at a particular *point* as opposed to its average rate of change over an *interval.*

To motivate this definition we shall consider an example from the physical sciences.

Example 27 It is a known physical fact that a rock dropped from a height falls toward the earth so that the distance fallen (in feet) after t seconds is approximately

$$f(t) = 16t^2.$$

(Figure 11.19). Let us try to calculate the instantaneous speed of the rock precisely 3 seconds after it has been released. Since we are not cal-

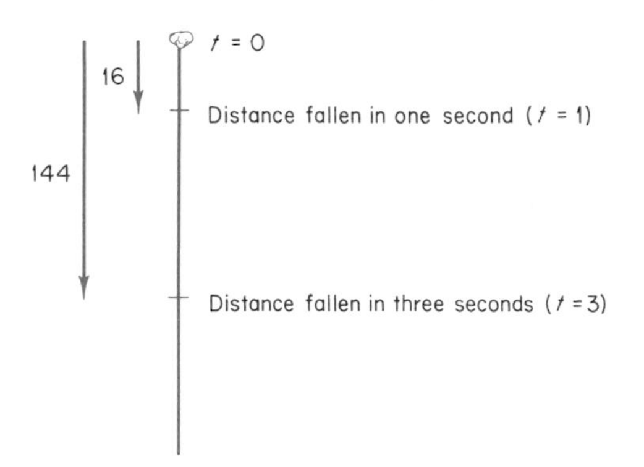

Figure 11.19

culating an average speed, we cannot simply divide distance traveled by the time elapsed. However, we can argue as follows. Over a very short interval of time, say $\frac{1}{10}$ second, the speed of the rock cannot vary much. Thus the error will be small if we *approximate* the instantaneous speed when $t = 3$ by the average speed over the interval from $t = 3$ to $t = 3.1$. The calculation is as follows:

$$\begin{aligned}\text{instantaneous speed when } t = 3 \\ &\approx \text{average speed between } t = 3 \text{ and } t = 3.1 \\ &= \frac{f(3.1) - f(3)}{3.1 - 3} = \frac{16(3.1)^2 - 16(3)^2}{0.1} \\ &= 97.6 \quad \text{ft/sec.}\end{aligned}$$

To improve this approximation we might calculate the average speed over an even smaller interval, say $t = 3$ to $t = 3.01$ or $t = 3$ to $t = 3.001$. The smaller the interval, the better the approximation. In Table 11.4

Table 11.4

Time interval (sec)	Average speed (ft/sec)
$t = 3$ to $t = 3.1$	97.6
$t = 3$ to $t = 3.01$	96.16
$t = 3$ to $t = 3.001$	96.016
$t = 3$ to $t = 3.0001$	96.0016
$t = 3$ to $t = 3.00001$	96.00016

we have summarized the results of several such calculations. Since the average speeds tend toward 96 ft/sec as the interval size decreases, we can reasonably conclude that 96 ft/sec is the instantaneous speed when $t = 3$. We can obtain this same result in another (more precise) way. On each line of Table 11.4, the left-hand side was some interval of the form

$$t = 3 \qquad \text{to} \qquad t = 3 + h$$

(e. g., $h = 0.1$ on line one, $h = 0.01$ on line two, and so forth.) On the right-hand side of the table we computed the average speed over this interval

$$\begin{aligned} \text{average speed} &= \frac{f(3+h) - f(3)}{(3+h) - 3} = \frac{16(3+h)^2 - 16(3^2)}{h} \\ &= \frac{16(9 + 6h + h^2) - 16(9)}{h} \\ &= \frac{144 + 96h + 16h^2 - 144}{h} \\ &= \frac{96h + 16h^2}{h} \\ &= 96 + 16h. \end{aligned}$$

We then observed what happened to this average speed as the interval size decreased, that is, as $h \to 0$

$$\lim_{h \to 0} (96 + 16h) = 96.$$

This agrees with the result conjectured from Table 11.4.

Motivated by this example, we make the following definition.

Definition If a quantity y is a function of a quantity x, say $y = f(x)$, then the ***instantaneous rate of change*** of y with respect to x at the point x_0 is

$$\lim_{h \to 0} \frac{f(x_0 + h) - f(x_0)}{h} \tag{3}$$

Note that

$$\frac{f(x_0 + h) - f(x_0)}{h}$$

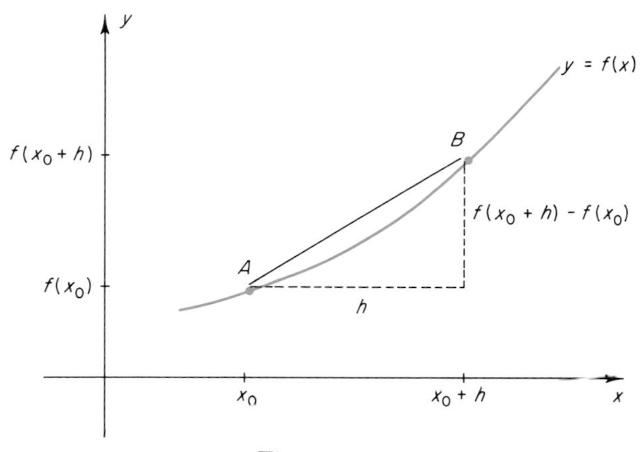

Figure 11.20

is just the average rate of change of y with respect to x over the interval from $x = x_0$ to $x = x_0 + h$ (Figure 11.20). Thus we are defining the instantaneous rate of change as a limit of average rates of change over smaller and smaller intervals.

Instantaneous rate of change has an important geometric interpretation. Since

$$\frac{f(x_0 + h) - f(x_0)}{h}$$

represents the slope of the secant line joining $A(x_0, f(x_0))$ and $B(x_0 + h, f(x_0 + h))$ (Figure 11.20), the instantaneous rate of change at x_0

$$\lim_{h \to 0} \frac{f(x_0 + h) - f(x_0)}{h}$$

can be interpreted as the limit of these slopes as $h \to 0$. However, as $h \to 0$, the point B moves along the graph of f toward A, and the secant lines from A to B tend toward a line called the ***tangent*** line to the graph of f at A (Figure 11.21). Thus we can interpret (3) as the slope of the line tangent to the graph of f at the point $(x_0, f(x_0))$. We call the slope of this tangent, the ***slope of the curve*** $y = f(x)$ at $(x_0, f(x_0))$, or sometimes simply the ***slope*** of $y = f(x)$ at x_0.

Example 28 Let us consider again the manufacturer's problem discussed in Example 25. For each x gallons of liquid produced, the manufacturer's cost is

$$C(x) = x^2 - 2x + 5.$$

As the manufacturer increases the production, the costs change. Let us

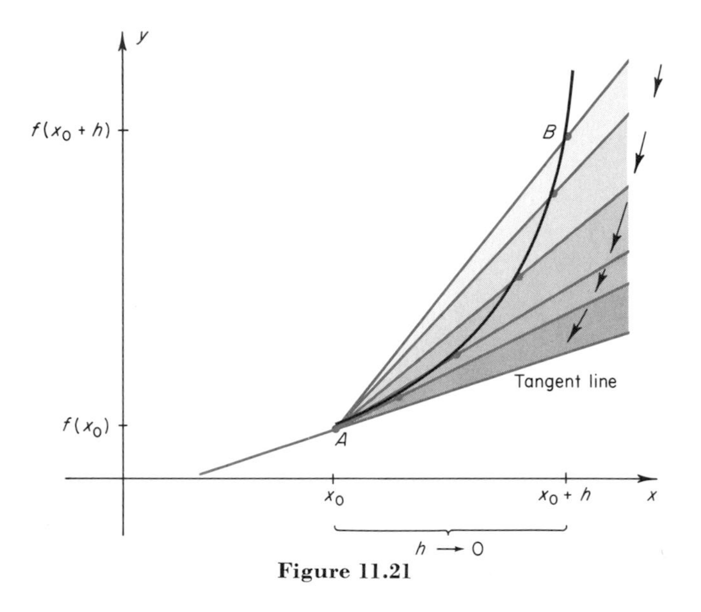

Figure 11.21

try to determine the rate at which the cost will be changing at the instant a production level of $x_0 = 5$ gallons is reached.

Solution From (3) the instantaneous rate of change in cost at a production level of x_0 is

$$\lim_{h \to 0} \frac{C(x_0 + h) - C(x_0)}{h}$$

$$= \lim_{h \to 0} \frac{[(x_0 + h)^2 - 2(x_0 + h) + 5] - [x_0^2 - 2x_0 + 5]}{h}$$

$$= \lim_{h \to 0} \frac{[x_0^2 + 2x_0h + h^2 - 2x_0 - 2h + 5] - [x_0^2 - 2x_0 + 5]}{h}$$

$$= \lim_{h \to 0} \frac{2x_0h + h^2 - 2h}{h}$$

$$= \lim_{h \to 0} (2x_0 + h - 2)$$

$$= 2x_0 - 2.$$

In particular, at a production level of $x_0 = 5$ the cost is increasing at a rate of

$$2x_0 - 2 = 2(5) - 2 = 8 \quad \text{dollars/gallon.}$$

Geometrically, the graph of the cost function $C(x) = x^2 - 2x + 5$ has a tangent of slope 8 at the point where $x_0 = 5$ (Figure 11.22).

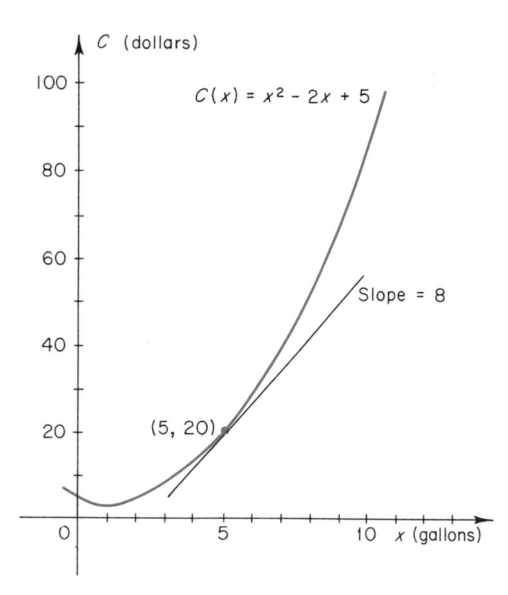

Figure 11.22

EXERCISE SET 11.5

1. Suppose

$$y = 3x + 4.$$

(a) Find the average rate of change of y with respect to x between $x = 0$ and $x = 1$.

(b) Find the average rate of change of y with respect to x between $x = -3$ and $x = 4$.

(c) Find the average rate of change of y with respect to x between any two values $x = a$ and $x = b$.

(d) Explain the result in (c) by graphing $y = 3x + 4$.

2. Suppose

$$y = \tfrac{1}{2}x^2.$$

(a) Find the average rate of change of y with respect to x between $x = 1$ and $x = 4$.

(b) Sketch the graph of the curve and draw a secant line whose slope is the average rate of change obtained in (a).

In Exercises 3–7 find the average rate of change of y with respect to x between x_1 and x_2.

3. $y = 3x - 1;\quad x_1 = 4,\ x_2 = 5$

4. $y = 2x^2 + 3;\quad x_1 = -2,\ x_2 = 6$

5. $y = 3;\quad x_1 = 2,\ x_2 = 4$

6. $y = \dfrac{1}{x};\quad x_1 = -3,\ x_2 = -1$

7. $y = \dfrac{3x^2 - 2}{x - 1};\quad x_1 = 2,\ x_2 = 4.$

8. For the function graphed in Figure 11.23, find the average rate of change of y with respect to x between:

(a) $x_1 = 2$ and $x_2 = 5$ (b) $x_1 = 3$ and $x_2 = 8$

(c) $x_1 = 2$ and $x_2 = 10$ (d) $x_1 = 11$ and $x_2 = 12.$

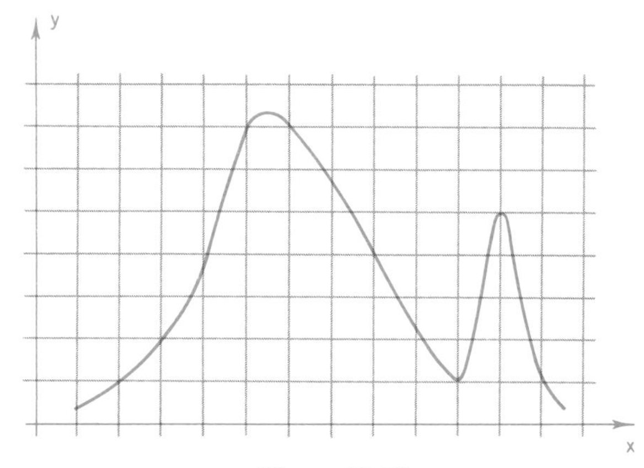

Figure 11.23

In Exercises 9–12 find the instantaneous rate of change of y with respect to x at the given value of x_0.

9. $y = 2x;\quad x_0 = 2$

10. $y = x^2;\quad x_0 = 3$

11. $y = x^2 + 1;\quad x_0 = 0$

12. $y = 3x^2 - 2;\quad x_0 = 4.$

13. As discussed in Example 27, a rock dropped from a height falls to earth so that the distance fallen (in feet) after t seconds is approximately

$$f(t) = 16t^2.$$

(a) Find the distance traveled by the rock between the third and sixth seconds after release (that is, between $t = 3$ and $t = 6$).

(b) Find the average speed during this time interval.

(c) Find the instantaneous speed at $t = 3$.

(d) Find the instantaneous speed at $t = 4$.

14. **(Manufacturing)** Consider the manufacturing problem in Example 28.

(a) Find the average rate of change of cost when the production level increases from 5 to 10 gallons.

(b) On the graph of $C(x)$, draw a secant line whose slope is the average rate of change found in part (a).

(c) Find the instantaneous rate of change in cost at a production level of $x_0 = 8$ gallons.

(d) On the graph of $C(x)$, draw a tangent line whose slope is the instantaneous rate of change found in part (b).

15. **(Population Growth)** The following table lists population statistics for the city of San Francisco, California (based on data from the *World Almanac*).

Year	Population
1900	342,782
1950	775,357
1960	740,316
1970	715,674

(a) What is the average rate of change of population with respect to time between the years 1900 and 1970?

(b) What is the average rate of change of population with respect to time between the years 1960 and 1970?

(c) On the average, was the population declining more rapidly over the ten year span between 1960 and 1970 or over the twenty year span between 1950 and 1970?

16. **(Epidemics)** In an outbreak of σ-type malaria, the total number of cases reported over a 10-day period is indicated in the following table.

Day	Total number of cases reported to date	Day	Total number of cases reported to date
1	20	6	245
2	80	7	247
3	140	8	250
4	220	9	250
5	230	10	250

(a) Make a graph to describe these data.

(b) What is the average rate of change in the total number of cases reported with respect to time between day 2 and day 4 inclusive?

(c) Over which consecutive two days does the largest average rate of increase in the number of cases reported occur?

(d) Over the last two days, what is the average rate of change in total number of cases reported?

(e) What does the result in (d) tell you?

17. (**Advertising**) The graph in Figure 11.24 illustrates a response curve for a college fund-raising campaign. With an amount x spent on advertising, the college receives donations amounting to $D(x)$.

(a) Find the average rate of change of D as x changes from

0	to	10	20	to	30
10	to	20	30	to	40

(b) Can you explain, from a practical point of view why one would anticipate these rates of change to be decreasing?

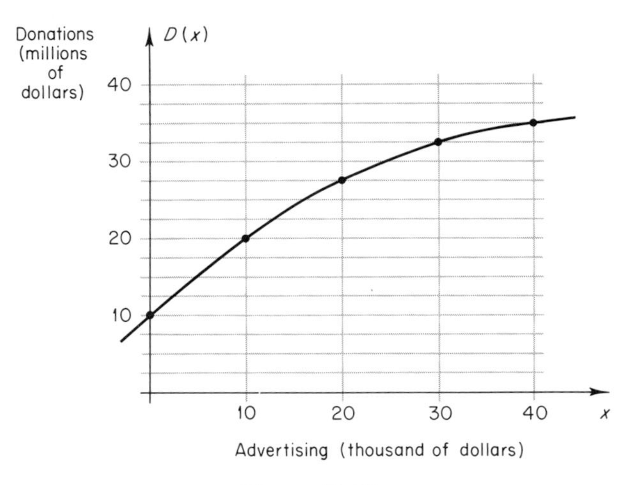

Figure 11.24

12
THE DERIVATIVE

12.1 THE DERIVATIVE

The notion of instantaneous rate of change leads to one of the cornerstones of calculus, the derivative. In this section we introduce this idea and develop some more efficient techniques for computing instantaneous rates of change and slopes of tangents.

In the last section of Chapter 11 we saw that the expression

$$\lim_{h\to 0}\frac{f(x_0+h)-f(x_0)}{h} \tag{1}$$

could be interpreted either as the slope of the tangent to $y = f(x)$ at the point $(x_0, f(x_0))$ or as the instantaneous rate of change of y with respect to x at this point. If we replace the fixed value x_0 in this expression by a variable x, we obtain an important function:

Definition The ***derivative*** of a function f is denoted by $f'(x)$ and is defined by

$$f'(x) = \lim_{h\to 0}\frac{f(x+h)-f(x)}{h} \tag{2}$$

To compute the derivative of f is to ***differentiate*** f, and the process of obtaining a derivative is called ***differentiation.***

Note that if we substitute a specific value for x in (2), say $x = x_0$, we obtain

$$f'(x_0) = \lim_{h \to 0} \frac{f(x_0 + h) - f(x_0)}{h} = \text{slope of tangent to } y = f(x) \text{ at } x_0$$

Thus we may think of the derivative $f'(x)$ as a function whose value at a point x_0 is the slope of the curve $y = f(x)$ at x_0.

Example 1 Let f be defined by $f(x) = x^2$. Then the derivative of f is

$$\begin{aligned} f'(x) = \lim_{h \to 0} \frac{f(x+h) - f(x)}{h} &= \lim_{h \to 0} \frac{(x+h)^2 - x^2}{h} \\ &= \lim_{h \to 0} \frac{x^2 + 2xh + h^2 - x^2}{h} \\ &= \lim_{h \to 0} (2x + h) \\ &= 2x. \end{aligned}$$

By substituting various specific values for x in the derivative

$$f'(x) = 2x$$

we can obtain the slope of the tangent to the graph of $f(x) = x^2$ at various points. For example, the slopes of the tangent lines to $f(x) = x^2$ at $x = 2$, $x = 0$, and $x = -2$ are

$$\begin{aligned} f'(2) &= 2(2) = 4 \\ f'(0) &= 2(0) = 0 \\ f'(-2) &= 2(-2) = -4 \end{aligned}$$

(Figure 12.1).

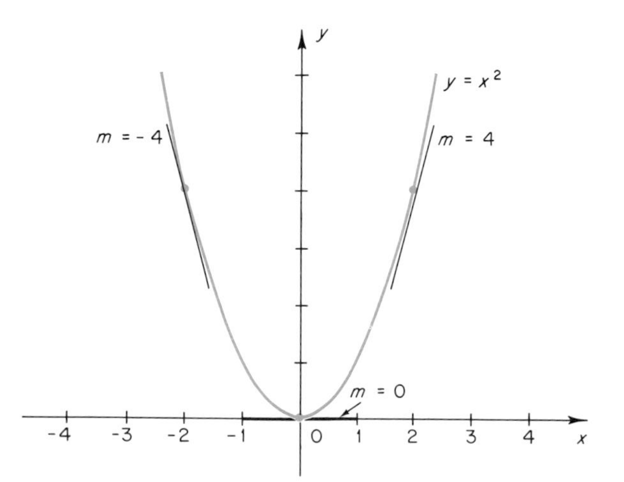

Figure 12.1

Some other commonly used notations for the derivative $f'(x)$ are

$$D_x(f(x)) \qquad \text{and} \qquad \frac{d}{dx}(f(x)).$$

For example, we showed earlier that $f'(x) = 2x$ if $f(x) = x^2$. In the above notations we could write

$$D_x(x^2) = 2x \qquad \text{or} \qquad \frac{d}{dx}(x^2) = 2x.$$

If we write $y = f(x)$, then $f'(x)$ can be written

$$\frac{dy}{dx} \qquad \text{or} \qquad y'.$$

Thus, if $y = x^2$, then we might write

$$y' = 2x \qquad \text{or} \qquad \frac{dy}{dx} = 2x.$$

The last notation is especially useful when variables other than x and y are used. For example, if $u = f(t)$, the derivative of f might be denoted

$$\frac{du}{dt};$$

and if $A = h(r)$, the derivative of h might be denoted

$$\frac{dA}{dr}.$$

For example, if $u = t^2$, then

$$\frac{du}{dt} = 2t.$$

Before proceeding with more examples, let us summarize the procedure for calculating a derivative $f'(x)$.

Step 1 Calculate $f(x + h)$.

Step 2 Form the ratio

$$\frac{f(x+h) - f(x)}{h}$$

(sometimes called the ***difference quotient***).

Step 3 Find

$$\lim_{h \to 0} \frac{f(x+h) - f(x)}{h}$$

if it exists.

Example 2 It is obvious geometrically that a tangent at any point to a line $y = ax + b$ will coincide with the line and will therefore have slope a (Figure 12.2). As a result, we should suspect that the derivative of

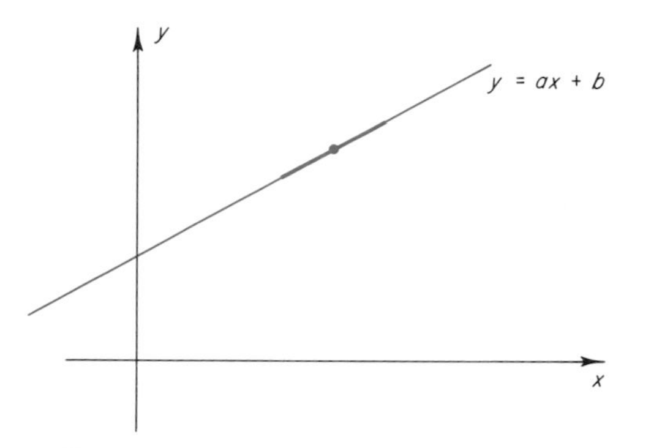

Figure 12.2 Each tangent coincides with the line.

$f(x) = ax + b$ will be constant with value a at each point. Let us see that this is in fact the case:

$$\begin{aligned}
f'(x) &= \lim_{h \to 0} \frac{f(x+h) - f(x)}{h} \\
&= \lim_{h \to 0} \frac{[a(x+h) + b] - [ax + b]}{h} \\
&= \lim_{h \to 0} \frac{ax + ah + b - ax - b}{h} \\
&= \lim_{h \to 0} \frac{ah}{h} \\
&= \lim_{h \to 0} a \\
&= a \qquad \text{(the limit of a constant).}
\end{aligned}$$

Thus $f'(x) = a$ as we suspected.

Example 3 If f is a constant function, say $f(x) = b$, then the graph of f is a horizontal line (slope 0). Thus from the last example

$$f'(x) = 0.$$

In brief,

> The derivative of a constant is zero.

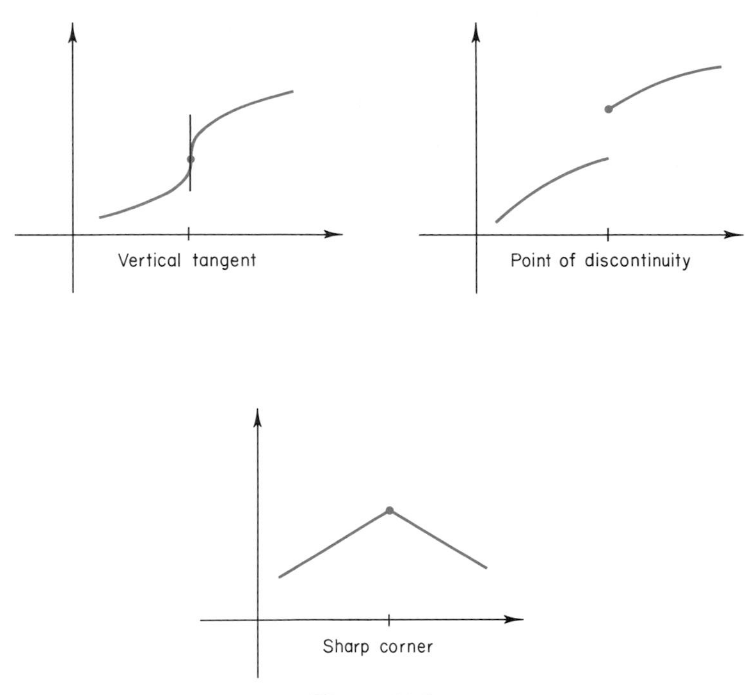

Figure 12.3

It is possible that the limit in the derivative definition may not exist for certain values of x, in which case the derivative of f will not be defined at those points. Some examples of such points are (Figure 12.3):

(a) points where f has a vertical tangent (so that the tangent has infinite slope),

(b) points where f is discontinuous,

(c) points where the graph of f has a sharp corner.

A few exercises at the end of this section deal with this topic. As a matter of terminology, a function f is called ***differentiable at a*** if the derivative of f exists at $x = a$. A function that is differentiable at every point of its domain is called a ***differentiable function.***

In the remainder of this section we shall discuss five rules that will simplify the calculation of derivatives. We shall not prove these rules, but rather concentrate on examples illustrating how they are used.

The Power Rule For any real value of r,

$$\frac{d}{dx}(x^r) = rx^{r-1}.$$

In other words, the derivative of x raised to a power is the power times x raised to the power less 1.

Example 4

$$\frac{d}{dx}(x^6) = 6x^{6-1} = 6x^5.$$

$$\frac{d}{dx}(x^2) = 2x^{2-1} = 2x^1 = 2x \qquad \text{(see Example 1).}$$

$$\frac{d}{dx}(x) = \frac{d}{dx}(x^1) = 1x^{1-1} = 1x^0 = 1 \cdot 1 = 1.$$

$$\frac{d}{dx}(\sqrt{x}) = \frac{d}{dx}(x^{1/2}) = \frac{1}{2}x^{(1/2)-1} = \frac{1}{2}x^{-1/2} = \frac{1}{2\sqrt{x}}.$$

$$\frac{d}{dx}(x^{-2/3}) = -\frac{2}{3}x^{-(2/3)-1} = -\frac{2}{3}x^{-5/3}.$$

The Constant Times a Function Rule For any constant k,

$$\frac{d}{dx}[kf(x)] = k\frac{d}{dx}[f(x)].$$

In other words, a constant k can be "moved" through the derivative sign.

Example 5

$$\frac{d}{dx}(6x^4) = 6\frac{d}{dx}(x^4) = 6 \cdot 4x^3 = 24x^3.$$

$$\frac{d}{dx}\left(-\frac{1}{7}x^2\right) = -\frac{1}{7}\frac{d}{dx}(x^2) = -\frac{1}{7} \cdot 2x = -\frac{2}{7}x.$$

$$\frac{d}{dx}\left(-\frac{4}{x^{3/2}}\right) = \frac{d}{dx}(-4x^{-3/2}) = -4 \cdot \frac{d}{dx}(x^{-3/2})$$

$$= -4 \cdot \left(-\frac{3}{2}x^{-3/2-1}\right) = 6x^{-5/2}.$$

$$\frac{d}{dx}(\pi x) = \pi \cdot \frac{d}{dx}(x) = \pi \cdot 1 = \pi.$$

$$\frac{d}{dx}(-x^3) = \frac{d}{dx}(-1 \cdot x^3) = -1 \cdot \frac{d}{dx}(x^3) = -3x^2.$$

The Sum and Difference Rule

$$\frac{d}{dx}[f(x) + g(x)] = \frac{d}{dx}[f(x)] + \frac{d}{dx}[g(x)],$$

$$\frac{d}{dx}[f(x) - g(x)] = \frac{d}{dx}[f(x)] - \frac{d}{dx}[g(x)].$$

In other words, the derivative of a sum is the sum of the derivatives and the derivative of a difference is the difference of the derivatives.

Example 6

$$\frac{d}{dx}(x^5 + x^4) = \frac{d}{dx}(x^5) + \frac{d}{dx}(x^4) = 5x^4 + 4x^3.$$

$$\frac{d}{dx}(1 - x^3) = \frac{d}{dx}(1) - \frac{d}{dx}(x^3) = 0 - 3x^2 = -3x^2.$$

$$\begin{aligned}\frac{d}{dx}(3x^4 - 2x^3 + 4x^2 - 3x + 2) &= 3 \cdot 4x^3 - 2 \cdot 3x^2 + 4 \cdot 2x - 3 + 0 \\ &= 12x^3 - 6x^2 + 8x - 3.\end{aligned}$$

In light of the sum and difference rule, one might conjecture that the derivative of a product of two functions is the product of their derivatives. However, this is false in general. For example, we have

$$x^6 = x^4 \cdot x^2$$

yet

$$\frac{d}{dx}(x^6) \neq \frac{d}{dx}(x^4) \cdot \frac{d}{dx}(x^2)$$

since

$$\frac{d}{dx}(x^6) = 6x^5$$

while

$$\frac{d}{dx}(x^4) \cdot \frac{d}{dx}(x^2) = 4x^3 \cdot 2x = 8x^4.$$

The correct rule is more complicated.

The Product Rule

$$\frac{d}{dx}[f(x) \cdot g(x)] = f(x) \cdot \frac{d}{dx}[g(x)] + g(x) \cdot \frac{d}{dx}[f(x)].$$

In other words, the derivative of a product of two functions is the first function times the derivative of the second plus the second function times the derivative of the first.

Example 7 Since $x^6 = x^4 \cdot x^2$, we should obtain the same result whether we differentiate x^6 by the power rule or differentiate $x^4 \cdot x^2$ by the product rule. This is the case since

$$\begin{aligned}\frac{d}{dx}(x^4 \cdot x^2) &= x^4 \frac{d}{dx}(x^2) + x^2 \frac{d}{dx}(x^4) \\ &= x^4(2x) + x^2(4x^3) \\ &= 2x^5 + 4x^5 \\ &= 6x^5 \\ &= \frac{d}{dx}(x^6).\end{aligned}$$

Example 8

$$\begin{aligned}&\frac{d}{dx}[(x^5 + 2x^3 - 5)(3x^4 - 2x^2 + 8)] \\ &\quad = (x^5 + 2x^3 - 5) \cdot \frac{d}{dx}(3x^4 - 2x^2 + 8) \\ &\qquad\qquad + (3x^4 - 2x^2 + 8) \cdot \frac{d}{dx}(x^5 + 2x^3 - 5) \\ &\quad = (x^5 + 2x^3 - 5)(12x^3 - 4x) + (3x^4 - 2x^2 + 8)(5x^4 + 6x^2).\end{aligned}$$

If desired, we could rewrite this derivative by multiplying out and collecting terms. Also, we could have obtained the derivative by multiplying out the factors $x^5 + 2x^3 - 5$ and $3x^4 - 2x^2 + 8$ before differentiating. However, this is unnecessary extra work.

Just as the derivative of a product is not the product of the derivatives, so the derivative of a quotient is not the quotient of the derivatives. The correct rule is as follows.

The Quotient Rule

$$\frac{d}{dx}\left[\frac{f(x)}{g(x)}\right] = \frac{g(x) \cdot \frac{d}{dx}[f(x)] - f(x) \cdot \frac{d}{dx}[g(x)]}{[g(x)]^2},$$

In other words, the derivative of a quotient is the denominator times the derivative of the numerator minus the numerator times the derivative of the denominator, all divided by the square of the denominator.

Example 9 Since $x^4 = x^6/x^2$ we should obtain the same result whether we differentiate x^4 by the power rule or differentiate x^6/x^2 by the quotient rule. This is the case since

$$\begin{aligned}\frac{d}{dx}\left(\frac{x^6}{x^2}\right) &= \frac{x^2 \cdot \frac{d}{dx}(x^6) - x^6 \cdot \frac{d}{dx}(x^2)}{(x^2)^2} \\ &= \frac{x^2(6x^5) - x^6(2x)}{x^4} \\ &= \frac{6x^7 - 2x^7}{x^4} \\ &= \frac{4x^7}{x^4} \\ &= 4x^3 \\ &= \frac{d}{dx}(x^4).\end{aligned}$$

Example 10 If

$$y = \frac{3x^2 - 1}{x + 2},$$

then by the quotient rule

$$\begin{aligned}\frac{dy}{dx} &= \frac{(x + 2) \cdot \frac{d}{dx}(3x^2 - 1) - (3x^2 - 1) \cdot \frac{d}{dx}(x + 2)}{(x + 2)^2} \\ &= \frac{(x + 2)(6x) - (3x^2 - 1)(1)}{(x + 2)^2};\end{aligned}$$

or if desired we could multiply out and write

$$\frac{dy}{dx} = \frac{3x^2 + 12x + 1}{x^2 + 4x + 4}.$$

EXERCISE SET 12.1

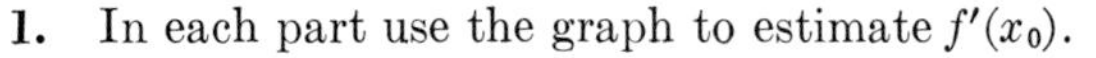

1. In each part use the graph to estimate $f'(x_0)$.

(a)

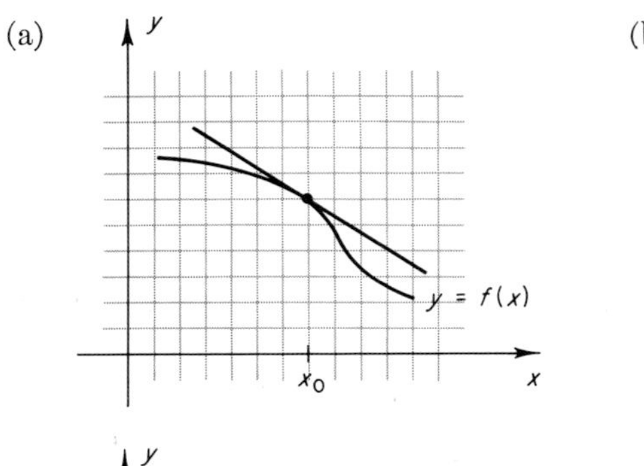

(b)

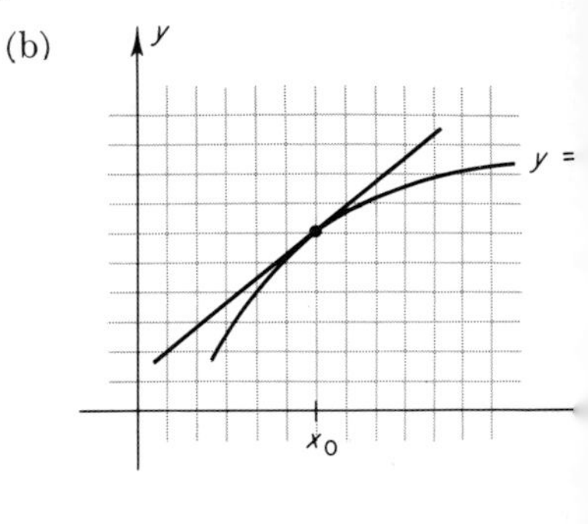

(c)

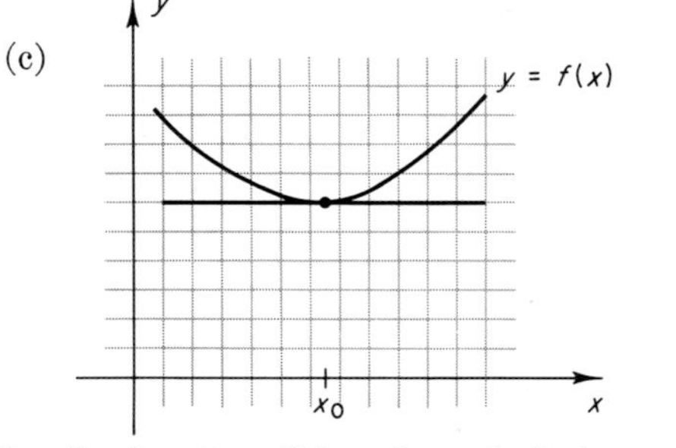

(d)

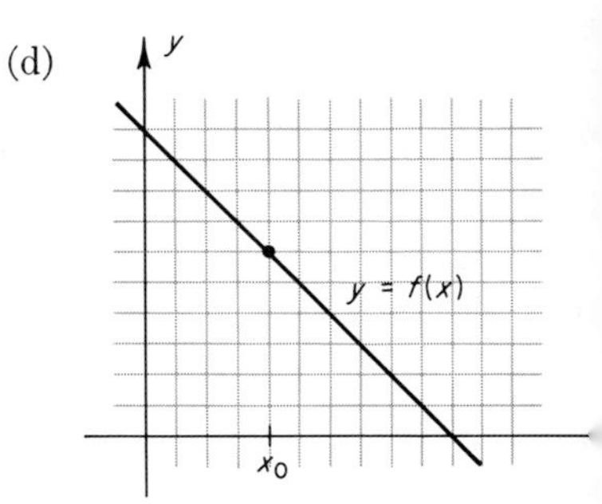

2. For the function $f(x) = 2x - 3$, find:

(a) $\dfrac{f(x+h) - f(x)}{h}$

(b) $\lim\limits_{h\to 0} \dfrac{f(x+h) - f(x)}{h}$

(c) $f'(-1)$, $f'(0)$, and $f'(2)$.

3. For the function $f(x) = \frac{1}{2}x^2$ find:

(a) $\dfrac{f(x+h) - f(x)}{h}$.

(b) $f'(x)$.

(c) $f'(-2)$, $f'(0)$, and $f'(4)$.

(d) Using graph paper, carefully plot the graph of f and draw approximate tangent lines at the points where $x = -2$, $x = 0$, and $x = 4$.

(e) Are the slopes of your tangent lines in reasonable agreement with the calculations in part (c)?

In Exercises 4–7 calculate the difference quotient

$$\frac{f(x+h) - f(x)}{h}$$

and then obtain $f'(x)$ by computing the limit of the difference quotient as $h \to 0$.

4. $f(x) = 2x + 1$

5. $f(x) = 2x^2 + 3$

6. $f(x) = 6$

7. $f(x) = \dfrac{1}{x}$.

In Exercises 8–29 use the rules of differentiation discussed in this section to find $f'(x)$.

8. $f(x) = x^6$.

9. $f(x) = 2x + 3$

10. $f(x) = 3x^4$

11. $f(x) = -2$

12. $f(x) = 3x^3 - 2x^2 + x - 8$

13. $f(x) = 5\sqrt{x}$

14. $f(x) = \dfrac{2}{x^2} + 3x^3$

15. $f(x) = x^{1/3} - x^{-1/2}$

16. $f(x) = 1 + x^{0.08}$

17. $f(x) = \dfrac{3}{\sqrt[4]{x}}$

18. $f(x) = x^2 \cdot x^9$ (two ways)

19. $f(x) = (x - 2)(x + 1)$ (two ways)

20. $f(x) = (3x^2 - 2x)(2x^3 + 1)$

21. $f(x) = (x^2 + 5)(x^2 - 2x + 3)$

22. $f(x) = (2x + \sqrt{x})(x^3 + 1)$

23. $f(x) = (x^{1/3} + 7x)(x^{10} - 9x^8 + 1)$

24. $f(x) = \dfrac{x^{11}}{x^5}$ (two ways)

25. $f(x) = \dfrac{1}{x}$ (two ways)

26. $f(x) = \dfrac{x}{100 - x}$

27. $f(x) = \dfrac{x^2 + 1}{x^2 - 1}$

28. $f(x) = \dfrac{3x^2 - 2x + 1}{2x^3 + 1}$

29. $f(x) = \dfrac{\sqrt{x} + x}{2 + 3\sqrt{x}}$.

30. Suppose $f(x) = 3x^2 - 2x + 1$. Compute $f'(2)$, $f'(0)$, $f'(-2)$.

31. Suppose

$$f(x) = \frac{3x - 1}{2x + 5}.$$

Compute $f'(0)$, $f'(1)$, $f'(-2)$.

32. Find the slope of the tangent to the graph of $f(x) = x^2 + x$ at the point where:

(a) $x = 0$ (b) $x = 1$ (c) $x = -1$ (d) $x = 2$.

33. Find the slope of the tangent to the graph of $f(x) = 1/x$ at the point where:

(a) $x = -1$ (b) $x = 1$ (c) $x = 3$ (d) $x = -3$.

34. Suppose f is defined by $f(x) = 2x^2 - 3x + 4$. Find the point on the graph of f where the slope of the tangent line is 9.

35. Suppose f is defined by $f(x) = 3x^2 - 2x - 5$. Find the point on the graph of f where the tangent line is parallel to the line $y = x + 2$.

36. Suppose f is defined by $f(x) = 2x^3 + 3x^2 - 12x + 5$. Find the points on the graph of f where the tangent line is horizontal.

37. (a) Suppose $A = 3r^2$; find $\dfrac{dA}{dr}$.

(b) Find $\dfrac{d}{dt}(3t^2)$.

(c) Find y' if $y = 3x^2$.

(d) Find $\dfrac{d}{ds}(3s^2)$.

(e) Find $\dfrac{du}{dt}$ if $u = 3t^2$.

(f) Find $\dfrac{dC}{dx}$ if $C(x) = x^2$.

(g) Find $D_x(x^7 + 1)$.

38. Find dy/dx if

(a) $y = 3x^3 + x^2 - 2x + 4$ (b) $y = (2x^2 - 1)(3x^3 + 2x)$.

39. Find dy/dx if

(a) $y = 3\sqrt{x} + 2x^3$ (b) $y = \dfrac{3}{x^2} + 2x$.

40. List the points where the function graphed in Figure 12.4 is not differentiable.

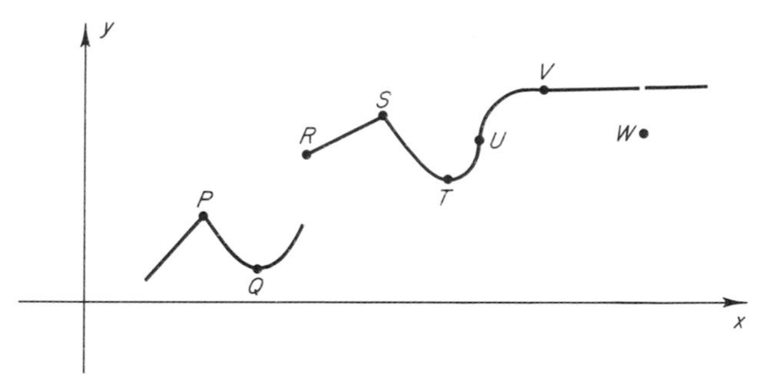

Figure 12.4

41. Graph the following functions and find the values of x where the functions are not differentiable.

(a) $f(x) = |x - 2|$ (b) $f(x) = \dfrac{1}{x}$

(c) $f(x) = \dfrac{|x|}{x}$ (d) $f(x) = \begin{cases} 3x + 1 & \text{if } x \geq 1 \\ 4x & \text{if } x < 1. \end{cases}$

42. For transporting a load of goods, a trucking firm charges a service fee of

\$100 and a mileage fee of \$3 per mile for the first 100 miles or less and \$2 for each additional mile.

(a) Graph the cost $C(x)$ of shipping a load of goods x miles.

(b) Is $C(x)$ a continuous function?

(c) Is $C(x)$ a differentiable function?

12.2 THE CHAIN RULE

If we were interested in the derivative of

$$y = (1 + x)^3,$$

we could expand the right-hand side to obtain

$$y = 1 + 3x + 3x^2 + x^3$$

and then compute

$$\frac{dy}{dx} = 3 + 6x + 3x^2. \tag{1}$$

However, this approach would not be practical if we want to differentiate $y = (1 + x)^{500}$. We need other methods. Another simple function we cannot yet differentiate unless we resort to the derivative definition is

$$f(x) = (1 + x)^{1/2} = \sqrt{1 + x}.$$

In this section we introduce an important result, called the "chain rule," that will greatly expand the class of functions we can differentiate.

The basic idea behind the chain rule is to "decompose" functions we cannot differentiate into functions we can differentiate. To illustrate what we have in mind, let us reexamine the problem of obtaining the derivative dy/dx of

$$y = (1 + x)^3. \tag{2}$$

Let us define a new variable u by

$$u = 1 + x. \tag{3}$$

From (2) the expression for y in terms of u is

$$y = u^3. \tag{4}$$

But expressions (3) and (4) are easy to differentiate:

$$\frac{du}{dx} = 1 \qquad \text{and} \qquad \frac{dy}{du} = 3u^2.$$

We now ask if there is some way to relate these easy derivatives to the harder derivative dy/dx. The basic idea is simple: the derivative du/dx

is the rate at which u varies with x, and the derivative dy/du is the rate at which y varies with u; thus the rate dy/dx at which y varies with x is

$$\frac{dy}{dx} = \frac{dy}{du} \cdot \frac{du}{dx} \tag{5}$$

(For example, if u varies twice as fast as x and y varies three times as fast as u, then y varies $3 \cdot 2 = 6$ times as fast as x.) Formula (5) is called the ***chain rule.***

If we apply the chain rule to our problem above we obtain

$$\frac{dy}{dx} = \frac{dy}{du} \cdot \frac{du}{dx} = 3u^2 \cdot 1 = 3u^2$$

or using (3) to express the answer in terms of x

$$\frac{dy}{dx} = 3(1 + x)^2 = 3(1 + 2x + x^2) = 3 + 6x + 3x^2.$$

Note that this agrees with result (1) obtained by first expanding $(1 + x)^3$.

Example 11 Find dy/dx if

$$y = (1 + x^2)^{100}.$$

Solution Let

$$u = 1 + x^2$$

so that

$$y = u^{100}.$$

We have

$$\frac{du}{dx} = \frac{d}{dx}(1 + x^2) = 2x$$

$$\frac{dy}{du} = \frac{d}{du}(u^{100}) = 100u^{99} = 100(1 + x^2)^{99}$$

so that from the chain rule

$$\frac{dy}{dx} = \frac{dy}{du} \cdot \frac{du}{dx} = 100(1 + x^2)^{99} \cdot 2x = 200x(1 + x^2)^{99}.$$

Let us summarize the steps involved in using the chain rule to differentiate $y = f(x)$.

Step 1	Try to introduce a new variable u (depending on x) in such a way that dy/du and du/dx are easy to calculate.
Step 2	Calculate dy/du and du/dx.
Step 3	Express dy/du in terms of x.
Step 4	Calculate dy/dx from the chain rule.

Example 12 Find

$$\frac{d}{dx}(\sqrt{x+1}).$$

Solution Let

$$y = \sqrt{x+1}$$

and define

$$u = x + 1$$

so that

$$y = \sqrt{u} = u^{1/2}.$$

Thus

$$\frac{du}{dx} = \frac{d}{dx}(x+1) = 1$$

$$\frac{dy}{du} = \frac{d}{du}(u^{1/2}) = \frac{1}{2}u^{-1/2} = \frac{1}{2\sqrt{u}} = \frac{1}{2\sqrt{x+1}}.$$

Applying the chain rule gives

$$\frac{d}{dx}(\sqrt{x+1}) = \frac{dy}{dx} = \frac{dy}{du}\cdot\frac{du}{dx} = \frac{1}{2\sqrt{x+1}}\cdot 1 = \frac{1}{2\sqrt{x+1}}.$$

Example 13 Find $h'(x)$ if

$$h(x) = \left(\frac{x}{1+x}\right)^{1/3}.$$

Solution Let $y = h(x)$ and define

$$u = \frac{x}{1+x},$$

so that

$$y = u^{1/3}$$

Thus

$$\frac{du}{dx} = \frac{d}{dx}\left(\frac{x}{1+x}\right) = \frac{(1+x)(1) - x(1)}{(1+x)^2} \qquad \text{(quotient rule)}$$

$$= \frac{1}{(1+x)^2}$$

and

$$\frac{dy}{du} = \frac{d}{du}(u^{1/3}) = \frac{1}{3}u^{-2/3} = \frac{1}{3}\left(\frac{x}{1+x}\right)^{-2/3}$$

Therefore, from the chain rule

$$h'(x) = \frac{dy}{dx} = \frac{dy}{du}\cdot\frac{du}{dx} = \frac{1}{3}\left(\frac{x}{1+x}\right)^{-2/3}\cdot\frac{1}{(1+x)^2}.$$

In the above examples we used the chain rule to differentiate powers of functions, specifically

$$(1+x^2)^{100}, \qquad \sqrt{x+1} = (x+1)^{1/2}, \qquad \text{and} \qquad \left(\frac{x}{1+x}\right)^{1/3}$$

Rather than apply the chain rule every time we want to differentiate a power of a function, it is more efficient to obtain, once and for all, a formula for differentiating functions of the form

$$[g(x)]^k.$$

If we let $y = [g(x)]^k$ and define $u = g(x)$, then we obtain

$$y = u^k.$$

Thus, by the chain rule

$$\frac{d}{dx}[g(x)]^k = \frac{dy}{dx} = \frac{dy}{du}\cdot\frac{du}{dx}. \tag{6}$$

But

$$\frac{dy}{du} = \frac{d}{du}(u^k) = ku^{k-1} = k[g(x)]^{k-1}$$

and

$$\frac{du}{dx} = g'(x)$$

so that (6) becomes

$$\frac{d}{dx}[g(x)]^k = k[g(x)]^{k-1}\cdot g'(x) \tag{7}$$

In other words, the derivative of a function to a power is the power times the function to the power less one, times the derivative of the function.

Example 14 From (7)

$$\frac{d}{dx}(3 + x - x^3)^{17} = 17(3 + x - x^3)^{16} \cdot \frac{d}{dx}(3 + x - x^3)$$
$$= 17(3 + x - x^3)^{16} \cdot (1 - 3x^2).$$

If variables other than y, u, and x are used, then the chain rule formula must be altered appropriately. For example, if w is a function of v and v is a function of t, then w is a function of t and the chain rule becomes

$$\frac{dw}{dt} = \frac{dw}{dv} \cdot \frac{dv}{dt}.$$

Similarly for other variables.

Example 15 Suppose a manufacturer finds that the cost (in dollars) of making x units of a product is given by

$$C(x) = (5\sqrt{x} + 10)^2.$$

If the rate of production is maintained at a constant three units per day, find the rate at which the cost is changing with time when the level of production is $x = 36$ units.

Solution Since x is a function of the time t and C is a function of x, it follows that C is a function of t. Thus, using the chain rule, we have

$$\frac{dC}{dt} = \frac{dC}{dx} \cdot \frac{dx}{dt}$$

$$= \frac{d}{dx}[(5\sqrt{x} + 10)^2] \cdot \frac{dx}{dt}$$

$$= 2(5\sqrt{x} + 10) \cdot \frac{d}{dx}(5\sqrt{x} + 10) \cdot \frac{dx}{dt}$$

$$= 2(5\sqrt{x} + 10) \cdot \frac{5}{2\sqrt{x}} \cdot \frac{dx}{dt}$$

$$= \frac{25\sqrt{x} + 50}{\sqrt{x}} \cdot \frac{dx}{dt}. \tag{8}$$

We are given that

$$\frac{dx}{dt} = 3$$

so that when $x = 36$ we obtain from (8)

$$\frac{dC}{dt} = \frac{25\sqrt{36} + 50}{\sqrt{36}} \cdot 3 = 100 \quad \text{(dollars per day)}.$$

Additional applications of the chain rule are given in Section 12.3.

EXERCISE SET 12.2

In Exercises 1–12 find dy/dx.

1. $y = (1 + 2x)^{18}$

2. $y = (2 - 3x^2)^{25}$

3. $y = \sqrt{2 - x}$

4. $y = (3 - 2x^2)^{-20}$

5. $y = \dfrac{1}{(2x^2 - 3x)^6}$

6. $y = \sqrt{x^3 + 2x + 1}$

7. $y = (7x^3 - 6x^2 + 2)^{1/3}$

8. $y = x\sqrt{x + 2}$

9. $y = (3x^4 + 2)^{10}(2x^3 - 3)^{20}$

10. $y = \sqrt{\dfrac{x - 4}{x + 11}}$

11. $y = \dfrac{2}{(3x^2 + 5)^{2/3}}$

12. $y = x^2\sqrt{1 - x}$.

13. (a) Suppose r is a function of s and s is a function of t. Use the chain rule to relate

$$\frac{dr}{ds}, \frac{dr}{dt} \quad \text{and} \quad \frac{ds}{dt}.$$

(b) Suppose C is a function of z and z is a function of w. Use the chain rule to relate

$$\frac{dz}{dw}, \frac{dC}{dz} \quad \text{and} \quad \frac{dC}{dw}.$$

14. **(Manufacturing)** Suppose a manufacturer finds that the cost (in dollars) of manufacturing x units of a product is given by

$$C(x) = (100 + 3\sqrt{x})^3.$$

If the rate of production is maintained at a constant 50 units per day, find the rate at which cost is changing with time when the level of production is 81 units.

15. If $y = (x^2 + 4)^{3/2}$, find all values of x at which $dy/dx = 0$.

16. **(Crop Yield)** Assume that soybean yield y (pounds per square yard) is related to the amount x of high-phosphorus fertilizer applied (pounds per square yard) by the "restricted growth" equation

$$y = 50[1 - 10(x^2 + 2)^{-3}].$$

Find the instantaneous rate of change of y with respect to x when $x = 1$.

12.3 APPLICATIONS OF THE DERIVATIVE

In this section we discuss several applications of the derivative: other applications appear in exercises and in later sections of the text.

Equations of Tangent Lines

Example 16 Let f be defined by

$$f(x) = x^2 + 3x - 4.$$

Find the equation of the tangent to the curve $y = f(x)$ at the point

(a) (x_0, y_0)
(b) $(-2, -6)$.

Solution (a) As discussed earlier, the slope of the tangent to $y = f(x)$ at the point (x_0, y_0) is $f'(x_0)$. In this case

$$f'(x) = 2x + 3$$

so that

$$f'(x_0) = 2x_0 + 3.$$

Thus the tangent has slope $m = 2x_0 + 3$ and passes through (x_0, y_0), so that the point–slope form of its equation is (see Section 2.2):

$$y = y_0 + (2x_0 + 3)(x - x_0). \tag{1}$$

(b) In particular, the equation of the tangent line at $(-2, -6)$ can be obtained by substituting $x_0 = -2$ and $y_0 = -6$ in (1). This yields

$$y = -6 + (-1)(x + 2)$$

or

$$y = -x - 8.$$

An Application in Ecology

Example 17 A manufacturer proposes to begin dumping a biodegradable liquid waste product into a nearby lake. The production will be such that the amount of waste dumped at time t will be

$$A = 3t^{3/2}$$

(where A is in gallons and t is in weeks elapsed after the start of dumping). If the liquid decomposes at a constant rate of 27 gallons per week, how long will it take until the manufacturer is dumping liquid more rapidly than it is decomposing?

Solution The rate at which the liquid is being dumped is

$$\frac{dA}{dt} = \frac{d}{dt}(3t^{3/2}) = 3\left(\frac{3}{2}\right)t^{1/2} = \frac{9}{2}\sqrt{t} \quad \text{gallons/week.}$$

Thus the manufacturer will be dumping at the *same* rate that the material is decomposing when

$$\frac{9}{2}\sqrt{t} = 27$$

or

$$\sqrt{t} = \frac{2}{9}(27) = 6$$

or

$$t = 36.$$

Therefore, after 36 weeks the manufacturer will be dumping more rapidly than the material is decomposing (Figure 12.5).

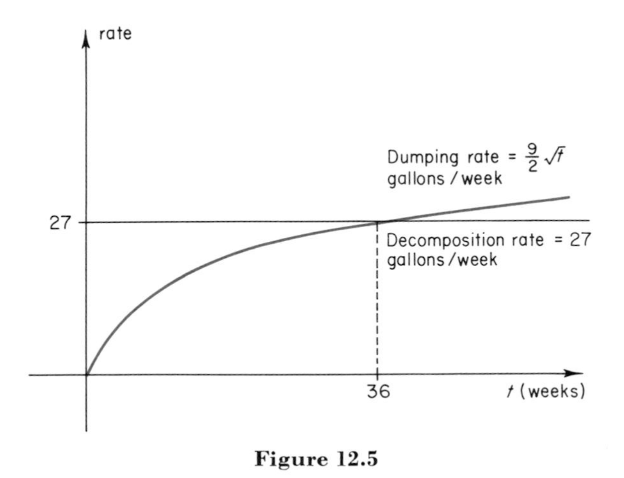

Figure 12.5

Marginal Analysis

Economists and business people are frequently interested in how changes in such parameters as inventory, production, supply, advertising, and price affect other parameters like profit, inflation, demand, and employment. Such problems are often studied using ***marginal analysis.*** The word ***marginal*** is the economist's terminology for rate of change or derivative. Thus marginal analysis is concerned with the rate at which one economic variable changes relative to another.

Three functions of importance to a manufacturer are:

$C(x)$ = total cost of producing x units of the product.

$R(x)$ = total revenue received from selling x units of the product.

$P(x)$ = total profit obtained from selling x units of the product.

These are called respectively the ***cost function, revenue function,*** and ***profit function.*** If all units produced are sold, then these are related by

$$\begin{aligned} P(x) \quad &= \quad R(x) \quad - \quad C(x) \\ (\text{profit}) \quad &= \quad (\text{revenue}) \quad - \quad (\text{cost}) \end{aligned} \tag{2}$$

The derivatives of $C(x)$, $R(x)$, and $P(x)$ are called, respectively, the ***marginal cost*** (MC), the ***marginal revenue*** (MR), and the ***marginal profit*** (MP):

$C'(x)$ = marginal cost = rate of change of cost with respect to the number x of units produced.

$R'(x)$ = marginal revenue = rate of change of revenue with respect to the number x of units sold.

$P'(x)$ = marginal profit = rate of change of profit with respect to the number x of units sold.

If all units produced are sold, then the relationship between marginal profit, marginal cost, and marginal revenue is obtained by differentiating both sides of (2):

$$P'(x) \quad = \quad R'(x) \quad - \quad C'(x) \tag{2a}$$

$$\begin{pmatrix}\text{marginal}\\ \text{profit}\end{pmatrix} = \begin{pmatrix}\text{marginal}\\ \text{revenue}\end{pmatrix} - \begin{pmatrix}\text{marginal}\\ \text{cost}\end{pmatrix}$$

or

$$\text{MP} = \text{MR} - \text{MC}.$$

Example 18 Suppose that a manufacturer's cost and revenue functions are

$$C(x) = 3000 + 200x + \frac{x^2}{5} \tag{3}$$

$$R(x) = 350x + \frac{x^2}{20} \tag{4}$$

where $C(x)$ and $R(x)$ are in dollars. Then the marginal cost and marginal revenue are

$$C'(x) = 200 + \frac{2x}{5} \tag{5}$$

$$R'(x) = 350 + \frac{x}{10} \tag{6}$$

and the marginal profit is

$$\begin{aligned} P'(x) &= R'(x) - C'(x) \\ &= \left(350 + \frac{x}{10}\right) - \left(200 + \frac{2x}{5}\right) \\ &= 150 - \frac{3x}{10}. \end{aligned} \tag{7}$$

If $x = 20$ (units), then from (3) the total cost for manufacturing 20 units is

$$C(20) = 3000 + 200(20) + \frac{(20)^2}{5} = 7080 \quad \text{dollars.}$$

From (5), the marginal cost when $x = 20$ units is

$$C'(20) = 200 + \frac{2(20)}{5} = 208 \quad \text{dollars/unit,}$$

which means that cost is increasing at a rate of 208 dollars per unit at the time the sales level reaches $x = 20$ units. Similarly, from (4) the total revenue received from selling $x = 20$ units is

$$R(20) = 350(20) + \frac{(20)^2}{20} = 7020 \quad \text{dollars,}$$

and the marginal revenue at this production level is

$$R'(20) = 350 + \frac{20}{10} = 352 \quad \text{dollars/unit.}$$

Thus the total profit on $x = 20$ units is

$$P(20) = R(20) - C(20) = 7020 - 7080 = -60 \quad \text{dollars,}$$

which means that the manufacturer has lost \$60 at this stage of production; however, the marginal profit at this production level is

$$P'(20) = R'(20) - C'(20) = 352 - 208 = 144 \quad \text{dollars/unit,}$$

which means that the total profit is increasing at a rate of \$144 per unit when the production level is $x = 20$ units.

Frequently, the marginal cost $C'(x)$ is interpreted as the cost of producing the $(x + 1)$th unit. As an illustration, in Example 18 we found $C'(20) = 208$ dollars/unit, which told us that the *total* cost was increasing at a rate of \$208 per unit produced when $x = 20$. This can be interpreted to mean that it will cost an additional \$208 to produce the twenty-first unit. Strictly speaking, this is not quite correct since the rate of increase $C'(20) = 208$ applies only at the point where the production level is $x = 20$ units. Usually, this rate will vary slightly during the time period between the production level $x = 20$ and the level $x = 21$, so that the cost may not be precisely \$208. However, if the variation in the marginal cost is not too great, this figure should be reasonably accurate. Thus we have the following common approximations:

$C'(x)$ = marginal cost $\approx$ cost of producing the $(x + 1)$th unit.

$R'(x)$ = marginal revenue $\approx$ added revenue obtained by selling the $(x + 1)$th unit.

$P'(x)$ = marginal profit $\approx$ added profit (loss) incurred by selling the $(x + 1)$th unit.

As an illustration, we found in Example 18 that $P(20) = -60$, while $P'(20) = 144$. Thus the manufacturer has lost \$60 when his production level has reached $x = 20$ units, yet at this stage of production he will make a profit of approximately \$144 on his twenty-first unit.

We conclude with some comments about marginal cost. Manufacturing costs can be classified as ***fixed costs*** or ***variable costs.*** Fixed costs include such items as rent, insurance, and administrative wages. They are incurred even if no items are produced. Variable costs include such items as cost of materials, cost of transporting manufactured goods, and production wages. These costs vary with the number x of items manufactured. For example, for the cost function $C(x) = x^2 + 3x + 1000$ (dollars), the fixed and variable costs break down as follows:

$$C(x) = \underbrace{\$x^2 + 3x}_{\text{variable costs}} + \underbrace{\$1000}_{\text{fixed costs}}.$$

More generally, any cost function can be written in the form

$$C(x) = \underset{\text{variable costs}}{f(x)} + \underset{\text{fixed costs}}{k}.$$

Because the derivative of a constant is zero, it follows that the marginal cost is

$$C'(x) = f'(x) + 0 = f'(x),$$

which demonstrates that fixed costs do not affect the marginal cost. This is a basic principle of economics.

Related Rates

In a *related rates* problem one is concerned with the rates of change of two or more variables (usually with respect to time t). In a typical problem, one knows the rate dx/dt at which x changes with time, and the problem is to determine the rate dy/dt at which y changes with time, where y is a known function of x, say $y = f(x)$. The key to solving such problems is the chain rule:

$$\frac{dy}{dt} = \frac{dy}{dx} \cdot \frac{dx}{dt}.$$

Example 19 A stone dropped in a lake produces a circular ripple whose radius increases at a constant rate of 2 feet per second. How rapidly is the area of the ripple increasing when the radius is 10 feet?

Solution Let

$$\begin{aligned} r &= \text{the radius of the ripple in feet,} \\ A &= \text{the area of the ripple in square feet,} \\ t &= \text{time in seconds.} \end{aligned}$$

We are given that

$$\frac{dr}{dt} = 2 \quad \text{ft/sec} \tag{8}$$

We want to find dA/dt when $r = 10$.

From the area formula for a circle we obtain the relationship

$$A = \pi r^2$$

and from the chain rule

$$\frac{dA}{dt} = \frac{dA}{dr} \cdot \frac{dr}{dt}$$

or

$$\frac{dA}{dt} = 2\pi r \cdot \frac{dr}{dt}$$

or from (8)

$$\frac{dA}{dt} = 4\pi r.$$

Thus when $r = 10$

$$\frac{dA}{dt} = 40\pi \approx 126.$$

Thus the area is increasing at an approximate rate of 126 square feet per second.

Example 20 The wholesale price of bananas in a local market is related to the supply (in crates) by

$$P = 3.00 + \frac{30}{S}$$

where P = price per pound (in dollars) and S = supply (in crates.) At what rate will the price be changing if 100 crates are available, but the number of crates is decreasing at a rate of 10 crates per day?

Solution Let t = time in days. Since S is *decreasing* at a rate of 10 crates per day, we have

$$\frac{dS}{dt} = -10.$$

By the chain rule

$$\frac{dP}{dt} = \frac{dP}{dS} \cdot \frac{dS}{dt}$$

or

$$\frac{dP}{dt} = \frac{d}{dS}\left(3 + \frac{30}{S}\right) \cdot \frac{dS}{dt}$$

or

$$\frac{dP}{dt} = \left(-\frac{30}{S^2}\right)(-10) = \frac{300}{S^2}.$$

Thus when $S = 100$ we have

$$\frac{dP}{dt} = \frac{300}{(100)^2} = 0.03 \quad \text{dollars per day}$$

Thus the price per pound is increasing at a rate of 3 cents per day.

EXERCISE SET 12.3

In Exercises 1–4 find the equation of the tangent to the curve $y = f(x)$ at the indicated point. Graph the function and the tangent line.

1. $f(x) = x^2$; $(2, 4)$

2. $f(x) = x^2 + 1$; $(0, 1)$

3. $f(x) = 1/x$; $(2, \frac{1}{2})$

4. $f(x) = -2x + 3$; $(1, 1)$.

5. Find the equation of the tangent to

$$y = \frac{x - 1}{x + 2}$$

at the point where $x = 3$.

6. Find the equation of the tangent to

$$y = (x^2 + 2x)^{18}$$

at the point where $x = -1$.

7. An object travels in such a way that its distance in miles from the starting point after t hours is

$$s = 10\sqrt{t}.$$

Find the instantaneous velocity at the end of $t = 4$ hours.

8. The volume V of a spherical malignant tumor is given by

$$V = \tfrac{4}{3}\pi r^3$$

(a) Find the rate of change of the volume with respect to radius.

(b) At what rate is V changing relative to r when $r = 0.5$ centimeters?

9. Referring to Exercise 8, at what rate with respect to time will the volume V be changing if the radius r increases at a constant rate of 0.1 centimeters per month?

10. A bacteria colony has an initial population of 20,000. After t hours the colony increases to a population $P(t)$ given by

$$P(t) = 20{,}000(1 + 0.5t + t^2).$$

(a) Find the rate of change of the population P with respect to time (this is called the *growth* rate).

(b) Find the number of bacteria present after $t = 10$ hours and find the growth rate at that moment.

11. (**Advertising**) A manufacturer estimates that the firm can sell Q units of its product with an advertising expenditure of x thousand dollars where

$$Q(x) = -x^2 + 500x + 8.$$

(a) What is the rate of change of the number of units sold with respect to the amount spent on advertising?

(b) What is this rate of change when $x = 20$?

12. (**Manufacturing**) The total cost in dollars of refining x units of raw sugar is given by $C(x) = x^2 - \ln x$. If sugar is refined at a constant rate of 10 units per hour, determine the rate at which the cost is changing when 40 units have been produced.

13. (**Ecology**) In a wildlife reserve, the population P of gray hawks depends on the population x of its basic food supply, rabbits. Research suggests the relationship is approximately

$$P = 0.002x + 0.0005x^2.$$

If the number of rabbits is allowed to increase at a constant controlled rate of 500 per year, how rapidly will the hawk population be increasing at a time when the reserve contains 400 rabbits?

14. (**Ecology**) A manufacturer proposes to begin dumping a biodegradable liquid waste product into a nearby stream. The production is such that the amount A (in gallons) of waste dumped at time t (in weeks elapsed after the start of dumping) will be

$$A = (t + 1)^{5/2}.$$

If the liquid decomposes at a constant rate of 20 gallons per week, how long will it take until the manufacturer is dumping liquid more rapidly than it is decomposing?

15. (**Marginal Analysis**) A manufacturer's cost and revenue functions are

$$C(x) = 1000 + 50x + \frac{x^2}{10}$$

$$R(x) = 400 + \frac{x}{20}.$$

(a) Find the marginal cost, marginal revenue, and marginal profit.

(b) At what rate is cost changing when the production level is $x = 25$ units?

(c) At what rate is revenue changing when the production level is $x = 25$ units?

(d) At what rate is the profit changing when $x = 25$ units?

(e) Estimate the cost of producing the twenty-sixth unit.

16. (**Marginal Analysis**) Repeat the directions of Exercise 15 given that

$$C(x) = \tfrac{1}{2}x^2 + 8x + 24$$

$$P(x) = -0.01x^2 + 25x - 50.$$

17. (**Marginal Analysis**) A manufacturer determines that to sell x units, the price P in dollars must be

$$P(x) = 200 - \tfrac{1}{2}x.$$

The cost in dollars of producing x units is

$$C(x) = 3000 + \tfrac{1}{4}x^2.$$

(a) Find the revenue and profit functions.

(b) Find the marginal revenue, cost, and profit functions.

(c) Estimate the added revenue obtained by selling the twenty-sixth unit.

(d) Estimate added profit (loss) incurred by selling the twenty-sixth unit.

18. A cubical block of ice melts so that each side decreases at a rate of 1 inch per hour. How fast is the volume of the block decreasing when the sides are 6 inches?

12.4 DERIVATIVES OF THE EXPONENTIAL AND LOGARITHMIC FUNCTIONS†

Our primary objective in this section is to obtain the derivatives of the *natural logarithm function* $\ln x$ and the *natural exponential function* e^x. As

† Readers are advised to review the material in Section 9.1 before starting this section.

we shall see later in the text, these functions have numerous important applications.

To obtain the derivative of

$$f(x) = e^x$$

we shall use the derivative definition. Thus

$$\begin{aligned} f'(x) &= \lim_{h \to 0} \frac{f(x+h) - f(x)}{h} && \text{(derivative definition)} \\ &= \lim_{h \to 0} \frac{e^{x+h} - e^x}{h} && \text{(substitution)} \\ &= \lim_{h \to 0} \frac{e^x e^h - e^x}{h} && \text{(property of exponents)} \\ &= \lim_{h \to 0} e^x \cdot \frac{e^h - 1}{h} && \text{(factoring).} \end{aligned}$$

Because the factor e^x does not involve h it remains *constant* as $h \to 0$. Therefore, since we are allowed to move a constant through a limit sign, we can write

$$f'(x) = e^x \lim_{h \to 0} \frac{e^h - 1}{h}. \tag{1}$$

Although we shall omit the formal proof, it is not hard to see with the aid of a hand calculator that

$$\lim_{h \to 0} \frac{e^h - 1}{h} = 1$$

(Table 12.1). Thus (1) becomes

$$f'(x) = e^x \cdot 1 = e^x;$$

or equivalently,

$$\frac{d}{dx}[e^x] = e^x \tag{2}$$

In other words, the derivative of the natural exponential function is the exponential function itself.

To differentiate functions like

$$e^{2x}, \qquad e^{x^2}, \qquad e^{\sqrt{x}}, \qquad e^{5x^3 - x}$$

we must use the chain rule. However, rather than apply the chain rule

Table 12.1

Approaching 0 from the left						
h	-1	-0.5	-0.25	-0.1	-0.01	-0.001
$\dfrac{e^h - 1}{h}$	0.632	0.787	0.885	0.951	0.995	0.9995

Approaching 0 from the right						
h	0.001	0.01	0.1	0.25	0.5	1
$\dfrac{e^h - 1}{h}$	1.0005	1.005	1.052	1.136	1.297	1.718

every time we encounter such functions, it will be helpful to obtain a derivative formula applicable to any function of the form

$$e^{g(x)}.$$

In Exercise 27 we ask the reader to apply the chain rule to derive the formula

$$\frac{d}{dx}[e^{g(x)}] = e^{g(x)} \cdot g'(x) \tag{3}$$

Example 21 From formula (3)

$$\frac{d}{dx}[e^{5x}] = e^{5x} \cdot \frac{d}{dx}(5x) = e^{5x} \cdot 5 = 5e^{5x}.$$

$$\frac{d}{dx}[e^{x^2}] = e^{x^2} \cdot \frac{d}{dx}(x^2) = e^{x^2} \cdot 2x = 2xe^{x^2}.$$

$$\frac{d}{dx}[e^{-\sqrt{x}}] = e^{-\sqrt{x}} \cdot \frac{d}{dx}(-\sqrt{x}) = e^{-\sqrt{x}} \frac{d}{dx}(-x^{1/2})$$

$$= e^{-\sqrt{x}}(-\tfrac{1}{2}x^{-1/2}) = -\tfrac{1}{2}\, e^{-\sqrt{x}}x^{-1/2}.$$

Example 22 In chapter 13 we shall encounter expressions of the form

$$Q = Q_0e^{kt} \tag{4}$$

where Q_0 and k are constants. Show that Q satisfies the condition

$$\frac{dQ}{dt} = kQ.$$

(This relationship states that the rate of change of Q is proportional to Q.)

Solution

$$\begin{aligned}\frac{dQ}{dt} &= \frac{d}{dt}(Q_0e^{kt}) \\ &= Q_0\frac{d}{dt}(e^{kt}) \qquad (Q_0 \text{ is constant}) \\ &= Q_0e^{kt}\frac{d}{dt}(kt) \qquad (\text{formula } (3)) \\ &= Q_0e^{kt}\cdot k \\ &= k(Q_0e^{kt}) \\ &= kQ.\end{aligned}$$

Thus

$$\frac{dQ}{dt} = kQ.$$

Example 23 What is the instantaneous rate of change of

$$h(x) = \frac{x}{1+e^x}$$

when $x = 0$?

Solution By the quotient rule

$$\begin{aligned}h'(x) &= \frac{(1+e^x)\cdot\frac{d}{dx}(x) - x\cdot\frac{d}{dx}(1+e^x)}{(1+e^x)^2} \\ &= \frac{(1+e^x)\cdot 1 - x\cdot(e^x)}{(1+e^x)^2} \\ &= \frac{1+e^x-xe^x}{(1+e^x)^2}.\end{aligned}$$

Thus the instantaneous rate of change when $x = 0$ is

$$h'(0) = \frac{1+e^0-0e^0}{(1+e^0)^2} = \frac{2}{(2)^2} = \frac{1}{2}.$$

We now turn to the derivative of the logarithm function

$$g(x) = \ln x. \tag{5}$$

As discussed in Section 9.1, we can rewrite (5) as

$$x = e^{g(x)} \tag{6}$$

so that

$$\frac{d}{dx}[x] = \frac{d}{dx}[e^{g(x)}]$$

or from (3)

$$1 = e^{g(x)} \cdot g'(x)$$

or from (6)

$$1 = x \cdot g'(x)$$

or

$$g'(x) = \frac{1}{x}.$$

In other words

$$\frac{d}{dx}[\ln x] = \frac{1}{x}.$$

To differentiate functions like

$$\ln(2x + 1), \qquad \ln \frac{1}{1 + x}, \qquad \ln(e^x - 1)$$

we need the chain rule. In Exercise 28 we ask the reader to use the chain rule to derive the formula

$$\frac{d}{dx}[\ln(g(x))] = \frac{1}{g(x)} \cdot g'(x). \tag{7}$$

Example 24 The function

$$h(x) = \ln(x^2)$$

can be differentiated in two ways. We can apply formula (7), in which case we get

$$h'(x) = \frac{d}{dx}[\ln(x^2)] = \frac{1}{x^2} \cdot \frac{d}{dx}[x^2] = \frac{1}{x^2} \cdot 2x = \frac{2}{x},$$

or we can use properties of the natural logarithm to write

$$h(x) = \ln(x^2) = 2 \ln x$$

so that

$$h'(x) = \frac{d}{dx}[2 \ln x] = 2\frac{d}{dx}[\ln x] = 2 \cdot \frac{1}{x} = \frac{2}{x}.$$

Example 25 Using formula (7) we obtain

$$\frac{d}{dx}[\ln(5x)] = \frac{1}{5x} \cdot \frac{d}{dx}[5x] = \frac{1}{5x} \cdot 5 = \frac{1}{x},$$

$$\frac{d}{dx}[\ln(x^3 + 3x + 2)] = \frac{1}{x^3 + 3x + 2} \cdot \frac{d}{dx}(x^3 + 3x + 2) = \frac{3x^2 + 3}{x^3 + 3x + 2}.$$

Example 26 Find the slope of the tangent to the curve

$$y = x \ln x$$

at the point where $x = 1$.

Solution The slope of the tangent is the value of dy/dx at $x = 1$. By the product rule

$$\begin{aligned} \frac{dy}{dx} &= x \cdot \frac{d}{dx}[\ln x] + \ln x \cdot \frac{d}{dx}[x] \\ &= x \cdot \frac{1}{x} + (\ln x) \cdot 1 \\ &= 1 + \ln x. \end{aligned}$$

Thus when $x = 1$,

$$\frac{dy}{dx} = 1 + \ln(1) = 1 + 0 = 1.$$

EXERCISE SET 12.4

In Exercises 1–22 find dy/dx.

1. $y = e^{2x}$

2. $y = e^{-x}$

3. $y = 3e^{4x^2}$

4. $y = e^{3\sqrt{x}}$

5. $y = 2x - e^{4x} + 3e^{x^3}$

6. $y = xe^x$

7. $y = \dfrac{e^x}{x}$

8. $y = \dfrac{1}{1 + e^x}$

9. $y = \dfrac{1}{5x} e^{-x^2}$

10. $y = x^4 e^{-x}$

11. $y = (e^x)^2$

12. $y = e^{\sqrt{x+1}}$

13. $y = e^{(1-x)^7}$

14. $y = \sqrt{e^x + 1}$

15. $y = -3 \ln x$

16. $y = x^2 \ln x$

17. $y = \dfrac{\ln x}{x^3}$

18. $y = (\ln x)^5$

19. $y = \ln (e^x + x)$

20. $y = \ln (7x^3 + 5)$

21. $y = e^x \ln x$

22. $y = \ln (x^{100})$.

23. Find the equation of the tangent line to $y = e^x$ at $(0, 1)$.

24. Find the equation of the tangent line to $y = \ln x$ at the point where $x = 1$.

25. Let $y = e^x \ln (x + 1)$. Find the instantaneous rate of change of y with respect to x when $x = 0$.

26. Let $w = t \ln t$. Find the instantaneous rate of change of w with respect to t when $t = e$.

27. Use the chain rule to derive the formula

$$\frac{d}{dx}\left[e^{g(x)}\right] = e^{g(x)} \cdot g'(x).$$

28. Use the chain rule to derive the formula

$$\frac{d}{dx}\left[\ln (g(x))\right] = \frac{1}{g(x)} \cdot g'(x).$$

29. **(Ecology)** A lake is stocked with 500 steelhead trout. If the lake can support a maximum of 5500 such trout, then according to a standard model (the *inhibited growth model*) the number $P(t)$ of trout in the lake after t months will have a formula of the form

$$P(t) = \frac{5000}{1 + 10e^{-kt}}$$

where k is a constant.

(a) Estimate the constant k if the trout population after six months is 800. (Use Appendix Table V)

(b) Find dP/dt.

(c) How rapidly is the trout population growing after 10 months?

30. (**Learning**) It is determined by experimentation that in a certain assembly line operation, a typical experienced worker can produce at most 100 units per day. If a new worker is placed on the assembly line, then according to a standard model (the *Hullian* or *learning* model) the worker's daily output $P(t)$ after t days of training will have a formula of the form

$$P(t) = 100(1 - e^{-kt})$$

where k is a constant that varies with the individual.

(a) Estimate the constant k for an individual who produces 50 units per day after one day of training.

(b) Find dP/dt.

(c) Find the rate at which $P(t)$ is increasing with time after five days of training.

13 APPLICATIONS OF DIFFERENTIATION

13.1 APPLICATIONS TO EXPONENTIAL GROWTH AND DECAY MODELS

Many physical quantities increase or decrease with time in proportion to the amount of the quantity present. Some typical examples are human population, certain kinds of investment interest, radioactivity, alcohol level in the blood after drinking, and bacteria in a culture. In this section we shall use the derivative to study the growth of such quantities.

Exponential Growth Models A physical quantity is said to have an ***exponential growth model*** if at each instant of time its rate of growth is proportional to the amount of the quantity present.

Let us consider a quantity with an exponential growth model. At each instant of time we shall denote by $Q = Q(t)$ the amount of the quantity present at time t, where t represents the *time elapsed* from some initial observation. For example, if time is measured in seconds, $t = 1$ means 1 second after the initial observation, $t = 7.6$ means 7.6 seconds after the initial observation, and $t = 0$ is the value of t at the initial observation.

Because the quantity Q has an exponential growth model, its growth rate dQ/dt is proportional to the amount Q present at each instant. Thus dQ/dt and Q are related by

$$\frac{dQ}{dt} = kQ \tag{1}$$

where k is a constant of proportionality. Equation (1) is an example of a ***differential equation***; that is, an equation involving the derivative of an unknown function. However, we discovered in Example 22 of Chapter 12 that this equation is satisfied when

$$Q = Q_0 e^{kt} \tag{2}$$

where Q_0 is a constant. From this formula, the value of Q present initially $(t = 0)$ is

$$Q = Q_0 e^{k(0)} = Q_0 e^0 = Q_0 \cdot 1 = Q_0.$$

Thus the constant Q_0 in formula (2) is the amount present initially. The constant k in formula (2) is called the ***growth constant*** since it determines how rapidly Q grows with time. At each instant the quantity Q grows at a rate given by

$$\text{growth rate} = k \cdot (\text{amount } Q \text{ present}). \tag{3}$$

Thus, the larger the value of k, the larger the growth rate at each instant.

Example 1 Suppose a quantity Q has an exponential growth model described by the formula

$$Q = 60e^{0.05t} \tag{4}$$

where t is measured in hours. For this model the amount present initially is

$$Q_0 = 60 \quad \text{(units)}$$

and the growth constant is

$$k = 0.05.$$

From formulas (3) and (4) the initial growth rate (at $t = 0$ or $Q_0 = 60$) is

$$\text{growth rate} = kQ_0 = (0.05)(60) = 3 \text{ units/hour}$$

and the growth rates when $Q_0 = 300$ and $Q_0 = 1000$ are

$$\text{growth rate} = kQ_0 = (0.05)(300) = 15 \text{ units/hour} \qquad \text{when } Q_0 = 300$$

and

$$\text{growth rate} = kQ_0 = (0.05)(1000) = 50 \text{ units/hour} \qquad \text{when } Q_0 = 1000.$$

Observe how the growth *rate* increases with time in an exponential growth model. This is quite different from a linear model where the growth rate always remains constant. Figure 13.1 shows a typical growth pattern for a quantity Q with an exponential growth model.

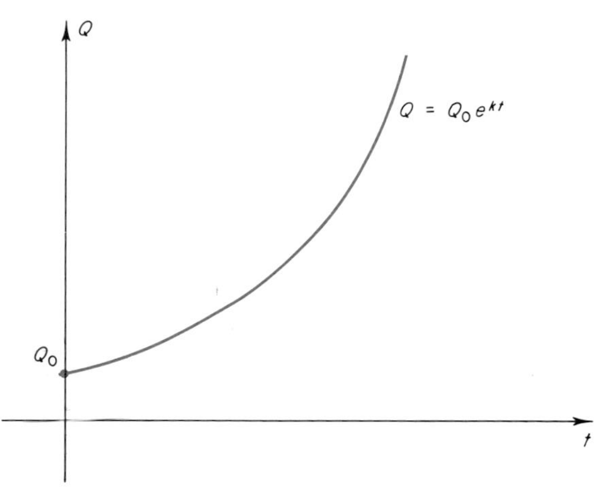

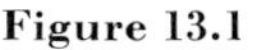
Figure 13.1

> **Exponential Decay Models** A physical quantity is said to have an ***exponential decay model*** if at each instant of time its rate of decrease (or decay) is proportional to the amount present.

If Q has an exponential decay model then

$$\frac{dQ}{dt} = -kQ,$$

where the minus indicates that Q *decreases* at a rate proportional to Q. This differential equation is satisfied by (verify):

$$Q = Q_0 e^{-kt}$$

where Q_0 is the amount present initially, t is the time elapsed, and k is the *decay* constant. At each instant the quantity Q *decays* at the rate

$$\text{decay rate} = k \cdot (\text{amount } Q \text{ present}). \tag{5}$$

Example 2 If Q has an exponential decay model described by the formula

$$Q = 100e^{-0.05t}$$

(t in hours), then the amount present initially is

$$Q_0 = 100$$

and the decay constant is $k = 0.05$ so that initially Q is decreasing at the rate

$$\text{decay rate} = kQ_0 = (0.05)(100) = 5 \text{ units/hour when } Q_0 = 100.$$

Figure 13.2 shows a typical decay pattern for a quantity with an exponential decay model.

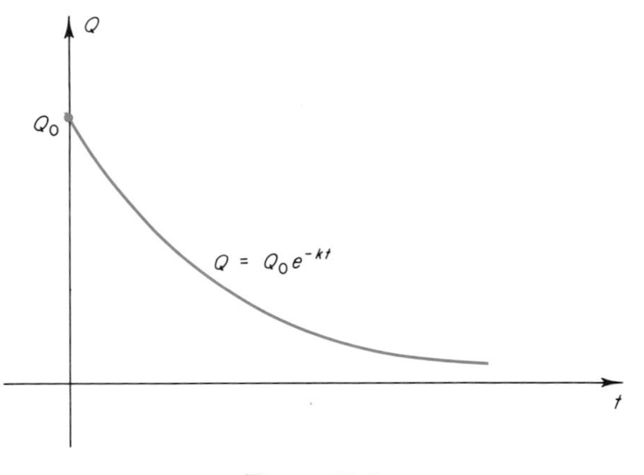

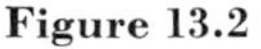

Figure 13.2

Remark In exponential growth and decay models the growth (or decay) constant k is often expressed as a percentage. Thus a growth constant of 2% would mean $k = 0.02$ and a growth constant of 300% would mean $k = 3$. We also note that the growth (decay) constant for an exponential model is often called the ***growth rate*** (***decay rate***). Strictly speaking, this is not correct. As indicated in formula (1), the growth or decay rate dQ/dt is not k, but rather kQ, that is, k times the amount present. However, the description of k as a growth (decay) rate is so common that we shall adopt this standard terminology.

We now turn to some additional examples of exponential growth and decay models.

Global Population Growth

In 1798 the controversial English social philosopher Thomas Malthus† published a book entitled, *An Essay on the Principle of Population as it Affects the Future Improvement of Society*. In this work Malthus proposed

† *Thomas Robert Malthus* (1766–1834)—English social philosopher. Malthus was born to a prosperous middle class family. His theory that population always tends to outrun food supply led him to advocate limits on human reproduction. Malthus spent much of his life as a professor of history and economics for the East India Company's Haileybury College.

that human population has an exponential growth model and that eventually population must exceed the food supply. History has shown the Malthusian model of population growth to be inaccurate for technologically developed countries. However, many modern demographers (population scientists) feel that exponential growth models are suitable for lesser developed countries and for the global population as a whole, at least over a period of thirty or forty years.

Example 3 According to data published by the United Nations, the world population at the beginning of 1975 was approximately 4 billion and growing at a rate of approximately 2%. Assuming an exponential growth model, estimate the world population for the year 2000.

Solution Let us measure time t in years and population Q in billions. If we take the beginning of 1975 as the initial observation, then the initial value of Q is

$$Q_0 = 4 \quad \text{(billion)}.$$

Since the growth rate is 2%, we have

$$k = 0.02$$

so that the equation

$$Q = Q_0 e^{kt}$$

becomes

$$Q = 4e^{0.02t}.$$

By the year 2000 the time t elapsed from the initial observation in 1975 will be $t = 25$ (years). Thus from the computation above the population by the year 2000 will be

$$Q = 4e^{0.02(25)} = 4e^{0.5}$$

or from Appendix Table VI

$$Q = 4(1.6487) = 6.5948$$

which is approximately 6.6 billion.

In Figure 13.3 we have sketched a graph of global population over time based on data from the United Nations. Observe that the shape of this curve is similar to that of the general exponential growth curve shown in Figure 13.1.

Example 4 The following statement appears in the *Encyclopedia Americana*,† "Average annual growth of world population is about 2% which means

† *Encyclopedia Americana*, Vol. 22, Americana Corporation, Grolier, 1976.

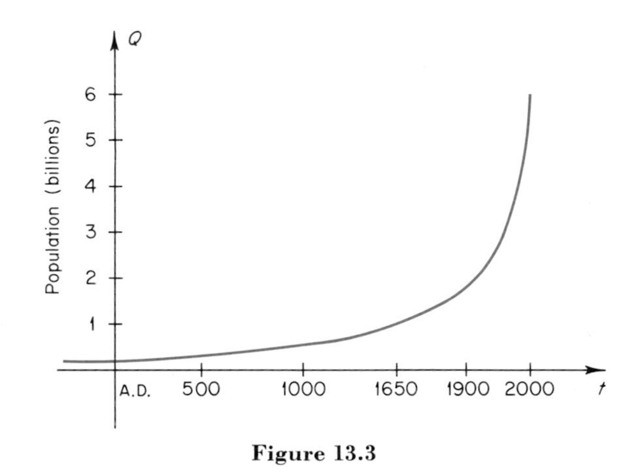

Figure 13.3

doubling of the world's population every 35 years." Show that this statement is correct, assuming an exponential growth model.

Solution Let Q_0 be the world population at any point in time. Since $k = 0.02$, the population at any later time will be

$$Q = Q_0 e^{kt} = Q_0 e^{0.02t}. \tag{6}$$

The population will have doubled when $Q = 2Q_0$. Thus from (6) the time t required for doubling will satisfy

$$2Q_0 = Q_0 e^{0.02t}.$$

Dividing through by Q_0 yields

$$2 = e^{0.02t}$$

and taking the natural logarithm of both sides gives

$$\ln 2 = \ln e^{0.02t} = (0.02t) \ln e = .02t$$

Thus the doubling time is

$$t = \frac{1}{0.02} \ln 2$$

or from Appendix Table V

$$t = \frac{1}{0.02} (0.6931) = 34.655$$

or approximately 35 years.

It is interesting to note that the doubling time of 35 years does not depend on the quantity Q_0 present initially. At any point in time the population will double in the subsequent 35 years (for a 2% growth rate.)

Thus, with a continued 2% growth rate the global population of 4 billion in 1975 will double to 8 billion by the year 2010 and will double again to 16 billion by 2045. Therefore, between 2010 and 2045 the earth will add to its *existing* population 8 billion new people—twice the present global population!

It was not accidental in the previous example that the time required for doubling was independent of the initial population. This is a property of all exponential growth models. To see this let Q be any quantity with an exponential growth model. From any initial value Q_0 the time required to double will satisfy

$$2Q_0 = Q_0 e^{kt}$$

or on dividing by Q_0

$$2 = e^{kt}$$

so that

$$\ln 2 = \ln e^{kt}$$

or

$$\ln 2 = kt \ln e = kt$$

or

$$t = \frac{\ln 2}{k}$$

which is independent of Q_0. We shall call

$$T = \frac{\ln 2}{k} \qquad (7)$$

the ***doubling time*** for the growth model. In the case of an exponential decay model the time required for Q to reduce by half is given by the same formula (Exercise 16). For such models we will call T the ***halving time*** or sometimes the ***half-life.*** Doubling and halving times are illustrated graphically in Figure 13.4.

Radioactive Decay

Radioactive elements continually undergo a process of disintegration called ***radioactive decay.*** It is a physical fact that the rate of decay is proportional to the amount of the element present. As a consequence,

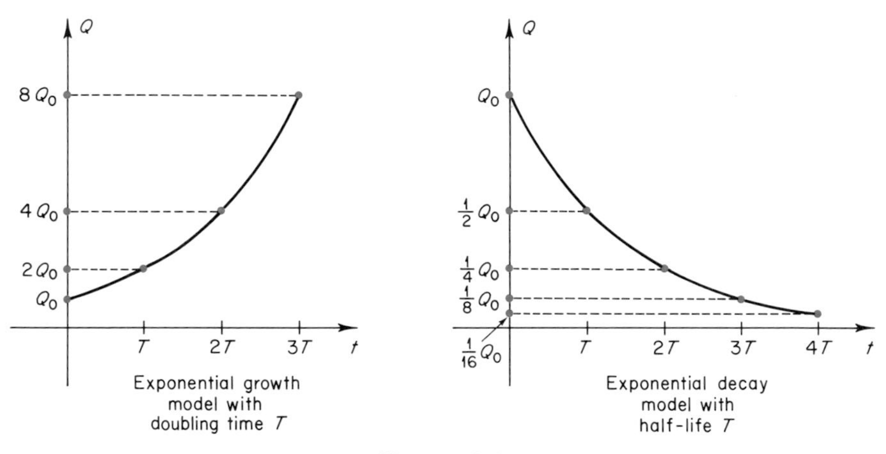

Figure 13.4

the quantity Q of any radioactive substance has an exponential decay model, that is, Q is given by

$$Q = Q_0 e^{-kt}$$

where Q_0 is the amount present initially, t is the time elapsed, and k is the rate of decay.

Example 5 Potassium-42 has a decay rate of approximately 5.5% per hour.

(a) If 1000 grams of potassium-42 are present initially, how much will be left after two hours?

(b) What is the half-life of potassium-42?

Solution (a) From the given data, $Q_0 = 1000$ and $k = 0.055$, so that the quantity Q left after $t = 2$ hours is

$$Q = Q_0 e^{-kt} = 1000e^{-(0.055)2} = 1000e^{-0.11}$$

or from Appendix Table VI

$$Q = 1000(0.8958) = 895.8 \quad \text{grams}$$

(b) From formula (7) the half-life T is

$$T = \frac{\ln 2}{k} = \frac{0.6931}{0.055} = 12.6 \quad \text{hours.}$$

Example 6 Carbon-14 is a radioactive carbon isotope with a half-life of 5750 years. What is the decay rate?

Solution Formula (7) can be written

$$k = \frac{\ln 2}{T}.$$

Substituting the half-life $T = 5750$ and the value $\ln 2 = 0.6931$, yields the decay rate

$$k = \frac{0.6931}{5750} = 0.0001205$$

or approximately

$$k = 0.012\% \quad \text{per year.}$$

Carbon Dating of Fossils

When the nitrogen in the earth's upper atmosphere is bombarded by cosmic radiation, the radioactive element carbon-14 is produced. This carbon-14 combines with oxygen to form carbon dioxide, which is ingested by plants, which in turn are eaten by animals. In this way all living plants and animals absorb quantities of radioactive carbon-14. In 1947 the American nuclear scientist W. F. Libby† proposed the theory that the percentage of carbon-14 in the atmosphere and in living tissues of plants and animals is the same. When a plant or animal dies, the carbon-14 in the tissue begins to decay. Thus, the age of a plant or animal fossil can be estimated by determining how much of its carbon-14 content has decayed. Libby won the Nobel prize in 1960 for his discovery.

In the years 1950 and 1951 a research team from the Texas Memorial Museum unearthed charred bison bones and the so-called "Folsom points," which were projectile tips probably used for darts. It was clear from the evidence that the bones came from a bison cooked by the makers of the points. Thus, by carbon-14 dating of the bones, the research team was able to clearly establish that hunters ("Folsom man") roamed North America some 10,000 years ago. The following example illustrates how such calculations are made by anthropologists.

Example 7 (Carbon Dating and the "Folsom Points")‡ Assuming that chemical tests show the Folsom bison bones to have lost 70% of their carbon-14 estimate the age of the bones.

Solution From Example 6 the decay rate for carbon-14 is $k = 0.012\%$ per year, so that the amount Q of carbon-14 that remains after t years from an initial amount Q_0 is

$$Q = Q_0 e^{-0.00012t}.$$

† W. F. Libby, "Radiocarbon dating," *American Scientist*, **44,** (1956), 98–112.

‡ The data in this problem are based on results reported by E. H. Sellards, "Age of Folsom Man," *Science* **115** (1952), 98.

Since the bones lost 70% of their carbon-14, the amount that remains is 30% of the original quantity Q_0. Thus, to find the age t we solve

$$30\% Q_0 = Q_0 e^{-0.00012t}$$

or

$$\begin{aligned} 0.3Q_0 &= Q_0 e^{-0.00012t} \\ 0.3 &= e^{-0.00012t} \\ \ln 0.3 &= \ln e^{-0.00012t} \\ \ln 0.3 &= -0.00012t \\ -1.2040 &= -0.00012t \qquad \text{(Appendix Table V)} \\ t &= \frac{1.2040}{0.00012} = 10{,}033. \end{aligned}$$

Thus, the bison bones are approximately 10,033 years old.

Estimating Future U.S. Gasoline Production

Past data on U.S. production of gasoline are shown graphically in Figure 13.5 (based on figures reported by the United States Bureau of

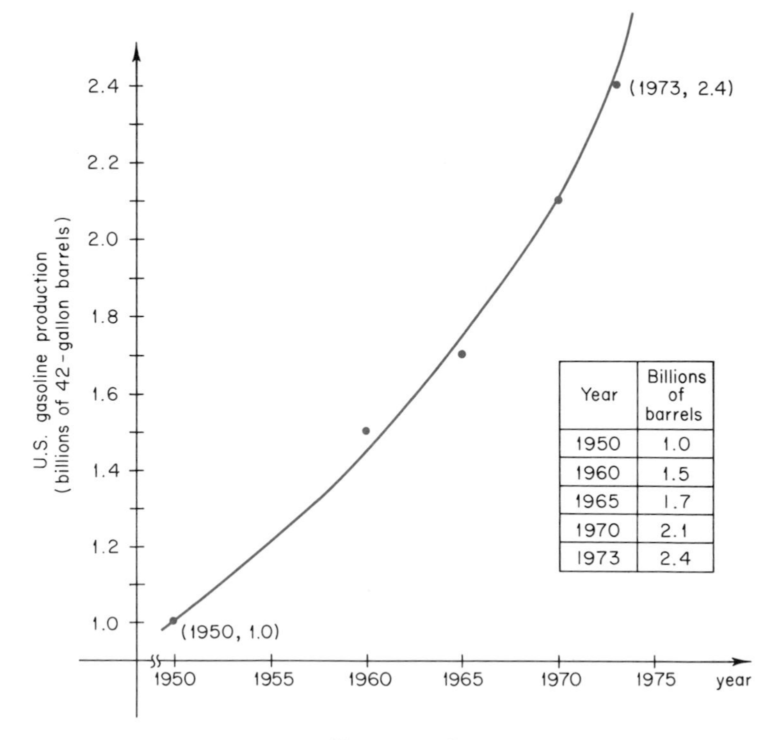

Year	Billions of barrels
1950	1.0
1960	1.5
1965	1.7
1970	2.1
1973	2.4

Figure 13.5

Mines). The shape of this graph strongly suggests the possibility of describing U.S. gasoline production by an exponential growth model.

Example 8 Assuming an exponential growth model, find

(a) the growth rate k using the data from the years 1950 and 1973,

(b) the estimated gasoline production for the year 2000.

Solution (a) If we take $t = 0$ to be the year 1950, then the production Q_0 for this initial year is

$$Q_0 = 1.0 \quad \text{(billion barrels)}$$

which means that the production t years later will be

$$Q = Q_0e^{kt} = e^{kt}. \tag{8}$$

Since the year 1973 corresponds to $t = 23$ $(1973 - 1950 = 23)$ and since the production for 1973 is

$$Q = 2.4,$$

we obtain from (8)

$$2.4 = e^{k(23)} = e^{23k}$$

$$\ln 2.4 = \ln e^{23k} = 23k$$

$$k = \frac{\ln 2.4}{23} = \frac{0.8755}{23} \approx 0.0381.$$

Thus, the growth rate is about 3.81%.

(b) From the calculation in (a) the growth equation for this model is

$$Q = Q_0e^{kt} = e^{0.0381t}.$$

Since the year 2000 corresponds to $t = 50$ $(2000 - 1950 = 50)$, we obtain

$$Q = e^{0.0381(50)} \approx e^{1.9} = 6.686 \qquad \text{(Appendix Table VI)}$$

Thus, the estimated production level for the year 2000 is 6.686 billion barrels.

EXERCISE SET 13.1

1. Consider the exponential growth model given by

$$Q = 500e^{0.02t}$$

(a) Find the growth constant.

(b) What is the initial value of Q?

(c) Use Appendix Table VI to complete the following table

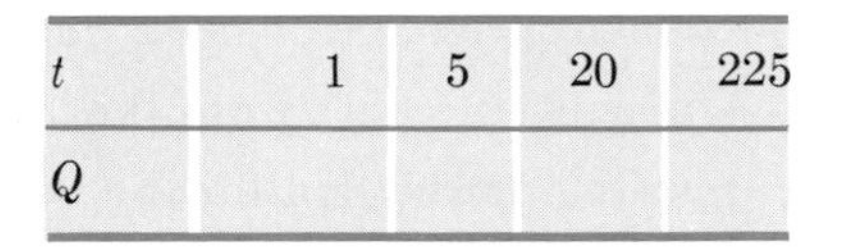

t	1	5	20	225
Q				

(d) At what rate is Q growing initially? (Use formula (3).)

(e) At what rate is Q growing when $t = 3$? (Use formula (3).)

2. Consider the exponential decay model given by

$$Q = 3000e^{-0.005t}.$$

(a) Find the decay constant.

(b) What is the initial value of Q?

(c) Use Appendix Table VI to complete the following chart.

t	10	30	200	3000
Q				

(d) At what rate is Q decaying initially? (Use formula (5).)

(e) At what rate is Q decaying when $Q = 100$? (Use formula (5).)

3. For the growth model in Exercise 1, use Appendix Table V to complete the following chart.

Q	500	1250	2300
t			

4. For the decay model in Exercise 2, use Appendix Table V to complete the following chart.

Q	3000	1500	600
t			

5. Consider the exponential growth model given by

$$Q = Q_0e^{0.4t}$$

where t is in seconds.

(a) How long does it take for Q to double in value?

(b) How long to quadruple in value?

6. Consider the exponential decay model given by

$$Q = Q_0e^{-0.6t}$$

where t is in years.

(a) How long does it take for Q to decay to half of its value?

(b) How long does it take to decay to one-fourth of its value?

7. (a) Find the growth rate of an exponential growth model with a doubling time of five years.

(b) Find the decay rate of an exponential decay model with a half-life of 100 hours.

8. Radium-228 has a half-life of 6.7 years. How much time is required for 80% of the radium to decay?

9. The decay rate of krypton-85 is 6.3% per year. What is its half-life?

10. Find a formula relating the growth rate and *tripling time* for an exponential model.

11. **(Global Population Growth)** From the United Nations data in Example 4, the world population at the beginning of 1975 was approximately 4 billion and growing at a rate of approximately 2% per year.

(a) Using this data and assuming an exponential growth model complete the following chart.

Year	1975	1980	1985	1990	1995	2000	2005	2010
World population (in billions)								

(b) Sketch a graph of these data.

12. **(Anesthesiology)** Sodium pentobarbital, commonly used for surgical anesthesia, is absorbed by the body organs at a rate proportional to its concentration in the bloodstream. As a result, the bloodstream concentration follows an exponential decay model. Assume that a surgical patient requires 25 mg of sodium pentobarbital in the bloodstream per kilogram of body weight to maintain a proper level of anesthesia for surgery. How many milligrams of the drug must be administered to a 60-kg patient to maintain a proper level of anesthesia for a $\frac{1}{2}$-hour operation assuming the drug bloodstream concentration decays at a rate of 14% per hour?

13. **(Carbon Dating)** Fossil remains of a human skeleton have $\frac{1}{20}$ of the original carbon-14 content. Estimate the age of the fossil.

14. **(Nuclear Energy Supply)** The power supply for an experimental lunar sensor uses a radioisotope whose power output P in watts decreases with time according to the model

$$P = 40e^{-0.006t}$$

where t is in days.

(a) What is the power output after 200 days of operation?

(b) What is the percentage decrease in power output over this 200-day period?

(c) What is the half-life of the power supply?

(d) If the sensing device requires 16 watts of power to operate, how long will it stay in operation?

15. **(Modeling Business Expansion)** To plan for future expansion, a manufacturer wants to estimate the projected sales volume for the year 1980. The firm's past sales record is shown in Figure 13.6.

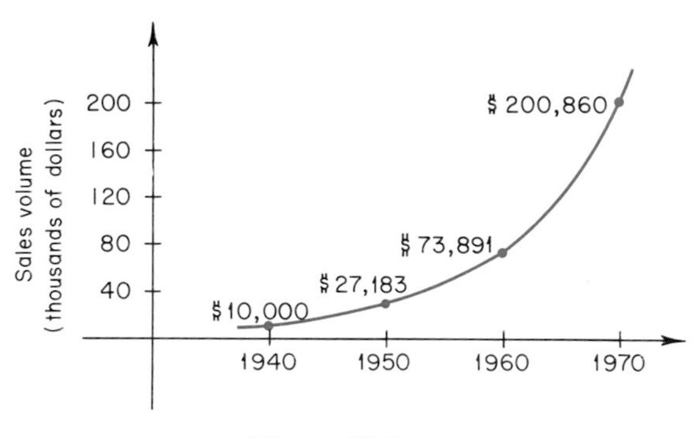

Figure 13.6

Assume an exponential growth model and follow the method of Example 8 to find:

(a) the growth rate k based on the data for the year 1970;

(b) the estimated sales volume for the year 1980.

16. Show that for an exponential decay model, the halving life is given by formula (7).

13.2 APPLICATIONS OF DIFFERENTIATION TO OPTIMIZATION

Many practical problems reduce to maximizing or minimizing some quantity: a businessman wants to maximize his profits and minimize his costs, an engineer wants to maximize the strength of a bridge structure, an airline wants to minimize the flying time for a certain trip, and so forth. In this section we shall show how differentiation can be used to solve certain kinds of maximization and minimization problems. We begin with an example.

Example 9 A farmer wants to fence in a rectangular field bordering on a straight stream (Figure 13.7). He will not fence the side along the stream. If he

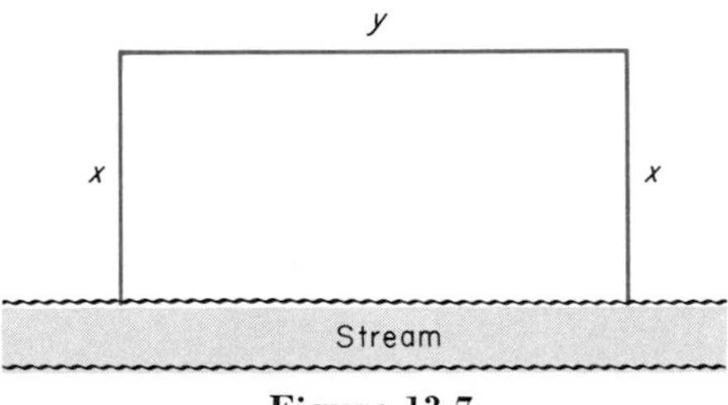

Figure 13.7

has 1000 feet of fence to work with, what is the maximum area he can enclose?

Solution To attack this problem, let us denote the lengths of the sides by x and y (as shown in Figure 13.7), so that the area A to be maximized is

$$A = xy \tag{1}$$

Since the farmer has 1000 feet of fence, we have the following relationship between x and y:

$$1000 = 2x + y$$

or

$$y = 1000 - 2x. \tag{2}$$

Thus we can rewrite (1) as

$$A = x(1000 - 2x) = 1000x - 2x^2. \tag{3}$$

In this formula the variable x is subject to certain physical restrictions. For example, since x represents a length, we must have $x \geq 0$. Moreover, there is only 1000 feet of fence available, so we cannot use more than this amount on the two sides of length x; thus $2x \leq 1000$ or $x \leq 500$.

In light of these physical restrictions and (3), we can formulate our problem as follows:

> Find the maximum value of
>
> $$A = 1000x - 2x^2 \tag{4}$$
>
> where x satisfies $0 \leq x \leq 500$.

One way to attack this problem is to graph Equation (4) and try to determine the maximum value of A from the graph. In Figure 13.8 we

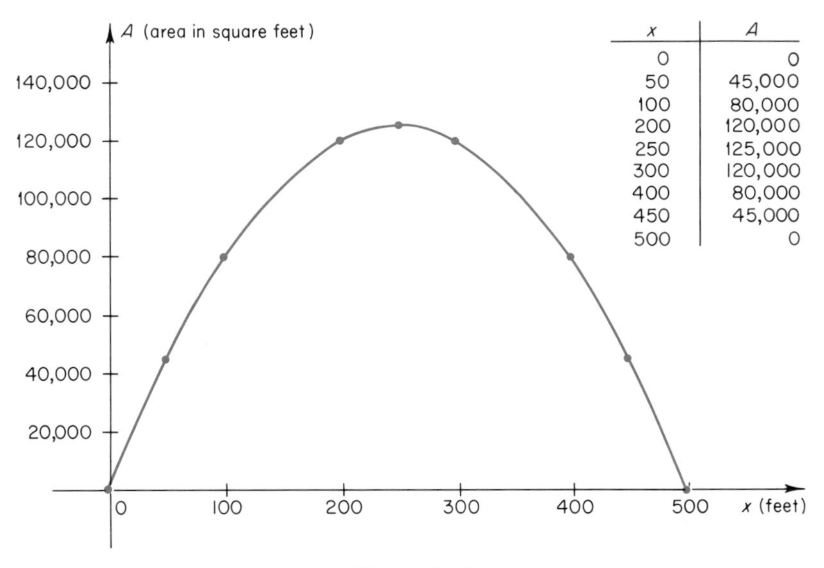

Figure 13.8

have tabulated and plotted some points on this graph. In light of this figure it appears that maximum area is

$$A = 125{,}000 \text{ square feet}$$

which occurs when

$$x = 250 \text{ feet}$$

and from (2)

$$y = 1000 - 2x = 1000 - 2(250) = 500 \text{ feet}$$

Although this graphical solution is simple, it suffers from a major defect. It is possible that some value of x different from any of those tabulated might produce a value for the area surpassing any that we have recorded and plotted. Moreover, it will do no good to plot more points since we might still omit the very value of x that produces the largest area. What we need is a definitive way to locate the highest point on the graph of (4). This is where the derivative comes into play.

Observe that the highest point on the graph divides the region where the curve is rising from the region where it is falling (Figure 13.9). At this point the tangent to the curve is horizontal, that is, has slope zero. Thus, to determine where the curve has its peak we need only determine where the tangent has slope zero. The computations are as follows: At each point of the curve

$$A = 1000x - 2x^2$$

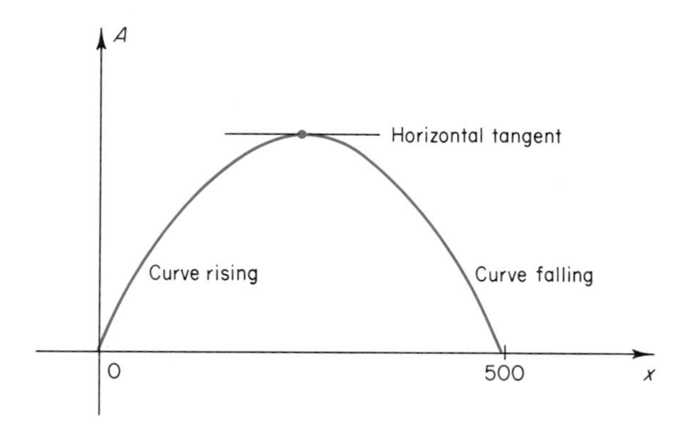

Figure 13.9

the slope of the tangent is

$$\frac{dA}{dx} = 1000 - 4x.$$

Thus the tangent has slope zero if

$$1000 - 4x = 0$$

or

$$x = 250,$$

which demonstrates definitively that the maximum area does indeed occur when $x = 250$ ft.

Before we discuss any additional examples, it will be helpful to introduce some terminology and notation.

By an ***interval*** we mean a set of real numbers that forms a line segment on a coordinate axis. The interval may include or exclude one or both of the endpoints. If a and b are real numbers such that $a < b$, then the ***closed interval*** from a to b, denoted $[a, b]$, is the set of points

$$\{x \mid a \leq x \leq b\}$$

and the ***open interval*** from a to b, denoted (a, b), is the set

$$\{x \mid a < x < b\}$$

These sets are shown in Figure 13.10. We shall use a square bracket

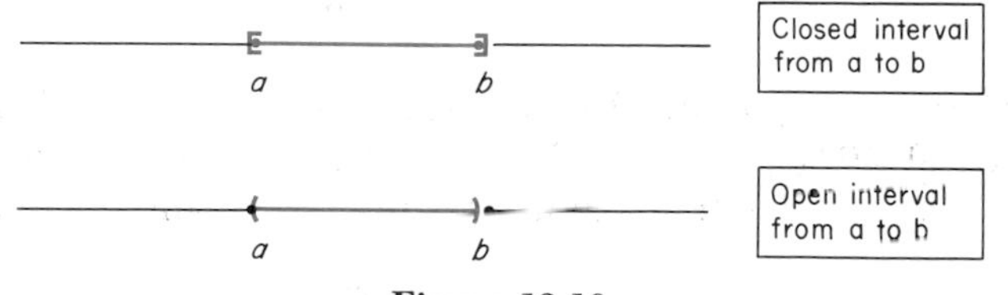

Figure 13.10

[or] to indicate an endpoint that is included in the interval and a rounded bracket (or) to indicate an endpoint that is excluded. Thus

$$[a, b) = \{x \mid a \leq x < b\}$$

is the interval

which includes a but excludes b.

We shall also need intervals that extend indefinitely in one or both directions. Some intervals of this type and their notation are

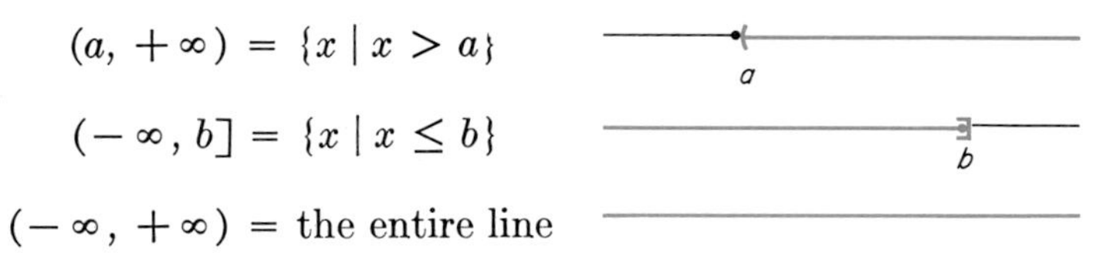

$$(a, +\infty) = \{x \mid x > a\}$$

$$(-\infty, b] = \{x \mid x \leq b\}$$

$$(-\infty, +\infty) = \text{the entire line}$$

In this notation, the symbol $+\infty$ (read "plus infinity") indicates that the interval extends indefinitely in the positive direction, and $-\infty$ ("minus infinity") indicates that the interval extends indefinitely in the negative direction.

Example 10

Interval

$(-1, 2)$

$[1, 3]$

$(0, +\infty)$

$(-\infty, 2]$

In Example 9 above, the physical problem of fencing a field with maximum area reduced to the mathematical problem of maximizing

$$A = 1000x - 2x^2$$

where x was required by physical considerations to lie in the closed interval $[0, 500]$ (see (4)). This is typical of many optimization problems; the objective is to maximize (or minimize) some function

$$f(x)$$

where x is required to lie in some specified interval. There is some terminology associated with such problems.

Definition On a specified interval, a function f is said to have an ***absolute maximum*** at $x = c$ if $f(c)$ is the largest value of $f(x)$ for x in the interval. Similarly, f has an ***absolute minimum*** at $x = c$ if $f(c)$ is the smallest value of $f(x)$ for x in the interval.

Example 11 In Figure 13.11a the function f has an absolute maximum at c_1 and an absolute minimum at c_2. In Figure 13.11b f has an absolute maximum at the endpoint a and an absolute minimum at the endpoint b.

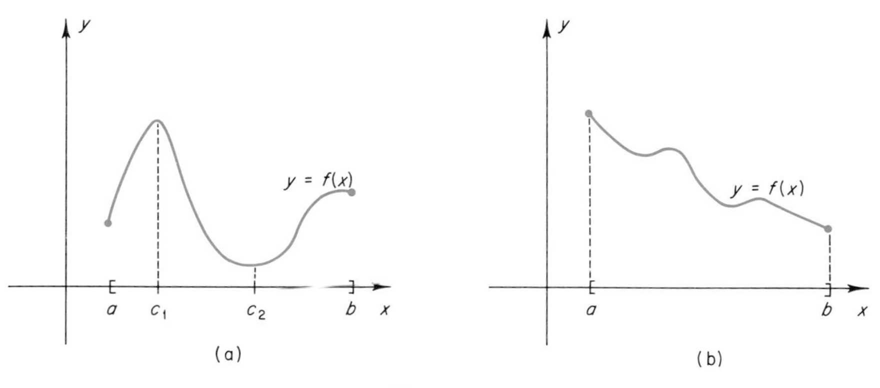

Figure 13.11

If we are given a closed interval $[a, b]$ and if we draw the graph of a continuous function by starting above the left-hand endpoint a and not

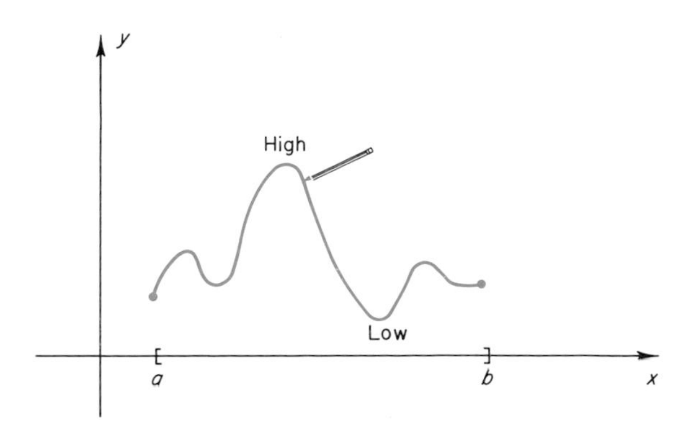

Figure 13.12

removing our pencil from the paper until we are above the right-hand endpoint b, then it is intuitively clear that our pencil will reach a highest point and a lowest point somewhere along the curve (Figure 13.12). This suggests the following result, which we state without proof.

Extreme Value Theorem If a function f is continuous at each point of a closed interval $[a, b]$, then f has an absolute maximum and an absolute minimum on $[a, b]$.

In this theorem the requirements that f be continuous and that the interval be closed are essential. For example, the discontinuous function in Figure 13.13a has no absolute maximum on $[a, b]$, and the continuous

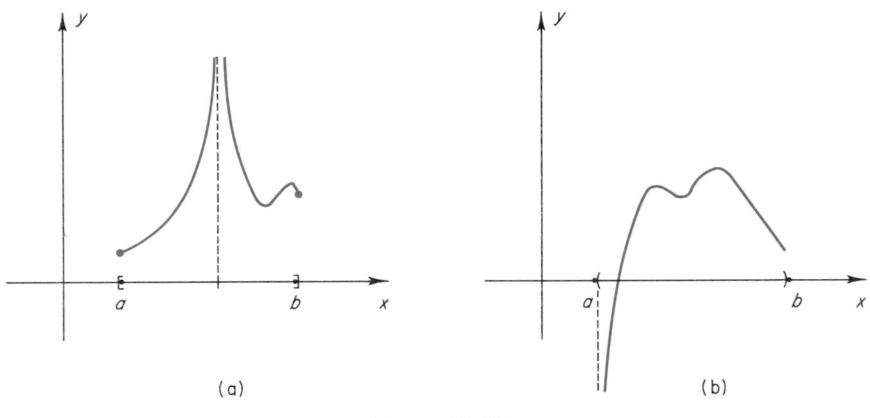

Figure 13.13

function in Figure 13.13b has no absolute minimum on the open interval (a, b).

If f is differentiable at each point of a closed interval $[a, b]$, then there are two possibilities for the location of an absolute maximum; it can occur at one of the endpoints a or b (Figure 13.14a), or it can occur at a point where the derivative of f is zero (Figure 13.14b). The same possibilities hold for an absolute minimum.

There is some standard terminology that will be helpful to fix these

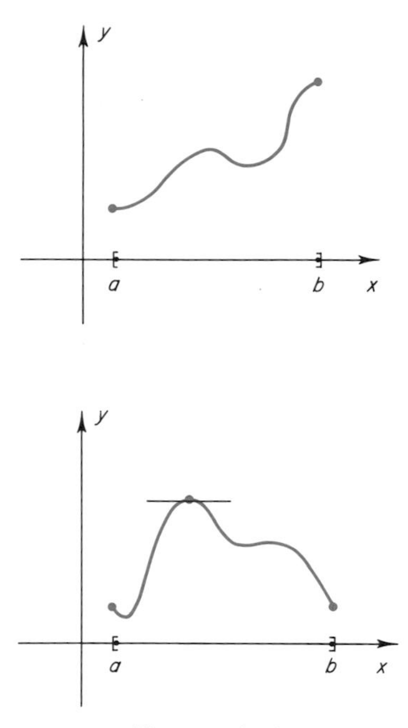

Figure 13.14

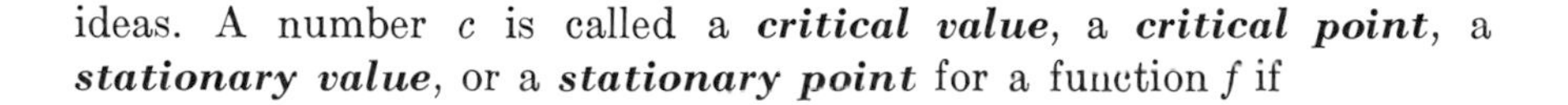

ideas. A number c is called a ***critical value***, a ***critical point***, a ***stationary value***, or a ***stationary point*** for a function f if

$$f'(c) = 0$$

Using this terminology we can summarize the results above as follows:

> **Theorem** If f is differentiable at each point of a given interval and if f has an absolute maximum on the interval, then this maximum must occur at an endpoint (if there is one) or at a critical point. Similarly for an absolute minimum.

In the case where we are maximizing or minimizing a continuous function over a closed interval $[a, b]$, we need not worry about the existence of this maximum or minimum; the existence is assured by the extreme value theorem.

Procedure for Finding an Absolute Maximum or Minimum of a Differentiable Function on a Closed Interval $[a, b]$.

Step 1 Find the critical values of f on $[a, b]$.

Step 2 Evaluate f at the critical values and endpoints a and b.

Step 3 The largest of the values obtained in Step 2 is the absolute maximum for f, and the smallest is the absolute minimum.

When we are concerned with maximum or minimum values of f over other kinds of intervals, we must worry about the possibility that the maximum or minimum may not exist. We consider such problems in the next section.

Example 12 Find the absolute minimum and maximum of the function defined by

$$f(x) = 2x^3 - 3x^2 - 36x + 5$$

on the interval $[-5, 5]$.

Solution ***Step 1.*** We can find the critical values of f on $[a, b]$ by setting $f'(x) = 0$. We obtain

$$f'(x) = 6x^2 - 6x - 36 = 0$$

$$6(x^2 - x - 6) = 0$$

$$6(x - 3)(x + 2) = 0,$$

so that the critical values of f are

$$x = -2 \quad \text{and} \quad x = 3.$$

Step 2. We evaluate f at the critical values and at the endpoints. This yields

$$f(-2) = 2(-2)^3 - 3(-2)^2 - 36(-2) + 5 = 49$$

$$f(3) = 2(3)^3 - 3(3)^2 - 36(3) + 5 = -76$$

$$f(-5) = 2(-5)^3 - 3(-5)^2 - 36(-5) + 5 = -140$$

$$f(5) = 2(5)^3 - 3(5)^2 - 36(5) + 5 = 0.$$

Hence, the absolute minimum value of f is -140 and it occurs at $x = -5$; the absolute maximum value of f is 49 and it occurs at $x = -2$.

Example 13 A thin rectangular sheet of cardboard 16 cm × 30 cm will be used to make a box by cutting a square from each corner and folding up the sides (Figure 13.15). What size square should be cut from each corner

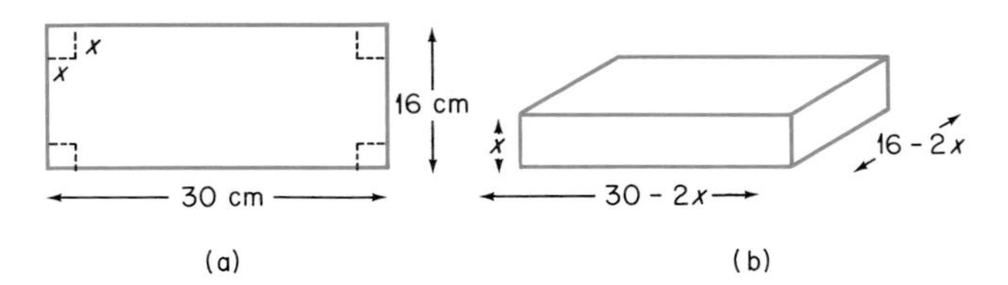

Figure 13.15

to yield a box of maximum possible volume? What is the maximum volume?

Solution The volume V of the box is the product of its length, width, and height so that

$$V = (30 - 2x)(16 - 2x)x = 4x^3 - 92x^2 + 480x \tag{5}$$

(Figure 13.15*b*). The variable x in this formula satisfies certain physical restrictions. We must have $x \geq 0$ since x is a length; and since the minimum dimension of the cardboard is 16 cm, we must have $2x \leq 16$ or $x \leq 8$ (Figure 13.15*a*). Thus

$$0 \leq x \leq 8.$$

Therefore, our problem reduces to maximizing (5) over the closed interval [0, 8]. We begin by locating the critical points of V that lie in this interval. Since

$$\frac{dV}{dx} = 12x^2 - 184x + 480$$

the critical points of V occur when $dV/dx = 0$ or

$$12x^2 - 184x + 480 = 0$$

or on dividing by 4

$$3x^2 - 46x + 120 = 0.$$

Using the quadratic formula to solve this equation, we obtain

$$x = \frac{-b \pm \sqrt{b^2 - 4ac}}{2a} = \frac{46 \pm \sqrt{(46)^2 - 4(3)(120)}}{2(3)} = \frac{46 \pm 26}{6},$$

which yields the critical values

$$x = \tfrac{10}{3} \qquad \text{and} \qquad x = 12.$$

Since the value $x = 12$ is outside the interval $[0, 8]$, we need only check the value of V at the endpoints of $[0, 8]$ and at $x = \tfrac{10}{3}$. We obtain

x	0	$\frac{10}{3}$	8
$V = (30 - 2x)(16 - 2x)x$	0	$\frac{19600}{27}$	0

Thus a maximum volume of

$$V = \frac{19600}{27} \approx 726 \quad \text{cm}^3$$

occurs when

$$x = \frac{10}{3} \quad \text{cm.}$$

Example 14 (Profit Analysis) Suppose a manufacturer is limited by the firm's production facilities to a daily output of at most 80 units and suppose that the daily cost and revenue functions are

$$C(x) = x^2 + 4x + 200 \quad \text{(dollars)}$$

$$R(x) = 108x - x^2 \quad \text{(dollars)}$$

(see Marginal Analysis, Section 12.3). Assuming that all units produced are sold, how many units should be manufactured daily to maximize the profit?

Solution We want to determine a value of x in the interval $[0, 80]$ at which the profit function

$$\begin{aligned} P(x) = R(x) - C(x) &= (108x - x^2) - (x^2 + 4x + 200) \\ &= -2x^2 + 104x - 200 \end{aligned}$$

has its maximum value. Since

$$P'(x) = -4x + 104$$

the critical value of $P(x)$ occurs when

$$-4x + 104 = 0$$

or

$$x = 26.$$

Evaluating $P(x)$ at the endpoints of [0, 80] and at this critical point we obtain:

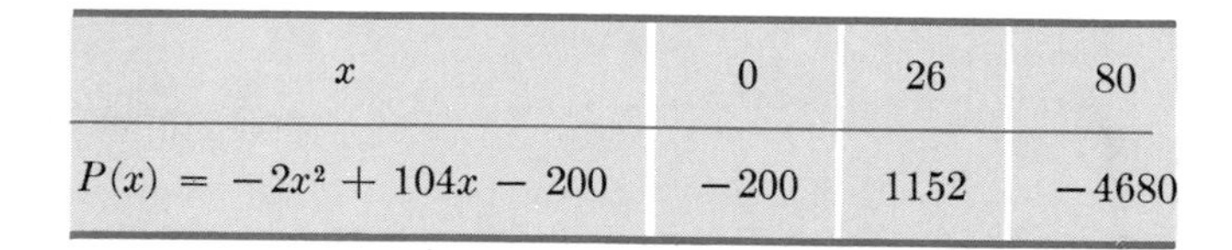

x	0	26	80
$P(x) = -2x^2 + 104x - 200$	-200	1152	-4680

Thus a maximum daily profit of $1152 is achieved by producing and selling 26 units per day.

Inventory Control

One of the important problems in any retail business is ***inventory control.*** On the one hand there must be enough inventory to meet demand, and on the other hand the business must avoid excess inventory since this results in unnecessary storage, insurance, and management costs. Moreover, money not committed to inventory can be earning interest elsewhere. As an example, suppose a retailer of automobile tires expects to sell 8000 tires during the year with sales occurring at a fairly constant rate. Although the retailer could order the 8000 tires all at once, this would result in high ***holding costs*** (insurance, storage rental, security, and so forth). To reduce the holding costs the retailer might, instead, make many smaller orders during the year. However, this results in high ***reorder costs*** (delivery charges, paperwork, loading and unloading, and so forth). Thus we are led to the problem of determining an ordering strategy that will strike an optimal balance between holding costs and reorder costs. More precisely, the problem is to minimize

$$\text{total annual inventory cost} = \text{annual holding cost} + \text{annual reorder cost}$$

by choosing an appropriate ***lot size*** (amount to be reordered each time). The lot size that minimizes the total annual inventory cost is called the ***economic ordering quantity*** (EOQ).

Example 15 A tire dealer expects to sell 8000 tires during the year with sales occurring at a relatively constant rate. The annual holding cost is $8.00 per tire, and the contract with the wholesaler calls for a flat fee of $80 per reorder, regardless of size. How many times per year and in what lot size should the dealer reorder to minimize the total annual inventory cost?

Solution We shall assume that the lot size x is the same for each order and that each order is received just as the inventory on hand falls to zero (Figure 13.16). With these assumptions, the largest number of tires on hand at any one time is x and since sales are assumed to occur at a constant rate it is reasonable that the ***average inventory*** during the year

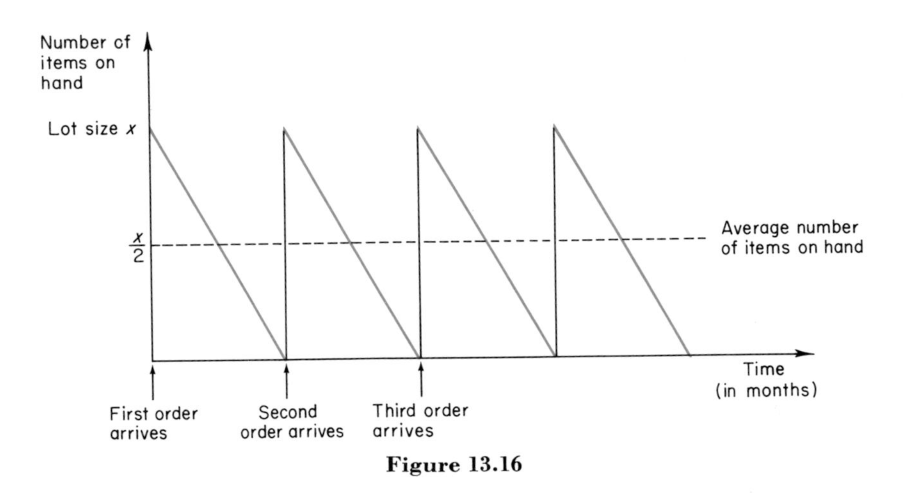

Figure 13.16

will be $x/2$ (Figure 13.16). Thus the annual holding cost $H(x)$ in dollars will be

$$H(x) = \begin{pmatrix}\text{annual holding}\\ \text{cost per tire}\end{pmatrix} \cdot \begin{pmatrix}\text{average number}\\ \text{of tires}\end{pmatrix} = 8\left(\frac{x}{2}\right) = 4x.$$

To determine the annual reorder cost we argue as follows. The number of orders is

$$\begin{pmatrix}\text{number of}\\ \text{orders}\end{pmatrix} = \frac{(\text{annual sales})}{(\text{lot size})} = \frac{8000}{x}$$

so that the annual reorder cost $O(x)$ in dollars is

$$O(x) = \begin{pmatrix}\text{cost of}\\ \text{each order}\end{pmatrix} \cdot \begin{pmatrix}\text{number of}\\ \text{orders}\end{pmatrix} = 80\left(\frac{8000}{x}\right) = \frac{640{,}000}{x}.$$

Thus the total annual inventory cost $K(x)$ is

$$K(x) = H(x) + O(x) = 4x + \frac{640{,}000}{x} \tag{6}$$

Because the total annual sales are 8000 tires, the lot size x has a minimum value of $x = 1$ and a maximum value of 8000. Thus the problem reduces to locating the minimum value of $K(x)$ for x in the interval $[1, 8000]$. We proceed as follows. From (6)

$$K'(x) = 4 - \frac{640{,}000}{x^2}.$$

Setting $K'(x) = 0$ to obtain the critical values, we have

$$4 - \frac{640{,}000}{x^2} = 0$$

or

$$4x^2 = 640{,}000$$

$$x^2 = 160{,}000.$$

Thus

$$x = 400 \quad \text{and} \quad x = -400$$

so that $x = 400$ is the only critical point in the interval [1, 8000]. Evaluating $K(x)$ at the endpoints and critical point we obtain

x	1	400	8000
$K(x) = 4x + \dfrac{640{,}000}{x}$	640,004	3200	32,080

Thus a minimum total annual inventory cost of $3200 is achieved by a lot size (EOQ) of 400 tires, which means that the dealer should meet his annual demand of 8000 tires with

$$\frac{8000}{400} = 20 \quad \text{orders}$$

of 400 tires each.

EXERCISE SET 13.2

1. Sketch the following intervals:

(a) (1, 3) (b) [1, 3] (c) [1, 3) (d) (1, 3]
(e) $(-\infty, 1)$ (f) $(-\infty, 1]$ (g) $(3, +\infty)$ (h) $[3, +\infty)$.

2. Sketch the following intervals:

(a) $(-3, -1)$ (b) $(-\infty, -4]$ (c) $[-5, +\infty)$ (d) $(0, +\infty)$
(e) $(-\infty, 0]$ (f) [0, 3] (g) $[-2, 2]$ (h) $(-4, 5)$.

In Exercises 3–10 find the absolute maximum and absolute minimum values for $f(x)$ on the given interval, and specify the values of x where the maximum and minimum occur.

3. $f(x) = 2x + 6$ on $[-1, 6]$

4. $f(x) = x^2 + 2x - 1$ on $[-2, 3]$

5. $f(x) = x - x^2$ on $[0, 2]$

6. $f(x) = x^3 - x^2$ on $[-1, 4]$

7. $f(x) = x^4 - 2x^3$ on $[0, 2]$

8. $f(x) = \dfrac{1}{x}$ on $[\frac{1}{2}, 5]$

9. $f(x) = \dfrac{2}{1 + x^2}$ on $[-2, 2]$

10. $f(x) = e^x - e^{-x}$ on $[-1, 1]$.

11. Suppose that 240 feet of fencing will be used to enclose a rectangular field. Find the dimensions of the rectangle that will yield the largest possible area.

12. A garden store wants to build a rectangular enclosure to display its shrubs. Three sides will be built from chain-link fence costing \$20 per running-foot and the remaining side will be built from cedar fence costing \$10 per running foot (Figure 13.17). Find the dimensions of the enclosure of largest possible area that can be built with \$2000 worth of fence.

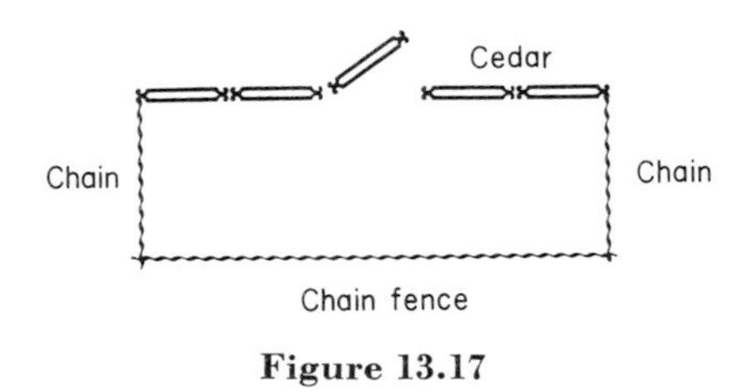

Figure 13.17

13. An agriculturalist wants to enclose two equal rectangular areas for experimentation as shown in Figure 13.18. If 360 yards of fencing are available, what is the largest total area that can be enclosed? (*Hint:* See the dimensions indicated in the figure.)

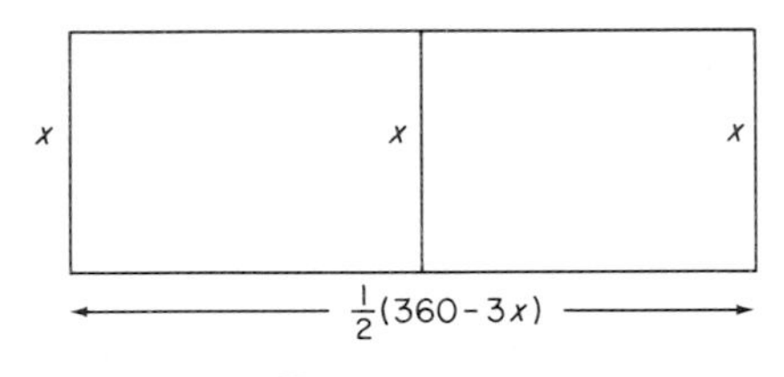

Figure 13.18

14. A square piece of cardboard 12 in. by 12 in. will be used to make an open box by cutting a square from each corner and folding up the sides (see Example 13).

(a) What size square should be cut from each corner to yield a box of maximum volume?

(b) What is the maximum volume?

15. A manufacturer of electronic components needs two wire elements: one is a circle, the other is a square. These are made by cutting a piece of wire 20 cm long into two pieces, and bending one piece into a square and the other into a circle. How long should each piece of wire be to maximize the sum of areas? (*Hint:* See Figure 13.19.)

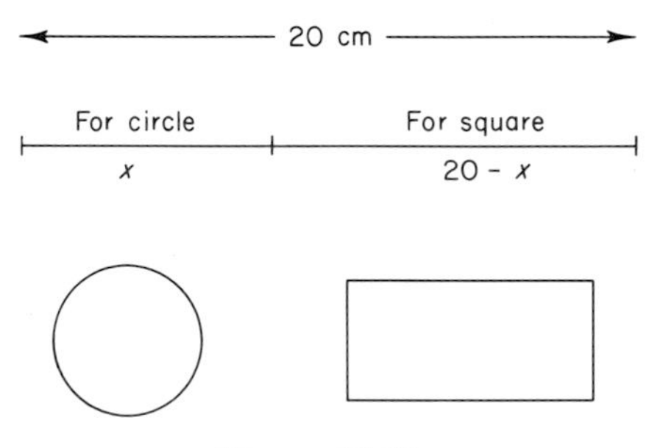

Figure 13.19

16. **(Advertising)** A manufacturer believes that the firm's yearly profit $P(x)$ (in thousands of dollars) is related to its yearly advertising expenditure x (in thousands of dollars) by $P(x) = 10 + 46x - \frac{1}{2}x^2$. Assuming the firm can afford at most \$50,000 a year for advertising, how much should be spent to maximize the profit?

17. **(Profit Analysis)** A manufacturer is limited by the firm's production facilities to an output of at most 125 units per day. The daily cost and revenue functions for producing x units are

$$C(x) = 3x^2 - 750x + 100$$

$$R(x) = 50x - x^2.$$

Assuming that all units produced are sold, how many units should be manufactured daily to maximize the profit?

18. **(Inventory Control)** A camera wholesaler expects to sell 1000 cameras during the year with sales occurring at a relatively constant rate. The annual holding cost is \$7.00 per camera and the contract with the wholesaler calls for a flat fee of \$35 for each order, regardless of size. How many times per year and in what lot size should the dealer reorder to minimize the total annual inventory cost?

19. **(Inventory Control)** A department store sells 2500 top-of-the-line refrigerators per year, with sales occurring at a relatively constant rate. The annual holding cost per refrigerator is \$10. To reorder there is a basic \$20 service fee per reorder plus a \$9 insurance charge for each refrigerator ordered. How many times per year and in what lot size should the store reorder to minimize the total annual inventory cost?

20. The U.S. Postal service has the following restriction on mailing a fourth-class parcel in the form of a rectangular box. The perimeter of one end plus the length of the box must be no more than 100 inches. What is the largest volume of a permissible rectangular parcel whose ends are squares?

21. In designing a department store it is estimated that each sales counter will produce an average daily profit of \$1000 provided the number of counters is between 0 and 20. If the number of counters is above 20, the average profit on every counter will be reduced by \$50 for each counter above 20.

How many counters should be installed for maximum average daily profit if the store has room for at most 40 counters?

22. A truck traveling at x miles per hour consumes gasoline at the rate of

$$G(x) = \frac{1}{1000}\left(\frac{2000}{x} + x\right)$$

gallons per mile.

(a) If gasoline costs 60 cents per gallon, how much will it cost for gasoline (in dollars) to travel 100 miles at x miles per hour?

(b) If a 100-mile stretch of open road has a minimum speed limit of 30 miles per hour and a maximum limit of 60 miles per hour, what is the most economical speed for the truck?

(c) What is the most economical speed if the driver is paid \$25 per hour for the 100 mile trip?

13.3 FIRST AND SECOND DERIVATIVE TESTS

In this section we develop some additional mathematical tools for solving maximization and minimization problems.

At points where $f'(x) > 0$ the curve $y = f(x)$ has a tangent with positive slope, and at points where $f'(x) < 0$ the curve has a tangent with negative slope. Thus it follows that the curve $y = f(x)$ will be *increasing* (rising) on any interval where $f'(x) > 0$ for all x in the interval and *decreasing* (falling) on any interval where $f'(x) < 0$ for all x in the interval (Figure 13.20). Recall that a value of x where $f'(x) = 0$ is called a critical value or *stationary* value for f. At such points, the curve $y = f(x)$ is neither rising nor falling.

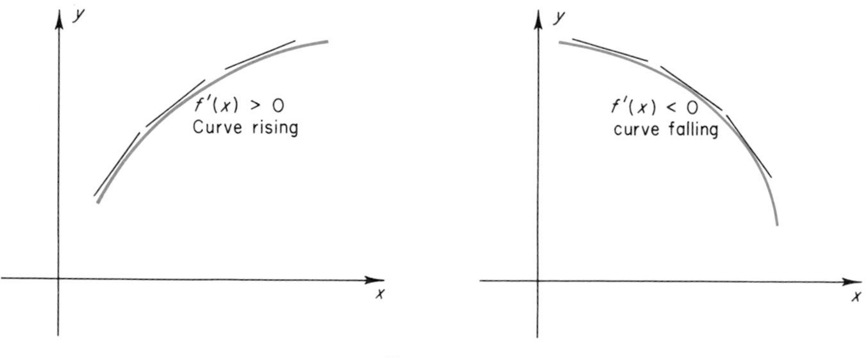

Figure 13.20

Example 16 Let

$$f(x) = x^2 - 2x + 6.$$

Where is $y = f(x)$ increasing? Decreasing? What are the stationary values for f?

Solution Since

$$f'(x) = 2x - 2,$$

the stationary value of f occurs when

$$2x - 2 = 0.$$

Solving for x yields $x = 1$ as the only stationary value.

Since we can write

$$f'(x) = 2x - 2 = 2(x - 1),$$

we see that $f'(x) > 0$ if $x > 1$ and $f'(x) < 0$ if $x < 1$. Thus, $y = f(x)$ is increasing when $x > 1$ and decreasing when $x < 1$. The graph of f is shown in Figure 13.21.

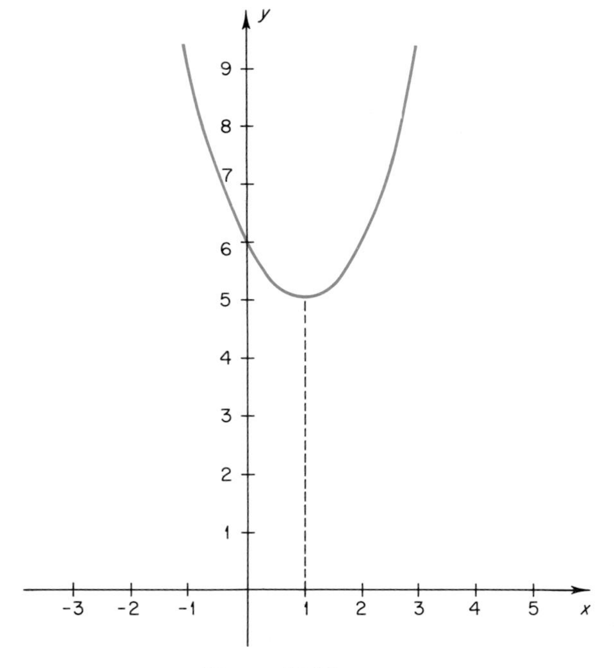

Figure 13.21

Example 17 Let

$$f(x) = 2x^3 - 3x^2 - 12x + 2.$$

Where is $y = f(x)$ increasing? Decreasing? What are the stationary values for f?

Solution Since

$$f'(x) = 6x^2 - 6x - 12 = 6(x^2 - x - 2) = 6(x - 2)(x + 1),$$

the stationary values of f occur when

$$6(x - 2)(x + 1) = 0.$$

Thus the stationary values are $x = 2$ and $x = -1$.

Since

$$f'(x) = 6(x - 2)(x + 1),$$

we have $f'(x) > 0$ when the factors $x - 2$ and $x + 1$ have the same sign, and we have $f'(x) < 0$ when the signs are opposite. From the analysis in Figure 13.22 we see that $y = f(x)$ is increasing on the intervals $(-\infty, -1)$

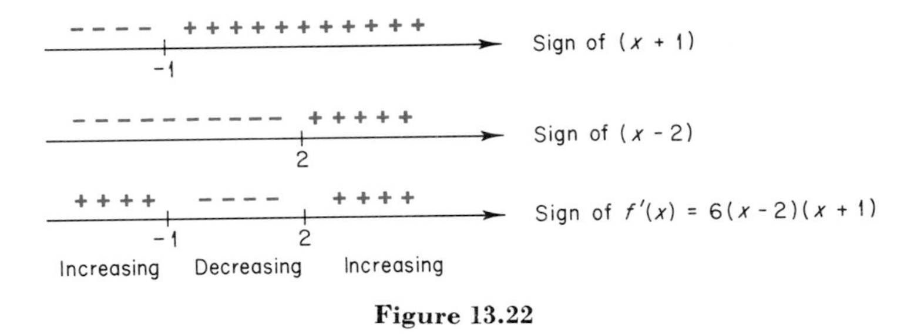

Figure 13.22

and $(2, +\infty)$ and decreasing on the interval $(-1, 2)$. The graph of f is shown in Figure 13.23.

Geometrically, the stationary points of f are the values of x at which the graph of f has a horizontal tangent. We can distinguish three types of stationary points: "peaks," "troughs," and "inflection points" according to the shape of the graph of f about the point (Figure 13.24). If a peak occurs at a stationary point of f, then f is said to have a ***relative maximum*** (or ***local maximum***) at the point; and if a trough occurs, then f is said to have a ***relative minimum*** (or ***local minimum***) at the point.

Example 18 A relative maximum or minimum may or may not be an absolute maximum or minimum for a function. For example, in Figure 13.25*a*, f has both a relative maximum and an absolute maximum at c_1, while at c_2 there is a relative minimum but not an absolute minimum.

As indicated in Figure 13.25*b*, the relative maxima and minima of a function can be identified by the behavior of the derivative in the vicinity of the point. At a relative maximum the graph of f is increasing on an interval extending left from the point ($f'(x) > 0$) and decreasing ($f'(x) < 0$) on an interval extending right from the point; at a relative minimum, the graph is decreasing on the left and increasing on the right. For inflection points the graph is either decreasing on the left and right or increasing on the left and right. In summary, we have the following important result.

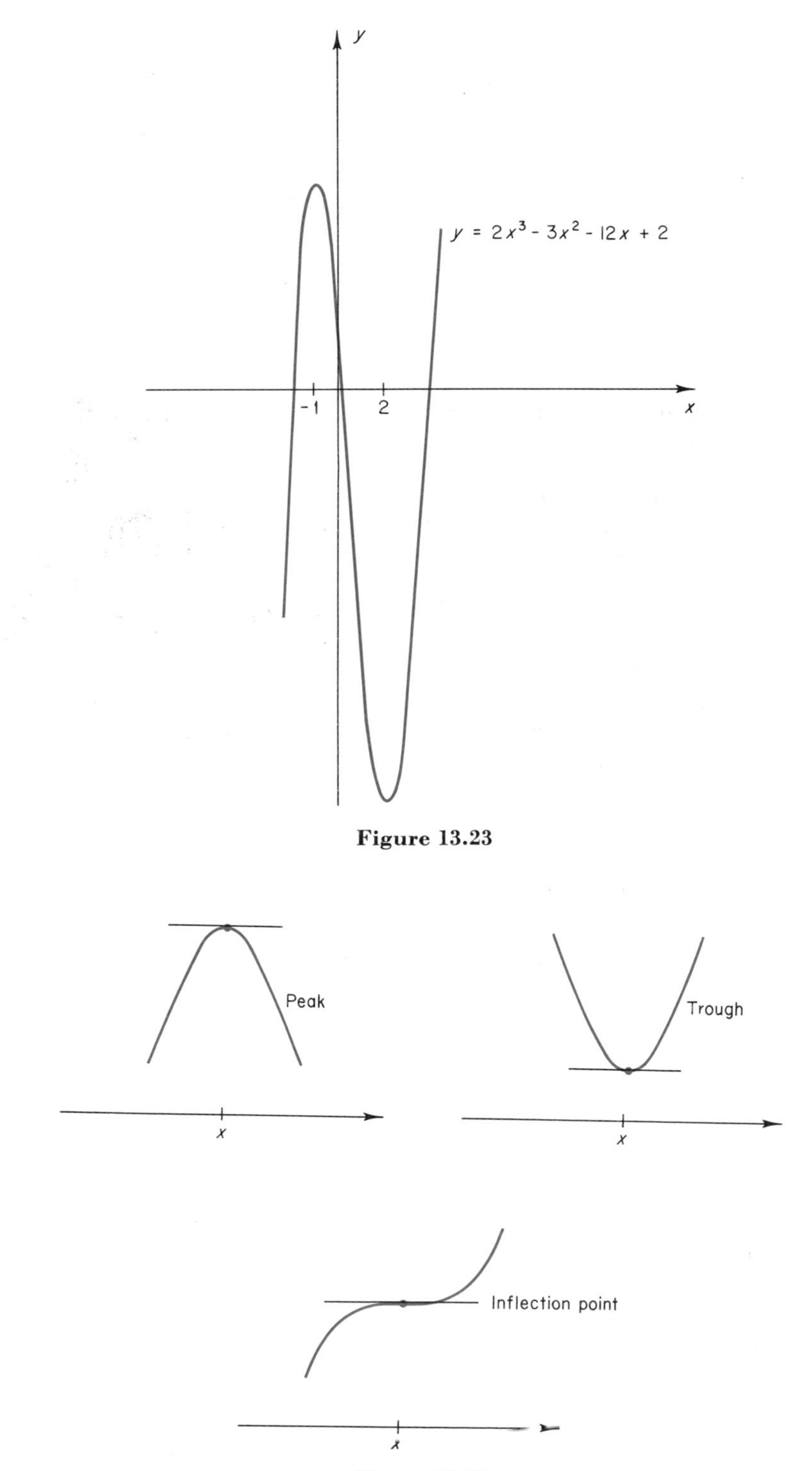

Figure 13.23

Figure 13.24

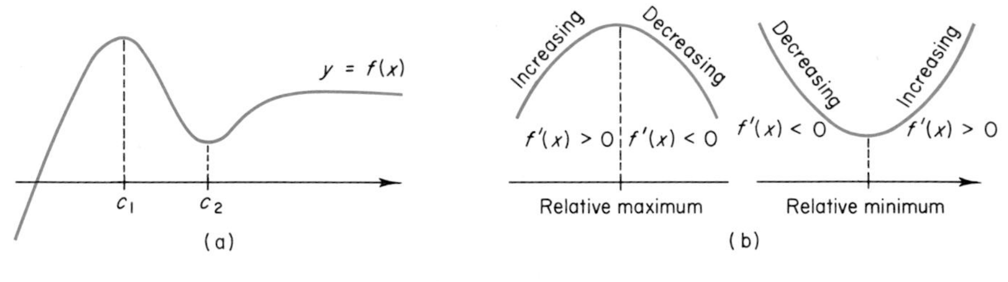

Figure 13.25

The First Derivative Test If c is a stationary point for a function f, then:

(a) f has a ***relative maximum*** at c if $f'(x) > 0$ on an interval extending left from c and $f'(x) < 0$ on an interval extending right from c.

(b) f has a ***relative minimum*** at c if $f'(x) < 0$ on an interval extending left from c and $f'(x) > 0$ on an interval extending right from c.

(c) f has neither a relative maximum nor a relative minimum at c if $f'(x) > 0$ on an interval extending left from c and on an interval extending right from c or if $f'(x) < 0$ on such intervals.

To paraphrase this result loosely, a relative maximum occurs if the sign of f' changes from $+$ to $-$ at c, a relative minimum occurs if the sign of f' changes from $-$ to $+$ at c, and neither occurs if f' does not change sign at c.

Example 19 Classify the stationary points of $f(x) = \frac{1}{3}x^3 - \frac{1}{2}x^2$.

Solution Differentiating yields

$$f'(x) = x^2 - x = x(x - 1),$$

and on setting $x(x - 1) = 0$ we see that $x = 0$ and $x = 1$ are the stationary points.

From the analysis in Figure 13.26, the sign of f' changes from $+$ to $-$ at 0, so that a relative maximum occurs at $x = 0$; also the sign of f' changes from $-$ to $+$ at 1 so that a relative minimum occurs at $x = 1$.

There is another test for relative maxima and minima that is often easier to apply than the first derivative test. This test uses the "second

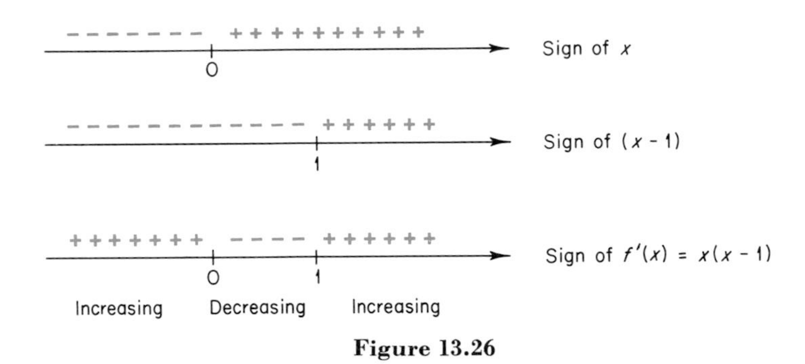

Figure 13.26

derivative" of the function, that is, the derivative of the derivative. More precisely, the ***second derivative*** of a function f is

$$\frac{d}{dx}[f'(x)]$$

and is denoted by $f''(x)$. For example, if $f(x) = x^3$, then

$$f'(x) = \frac{d}{dx}[x^3] = 3x^2$$

$$f''(x) = \frac{d}{dx}[3x^2] = 6x.$$

If $y = f(x)$, then the second derivative is also denoted by

$$y'' \quad \text{or} \quad \frac{d^2y}{dx^2} \quad \text{or} \quad \frac{d^2}{dx^2}[f(x)].$$

To see the significance of the second derivative, suppose $f''(x) > 0$ at each point x in some interval (a, b). This would indicate that the function $f'(x)$ is increasing on this interval. Geometrically, this means that slopes of the tangents to $y = f(x)$ are increasing as we travel left to right, indicating that the curve $y = f(x)$ bends upward over the interval (a, b) (Figure 13.27a). Such a curve is said to be ***concave up*** over this interval.

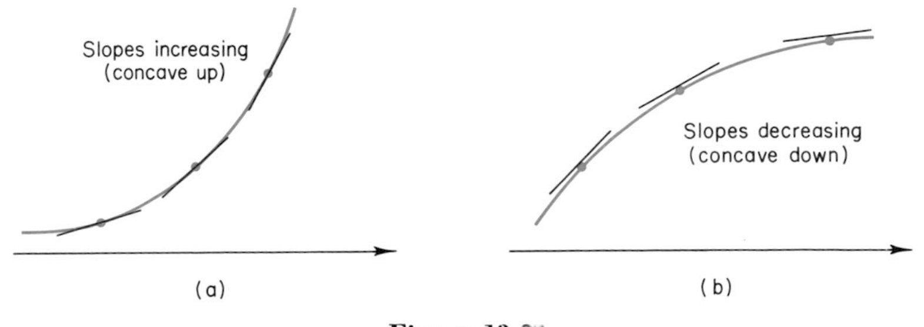

Figure 13.27

Similarly, if $f''(x) < 0$ for each x in (a, b), then the slopes of the tangents, given by $f'(x)$, will be decreasing as we travel left to right, indicating a downward bend in $y = f(x)$ over (a, b) (Figure 13.27b). Such a curve is said to be ***concave down*** over the interval.

The following theorem summarizes these results.

> **Theorem** On a given interval, the curve $y = f(x)$ is
>
> (a) concave up if $f''(x) > 0$ at each point in the interval;
>
> (b) concave down if $f''(x) < 0$ at each point in the interval.

Some people think of a curve that is concave up as one that "holds" water and a curve that is concave down as one that "spills" water. With this in mind, Figure 13.28 may help to remember the above theorem.

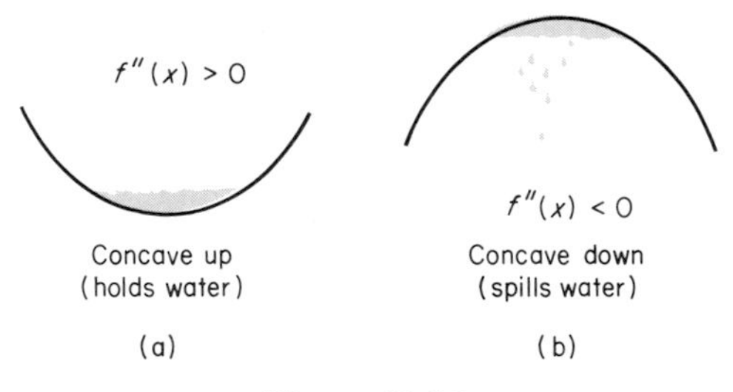

Figure 13.28

Example 20 Let

$$f(x) = 2x^3 - 3x^2 - 12x + 2.$$

Where is $y = f(x)$ concave up? Concave down?

Solution We have

$$f'(x) = 6x^2 - 6x - 12$$

$$f''(x) = 12x - 6 = 6(2x - 1).$$

Thus $f''(x) > 0$ if

$$2x - 1 > 0 \quad \text{or} \quad 2x > 1 \quad \text{or} \quad x > \tfrac{1}{2},$$

and $f''(x) < 0$ if

$$2x - 1 < 0 \quad \text{or} \quad 2x < 1 \quad \text{or} \quad x < \tfrac{1}{2}.$$

Therefore the graph is concave up if $x > \frac{1}{2}$ and concave down if $x < \frac{1}{2}$ (see Figure 13.29).

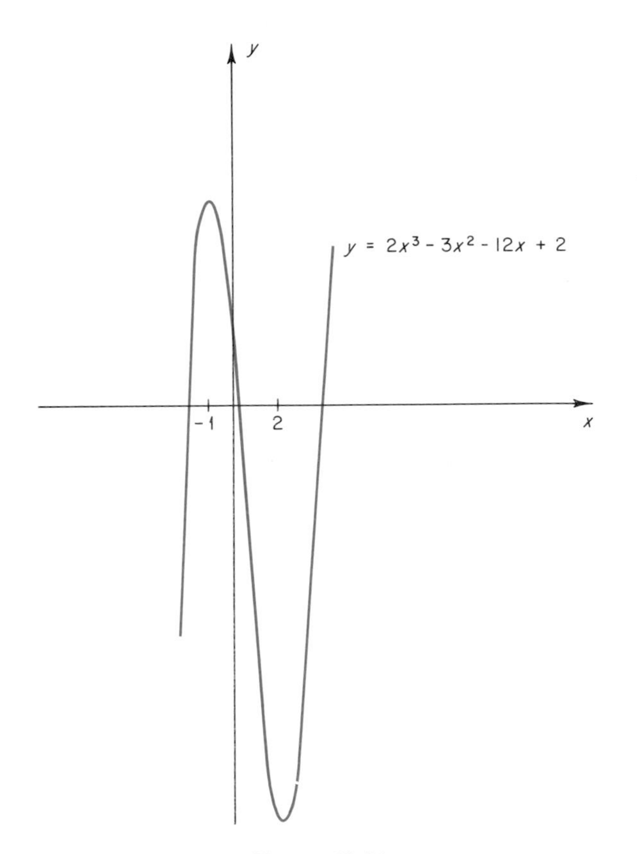

Figure 13.29

In the above example, the curve changed from concave down to concave up at $x = \frac{1}{2}$; a value of x where a curve changes from concave down to concave up or vice versa is called an ***inflection point*** for the curve.

Analysis of Unemployment

Example 21 In times of economic recession or depression, unemployment tends to grow at an ever increasing rate until it is brought under control by appropriate economic policy and pressures from the marketplace. Figure 13.30 shows a typical unemployment curve. The inflection point in July indicates that unemployment is coming under control. After this point is reached the curve is concave down, indicating a declining growth rate in unemployment. One can then anticipate that unemployment will soon "peak out" and begin to decline. In everyday language one might say that the economy "turned the corner" on unemployment in July.

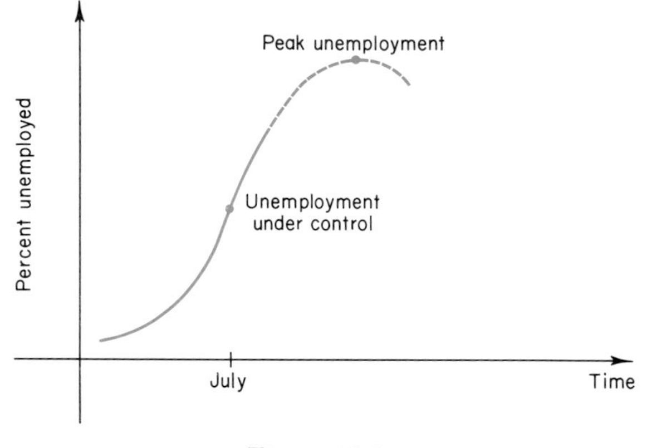

Figure 13.30

We now show how the second derivative can be used to "test" stationary points to determine whether they correspond to relative maxima or to relative minima. Assume that c is a critical point for f. If $f''(c) > 0$, then the curve $y = f(x)$ is concave up about c (Figure 13.28*a*) indicating a relative minimum at c. On the other hand, if $f''(c) < 0$, then $y = f(x)$ is concave down about c (Figure 13.28*b*), indicating a relative maximum at c. This suggests the following result.

The Second Derivative Test

(a) If $f'(c) = 0$ and $f''(c) > 0$, then f has a relative minimum at c.

(b) If $f'(c) = 0$ and $f''(c) < 0$, then f has a relative maximum at c.

Example 22 Use the second derivative test to locate the relative maxima and relative minima of $f(x) = \frac{1}{3}x^3 - \frac{1}{2}x^2$.

Solution We have

$$f'(x) = x^2 - x$$

$$f''(x) = 2x - 1.$$

On setting $f'(x) = x^2 - x = x(x - 1) = 0$, we obtain the stationary points $x = 0$ and $x = 1$. Since

$$f''(0) = 2(0) - 1 = -1 < 0,$$

there is a relative maximum at $x = 0$; and since

$$f''(1) = 2(1) - 1 = 1 > 0,$$

there is a relative minimum at $x = 1$. This agrees with the results obtained in Example 19, using the first derivative test.

The reader should note that the second derivative test does not specify the nature of the stationary point when $f''(c) = 0$. In this case there may be a relative maximum, a relative minimum, or neither at c. To determine what happens at such a point, we must rely on the first derivative test.

EXERCISE SET 13.3

Exercises 1–3 refer to Figure 13.31.

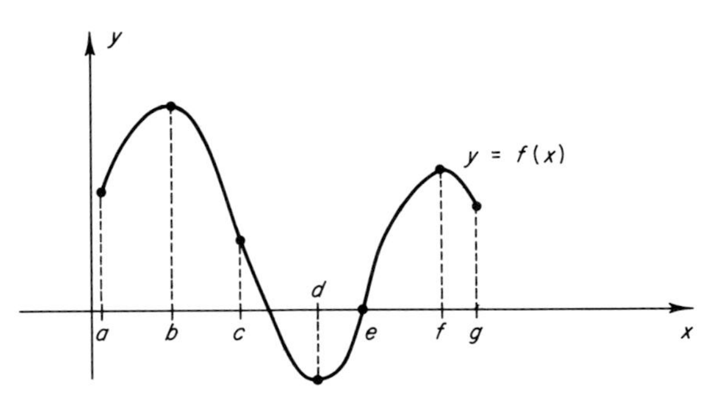

Figure 13.31

1. Over what open intervals is $y = f(x)$
(a) increasing? (b) decreasing?
(c) concave up? (d) concave down?

2. (a) Where is $f'(x) = 0$?
(b) Over what open intervals is $f'(x) > 0$?
(c) Over what open intervals is $f'(x) < 0$?
(d) Over what open intervals is $f''(x) > 0$?
(e) Over what open intervals is $f''(x) < 0$?

3. At what points does f have
(a) a relative maximum? (b) a relative minimum?
(c) an inflection point? (d) a stationary value?

Exercises 4–6 refer to Figure 13.32.

4. Over what open intervals is $y = f(x)$
(a) increasing? (b) decreasing?
(c) concave up? (d) concave down?

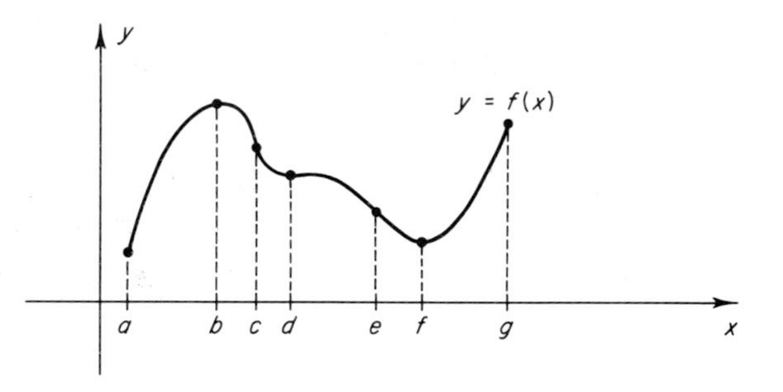

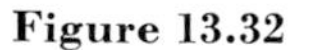

Figure 13.32

5. (a) Where is $f'(x) = 0$?

(b) Over what open intervals is $f'(x) > 0$?

(c) Over what open intervals is $f'(x) < 0$?

(d) Over what open intervals is $f''(x) > 0$?

(e) Over what open intervals is $f''(x) < 0$?

6. At what points does f have

(a) a relative maximum? (b) a relative minimum?

(c) an inflection point? (d) a stationary value?

In Exercises 7–16 where is $y = f(x)$ increasing? Decreasing? Locate and classify the stationary values as relative maxima, relative minima, or neither.

7. $f(x) = x^2 - 6x + 3$

8. $f(x) = 2x^2 + 12x - 3$

9. $f(x) = -2x + 1$

10. $f(x) = 4x - 2$

11. $f(x) = 4x^2 + 4$

12. $f(x) = \frac{5}{2}x^2 - \frac{10}{3}x^3$

13. $f(x) = x^3 + 3x^2 - 9x + 2$

14. $f(x) = 2x^3 + 3x^2 - 12x + 8$

15. $f(x) = e^{x^2}$

16. $f(x) = x \ln x$.

In Exercises 17–22 find $f''(x)$.

17. $f(x) = 3x^2 + 2x + 1$

18. $f(x) = 4x^5 - 2x^4$

19. $f(x) = 3x + 5$

20. $f(x) = xe^x$

21. $f(x) = x \ln x$

22. $f(x) = 1/x$.

In Exercises 23–28 where is the curve concave up? Concave down?

23. $f(x) = x^2 + 3x - 10$

24. $f(x) = 2x^3 + 3x^2 - 36x + 5$

25. $f(x) = x^3 - 6x^2 - 36x + 15$

26. $f(x) = x^4 - 4x^3 + 8$

27. $f(x) = (x - 3)^5$

28. $f(x) = xe^{-x}$.

29. Let $f(x) = x^3 - 3x + 9$. Locate the relative maxima and minima using:

(a) the first derivative test,

(b) the second derivative test.

30. Let $f(x) = 2x^3 - 3x^2 - 12x + 18$. Locate the relative maxima and minima using:

(a) the first derivative test,

(b) the second derivative test.

In Exercises 31–36 use any method to locate the relative maxima and relative minima, if any.

31. $f(x) = x^4 - 2x^3$

32. $f(x) = x^4$

33. $f(x) = x^3$

34. $f(x) = \frac{1}{4}x^4 - \frac{1}{2}x^2 + 2$

35. $f(x) = xe^x$

36. $f(x) = \dfrac{1}{x^2 + 1}$

37. Determine where the curve is increasing, decreasing, concave up, and concave down. Locate all relative maxima, minima, and inflection points, and then use all this information to help sketch the curve

(a) $y = x^2 + x + 1$ (b) $y = x^3 - 12x^2 + 36x$.

38. Follow the directions of Exercise 37 for

(a) $y = xe^x$ (b) $y = x + \dfrac{1}{x}$.

39. **(Learning)** Learning of most skills starts at a rapid rate and then slows down. A psychologist measures the learning performance of a laboratory rat by a numerical score on a standardized test. Assume the rat's score $P(t)$ after t weeks of learning is

$$P(t) = 15t^2 - t^3.$$

At what point in time does the rat's *rate* of learning begin to decline?

40. **(Marginal Analysis)** Suppose a manufacturer's profit function $P(x)$ has a stationary value at a production level of x_0 units. Show that the marginal cost and marginal revenue at this production level are equal.

13.4 MORE APPLICATIONS TO OPTIMIZATION

In this section we consider some maximization and minimization problems to which the extreme value theorem does not apply. We begin with an example.

Example 23 A manufacturer wants to design a closed can in the shape of a right circular cylinder, having a storage capacity (volume) of 100 cubic inches. What dimensions should be chosen to minimize the amount of metal needed for its manufacture?

Solution The dimensions of the can may be specified by its radius r and height h (Figure 13.33). From geometry we know that the volume V of such a

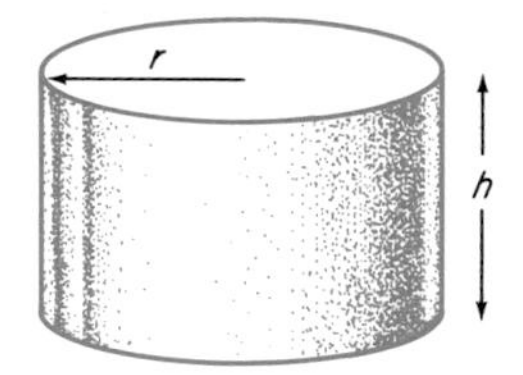

Figure 13.33

cylinder is

$$V = \pi r^2 h$$

and since the volume V is required to be 100 in.^{3}, r and h must satisfy

$$100 = \pi r^2 h$$

or

$$h = \frac{100}{\pi r^2}. \tag{1}$$

Thus, once the optimal value for r is obtained, the optimal value of h is automatically determined by Equation (1). If we assume the can is made from sheet metal of uniform thickness, then the amount of metal required is determined by the surface area of the can. The area is made up of a circular top and bottom, and a side made by bending and welding a rectangular sheet (Figure 13.34). As indicated in the figure, the height

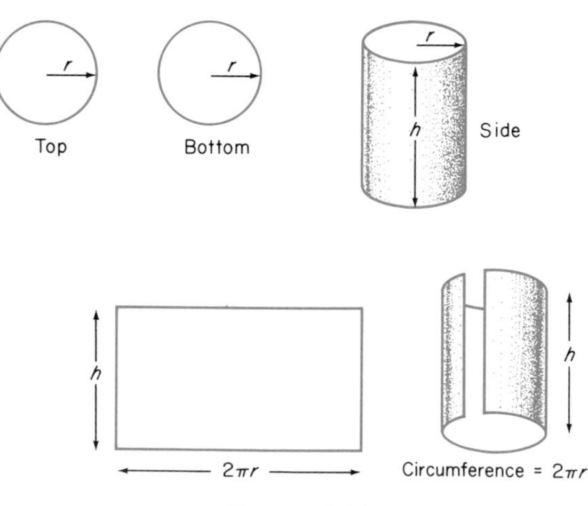

Figure 13.34

of the rectangle will be the height of the can and the length of the rectangle will be the circumference of the can (which is $2\pi r$, as the reader may recall from geometry). Thus the surface area S will be

$$S = \underbrace{\pi r^2}_{\substack{\text{area}\\\text{of}\\\text{top}}} + \underbrace{\pi r^2}_{\substack{\text{area}\\\text{of}\\\text{bottom}}} + \underbrace{2\pi rh}_{\substack{\text{area}\\\text{of}\\\text{side}}} = 2\pi r^2 + 2\pi rh.$$

Since the height of the can must satisfy Equation (1), we can rewrite the surface area S in terms of r alone; we obtain

$$S = 2\pi r^2 + 2\pi r\left(\frac{100}{\pi r^2}\right) = 2\pi r^2 + \frac{200}{r}. \tag{2}$$

What are the physical restrictions on r? Because a can cannot have a negative radius and because we cannot build a can of zero radius, we must have

$$r > 0$$

that is, r must lie in the interval $(0, +\infty)$. Since this is not a closed interval $[a, b]$, the extreme value theorem does not apply; thus we have no assurance that S actually has a minimum value for $r > 0$. However, these doubts can be removed by graphing the equation

$$S = 2\pi r^2 + \frac{200}{r}$$

(Figure 13.35). As before, we can find the low point on this graph by

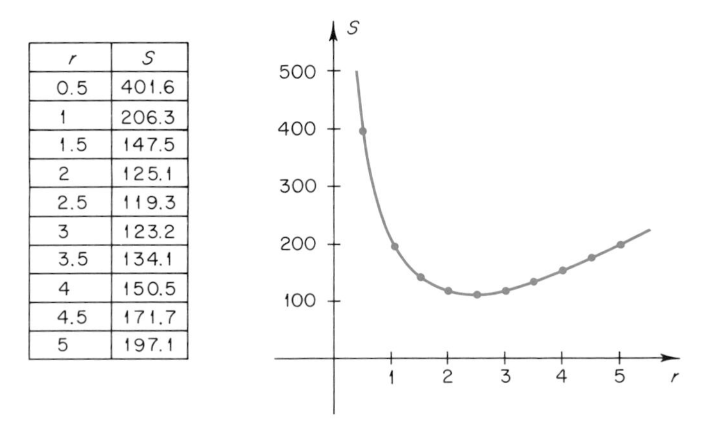

r	S
0.5	401.6
1	206.3
1.5	147.5
2	125.1
2.5	119.3
3	123.2
3.5	134.1
4	150.5
4.5	171.7
5	197.1

Figure 13.35

determining where the tangent is horizontal. The calculations are as follows:

$$\begin{aligned}\frac{dS}{dr} &= \frac{d}{dr}\left[2\pi r^2 + \frac{200}{r}\right] \\ &= 2\pi \frac{d}{dr}[r^2] + 200 \frac{d}{dr}\left[\frac{1}{r}\right] \\ &= 2\pi(2r) + 200\left(-\frac{1}{r^2}\right) \\ &= 4\pi r - \frac{200}{r^2}.\end{aligned}$$

Setting $dS/dr = 0$, we obtain

$$4\pi r - \frac{200}{r^2} = 0,$$

or on multiplying through by r^2

$$4\pi r^3 - 200 = 0$$

or

$$r = \sqrt[3]{\frac{200}{4\pi}} = \sqrt[3]{\frac{50}{\pi}}.$$

Thus the minimum amount of metal will be used when the radius is

$$r = \sqrt[3]{\frac{50}{\pi}} \approx 2.52 \quad \text{in.}$$

and the height is (see (1))

$$h = \frac{100}{\pi r^2} = \frac{100}{\pi(\sqrt[3]{50/\pi})^2} \approx 5.03 \quad \text{in.}$$

(The approximate values for r and h were obtained using a hand-held calculator. The approximate value for r is in agreement with Figure 13.35.)

It is interesting to note that the surface area S does not have a maximum value (Figure 13.35). Thus, had we been interested in choosing r and h to maximize S, our problem would have had no solution.

The following result is useful in optimization problems where there is only one stationary point (as in the example above).

Theorem Suppose a differentiable function f has *only one* stationary point c in a certain interval.

(a) If f has a relative maximum at c, then $f(c)$ is the maximum value of f on the interval (Figure 13.36a);

(b) If f has a relative minimum at c, then $f(c)$ is the minimum value of f on the interval (Figure 13.36b).

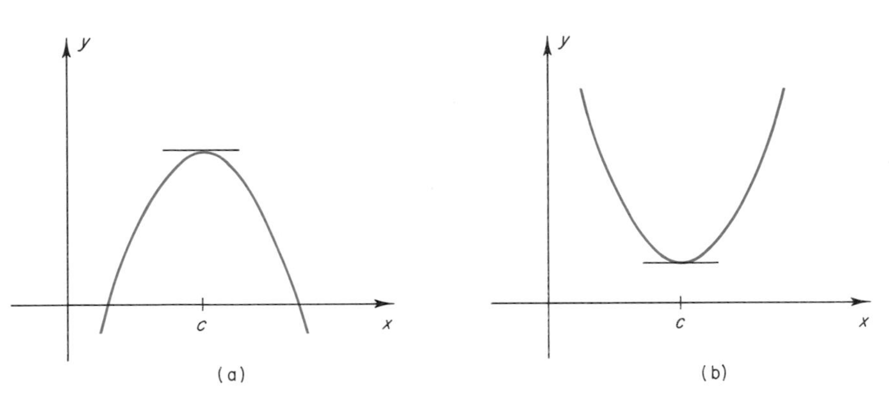

Figure 13.36

Example 24 In Example 23 we had to graph the surface area function in order to deduce that the surface area could be minimized. This tedious graphing can be avoided by using the preceding theorem and the second derivative test. We saw in Example 23 that

$$r = \sqrt[3]{\frac{50}{\pi}}$$

is the only stationary point in the interval $(0, +\infty)$ for the surface area function S. Since the second derivative

$$\frac{d^2S}{dr^2} = 4\pi + \frac{400}{r^3}$$

evaluated at the above stationary point is positive (verify), S has a relative minimum and thus an absolute minimum at this point.

Example 25 (Hydrocarbon Emissions) The rate at which hydrocarbons are emitted by an automobile engine depends on the speed of the automobile. Suppose the emission rate $R(x)$ in milligrams per minute is related to the speed x in miles per hour by

$$R(x) = xe^{-x/30}.$$

At what speed is the emission rate $R(x)$ a maximum?

Solution Since x represents speed, we must have $x \geq 0$; so the problem is to maximize $R(x)$ over the interval $[0, +\infty)$. We first find the stationary values of $R(x)$. Using the product rule for derivatives, we obtain

$$R'(x) = x\left(-\tfrac{1}{30}e^{-x/30}\right) + e^{-x/30} = e^{-x/30}\left[1 - \tfrac{1}{30}x\right].$$

Setting $R'(x) = 0$ and dividing by $e^{-x/30}$ yields

$$1 - \tfrac{1}{30}x = 0.$$

Solving gives $x = 30$ as the only stationary point. To determine the nature of this point, we apply the second derivative test. Differentiating $R'(x)$ above yields

$$R''(x) = e^{-x/30}\left(-\tfrac{1}{30}\right) + \left[1 - \tfrac{1}{30}x\right]\left(-\tfrac{1}{30}e^{-x/30}\right)$$

so that

$$R''(30) = e^{-1}\left(-\frac{1}{30}\right) + 0 = -\frac{1}{30e}.$$

Thus $R''(30) < 0$, which means that $R(x)$ has a relative maximum and an absolute maximum at $x = 30$. Therefore the maximum rate of hydrocarbon emission occurs when the automobile is driven at 30 miles per hour.

Example 26 (Oil Pricing Policies) An oil producing country is selling 1,000,000 barrels per day at \$10 per barrel. A price increase is contemplated, but it is estimated that each 10 cent per barrel increase in price will result in 5000 fewer barrels sold per day. How much of an increase, if any, should be made to maximize the daily revenue?

Solution If we denote the price increase (in dollars) by x, then

$$10x$$

denotes the number of 10 cent price increases and

$$(5000)(10x) = (50{,}000)x$$

represents the decrease in the number of barrels sold per day. Thus with a price increase of x dollars per barrel, the number of barrels sold per day will be

$$1{,}000{,}000 - 50{,}000x$$

and the daily revenue $R(x)$ in dollars at the new selling price of $10 + x$ dollars will be

$$R(x) = \begin{pmatrix}\text{number of}\\ \text{barrels sold}\end{pmatrix} \cdot \begin{pmatrix}\text{price per}\\ \text{barrel}\end{pmatrix} = (1{,}000{,}000 - 50{,}000x)(10 + x)$$

or

$$R(x) = 10{,}000{,}000 + (500{,}000)x - (50{,}000)x^2.$$

Since the price increase x can be any nonnegative number, the problem reduces to maximizing $R(x)$ over the interval $[0, +\infty)$. We begin by finding the stationary points:

$$R'(x) = 500{,}000 - 100{,}000x = 0,$$

so

$$x = 5$$

is the only critical point. Since

$$R''(x) = -100{,}000$$

we have

$$R''(5) = -100{,}000 < 0,$$

so $R(x)$ has a relative maximum and an absolute maximum at $x = 5$. Thus the oil producer should raise the price by \$5.00, to \$15.00 per barrel, to maximize the revenue.

Example 27 A utility company is planning to run a power line from a generator located at a point A on a straight shoreline to an offshore oil rig located at a point B, which is 2 miles from the closest shore point C (Figure 13.37). The point C is 8 miles down the coast from A. The line is to run

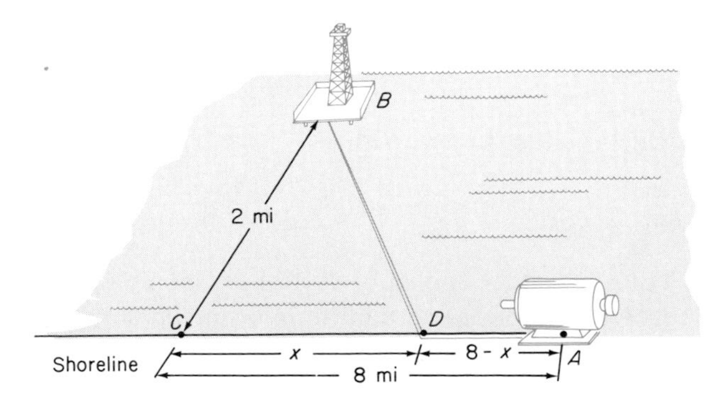

Figure 13.37

from A along the shore to a point D between A and C and then straight to B. If it costs \$1000 per mile to run line under water and \$800 per mile under ground, where should the point D be located to minimize the total cost?

Solution Let x denote the length of CD, so that x satisfies $0 \leq x \leq 8$. The length of the power line running under water is the length of the line segment BD, which is, by the Pythagorean theorem,

$$\sqrt{x^2 + 2^2} = \sqrt{x^2 + 4}.$$

The length of the power line under ground is the length of AD, which is $8 - x$. Thus the total cost is given by

$$C(x) = 800(8 - x) + 1000\sqrt{x^2 + 4}.$$

To find the minimum value of $C(x)$, we find its stationary points:

$$C'(x) = -800 + 1000 \cdot \frac{1}{2} \cdot \frac{2x}{\sqrt{x^2 + 4}} = 0$$

or

$$-800 + \frac{1000x}{\sqrt{x^2 + 4}} = 0$$

or

$$4\sqrt{x^2 + 4} = 5x.$$

Squaring both sides and simplifying, we obtain

$$16(x^2 + 4) = 25x^2$$

or

$$9x^2 = 64$$

which yields the two values

$$x = \tfrac{8}{3} \qquad \text{and} \qquad x = -\tfrac{8}{3}.$$

Thus $x = \frac{8}{3}$ is the only stationary point in the interval $[0, 8]$. We can determine whether a minimum occurs at $x = \frac{8}{3}$ by using the second derivative test, or since $[0, 8]$ is a closed interval, by evaluating $C(x)$ at the end points and the critical point. Because the second derivative of $C(x)$ is a little messy to calculate, we shall use the second approach.

x	0	$\frac{8}{3}$	8
$C(x) = 800(8 - x) + 1000\sqrt{x^2 + 4}$	\$8400	\$7600	$1000\sqrt{68} \approx$ \$8246.21

Thus a minimum cost of \$7600 is achieved by choosing $D = \frac{8}{3}$ mile from C.

EXERCISE SET 13.4

In Exercises 1–4 find the maximum and minimum values of f on the stated interval. If a maximum or minimum does not exist, say so.

1. $f(x) = x^2 - 3x + 2$ on $(-\infty, +\infty)$

2. $f(x) = 1 + 4x - \frac{1}{3}x^3$ on $(0, +\infty)$

3. $f(x) = 4x^2 + (1/x)$ on $(0, 1)$

4. $f(x) = xe^{-x}$ on $(-\infty, +\infty)$.

5. Suppose that x hours after a drug is administered to a patient, the drug's concentration in the body is given by

$$k(x) = \frac{2x}{x^2 + 4}.$$

How long after the drug has been administered is the concentration at its maximum value?

6. **(Epidemic)** Suppose that the number of people in a certain city who are affected by an epidemic is given by

$$P(t) = -t^2 + 80t + 10$$

where t is the number of days after the disease has been detected. On what day will the maximum number of people be affected, and how many people will be affected on this day?

7. Find the minimum value of $x^2 + 432/x$ for x in the interval $(0, +\infty)$.

8. It costs a pharmaceutical firm $x^3 + 200x + 1000$ dollars to manufacture x gallons of penicillin that it sells for 500 dollars a gallon.
(a) What size batch should it produce to achieve the maximum profit on the batch?
(b) What is the maximum profit?

9. You are assigned the job of designing a rectangular box with a square bottom and open top. If the box is required to have a volume of 62.5 cubic inches, what dimensions would you use to obtain a box with minimum surface area? What is the minimum surface area?

10. A shipper needs a closed rectangular container with a volume of 96 cubic feet and a square bottom. The heavy-duty plastic needed for the top and bottom costs \$3 per square foot and the standard plastic for the sides costs

$2 per square foot. What dimensions yield a container of minimum cost? What is the minimum cost?

11. If the marginal revenue and marginal cost for producing x units are MR $= 100 - \frac{1}{20}x^2$ and MC $= 10 + \frac{1}{20}x^2$, how many items should be produced to maximize the profit?

12. An airline finds that when its cargo plane flies a full load at x miles per hour, it consumes fuel at the rate of

$$F(x) = \frac{1}{500}\left(\frac{10{,}000}{x} + x\right)$$

gallons per mile. If the cost of jet fuel is $2 per gallon, what is the most economical speed for the plane to make a 1000-mile trip? What is the minimum fuel cost for the trip?

13. It costs $5 + (x/100)$ dollars per mile to drive a truck x miles per hour. If the driver receives 16 dollars per hour, how fast should the truck be driven to minimize the cost per mile?

14. A manufacturer of bicycles finds that when x bicycles per day are produced the following costs are incurred: a fixed cost of $1000, a labor cost of $10 per bicycle, and a cost of $\$25{,}000/x$ for advertising. How many bicycles should be produced daily to minimize the total cost?

15. A bacteria colony consisting of 10,000 bacteria is treated with a drug. After t hours have elapsed, the number of bacteria present is given by

$$N(t) = 10{,}000\left[\frac{t}{50} - \ln(1 + t)\right].$$

When is the number of bacteria present a minimum?

16. A manufacturer wants to design an open can (no top) in the shape of a right circular cylinder having a storage capacity of 1000 cubic inches. What dimensions should be used to minimize the amount of metal needed for the can?

17. A tour operator charters an airplane for a trip to Europe. If 200 passengers agree to go, the price will be $400 per passenger. For each passenger under 200, the price will rise $20 per passenger. How many passengers must go to give the operator the greatest revenue? In this case what will each passenger pay for the fare? (*Hint:* Let x = the number of passengers under 200.)

18. A motel owner finds that if a rent of $20 per room is charged, then 100 rooms per night can be rented. For each $1 increase in the rate per night, two fewer rooms are rented. How much of an increase, if any, should be made to maximize the daily revenue?

19. An orange grower finds that the average yield is 120 bushels per tree when 50 trees are planted on his grove. For each additional tree planted, the average yield per tree decreases by two bushels. How many trees should be planted to maximize the yield?

20. A utility company wants to run a power line from a generator located at a point A on a straight shoreline to an offshore oil rig located at a point B

which is 6 miles from the closest shorepoint C. The point C is 7 miles down the coast from A. The line is to run from A along the shore to a point D between A and C and then straight to B. If it costs \$5000 per mile to run line under water and \$4000 per mile over land, where should the point D be located to minimize the total cost? (*Hint:* See Example 27.)

13.5 PARTIAL DERIVATIVES (Optional)

The only functions studied so far in this text have been functions of one independent variable. However, many functions in applied problems depend on more than one variable. For example, the familiar formula for the area of a rectangle

$$A = lw$$

expresses the area A as a function of two variables, the length l and the width w; and the volume V of a box:

$$V = lwh$$

is a function of the three variables length, width, and height. More generally, one can conceive of functions that depend on four, five, or even more variables. For simplicity, we shall limit the discussion in this section to functions

$$f(x, y)$$

involving two independent variables x and y.

Let $f(x, y)$ be a function of two variables, and let us introduce a dependent variable z as

$$z = f(x, y).$$

We shall be interested in studying the rate at which z changes as we vary the independent variables x and y. Of particular importance are the rate of change of z with respect to x when the variable y remains constant, and the rate of change of z with respect to y when the variable x remains constant. This leads us to the following definitions:

If $z = f(x, y)$, then the ***partial derivative of z with respect to x*** is the derivative of z with respect to x with y treated as a constant. This partial derivative is denoted by

$$\frac{\partial z}{\partial x} \quad \text{or} \quad \frac{\partial f}{\partial x} \quad \text{or} \quad f_x$$

If $z = f(x, y)$, then the ***partial derivative of z with respect to y*** is the derivative of z with respect to y with x treated as a constant. This partial derivative is denoted by

$$\frac{\partial z}{\partial y} \quad \text{or} \quad \frac{\partial f}{\partial y} \quad \text{or} \quad f_y$$

The symbols $\partial z/\partial x$ and $\partial z/\partial y$ are used instead of dz/dx and dz/dy to indicate that we are differentiating a function having more than one variable.

Example 28 Let

$$z = f(x, y) = x^2 + xy + y^2.$$

Find the partial derivative of z with respect to x and the partial derivative of z with respect to y.

Solution To find the partial derivative of z with respect to x, we treat y as a constant and differentiate with respect to x, obtaining

$$\begin{aligned}\frac{\partial z}{\partial x} &= \frac{\partial}{\partial x}(x^2 + xy + y^2)\\ &= \frac{\partial}{\partial x}(x^2) + y\frac{\partial}{\partial x}(x) + \frac{\partial}{\partial x}(y^2)\\ &= 2x + y + 0\\ &= 2x + y.\end{aligned}$$

Alternatively, we can express this result in the notation

$$f_x(x, y) = 2x + y.$$

To find the partial derivative of z with respect to y we treat x as a constant and differentiate with respect to y, obtaining

$$\begin{aligned}\frac{\partial z}{\partial y} &= \frac{\partial}{\partial y}(x^2 + xy + y^2)\\ &= \frac{\partial}{\partial y}(x^2) + x\frac{\partial}{\partial y}(y) + \frac{\partial}{\partial y}(y^2)\\ &= 0 + x + 2y\\ &= x + 2y.\end{aligned}$$

Alternatively, we can express this result in the notation

$$f_y(x, y) = x + 2y.$$

If $z = f(x, y)$, then the values of the partial derivatives when $x = x_0$ and $y = y_0$ are denoted by

$$\left.\frac{\partial z}{\partial x}\right|_{(x_0,y_0)} \quad \text{and} \quad \left.\frac{\partial z}{\partial y}\right|_{(x_0,y_0)}$$

or

$$f_x(x_0, y_0) \quad \text{and} \quad f_y(x_0, y_0).$$

Example 29 Let

$$z = f(x, y) = x^2 + xy + y^2.$$

Then from the previous example

$$\frac{\partial z}{\partial x} = f_x(x, y) = 2x + y$$

$$\frac{\partial z}{\partial y} = f_y(x, y) = x + 2y.$$

Thus, if $x = 1$ and $y = 3$, we obtain

$$\left.\frac{\partial z}{\partial x}\right|_{(1,3)} = f_x(1, 3) = 2(1) + 3 = 5.$$

With y held constant, this represents the instantaneous rate of change of z with respect to x when $x = 1$, $y = 3$.

Similarly,

$$\left.\frac{\partial z}{\partial y}\right|_{(1,3)} = f_y(1, 3) = 1 + 2(3) = 7$$

represents the instantaneous rate of change of z with respect to y when $x = 1$, $y = 3$, and x remains constant.

Example 30 Suppose the cost C (in cents) of manufacturing one cardboard box of paperclips is given by

$$C = 3x + 10y + 2$$

where

x = the unit cost of cardboard (in cents)

y = the unit cost of metal (in cents).

Then

$$\frac{\partial C}{\partial x} = 3 \qquad \text{and} \qquad \frac{\partial C}{\partial y} = 10.$$

The first result tells us that if the cost y of metal remains constant, the manufacturing cost C will increase at a constant rate of 3 cents a box for each 1-cent increase in the cost of cardboard. Similarly, if the cost x of cardboard remains fixed, the cost C will increase at a constant rate of 10 cents per box for each 1-cent increase in the cost of metal.

If $z = f(x, y)$, then

$$f_x = \frac{\partial z}{\partial x} \qquad \text{and} \qquad f_y = \frac{\partial z}{\partial y}$$

are both functions of x and y, which means we can form partial derivatives of them, thereby obtaining what are called the ***second order partial derivatives***

$$\frac{\partial}{\partial x}\left(\frac{\partial z}{\partial x}\right) = \frac{\partial^2 z}{\partial x^2} \qquad \text{or} \qquad (f_x)_x = f_{xx}$$

$$\frac{\partial}{\partial y}\left(\frac{\partial z}{\partial x}\right) = \frac{\partial^2 z}{\partial y\, \partial x} \qquad \text{or} \qquad (f_x)_y = f_{xy}$$

$$\frac{\partial}{\partial x}\left(\frac{\partial z}{\partial y}\right) = \frac{\partial^2 z}{\partial x\, \partial y} \qquad \text{or} \qquad (f_y)_x = f_{yx}$$

$$\frac{\partial}{\partial y}\left(\frac{\partial z}{\partial y}\right) = \frac{\partial^2 y}{\partial y^2} \qquad \text{or} \qquad (f_y)_y = f_{yy}.$$

Observe that in the above notation, f_{xy} means that we first differentiate with respect to x, then with respect to y, while

$$\frac{\partial^2 z}{\partial x\, \partial y} = \frac{\partial}{\partial x}\left(\frac{\partial z}{\partial y}\right)$$

means that we first differentiate with respect to y and then with respect to x.

Example 31 If

$$f(x, y) = x^2 e^y + y e^x + x y^2,$$

find f_{xx}, f_{xy}, f_{yx}, f_{yy}.

Solution First we form the partial derivatives:

$$f_x = 2x e^y + y e^x + y^2$$

$$f_y = x^2 e^y + e^x + 2xy.$$

Then we obtain the second order partial derivatives:

$$f_{xx} = \frac{\partial}{\partial x}(f_x) = \frac{\partial}{\partial x}(2xe^y + ye^x + y^2) = 2e^y + ye^x$$

$$f_{yy} = \frac{\partial}{\partial y}(f_y) = \frac{\partial}{\partial y}(x^2e^y + e^x + 2xy) = x^2e^y + 2x$$

$$f_{xy} = \frac{\partial}{\partial y}(f_x) = \frac{\partial}{\partial y}(2xe^y + ye^x + y^2) = 2xe^y + e^x + 2y$$

$$f_{yx} = \frac{\partial}{\partial x}(f_y) = \frac{\partial}{\partial x}(x^2e^y + e^x + 2xy) = 2xe^y + e^x + 2y.$$

In the previous example it turned out that

$$f_{xy} = f_{yx}.$$

Although this equation does not always hold, it is true for many of the commonly encountered functions.

In Section 13.2 we studied the problem of maximizing or minimizing a function $f(x)$ of one variable. We now consider the problem of maximizing and minimizing functions $f(x, y)$ involving two variables. For functions of one variable, a point a such that $f'(a) = 0$ was called a critical point or stationary point for $f(x)$. Similarly, a point (a, b) is called a ***critical point*** or ***stationary point*** for $f(x, y)$ if

$$f_x(a, b) = 0 \qquad \text{and} \qquad f_y(a, b) = 0.$$

Example 32 Find the critical points of

$$f(x, y) = 2x^2 + xy + y^2 + 3y - 2x.$$

Solution

$$f_x(x, y) = 4x + y - 2$$

$$f_y(x, y) = x + 2y + 3.$$

To find the critical points we must solve the system of equations

$$\begin{aligned} 4x + y - 2 &= 0 \\ x + 2y + 3 &= 0 \end{aligned} \qquad \text{or} \qquad \begin{aligned} 4x + y &= 2 \\ x + 2y &= -3. \end{aligned}$$

This system of *linear equations* can be solved as discussed in Section 2.2 to obtain

$$x = 1, \qquad y = -2.$$

Thus $(1, -2)$ is the only critical point of f.

Example 33 Find the critical points of

$$f(x, y) = 2xy + x^2 - y^3.$$

Solution

$$f_x(x, y) = 2y + 2x$$
$$f_y(x, y) = 2x - 3y^2.$$

To find the critical points we must solve the system of equations

$$2y + 2x = 0$$
$$2x - 3y^2 = 0.$$

These are not linear equations, so the methods of Section 2.2 do not apply. To solve them, we solve the first equation for y in terms of x, obtaining

$$y = -x; \qquad (1)$$

then substitute this value of y in the second equation, obtaining

$$2x - 3x^2 = 0$$

or

$$x(2 - 3x) = 0$$

or

$$x = 0, \qquad x = \tfrac{2}{3}.$$

From (1) the corresponding y values are

$$y = 0, \qquad y = -\tfrac{2}{3}$$

which yields two critical points

$$(0, 0) \qquad \text{and} \qquad (\tfrac{2}{3}, -\tfrac{2}{3}).$$

Recall that for functions of one variable we distinguish between relative maxima and absolute maxima. A function $f(x)$ is said to have an absolute maximum at a if $f(a)$ is the largest value for $f(x)$. On the other hand, to have a relative maximum at a, $f(a)$ need only be the largest value for $f(x)$ in the "immediate vicinity" of a (Figure 13.38). A relative

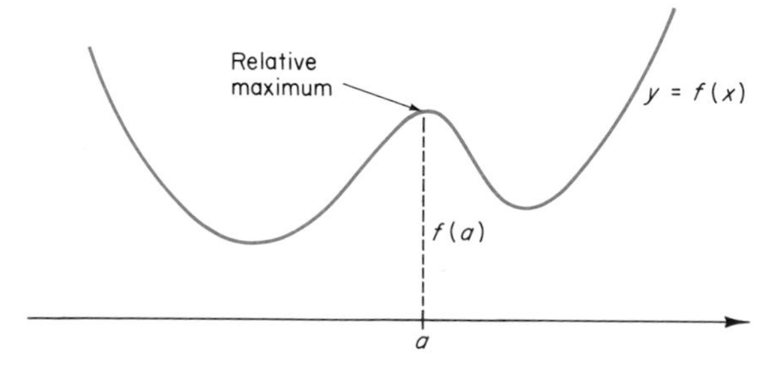

Figure 13.38

maximum may or may not be an absolute maximum for f. Similarly, a relative minimum may or may not be an absolute minimum for f.

For functions $f(x, y)$ of two variables, there are analogous notions. The function f is said to have an ***absolute maximum*** at (a, b) if $f(a, b)$ is the largest value that $f(x, y)$ can have, and f is said to have a ***relative maximum*** at (a, b) if $f(a, b)$ is the largest value for $f(x, y)$ for x "near" a and y "near" b. The terms ***absolute minimum*** and ***relative minimum*** are defined similarly. (The term *near* is admittedly vague, but the reader's intuitive interpretation of this term will be sufficient for our purposes.)

For functions of one variable, we were able to classify critical points as relative maxima or minima by using the second derivative test. The analogous test for functions of two variables is more complicated.

Test for Relative Maxima and Minima Let (a, b) be a critical point for $f(x, y)$ and let

$$M = f_{xx}(a, b) \cdot f_{yy}(a, b) - [f_{xy}(a, b)]^2;$$

then:

(a) f has a *relative maximum* at (a, b) if $M > 0$ and $f_{xx}(a, b) < 0$;

(b) f has a *relative minimum* at (a, b) if $M > 0$ and $f_{xx}(a, b) > 0$;

(c) f has neither a relative maximum nor a relative minimum at (a, b) if $M < 0$;

(d) The test does not apply if $M = 0$.

The proof or even an intuitive motivation for this result is beyond the scope of this textbook. However, we can give some examples.

Example 34 In Example 32 we showed $(1, -2)$ to be a critical point of

$$f(x, y) = 2x^2 + xy + y^2 + 3y - 2x.$$

To apply the above test we must calculate

$$f_{xx}(1, -2), \qquad f_{yy}(1, -2), \qquad \text{and} \qquad f_{xy}(1, -2).$$

But

$$f_x(x, y) = 4x + y - 2$$

$$f_y(x, y) = x + 2y + 3$$

and

$$f_{xx}(x, y) = 4, \quad \text{so} \quad f_{xx}(1, -2) = 4$$
$$f_{xy}(x, y) = 1, \quad \text{so} \quad f_{xy}(1, -2) = 1$$
$$f_{yy}(x, y) = 2, \quad \text{so} \quad f_{yy}(1, -2) = 2.$$

Thus

$$\begin{aligned} M &= f_{xx}(1, -2)f_{yy}(1, -2) - [f_{xy}(1, -2)]^2 \\ &= (4)(2) - (1)^2 = 7. \end{aligned}$$

Since $M > 0$ and $f_{xx}(1, -2) = 4 > 0$, it follows that $f(x, y)$ has a relative minimum at $(1, -2)$.

Example 35 In Example 33 we showed $(0, 0)$ to be a critical point of

$$f(x, y) = 2xy + x^2 - y^3.$$

We leave it for the reader to show

$$f_{xx}(x, y) = 2, \qquad f_{yy}(x, y) = -6y, \qquad f_{xy}(x, y) = 2.$$

Thus at the critical point $(0, 0)$

$$\begin{aligned} M &= f_{xx}(0, 0)f_{yy}(0, 0) - [f_{xy}(0, 0)]^2 \\ &= (2)(0) - (2)^2 \\ &= -4. \end{aligned}$$

Since $M < 0$, there is neither a relative maximum nor a relative minimum at $(0, 0)$.

Example 36 An automobile manufacturer determines that the profit P (in millions of dollars) is related to the advertising expenditures by

$$P(x, y) = -3x^2 - 2y^2 + x + 2y + 4xy + 150$$

where

x = amount (in millions of dollars) spent on television advertising

y = amount (in millions of dollars) spent on magazine advertising.

Find the values of x and y that maximize the profit and determine the maximum profit.

Solution To find the critical points solve the equations

$$P_x(x, y) = 0$$
$$P_y(x, y) = 0.$$

Thus we have

$$P_x(x, y) = -6x + 1 + 4y = 0$$
$$P_y(x, y) = -4y + 2 + 4x = 0$$

Solving this linear system we obtain (verify):

$$x = \tfrac{3}{2}, \qquad y = 2.$$

We leave it for the reader to verify that

$$M = P_{xx}(\tfrac{3}{2}, 2) \cdot P_{yy}(\tfrac{3}{2}, 2) - [P_{xy}(\tfrac{3}{2}, 2)]^2 = (-6) \cdot (-4) - (4)^2$$
$$= 8 > 0$$

and

$$P_{xx}(\tfrac{3}{2}, 2) = -6 < 0;$$

so $P(x, y)$ has a relative maximum $x = \frac{3}{2}$, $y = 2$. That is, \$1,500,000 ($\frac{3}{2}$ million) should be spent on television advertising and \$2 million on magazine advertising. The resulting profit P is

$$P = -3(\tfrac{3}{2})^2 - 2(2)^2 + \tfrac{3}{2} + 2(2) + 4(\tfrac{3}{2})2 + 150$$
$$= \tfrac{611}{4} \text{ millions of dollars}$$
$$= \$152{,}750{,}000.$$

EXERCISE SET 13.5

In Exercises 1–8 find f_x, f_y, and f_{yx}.

1. $f(x, y) = x^3 + xy + 2y^2$

2. $f(x, y) = x^2y - y^2x + y^3$

3. $f(x, y) = xe^y + y^2e^x - y^3$

4. $f(x, y) = xe^{3x+2y}$

5. $f(x, y) = e^{xy}$

6. $f(x, y) = \dfrac{y}{x} - \dfrac{x}{y}$

7. $f(x, y) = \ln(xy)$

8. $f(x, y) = (x + y) \ln(xy)$.

In Exercises 9–12 calculate $f_x(x_0, y_0)$, $f_y(x_0, y_0)$, and $f_{xy}(x_0, y_0)$.

9. $f(x, y) = xy^2 + 2xy + y^3$; $\quad (x_0, y_0) = (-1, 2)$

10. $f(x, y) = x^3 + xy^3 - x^2y + y^4$; $\quad (x_0, y_0) = (1, 2)$

11. $f(x, y) = xe^y + y^2e^x$; $\quad (x_0, y_0) = (0, 1)$

12. $f(x, y) = x \ln(xy)$; $\quad (x_0, y_0) = (1, 1)$.

In Exercises 13–18 calculate f_{xx}, f_{yy}, f_{xy}, and f_{yx}.

13. $f(x, y) = 2x^2 + xy - y^2$

14. $f(x, y) = 3x^2 - 2xy + y^4x^2$

15. $f(x, y) = 2x + 5y$

16. $f(x, y) = e^{xy}$

17. $f(x, y) = xe^{x+y}$

18. $f(x, y) = y \ln x$.

In Exercises 19–22 evaluate f_{xx}, f_{yy}, f_{xy}, and f_{yx} at the given point (x_0, y_0).

19. $f(x, y) = 3x^2y + yx - 2y^2$; $(x_0, y_0) = (-1, 1)$

20. $f(x, y) = 2x^2y + y^3$; $(x_0, y_0) = (-2, 3)$

21. $f(x, y) = e^{2x+3y}$; $(x_0, y_0) = (0, 1)$

22. $f(x, y) = y \ln x$; $(x_0, y_0) = (2, 3)$.

In Exercises 23–31 find the relative maxima and minima, if any.

23. $f(x, y) = x^2 + xy + y^2 - 4x - 5y$

24. $f(x, y) = x^2 + 2x + y^2 + 6y + 8$

25. $f(x, y) = x^2 + y^2$

26. $f(x, y) = x^2 - y^2$

27. $f(x, y) = -2x^2 - 2xy - y^2 + 4x + 2y$

28. $f(x, y) = 2x^3 - 6xy + 6x + 3y^2 - 18y$

29. $f(x, y) = x^2 + y^2 - 2xy$

30. $f(x, y) = -x^2 + 6x - y^2 + 2y + 12$

31. $f(x, y) = x^3 + 2xy - y^3$

32. Let $f(x, y) = \ln(x^2 + y^2)$. Show that $f_{xx} + f_{yy} = 0$.

33. A manufacturer's cost C depends on the number of employees x and on the cost of materials and overhead y (in thousands of dollars).

(a) If the relationship is

$$C(x, y) = 20 + 3x^2 + 4y$$

find $\partial C/\partial x$ and $\partial C/\partial y$.

(b) Evaluate $\partial C/\partial x$ and $\partial C/\partial y$ with $x = 10$ and $y = 3$.

(c) In words, explain the meaning of the values obtained in (b).

34. The pollution index I for a certain lake is defined to be

$$I = \ln x + \ln y + 3xy$$

where x is the number of milligrams of detergent and y is the number of milligrams of metallic salts in a standard sample.

(a) Find $I_x(1, 2)$ and $I_y(1, 2)$.

(b) In words, explain the meaning of the values obtained in (a).

35. **(Biomedical Engineering)** In certain fields of engineering design it becomes important to know how the average person's body surface area varies with height and weight (e.g., the body surface area affects the amount of

moisture evaporating in a fixed time period). A commonly used empirical formula relating body surface area A (square meters) to weight W (kilograms) and height H (meters) is

$$A = 2.024W^{0.425}H^{0.725}$$

(a) Find $\partial A/\partial W$, $\partial A/\partial H$ when $W = 91$ and $H = 2$ (use the approximations

$$91^{0.425} \approx 6.80, \qquad 91^{-0.575} \approx 0.075, \qquad 2^{0.725} \approx 1.65, \qquad 2^{-0.725} \approx 0.826$$

(b) Give a physical interpretation of the results in (a).

36. A company manufactures two items, I and II, that sell for \$40 and \$60, respectively. The cost of producing x units of I and y units of II is

$$600 + 6x + 4y + .02(6x^2 + 2xy + 6y^2).$$

Assuming that profit = revenue − cost, find the values of x and y that maximize the profit.

37. A department store's daily profit P in dollars depends on the number of salespeople x and on the number of departments y. The relationship is

$$P = [8000 - (4 - x)^2 - (8 - y)^2].$$

(a) What values of x and y will maximize the profit?

(b) What is the maximum profit?

38. Find the dimensions of a closed rectangular box of least surface area and having a volume of 50 cubic inches. (*Hint:* Let the dimensions be x, y, and z and begin by expressing V in terms of x and y alone.)

39. An electronics firm finds that if it spends x million dollars on research and y million dollars on development, then its total expenditure $E(x, y)$ (in millions of dollars) is given by

$$E(x, y) = 2{,}000 - \tfrac{1}{2}x^2 + 50x - \tfrac{1}{3}y^3 + 16y.$$

How much money should be spent on research and how much on development to minimize the total expenditure?

40. **(Least Squares)** In Section 2.3, the method of least squares was discussed. In Example 26 of that section, we found the line of best fit for the data

$$(1, 3), \quad (2, 1), \quad (3, 4), \quad (4, 3).$$

We shall show how to obtain this result using calculus. Let the line of best fit be denoted by

$$y = mx + b.$$

Recall that the method of least squares seeks to minimize

$$d_1^2 + d_2^2 + d_3^2 + d_4^2.$$

(a) Using Figures 2.22b and 2.23, show that

$$d_1 = 3 - (1m + b) \qquad d_2 = (2m - b) + 1$$

$$d_3 = 4 - (3m + b) \qquad d_4 = (4m + b) - 3.$$

(b) Use the calculus to obtain the values of m and b that minimize

$$\begin{aligned} E(m, b) &= [3 - (m + b)]^2 + [(2m + b) - 1]^2 + [4 - (3m + b)]^2 \\ &\quad + [(4m + b) - 3]^2 \\ &= d_1^2 + d_2^2 + d_3^2 + d_4^2 \end{aligned}$$

and check that your result agrees with that in Example 26 of Chapter 2.

14 INTEGRATION

Traditionally, calculus is divided into two main areas: ***differential calculus,*** which is concerned with tangents to curves and their applications, and ***integral calculus,*** which is concerned with areas under curves and their applications. In differential calculus the fundamental mathematical tool is the derivative; in integral calculus the basic tool is the "integral," which we shall study in this chapter. As we shall see, there is a close relationship between the problem of finding tangents and the problem of finding areas, so that the distinction between differential calculus and integral calculus is not always clear-cut.

14.1 Antiderivatives and Indefinite Integrals

In Chapter 12 we saw that the instantaneous rate of change of a function f is given at each point by its derivative f'. In this section we study the reverse problem: given the instantaneous rate of change f' at each point, can we find f itself? For example, given the instantaneous speed of an object at each instant, can we determine its position at each instant? Or, given a marginal cost function, can we determine the cost function itself?

We begin with an example. Let us try to find $F(x)$ given that

$$F'(x) = 2x \tag{1}$$

From our experience with derivatives we can guess the answer

$$F(x) = x^2$$

almost immediately. However, there are other answers as well; for example

$$F(x) = x^2 + 1, \qquad F(x) = x^2 - \tfrac{1}{3}, \qquad F(x) = x^2 + \sqrt{2}.$$

In fact, for any constant C,

$$F(x) = x^2 + C \tag{2}$$

will have derivative $F'(x) = 2x$.

The process of obtaining a function from its derivative is called ***anti-differentiation*** or ***integration***, and a function F such that $F'(x) = f(x)$ is called an ***antiderivative*** of f. Thus we have shown above that any function of the form $F(x) = x^2 + C$ is an antiderivative of $f(x) = 2x$. One might reasonably ask whether there are any antiderivatives of $f(x) = 2x$ that we have missed; that is, antiderivatives that cannot be obtained by substituting a value for C in (2). The following result shows that the answer is no; we have indeed obtained all the antiderivatives.

The Equal Derivative Principle If $F'(x) = G'(x)$ at each point of an interval, then $F(x) - G(x)$ is constant over that interval.

Although we omit the proof, the result is easy to visualize; since $F'(x) = G'(x)$ at each point, the tangents to the graphs of F and G have the same slope, hence are parallel at each x (Figure 14.1). Thus the graphs of F and G are themselves "parallel"; that is, $F(x) - G(x)$ is constant. In light of the equal derivative principle, if $F(x)$ is any antiderivative of

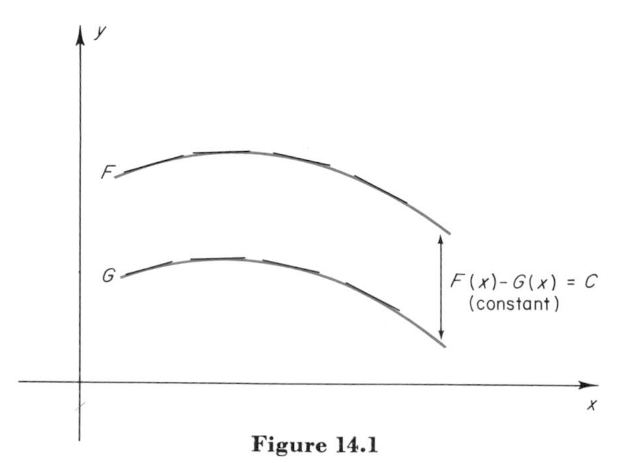

Figure 14.1

$f(x) = 2x$, then $F(x) - x^2$ is constant since $F(x)$ and x^2 both have derivative $2x$. Thus for some constant C,

$$F(x) - x^2 = C$$

or

$$F(x) = x^2 + C,$$

which proves that (2) describes *all* antiderivatives of x^2.

In general, if $F(x)$ is any antiderivative of $f(x)$, then the set of all antiderivatives is specified by the formula $F(x) + C$ and we write

$$\int f(x)\, dx = F(x) + C$$

Thus, we can write

$$\int 2x\, dx = x^2 + C.$$

Just as the symbol d/dx represents the operation of differentiation, so the symbol $\int dx$ represents the operation of antidifferentiation. The symbol $\int$ is called an ***integral sign*** and the symbol $\int f(x)\, dx$ is called the ***indefinite integral*** of $f(x)$. The arbitrary constant C is called the ***constant of integration,*** and the function $f(x)$ is called the ***integrand.***

Since integration is the reverse of differentiation, every differentiation formula has a companion integration formula. For example, if k is a constant, then we can write:

Differentiation Formula	**Corresponding Integration Formula**	
$\dfrac{d}{dx}[kx] = k$	$\displaystyle\int k\, dx = kx + C$	(3)
$\dfrac{d}{dx}[e^x] = e^x$	$\displaystyle\int e^x\, dx = e^x + C$	(4)
$\dfrac{d}{dx}[\ln x] = \dfrac{1}{x}$	$\displaystyle\int \frac{1}{x}\, dx = \ln x + C$	(5)
$\dfrac{d}{dx}\left[\dfrac{x^{k+1}}{k+1}\right] = x^k$ $(k \neq -1)$	$\displaystyle\int x^k\, dx = \frac{x^{k+1}}{k+1} + C$ $(k \neq -1)$	(6)

Remark Because $\ln x$ is only defined when $x > 0$, formula (5) only applies for such x.

Integration formula (6) can be paraphrased: *to integrate a power of x (other than -1), increase the power on x by 1 and divide by the increased power.*

Example 1 From formula (6)

$$\int x^2\,dx = \frac{x^3}{3} + C.$$

$$\int x^3\,dx = \frac{x^4}{4} + C.$$

$$\int \sqrt{x}\,dx = \int x^{1/2}\,dx = \frac{x^{1+(1/2)}}{1+\frac{1}{2}} + C = \tfrac{2}{3}x^{3/2} + C.$$

$$\int x^{-5}\,dx = \frac{x^{-5+1}}{-5+1} + C = -\frac{x^{-4}}{4} + C.$$

Example 2 From formula (3)

$$\int 2\,dx = 2x + C.$$

$$\int dx = \int 1\,dx = 1x + C = x + C.$$

The following rules are obtained by reversing the corresponding differentiation rules of Section 12.1.

The Constant Times a Function Rule If k is a constant, than

$$\int kf(x)\,dx = k\int f(x)\,dx.$$

In other words, a *constant* can be moved past an integral sign.

The Sum and Difference Rules

$$\int [f(x) + g(x)]\,dx = \int f(x)\,dx + \int g(x)\,dx.$$

$$\int [f(x) - g(x)]\,dx = \int f(x)\,dx - \int g(x)\,dx.$$

In other words, the integral of a sum (difference) is the sum (difference) of the integrals.

Example 3

$$\int 3x^7\,dx = 3\int x^7\,dx = 3\,\frac{x^8}{8} + C = \frac{3}{8}x^8 + C.$$

$$\int (x + e^x)\,dx = \int x\,dx + \int e^x\,dx = \frac{x^2}{2} + e^x + C.$$

$$\begin{aligned}
\int (2x^3 - 4x^2 + 1)\,dx &= \int 2x^3\,dx - \int 4x^2\,dx + \int 1\,dx \\
&= 2\int x^3\,dx - 4\int x^2\,dx + \int 1\,dx \\
&= 2\left(\frac{x^4}{4}\right) - 4\left(\frac{x^3}{3}\right) + x + C \\
&= \frac{1}{2}x^4 - \frac{4}{3}x^3 + x + C.
\end{aligned}$$

The result in any integration problem can be checked by differentiation. Thus, the last result in Example 3 can be checked as follows:

$$\begin{aligned}
\frac{d}{dx}\left(\frac{1}{2}x^4 - \frac{4}{3}x^3 + x + C\right) &= \frac{1}{2}(4x^3) - \frac{4}{3}(3x^2) + 1 \\
&= 2x^3 - 4x^2 + 1.
\end{aligned}$$

Example 4

$$\begin{aligned}
\int \left(\frac{1}{x} + \frac{2}{x^3}\right) dx &= \int \frac{1}{x}\,dx + \int \frac{2}{x^3}\,dx \\
&= \int \frac{1}{x}\,dx + 2\int x^{-3}\,dx \\
&= \ln x + 2\left(\frac{x^{-2}}{-2}\right) + C \\
&= \ln x - \frac{1}{x^2} + C.
\end{aligned}$$

Example 5 When a letter other than x is used for the independent variable of a function, we must make the appropriate notational changes in the integration formulas:

$$\int t^3\,dt = \frac{t^4}{4} + C.$$

$$\int 3A^2\,dA = 3\int A^2\,dA = 3\,\frac{A^3}{3} + C = A^3 + C.$$

$$\int e^u\,du = e^u + C.$$

In applied problems, there are often conditions that determine a specific value for the constant of integration.

Example 6 A manufacturer determines that the marginal cost in dollars is given by

$$M(x) = x^2 + 3x.$$

Find the cost function $C(x)$, assuming that the fixed cost (cost when $x = 0$ units are produced) is \$30.

Solution Since the marginal cost $M(x) = x^2 + 3x$ is the derivative of the cost function, we have

$$C'(x) = x^2 + 3x.$$

Thus the unknown cost function $C(x)$ is an antiderivative of $x^2 + 3x$; consequently, it can be determined by the integration

$$\begin{aligned} C(x) &= \int (x^2 + 3x)\,dx \\ &= \int x^2\,dx + 3\int x\,dx \\ &= \tfrac{1}{3}x^3 + \tfrac{3}{2}x^2 + K. \end{aligned}$$

(We have used K for the constant of integration to avoid confusion with the cost C.) Since $C = \$30$ when $x = 0$, we have

$$C(0) = 30 = \tfrac{1}{3}(0) + \tfrac{3}{2}(0) + K$$

or $K = 30$. Thus the cost function is

$$C(x) = \tfrac{1}{3}x^3 + \tfrac{3}{2}x^2 + 30.$$

EXERCISE SET 14.1

In Exercises 1–18 find the indefinite integral.

1. $\int 5\,dx$ **2.** $\int 3\,dx$ **3.** $\int t^5\,dt$

4. $\int x^8\,dx$ **5.** $\int x^{1/4}\,dx$ **6.** $\int t^{2/3}\,dt$

7. $\int \frac{6}{x^2}\,dx$ **8.** $\int \frac{3}{t^5}\,dt$ **9.** $\int 2e^u\,du$

10. $\int 7e^x\,dx$ **11.** $\int \frac{50}{x}\,dx$ **12.** $\int \frac{30}{t}\,dt$

13. $\int \sqrt{x}\,dx$ **14.** $\int (3A^2 + 2A - 1)\,dA$

15. $\int (3x^2 - 5x^{3/4} + 4)\, dx$

16. $\int \left(t^2 + e^t - \frac{1}{t}\right) dt$

17. $\int \left(2x^{2/3} - 3e^x + \frac{2}{x}\right) dx$

18. $\int \left(3x^2 - 2x^{3/2} + \frac{4}{x^3}\right) dx.$

In Exercises 19–24 find f from the given information.

19. $f'(x) = x + 2, \quad f(3) = 5$

20. $f'(x) = x - 5, \quad f(2) = 4$

21. $f'(x) = x^2 + 3, \quad f(-1) = 2$

22. $f'(x) = x^2 - 5, \quad f(0) = 3$

23. $f'(x) = 2 + 3e^x, \quad f(0) = 5$

24. $f'(x) = 2/x, \quad f(1) = -3.$

25. Given that the point $(-2, 1)$ lies on the curve $y = f(x)$ and given that $f'(x) = 3x^2$, find the curve $y = f(x)$.

26. Given that the point $(0, 8)$ lies on the curve $w = g(t)$ and given that $dw/dt = 7e^t$, find the curve $w = g(t)$.

27. Suppose that the rate of change of a certain population $P(t)$ with respect to time is given by

$$P'(t) = 25{,}000 + 4t^{2/5},$$

and at time $t = 0$, the population is $P(0) = 50{,}000$.

(a) Find an expression for $P(t)$

(b) What will the population be when $t = 20$?

28. An automobile moves along a straight track in such a way that its velocity $v(t)$ after t seconds is $v(t) = \sqrt{t}$ (feet per second). Find the total distance traveled by the car at the end of 5 seconds ($t = 5$). (*Hint*: Let $s(t)$ denote the distance traveled after t seconds. Thus $s(0) = 0$ and $ds/dt = v(t)$.)

29. (a) Show by differentiating that any function of the form $F(x) = x + C$ is an antiderivative of the constant function 1.

(b) From (a) and the equal derivative principle, the functions of the form $F(x) = x + C$ should have "parallel" graphs. Graph $F(x)$ in the cases $C = -1$, $C = 0$, $C = 1$, and $C = 2$.

30. (**Marginal Analysis**) A company determines that its marginal cost is

$$C'(x) = x^2 - 3x.$$

Find the cost function $C(x)$ if the fixed cost (cost of producing zero units) is \$1000.

14.2 THE DEFINITE INTEGRAL

As we shall see, many applied problems reduce to finding the area under a curve $y = f(x)$ over an interval $[a, b]$ (Figure 14.2). In this section we discuss this problem.

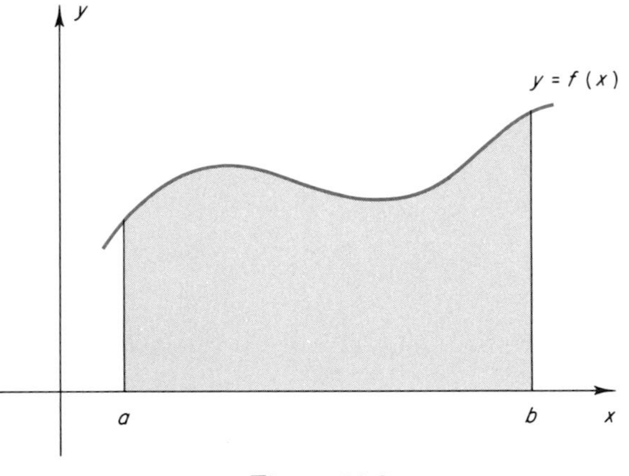

Figure 14.2

Let f be a continuous function whose graph does not dip below the x axis over an interval $[a, b]$. Surprising as it may seem, the key to finding the area under $y = f(x)$ over the fixed interval $[a, b]$ is to first study how the area under this curve *varies* as we change the right-hand endpoint b. For this purpose, we replace the fixed right-hand endpoint b by a variable endpoint x, and we denote by $A(x)$ the area under the curve over the interval $[a, x]$ (Figure 14.3). We call $A(x)$ the ***area function*** for f starting from a.

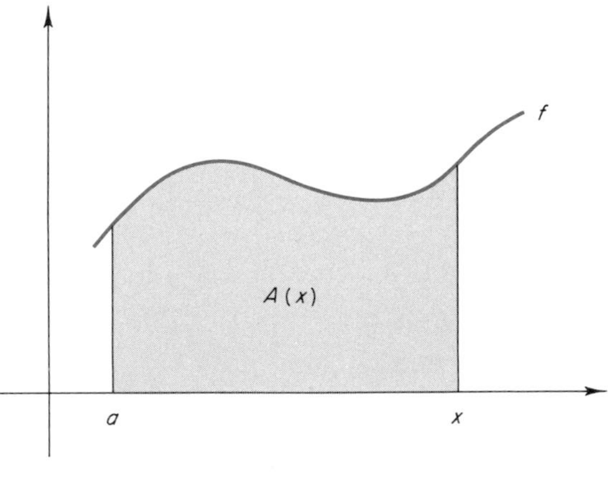

Figure 14.3

Example 7 Let f be the constant function defined by $f(x) = 2$. Since the graph of f is the horizontal line $y = 2$ (Figure 14.4), the area under the graph of f over the interval $[0, x]$ is the area of a rectangle of height 2 and base x. Thus the area function for f starting from 0 is

$$A(x) = 2x.$$

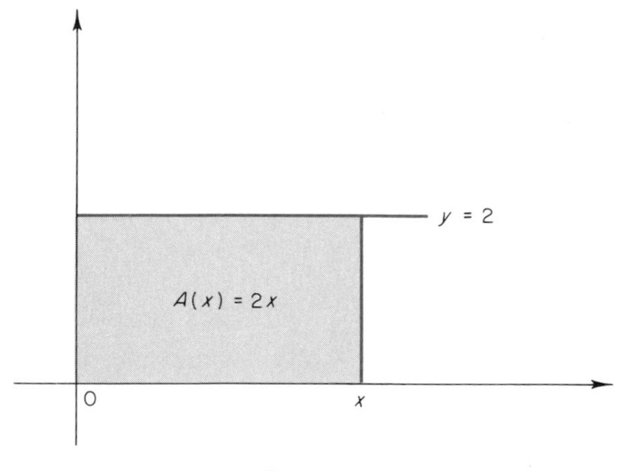

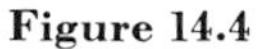
Figure 14.4

Example 8 Let

$$f(x) = x - 1.$$

The graph of f is the line $y = x - 1$. As indicated in Figure 14.5, the area under this graph over the interval $[1, x]$ is the area of a triangle having base $x - 1$ and height $x - 1$. Thus the area function for f starting from 1 is

$$A(x) = \tfrac{1}{2}(x - 1)(x - 1) = \tfrac{1}{2}(x^2 - 2x + 1) = \tfrac{1}{2}x^2 - x + \tfrac{1}{2}.$$

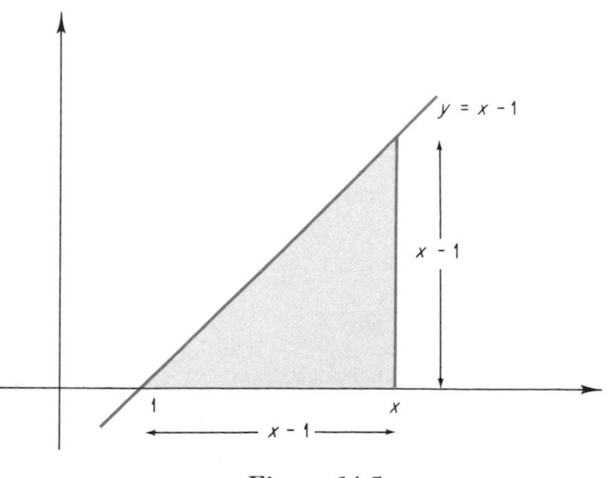

Figure 14.5

In these two examples we were able to obtain explicit formulas for the area functions because the curves were simple enough that we could use area formulas from geometry. However, even for many simple curves there are no formulas from elementary geometry to help us determine the area function. For example, no basic geometry formula will help us find the area under the curve

$$y = x^2$$

over the interval $[0, x]$ (Figure 14.6). What is surprising, however, is that the *derivative* of the area function is always easy to obtain, no matter

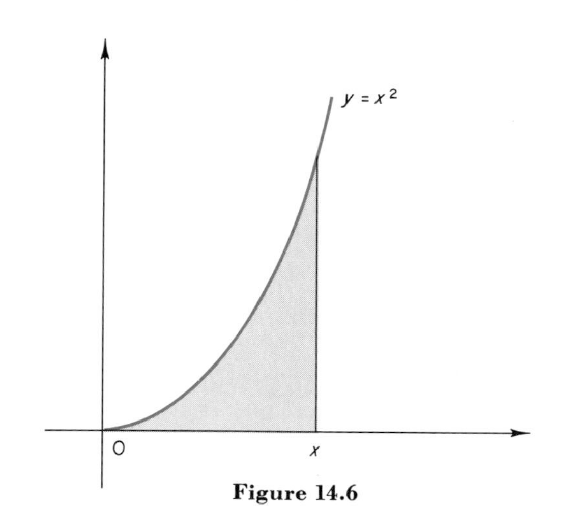

Figure 14.6

how complicated the function f (as long as f is continuous). To see why, let us examine the derivatives of the area functions in the examples above:

	Function	Area function	Derivative of area function
Example 7	$f(x) = 2$	$A(x) = 2x$	$A'(x) = 2$
Example 8	$f(x) = x - 1$	$A(x) = \frac{1}{2}x^2 - x + \frac{1}{2}$	$A'(x) = x - 1$

In each case, the derivative of the area function turned out to be the same as the original function f. This is not accidental, it is a consequence of the following major result (see Exercise 30):

> ***The Fundamental Theorem of Calculus*** If f is a function that is continuous and has nonnegative values on an interval containing the point a, and if $A(x)$ is the area function for f starting from a, then at each point of the interval
>
> $$A'(x) = f(x).$$

Example 9 Find the area function starting from $a = 0$ for $f(x) = x^2$.

Solution From the fundamental theorem of calculus, the area function $A(x)$ satisfies

$$A'(x) = x^2$$

which states that $A(x)$ is an antiderivative of x^2. Thus

$$A(x) = \int x^2\, dx = \frac{x^3}{3} + C \tag{1}$$

where the constant of integration is yet undetermined. However, if $x = 0$ the interval $[0, x]$ reduces to a point, in which case the area under the curve over the interval is 0, that is, $A(x) = 0$ if $x = 0$. Thus, from (1)

$$0 = A(0) = C$$

so that the area function is

$$A(x) = \frac{x^3}{3}.$$

Example 10 Find the area under the curve $y = x^2$ over the interval $[0, 2]$ (Figure 14.7).

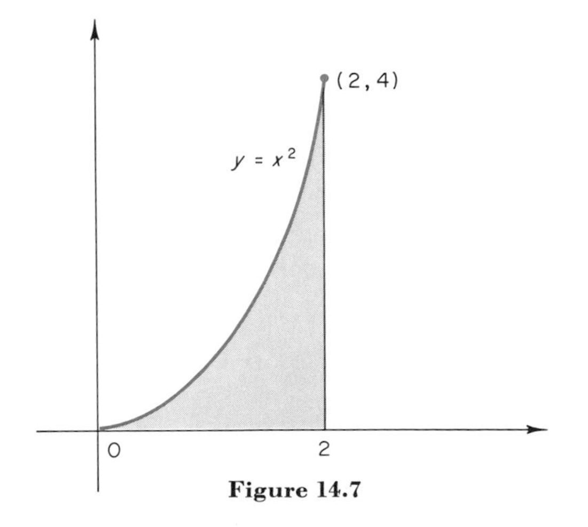

Figure 14.7

Solution From Example 9 the area under the curve over $[0, x]$ is

$$A(x) = \frac{x^3}{3}.$$

Letting $x = 2$ will yield the area under the curve over $[0, 2]$. This area is

$$A(2) = \frac{(2)^3}{3} = \frac{8}{3} \quad \text{(square units).}$$

Example 9 illustrates a point worth noting. If $A(x)$ is the area function for f starting from a, then $A(x) = 0$ when $x = a$; that is,

$$A(a) = 0 \tag{2}$$

As shown in Example 9, this condition can be used to single out the area function $A(x)$ for f from the many antiderivatives of $f(x)$. However, for many purposes this is unnecessary work. To see why, let $F(x)$ be *any* antiderivative of $f(x)$. From the equal derivative principle, $F(x)$ and $A(x)$ differ only by some constant C; that is,

$$F(x) = A(x) + C.$$

Thus

$$\begin{aligned} F(b) - F(a) &= [A(b) + C] - [A(a) + C] \\ &= A(b) - A(a), \\ &= A(b) \qquad \text{(from (2))}. \end{aligned}$$

Thus

$$F(b) - F(a) = A(b) = \text{area under } y = f(x) \text{ over } [a, b].$$

In other words, the difference

$$F(b) - F(a)$$

has the same value for every antiderivative $F(x)$ of $f(x)$ and this difference is the area under $y = f(x)$ over $[a, b]$. This important result leads to the following definition.

Definition If $f(x)$ is a continuous function on the interval $[a, b]$, then the ***definite integral*** of f from a to b, written $\int_a^b f(x)\,dx$, is defined by

$$\int_a^b f(x)\,dx = F(b) - F(a) \tag{3}$$

where F is any antiderivative of f.

The numbers a and b in (3) are called the ***limits of integration*** and the expression on the righthand side of (3) is often written

$$F(x)\Big]_a^b$$

which means, subtract $F(a)$ from $F(b)$. For example

$$x^2\Big]_1^3 = (3)^2 - (1)^2 = 8.$$

Example 11 Evaluate the definite integral $\int_2^3 x^3\,dx$.

Solution

$$\int x^3\,dx = \frac{x^4}{4} + C.$$

Since any antiderivative of x^3 will suffice to evaluate the given definite integral, we take $C = 0$, so that $F(x) = x^4/4$. Thus

$$\int_2^3 x^3\,dx = \frac{x^4}{4}\bigg]_2^3 = \left(\frac{3^4}{4}\right) - \left(\frac{2^4}{4}\right) = \frac{81}{4} - \frac{16}{4} = \frac{65}{4}.$$

The computations in the last example can be arranged more compactly by writing

$$\int_2^3 x^3\,dx = \int x^3\,dx\bigg]_2^3 = \frac{x^4}{4}\bigg]_2^3 = \frac{81}{4} - \frac{16}{4} = \frac{65}{4}.$$

Example 12 Use the definite integral to find the area under $y = x^2$ over the interval $[0, 2]$.

Solution The area A is

$$A = \int_0^2 x^2\,dx = \int x^2\,dx\bigg]_0^2 = \frac{x^3}{3}\bigg]_0^2 = \frac{8}{3} - 0 = \frac{8}{3}.$$

This agrees with the result obtained in Example 10 where we solved this same problem (less efficiently) using the area function.

Example 13 Find the area under the curve $y = e^x$ over the interval $[0, 2]$ (Figure 14.8).

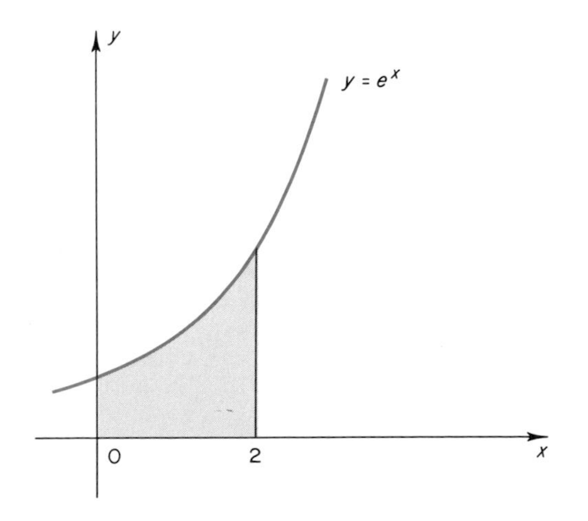

Figure 14.8

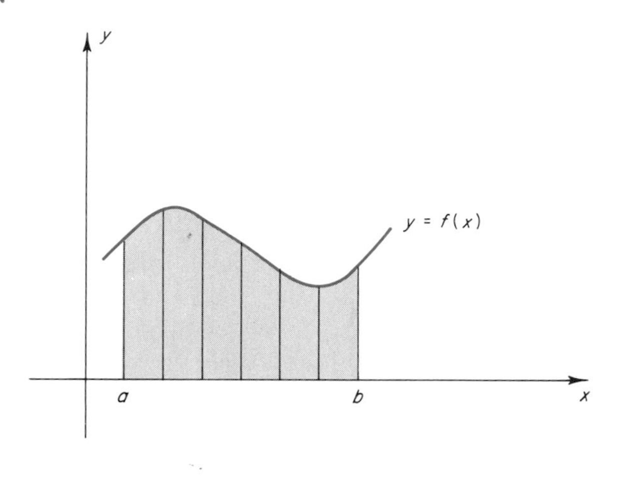

Figure 14.9

Solution The area A we want is

$$A = \int_0^2 e^x \, dx = \int e^x \, dx \Big]_0^2 = e^x \Big]_0^2 = e^2 - e^0 = e^2 - 1.$$

From Appendix Table VI this area is approximately 6.3891.

Riemann Sums (Optional)

There is another approach to the definition of the definite integral due to the mathematician Bernhard Riemann.† To motivate the idea we consider a function f that is continuous on an interval $[a, b]$ and such that $f(x) \geq 0$ for all x in the interval. Thus the graph of f over $[a, b]$ does not go below the x axis. As we have already seen, the definite integral

$$\int_a^b f(x) \, dx$$

represents the area under the curve $y = f(x)$ over the interval $[a, b]$.

However, this area can also be obtained by the following procedure. Divide the interval $[a, b]$ into a fixed number of intervals of equal length. For example, in Figure 14.9 we divided the interval $[a, b]$ into six subintervals. Using these subintervals, we can decompose the area under

† *Georg Friedrich Bernhard Riemann* (1826–1866)—Riemann was a German mathematician. Bernhard Riemann, as he is commonly known, was the son of a Protestant minister. He obtained his elementary education from his father and showed brilliance in mathematics at an early age. In college, at Göttingen University, he studied theology and philology, but eventually transferred to mathematics and studied under Gauss. In 1862 Riemann contracted pleuritis and was seriously ill for the rest of his life. He died in 1866 at the age of thirty-nine. Riemann's early death was unfortunate since his mathematical work was brilliant and of fundamental importance. His work in geometry was used by Albert Einstein some 50 years later in formulating the theory of relativity.

$y = f(x)$ over $[a, b]$ into strips, as shown in Figure 14.9. In general, there will be no simple formula for the areas of these strips because of their curved upper boundaries. However, we can approximate each strip by a rectangle whose base is the subinterval and whose height is the value of $f(x)$ at any point x chosen in the subinterval (Figure 14.10).

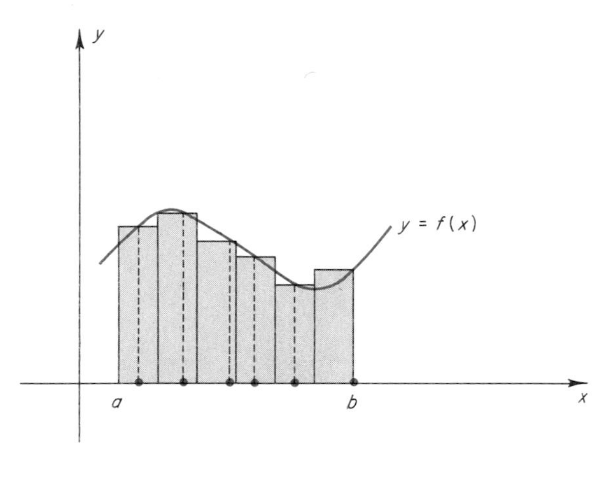

Figure 14.10

The sum of these rectangular areas, called a ***Riemann sum***, serves as an approximation to the entire area under $y = f(x)$ over $[a, b]$. Now we come to the crucial observation. As we let the number of subintervals get larger and larger, the widths of the approximating rectangles decrease and the *error* in approximating the total area by the sum of the rectangular areas gets smaller and smaller (Figure 14.11). Thus, the *exact* area $\int_a^b f(x)\, dx$ is obtained as a limit of Riemann sums. The following example illustrates this idea.

Example 14 If $f(x) = x$, then the area under $y = f(x)$ over the interval $[0, 2]$ is given by the definite integral

$$\int_0^2 x\, dx$$

whose value is

$$\int_0^2 x\, dx = \frac{x^2}{2}\Big]_0^2 = 2 - 0 = 2. \tag{4}$$

To obtain this same result using Riemann sums, we divide the interval $[0, 2]$ into n equal subintervals. (It is desirable not to specify a numerical value for n since we shall want to let the number of subintervals increase.) As indicated in Figure 14.12, each of the n subintervals will have length $2/n$.

Over each subinterval, we construct a rectangle of height $f(x)$ where x

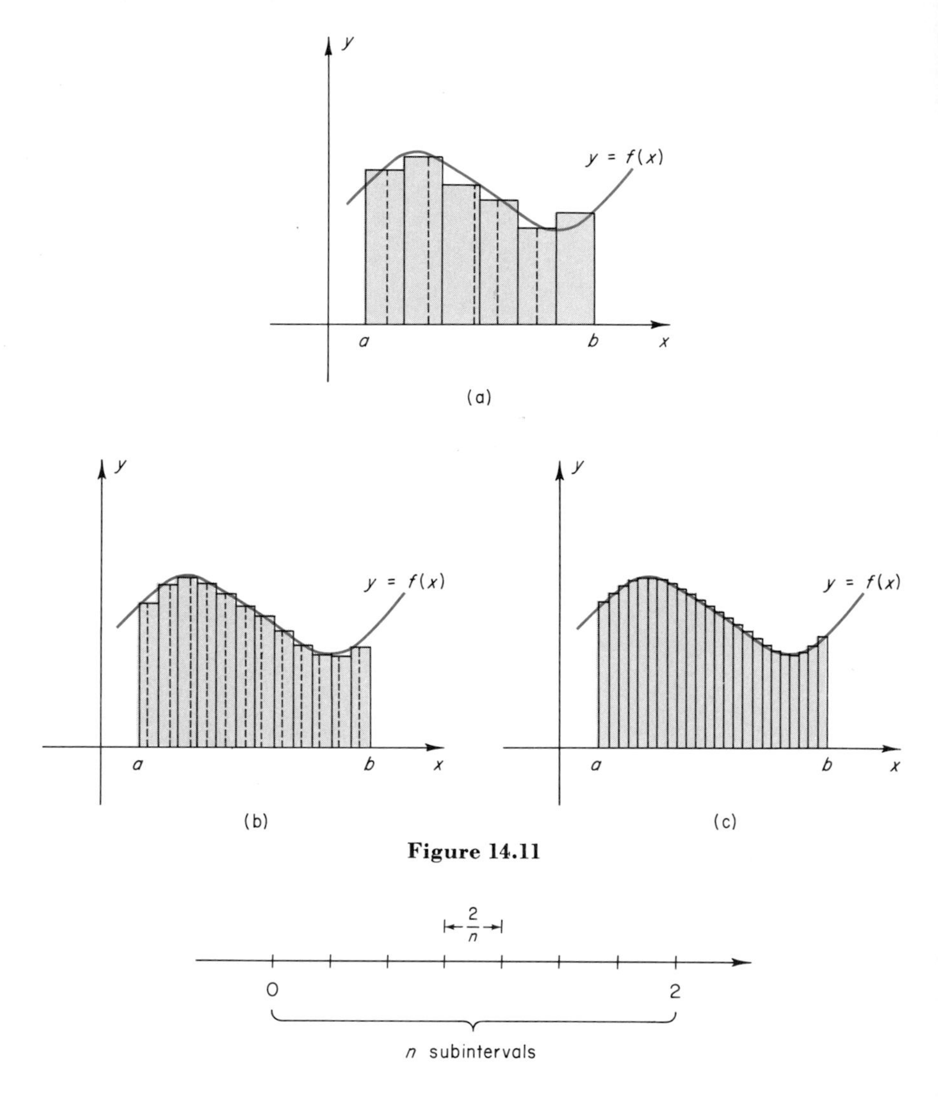

Figure 14.11

Figure 14.12

is any point in the subinterval. To be specific, we shall use the right-hand endpoint of each subinterval (Figure 14.13). With this choice, the heights of the rectangles will be

$$\frac{2}{n}, \quad \frac{4}{n}, \quad \frac{6}{n}, \quad \ldots, \quad \frac{2n}{n} \ (= 2)$$

and the sum of their areas will be

$$\left(\frac{2}{n}\right)\left(\frac{2}{n}\right) + \left(\frac{4}{n}\right)\left(\frac{2}{n}\right) + \left(\frac{6}{n}\right)\left(\frac{2}{n}\right) + \cdots + \left(\frac{2n}{n}\right)\left(\frac{2}{n}\right)$$

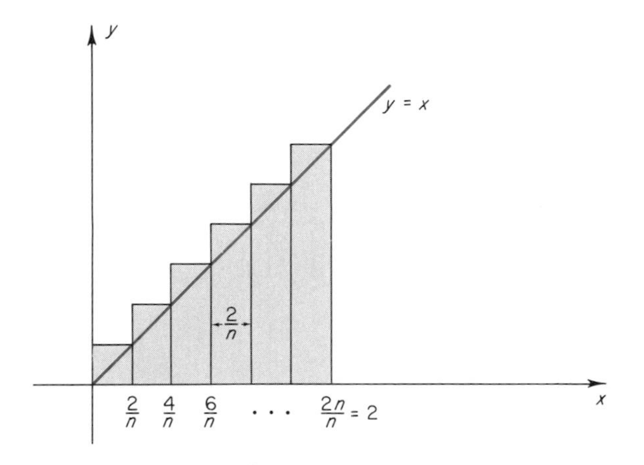

Figure 14.13

or

$$\frac{2}{n}\left[\frac{2}{n}+\frac{4}{n}+\frac{6}{n}+\cdots+\frac{2n}{n}\right]$$

and upon factoring out another $2/n$, we obtain the Riemann sum

$$\frac{4}{n^2}[1+2+3+\cdots+n]. \tag{5}$$

For example, with five subintervals ($n = 5$) this formula would yield

$$\frac{4}{5^2}[1+2+3+4+5]=\frac{60}{25}=\frac{12}{5}=2.4.$$

As we let n get larger and larger, the Riemann sum in (5) should approach the exact area under the curve.

To see that this is in fact the case, we have used a hand calculator to evaluate the Riemann sums in (5) for various n (Table 14.1). This table

Table 14.1

n Number of Subintervals	Value of the Riemann Sum (formula (5))
3	2.6667
5	2.4000
8	2.2500
10	2.2000
20	2.1000
50	2.0400
100	2.0200
1,000	2.0020
10,000	2.0002

suggests that the Riemann sums approach the value 2 as n increases. This agrees with the result obtained by integration (see (4)).

EXERCISE SET 14.2

In Exercises 1–14 use a definite integral to find the area under the given curve over the indicated interval.

1. $y = 3$, $[-1, 2]$

2. $y = 4$, $[2, 5]$

3. $y = 3x$, $[1, 4]$

4. $y = 2x$, $[2, 5]$

5. $y = x^2$, $[1, 3]$

6. $y = x^2 - 2$, $[2, 4]$

7. $y = 3 - x^2$, $[-1, 1]$

8. $y = 4 - 2x^2$, $[0, 2]$

9. $y = 3x - x^2$, $[0, 3]$

10. $y = \sqrt{x}$, $[1, 4]$

11. $y = e^x$, $[0, 1]$

12. $y = e^x$, $[0, 5]$

13. $y = 2/x$, $[1, 2]$

14. $y = 3/x$, $[2, 5]$

In Exercises 15–22 evaluate the definite integral.

15. $\int_{-1}^{2} dx$

16. $\int_{2}^{5} dx$

17. $\int_{3}^{5} x^3\, dx$

18. $\int_{-1}^{2} x^5\, dx$

19. $\int_{0}^{2} e^t\, dt$

20. $\int_{-1}^{0} e^s\, ds$

21. $\int_{1}^{3} \frac{1}{x}\, dx$

22. $\int_{2}^{8} \frac{1}{t}\, dt.$

In Exercises 23–28 give a definite integral whose value is the shaded area; then calculate the area from your integral. (Use Appendix tables, where needed.)

23. **24.**

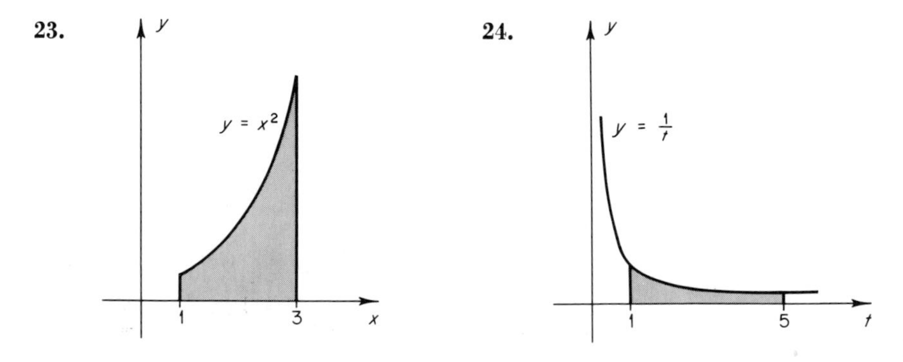

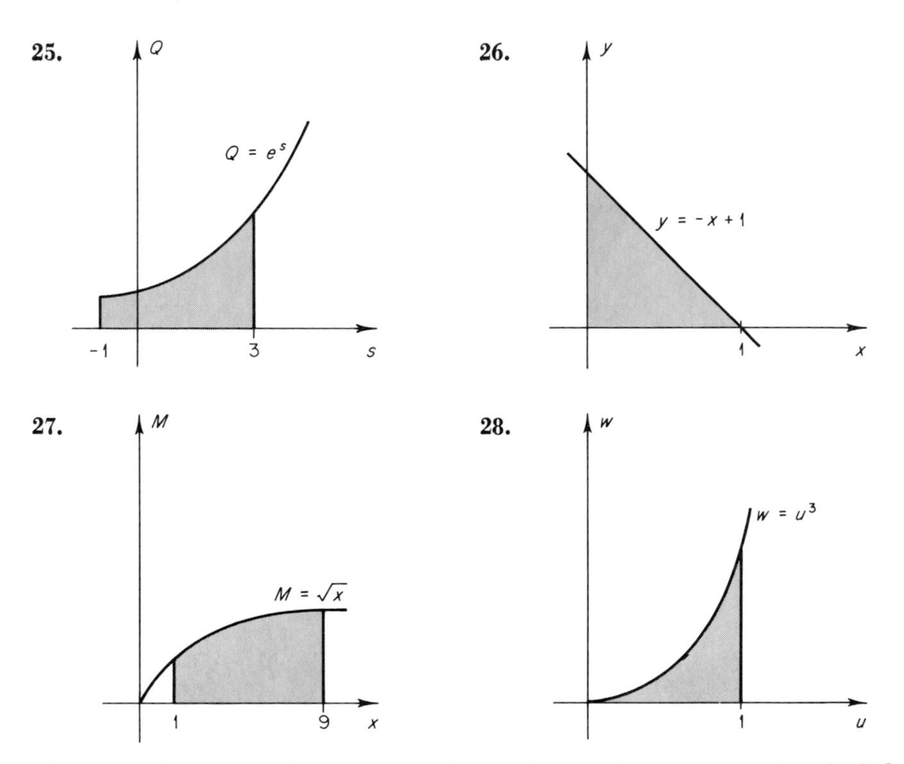

29. Find the area of the shaded region in Figure 14.14. (*Hint*: The shaded area forms part of a rectangle. Find the area of the remaining portion of the rectangle first.)

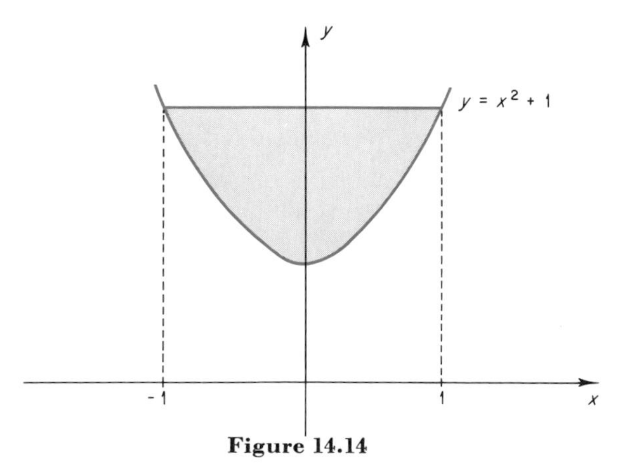

Figure 14.14

30. **(The Fundamental Theorem of Calculus)** This exercise is intended to *motivate* the fundamental theorem of calculus. It is not intended as a formal proof.

(a) Let $A(x)$ be the area function for f starting from a. In Figure 14.15 shade the area $A(x_0 + h) - A(x_0)$.

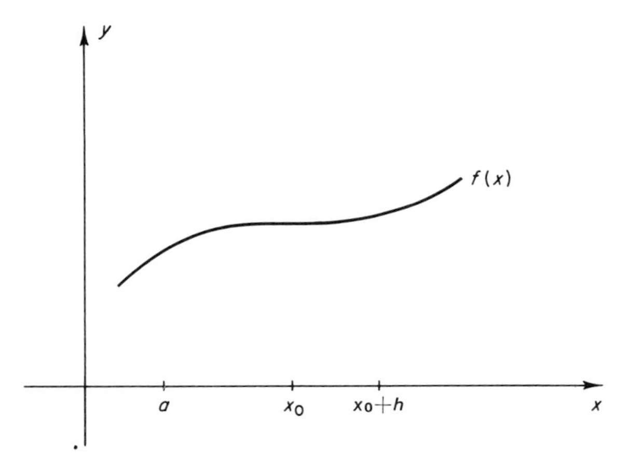

Figure 14.15

(b) Explain why one might expect the approximation

$$A(x_0 + h) - A(x_0) \approx h \cdot f(x_0)$$

to get better as h gets small.

(c) Let us rewrite the approximation in (b) as

$$\frac{A(x_0 + h) - A(x_0)}{h} \approx f(x_0).$$

By taking an appropriate limit of both sides, deduce

$$A'(x_0) = f(x_0)$$

which is just the statement of the fundamental theorem of calculus at the point $x = x_0$.

31. (Optional) Let $f(x) = 4x$ and consider the interval $[0, 4]$.

(a) Divide the interval into n equal subintervals and using the midpoint of each subinterval, obtain the Riemann sum that approximates the area under $y = 4x$ over $[0, 4]$.

(b) Evaluate the Riemann sum obtained in (a) for $n = 3$, $n = 5$, and $n = 10$.

(c) Find the exact area by integration and determine the error for each of the approximations in (b).

32. (Optional) Let $f(x) = 1/x$ and consider the interval $[1, 3]$.

(a) Divide the interval $[1, 3]$ into n equal subintervals. Using the right-hand endpoint of each subinterval, obtain a Riemann sum that approximates the area under $y = 1/x$ over $[1, 3]$.

(b) Evaluate the Riemann sum obtained in (a) for $n = 3$, $n = 4$, and $n = 6$. (It can be shown that to four decimal places the area is 1.0986.)

14.3 PROPERTIES OF THE DEFINITE INTEGRAL

In the previous section we saw that the definite integral $\int_a^b f(x)\,dx$ of a function f that is continuous and nonnegative on $[a, b]$ can be interpreted as the area under the graph of f over the interval $[a, b]$ (Figure 14.16). However, if the reader will take a moment to reread the definition

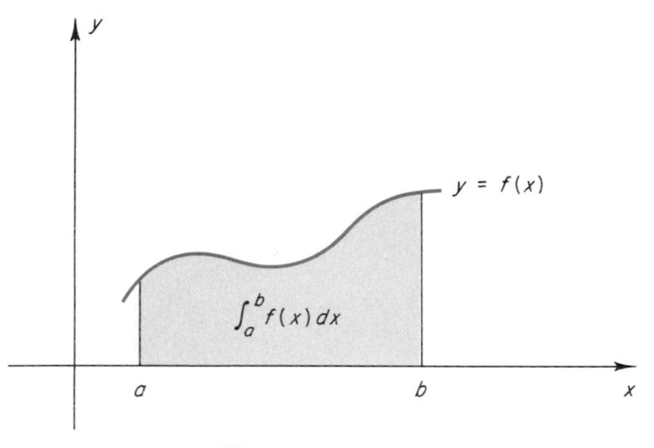

Figure 14.16

of the definite integral, it will be seen that we did not require f to be nonnegative on $[a, b]$; indeed, the defining equation

$$\int_a^b f(x)\,dx = F(b) - F(a)$$

makes perfectly good sense, even if f assumes both positive and negative values on $[a, b]$. For example, the function

$$f(x) = 1 - x$$

has both positive and negative values for x in the interval $[0, 2]$ (Figure 14.17), yet we can still write

$$\int_0^2 (1 - x)\,dx = \int (1 - x)\,dx\Big]_0^2 = x - \frac{x^2}{2}\Big]_0^2 = \left[2 - \frac{4}{2}\right] - [0] = 0.$$

What does the integral

$$\int_a^b f(x)\,dx$$

represent when $f(x)$ assumes both positive and negative values on $[a, b]$? Clearly, the integral *cannot* represent the total area between the graph of f and the interval $[a, b]$ since we saw above that

$$\int_0^2 (1 - x)\,dx = 0,$$

yet from Figure 14.17 the total area between the graph of $f(x) = 1 - x$ and the interval $[0, 2]$ is certainly not zero. Before we attempt to answer this question, it will be helpful to summarize some basic properties of definite integrals.

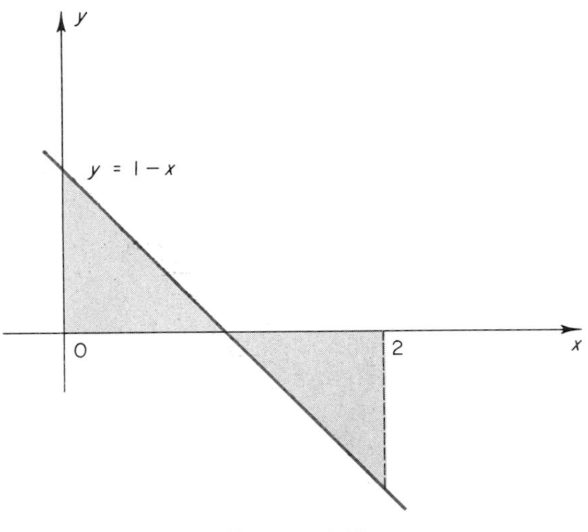

Figure 14.17

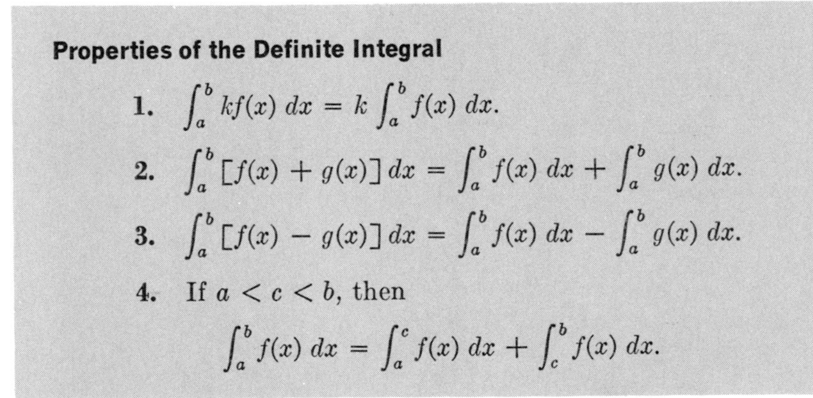

Properties of the Definite Integral

1. $\int_a^b kf(x)\,dx = k\int_a^b f(x)\,dx.$

2. $\int_a^b [f(x) + g(x)]\,dx = \int_a^b f(x)\,dx + \int_a^b g(x)\,dx.$

3. $\int_a^b [f(x) - g(x)]\,dx = \int_a^b f(x)\,dx - \int_a^b g(x)\,dx.$

4. If $a < c < b$, then

$$\int_a^b f(x)\,dx = \int_a^c f(x)\,dx + \int_c^b f(x)\,dx.$$

Properties 1, 2, and 3 follow from the corresponding properties for indefinite integrals. To obtain Property 4 we need only observe that if F is an antiderivative for f, then

$$\begin{aligned}
\int_a^c f(x)\,dx + \int_c^b f(x)\,dx &= \int f(x)\,dx\Big]_a^c + \int f(x)\,dx\Big]_c^b \\
&= [F(c) - F(a)] + [F(b) - F(c)] \\
&= F(b) - F(a) \\
&= \int_a^b f(x)\,dx.
\end{aligned}$$

When f is a nonnegative function on $[a, b]$, Property 4 states (Figure 14.18) that the area under $y = f(x)$ from a to b is the area from a to c plus the area from c to b.

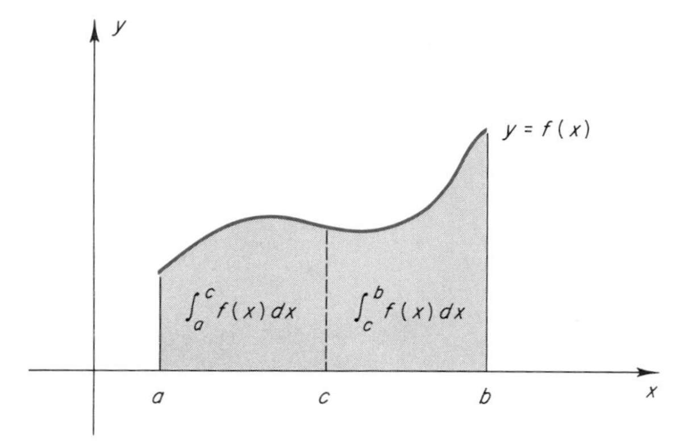

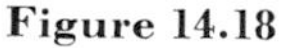
Figure 14.18

Example 15

$$\int_2^5 \left(1 + 3x^2 - \frac{1}{x}\right) dx = \int_2^5 dx + 3\int_2^5 x^2\,dx - \int_2^5 \frac{1}{x}\,dx$$

$$= x\Big]_2^5 + 3\,\frac{x^3}{3}\Big]_2^5 - \ln x\Big]_2^5$$

$$= (5 - 2) + 3\left(\frac{125}{3} - \frac{8}{3}\right) - (\ln 5 - \ln 2)$$

$$= 3 + 117 + \ln 2 - \ln 5$$

$$= 120 + \ln\left(\frac{2}{5}\right).$$

We now turn to the problem of interpreting $\int_a^b f(x)\,dx$ when f may have negative values on $[a, b]$. We begin by considering the curves

$$y = f(x) \qquad \text{and} \qquad y = -f(x)$$

where $f(x) \leq 0$ for all x in $[a, b]$. In this case $-f(x) \geq 0$ for all x in $[a, b]$ (Figure 14.19). As indicated in Figure 14.19, the curves $y = f(x)$ and $y = -f(x)$ are reflections of one another in the x axis, and the shaded regions in the figure have the same area A.

Because $-f(x)$ is nonnegative on the interval $[a, b]$, the area A under $y = -f(x)$ over $[a, b]$ is given by the definite integral

$$A = \int_a^b -f(x)\,dx = -\int_a^b f(x)\,dx$$

or

$$\int_a^b f(x)\,dx = -A$$

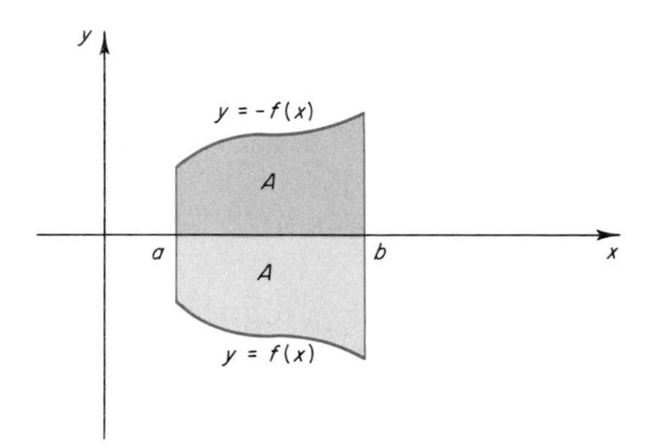

Figure 14.19

This illustrates that if f assumes only negative values (or zero) on $[a, b]$, then

$$\int_a^b f(x)\,dx$$

represents the *negative* of the area between $y = f(x)$ and the interval $[a, b]$.

To interpret $\int_a^b f(x)\,dx$ in the case where $f(x)$ has both positive and negative values on $[a, b]$, consider the curve $y = f(x)$ in Figure 14.20.

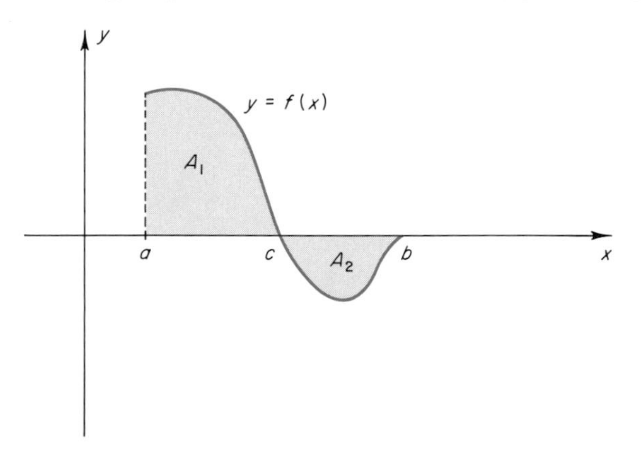

Figure 14.20

According to Property 4 for definite integrals, we can evaluate $\int_a^b f(x)\,dx$ by writing

$$\int_a^b f(x)\,dx = \int_a^c f(x)\,dx + \int_c^b f(x)\,dx. \tag{1}$$

But $f(x)$ is nonnegative on the interval $[a, c]$, so that $\int_a^c f(x)\,dx$ is the area A_1, and $f(x)$ is less than or equal to zero on $[c, b]$ so that $\int_c^b f(x)\,dx$ is the negative of area A_2. Thus (1) can be written

$$\int_a^b f(x)\,dx = A_1 + (-A_2) = A_1 - A_2.$$

This illustrates the following result:

A Geometric Interpretation of the Definite Integral If f is a continuous function on $[a, b]$, then the definite integral $\int_a^b f(x)\,dx$ represents the area *above* the interval $[a, b]$ and under $y = f(x)$ minus the area *below* the interval $[a, b]$ and above $y = f(x)$.

Example 16 In the beginning of this section we saw that

$$\int_0^2 (1 - x)\,dx = 0.$$

The explanation for this result should now be clear from Figure 14.17. The area of the triangular region above the x axis is equal to the triangular area below the x axis.

Example 17 We have

$$\begin{aligned}\int_{-2}^3 (x^2 - 4)\,dx = \int (x^2 - 4)\,dx\Big]_{-2}^3 &= \left(\frac{x^3}{3} - 4x\right)\Big]_{-2}^3\\ &= \left(\frac{27}{3} - 12\right) - \left(-\frac{8}{3} + 8\right)\\ &= -3 - \frac{16}{3} = -\frac{25}{3}.\end{aligned}$$

The negative value for the integral indicates that the portion of the area between the curve $y = x^2 - 4$ and the interval $[-2, 3]$ lying below the x axis is greater than the portion that lies above the x axis (Figure 14.21).

Example 18 Show that the result obtained in Example 17 is the area above the x axis under the curve minus the area below.

Solution We shall calculate separately the areas above and below the x axis and then subtract.

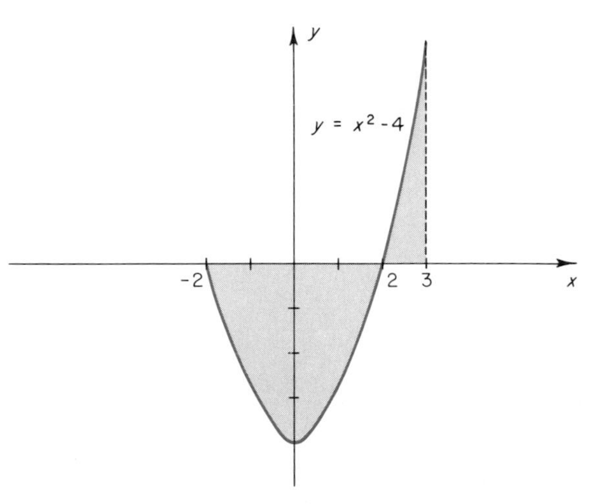

Figure 14.21

The area above the x axis is

$$\int_2^3 (x^2 - 4)\, dx = \int (x^2 - 4)\, dx \Big]_2^3 = \frac{x^3}{3} - 4x \Big]_2^3$$

$$= \left(\frac{27}{3} - 12\right) - \left(\frac{8}{3} - 8\right)$$

$$= -3 - \left(-\frac{16}{3}\right) = \frac{7}{3}.$$

The *negative* of the area below the x axis is

$$\int_{-2}^2 (x^2 - 4)\, dx = \int (x^2 - 4)\, dx \Big]_{-2}^2 = \frac{x^3}{3} - 4x \Big]_{-2}^2$$

$$= \left(\frac{8}{3} - 8\right) - \left(-\frac{8}{3} + 8\right)$$

$$= \frac{16}{3} - 16 = -\frac{32}{3}.$$

Thus the area below the x axis is $\frac{32}{3}$ and the area above minus the area below is

$$\frac{7}{3} - \frac{32}{3} = -\frac{25}{3}.$$

This agrees with the result in Example 17.

Suppose we want to find the area over the interval $[a, b]$ between the two curves $y = f(x)$ and $y = g(x)$ shown in Figure 14.22*a*. As indicated in parts (b) and (c) of the figure, the area A of this region can be written as

$$A = A_f - A_g \tag{2}$$

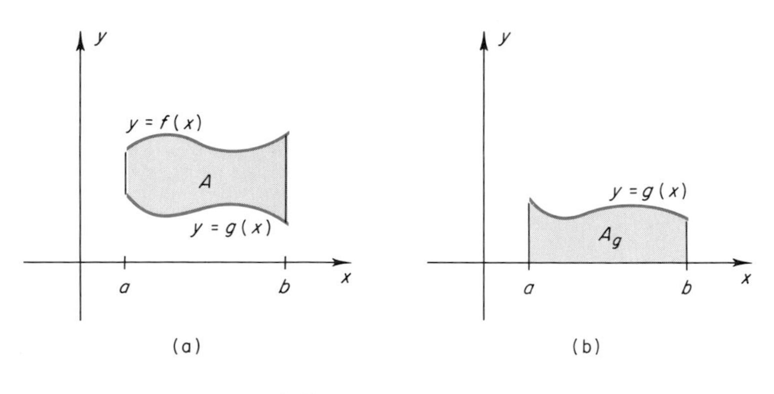

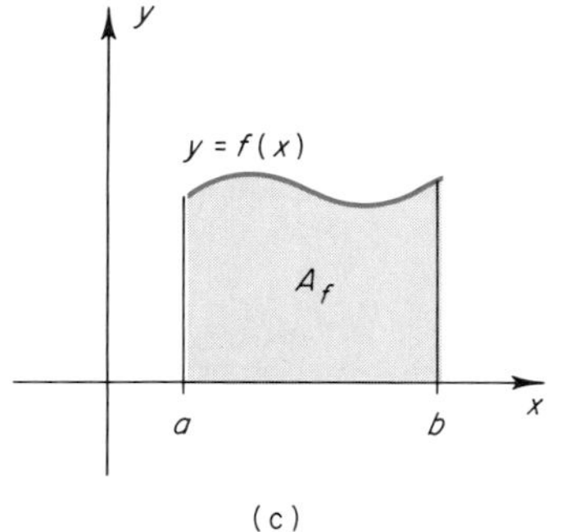

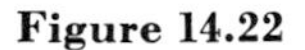
Figure 14.22

where A_f is the area under $y = f(x)$ over $[a, b]$ and A_g is the area under $y = g(x)$ over $[a, b]$. In terms of definite integrals (2) can be written

$$A = \int_a^b f(x)\,dx - \int_a^b g(x)\,dx = \int_a^b [f(x) - g(x)]\,dx.$$

This suggests the following result.

> If f and g are continuous functions with $f(x) \geq g(x)$ on $[a, b]$, then the area A between $y = f(x)$ and $y = g(x)$ over $[a, b]$ is
>
> $$A = \int_a^b [f(x) - g(x)]\,dx.$$

Example 19 Find the area of the region enclosed by the curves $y = x^2$ and $y = x$.

Solution We begin by sketching the curves to obtain a clear picture of the region involved (Figure 14.23). Next we determine the points of intersection, if any, by observing that any such point (x, y) must satisfy

$$y = x^2 \qquad \text{and} \qquad y = x$$

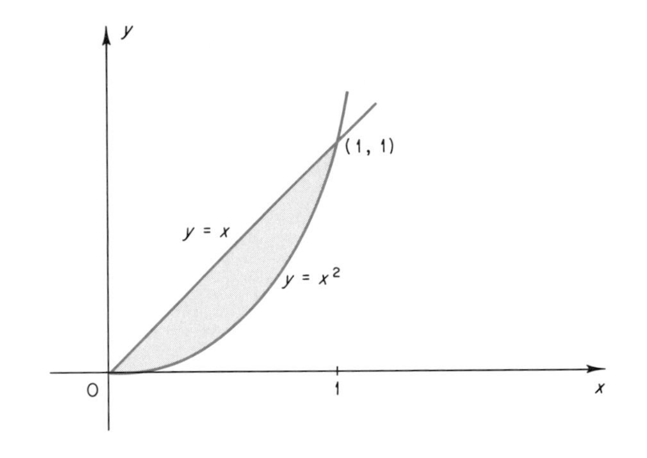

Figure 14.23

so that

$$x = x^2$$
$$x^2 - x = 0$$
$$x(x - 1) = 0$$

which yields

$$x = 0 \qquad \text{and} \qquad x = 1.$$

Thus the interval with which we are concerned is [0, 1] and the area is

$$\int_0^1 (x - x^2)\, dx = \frac{x^2}{2} - \frac{x^3}{3}\Big]_0^1 = \left(\frac{1}{2} - \frac{1}{3}\right) - 0 = \frac{1}{6}.$$

EXERCISE SET 14.3

1. (a) Evaluate the definite integral

$$\int_0^3 (2 - x)\, dx.$$

(b) Use the formula for the area of a triangle (area $= \frac{1}{2} \cdot$ base $\cdot$ height) and the graph of $y = 2 - x$ (Figure 14.24) to check the result in (a).

2. (a) Evaluate the definite integral

$$\int_{-1}^1 \tfrac{1}{2}x\, dx.$$

(b) Use the formula for the area of a triangle (area $= \frac{1}{2} \cdot$ base $\cdot$ height) and the graph of $y = 2x$ (Figure 14.25) to check the result in (a).

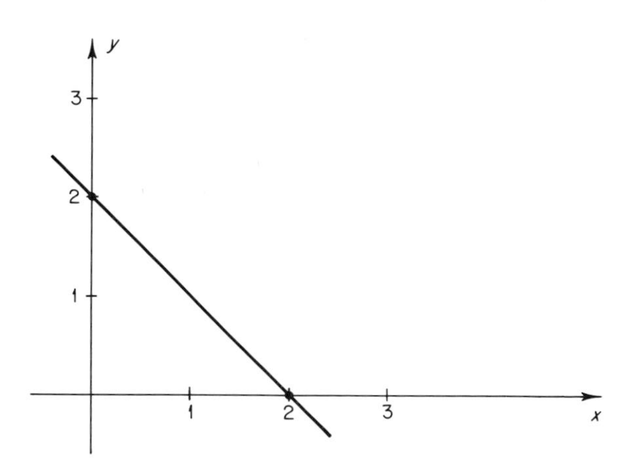

Figure 14.24

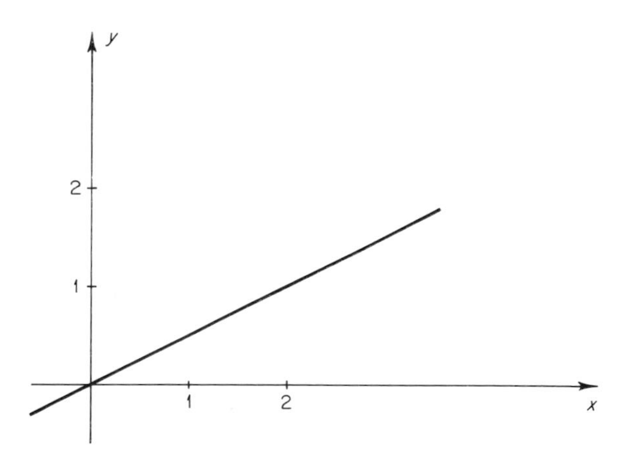

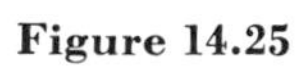
Figure 14.25

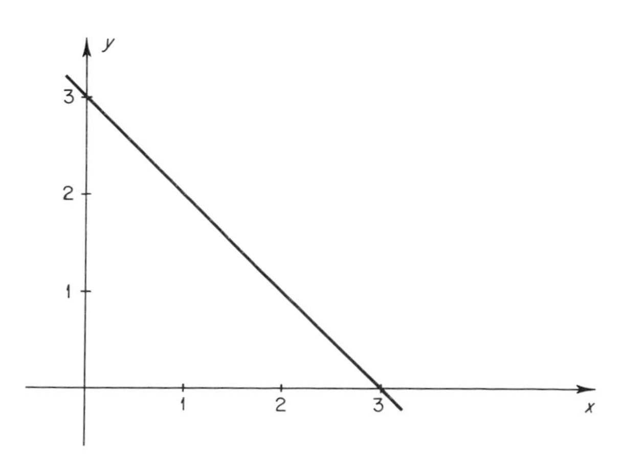

Figure 14.26

3. (a) Evaluate the definite integral

$$\int_2^3 (3 - x)\, dx.$$

(b) Use the formula for the area of a triangle (area $= \frac{1}{2} \cdot$ base $\cdot$ height) and the graph of $y = x - 3$ in Figure 14.26 to check the result in (a).

In Exercises 4–15 evaluate the definite integral.

4. $\int_2^4 3x\, dx$

5. $\int_2^5 5x^2\, dx$

6. $\int_0^1 (2x^2 + 3x^4)\, dx$

7. $\int_1^2 (3x^3 - 7x^2)\, dx.$

8. $\int_{-1}^2 (4x^3 - 2x^2 + 2)\, dx$

9. $\int_9^{16} 5\sqrt{x}\, dx$

10. $\int_0^4 (2e^t - 4t^3)\, dt$

11. $\int_0^1 (2\sqrt{s} + s^2 - 2)\, ds$

12. $\int_1^2 \left(3x - \frac{2}{3x^2} + \frac{4}{x}\right) dx$

13. $\int_0^1 (2u^{-1/3} + 2u^3)\, du$

14. $\int_{-1}^1 (x^{1/5} - x^{1/3})\, dx$

15. $\int_2^2 \left(3x^4 + \frac{3}{4x^2} - 2\right) dx.$

16. (a) Evaluate

$$\int_0^1 x^2\, dx + \int_1^2 x^2\, dx$$

by integrating each term.

(b) Write the sum in (a) as a single definite integral and evaluate this integral.

17. (a) Evaluate

$$\int_{-1}^2 e^x\, dx + \int_2^3 e^x\, dx$$

by integrating each term.

(b) Write the sum in (a) as a single integral and evaluate this integral.

In Exercises 18–27 find the areas of the shaded regions.

18. **19.**

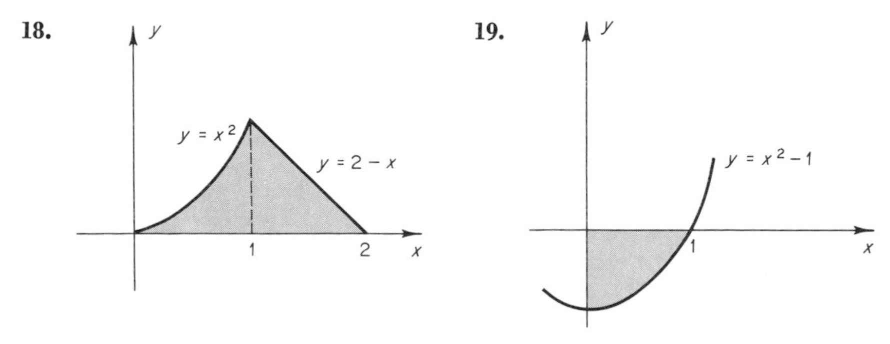

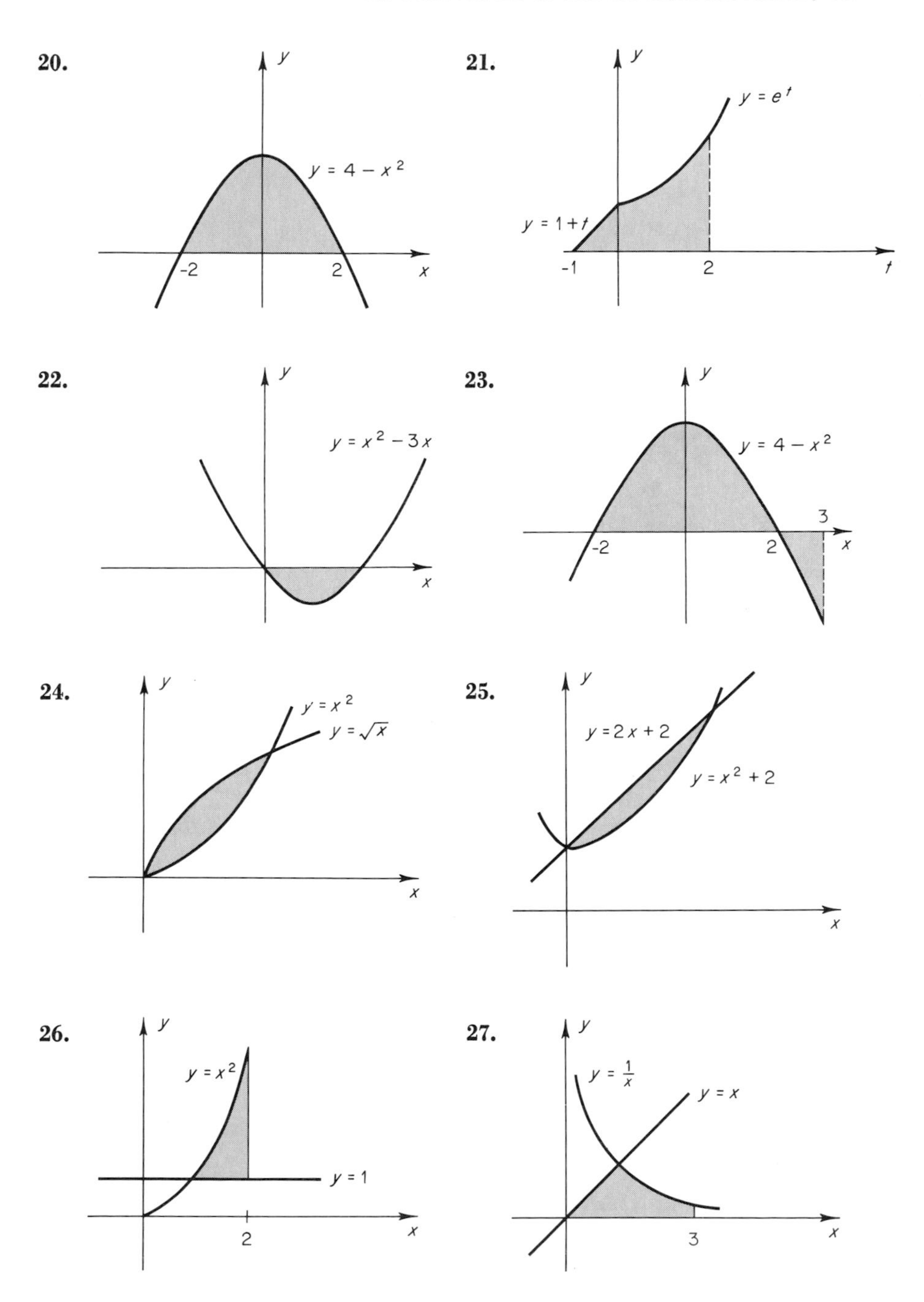

28. Find the area of the region enclosed by the curves $y = 3x^2$ and $y = 3x$.

29. Find the area of the region enclosed by the curves $y = x^2$ and $y = x + 2$.

30. Find the total area of the region bounded between the curve $y = x^2 - 4x$, the line $x = -3$, and the x axis.

14.4 TECHNIQUES OF INTEGRATION (Optional)

In this section we discuss three additional methods for evaluating integrals: integration by substitution, integration by parts, and integration using tables. It is not our objective to make the reader an expert in evaluating integrals, but rather to survey some of the basic ideas.

Integration by Substitution

The technique of substitution, which we now describe, can often be used to reduce an unfamiliar integration problem to a familiar one.

To illustrate the basic idea, consider the problem

$$\int (\tfrac{1}{2}x^2 + 4)^{80}\, x\, dx \tag{1}$$

and let us set

$$u = \tfrac{1}{2}x^2 + 4$$

so that

$$\frac{du}{dx} = x. \tag{2}$$

Thus integral (1) above can be written

$$\int (\tfrac{1}{2}x^2 + 4)^{80}\, x\, dx = \int u^{80} \frac{du}{dx}\, dx. \tag{3}$$

Up to now the symbol dx in the notation $\int f(x)\, dx$ has had no real purpose. However, without going into detail, let suffice to say that the dx is included so that in problems of substitution we can "cancel" the dx's in the expression

$$\frac{du}{dx}\, dx$$

and write

$$du = \frac{du}{dx}\, dx.$$

Accepting this, we obtain from (3)

$$\begin{aligned} \int (\tfrac{1}{2}x^2 + 4)^{80}\, x\, dx &= \int u^{80}\, du \\ &= \frac{1}{81} u^{81} + C \\ &= \frac{1}{81} (\tfrac{1}{2}x^2 + 4)^{81} + C. \end{aligned} \tag{4}$$

Usually these computations would be carried out by first rewriting (2) in the form

$$du = x\,dx$$

and then substituting

$$u = \tfrac{1}{2}x^2 + 4 \qquad \text{and} \qquad du = x\,dx$$

to obtain (4). The following example illustrates this idea.

Example 20 Evaluate

$$\int \frac{2x}{x^2+1}\,dx.$$

Solution Let $u = x^2 + 1$ so that

$$\frac{du}{dx} = 2x \qquad \text{or} \qquad du = 2x\,dx.$$

Substituting in the given integral we obtain

$$\begin{aligned}\int \frac{2x}{x^2+1}\,dx &= \int \frac{1}{x^2+1}\,2x\,dx = \int \frac{1}{u}\,du \\ &= \ln u + C \\ &= \ln(x^2+1) + C.\end{aligned}$$

In the above example the reader may have wondered how we "knew" to let $u = x^2 + 1$ rather than $u = 2x$, $u = 1/(1+x^2)$, or some other choice. The answer is that we *did not* really know in advance that the substitution $u = x^2 + 1$ would work; substitution is a "trial and error" technique, so that sometimes we might have to try several different substitutions before we find one that works or we reach the conclusion that there is no substitution that works.

Example 21 Evaluate

$$\int x\sqrt{x^2+9}\,dx.$$

Solution Let

$$u = x^2 + 9$$

so that

$$\frac{du}{dx} = 2x \qquad \text{or} \qquad du = 2x\,dx.$$

In the given integral we do not quite have the $2x\,dx$ needed for du; we have only $x\,dx$. To remedy this, we simply introduce the missing factor of 2, where needed, and correct for it by multiplying by $\frac{1}{2}$ as follows:

$$\begin{aligned}\int x\sqrt{x^2+9}\,dx &= \frac{1}{2}\int \sqrt{x^2+9}\cdot 2x\,dx \\ &= \frac{1}{2}\int \sqrt{u}\,du = \frac{1}{2}\int u^{1/2}\,du \\ &= \frac{1}{2}\cdot\frac{2}{3}u^{3/2}+C \\ &= \frac{1}{3}(x^2+9)^{3/2}+C.\end{aligned}$$

Example 22 Evaluate

$$\int_0^1 \frac{3x^2}{(x^3+9)^2}\,dx.$$

Solution We first evaluate the indefinite integral

$$\int \frac{3x^2}{(x^3+9)^2}\,dx.$$

Suppose we try the substitution

$$u = x^3$$

so that

$$\frac{du}{dx} = 3x^2 \qquad \text{or} \qquad du = 3x^2\,dx.$$

This yields

$$\int \frac{3x^2}{(x^3+9)^2}\,dx = \int \frac{1}{(u+9)^2}\,du$$

which is not an integral we know how to evaluate directly. Let us try a different substitution, say

$$u = x^3+9$$

so that

$$\frac{du}{dx} = 3x^2 \qquad \text{or} \qquad du = 3x^2\,dx.$$

Thus

$$\int \frac{3x^2}{(x^3+9)^2}\,dx = \int \frac{1}{u^2}\,du = \int u^{-2}\,du$$

$$= -u^{-1} + C = -\frac{1}{u} + C$$

$$= -\frac{1}{x^3+9} + C.$$

Therefore

$$\int_0^1 \frac{3x^2}{(x^3+9)^2}\,dx = -\frac{1}{x^3+9}\Big]_0^1 = \left(-\frac{1}{10}\right) - \left(-\frac{1}{9}\right)$$

$$= \frac{1}{9} - \frac{1}{10} = \frac{1}{90}.$$

Example 23 Evaluate

$$\int e^{x^3} x^2\,dx.$$

Solution Let

$$u = x^3$$

so that

$$\frac{du}{dx} = 3x^2 \qquad \text{or} \qquad du = 3x^2\,dx.$$

Thus

$$\int e^{x^3} x^2\,dx = \frac{1}{3} \int e^{x^3} \cdot 3x^2\,dx$$

$$= \frac{1}{3} \int e^u\,du$$

$$= \frac{1}{3} e^u + C$$

$$= \frac{1}{3} e^{x^3} + C.$$

Example 24 Evaluate

$$\int \frac{e^x}{1+e^x}\,dx.$$

Solution Let

$$u = 1 + e^x$$

so that

$$\frac{du}{dx} = e^x \qquad \text{or} \qquad du = e^x\,dx.$$

Thus

$$\int \frac{e^x}{1+e^x}\,dx = \int \frac{du}{u}$$
$$= \ln u + C$$
$$= \ln(1 + e^x) + C.$$

Example 25 Evaluate

$$\int_1^e \frac{\ln x}{x}\,dx.$$

Solution In the indefinite integral

$$\int \frac{\ln x}{x}\,dx$$

let

$$u = \ln x$$

so that

$$\frac{du}{dx} = \frac{1}{x} \qquad \text{or} \qquad du = \frac{1}{x}\,dx.$$

Thus

$$\int \frac{\ln x}{x}\,dx = \int u\,du = \frac{u^2}{2} + C$$
$$= \frac{(\ln x)^2}{2} + C.$$

Therefore

$$\int_1^e \frac{\ln x}{x}\,dx = \left.\frac{(\ln x)^2}{2}\right]_1^e = \frac{(\ln e)^2}{2} - \frac{(\ln 1)^2}{2}$$
$$= \frac{1}{2} - 0 = \frac{1}{2}.$$

Integration by Parts

Another important integration method is based on the formula for the derivative of a product of two functions:

$$\frac{d}{dx}[f(x) \cdot g(x)] = f(x)g'(x) + g(x)f'(x). \tag{5}$$

If we integrate both sides of (5), we obtain

$$\int \frac{d}{dx}[f(x) \cdot g(x)]\, dx = \int f(x)g'(x)\, dx + \int g(x)f'(x)\, dx$$

or

$$f(x)g(x) = \int f(x)g'(x)\, dx + \int g(x)f'(x)\, dx$$

or

$$\int f(x)g'(x)\, dx = f(x)g(x) - \int g(x)f'(x)\, dx \tag{6}$$

Formula (6) is called the formula for ***integration by parts;*** it expresses one integral in terms of another. To use this formula for evaluating an integral

$$\int h(x)\, dx$$

one tries to factor the function h into the form

$$h(x) = f(x)g'(x)$$

so that $\int g(x)f'(x)\, dx$ is easier to evaluate than

$$\int h(x)\, dx = \int f(x)g'(x)\, dx.$$

Example 26 Evaluate

$$\int xe^x\, dx.$$

Solution Let

$$f(x) = x \qquad \text{and} \qquad g'(x) = e^x$$

so that

$$f'(x) = 1 \qquad \text{and} \qquad g(x) = \int e^x\,dx = e^x.$$

Thus, from (6),

$$\begin{aligned}\int xe^x\,dx &= \int f(x)g'(x)\,dx \\ &= f(x)g(x) - \int g(x)f'(x)\,dx \\ &= xe^x - \int e^x \cdot 1\,dx \\ &= xe^x - e^x + C.\end{aligned}$$

As with substitution, integration by parts is a "trial and error" method; its success hinges on making the right choice for $f(x)$ and $g'(x)$. For example, in the evaluation of $\int xe^x\,dx$ above, had we chosen

$$f(x) = e^x \qquad \text{and} \qquad g'(x) = x$$

so that

$$f'(x) = e^x \qquad \text{and} \qquad g(x) = \int x\,dx = \frac{x^2}{2}$$

we would have obtained from (6)

$$\begin{aligned}\int xe^x\,dx &= \int f(x)g'(x)\,dx \\ &= f(x)g(x) - \int g(x)f'(x)\,dx \\ &= \frac{x^2}{2}e^x - \int \frac{x^2}{2}e^x\,dx.\end{aligned}$$

However, the new integral $\int (x^2/2)e^x\,dx$ is more complicated than the original, so that this choice of $f(x)$ and $g'(x)$ is of no help.

Example 27 Evaluate

$$\int x \ln x\,dx.$$

Solution Let

$$f(x) = \ln x \qquad \text{and} \qquad g'(x) = x$$

so that

$$f'(x) = \frac{1}{x} \qquad \text{and} \qquad g(x) = \int x\,dx = \frac{x^2}{2}.$$

Thus

$$\begin{aligned}\int x \ln x\, dx &= \int f(x)g'(x)\, dx \\ &= f(x)g(x) - \int g(x)f'(x)\, dx \\ &= \frac{x^2}{2} \ln x - \int \left(\frac{x^2}{2}\right)\left(\frac{1}{x}\right) dx \\ &= \frac{x^2}{2} \ln x - \frac{1}{2}\int x\, dx \\ &= \frac{x^2}{2} \ln x - \frac{1}{2}\left(\frac{x^2}{2}\right) + C \\ &= \frac{x^2}{2} \ln x - \frac{x^2}{4} + C.\end{aligned}$$

Tables of Integrals

So far we have just touched on the problem of integration. There remain many integrals we cannot yet evaluate and many integration techniques we do not have time to discuss. However, even with a limited knowledge of integration, it is possible to solve very complicated integration problems using tables of integral formulas. There exist many such tables, some of which contain hundreds of formulas. Generally, these tables are arranged according to the form of the integrand and the user has to match the integral with one of the forms in the table. Sometimes the integrand will require a little algebraic manipulation before it matches one of the table forms. To illustrate the idea we will evaluate some integrals using the short table of integrals in Appendix Table VIII.

Example 28 Use the integral table in Appendix VIII to evaluate:

(a) $\displaystyle\int \frac{dx}{\sqrt{x^2 - 4}}$ (b) $\displaystyle\int \frac{dx}{x^2 - 5}$ (c) $\displaystyle\int \frac{dx}{\sqrt{4x^2 + 9}}$.

Solution (a) This integral matches with Formula 9 when $a = 2$. Thus

$$\int \frac{dx}{\sqrt{x^2 - 4}} = \ln(x + \sqrt{x^2 - 4}) + C.$$

(b) This integral matches with Formula 10 when $a = \sqrt{5}$ $(a^2 = 5)$. Thus

$$\int \frac{dx}{x^2 - 5} = \frac{1}{2\sqrt{5}} \ln\left(\frac{x - \sqrt{5}}{x + \sqrt{5}}\right) + C.$$

(c) This integral does not match any of the forms in the table exactly. However, if it were not for the factor of 4 multiplying the x^2, it would match Formula 8. We remedy this as follows:

$$\int \frac{dx}{\sqrt{4x^2+9}} = \int \frac{dx}{\sqrt{4(x^2+\frac{9}{4})}} = \frac{1}{2}\int \frac{dx}{\sqrt{x^2+\frac{9}{4}}}.$$

We now apply Formula 8 with $a^2 = \frac{9}{4}$ to obtain

$$\int \frac{dx}{\sqrt{4x^2+9}} = \frac{1}{2}\int \frac{dx}{\sqrt{x^2+\frac{9}{4}}} = \frac{1}{2}\ln\left(x+\sqrt{x^2+\frac{9}{4}}\right) + C.$$

Example 29 Evaluate

$$\int \frac{dx}{x(4-2x)}.$$

Solution From Formula 6 with $a = -2$ and $b = 4$ we obtain

$$\int \frac{dx}{x(4-2x)} = \frac{1}{4}\ln\left(\frac{x}{4-2x}\right) + C.$$

EXERCISE SET 14.4

In Exercises 1–16 evaluate the integral by the method of substitution.

1. $\int 3x^2(x^3+5)^{10}\,dx$

2. $\int \frac{3x^2}{x^3+1}\,dx$

3. $\int \frac{2x}{(x^2-1)^{15}}\,dx$

4. $\int x^3(x^4+1)\,dx$

5. $\int \frac{x}{(x^2+4)^3}\,dx$

6. $\int \frac{1}{x+2}\,dx$

7. $\int \frac{1}{2-x}\,dx$

8. $\int 3t^2\sqrt{t^3+5}\,dt$

9. $\int t^3\sqrt{t^4+2}\,dt$

10. $\int 2xe^{x^2}\,dx$

11. $\int e^{-2x}\,dx$

12. $\int (3x^2+2x)e^{(x^3+x^2)}\,dx$

13. $\int \frac{(\ln x)^2}{x}\,dx$

14. $\int \frac{\ln 3x}{x}\,dx$

15. $\int_0^1 e^{3t}\,dt$

16. $\int x(x^2-1)^{23}\,dx.$

In Exercises 17–22 evaluate the integral using integration by parts.

17. $\int 3xe^{3x}\,dx$

18. $\int xe^{-x}\,dx$

19. $\int x\sqrt{x+1}\,dx$

20. $\int x^2 \ln x\,dx$

21. $\int \sqrt{x} \ln x\,dx$

22. $\int_0^2 xe^{-3x}\,dx.$

23. Use integration by parts to evaluate $\int \ln x\,dx$. (*Hint*: Let $f(x) = \ln x$, $g'(x) = 1$ in the integration by parts formula.)

24. Use integration by parts twice to evaluate $\int x^2e^x\,dx$.

In Exercises 25–32 find the indicated integral by using Appendix Table VIII.

25. $\int \frac{x}{3x+6}\,dx$

26. $\int \frac{1}{\sqrt{x^2+25}}\,dx$

27. $\int \frac{1}{\sqrt{9x^2-25}}\,dx$

28. $\int \frac{1}{t(3t-2)^2}\,dt$

29. $\int \frac{1}{16t^2-9}\,dt$

30. $\int \frac{1}{(3-2x)^2}\,dx$

31. $\int \frac{dx}{x\sqrt{4+9x^2}}$

32. $\int \frac{1}{9-4x^2}\,dx.$

15 APPLICATIONS OF INTEGRATION

In this chapter we shall show how the definite integral relates the rate of change of a function to its total change in value between two points. We shall also discuss a variety of applications of this relationship.

Recall from Section 14.2 that the defining equation for the definite integral is

$$\int_a^b f(x)\,dx = F(b) - F(a)$$

where F is any antiderivative of the integrand. In particular, if we integrate the derivative $f'(x)$ of a function f, we obtain

$$\int_a^b f'(x)\,dx = f(b) - f(a) \tag{1}$$

since $f(x)$ is an antiderivative of the integrand $f'(x)$. This result has a useful geometric interpretation. The right-hand side of (1) represents the total change in the value of $f(x)$ as x varies from a to b (Figure 15.1). The left-hand side is the integral from a to b of the rate of change of f with respect to x. Thus

The definite integral $\int_a^b f'(x)\,dx$ of the rate of change of f with respect to x yields the total change in the value of f as x varies from a to b.

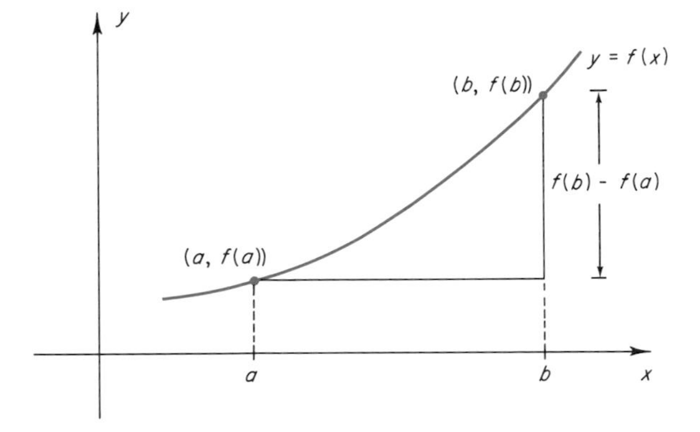

Figure 15.1

Application to Marginal Analysis

Recall from Section 12.3 that marginal cost (MC), marginal revenue (MR), and marginal profit (MP) are the derivatives of the cost function, revenue function, and profit function, respectively. That is,

$$\text{MC} = \text{marginal cost} = C'(x)$$
$$\text{MR} = \text{marginal revenue} = R'(x)$$
$$\text{MP} = \text{marginal profit} = P'(x).$$

Thus, if we integrate MC, MR, or MP over an interval $[a, b]$ we obtain from (1)

$$\int_a^b (\text{MC})\,dx = \int_a^b C'(x)\,dx = C(b) - C(a)$$
$$= \text{total change in cost as } x \text{ varies from } a \text{ to } b;$$
$$\int_a^b (\text{MR})\,dx = \int_a^b R'(x)\,dx = R(b) - R(a)$$
$$= \text{total change in revenue as } x \text{ varies from } a \text{ to } b;$$
$$\int_a^b (\text{MP})\,dx = \int_a^b P'(x)\,dx = P(b) - P(a)$$
$$= \text{total change in profit as } x \text{ varies from } a \text{ to } b.$$

Example 1 A manufacturer determines that the firm's marginal cost and marginal revenue functions are:

$$\text{MC} = C'(x) = 100 - 0.1x$$
$$\text{MR} = R'(x) = 100 + 0.1x.$$

(a) Find the added revenue that results when the production level is increased from 20 to 30 units.

(b) Find the total revenue resulting from the sale of 30 units.

(c) If the fixed cost (cost of producing $x = 0$ units) is \$400, find the total cost of producing 30 units.

Solution (a) As the production level increases from $x = 20$ to $x = 30$, the revenue changes an amount

$$\int_{20}^{30} (\text{MR})\,dx = \int_{20}^{30} (100 + 0.1x)\,dx = (100x + 0.05x^2)\Big]_{20}^{30}$$

$$= 3045 - 2020 = 1025.$$

Thus the added revenue received is \$1025.

(b) As the production level increases from $x = 0$ to $x = 30$, the change in revenue received is

$$\int_{0}^{30} (\text{MR})\,dx = \int_{0}^{30} (100 + 0.1x)\,dx = (100x + 0.05x^2)\Big]_{0}^{30}$$

$$= 3045 - 0 = 3045.$$

Thus, as production increases from $x = 0$ to $x = 30$ units, the revenue increases \$3045. If we assume that the revenue for manufacturing $x = 0$ units is zero, then the total revenue received for manufacturing 30 units will be \$3045.

(c) As the production level increases from $x = 0$ to $x = 30$, the cost changes an amount

$$\int_{0}^{30} (\text{MC})\,dx = \int_{0}^{30} (100 - 0.1x)\,dx = (100x - 0.05x^2)\Big]_{0}^{30}$$

$$= 2955 - 0 = 2955.$$

Thus the total cost for manufacturing 30 units will be the fixed cost plus the added cost as the production increases from $x = 0$ to $x = 30$ units; that is

$$\text{total cost} = \$400 + \$2955 = \$3355.$$

Application to Analysis of Natural Resources

The demand for many of the earth's natural resources (like coal, copper, zinc, oil, or gas) seems to follow an exponential growth pattern. More precisely, if we pick some reference point $t = 0$ in time and let $A(t)$ denote the amount of the product consumed during the time interval $[0, t]$, then frequently the consumption rate $R(t) = A'(t)$ will have an exponential model

$$R(t) = R_0 e^{kt}$$

where R_0 is the rate of consumption at time $t = 0$ and k is the growth constant for the model. Since $A(t) = 0$ when $t = 0$ (why?), it follows that the total consumption during a time interval $[0, T]$ is

$$\int_0^T R(t)\,dt = \int_0^T A'(t)\,dt = A(T) - A(0) = A(T).$$

Thus

$$\begin{aligned}\int_0^T R(t)\,dt &= A(T)\\ &= \text{total consumption during the time interval } [0, T].\end{aligned}$$

Example 2 In the beginning of 1974 ($t = 0$) zinc was being consumed at a rate of†

$$478{,}850 \quad \text{short tons}$$

per year and the consumption rate was increasing at 4.5% per year. Assuming an exponential model for the rate of consumption, estimate the total amount of zinc that will be used from 1974 to 1984.

Solution From the data, the growth rate is

$$R(t) = 478{,}850e^{0.045t}$$

Thus the total consumption between $t = 0$ (1974) and $t = 10$ (1984) will be

$$\begin{aligned}\int_0^{10} R(t)\,dt &= \int_0^{10} 478{,}850e^{0.045t}\,dt\\ &= \frac{478{,}850}{0.045}e^{0.045t}\Big]_0^{10}\\ &= 10{,}641{,}111(e^{0.45} - e^0)\\ &= 10{,}641{,}111(0.5683) \quad \text{(Appendix Table VI)}\\ &= 6{,}047{,}343 \quad \text{(short tons).}\end{aligned}$$

Application to the Study of Motion

Example 3 An automobile moves along a straight track in such a way that its speed $g(t)$ after t seconds is

$$g(t) = 3t^{1/2} \quad \text{feet per second.}$$

How far does the automobile travel during the first 100 seconds?

† *The World Almanac* (New York: Newspaper Enterprise Association, 1977).

Solution The speed $g(t)$ is the rate at which distance traveled changes with time; thus the total distance traveled during the first 100 seconds ($t = 0$ to $t = 100$) is

$$\int_0^{100} g(t)\,dt = \int_0^{100} 3t^{1/2}\,dt = 3 \cdot \frac{2}{3} t^{3/2} \Big]_0^{100}$$

$$= 2000 - 0$$

$$= 2000 \quad \text{feet.}$$

Application to Biology

Example 4 An experimental drug changes the average subject's body temperature at a rate

$$r(t) = -0.003t^2 + 0.01t$$

where $r(t)$ is in Fahrenheit degrees per hour and t is the number of hours elapsed after administration of the drug. How much of a temperature change will occur between the second and fifth hours ($t = 2$ to $t = 5$)?

Solution Since $r(t)$ is the *rate* at which temperature changes with time, the total change in temperature from $t = 2$ to $t = 5$ is

$$\int_2^5 r(t)\,dt = \int_2^5 (-0.003t^2 + 0.01t)\,dt$$

$$= (-0.001t^3 + 0.005t^2)\Big]_2^5 = 0 - 0.012 = -0.012.$$

Thus the temperature decreases 0.012 degrees Fahrenheit during the period $t = 2$ to $t = 5$.

Applications in Probability

A function $f(x)$ is called a ***density function*** if:

(1) $f(x) \geq 0$ for all x.
(2) The total area under the graph of f is one.

Some typical density functions are sketched in Figure 15.2. We have encountered some important density functions already, the ***Gaussian*** or ***normal*** curves of Section 7.6 and the chi-square curves of Section 7.7.

Density functions play an important role in probability problems involving continuous random variables, that is, random variables whose possible values form an entire interval of real numbers (Section 7.1).

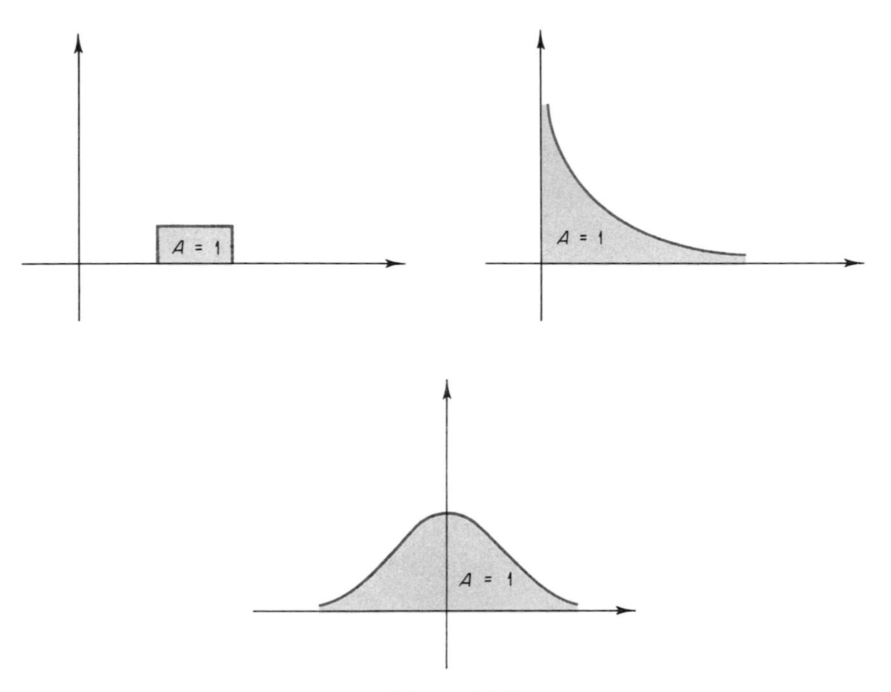

Figure 15.2

Given a continuous random variable X, it is often possible to find a density function $f(x)$ with the property that

$$P(a \leq X \leq b) = \int_a^b f(x)\, dx;$$

that is, the probability that X takes on a value between a and b is the area under $f(x)$ over the interval $[a, b]$ (Figure 15.3). In this case $f(x)$ is called the ***density function*** for X.

As an illustration, let X represent a real number chosen "at random"

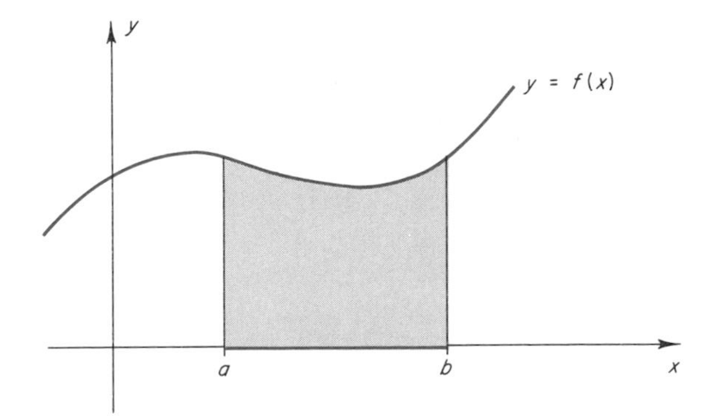

Figure 15.3 The shaded area is the probability that X takes on a value in the interval $[a, b]$.

from the interval [0, 10] and let us ask for the probability that X falls in the interval [2, 4] (Figure 15.4). Because the interval [2, 4] has

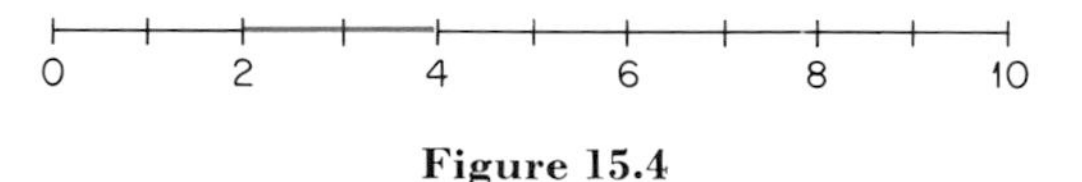

Figure 15.4

length 2, and because the entire interval of possible values for X has length 10, it is intuitively clear that

$$P(2 \leq X \leq 4) = \tfrac{2}{10} = \tfrac{1}{5}.$$

More generally, if we ask for the probability that X lies in a subinterval $[a, b]$, then we obtain

$$P(a \leq X \leq b) = \frac{\text{length of subinterval}}{\text{length of entire interval}} = \frac{b - a}{10}$$

(Figure 15.5). However, if we let f be the density function given by

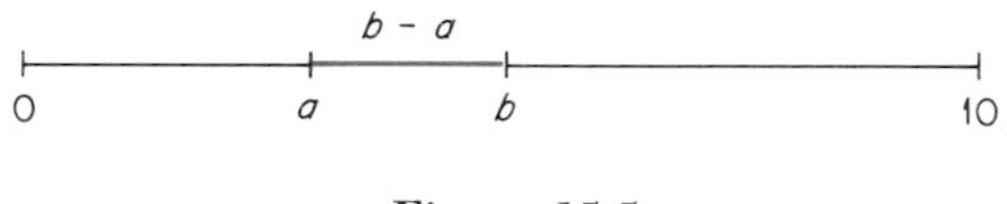

Figure 15.5

$$f(x) = \tfrac{1}{10} \qquad \text{where} \quad 0 \leq x \leq 10$$

(Figure 15.6a) then $(b - a)/10$ is also the area under $f(x)$ over the interval $[a, b]$ (Figure 15.6b), so that

$$P(a \leq X \leq b) = \int_a^b f(x)\,dx.$$

Thus f is a density function for X.

More generally, if X is a number chosen at random from an interval

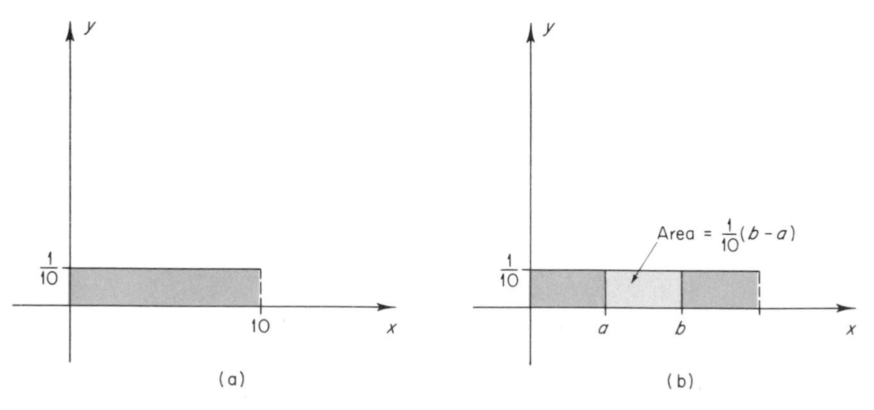

Figure 15.6

$[c, d]$, then X is said to be ***uniformly distributed*** over $[c, d]$. A density function for X is

$$f(x) = \frac{1}{d - c} \qquad \text{where} \quad c \leq x \leq d.$$

Example 5 Due to variations in air traffic and weather, the 10:00 A.M. flight from Philadelphia to Chicago takes off (at random) between 10:00 A.M. and 10:15 A.M. What is the probability that a passenger will experience a delay of 10 minutes or more?

Solution Let the random variable X be the delay (in minutes). Then X is uniformly distributed over the interval $[0, 15]$ and its density function is

$$f(x) = \frac{1}{15} \qquad \text{where} \quad 0 \leq x \leq 15.$$

Thus the probability of a delay of 10 minutes or more is

$$P(10 \leq X \leq 15) = \int_{10}^{15} \frac{1}{15}\, dx = \frac{1}{15}\, x \Big]_{10}^{15} = \frac{1}{15}(15 - 10) = \frac{1}{3}.$$

One of the most important density functions in applications is

$$f(x) = ke^{-kx}$$

where $x \geq 0$ and k is a positive constant (Figure 15.7). A random variable X with such a density function is said to be ***exponentially distributed.*** Although we shall omit the details, the constant k may be interpreted as

$$k = \frac{1}{\text{average value of } X}.$$

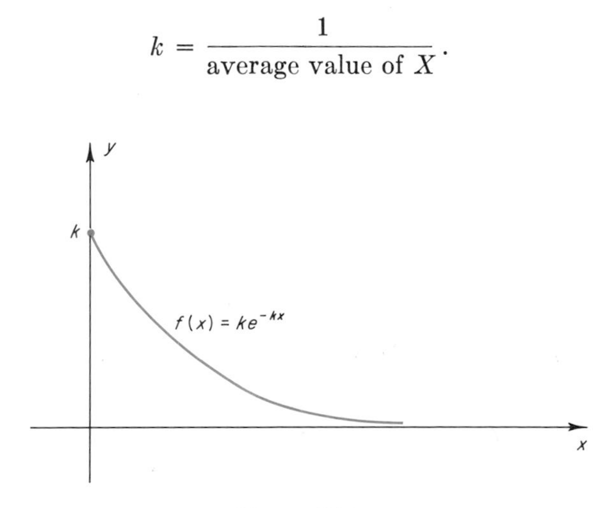

Figure 15.7

Examples of exponentially distributed random variables are:

the length of life of a manufactured item such as a lightbulb, a resistor, or a television;

the distance between two successive cars on a highway;

the duration of a phone call.

Example 6 Company records show the average length of a firm's business calls to be five minutes. What is the probability that a call will be three minutes or less?

Solution Let X be the length of the call (in minutes) and assume X to be exponentially distributed. Thus,

$$k = \frac{1}{\text{average value of } X} = \frac{1}{5} = 0.2,$$

so the density function for X is

$$f(x) = 0.2e^{-0.2x}$$

where $x \geq 0$. The probability that a call will be three minutes or less is

$$\begin{aligned} P(0 \leq X \leq 3) &= \int_0^3 0.2e^{-0.2x}\,dx \\ &= -e^{-0.2x}\Big]_0^3 \\ &= -e^{-0.6} + e^0 \\ &= 1 - e^{-0.6} \\ &= 1 - 0.5488 \qquad \text{(Appendix Table VI)} \\ &= 0.4512 \end{aligned}$$

In other words, about 45.12% of all calls will be three minutes or less.

EXERCISE SET 15

1. In each part, use formula (1) to compute $\int_0^2 f'(x)\,dx$.
   (a) $f'(x) = x^3$ (b) $f'(x) = e^{x^2/2}$.
2. Use the data in the graph in Figure 15.8 to find $\int_1^3 f'(x)\,dx$.
3. **(Marginal Analysis)** A manufacturer determines that the firm's marginal cost and marginal revenue functions are:

$$\text{MC} = C'(x) = 200 - 0.4x$$

$$\text{MR} = R'(x) = 200 + 0.2x.$$

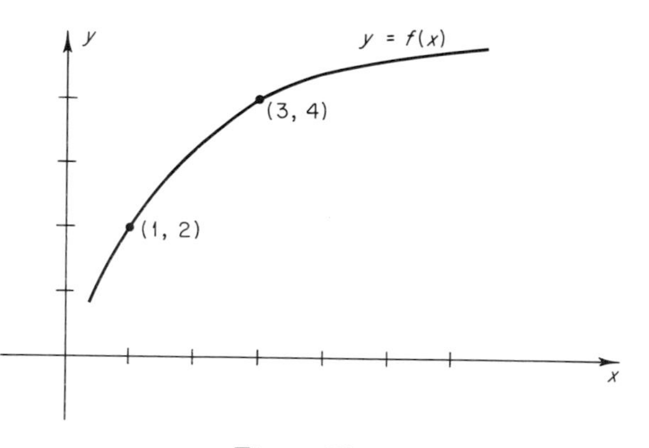

Figure 15.8

(a) Find the added revenue that results when the production level is increased from 10 to 50 units.

(b) Find the total revenue resulting from the sale of 70 units.

(c) If the fixed cost is \$1000, find the total cost of producing 70 units.

4. **(Marginal Analysis)** Consider the manufacturing problem of Exercise 3.

   (a) Find the total profit from manufacturing and selling 70 units. (*Hint:* Do not forget to take the fixed cost into account.)

   (b) Find the profit from manufacturing and selling k units.

5. **(Consumption of Natural Resources)** In 1975 the rate of energy consumption in the U.S. was about 72 quadrillion B.T.U.'s per year (*Time Magazine*, May, 1977) and growing at 4% per year. Assuming an exponential model for the rate of consumption, estimate the total amount of energy that will be used from 1975 ($t = 0$) to 1985 ($t = 10$).

6. **(Consumption of Natural Resources)** In 1974, U.S. natural gas reserves were being consumed at a rate of 22 million cubic feet per year (U.S. Transportation Department data).

   (a) Assuming a growth of 2% per year in the rate of consumption, estimate the amount of gas that will be used from 1974 ($t = 0$) to 1994 ($t = 20$).

   (b) The U.S. Geologic Survey estimates U.S. gas reserves at 655 trillion cubic feet. Starting from 1974 as $t = 0$, how long will U.S. gas reserves last at the rate of consumption in (a)?

7. A stone dropped from the top of a building falls so that its speed after t seconds is

$$v(t) = 32t \quad \text{feet per second.}$$

   (a) How far will the stone fall in 25 seconds?

   (b) How long will it take for the stone to fall 400 feet?

8. A processing plant begins dumping sewage into a stream at the rate of

$$R(t) = 300t^2 + 3t \quad \text{gallons per day.}$$

where t is the number of days elapsed after dumping begins.

(a) What is the amount of sewage dumped between the fifth and tenth days?

(b) Assuming no sewage is being dumped at time $t = 0$, what is the total amount of sewage dumped during the first five days ($t = 0$ to $t = 5$)?

9. As the result of a five-day sales campaign a company's rate of sales are expected to grow according to the curve shown in Figure 15.9, where $r(t)$

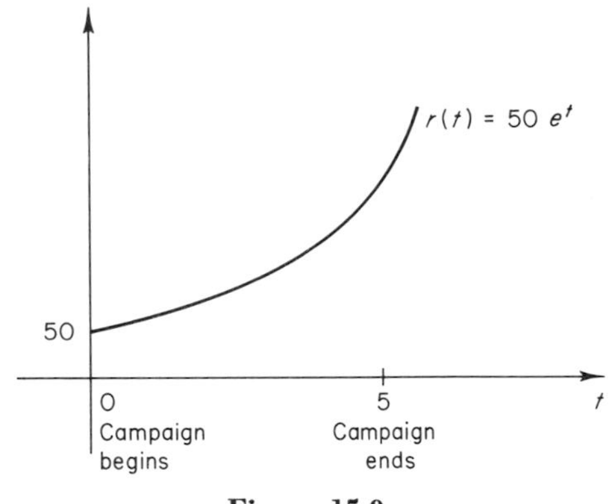

Figure 15.9

is in dollars per day and t is the number of days elapsed from the start of the campaign. Find the increase in total sales during the campaign period.

10. (**Biology**) An astronaut's reaction time changes at a rate

$$r(t) = 0.004t^2 - 0.02t$$

where $r(t)$ is in milliseconds per hour and t is the number of hours that the astronaut is weightless. How much of an increase in reaction time will occur between the fifth and tenth hours of weightlessness?

11. Let $f(x) = \frac{1}{4}x$ if $1 \le x \le 3$ and $f(x) = 0$ otherwise.

(a) Show that f is a density function.

(b) Let X be a continuous random variable whose density function is f. Find $P(1 \le X \le 2)$.

(c) For the random variable in (b), find $P(X \ge 0)$.

12. Let $f(x) = \frac{3}{2}x^2$ if $-1 \le x \le 1$ and $f(x) = 0$ otherwise.

(a) Show that f is a density function.

(b) Let X be a continuous random variable whose density function is f. Find $P(-\frac{1}{2} \le X \le \frac{1}{2})$.

(c) For the random variable in (b), find $P(X \le 0)$.

13. A number X is chosen at random from the interval $[5, 25]$.

(a) Find a density function for X.

(b) What is the probability that X satisfies $1 \le X \le 4$?

(c) What is the probability that $X \ge 3$?

14. The voltage in a 120-volt line varies randomly between 115 and 125 volts.

(a) What is the probability that the voltage will fall between 118 and 123 volts?

(b) If a photographic enlarger works improperly when the voltage is 118 volts or lower, what percentage of the time will the enlarger work properly?

15. The 9:00 A.M. train from Philadelphia to New York leaves at random between 9:00 and 9:10 A.M.

(a) What is the probability that a passenger boarding at 9:00 A.M. will wait longer than 5 minutes for the train to leave?

(b) What is the probability that the wait will be less than 10 minutes?

(c) If a person reaches the station at 9:07 A.M. what are the chances that the passenger will still catch the train?

16. Suppose a commuter train runs every half hour, and you arrive at the station at random.

(a) What is the probability you will wait at least 10 minutes?

(b) What is the probability that you will wait 15 minutes or less?

(c) What is the probability that you will wait at least 2 minutes, but no longer than 5 minutes?

17. Assume that the average wait in a dentist's office is 20 minutes and that the waiting time is exponentially distributed.

(a) What is the probability a patient will wait 10 minutes or less before seeing the dentist?

(b) What percentage of the patients wait longer than 10 minutes?

18. Suppose that the average distance between successive cars on a bridge is 150 feet and that the distance is an exponentially distributed random variable.

(a) What is the probability that the distance between two successive cars will be no more than 75 feet?

(b) Between 30 and 75 feet?

19. A certain brand of lightbulb is claimed to have an average life of 1000 hours. Assuming the life of such a bulb to be exponentially distributed, find

(a) the probability that a bulb will last 500 hours or less;

(b) the probability that a bulb will last more than 500 hours;

(c) the percentage of bulbs that will last between 1000 and 2000 hours.

(For readers familiar with Section 6.5.)

20. (a) Referring to Exercise 19, what is the probability that two bulbs selected independently will both burn out in 800 hours or less?

(b) If three bulbs are selected independently, what is the probability that at least two will last longer than 1000 hours?

APPENDIX/TABLES

Table I

Areas under the Standard Normal Curve

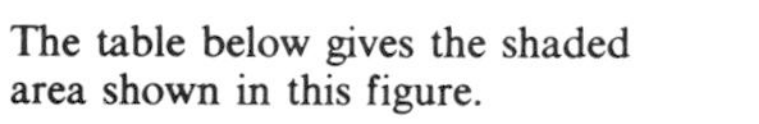

The table below gives the shaded area shown in this figure.

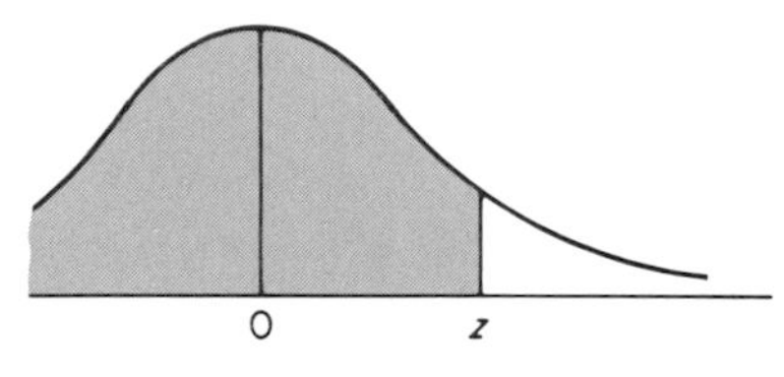

z	0	1	2	3	4	5	6	7	8	9
0.0	.5000	.5040	.5080	.5120	.5160	.5199	.5239	.5279	.5319	.5359
0.1	.5398	.5438	.5478	.5517	.5557	.5596	.5636	.5675	.5714	.5753
0.2	.5793	.5832	.5871	.5910	.5948	.5987	.6026	.6064	.6103	.6141
0.3	.6179	.6217	.6255	.6293	.6331	.6368	.6406	.6443	.6480	.6517
0.4	.6554	.6591	.6628	.6664	.6700	.6736	.6772	.6808	.6844	.6879
0.5	.6915	.6950	.6985	.7019	.7054	.7088	.7123	.7157	.7190	.7224
0.6	.7257	.7291	.7324	.7357	.7389	.7422	.7454	.7486	.7517	.7549
0.7	.7580	.7611	.7642	.7673	.7704	.7734	.7764	.7794	.7823	.7852
0.8	.7881	.7910	.7939	.7967	.7995	.8023	.8051	.8078	.8106	.8133
0.9	.8159	.8186	.8212	.8238	.8264	.8289	.8315	.8340	.8365	.8389
1.0	.8413	.8438	.8461	.8485	.8508	.8531	.8554	.8577	.8599	.8621
1.1	.8643	.8665	.8686	.8708	.8729	.8749	.8770	.8790	.8810	.8830
1.2	.8849	.8869	.8888	.8907	.8925	.8944	.8962	.8980	.8997	.9015
1.3	.9032	.9049	.9066	.9082	.9099	.9115	.9131	.9147	.9162	.9177
1.4	.9192	.9207	.9222	.9236	.9251	.9265	.9279	.9292	.9306	.9319
1.5	.9332	.9345	.9357	.9370	.9382	.9394	.9406	.9418	.9429	.9441
1.6	.9452	.9463	.9474	.9484	.9495	.9505	.9515	.9525	.9535	.9545
1.7	.9554	.9564	.9573	.9582	.9591	.9599	.9608	.9616	.9625	.9633
1.8	.9641	.9649	.9656	.9664	.9671	.9678	.9686	.9693	.9699	.9706
1.9	.9713	.9719	.9726	.9732	.9738	.9744	.9750	.9756	.9761	.9767
2.0	.9772	.9778	.9783	.9788	.9793	.9798	.9803	.9808	.9812	.9817
2.1	.9821	.9826	.9830	.9834	.9838	.9842	.9846	.9850	.9854	.9857
2.2	.9861	.9864	.9868	.9871	.9875	.9878	.9881	.9884	.9887	.9890
2.3	.9893	.9896	.9898	.9901	.9904	.9906	.9909	.9911	.9913	.9916
2.4	.9918	.9920	.9922	.9925	.9927	.9929	.9931	.9932	.9934	.9936
2.5	.9938	.9940	.9941	.9943	.9945	.9946	.9948	.9949	.9951	.9952
2.6	.9953	.9955	.9956	.9957	.9959	.9960	.9961	.9962	.9963	.9964
2.7	.9965	.9966	.9967	.9968	.9969	.9970	.9971	.9972	.9973	.9974
2.8	.9974	.9975	.9976	.9977	.9977	.9978	.9979	.9979	.9980	.9981
2.9	.9981	.9982	.9982	.9983	.9984	.9984	.9985	.9985	.9986	.9986
3.0	.9987	.9987	.9987	.9988	.9988	.9989	.9989	.9989	.9990	.9990
3.1	.9990	.9991	.9991	.9991	.9992	.9992	.9992	.9992	.9993	.9993
3.2	.9993	.9993	.9994	.9994	.9994	.9994	.9994	.9995	.9995	.9995
3.3	.9995	.9995	.9995	.9996	.9996	.9996	.9996	.9996	.9996	.9997
3.4	.9997	.9997	.9997	.9997	.9997	.9997	.9997	.9997	.9997	.9998

Table II

Table of Binomial Probabilities

n	x	.01	.05	.10	.20	.30	.40	.50	.60	.70	.80	.90	.95	.99	x
2	0	.980	.902	.810	.640	.490	.360	.250	.160	.090	.040	.010	.002	.000	0
	1	.020	.095	.180	.320	.420	.480	.500	.480	.420	.320	.180	.095	.020	1
	2	.000	.002	.010	.040	.090	.160	.250	.360	.490	.640	.810	.902	.980	2
3	0	.970	.857	.729	.512	.343	.216	.125	.064	.027	.008	.001	.000	.000	0
	1	.029	.135	.243	.384	.441	.432	.375	.288	.189	.096	.027	.007	.000	1
	2	.000	.007	.027	.096	.189	.288	.375	.432	.441	.384	.243	.135	.029	2
	3	.000	.000	.001	.008	.027	.064	.125	.216	.343	.512	.729	.857	.970	3
4	0	.961	.815	.656	.410	.240	.130	.062	.026	.008	.002	.000	.000	.000	0
	1	.039	.171	.292	.410	.412	.346	.250	.154	.076	.026	.004	.000	.000	1
	2	.001	.014	.049	.154	.265	.346	.375	.346	.265	.154	.049	.014	.001	2
	3	.000	.000	.004	.026	.076	.154	.250	.346	.412	.410	.292	.171	.039	3
	4	.000	.000	.000	.002	.008	.026	.062	.130	.240	.410	.656	.815	.961	4
5	0	.951	.774	.590	.328	.168	.078	.031	.010	.002	.000	.000	.000	.000	0
	1	.048	.204	.328	.410	.360	.259	.156	.077	.028	.006	.000	.000	.000	1
	2	.001	.021	.073	.205	.309	.346	.312	.230	.132	.051	.008	.001	.000	2
	3	.000	.001	.008	.051	.132	.230	.312	.346	.309	.205	.073	.021	.001	3
	4	.000	.000	.000	.006	.028	.077	.156	.259	.360	.410	.328	.204	.048	4
	5	.000	.000	.000	.000	.002	.010	.031	.078	.168	.328	.590	.774	.951	5
6	0	.941	.735	.531	.262	.118	.047	.016	.004	.001	.000	.000	.000	.000	0
	1	.057	.232	.354	.393	.303	.187	.094	.037	.010	.002	.000	.000	.000	1
	2	.001	.031	.098	.246	.324	.311	.234	.138	.060	.015	.001	.000	.000	2
	3	.000	.002	.015	.082	.185	.276	.312	.276	.185	.082	.015	.002	.000	3
	4	.000	.000	.001	.015	.060	.138	.234	.311	.324	.246	.098	.031	.001	4
	5	.000	.000	.000	.002	.010	.037	.094	.187	.303	.393	.354	.232	.057	5
	6	.000	.000	.000	.000	.001	.004	.016	.047	.118	.262	.531	.735	.941	6
7	0	.932	.698	.478	.210	.082	.028	.008	.002	.000	.000	.000	.000	.000	0
	1	.066	.257	.372	.367	.247	.131	.055	.017	.004	.000	.000	.000	.000	1
	2	.002	.041	.124	.275	.318	.261	.164	.077	.025	.004	.000	.000	.000	2
	3	.000	.004	.023	.115	.227	.290	.273	.194	.097	.029	.003	.000	.000	3
	4	.000	.000	.003	.029	.097	.194	.273	.290	.227	.115	.023	.004	.000	4
	5	.000	.000	.000	.004	.025	.077	.164	.261	.318	.275	.124	.041	.002	5
	6	.000	.000	.000	.000	.004	.017	.055	.131	.247	.367	.372	.257	.066	6
	7	.000	.000	.000	.000	.000	.002	.008	.028	.082	.210	.478	.698	.932	7
8	0	.923	.663	.430	.168	.058	.017	.004	.001	.000	.000	.000	.000	.000	0
	1	.075	.279	.383	.336	.198	.090	.031	.008	.001	.000	.000	.000	.000	1
	2	.003	.051	.149	.294	.296	.209	.109	.041	.010	.001	.000	.000	.000	2
	3	.000	.005	.033	.147	.254	.279	.219	.124	.047	.009	.000	.000	.000	3
	4	.000	.000	.005	.046	.136	.232	.273	.232	.136	.046	.005	.000	.000	4
	5	.000	.000	.000	.009	.047	.124	.219	.279	.254	.147	.033	.005	.000	5
	6	.000	.000	.000	.001	.010	.041	.109	.209	.296	.294	.149	.051	.003	6
	7	.000	.000	.000	.000	.001	.008	.031	.090	.198	.336	.383	.279	.075	7
	8	.000	.000	.000	.000	.000	.001	.004	.017	.058	.168	.430	.663	.923	8
9	0	.914	.630	.387	.134	.040	.010	.002	.000	.000	.000	.000	.000	.000	0
	1	.083	.299	.387	.302	.156	.060	.018	.004	.000	.000	.000	.000	.000	1
	2	.003	.063	.172	.302	.267	.161	.070	.021	.004	.000	.000	.000	.000	2
	3	.000	.008	.045	.176	.267	.251	.164	.074	.021	.003	.000	.000	.000	3
	4	.000	.001	.007	.066	.172	.251	.246	.167	.074	.017	.001	.000	.000	4
	5	.000	.000	.001	.017	.074	.167	.246	.251	.172	.066	.007	.001	.000	5
	6	.000	.000	.000	.003	.021	.074	.164	.251	.267	.176	.045	.008	.000	6
	7	.000	.000	.000	.000	.004	.021	.070	.161	.267	.302	.172	.063	.003	7
	8	.000	.000	.000	.000	.000	.004	.018	.060	.156	.302	.387	.299	.083	8
	9	.000	.000	.000	.000	.000	.000	.002	.010	.040	.134	.387	.630	.914	9

Table II (Continued)

Table of Binomial Probabilities

n	x	.01	.05	.10	.20	.30	.40	.50	.60	.70	.80	.90	.95	.99	x
10	0	.904	.599	.349	.107	.028	.006	.001	.000	.000	.000	.000	.000	.000	0
	1	.091	.315	.387	.268	.121	.040	.010	.002	.000	.000	.000	.000	.000	1
	2	.004	.075	.194	.302	.233	.121	.044	.011	.001	.000	.000	.000	.000	2
	3	.000	.010	.057	.201	.267	.215	.117	.042	.009	.001	.000	.000	.000	3
	4	.000	.001	.011	.088	.200	.251	.205	.111	.037	.006	.000	.000	.000	4
	5	.000	.000	.001	.026	.103	.201	.246	.201	.103	.026	.001	.000	.000	5
	6	.000	.000	.000	.006	.037	.111	.205	.251	.200	.088	.011	.001	.000	6
	7	.000	.000	.000	.001	.009	.042	.117	.215	.267	.201	.057	.010	.000	7
	8	.000	.000	.000	.000	.001	.011	.044	.121	.233	.302	.194	.075	.004	8
	9	.000	.000	.000	.000	.000	.002	.010	.040	.121	.268	.387	.315	.091	9
	10	.000	.000	.000	.000	.000	.000	.001	.006	.028	.107	.349	.599	.904	10
11	0	.895	.569	.314	.086	.020	.004	.000	.000	.000	.000	.000	.000	.000	0
	1	.099	.329	.384	.236	.093	.027	.005	.001	.000	.000	.000	.000	.000	1
	2	.005	.087	.213	.295	.200	.089	.027	.005	.001	.000	.000	.000	.000	2
	3	.000	.014	.071	.221	.257	.177	.081	.023	.004	.000	.000	.000	.000	3
	4	.000	.001	.016	.111	.220	.236	.161	.070	.017	.002	.000	.000	.000	4
	5	.000	.000	.002	.039	.132	.221	.226	.147	.057	.010	.000	.000	.000	5
	6	.000	.000	.000	.010	.057	.147	.226	.221	.132	.039	.002	.000	.000	6
	7	.000	.000	.000	.002	.017	.070	.161	.236	.220	.111	.016	.001	.000	7
	8	.000	.000	.000	.000	.004	.023	.081	.177	.257	.221	.071	.014	.000	8
	9	.000	.000	.000	.000	.001	.005	.027	.089	.200	.295	.213	.087	.005	9
	10	.000	.000	.000	.000	.000	.001	.005	.027	.093	.236	.384	.329	.099	10
	11	.000	.000	.000	.000	.000	.000	.000	.004	.020	.086	.314	.569	.895	11
12	0	.886	.540	.282	.069	.014	.002	.000	.000	.000	.000	.000	.000	.000	0
	1	.107	.341	.377	.206	.071	.017	.003	.000	.000	.000	.000	.000	.000	1
	2	.006	.099	.230	.283	.168	.064	.016	.002	.000	.000	.000	.000	.000	2
	3	.000	.017	.085	.236	.240	.142	.054	.012	.001	.000	.000	.000	.000	3
	4	.000	.002	.021	.133	.231	.213	.121	.042	.008	.001	.000	.000	.000	4
	5	.000	.000	.004	.053	.158	.227	.193	.101	.029	.003	.000	.000	.000	5
	6	.000	.000	.000	.016	.079	.177	.226	.177	.079	.016	.000	.000	.000	6
	7	.000	.000	.000	.003	.029	.101	.193	.227	.158	.053	.004	.000	.000	7
	8	.000	.000	.000	.001	.008	.042	.121	.213	.231	.133	.021	.002	.000	8
	9	.000	.000	.000	.000	.001	.012	.054	.142	.240	.236	.085	.017	.000	9
	10	.000	.000	.000	.000	.000	.002	.016	.064	.168	.283	.230	.099	.006	10
	11	.000	.000	.000	.000	.000	.000	.003	.017	.071	.206	.377	.341	.107	11
	12	.000	.000	.000	.000	.000	.000	.000	.002	.014	.069	.282	.540	.886	12

Table III

Table of 5% Critical Levels for χ^2-Curves

ν Degrees of Freedom	1	2	3	4	5	6	7	8	9	10
c 5% Critical Level	3.84	5.99	7.81	9.49	11.1	12.6	14.4	15.5	16.9	18.3

Table IV

Table of Square Roots and Cube Roots

n	$\sqrt{n}$	$\sqrt{10n}$	$\sqrt[3]{n}$	n	$\sqrt{n}$	$\sqrt{10n}$	$\sqrt[3]{n}$
1.0	1.000	3.162	1.000	5.5	2.345	7.416	1.765
1.1	1.049	3.317	1.032	5.6	2.366	7.483	1.776
1.2	1.095	3.464	1.063	5.7	2.387	7.550	1.786
1.3	1.140	3.606	1.091	5.8	2.408	7.616	1.797
1.4	1.183	3.742	1.119	5.9	2.429	7.681	1.807
1.5	1.225	3.873	1.145	6.0	2.449	7.746	1.817
1.6	1.265	4.000	1.170	6.1	2.470	7.810	1.827
1.7	1.304	4.123	1.193	6.2	2.490	7.874	1.837
1.8	1.342	4.243	1.216	6.3	2.510	7.937	1.847
1.9	1.378	4.359	1.239	6.4	2.530	8.000	1.857
2.0	1.414	4.472	1.260	6.5	2.550	8.062	1.866
2.1	1.449	4.583	1.281	6.6	2.569	8.124	1.876
2.2	1.483	4.690	1.301	6.7	2.588	8.185	1.885
2.3	1.517	4.796	1.320	6.8	2.608	8.246	1.895
2.4	1.549	4.899	1.339	6.9	2.627	8.307	1.904
2.5	1.581	5.000	1.357	7.0	2.646	8.367	1.913
2.6	1.612	5.099	1.375	7.1	2.665	8.426	1.922
2.7	1.643	5.196	1.392	7.2	2.683	8.485	1.931
2.8	1.673	5.292	1.409	7.3	2.702	8.544	1.940
2.9	1.703	5.385	1.426	7.4	2.720	8.602	1.949
3.0	1.732	5.477	1.442	7.5	2.739	8.660	1.957
3.1	1.761	5.568	1.458	7.6	2.757	8.718	1.966
3.2	1.789	5.657	1.474	7.7	2.775	8.775	1.975
3.3	1.817	5.745	1.489	7.8	2.793	8.832	1.983
3.4	1.844	5.831	1.504	7.9	2.811	8.888	1.992
3.5	1.871	5.916	1.518	8.0	2.828	8.944	2.000
3.6	1.897	6.000	1.533	8.1	2.846	9.000	2.008
3.7	1.924	6.083	1.547	8.2	2.864	9.055	2.017
3.8	1.949	6.164	1.560	8.3	2.881	9.110	2.025
3.9	1.975	6.245	1.574	8.4	2.898	9.165	2.033
4.0	2.000	6.325	1.587	8.5	2.915	9.220	2.041
4.1	2.025	6.403	1.601	8.6	2.933	9.274	2.049
4.2	2.049	6.481	1.613	8.7	2.950	9.327	2.057
4.3	2.074	6.557	1.626	8.8	2.966	9.381	2.065
4.4	2.098	6.633	1.639	8.9	2.983	9.434	2.072
4.5	2.121	6.708	1.651	9.0	3.000	9.487	2.080
4.6	2.145	6.782	1.663	9.1	3.017	9.539	2.088
4.7	2.168	6.856	1.675	9.2	3.033	9.592	2.095
4.8	2.191	6.928	1.687	9.3	3.050	9.644	2.103
4.9	2.214	7.000	1.698	9.4	3.066	9.695	2.110
5.0	2.236	7.071	1.710	9.5	3.082	9.747	2.118
5.1	2.258	7.141	1.721	9.6	3.098	9.798	2.125
5.2	2.280	7.211	1.732	9.7	3.114	9.849	2.133
5.3	2.302	7.280	1.744	9.8	3.130	9.899	2.140
5.4	2.324	7.348	1.754	9.9	3.146	9.950	2.147

Table V

Table of Natural Logarithms

n	$\log_e n$	n	$\log_e n$	n	$\log_e n$
		4.5	1.5041	9.0	2.1972
0.1	−2.3026	4.6	1.5261	9.1	2.2083
0.2	−1.6094	4.7	1.5476	9.2	2.2192
0.3	−1.2040	4.8	1.5686	9.3	2.2300
0.4	−0.9163	4.9	1.5892	9.4	2.2407
0.5	−0.6931	5.0	1.6094	9.5	2.2513
0.6	−0.5108	5.1	1.6292	9.6	2.2618
0.7	−0.3567	5.2	1.6487	9.7	2.2721
0.8	−0.2231	5.3	1.6677	9.8	2.2824
0.9	−0.1054	5.4	1.6864	9.9	2.2925
1.0	0.0000	5.5	1.7047	10	2.3026
1.1	0.0953	5.6	1.7228	11	2.3979
1.2	0.1823	5.7	1.7405	12	2.4849
1.3	0.2624	5.8	1.7579	13	2.5649
1.4	0.3365	5.9	1.7750	14	2.6391
1.5	0.4055	6.0	1.7918	15	2.7081
1.6	0.4700	6.1	1.8083	16	2.7726
1.7	0.5306	6.2	1.8245	17	2.8332
1.8	0.5878	6.3	1.8405	18	5.8904
1.9	0.6419	6.4	1.8563	19	2.9444
2.0	0.6931	6.5	1.8718	20	2.9957
2.1	0.7419	6.6	1.8871	25	3.2189
2.2	0.7885	6.7	1.9021	30	3.4012
2.3	0.8329	6.8	1.9169	35	3.5553
2.4	0.8755	6.9	1.9315	40	3.6889
2.5	0.9163	7.0	1.9459	45	3.8067
2.6	0.9555	7.1	1.9601	50	3.9120
2.7	0.9933	7.2	1.9741	55	4.0073
2.8	1.0296	7.3	1.9879	60	4.0943
2.9	1.0647	7.4	2.0015	65	4.1744
3.0	1.0986	7.5	2.0149	70	4.2485
3.1	1.1314	7.6	2.0281	75	4.3175
3.2	1.1632	7.7	2.0142	80	4.3820
3.3	1.1939	7.8	2.0541	85	4.4427
3.4	1.2238	7.9	2.0669	90	4.4998
3.5	1.2528	8.0	2.0794	95	4.5539
3.6	1.2809	8.1	2.0919	100	4.6052
3.7	1.3083	8.2	2.1041		
3.8	1.3350	8.3	2.1163		
3.9	1.3610	8.4	2.1282		
4.0	1.3863	8.5	2.1401		
4.1	1.4110	8.6	2.1518		
4.2	1.4351	8.7	2.1633		
4.3	1.4586	8.8	2.1748		
4.4	1.4816	8.9	2.1861		

Table VI

Table of Exponentials and Their Reciprocals

x	e^x	e^{-x}	x	e^x	e^{-x}
0.00	1.0000	1.0000	1.3	3.6693	0.2725
0.01	1.0101	0.9900	1.4	4.0552	0.2466
0.02	1.0202	0.9802	1.5	4.4817	0.2231
0.03	1.0305	0.9704	1.6	4.9530	0.2019
0.04	1.0408	0.9608	1.7	5.4739	0.1827
0.05	1.0513	0.9512	1.8	6.0496	0.1653
0.06	1.0618	0.9418	1.9	6.6859	0.1496
0.07	1.0725	0.9324	2.0	7.3891	0.1353
0.08	1.0833	0.9231	2.1	8.1662	0.1225
0.09	1.0942	0.9139	2.2	9.0250	0.1108
0.10	1.1052	0.9048	2.3	9.9742	0.1003
0.11	1.1163	0.8958	2.4	11.023	0.0907
0.12	1.1275	0.8869	2.5	12.182	0.0821
0.13	1.1388	0.8781	2.6	13.464	0.0743
0.14	1.1503	0.8694	2.7	14.880	0.0672
0.15	1.1618	0.8607	2.8	16.445	0.0608
0.16	1.1735	0.8521	2.9	18.174	0.0550
0.17	1.1853	0.8437	3.0	20.086	0.0498
0.18	1.1972	0.8353	3.1	22.198	0.0450
0.19	1.2092	0.8270	3.2	24.533	0.0408
0.20	1.2214	0.8187	3.3	27.113	0.0369
0.21	1.2337	0.8106	3.4	29.964	0.0334
0.22	1.2461	0.8025	3.5	33.115	0.0302
0.23	1.2586	0.7945	3.6	36.598	0.0273
0.24	1.2712	0.7866	3.7	40.447	0.0247
0.25	1.2840	0.7788	3.8	44.701	0.0224
0.26	1.2969	0.7711	3.9	49.402	0.0202
0.27	1.3100	0.7634	4.0	54.598	0.0183
0.28	1.3231	0.7558	4.1	60.340	0.0166
0.29	1.3364	0.7483	4.2	66.686	0.0150
0.30	1.3499	0.7408	4.3	73.700	0.0136
0.35	1.4191	0.7047	4.4	81.451	0.0123
0.40	1.4918	0.6703	4.5	90.017	0.0111
0.45	1.5683	0.6376	4.6	99.484	0.0101
0.50	1.6487	0.6065	4.7	109.95	0.0091
0.55	1.7333	0.5769	4.8	121.51	0.0082
0.60	1.8221	0.5488	4.9	134.29	0.0074
0.65	1.9155	0.5220	5	148.41	0.0067
0.70	2.0138	0.4966	6	403.43	0.0025
0.75	2.1170	0.4724	7	1096.6	0.0009
0.80	2.2255	0.4493	8	2981.0	0.0003
0.85	2.3396	0.4274	9	8103.1	0.0001
0.90	2.4596	0.4066	10	22026	0.00005
0.95	2.5857	0.3867	11	59874	0.00002
1.0	2.7183	0.3679	12	162,754	0.000006
1.1	3.0042	0.3329	13	442,413	0.000002
1.2	3.3201	0.3012	14	1,202,604	0.0000008
			15	3,269,017	0.0000003

Table VII

i = rate of interest per period $\quad n$ = number of periods $\quad i = \frac{1}{4}\%$

n	$(1+i)^n$	$(1+i)^{-n}$	$s_{\overline{n}\|i}$	$a_{\overline{n}\|i}$	$\frac{1}{s_{\overline{n}\|i}}$	$\frac{1}{a_{\overline{n}\|i}}$
1	1.0025 0000	0.9975 0623	1.0000 0000	0.9975 0623	1.0000 0000	1.0025 0000
2	1.0050 0625	0.9950 1869	2.0025 0000	1.9925 2492	0.4993 7578	0.5018 7578
3	1.0075 1877	0.9925 3734	3.0075 0625	2.9850 6227	0.3325 0139	0.3350 0139
4	1.0100 3756	0.9900 6219	4.0150 2502	3.9751 2446	0.2490 6445	0.2515 6445
5	1.0125 6266	0.9875 9321	5.0250 6258	4.9627 1766	0.1990 0250	0.2015 0250
6	1.0150 9406	0.9851 3038	6.0376 2523	5.9478 4804	0.1656 2803	0.1681 2803
7	1.0176 3180	0.9826 7370	7.0527 1930	6.9305 2174	0.1417 8928	0.1442 8928
8	1.0201 7588	0.9802 2314	8.0703 5110	7.9107 4487	0.1239 1035	0.1264 1035
9	1.0227 2632	0.9777 7869	9.0905 2697	8.8885 2357	0.1100 0462	0.1125 0462
10	1.0252 8313	0.9753 4034	10.1132 5329	9.8638 6391	0.0988 8015	0.1013 8015
11	1.0278 4634	0.9729 0807	11.1385 3642	10.8367 7198	0.0897 7840	0.0922 7840
12	1.0304 1596	0.9704 8187	12.1663 8277	11.8072 5384	0.0821 9370	0.0846 9370
13	1.0329 9200	0.9680 6171	13.1967 9872	12.7753 1555	0.0757 7595	0.0782 7595
14	1.0355 7448	0.9656 4759	14.2297 9072	13.7409 6314	0.0702 7510	0.0727 7510
15	1.0381 6341	0.9632 3949	15.2653 6520	14.7042 0264	0.0655 0777	0.0680 0777
16	1.0407 5882	0.9608 3740	16.3035 2861	15.6650 4004	0.0613 3642	0.0638 3642
17	1.0433 6072	0.9584 4130	17.3442 8743	16.6234 8133	0.0576 5587	0.0601 5587
18	1.0459 6912	0.9560 5117	18.3876 4815	17.5795 3250	0.0543 8433	0.0568 8433
19	1.0485 8404	0.9536 6700	19.4336 1727	18.5331 9950	0.0514 5722	0.0539 5722
20	1.0512 0550	0.9512 8878	20.4822 0131	19.4844 8828	0.0488 2288	0.0513 2288
21	1.0538 3352	0.9489 1649	21.5334 0682	20.4334 0477	0.0464 3947	0.0489 3947
22	1.0564 6810	0.9465 5011	22.5872 4033	21.3799 5488	0.0442 7278	0.0467 7278
23	1.0591 0927	0.9441 8964	23.6437 0843	22.3241 4452	0.0422 9455	0.0447 9455
24	1.0617 5704	0.9418 3505	24.7028 1770	23.2659 7957	0.0404 8121	0.0429 8121
25	1.0644 1144	0.9394 8634	25.7645 7475	24.2054 6591	0.0388 1298	0.0413 1298
26	1.0670 7247	0.9371 4348	26.8289 8619	25.1426 0939	0.0372 7312	0.0397 7312
27	1.0697 4015	0.9348 0646	27.8960 5865	26.0774 1585	0.0358 4736	0.0383 4736
28	1.0724 1450	0.9324 7527	28.9657 9880	27.0098 9112	0.0345 2347	0.0370 2347
29	1.0750 9553	0.9301 4990	30.0382 1330	27.9400 4102	0.0332 9093	0.0357 9093
30	1.0777 8327	0.9278 3032	31.1133 0883	28.8678 7134	0.0321 4059	0.0346 4059
31	1.0804 7773	0.9255 1653	32.1910 9210	29.7933 8787	0.0310 6449	0.0335 6449
32	1.0831 7892	0.9232 0851	33.2715 6983	30.7165 9638	0.0300 5569	0.0325 5569
33	1.0858 8687	0.9209 0624	34.3547 4876	31.6375 0262	0.0291 0806	0.0316 0806
34	1.0886 0159	0.9186 0972	35.4406 3563	32.5561 1234	0.0282 1620	0.0307 1620
35	1.0913 2309	0.9163 1892	36.5292 3722	33.4724 3126	0.0273 7533	0.0298 7533
36	1.0940 5140	0.9140 3384	37.6205 6031	34.3864 6510	0.0265 8121	0.0290 8121
37	1.0967 8653	0.9117 5445	38.7146 1171	35.2982 1955	0.0258 3004	0.0283 3004
38	1.0995 2850	0.9094 8075	39.8113 9824	36.2077 0030	0.0251 1843	0.0276 1843
39	1.1022 7732	0.9072 1272	40.9109 2673	37.1149 1302	0.0244 4335	0.0269 4335
40	1.1050 3301	0.9049 5034	42.0132 0405	38.0198 6336	0.0238 0204	0.0263 0204
41	1.1077 9559	0.9026 9361	43.1182 3706	38.9225 5697	0.0231 9204	0.0256 9204
42	1.1105 6508	0.9004 4250	44.2260 3265	39.8229 9947	0.0226 1112	0.0251 1112
43	1.1133 4149	0.8981 9701	45.3365 9774	40.7211 9648	0.0220 5724	0.0245 5724
44	1.1161 2485	0.8959 5712	46.4499 3923	41.6171 5359	0.0215 2855	0.0240 2855
45	1.1189 1516	0.8937 2281	47.5660 6408	42.5108 7640	0.0210 2339	0.0235 2339
46	1.1217 1245	0.8914 9407	48.6849 7924	43.4023 7047	0.0205 4022	0.0230 4022
47	1.1245 1673	0.8892 7090	49.8066 9169	44.2916 4137	0.0200 7762	0.0225 7762
48	1.1273 2802	0.8870 5326	50.9312 0842	45.1786 9463	0.0196 3433	0.0221 3433
49	1.1301 4634	0.8848 4116	52.0585 3644	46.0635 3580	0.0192 0915	0.0217 0915
50	1.1329 7171	0.8826 3457	53.1886 8278	46.9461 7037	0.0188 0099	0.0213 0099

$i = \frac{1}{4}\%$

n	$(1+i)^n$	$(1+i)^{-n}$	$s_{\overline{n}\rvert i}$	$a_{\overline{n}\rvert i}$	$\frac{1}{s_{\overline{n}\rvert i}}$	$\frac{1}{a_{\overline{n}\rvert i}}$
51	1.1358 0414	0.8804 3349	54.3216 5449	47.8266 0386	0.0184 0886	0.0209 0886
52	1.1386 4365	0.8782 3790	55.4574 5862	48.7048 4176	0.0180 3184	0.0205 3184
53	1.1414 9026	0.8760 4778	56.5961 0227	49.5808 8953	0.0176 6906	0.0201 6906
54	1.1443 4398	0.8738 6312	57.7375 9252	50.4547 5265	0.0173 1974	0.0198 1974
55	1.1472 0484	0.8716 8391	58.8819 3650	51.3264 3656	0.0169 8314	0.0194 8314
56	1.1500 7285	0.8695 1013	60.0291 4135	52.1959 4669	0.0166 5858	0.0191 5858
57	1.1529 4804	0.8673 4178	61.1792 1420	53.0632 8847	0.0163 4542	0.0188 4542
58	1.1558 3041	0.8651 7883	62.3321 6223	53.9284 6730	0.0160 4308	0.0185 4308
59	1.1587 1998	0.8630 2128	63.4879 9264	54.7914 8858	0.0157 5101	0.0182 5101
60	1.1616 1678	0.8608 6911	64.6467 1262	55.6523 5769	0.0154 6869	0.0179 6869
61	1.1645 2082	0.8587 2230	65.8083 2940	56.5110 7999	0.0151 9564	0.0176 9564
62	1.1674 3213	0.8565 8085	66.9728 5023	57.3676 6083	0.0149 3142	0.0174 3142
63	1.1703 5071	0.8544 4474	68.1402 8235	58.2221 0557	0.0146 7561	0.0171 7561
64	1.1732 7658	0.8523 1395	69.3106 3306	59.0744 1952	0.0144 2780	0.0169 2780
65	1.1762 0977	0.8501 8848	70.4839 0964	59.9246 0800	0.0141 8764	0.0166 8764
66	1.1791 5030	0.8480 6831	71.6601 1942	60.7726 7631	0.0139 5476	0.0164 5476
67	1.1820 9817	0.8459 5343	72.8392 6971	61.6186 2974	0.0137 2886	0.0162 2886
68	1.1850 5342	0.8438 4382	74.0213 6789	62.4624 7355	0.0135 0961	0.0160 0961
69	1.1880 1605	0.8417 3947	75.2064 2131	63.3042 1302	0.0132 9674	0.0157 9674
70	1.1909 8609	0.8396 4037	76.3944 3736	64.1438 5339	0.0130 8996	0.0155 8996
71	1.1939 6356	0.8375 4650	77.5854 2345	64.9813 9989	0.0128 8902	0.0153 8902
72	1.1969 4847	0.8354 5786	78.7793 8701	65.8168 5774	0.0126 9368	0.0151 9368
73	1.1999 4084	0.8333 7442	79.9763 3548	66.6502 3216	0.0125 0370	0.0150 0370
74	1.2029 4069	0.8312 9618	81.1762 7632	67.4815 2834	0.0123 1887	0.0148 1887
75	1.2059 4804	0.8292 2312	82.3792 1701	68.3107 5146	0.0121 3898	0.0146 3898
76	1.2089 6291	0.8271 5523	83.5851 6505	69.1379 0670	0.0119 6385	0.0144 6385
77	1.2119 8532	0.8250 9250	84.7941 2797	69.9629 9920	0.0117 9327	0.0142 9327
78	1.2150 1528	0.8230 3491	86.0061 1329	70.7860 3411	0.0116 2708	0.0141 2708
79	1.2180 5282	0.8209 8246	87.2211 2857	71.6070 1657	0.0114 6511	0.0139 6511
80	1.2210 9795	0.8189 3512	88.4391 8139	72.4259 5169	0.0113 0721	0.0138 0721
81	1.2241 5070	0.8168 9289	89.6602 7934	73.2428 4458	0.0111 5321	0.0136 5321
82	1.2272 1108	0.8148 5575	90.8844 3004	74.0577 0033	0.0110 0298	0.0135 0298
83	1.2302 7910	0.8128 2369	92.1116 4112	74.8705 2402	0.0108 5639	0.0133 5639
84	1.2333 5480	0.8107 9670	93.3419 2022	75.6813 2072	0.0107 1330	0.0132 1330
85	1.2364 3819	0.8087 7476	94.5752 7502	76.4900 9548	0.0105 7359	0.0130 7359
86	1.2395 2928	0.8067 5787	95.8117 1321	77.2968 5335	0.0104 3714	0.0129 3714
87	1.2426 2811	0.8047 4600	97.0512 4249	78.1015 9935	0.0103 0384	0.0128 0384
88	1.2457 3468	0.8027 3915	98.2938 7060	78.9043 3850	0.0101 7357	0.0126 7357
89	1.2488 4901	0.8007 3731	99.5396 0527	79.7050 7581	0.0100 4625	0.0125 4625
90	1.2519 7114	0.7987 4046	100.7884 5429	80.5038 1627	0.0099 2177	0.0124 2177
91	1.2551 0106	0.7967 4859	102.0404 2542	81.3005 6486	0.0098 0004	0.0123 0004
92	1.2582 3882	0.7947 6168	103.2955 2649	82.0953 2654	0.0096 8096	0.0121 8096
93	1.2613 8441	0.7927 7973	104.5537 6530	82.8881 0628	0.0095 6446	0.0120 6446
94	1.2645 3787	0.7908 0273	105.8151 4972	83.6789 0900	0.0094 5044	0.0119 5044
95	1.2676 9922	0.7888 3065	107.0796 8759	84.4677 3966	0.0093 3884	0.0118 3884
96	1.2708 6847	0.7868 6349	108.3473 8681	85.2546 0315	0.0092 2957	0.0117 2957
97	1.2740 4564	0.7849 0124	109.6182 5528	86.0395 0439	0.0091 2257	0.0116 2257
98	1.2772 3075	0.7829 4388	110.8923 0091	86.8224 4827	0.0090 1776	0.0115 1776
99	1.2804 2383	0.7809 9140	112.1695 3167	87.6034 3967	0.0089 1508	0.0114 1508
100	1.2836 2489	0.7790 4379	113.4499 5550	88.3824 8346	0.0088 1446	0.0113 1446

$i = \frac{1}{2}\%$

n	$(1+i)^n$	$(1+i)^{-n}$	$s_{\overline{n}\rvert i}$	$a_{\overline{n}\rvert i}$	$\frac{1}{s_{\overline{n}\rvert i}}$	$\frac{1}{a_{\overline{n}\rvert i}}$
1	1.0050 0000	0.9950 2488	1.0000 0000	0.9950 2488	1.0000 0000	1.0050 0000
2	1.0100 2500	0.9900 7450	2.0050 0000	1.9850 9938	0.4987 5312	0.5037 5312
3	1.0150 7513	0.9851 4876	3.0150 2500	2.9702 4814	0.3316 7221	0.3366 7221
4	1.0201 5050	0.9802 4752	4.0301 0013	3.9504 9566	0.2481 3279	0.2531 3279
5	1.0252 5125	0.9753 7067	5.0502 5063	4.9258 6633	0.1980 0997	0.2030 0997
6	1.0303 7751	0.9705 1808	6.0755 0188	5.8963 8441	0.1645 9546	0.1695 9546
7	1.0355 2940	0.9656 8963	7.1058 7939	6.8620 7404	0.1407 2854	0.1457 2854
8	1.0407 0704	0.9608 8520	8.1414 0879	7.8229 5924	0.1228 2886	0.1278 2886
9	1.0459 1058	0.9561 0468	9.1821 1583	8.7790 6392	0.1089 0736	0.1139 0736
10	1.0511 4013	0.9513 4794	10.2280 2641	9.7304 1186	0.0977 7057	0.1027 7057
11	1.0563 9583	0.9466 1487	11.2791 6654	10.6770 2673	0.0886 5903	0.0936 5903
12	1.0616 7781	0.9419 0534	12.3355 6237	11.6189 3207	0.0810 6643	0.0860 6643
13	1.0669 8620	0.9372 1924	13.3972 4018	12.5561 5131	0.0746 4224	0.0796 4224
14	1.0723 2113	0.9325 5646	14.4642 2639	13.4887 0777	0.0691 3609	0.0741 3609
15	1.0776 8274	0.9279 1688	15.5365 4752	14.4166 2465	0.0643 6436	0.0693 6436
16	1.0830 7115	0.9233 0037	16.6142 3026	15.3399 2502	0.0601 8937	0.0651 8937
17	1.0884 8651	0.9187 0684	17.6973 0141	16.2586 3186	0.0565 0579	0.0615 0579
18	1.0939 2894	0.9141 3616	18.7857 8791	17.1727 6802	0.0532 3173	0.0582 3173
19	1.0993 9858	0.9095 8822	19.8797 1685	18.0823 5624	0.0503 0253	0.0553 0253
20	1.1048 9558	0.9050 6290	20.9791 1544	18.9874 1915	0.0476 6645	0.0526 6645
21	1.1104 2006	0.9005 6010	22.0840 1101	19.8879 7925	0.0452 8163	0.0502 8163
22	1.1159 7216	0.8960 7971	23.1944 3107	20.7840 5896	0.0431 1380	0.0481 1380
23	1.1215 5202	0.8916 2160	24.3104 0322	21.6756 8055	0.0411 3465	0.0461 3465
24	1.1271 5978	0.8871 8567	25.4319 5524	22.5628 6622	0.0393 2061	0.0443 2061
25	1.1327 9558	0.8827 7181	26.5591 1502	23.4456 3803	0.0376 5186	0.0426 5186
26	1.1384 5955	0.8783 7991	27.6919 1059	24.3240 1794	0.0361 1163	0.0411 1163
27	1.1441 5185	0.8740 0986	28.8303 7015	25.1980 2780	0.0346 8565	0.0396 8565
28	1.1498 7261	0.8696 6155	29.9745 2200	26.0676 8936	0.0333 6167	0.0383 6167
29	1.1556 2197	0.8653 3488	31.1243 9461	26.9330 2423	0.0321 2914	0.0371 2914
30	1.1614 0008	0.8610 2973	32.2800 1658	27.7940 5397	0.0309 7892	0.0359 7892
31	1.1672 0708	0.8567 4600	33.4414 1666	28.6507 9997	0.0299 0304	0.0349 0304
32	1.1730 4312	0.8524 8358	34.6086 2375	29.5032 8355	0.0288 9453	0.0338 9453
33	1.1789 0833	0.8482 4237	35.7816 6686	30.3515 2592	0.0279 4727	0.0329 4727
34	1.1848 0288	0.8440 2226	36.9605 7520	31.1955 4818	0.0270 5586	0.0320 5586
35	1.1907 2689	0.8398 2314	38.1453 7807	32.0353 7132	0.0262 1550	0.0312 1550
36	1.1966 8052	0.8356 4492	39.3361 0496	32.8710 1624	0.0254 2194	0.0304 2194
37	1.2026 6393	0.8314 8748	40.5327 8549	33.7025 0372	0.0246 7139	0.0296 7139
38	1.2086 7725	0.8273 5073	41.7354 4942	34.5298 5445	0.0239 6045	0.0289 6045
39	1.2147 2063	0.8232 3455	42.9441 2666	35.3530 8900	0.0232 8607	0.0282 8607
40	1.2207 9424	0.8191 3886	44.1588 4730	36.1722 2786	0.0226 4552	0.0276 4552
41	1.2268 9821	0.8150 6354	45.3796 4153	36.9872 9141	0.0220 3631	0.0270 3631
42	1.2330 3270	0.8110 0850	46.6065 3974	37.7982 9991	0.0214 5622	0.0264 5622
43	1.2391 9786	0.8069 7363	47.8395 7244	38.6052 7354	0.0209 0320	0.0259 0320
44	1.2453 9385	0.8029 5884	49.0787 7030	39.4082 3238	0.0203 7541	0.0253 7541
45	1.2516 2082	0.7989 6402	50.3241 6415	40.2071 9640	0.0198 7117	0.0248 7117
46	1.2578 7892	0.7949 8907	51.5757 8497	41.0021 8547	0.0193 8894	0.0243 8894
47	1.2641 6832	0.7910 3390	52.8336 6390	41.7932 1937	0.0189 2733	0.0239 2733
48	1.2704 8916	0.7870 9841	54.0978 3222	42.5803 1778	0.0184 8503	0.0234 8503
49	1.2768 4161	0.7831 8250	55.3683 2138	43.3635 0028	0.0180 6087	0.0230 6087
50	1.2832 2581	0.7792 8607	56.6451 6299	44.1427 8635	0.0176 5376	0.0226 5376

$i = \frac{1}{2}\%$

n	$(1+i)^n$	$(1+i)^{-n}$	$s_{\overline{n}\mid i}$	$a_{\overline{n}\mid i}$	$\frac{1}{s_{\overline{n}\mid i}}$	$\frac{1}{a_{\overline{n}\mid i}}$
51	1.2896 4194	0.7754 0902	57.9283 8880	44.9181 9537	0.0172 6269	0.0222 6269
52	1.2960 9015	0.7715 5127	59.2180 3075	45.6897 4664	0.0168 8675	0.0218 8675
53	1.3025 7060	0.7677 1270	60.5141 2090	46.4574 5934	0.0165 2507	0.0215 2507
54	1.3090 8346	0.7638 9324	61.8166 9150	47.2213 5258	0.0161 7686	0.0211 7686
55	1.3156 2887	0.7600 9277	63.1257 7496	47.9814 4535	0.0158 4139	0.0208 4139
56	1.3222 0702	0.7563 1122	64.4414 0384	48.7377 5657	0.0155 1797	0.0205 1797
57	1.3288 1805	0.7525 4847	65.7636 1086	49.4903 0505	0.0152 0598	0.0202 0598
58	1.3354 6214	0.7488 0445	67.0924 2891	50.2391 0950	0.0149 0481	0.0199 0481
59	1.3421 3946	0.7450 7906	68.4278 9105	50.9841 8855	0.0146 1392	0.0196 1392
60	1.3488 5015	0.7413 7220	69.7700 3051	51.7255 6075	0.0143 3280	0.0193 3280
61	1.3555 9440	0.7376 8378	71.1188 8066	52.4632 4453	0.0140 6096	0.0190 6096
62	1.3623 7238	0.7340 1371	72.4744 7507	53.1972 5824	0.0137 9796	0.0187 9796
63	1.3691 8424	0.7303 6190	73.8368 4744	53.9276 2014	0.0135 4337	0.0185 4337
64	1.3760 3016	0.7267 2826	75.2060 3168	54.6543 4839	0.0132 9681	0.0182 9681
65	1.3829 1031	0.7231 1269	76.5820 6184	55.3774 6109	0.0130 5789	0.0180 5789
66	1.3898 2486	0.7195 1512	77.9649 7215	56.0969 7621	0.0128 2627	0.0178 2627
67	1.3967 7399	0.7159 3544	79.3547 9701	56.8129 1165	0.0126 0163	0.0176 0163
68	1.4037 5785	0.7123 7357	80.7515 7099	57.5252 8522	0.0123 8366	0.0173 8366
69	1.4107 7664	0.7088 2943	82.1553 2885	58.2341 1465	0.0121 7206	0.0171 7206
70	1.4178 3053	0.7053 0291	83.5661 0549	58.9394 1756	0.0119 6657	0.0169 6657
71	1.4249 1968	0.7017 9394	84.9839 3602	59.6412 1151	0.0117 6693	0.0167 6693
72	1.4320 4428	0.6983 0243	86.4088 5570	60.3395 1394	0.0115 7289	0.0165 7289
73	1.4392 0450	0.6948 2829	87.8408 9998	61.0343 4222	0.0113 8422	0.0163 8422
74	1.4464 0052	0.6913 7143	89.2801 0448	61.7257 1366	0.0112 0070	0.0162 0070
75	1.4536 3252	0.6879 3177	90.7265 0500	62.4136 4543	0.0110 2214	0.0160 2214
76	1.4609 0069	0.6845 0923	92.1801 3752	63.0981 5466	0.0108 4832	0.0158 4832
77	1.4682 0519	0.6811 0371	93.6410 3821	63.7792 5836	0.0106 7908	0.0156 7908
78	1.4755 4622	0.6777 1513	95.1092 4340	64.4569 7350	0.0105 1423	0.0155 1423
79	1.4829 2395	0.6743 4342	96.5847 8962	65.1313 1691	0.0103 5360	0.0153 5360
80	1.4903 3857	0.6709 8847	98.0677 1357	65.8023 0538	0.0101 9704	0.0151 9704
81	1.4977 9026	0.6676 5022	99.5580 5214	66.4699 5561	0.0100 4439	0.0150 4439
82	1.5052 7921	0.6643 2858	101.0558 4240	67.1342 8419	0.0098 9552	0.0148 9552
83	1.5128 0561	0.6610 2346	102.5611 2161	67.7953 0765	0.0097 5028	0.0147 5028
84	1.5203 6964	0.6577 3479	104.0739 2722	68.4530 4244	0.0096 0855	0.0146 0855
85	1.5279 7148	0.6544 6248	105.5942 9685	69.1075 0491	0.0094 7021	0.0144 7021
86	1.5356 1134	0.6512 0644	107.1222 6834	69.7587 1135	0.0093 3513	0.0143 3513
87	1.5432 8940	0.6479 6661	108.6578 7968	70.4066 7796	0.0092 0320	0.0142 0320
88	1.5510 0585	0.6447 4290	110.2011 6908	71.0514 2086	0.0090 7431	0.0140 7431
89	1.5587 6087	0.6415 3522	111.7521 7492	71.6929 5608	0.0089 4837	0.0139 4837
90	1.5665 5468	0.6383 4350	113.3109 3580	72.3312 9958	0.0088 2527	0.0138 2527
91	1.5743 8745	0.6351 6766	114.8774 9048	72.9664 6725	0.0087 0493	0.0137 0493
92	1.5822 5939	0.6320 0763	116.4518 7793	73.5984 7487	0.0085 8724	0.0135 8724
93	1.5901 7069	0.6288 6331	118.0341 3732	74.2273 3818	0.0084 7213	0.0134 7213
94	1.5981 2154	0.6257 3464	119.6243 0800	74.8530 7282	0.0083 5950	0.0133 5950
95	1.6061 1215	0.6226 2153	121.2224 2954	75.4756 9434	0.0082 4930	0.0132 4930
96	1.6141 4271	0.6195 2391	122.8285 4169	76.0952 1825	0.0081 4143	0.0131 4143
97	1.6222 1342	0.6164 4170	124.4426 8440	76.7116 5995	0.0080 3583	0.0130 3583
98	1.6303 2449	0.6133 7483	126.0648 9782	77.3250 3478	0.0079 3242	0.0129 3242
99	1.6384 7611	0.6103 2321	127.6952 2231	77.9353 5799	0.0078 3115	0.0128 3115
100	1.6466 6849	0.6072 8678	129.3336 9842	78.5426 4477	0.0077 3194	0.0127 3194

$i = \frac{3}{4}\%$

n	$(1+i)^n$	$(1+i)^{-n}$	$s_{\overline{n}\|i}$	$a_{\overline{n}\|i}$	$\frac{1}{s_{\overline{n}\|i}}$	$\frac{1}{a_{\overline{n}\|i}}$
1	1.0075 0000	0.9925 5583	1.0000 0000	0.9925 5583	1.0000 0000	1.0075 0000
2	1.0150 5625	0.9851 6708	2.0075 0000	1.9777 2291	0.4981 3200	0.5056 3200
3	1.0226 6917	0.9778 3333	3.0225 5625	2.9555 5624	0.3308 4579	0.3383 4579
4	1.0303 3919	0.9705 5417	4.0452 2542	3.9261 1041	0.2472 0501	0.2547 0501
5	1.0380 6673	0.9633 2920	5.0755 6461	4.8894 3961	0.1970 2242	0.2045 2242
6	1.0458 5224	0.9561 5802	6.1136 3135	5.8455 9763	0.1635 6891	0.1710 6891
7	1.0536 9613	0.9490 4022	7.1594 8358	6.7946 3785	0.1396 7488	0.1471 7488
8	1.0615 9885	0.9419 7540	8.2131 7971	7.7366 1325	0.1217 5552	0.1292 5552
9	1.0695 6084	0.9349 6318	9.2747 7856	8.6715 7642	0.1078 1929	0.1153 1929
10	1.0775 8255	0.9280 0315	10.3443 3940	9.5995 7958	0.0966 7123	0.1041 7123
11	1.0856 6441	0.9210 9494	11.4219 2194	10.5206 7452	0.0875 5094	0.0950 5094
12	1.0938 0690	0.9142 3815	12.5075 8636	11.4349 1267	0.0799 5148	0.0874 5148
13	1.1020 1045	0.9074 3241	13.6013 9325	12.3423 4508	0.0735 2188	0.0810 2188
14	1.1102 7553	0.9006 7733	14.7034 0370	13.2430 2242	0.0680 1146	0.0755 1146
15	1.1186 0259	0.8939 7254	15.8136 7923	14.1369 9495	0.0632 3639	0.0707 3639
16	1.1269 9211	0.8873 1766	16.9322 8183	15.0243 1261	0.0590 5879	0.0665 5879
17	1.1354 4455	0.8807 1231	18.0592 7394	15.9050 2492	0.0553 7321	0.0628 7321
18	1.1439 6039	0.8741 5614	19.1947 1849	16.7791 8107	0.0520 9766	0.0595 9766
19	1.1525 4009	0.8676 4878	20.3386 7888	17.6468 2984	0.0491 6740	0.0566 6740
20	1.1611 8414	0.8611 8985	21.4912 1897	18.5080 1969	0.0465 3063	0.0540 3063
21	1.1698 9302	0.8547 7901	22.6524 0312	19.3627 9870	0.0441 4543	0.0516 4543
22	1.1786 6722	0.8484 1589	23.8222 9614	20.2112 1459	0.0419 7748	0.0494 7748
23	1.1875 0723	0.8421 0014	25.0009 6336	21.0533 1473	0.0399 9846	0.0474 9846
24	1.1964 1353	0.8358 3140	26.1884 7059	21.8891 4614	0.0381 8474	0.0456 8474
25	1.2053 8663	0.8296 0933	27.3848 8412	22.7187 5547	0.0365 1650	0.0440 1650
26	1.2144 2703	0.8234 3358	28.5902 7075	23.5421 8905	0.0349 7693	0.0424 7693
27	1.2235 3523	0.8173 0380	29.8046 9778	24.3594 9286	0.0335 5176	0.0410 5176
28	1.2327 1175	0.8112 1966	31.0282 3301	25.1707 1251	0.0322 2871	0.0397 2871
29	1.2419 5709	0.8051 8080	32.2609 4476	25.9758 9331	0.0309 9723	0.0384 9723
30	1.2512 7176	0.7991 8690	33.5029 0184	26.7750 8021	0.0298 4816	0.0373 4816
31	1.2606 5630	0.7932 3762	34.7541 7361	27.5683 1783	0.0287 7352	0.0362 7352
32	1.2701 1122	0.7873 3262	36.0148 2991	28.3556 5045	0.0277 6634	0.0352 6634
33	1.2796 3706	0.7814 7158	37.2849 4113	29.1371 2203	0.0268 2048	0.0343 2048
34	1.2892 3434	0.7756 5418	38.5645 7819	29.9127 7621	0.0259 3053	0.0334 3053
35	1.2989 0359	0.7698 8008	39.8538 1253	30.6826 5629	0.0250 9170	0.0325 9170
36	1.3086 4537	0.7641 4896	41.1527 1612	31.4468 0525	0.0242 9973	0.0317 9973
37	1.3184 6021	0.7584 6051	42.4613 6149	32.2052 6576	0.0235 5082	0.0310 5082
38	1.3283 4866	0.7528 1440	43.7798 2170	32.9580 8016	0.0228 4157	0.0303 4157
39	1.3383 1128	0.7472 1032	45.1081 7037	33.7052 9048	0.0221 6893	0.0296 6893
40	1.3483 4861	0.7416 4796	46.4464 8164	34.4469 3844	0.0215 3016	0.0290 3016
41	1.3584 6123	0.7361 2701	47.7948 3026	35.1830 6545	0.0209 2276	0.0284 2276
42	1.3686 4969	0.7306 4716	49.1532 9148	35.9137 1260	0.0203 4452	0.0278 4452
43	1.3789 1456	0.7252 0809	50.5219 4117	36.6389 2070	0.0197 9338	0.0272 9338
44	1.3892 5642	0.7198 0952	51.9008 5573	37.3587 3022	0.0192 6751	0.0267 6751
45	1.3996 7584	0.7144 5114	53.2901 1215	38.0731 8136	0.0187 6521	0.0262 6521
46	1.4101 7341	0.7091 3264	54.6897 8799	38.7823 1401	0.0182 8495	0.0257 8495
47	1.4207 4971	0.7038 5374	56.0999 6140	39.4861 6775	0.0178 2532	0.0253 2532
48	1.4314 0533	0.6986 1414	57.5207 1111	40.1847 8189	0.0173 8504	0.0248 8504
49	1.4421 4087	0.6934 1353	58.9521 1644	40.8781 9542	0.0169 6292	0.0244 6292
50	1.4529 5693	0.6882 5165	60.3942 5732	41.5664 4707	0.0165 5787	0.0240 5787

$i = \frac{3}{4}\%$

n	$(1+i)^n$	$(1+i)^{-n}$	$s_{\overline{n}\rceil i}$	$a_{\overline{n}\rceil i}$	$\frac{1}{s_{\overline{n}\rceil i}}$	$\frac{1}{a_{\overline{n}\rceil i}}$
51	1.4638 5411	0.6831 2819	61.8472 1424	42.2495 7525	0.0161 6888	0.0236 6888
52	1.4748 3301	0.6780 4286	63.3110 6835	42.9276 1812	0.0157 9503	0.0232 9503
53	1.4858 9426	0.6729 9540	64.7859 0136	43.6006 1351	0.0154 3546	0.0229 3546
54	1.4970 3847	0.6679 8551	66.2717 9562	44.2685 9902	0.0150 8938	0.0225 8938
55	1.5082 6626	0.6630 1291	67.7688 3409	44.9316 1193	0.0147 5605	0.0222 5605
56	1.5195 7825	0.6580 7733	69.2771 0035	45.5896 8926	0.0144 3478	0.0219 3478
57	1.5309 7509	0.6531 7849	70.7966 7860	46.2428 6776	0.0141 2496	0.0216 2496
58	1.5424 5740	0.6483 1612	72.3276 5369	46.8911 8388	0.0138 2597	0.0213 2597
59	1.5540 2583	0.6434 8995	73.8701 1109	47.5346 7382	0.0135 3727	0.0210 3727
60	1.5656 8103	0.6386 9970	75.4241 3693	48.1733 7352	0.0132 5836	0.0207 5836
61	1.5774 2363	0.6339 4511	76.9898 1795	48.8073 1863	0.0129 8873	0.0204 8873
62	1.5892 5431	0.6292 2592	78.5672 4159	49.4365 4455	0.0127 2795	0.0202 2795
63	1.6011 7372	0.6245 4185	80.1564 9590	50.0610 8640	0.0124 7560	0.0199 7560
64	1.6131 8252	0.6198 9266	81.7576 6962	50.6809 7906	0.0122 3127	0.0197 3127
65	1.6252 8139	0.6152 7807	83.3708 5214	51.2962 5713	0.0119 9460	0.0194 9460
66	1.6374 7100	0.6106 9784	84.9961 3353	51.9069 5497	0.0117 6524	0.0192 6524
67	1.6497 5203	0.6061 5170	86.6336 0453	52.5131 0667	0.0115 4286	0.0190 4286
68	1.6621 2517	0.6016 3940	88.2833 5657	53.1147 4607	0.0113 2716	0.0188 2716
69	1.6745 9111	0.5971 6070	89.9454 8174	53.7119 0677	0.0111 1785	0.0186 1785
70	1.6871 5055	0.5927 1533	91.6200 7285	54.3046 2210	0.0109 1464	0.0184 1464
71	1.6998 0418	0.5883 0306	93.3072 2340	54.8929 2516	0.0107 1728	0.0182 1728
72	1.7125 5271	0.5839 2363	95.0070 2758	55.4768 4880	0.0105 2554	0.0180 2554
73	1.7253 9685	0.5795 7681	96.7195 8028	56.0564 2561	0.0103 3917	0.0178 3917
74	1.7383 3733	0.5752 6234	98.4449 7714	56.6316 8795	0.0101 5796	0.0176 5796
75	1.7513 7486	0.5709 7999	100.1833 1446	57.2026 6794	0.0099 8170	0.0174 8170
76	1.7645 1017	0.5667 2952	101.9346 8932	57.7693 9746	0.0098 1020	0.0173 1020
77	1.7777 4400	0.5625 1069	103.6991 9949	58.3319 0815	0.0096 4328	0.0171 4328
78	1.7910 7708	0.5583 2326	105.4769 4349	58.8902 3141	0.0094 8074	0.0169 8074
79	1.8045 1015	0.5541 6701	107.2680 2056	59.4443 9842	0.0093 2244	0.0168 2244
80	1.8180 4398	0.5500 4170	109.0725 3072	59.9944 4012	0.0091 6821	0.0166 6821
81	1.8316 7931	0.5459 4710	110.8905 7470	60.5403 8722	0.0090 1790	0.0165 1790
82	1.8454 1691	0.5418 8297	112.7222 5401	61.0822 7019	0.0088 7136	0.0163 7136
83	1.8592 5753	0.5378 4911	114.5676 7091	61.6201 1930	0.0087 2847	0.0162 2847
84	1.8732 0196	0.5338 4527	116.4269 2845	62.1539 6456	0.0085 8908	0.0160 8908
85	1.8872 5098	0.5298 7123	118.3001 3041	62.6838 3579	0.0084 5308	0.0159 5308
86	1.9014 0536	0.5259 2678	120.1873 8139	63.2097 6257	0.0083 2034	0.0158 2034
87	1.9156 6590	0.5220 1169	122.0887 8675	63.7317 7427	0.0081 9076	0.0156 9076
88	1.9300 3339	0.5181 2575	124.0044 5265	64.2499 0002	0.0080 6423	0.0155 6423
89	1.9445 0865	0.5142 6873	125.9344 8604	64.7641 6875	0.0079 4064	0.0154 4064
90	1.9590 9246	0.5104 4043	127.8789 9469	65.2746 0918	0.0078 1989	0.0153 1989
91	1.9737 8565	0.5066 4063	129.8380 8715	65.7812 4981	0.0077 0190	0.0152 0190
92	1.9885 8905	0.5028 6911	131.8118 7280	66.2841 1892	0.0075 8657	0.0150 8657
93	2.0035 0346	0.4991 2567	133.8004 6185	66.7832 4458	0.0074 7382	0.0149 7382
94	2.0185 2974	0.4954 1009	135.8039 6531	67.2786 5467	0.0073 6356	0.0148 6356
95	2.0336 6871	0.4917 2217	137.8224 9505	67.7703 7685	0.0072 5571	0.0147 5571
96	2.0489 2123	0.4880 6171	139.8561 6377	68.2584 3856	0.0071 5020	0.0146 5020
97	2.0642 8814	0.4844 2850	141.9050 8499	68.7428 6705	0.0070 4696	0.0145 4696
98	2.0797 7030	0.4808 2233	143.9693 7313	69.2236 8938	0.0069 4592	0.0144 4592
99	2.0953 6858	0.4772 4301	146.0491 4343	69.7009 3239	0.0068 4701	0.0143 4701
100	2.1110 8384	0.4736 9033	148.1445 1201	70.1746 2272	0.0067 5017	0.0142 5017

$i = 1\%$

n	$(1+i)^n$	$(1+i)^{-n}$	$s_{\overline{n}\rvert i}$	$a_{\overline{n}\rvert i}$	$\frac{1}{s_{\overline{n}\rvert i}}$	$\frac{1}{a_{\overline{n}\rvert i}}$
1	1.0100 0000	0.9900 9901	1.0000 0000	0.9900 9901	1.0000 0000	1.0100 0000
2	1.0201 0000	0.9802 9605	2.0100 0000	1.9703 9506	0.4975 1244	0.5075 1244
3	1.0303 0100	0.9705 9015	3.0301 0000	2.9409 8521	0.3300 2211	0.3400 2211
4	1.0406 0401	0.9609 8034	4.0604 0100	3.9019 6555	0.2462 8109	0.2562 8109
5	1.0510 1005	0.9514 6569	5.1010 0501	4.8534 3124	0.1960 3980	0.2060 3980
6	1.0615 2015	0.9420 4524	6.1520 1506	5.7954 7647	0.1625 4837	0.1725 4837
7	1.0721 3535	0.9327 1805	7.2135 3521	6.7281 9453	0.1386 2828	0.1486 2828
8	1.0828 5671	0.9234 8322	8.2856 7056	7.6516 7775	0.1206 9029	0.1306 9029
9	1.0936 8527	0.9143 3982	9.3685 2727	8.5660 1758	0.1067 4036	0.1167 4036
10	1.1046 2213	0.9052 8695	10.4622 1254	9.4713 0453	0.0955 8208	0.1055 8208
11	1.1156 6835	0.8963 2372	11.5668 3467	10.3676 2825	0.0864 5408	0.0964 5408
12	1.1268 2503	0.8874 4923	12.6825 0301	11.2550 7747	0.0788 4879	0.0888 4879
13	1.1380 9328	0.8786 6260	13.8093 2804	12.1337 4007	0.0724 1482	0.0824 1482
14	1.1494 7421	0.8699 6297	14.9474 2132	13.0037 0304	0.0669 0117	0.0769 0117
15	1.1609 6896	0.8613 4947	16.0968 9554	13.8650 5252	0.0621 2378	0.0721 2378
16	1.1725 7864	0.8528 2126	17.2578 6449	14.7178 7378	0.0579 4460	0.0679 4460
17	1.1843 0443	0.8443 7749	18.4304 4314	15.5622 5127	0.0542 5806	0.0642 5806
18	1.1961 4748	0.8360 1731	19.6147 4757	16.3982 6858	0.0509 8205	0.0609 8205
19	1.2081 0895	0.8277 3992	20.8108 9504	17.2260 0850	0.0480 5175	0.0580 5175
20	1.2201 9004	0.8195 4447	22.0190 0399	18.0455 5297	0.0454 1531	0.0554 1531
21	1.2323 9194	0.8114 3017	23.2391 9403	18.8569 8313	0.0430 3075	0.0530 3075
22	1.2447 1586	0.8033 9621	24.4715 8598	19.6603 7934	0.0408 6372	0.0508 6372
23	1.2571 6302	0.7954 4179	25.7163 0183	20.4558 2113	0.0388 8584	0.0488 8584
24	1.2697 3465	0.7875 6613	26.9734 6485	21.2433 8726	0.0370 7347	0.0470 7347
25	1.2824 3200	0.7797 6844	28.2431 9950	22.0231 5570	0.0354 0675	0.0454 0675
26	1.2952 5631	0.7720 4796	29.5256 3150	22.7952 0366	0.0338 6888	0.0438 6888
27	1.3082 0888	0.7644 0392	30.8208 8781	23.5596 0759	0.0324 4553	0.0424 4553
28	1.3212 9097	0.7568 3557	32.1290 9669	24.3164 4316	0.0311 2444	0.0411 2444
29	1.3345 0388	0.7493 4215	33.4503 8766	25.0657 8530	0.0298 9502	0.0398 9502
30	1.3478 4892	0.7419 2292	34.7848 9153	25.8077 0822	0.0287 4811	0.0387 4811
31	1.3613 2740	0.7345 7715	36.1327 4045	26.5422 8537	0.0276 7573	0.0376 7573
32	1.3749 4068	0.7273 0411	37.4940 6785	27.2695 8947	0.0266 7089	0.0366 7089
33	1.3886 9009	0.7201 0307	38.8690 0853	27.9896 9255	0.0257 2744	0.0357 2744
34	1.4025 7699	0.7129 7334	40.2576 9862	28.7026 6589	0.0248 3997	0.0348 3997
35	1.4166 0276	0.7059 1420	41.6602 7560	29.4085 8009	0.0240 0368	0.0340 0368
36	1.4307 6878	0.6989 2495	43.0768 7836	30.1075 0504	0.0232 1431	0.0332 1431
37	1.4450 7647	0.6920 0490	44.5076 4714	30.7995 0994	0.0224 6805	0.0324 6805
38	1.4595 2724	0.6851 5337	45.9527 2361	31.4846 6330	0.0217 6150	0.0317 6150
39	1.4741 2251	0.6783 6967	47.4122 5085	32.1630 3298	0.0210 9160	0.0310 9160
40	1.4888 6373	0.6716 5314	48.8863 7336	32.8346 8611	0.0204 5560	0.0304 5560
41	1.5037 5237	0.6650 0311	50.3752 3709	33.4996 8922	0.0198 5102	0.0298 5102
42	1.5187 8989	0.6584 1892	51.8789 8946	34.1581 0814	0.0192 7563	0.0292 7563
43	1.5339 7779	0.6518 9992	53.3977 7936	34.8100 0806	0.0187 2737	0.0287 2737
44	1.5493 1757	0.6454 4546	54.9317 5715	35.4554 5352	0.0182 0441	0.0282 0441
45	1.5648 1075	0.6390 5492	56.4810 7472	36.0945 0844	0.0177 0505	0.0277 0505
46	1.5804 5885	0.6327 2764	58.0458 8547	36.7272 3608	0.0172 2775	0.0272 2775
47	1.5962 6344	0.6264 6301	59.6263 4432	37.3536 9909	0.0167 7111	0.0267 7111
48	1.6122 2608	0.6202 6041	61.2226 0777	37.9739 5949	0.0163 3384	0.0263 3384
49	1.6283 4834	0.6141 1921	62.8348 3385	38.5880 7871	0.0159 1474	0.0259 1474
50	1.6446 3182	0.6080 3882	64.4631 8218	39.1961 1753	0.0155 1273	0.0255 1273

$i = 1\%$

n	$(1+i)^n$	$(1+i)^{-n}$	$s_{\overline{n}\mid i}$	$a_{\overline{n}\mid i}$	$\frac{1}{s_{\overline{n}\mid i}}$	$\frac{1}{a_{\overline{n}\mid i}}$
51	1.6610 7814	0.6020 1864	66.1078 1401	39.7981 3617	0.0151 2680	0.0251 2680
52	1.6776 8892	0.5960 5806	67.7688 9215	40.3941 9423	0.0147 5603	0.0247 5603
53	1.6944 6581	0.5901 5649	69.4465 8107	40.9843 5072	0.0143 9956	0.0243 9956
54	1.7114 1047	0.5843 1336	71.1410 4688	41.5686 6408	0.0140 5658	0.0240 5658
55	1.7285 2457	0.5785 2808	72.8524 5735	42.1471 9216	0.0137 2637	0.0237 2637
56	1.7458 0982	0.5728 0008	74.5809 8192	42.7199 9224	0.0134 0824	0.0234 0824
57	1.7632 6792	0.5671 2879	76.3267 9174	43.2871 2102	0.0131 0156	0.0231 0156
58	1.7809 0060	0.5615 1365	78.0900 5966	43.8486 3468	0.0128 0573	0.0228 0573
59	1.7987 0960	0.5559 5411	79.8709 6025	44.4045 8879	0.0125 2020	0.0225 2020
60	1.8166 9670	0.5504 4962	81.6696 6986	44.9550 3841	0.0122 4445	0.0222 4445
61	1.8348 6367	0.5449 9962	83.4863 6655	45.5000 3803	0.0119 7800	0.0219 7800
62	1.8532 1230	0.5396 0358	85.3212 3022	46.0396 4161	0.0117 2041	0.0217 2041
63	1.8717 4443	0.5342 6097	87.1744 4252	46.5739 0258	0.0114 7125	0.0214 7125
64	1.8904 6187	0.5289 7126	89.0461 8695	47.1028 7385	0.0112 3013	0.0212 3013
65	1.9093 6649	0.5237 3392	90.9366 4882	47.6266 0777	0.0109 9667	0.0209 9667
66	1.9284 6015	0.5185 4844	92.8460 1531	48.1451 5621	0.0107 7052	0.0207 7052
67	1.9477 4475	0.5134 1429	94.7744 7546	48.6585 7050	0.0105 5136	0.0205 5136
68	1.9672 2220	0.5083 3099	96.7222 2021	49.1669 0149	0.0103 3889	0.0203 3889
69	1.9868 9442	0.5032 9801	98.6894 4242	49.6701 9949	0.0101 3280	0.0201 3280
70	2.0067 6337	0.4983 1486	100.6763 3684	50.1685 1435	0.0099 3282	0.0199 3282
71	2.0268 3100	0.4933 8105	102.6831 0021	50.6618 9539	0.0097 3870	0.0197 3870
72	2.0470 9931	0.4884 9609	104.7099 3121	51.1503 9148	0.0095 5019	0.0195 5019
73	2.0675 7031	0.4836 5949	106.7570 3052	51.6340 5097	0.0093 6706	0.0193 6706
74	2.0882 4601	0.4788 7078	108.8246 0083	52.1129 2175	0.0091 8910	0.0191 8910
75	2.1091 2847	0.4741 2949	110.9128 4684	52.5870 5124	0.0090 1609	0.0190 1609
76	2.1302 1975	0.4694 3514	113.0219 7530	53.0564 8638	0.0088 4784	0.0188 4784
77	2.1515 2195	0.4647 8726	115.1521 9506	53.5212 7364	0.0086 8416	0.0186 8416
78	2.1730 3717	0.4601 8541	117.3037 1701	53.9814 5905	0.0085 2488	0.0185 2488
79	2.1947 6754	0.4556 2912	119.4767 5418	54.4370 8817	0.0083 6983	0.0183 6983
80	2.2167 1522	0.4511 1794	121.6715 2172	54.8882 0611	0.0082 1885	0.0182 1885
81	2.2388 8237	0.4466 5142	123.8882 3694	55.3348 5753	0.0080 7179	0.0180 7179
82	2.2612 7119	0.4422 2913	126.1271 1931	55.7770 8666	0.0079 2851	0.0179 2851
83	2.2838 8390	0.4378 5063	128.3883 9050	56.2149 3729	0.0077 8887	0.0177 8887
84	2.3067 2274	0.4335 1547	130.6722 7440	56.6484 5276	0.0076 5273	0.0176 5273
85	2.3297 8997	0.4292 2324	132.9789 9715	57.0776 7600	0.0075 1998	0.0175 1998
86	2.3530 8787	0.4249 7350	135.3087 8712	57.5026 4951	0.0073 9050	0.0173 9050
87	2.3766 1875	0.4207 6585	137.6618 7499	57.9234 1535	0.0072 6418	0.0172 6418
88	2.4003 8494	0.4165 9985	140.0384 9374	58.3400 1520	0.0071 4089	0.0171 4089
89	2.4243 8879	0.4124 7510	142.4388 7868	58.7524 9030	0.0070 2056	0.0170 2056
90	2.4486 3267	0.4083 9119	144.8632 6746	59.1608 8148	0.0069 0306	0.0169 0306
91	2.4731 1900	0.4043 4771	147.3119 0014	59.5652 2919	0.0067 8832	0.0167 8832
92	2.4978 5019	0.4003 4427	149.7850 1914	59.9655 7346	0.0066 7624	0.0166 7624
93	2.5228 2869	0.3963 8046	152.2828 6933	60.3619 5392	0.0065 6673	0.0165 6673
94	2.5480 5698	0.3924 5590	154.8056 9803	60.7544 0982	0.0064 5971	0.0164 5971
95	2.5735 3755	0.3885 7020	157.3537 5501	61.1429 8002	0.0063 5511	0.0163 5511
96	2.5992 7293	0.3847 2297	159.9272 9256	61.5277 0299	0.0062 5284	0.0162 5284
97	2.6252 6565	0.3809 1383	162.5265 6548	61.9086 1682	0.0061 5284	0.0161 5284
98	2.6515 1831	0.3771 4241	165.1518 3114	62.2857 5923	0.0060 5503	0.0160 5503
99	2.6780 3349	0.3734 0832	167.8033 4945	62.6591 6755	0.0059 5936	0.0159 5936
100	2.7048 1383	0.3697 1121	170.4813 8294	63.0288 7877	0.0058 6574	0.0158 6574

$i = 1\frac{1}{4}\%$

n	$(1+i)^n$	$(1+i)^{-n}$	$s_{\overline{n}\rceil i}$	$a_{\overline{n}\rceil i}$	$\frac{1}{s_{\overline{n}\rceil i}}$	$\frac{1}{a_{\overline{n}\rceil i}}$
1	1.0125 0000	0.9876 5432	1.0000 0000	0.9876 5432	1.0000 0000	1.0125 0000
2	1.0251 5625	0.9754 6106	2.0125 0000	1.9631 1538	0.4968 9441	0.5093 9441
3	1.0379 7070	0.9634 1833	3.0376 5625	2.9265 3371	0.3292 0117	0.3417 0117
4	1.0509 4534	0.9515 2428	4.0756 2695	3.8780 5798	0.2453 6102	0.2578 6102
5	1.0640 8215	0.9397 7706	5.1265 7229	4.8178 3504	0.1950 6211	0.2075 6211
6	1.0773 8318	0.9281 7488	6.1906 5444	5.7460 0992	0.1615 3381	0.1740 3381
7	1.0908 5047	0.9167 1593	7.2680 3762	6.6627 2585	0.1375 8872	0.1500 8872
8	1.1044 8610	0.9053 9845	8.3588 8809	7.5681 2429	0.1196 3314	0.1321 3314
9	1.1182 9218	0.8942 2069	9.4633 7420	8.4623 4498	0.1056 7055	0.1181 7055
10	1.1322 7083	0.8831 8093	10.5816 6637	9.3455 2591	0.0945 0307	0.1070 0307
11	1.1464 2422	0.8722 7746	11.7139 3720	10.2178 0337	0.0853 6839	0.0978 6839
12	1.1607 5452	0.8615 0860	12.8603 6142	11.0793 1197	0.0777 5831	0.0902 5831
13	1.1752 6395	0.8508 7269	14.0211 1594	11.9301 8466	0.0713 2100	0.0838 2100
14	1.1899 5475	0.8403 6809	15.1963 7988	12.7705 5275	0.0658 0515	0.0783 0515
15	1.2048 2918	0.8299 9318	16.3863 3463	13.6005 4592	0.0610 2646	0.0735 2646
16	1.2198 8955	0.8197 4635	17.5911 6382	14.4202 9227	0.0568 4672	0.0693 4672
17	1.2351 3817	0.8096 2602	18.8110 5336	15.2299 1829	0.0531 6023	0.0656 6023
18	1.2505 7739	0.7996 3064	20.0461 9153	16.0295 4893	0.0498 8479	0.0623 8479
19	1.2662 0961	0.7897 5866	21.2967 6893	16.8193 0759	0.0469 5548	0.0594 5548
20	1.2820 3723	0.7800 0855	22.5629 7854	17.5993 1613	0.0443 2039	0.0568 2039
21	1.2980 6270	0.7703 7881	23.8450 1577	18.3696 9495	0.0419 3749	0.0544 3749
22	1.3142 8848	0.7608 6796	25.1430 7847	19.1305 6291	0.0397 7238	0.0522 7238
23	1.3307 1709	0.7514 7453	26.4573 6695	19.8820 3744	0.0377 9666	0.0502 9666
24	1.3473 5105	0.7421 9707	27.7880 8403	20.6242 3451	0.0359 8665	0.0484 8665
25	1.3641 9294	0.7330 3414	29.1354 3508	21.3572 6865	0.0343 2247	0.0468 2247
26	1.3812 4535	0.7239 8434	30.4996 2802	22.0812 5299	0.0327 8729	0.0452 8729
27	1.3985 1092	0.7150 4626	31.8808 7337	22.7962 9925	0.0313 6677	0.0438 6677
28	1.4159 9230	0.7062 1853	33.2793 8429	23.5025 1778	0.0300 4863	0.0425 4863
29	1.4336 9221	0.6974 9978	34.6953 7659	24.2000 1756	0.0288 2228	0.0413 2228
30	1.4516 1336	0.6888 8867	36.1290 6880	24.8889 0623	0.0276 7854	0.0401 7854
31	1.4697 5853	0.6803 8387	37.5806 8216	25.5692 9010	0.0266 0942	0.0391 0942
32	1.4881 3051	0.6719 8407	39.0504 4069	26.2412 7418	0.0256 0791	0.0381 0791
33	1.5067 3214	0.6636 8797	40.5385 7120	26.9049 6215	0.0246 6786	0.0371 6786
34	1.5255 6629	0.6554 9429	42.0453 0334	27.5604 5644	0.0237 8387	0.0362 8387
35	1.5446 3587	0.6474 0177	43.5708 6963	28.2078 5822	0.0229 5111	0.0354 5111
36	1.5639 4382	0.6394 0916	45.1155 0550	28.8472 6737	0.0221 6533	0.0346 6533
37	1.5834 9312	0.6315 1522	46.6794 4932	29.4787 8259	0.0214 2270	0.0339 2270
38	1.6032 8678	0.6237 1873	48.2629 4243	30.1025 0133	0.0207 1983	0.0332 1983
39	1.6233 2787	0.6160 1850	49.8662 2921	30.7185 1983	0.0200 5365	0.0325 5365
40	1.6436 1946	0.6084 1334	51.4895 5708	31.3269 3316	0.0194 2141	0.0319 2141
41	1.6641 6471	0.6009 0206	53.1331 7654	31.9278 3522	0.0188 2063	0.0313 2063
42	1.6849 6677	0.5934 8352	54.7973 4125	32.5213 1874	0.0182 4906	0.0307 4906
43	1.7060 2885	0.5861 5656	56.4823 0801	33.1074 7530	0.0177 0466	0.0302 0466
44	1.7273 5421	0.5789 2006	58.1883 3687	33.6863 9536	0.0171 8557	0.0296 8557
45	1.7489 4614	0.5717 7290	59.9156 9108	34.2581 6825	0.0166 9012	0.0291 9012
46	1.7708 0797	0.5647 1397	61.6646 3721	34.8228 8222	0.0162 1675	0.0287 1675
47	1.7929 4306	0.5577 4219	63.4354 4518	35.3806 2442	0.0157 6406	0.0282 6406
48	1.8153 5485	0.5508 5649	65.2283 8824	35.9314 8091	0.0153 3075	0.0278 3075
49	1.8380 4679	0.5440 5579	67.0437 4310	36.4755 3670	0.0149 1563	0.0274 1563
50	1.8610 2237	0.5373 3905	68.8817 8989	37.0128 7575	0.0145 1763	0.0270 1763

$i = 1\frac{1}{4}\%$

n	$(1+i)^n$	$(1+i)^{-n}$	$s_{\overline{n}\rvert i}$	$a_{\overline{n}\rvert i}$	$\frac{1}{s_{\overline{n}\rvert i}}$	$\frac{1}{a_{\overline{n}\rvert i}}$
51	1.8842 8515	0.5307 0524	70.7428 1226	37.5435 8099	0.0141 3571	0.0266 3571
52	1.9078 3872	0.5241 5332	72.6270 9741	38.0677 3431	0.0137 6897	0.0262 6897
53	1.9316 8670	0.5176 8229	74.5349 3613	38.5854 1660	0.0134 1653	0.0259 1653
54	1.9558 3279	0.5112 9115	76.4666 2283	39.0967 0776	0.0130 7760	0.0255 7760
55	1.9802 8070	0.5049 7892	78.4224 5562	39.6016 8667	0.0127 5145	0.0252 5145
56	2.0050 3420	0.4987 4461	80.4027 3631	40.1004 3128	0.0124 3739	0.0249 3739
57	2.0300 9713	0.4925 8727	82.4077 7052	40.5930 1855	0.0121 3478	0.0246 3478
58	2.0554 7335	0.4865 0594	84.4378 6765	41.0795 2449	0.0118 4303	0.0243 4303
59	2.0811 6676	0.4804 9970	86.4933 4099	41.5600 2419	0.0115 6158	0.0240 6158
60	2.1071 8135	0.4745 6760	88.5745 0776	42.0345 9179	0.0112 8993	0.0237 8993
61	2.1335 2111	0.4687 0874	90.6816 8910	42.5033 0054	0.0110 2758	0.0235 2758
62	2.1601 9013	0.4629 2222	92.8152 1022	42.9662 2275	0.0107 7410	0.0232 7410
63	2.1871 9250	0.4572 0713	94.9754 0034	43.4234 2988	0.0105 2904	0.0230 2904
64	2.2145 3241	0.4515 6259	97.1625 9285	43.8749 9247	0.0102 9203	0.0227 9203
65	2.2422 1407	0.4459 8775	99.3771 2526	44.3209 8022	0.0100 6268	0.0225 6268
66	2.2702 4174	0.4404 8173	101.6193 3933	44.7614 6195	0.0098 4065	0.0223 4065
67	2.2986 1976	0.4350 4368	103.8895 8107	45.1965 0563	0.0096 2560	0.0221 2560
68	2.3273 5251	0.4296 7277	106.1882 0083	45.6261 7840	0.0094 1724	0.0219 1724
69	2.3564 4442	0.4243 6817	108.5155 5334	46.0505 4656	0.0092 1527	0.0217 1527
70	2.3858 9997	0.4191 2905	110.8719 9776	46.4696 7562	0.0090 1941	0.0215 1941
71	2.4157 2372	0.4139 5462	113.2578 9773	46.8836 3024	0.0088 2941	0.0213 2941
72	2.4459 2027	0.4088 4407	115.6736 2145	47.2924 7431	0.0086 4501	0.0211 4501
73	2.4764 9427	0.4037 9661	118.1195 4172	47.6962 7093	0.0084 6600	0.0209 6600
74	2.5074 5045	0.3988 1147	120.5960 3599	48.0950 8240	0.0082 9215	0.0207 9215
75	2.5387 9358	0.3938 8787	123.1034 8644	48.4889 7027	0.0081 2325	0.0206 2325
76	2.5705 2850	0.3890 2506	125.6422 8002	48.8779 9533	0.0079 5910	0.0204 5910
77	2.6026 6011	0.3842 2228	128.2128 0852	49.2622 1761	0.0077 9953	0.0202 9953
78	2.6351 9336	0.3794 7879	130.8154 6863	49.6416 9640	0.0076 4436	0.0201 4436
79	2.6681 3327	0.3747 9387	133.4506 6199	50.0164 9027	0.0074 9341	0.0199 9341
80	2.7014 8494	0.3701 6679	136.1187 9526	50.3866 5706	0.0073 4652	0.0198 4652
81	2.7352 5350	0.3655 9683	138.8202 8020	50.7522 5389	0.0072 0356	0.0197 0356
82	2.7694 4417	0.3610 8329	141.5555 3370	51.1133 3717	0.0070 6437	0.0195 6437
83	2.8040 6222	0.3566 2547	144.3249 7787	51.4699 6264	0.0069 2881	0.0194 2881
84	2.8391 1300	0.3522 2268	147.1290 4010	51.8221 8532	0.0067 9675	0.0192 9675
85	2.8746 0191	0.3478 7426	149.9681 5310	52.1700 5958	0.0066 6808	0.0191 6808
86	2.9105 3444	0.3435 7951	152.8427 5501	52.5136 3909	0.0065 4267	0.0190 4267
87	2.9469 1612	0.3393 3779	155.7532 8945	52.8529 7688	0.0064 2041	0.0189 2041
88	2.9837 5257	0.3351 4843	158.7002 0557	53.1881 2531	0.0063 0119	0.0188 0119
89	3.0210 4948	0.3310 1080	161.6839 5814	53.5191 3611	0.0061 8491	0.0186 8491
90	3.0588 1260	0.3269 2425	164.7050 0762	53.8460 6036	0.0060 7146	0.0185 7146
91	3.0970 4775	0.3228 8814	167.7638 2021	54.1689 4850	0.0059 6076	0.0184 6076
92	3.1357 6085	0.3189 0187	170.8608 6796	54.4878 5037	0.0058 5272	0.0183 5272
93	3.1749 5786	0.3149 6481	173.9966 2881	54.8028 1518	0.0057 4724	0.0182 4724
94	3.2146 4483	0.3110 7636	177.1715 8667	55.1138 9154	0.0056 4425	0.0181 4425
95	3.2548 2789	0.3072 3591	180.3862 3151	55.4211 2744	0.0055 4366	0.0180 4366
96	3.2955 1324	0.3034 4287	183.6410 5940	55.7245 7031	0.0054 4541	0.0179 4541
97	3.3367 0716	0.2996 9666	186.9365 7264	56.0242 6698	0.0053 4941	0.0178 4941
98	3.3784 1600	0.2959 9670	190.2732 7980	56.3202 6368	0.0052 5560	0.0177 5560
99	3.4206 4620	0.2923 4242	193.6516 9580	56.6126 0610	0.0051 6391	0.0176 6391
100	3.4634 0427	0.2887 3326	197.0723 4200	56.9013 3936	0.0050 7428	0.0175 7428

$i = 1\frac{1}{2}\%$

n	$(1+i)^n$	$(1+i)^{-n}$	$s_{\overline{n}\rvert i}$	$a_{\overline{n}\rvert i}$	$\frac{1}{s_{\overline{n}\rvert i}}$	$\frac{1}{a_{\overline{n}\rvert i}}$
1	1.0150 0000	0.9852 2167	1.0000 0000	0.9852 2167	1.0000 0000	1.0150 0000
2	1.0302 2500	0.9706 6175	2.0150 0000	1.9558 8342	0.4962 7792	0.5112 7792
3	1.0456 7838	0.9563 1699	3.0452 2500	2.9122 0042	0.3283 8296	0.3433 8296
4	1.0613 6355	0.9421 8423	4.0909 0338	3.8543 8465	0.2444 4479	0.2594 4479
5	1.0772 8400	0.9282 6033	5.1522 6693	4.7826 4497	0.1940 8932	0.2090 8932
6	1.0934 4326	0.9145 4219	6.2295 5093	5.6971 8717	0.1605 2521	0.1755 2521
7	1.1098 4491	0.9010 2679	7.3229 9419	6.5982 1396	0.1365 5616	0.1515 5616
8	1.1264 9259	0.8877 1112	8.4328 3911	7.4859 2508	0.1185 8402	0.1335 8402
9	1.1433 8998	0.8745 9224	9.5593 3169	8.3605 1732	0.1046 0982	0.1196 0982
10	1.1605 4083	0.8616 6723	10.7027 2167	9.2221 8455	0.0934 3418	0.1084 3418
11	1.1779 4894	0.8489 3323	11.8632 6249	10.0711 1779	0.0842 9384	0.0992 9384
12	1.1956 1817	0.8363 8742	13.0412 1143	10.9075 0521	0.0766 7999	0.0916 7999
13	1.2135 5244	0.8240 2702	14.2368 2960	11.7315 3222	0.0702 4036	0.0852 4036
14	1.2317 5573	0.8118 4928	15.4503 8205	12.5433 8150	0.0647 2332	0.0797 2332
15	1.2502 3207	0.7998 5150	16.6821 3778	13.3432 3301	0.0599 4436	0.0749 4436
16	1.2689 8555	0.7880 3104	17.9323 6984	14.1312 6405	0.0557 6508	0.0707 6508
17	1.2880 2033	0.7763 8526	19.2013 5539	14.9076 4931	0.0520 7966	0.0670 7966
18	1.3073 4064	0.7649 1159	20.4893 7572	15.6725 6089	0.0488 0578	0.0638 0578
19	1.3269 5075	0.7536 0747	21.7967 1636	16.4261 6837	0.0458 7847	0.0608 7847
20	1.3468 5501	0.7424 7042	23.1236 6710	17.1686 3879	0.0432 4574	0.0582 4574
21	1.3670 5783	0.7314 9795	24.4705 2211	17.9001 3673	0.0408 6550	0.0558 6550
22	1.3875 6370	0.7206 8763	25.8375 7994	18.6208 2437	0.0387 0332	0.0537 0332
23	1.4083 7715	0.7100 3708	27.2251 4364	19.3308 6145	0.0367 3075	0.0517 3075
24	1.4295 0281	0.6995 4392	28.6335 2080	20.0304 0537	0.0349 2410	0.0499 2410
25	1.4509 4535	0.6892 0583	30.0630 2361	20.7196 1120	0.0332 6345	0.0482 6345
26	1.4727 0953	0.6790 2052	31.5139 6896	21.3986 3172	0.0317 3196	0.0467 3196
27	1.4948 0018	0.6689 8574	32.9866 7850	22.0676 1746	0.0303 1527	0.0453 1527
28	1.5172 2218	0.6590 9925	34.4814 7867	22.7267 1671	0.0290 0108	0.0440 0108
29	1.5399 8051	0.6493 5887	35.9987 0085	23.3760 7558	0.0277 7878	0.0427 7878
30	1.5630 8022	0.6397 6243	37.5386 8137	24.0158 3801	0.0266 3919	0.0416 3919
31	1.5865 2642	0.6303 0781	39.1017 6159	24.6461 4582	0.0255 7430	0.0405 7430
32	1.6103 2432	0.6209 9292	40.6882 8801	25.2671 3874	0.0245 7710	0.0395 7710
33	1.6344 7918	0.6118 1568	42.2986 1233	25.8789 5442	0.0236 4144	0.0386 4144
34	1.6589 9637	0.6027 7407	43.9330 9152	26.4817 2849	0.0227 6189	0.0377 6189
35	1.6838 8132	0.5938 6608	45.5920 8789	27.0755 9458	0.0219 3363	0.0369 3363
36	1.7091 3954	0.5850 8974	47.2759 6921	27.6606 8431	0.0211 5240	0.0361 5240
37	1.7347 7663	0.5764 4309	48.9851 0874	28.2371 2740	0.0204 1437	0.0354 1437
38	1.7607 9828	0.5679 2423	50.7198 8538	28.8050 5163	0.0197 1613	0.0347 1613
39	1.7872 1025	0.5595 3126	52.4806 8366	29.3645 8288	0.0190 5463	0.0340 5463
40	1.8140 1841	0.5512 6232	54.2678 9391	29.9158 4520	0.0184 2710	0.0334 2710
41	1.8412 2868	0.5431 1559	56.0819 1232	30.4589 6079	0.0178 3106	0.0328 3106
42	1.8688 4712	0.5350 8925	57.9231 4100	30.9940 5004	0.0172 6426	0.0322 6426
43	1.8968 7982	0.5271 8153	59.7919 8812	31.5212 3157	0.0167 2465	0.0317 2465
44	1.9253 3302	0.5193 9067	61.6888 6794	32.0406 2223	0.0162 1038	0.0312 1038
45	1.9542 1301	0.5117 1494	63.6142 0096	32.5523 3718	0.0157 1976	0.0307 1976
46	1.9835 2621	0.5041 5265	65.5684 1398	33.0564 8983	0.0152 5125	0.0302 5125
47	2.0132 7910	0.4967 0212	67.5519 4018	33.5531 9195	0.0148 0342	0.0298 0342
48	2.0434 7829	0.4893 6170	69.5652 1929	34.0425 5365	0.0143 7500	0.0293 7500
49	2.0741 3046	0.4821 2975	71.6086 9758	34.5246 8339	0.0139 6478	0.0289 6478
50	2.1052 4242	0.4750 0468	73.6828 2804	34.9996 8807	0.0135 7168	0.0285 7168

$i = 1\frac{1}{2}\%$

n	$(1+i)^n$	$(1+i)^{-n}$	$s_{\overline{n}\rvert i}$	$a_{\overline{n}\rvert i}$	$\frac{1}{s_{\overline{n}\rvert i}}$	$\frac{1}{a_{\overline{n}\rvert i}}$
51	2.1368 2106	0.4679 8491	75.7880 7046	35.4676 7298	0.0131 9469	0.0281 9469
52	2.1688 7337	0.4610 6887	77.9248 9152	35.9287 4185	0.0128 3287	0.0278 3287
53	2.2014 0647	0.4542 5505	80.0937 6489	36.3829 9690	0.0124 8537	0.0274 8537
54	2.2344 2757	0.4475 4192	82.2951 7136	36.8305 3882	0.0121 5138	0.0271 5138
55	2.2679 4398	0.4409 2800	84.5295 9893	37.2714 6681	0.0118 3018	0.0268 3018
56	2.3019 6314	0.4344 1182	86.7975 4292	37.7058 7863	0.0115 2106	0.0265 2106
57	2.3364 9259	0.4279 9194	89.0995 0606	38.1338 7058	0.0112 2341	0.0262 2341
58	2.3715 3998	0.4216 6694	91.4359 9865	38.5555 3751	0.0109 3661	0.0259 3661
59	2.4071 1308	0.4154 3541	93.8075 3863	38.9709 7292	0.0106 6012	0.0256 6012
60	2.4432 1978	0.4092 9597	96.2146 5171	39.3802 6889	0.0103 9343	0.0253 9343
61	2.4798 6807	0.4032 4726	98.6578 7149	39.7835 1614	0.0101 3604	0.0251 3604
62	2.5170 6609	0.3972 8794	101.1377 3956	40.1808 0408	0.0098 8751	0.0248 8751
63	2.5548 2208	0.3914 1669	103.6548 0565	40.5722 2077	0.0096 4741	0.0246 4741
64	2.5931 4442	0.3856 3221	106.2096 2774	40.9578 5298	0.0094 1534	0.0244 1534
65	2.6320 4158	0.3799 3321	108.8027 7215	41.3377 8618	0.0091 9094	0.0241 9094
66	2.6715 2221	0.3743 1843	111.4348 1374	41.7121 0461	0.0089 7386	0.0239 7386
67	2.7115 9504	0.3687 8663	114.1063 3594	42.0808 9125	0.0087 6376	0.0237 6376
68	2.7522 6896	0.3633 3658	116.8179 3098	42.4442 2783	0.0085 6033	0.0235 6033
69	2.7935 5300	0.3579 6708	119.5701 9995	42.8021 9490	0.0083 6329	0.0233 6329
70	2.8354 5629	0.3526 7692	122.3637 5295	43.1548 7183	0.0081 7235	0.0231 7235
71	2.8779 8814	0.3474 6495	125.1992 0924	43.5023 3678	0.0079 8727	0.0229 8727
72	2.9211 5796	0.3423 3000	128.0771 9738	43.8446 6677	0.0078 0779	0.0228 0779
73	2.9649 7533	0.3372 7093	130.9983 5534	44.1819 3771	0.0076 3368	0.0226 3368
74	3.0094 4996	0.3322 8663	133.9633 3067	44.5142 2434	0.0074 6473	0.0224 6473
75	3.0545 9171	0.3273 7599	136.9727 8063	44.8416 0034	0.0073 0072	0.0223 0072
76	3.1004 1059	0.3225 3793	140.0273 7234	45.1641 3826	0.0071 4146	0.0221 4146
77	3.1469 1674	0.3177 7136	143.1277 8292	45.4819 0962	0.0069 8676	0.0219 8676
78	3.1941 2050	0.3130 7523	146.2746 9967	45.7949 8485	0.0068 3645	0.0218 3645
79	3.2420 3230	0.3084 4850	149.4688 2016	46.1034 3335	0.0066 9036	0.0216 9036
80	3.2906 6279	0.3038 9015	152.7108 5247	46.4073 2349	0.0065 4832	0.0215 4832
81	3.3400 2273	0.2993 9916	156.0015 1525	46.7067 2265	0.0064 1019	0.0214 1019
82	3.3901 2307	0.2949 7454	159.3415 3798	47.0016 9720	0.0062 7583	0.0212 7583
83	3.4409 7492	0.2906 1531	162.7316 6105	47.2923 1251	0.0061 4509	0.0211 4509
84	3.4925 8954	0.2863 2050	166.1726 3597	47.5786 3301	0.0060 1784	0.0210 1784
85	3.5449 7838	0.2820 8917	169.6652 2551	47.8607 2218	0.0058 9396	0.0208 9396
86	3.5981 5306	0.2779 2036	173.2102 0389	48.1386 4254	0.0057 7333	0.0207 7333
87	3.6521 2535	0.2738 1316	176.8083 5695	48.4124 5571	0.0056 5584	0.0206 5584
88	3.7069 0723	0.2697 6666	180.4604 8230	48.6822 2237	0.0055 4138	0.0205 4138
89	3.7625 1084	0.2657 7996	184.1673 8954	48.9480 0234	0.0054 2984	0.0204 2984
90	3.8189 4851	0.2618 5218	187.9299 0038	49.2098 5452	0.0053 2113	0.0203 2113
91	3.8762 3273	0.2579 8245	191.7488 4889	49.4678 3696	0.0052 1516	0.0202 1516
92	3.9343 7622	0.2541 6990	195.6250 8162	49.7220 0686	0.0051 1182	0.0201 1182
93	3.9933 9187	0.2504 1369	199.5594 5784	49.9724 2055	0.0050 1104	0.0200 1104
94	4.0532 9275	0.2467 1300	203.5528 4971	50.2191 3355	0.0049 1273	0.0199 1273
95	4.1140 9214	0.2430 6699	207.6061 4246	50.4622 0054	0.0048 1681	0.0198 1681
96	4.1758 0352	0.2394 7487	211.7202 3459	50.7016 7541	0.0047 2321	0.0197 2321
97	4.2384 4057	0.2359 3583	215.8960 3811	50.9376 1124	0.0046 3186	0.0196 3186
98	4.3020 1718	0.2324 4909	220.1344 7868	51.1700 6034	0.0045 4268	0.0195 4268
99	4.3665 4744	0.2290 1389	224.4364 9586	51.3990 7422	0.0044 5560	0.0194 5560
100	4.4320 4565	0.2256 2944	228.8030 4330	51.6247 0367	0.0043 7057	0.0193 7057

$i = 1\frac{3}{4}\%$

n	$(1+i)^n$	$(1+i)^{-n}$	$s_{\overline{n}\mid i}$	$a_{\overline{n}\mid i}$	$\frac{1}{s_{\overline{n}\mid i}}$	$\frac{1}{a_{\overline{n}\mid i}}$
1	1.0175 0000	0.9828 0098	1.0000 0000	0.9828 0098	1.0000 0000	1.0175 0000
2	1.0353 0625	0.9658 9777	2.0175 0000	1.9486 9875	0.4956 6295	0.5131 6295
3	1.0534 2411	0.9492 8528	3.0528 0625	2.8979 8403	0.3275 6746	0.3450 6746
4	1.0718 5903	0.9329 5851	4.1062 3036	3.8309 4254	0.2435 3237	0.2610 3237
5	1.0906 1656	0.9169 1254	5.1780 8939	4.7478 5508	0.1931 2142	0.2106 2142
6	1.1097 0235	0.9011 4254	6.2687 0596	5.6489 9762	0.1595 2256	0.1770 2256
7	1.1291 2215	0.8856 4378	7.3784 0831	6.5346 4139	0.1355 3059	0.1530 3059
8	1.1488 8178	0.8704 1157	8.5075 3045	7.4050 5297	0.1175 4292	0.1350 4292
9	1.1689 8721	0.8554 4135	9.6564 1224	8.2604 9432	0.1035 5813	0.1210 5813
10	1.1894 4449	0.8407 2860	10.8253 9945	9.1012 2291	0.0923 7534	0.1098 7534
11	1.2102 5977	0.8262 6889	12.0148 4394	9.9274 9181	0.0832 3038	0.1007 3038
12	1.2314 3931	0.8120 5788	13.2251 0371	10.7395 4969	0.0756 1377	0.0931 1377
13	1.2529 8950	0.7980 9128	14.4565 4303	11.5376 4097	0.0691 7283	0.0866 7283
14	1.2749 1682	0.7843 6490	15.7095 3253	12.3220 0587	0.0636 5562	0.0811 5562
15	1.2972 2786	0.7708 7459	16.9844 4935	13.0928 8046	0.0588 7739	0.0763 7739
16	1.3199 2935	0.7576 1631	18.2816 7721	13.8504 9677	0.0546 9958	0.0721 9958
17	1.3430 2811	0.7445 8605	19.6016 0656	14.5950 8282	0.0510 1623	0.0685 1623
18	1.3665 3111	0.7317 7990	20.9446 3468	15.3268 6272	0.0477 4492	0.0652 4492
19	1.3904 4540	0.7191 9401	22.3111 6578	16.0460 5673	0.0448 2061	0.0623 2061
20	1.4147 7820	0.7068 2458	23.7016 1119	16.7528 8130	0.0421 9122	0.0596 9122
21	1.4395 3681	0.6946 6789	25.1163 8938	17.4475 4919	0.0398 1464	0.0573 1464
22	1.4647 2871	0.6827 2028	26.5559 2620	18.1302 6948	0.0376 5638	0.0551 5638
23	1.4903 6146	0.6709 7817	28.0206 5490	18.8012 4764	0.0356 8796	0.0531 8796
24	1.5164 4279	0.6594 3800	29.5110 1637	19.4606 8565	0.0338 8565	0.0513 8565
25	1.5429 8054	0.6480 9632	31.0274 5915	20.1087 8196	0.0322 2952	0.0497 2952
26	1.5699 8269	0.6369 4970	32.5704 3969	20.7457 3166	0.0307 0269	0.0482 0269
27	1.5974 5739	0.6259 9479	34.1404 2238	21.3717 2644	0.0292 9079	0.0467 9079
28	1.6254 1290	0.6152 2829	35.7378 7977	21.9869 5474	0.0279 8151	0.0454 8151
29	1.6538 5762	0.6046 4697	37.3632 9267	22.5916 0171	0.0267 6424	0.0442 6424
30	1.6828 0013	0.5942 4764	39.0171 5029	23.1858 4934	0.0256 2975	0.0431 2975
31	1.7122 4913	0.5840 2716	40.6999 5042	23.7698 7650	0.0245 7005	0.0420 7005
32	1.7422 1349	0.5739 8247	42.4121 9955	24.3438 5897	0.0235 7812	0.0410 7812
33	1.7727 0223	0.5641 1053	44.1544 1305	24.9079 6951	0.0226 4779	0.0401 4779
34	1.8037 2452	0.5544 0839	45.9271 1527	25.4623 7789	0.0217 7363	0.0392 7363
35	1.8352 8970	0.5448 7311	47.7308 3979	26.0072 5100	0.0209 5082	0.0384 5082
36	1.8674 0727	0.5355 0183	49.5661 2949	26.5427 5283	0.0201 7507	0.0376 7507
37	1.9000 8689	0.5262 9172	51.4335 3675	27.0690 4455	0.0194 4257	0.0369 4257
38	1.9333 3841	0.5172 4002	53.3336 2365	27.5862 8457	0.0187 4990	0.0362 4990
39	1.9671 7184	0.5083 4400	55.2669 6206	28.0946 2857	0.0180 9399	0.0355 9399
40	2.0015 9734	0.4996 0098	57.2341 3390	28.5942 2955	0.0174 7209	0.0349 7209
41	2.0366 2530	0.4910 0834	59.2357 3124	29.0852 3789	0.0168 8170	0.0343 8170
42	2.0722 6624	0.4825 6348	61.2723 5654	29.5678 0136	0.0163 2057	0.0338 2057
43	2.1085 3090	0.4742 6386	63.3446 2278	30.0420 6522	0.0157 8666	0.0332 8666
44	2.1454 3019	0.4661 0699	65.4531 5367	30.5081 7221	0.0152 7810	0.0327 7810
45	2.1829 7522	0.4580 9040	67.5985 8386	30.9662 6261	0.0147 9321	0.0322 9321
46	2.2211 7728	0.4502 1170	69.7815 5908	31.4164 7431	0.0143 3043	0.0318 3043
47	2.2600 4789	0.4424 6850	72.0027 3637	31.8589 4281	0.0138 8836	0.0313 8836
48	2.2995 9872	0.4348 5848	74.2627 8425	32.2938 0129	0.0134 6569	0.0309 6569
49	2.3398 4170	0.4273 7934	76.5623 8298	32.7211 8063	0.0130 6124	0.0305 6124
50	2.3807 8893	0.4200 2883	78.9022 2468	33.1412 0946	0.0126 7391	0.0301 7391

$i = 1\frac{3}{4}\%$

n	$(1+i)^n$	$(1+i)^{-n}$	$s_{\overline{n}\mid i}$	$a_{\overline{n}\mid i}$	$\frac{1}{s_{\overline{n}\mid i}}$	$\frac{1}{a_{\overline{n}\mid i}}$
51	2.4224 5274	0.4128 0475	81.2830 1361	33.5540 1421	0.0123 0269	0.0298 0269
52	2.4648 4566	0.4057 0492	83.7054 6635	33.9597 1913	0.0119 4665	0.0294 4665
53	2.5079 8046	0.3987 2719	86.1703 1201	34.3584 4632	0.0116 0492	0.0291 0492
54	2.5518 7012	0.3918 6947	88.6782 9247	34.7503 1579	0.0112 7672	0.0287 7672
55	2.5965 2785	0.3851 2970	91.2301 6259	35.1354 4550	0.0109 6129	0.0284 6129
56	2.6419 6708	0.3785 0585	93.8266 9043	35.5139 5135	0.0106 5795	0.0281 5795
57	2.6882 0151	0.3719 9592	96.4686 5752	35.8859 4727	0.0103 6606	0.0278 6606
58	2.7352 4503	0.3655 9796	99.1568 5902	36.2515 4523	0.0100 8503	0.0275 8503
59	2.7831 1182	0.3593 1003	101.8921 0405	36.6108 5526	0.0098 1430	0.0273 1430
60	2.8318 1628	0.3531 3025	104.6752 1588	36.9639 8552	0.0095 5336	0.0270 5336
61	2.8813 7306	0.3470 5676	107.5070 3215	37.3110 4228	0.0093 0172	0.0268 0172
62	2.9317 9709	0.3410 8772	110.3884 0522	37.6521 3000	0.0090 5892	0.0265 5892
63	2.9831 0354	0.3352 2135	113.3202 0231	37.9873 5135	0.0088 2455	0.0263 2455
64	3.0353 0785	0.3294 5587	116.3033 0585	38.3168 0723	0.0085 9821	0.0260 9821
65	3.0884 2574	0.3237 8956	119.3386 1370	38.6405 9678	0.0083 7952	0.0258 7952
66	3.1424 7319	0.3182 2069	122.4270 3944	38.9588 1748	0.0081 6813	0.0256 6813
67	3.1974 6647	0.3127 4761	125.5695 1263	39.2715 6509	0.0079 6372	0.0254 6372
68	3.2534 2213	0.3073 6866	128.7669 7910	39.5789 3375	0.0077 6597	0.0252 6597
69	3.3103 5702	0.3020 8222	132.0204 0124	39.8810 1597	0.0075 7459	0.0250 7459
70	3.3682 8827	0.2968 8670	135.3307 5826	40.1779 0267	0.0073 8930	0.0248 8930
71	3.4272 3331	0.2917 8054	138.6990 4653	40.4696 8321	0.0072 0985	0.0247 0985
72	3.4872 0990	0.2867 6221	142.1262 7984	40.7564 4542	0.0070 3600	0.0245 3600
73	3.5482 3607	0.2818 3018	145.6134 8974	41.0382 7560	0.0068 6750	0.0243 6750
74	3.6103 3020	0.2769 8298	149.1617 2581	41.3152 5857	0.0067 0413	0.0242 0413
75	3.6735 1098	0.2722 1914	152.7720 5601	41.5874 7771	0.0065 4570	0.0240 4570
76	3.7377 9742	0.2675 3724	156.4455 6699	41.8550 1495	0.0063 9200	0.0238 9200
77	3.8032 0888	0.2629 3586	160.1833 6441	42.1179 5081	0.0062 4285	0.0237 4285
78	3.8697 6503	0.2584 1362	163.9865 7329	42.3763 6443	0.0060 9806	0.0235 9806
79	3.9374 8592	0.2539 6916	167.8563 3832	42.6303 3359	0.0059 5748	0.0234 5748
80	4.0063 9192	0.2496 0114	171.7938 2424	42.8799 3474	0.0058 2093	0.0233 2093
81	4.0765 0378	0.2453 0825	175.8002 1617	43.1252 4298	0.0056 8828	0.0231 8828
82	4.1478 4260	0.2410 8919	179.8767 1995	43.3663 3217	0.0055 5936	0.0230 5936
83	4.2204 2984	0.2369 4269	184.0245 6255	43.6032 7486	0.0054 3406	0.0229 3406
84	4.2942 8737	0.2328 6751	188.2449 9239	43.8361 4237	0.0053 1223	0.0228 1223
85	4.3694 3740	0.2288 6242	192.5392 7976	44.0650 0479	0.0051 9375	0.0226 9375
86	4.4459 0255	0.2249 2621	196.9087 1716	44.2899 3099	0.0050 7850	0.0225 7850
87	4.5237 0584	0.2210 5770	201.3546 1971	44.5109 8869	0.0049 6636	0.0224 6636
88	4.6028 7070	0.2172 5572	205.8783 2555	44.7282 4441	0.0048 5724	0.0223 5724
89	4.6834 2093	0.2135 1914	210.4811 9625	44.9417 6355	0.0047 5102	0.0222 5102
90	4.7653 8080	0.2098 4682	215.1646 1718	45.1516 1037	0.0046 4760	0.0221 4760
91	4.8487 7496	0.2062 3766	219.9299 9798	45.3578 4803	0.0045 4690	0.0220 4690
92	4.9336 2853	0.2026 9057	224.7787 7295	45.5605 3860	0.0044 4882	0.0219 4882
93	5.0199 6703	0.1992 0450	229.7124 0148	45.7597 4310	0.0043 5327	0.0218 5327
94	5.1078 1645	0.1957 7837	234.7323 6850	45.9555 2147	0.0042 6017	0.0217 6017
95	5.1972 0324	0.1924 1118	239.8401 8495	46.1479 3265	0.0041 6944	0.0216 6944
96	5.2881 5429	0.1891 0190	245.0373 8819	46.3370 3455	0.0040 8101	0.0215 8101
97	5.3806 9699	0.1858 4953	250.3255 4248	46.5228 8408	0.0039 9480	0.0214 9480
98	5.4748 5919	0.1826 5310	255.7062 3947	46.7055 3718	0.0039 1074	0.0214 1074
99	5.5706 6923	0.1795 1165	261.1810 9866	46.8850 4882	0.0038 2876	0.0213 2876
100	5.6681 5594	0.1764 2422	266.7517 6789	47.0614 7304	0.0037 4880	0.0212 4880

$i = 2\%$

n	$(1+i)^n$	$(1+i)^{-n}$	$s_{\overline{n}\rceil i}$	$a_{\overline{n}\rceil i}$	$\frac{1}{s_{\overline{n}\rceil i}}$	$\frac{1}{a_{\overline{n}\rceil i}}$
1	1.0200 0000	0.9803 9216	1.0000 0000	0.9803 9216	1.0000 0000	1.0200 0000
2	1.0404 0000	0.9611 6878	2.0200 0000	1.9415 6094	0.4950 4950	0.5150 4950
3	1.0612 0800	0.9423 2233	3.0604 0000	2.8838 8327	0.3267 5467	0.3467 5467
4	1.0824 3216	0.9238 4543	4.1216 0800	3.8077 2870	0.2426 2375	0.2626 2375
5	1.1040 8080	0.9057 3081	5.2040 4016	4.7134 5951	0.1921 5839	0.2121 5839
6	1.1261 6242	0.8879 7138	6.3081 2096	5.6014 3089	0.1585 2581	0.1785 2581
7	1.1486 8567	0.8705 6018	7.4342 8338	6.4719 9107	0.1345 1196	0.1545 1196
8	1.1716 5938	0.8534 9037	8.5829 6905	7.3254 8144	0.1165 0980	0.1365 0980
9	1.1950 9257	0.8367 5527	9.7546 2843	8.1622 3671	0.1025 1544	0.1225 1544
10	1.2189 9442	0.8203 4830	10.9497 2100	8.9825 8501	0.0913 2653	0.1113 2653
11	1.2433 7431	0.8042 6304	12.1687 1542	9.7868 4805	0.0821 7794	0.1021 7794
12	1.2682 4179	0.7884 9318	13.4120 8973	10.5753 4122	0.0745 5960	0.0945 5960
13	1.2936 0663	0.7730 3253	14.6803 3152	11.3483 7375	0.0681 1835	0.0881 1835
14	1.3194 7876	0.7578 7502	15.9739 3815	12.1062 4877	0.0626 0197	0.0826 0197
15	1.3458 6834	0.7430 1473	17.2934 1692	12.8492 6350	0.0578 2547	0.0778 2547
16	1.3727 8571	0.7284 4581	18.6392 8525	13.5777 0931	0.0536 5013	0.0736 5013
17	1.4002 4142	0.7141 6256	20.0120 7096	14.2918 7188	0.0499 6984	0.0699 6984
18	1.4282 4625	0.7001 5937	21.4123 1238	14.9920 3125	0.0467 0210	0.0667 0210
19	1.4568 1117	0.6864 3076	22.8405 5863	15.6784 6201	0.0437 8177	0.0637 8177
20	1.4859 4740	0.6729 7133	24.2973 6980	16.3514 3334	0.0411 5672	0.0611 5672
21	1.5156 6634	0.6597 7582	25.7833 1719	17.0112 0916	0.0387 8477	0.0587 8477
22	1.5459 7967	0.6468 3904	27.2989 8354	17.6580 4820	0.0366 3140	0.0566 3140
23	1.5768 9926	0.6341 5592	28.8449 6321	18.2922 0412	0.0346 6810	0.0546 6810
24	1.6084 3725	0.6217 2149	30.4218 6247	18.9139 2560	0.0328 7110	0.0528 7110
25	1.6406 0599	0.6095 3087	32.0302 9972	19.5234 5647	0.0312 2044	0.0512 2044
26	1.6734 1811	0.5975 7928	33.6709 0572	20.1210 3576	0.0296 9923	0.0496 9923
27	1.7068 8648	0.5858 6204	35.3443 2383	20.7068 9780	0.0282 9309	0.0482 9309
28	1.7410 2421	0.5743 7455	37.0512 1031	21.2812 7236	0.0269 8967	0.0469 8967
29	1.7758 4469	0.5631 1231	38.7922 3451	21.8443 8466	0.0257 7836	0.0457 7836
30	1.8113 6158	0.5520 7089	40.5680 7921	22.3964 5555	0.0246 4992	0.0446 4992
31	1.8475 8882	0.5412 4597	42.3794 4079	22.9377 0152	0.0235 9635	0.0435 9635
32	1.8845 4059	0.5306 3330	44.2270 2961	23.4683 3482	0.0226 1061	0.0426 1061
33	1.9222 3140	0.5202 2873	46.1115 7020	23.9885 6355	0.0216 8653	0.0416 8653
34	1.9606 7603	0.5100 2817	48.0338 0160	24.4985 9172	0.0208 1867	0.0408 1867
35	1.9998 8955	0.5000 2761	49.9944 7763	24.9986 1933	0.0200 0221	0.0400 0221
36	2.0398 8734	0.4902 2315	51.9943 6719	25.4888 4248	0.0192 3285	0.0392 3285
37	2.0806 8509	0.4806 1093	54.0342 5453	25.9694 5341	0.0185 0678	0.0385 0678
38	2.1222 9879	0.4711 8719	56.1149 3962	26.4406 4060	0.0178 2057	0.0378 2057
39	2.1647 4477	0.4619 4822	58.2372 3841	26.9025 8883	0.0171 7114	0.0371 7114
40	2.2080 3966	0.4528 9042	60.4019 8318	27.3554 7924	0.0165 5575	0.0365 5575
41	2.2522 0046	0.4440 1021	62.6100 2284	27.7994 8945	0.0159 7188	0.0359 7188
42	2.2972 4447	0.4353 0413	64.8622 2330	28.2347 9358	0.0154 1729	0.0354 1729
43	2.3431 8936	0.4267 6875	67.1594 6777	28.6615 6233	0.0148 8993	0.0348 8993
44	2.3900 5314	0.4184 0074	69.5026 5712	29.0799 6307	0.0143 8794	0.0343 8794
45	2.4378 5421	0.4101 9680	71.8927 1027	29.4901 5987	0.0139 0962	0.0339 0962
46	2.4866 1129	0.4021 5373	74.3305 6447	29.8923 1360	0.0134 5342	0.0334 5342
47	2.5363 4352	0.3942 6836	76.8171 7576	30.2865 8196	0.0130 1792	0.0330 1792
48	2.5870 7039	0.3865 3761	79.3535 1927	30.6731 1957	0.0126 0184	0.0326 0184
49	2.6388 1179	0.3789 5844	81.9405 8966	31.0520 7801	0.0122 0396	0.0322 0396
50	2.6915 8803	0.3715 2788	84.5794 0145	31.4236 0589	0.0118 2321	0.0318 2321

$i = 2\%$

n	$(1+i)^n$	$(1+i)^{-n}$	$s_{\overline{n}\vert i}$	$a_{\overline{n}\vert i}$	$\frac{1}{s_{\overline{n}\vert i}}$	$\frac{1}{a_{\overline{n}\vert i}}$
51	2.7454 1979	0.3642 4302	87.2709 8948	31.7878 4892	0.0114 5856	0.0314 5856
52	2.8003 2819	0.3571 0100	90.0164 0927	32.1449 4992	0.0111 0909	0.0311 0909
53	2.8563 3475	0.3500 9902	92.8167 3746	32.4950 4894	0.0107 7392	0.0307 7392
54	2.9134 6144	0.3432 3433	95.6730 7221	32.8382 8327	0.0104 5226	0.0304 5226
55	2.9717 3067	0.3365 0425	98.5865 3365	33.1747 8752	0.0101 4337	0.0301 4337
56	3.0311 6529	0.3299 0613	101.5582 6432	33.5046 9365	0.0098 4656	0.0298 4656
57	3.0917 8859	0.3234 3738	104.5894 2961	33.8281 3103	0.0095 6120	0.0295 6120
58	3.1536 2436	0.3170 9547	107.6812 1820	34.1452 2650	0.0092 8667	0.0292 8667
59	3.2166 9685	0.3108 7791	110.8348 4257	34.4561 0441	0.0090 2243	0.0290 2243
60	3.2810 3079	0.3047 8227	114.0515 3942	34.7608 8668	0.0087 6797	0.0287 6797
61	3.3466 5140	0.2988 0614	117.3325 7021	35.0596 9282	0.0085 2278	0.0285 2278
62	3.4135 8443	0.2929 4720	120.6792 2161	35.3526 4002	0.0082 8643	0.0282 8643
63	3.4818 5612	0.2872 0314	124.0928 0604	35.6398 4316	0.0080 5848	0.0280 5848
64	3.5514 9324	0.2815 7170	127.5746 6216	35.9214 1486	0.0078 3855	0.0278 3855
65	3.6225 2311	0.2760 5069	131.1261 5541	36.1974 6555	0.0076 2624	0.0276 2624
66	3.6949 7357	0.2706 3793	134.7486 7852	36.4681 0348	0.0074 2122	0.0274 2122
67	3.7688 7304	0.2653 3130	138.4436 5209	36.7334 3478	0.0072 2316	0.0272 2316
68	3.8442 5050	0.2601 2873	142.2125 2513	36.9935 6351	0.0070 3173	0.0270 3173
69	3.9211 3551	0.2550 2817	146.0567 7563	37.2485 9168	0.0068 4665	0.0268 4665
70	3.9995 5822	0.2500 2761	149.9779 1114	37.4986 1929	0.0066 6765	0.0266 6765
71	4.0795 4939	0.2451 2511	153.9774 6937	37.7437 4441	0.0064 9446	0.0264 9446
72	4.1611 4038	0.2403 1874	158.0570 1875	37.9840 6314	0.0063 2683	0.0263 2683
73	4.2443 6318	0.2356 0661	162.2181 5913	38.2196 6975	0.0061 6454	0.0261 6454
74	4.3292 5045	0.2309 8687	166.4625 2231	38.4506 5662	0.0060 0736	0.0260 0736
75	4.4158 3546	0.2264 5771	170.7917 7276	38.6771 1433	0.0058 5508	0.0258 5508
76	4.5041 5216	0.2220 1737	175.2076 0821	38.8991 3170	0.0057 0751	0.0257 0751
77	4.5942 3521	0.2176 6408	179.7117 6038	39.1167 9578	0.0055 6447	0.0255 6447
78	4.6861 1991	0.2133 9616	184.3059 9558	39.3301 9194	0.0054 2576	0.0254 2576
79	4.7798 4231	0.2092 1192	188.9921 1549	39.5394 0386	0.0052 9123	0.0252 9123
80	4.8754 3916	0.2051 0973	193.7719 5780	39.7445 1359	0.0051 6071	0.0251 6071
81	4.9729 4794	0.2010 8797	198.6473 9696	39.9456 0156	0.0050 3405	0.0250 3405
82	5.0724 0690	0.1971 4507	203.6203 4490	40.1427 4663	0.0049 1110	0.0249 1110
83	5.1738 5504	0.1932 7948	208.6927 5180	40.3360 2611	0.0047 9173	0.0247 9173
84	5.2773 3214	0.1894 8968	213.8666 0683	40.5255 1579	0.0046 7581	0.0246 7581
85	5.3828 7878	0.1857 7420	219.1439 3897	40.7112 8999	0.0045 6321	0.0245 6321
86	5.4905 3636	0.1821 3157	224.5268 1775	40.8934 2156	0.0044 5381	0.0244 5381
87	5.6003 4708	0.1785 6036	230.0173 5411	41.0719 8192	0.0043 4750	0.0243 4750
88	5.7123 5402	0.1750 5918	235.6177 0119	41.2470 4110	0.0042 4416	0.0242 4416
89	5.8266 0110	0.1716 2665	241.3300 5521	41.4186 6774	0.0041 4370	0.0241 4370
90	5.9431 3313	0.1682 6142	247.1566 5632	41.5869 2916	0.0040 4602	0.0240 4602
91	6.0619 9579	0.1649 6217	253.0997 8944	41.7518 9133	0.0039 5101	0.0239 5101
92	6.1832 3570	0.1617 2762	259.1617 8523	41.9136 1895	0.0038 5859	0.0238 5859
93	6.3069 0042	0.1585 5649	265.3450 2093	42.0721 7545	0.0037 6868	0.0237 6868
94	6.4330 3843	0.1554 4754	271.6519 2135	42.2276 2299	0.0036 8118	0.0236 8118
95	6.5616 9920	0.1523 9955	278.0849 5978	42.3800 2254	0.0035 9602	0.0235 9602
96	6.6929 3318	0.1494 1132	284.6466 5898	42.5294 3386	0.0035 1313	0.0235 1313
97	6.8267 9184	0.1464 8169	291.3395 9216	42.6759 1555	0.0034 3242	0.0234 3242
98	6.9633 2768	0.1436 0950	298.1663 8400	42.8195 2505	0.0033 5383	0.0233 5383
99	7.1025 9423	0.1407 9363	305.1297 1168	42.9603 1867	0.0032 7729	0.0232 7729
100	7.2446 4612	0.1380 3297	312.2323 0591	43.0983 5164	0.0032 0274	0.0232 0274

$i = 2\frac{1}{4}\%$

n	$(1+i)^n$	$(1+i)^{-n}$	$s_{\overline{n}\rceil i}$	$a_{\overline{n}\rceil i}$	$\frac{1}{s_{\overline{n}\rceil i}}$	$\frac{1}{a_{\overline{n}\rceil i}}$
1	1.0225 0000	0.9779 9511	1.0000 0000	0.9779 9511	1.0000 0000	1.0225 0000
2	1.0455 0625	0.9564 7444	2.0225 0000	1.9344 6955	0.4944 3758	0.5169 3758
3	1.0690 3014	0.9354 2732	3.0680 0625	2.8698 9687	0.3259 4458	0.3484 4458
4	1.0930 8332	0.9148 4335	4.1370 3639	3.7847 4021	0.2417 1893	0.2642 1893
5	1.1176 7769	0.8947 1232	5.2301 1971	4.6794 5253	0.1912 0021	0.2137 0021
6	1.1428 2544	0.8750 2427	6.3477 9740	5.5544 7680	0.1575 3496	0.1800 3496
7	1.1685 3901	0.8557 6946	7.4906 2284	6.4102 4626	0.1335 0025	0.1560 0025
8	1.1948 3114	0.8369 3835	8.6591 6186	7.2471 8461	0.1154 8462	0.1379 8462
9	1.2217 1484	0.8185 2161	9.8539 9300	8.0657 0622	0.1014 8170	0.1239 8170
10	1.2492 0343	0.8005 1013	11.0757 0784	8.8662 1635	0.0902 8768	0.1127 8768
11	1.2773 1050	0.7828 9499	12.3249 1127	9.6491 1134	0.0811 3649	0.1036 3649
12	1.3060 4999	0.7656 6748	13.6022 2177	10.4147 7882	0.0735 1740	0.0960 1740
13	1.3354 3611	0.7488 1905	14.9082 7176	11.1635 9787	0.0670 7686	0.0895 7686
14	1.3654 8343	0.7323 4137	16.2437 0788	11.8959 3924	0.0615 6230	0.0840 6230
15	1.3962 0680	0.7162 2628	17.6091 9130	12.6121 6551	0.0567 8852	0.0792 8852
16	1.4276 2146	0.7004 6580	19.0053 9811	13.3126 3131	0.0526 1663	0.0751 1663
17	1.4597 4294	0.6850 5212	20.4330 1957	13.9976 8343	0.0489 4039	0.0714 4039
18	1.4925 8716	0.6699 7763	21.8927 6251	14.6676 6106	0.0456 7720	0.0681 7720
19	1.5261 7037	0.6552 3484	23.3853 4966	15.3228 9590	0.0427 6182	0.0652 6182
20	1.5605 0920	0.6408 1647	24.9115 2003	15.9637 1237	0.0401 4207	0.0626 4207
21	1.5956 2066	0.6267 1538	26.4720 2923	16.5904 2775	0.0377 7572	0.0602 7572
22	1.6315 2212	0.6129 2457	28.0676 4989	17.2033 5232	0.0356 2821	0.0581 2821
23	1.6682 3137	0.5994 3724	29.6991 7201	17.8027 8955	0.0336 7097	0.0561 7097
24	1.7057 6658	0.5862 4668	31.3674 0338	18.3890 3624	0.0318 8023	0.0543 8023
25	1.7441 4632	0.5733 4639	33.0731 6996	18.9623 8263	0.0302 3599	0.0527 3599
26	1.7833 8962	0.5607 2997	34.8173 1628	19.5231 1260	0.0287 2134	0.0512 2134
27	1.8235 1588	0.5483 9117	36.6007 0590	20.0715 0376	0.0273 2188	0.0498 2188
28	1.8645 4499	0.5363 2388	38.4242 2178	20.6078 2764	0.0260 2525	0.0485 2525
29	1.9064 9725	0.5245 2213	40.2887 6677	21.1323 4977	0.0248 2081	0.0473 2081
30	1.9493 9344	0.5129 8008	42.1952 6402	21.6453 2985	0.0236 9934	0.0461 9934
31	1.9932 5479	0.5016 9201	44.1446 5746	22.1470 2186	0.0226 5280	0.0451 5280
32	2.0381 0303	0.4906 5233	46.1379 1226	22.6376 7419	0.0216 7415	0.0441 7415
33	2.0839 6034	0.4798 5558	48.1760 1528	23.1175 2977	0.0207 5722	0.0432 5722
34	2.1308 4945	0.4692 9641	50.2599 7563	23.5868 2618	0.0198 9655	0.0423 9655
35	2.1787 9356	0.4589 6960	52.3908 2508	24.0457 9577	0.0190 8731	0.0415 8731
36	2.2278 1642	0.4488 7002	54.5696 1864	24.4946 6579	0.0183 2522	0.0408 2522
37	2.2779 4229	0.4389 9268	56.7974 3506	24.9336 5848	0.0176 0643	0.0401 0643
38	2.3291 9599	0.4293 3270	59.0753 7735	25.3629 9118	0.0169 2753	0.0394 2753
39	2.3816 0290	0.4198 8528	61.4045 7334	25.7828 7646	0.0162 8543	0.0387 8543
40	2.4351 8897	0.4106 4575	63.7861 7624	26.1935 2221	0.0156 7738	0.0381 7738
41	2.4899 8072	0.4016 0954	66.2213 6521	26.5951 3174	0.0151 0087	0.0376 0087
42	2.5460 0528	0.3927 7216	68.7113 4592	26.9879 0390	0.0145 5364	0.0370 5364
43	2.6032 9040	0.3841 2925	71.2573 5121	27.3720 3316	0.0140 3364	0.0365 3364
44	2.6618 6444	0.3756 7653	73.8606 4161	27.7477 0969	0.0135 3901	0.0360 3901
45	2.7217 5639	0.3674 0981	76.5225 0605	28.1151 1950	0.0130 6805	0.0355 6805
46	2.7829 9590	0.3593 2500	79.2442 6243	28.4744 4450	0.0126 1921	0.0351 1921
47	2.8456 1331	0.3514 1809	82.0272 5834	28.8258 6259	0.0121 9107	0.0346 9107
48	2.9096 3961	0.3436 8518	84.8728 7165	29.1695 4777	0.0117 8233	0.0342 8233
49	2.9751 0650	0.3361 2242	87.7825 1126	29.5056 7019	0.0113 9179	0.0338 9179
50	3.0420 4640	0.3287 2608	90.7576 1776	29.8343 9627	0.0110 1836	0.0335 1836

$i = 2\frac{1}{4}\%$

n	$(1+i)^n$	$(1+i)^{-n}$	$s_{\overline{n}\rvert i}$	$a_{\overline{n}\rvert i}$	$\frac{1}{s_{\overline{n}\rvert i}}$	$\frac{1}{a_{\overline{n}\rvert i}}$
51	3.1104 9244	0.3214 9250	93.7996 6416	30.1558 8877	0.0106 6102	0.0331 6102
52	3.1804 7852	0.3144 1810	96.9101 5661	30.4703 0687	0.0103 1884	0.0328 1884
53	3.2520 3929	0.3074 9936	100.0906 3513	30.7778 0623	0.0099 9094	0.0324 9094
54	3.3252 1017	0.3007 3287	103.3426 7442	31.0785 3910	0.0096 7654	0.0321 7654
55	3.4000 2740	0.2941 1528	106.6678 8460	31.3726 5438	0.0093 7489	0.0318 7489
56	3.4765 2802	0.2876 4330	110.0679 1200	31.6602 9768	0.0090 8530	0.0315 8530
57	3.5547 4990	0.2813 1374	113.5444 4002	31.9416 1142	0.0088 0712	0.0313 0712
58	3.6347 3177	0.2751 2347	117.0991 8992	32.2167 3489	0.0085 3977	0.0310 3977
59	3.7165 1324	0.2690 6940	120.7339 2169	32.4858 0429	0.0082 8268	0.0307 8268
60	3.8001 3479	0.2631 4856	124.4504 3493	32.7489 5285	0.0080 3533	0.0305 3533
61	3.8856 3782	0.2573 5801	128.2505 6972	33.0063 1086	0.0077 9724	0.0302 9724
62	3.9730 6467	0.2516 9487	132.1362 0754	33.2580 0573	0.0075 6795	0.0300 6795
63	4.0624 5862	0.2461 5635	136.1092 7221	33.5041 6208	0.0073 4704	0.0298 4704
64	4.1538 6394	0.2407 3971	140.1717 3083	33.7449 0179	0.0071 3411	0.0296 3411
65	4.2473 2588	0.2354 4226	144.3255 9477	33.9803 4405	0.0069 2878	0.0294 2878
66	4.3428 9071	0.2302 6138	148.5729 2066	34.2106 0543	0.0067 3070	0.0292 3070
67	4.4406 0576	0.2251 9450	152.9158 1137	34.4357 9993	0.0065 3955	0.0290 3955
68	4.5405 1939	0.2202 3912	157.3564 1713	34.6560 3905	0.0063 5500	0.0288 5500
69	4.6426 8107	0.2153 9278	161.8969 3651	34.8714 3183	0.0061 7677	0.0286 7677
70	4.7471 4140	0.2106 5309	166.5396 1758	35.0820 8492	0.0060 0458	0.0285 0458
71	4.8539 5208	0.2060 1769	171.2867 5898	35.2881 0261	0.0058 3816	0.0283 3816
72	4.9631 6600	0.2014 8429	176.1407 1106	35.4895 8691	0.0056 7728	0.0281 7728
73	5.0748 3723	0.1970 5065	181.1038 7705	35.6866 3756	0.0055 2169	0.0280 2169
74	5.1890 2107	0.1927 1458	186.1787 1429	35.8793 5214	0.0053 7118	0.0278 7118
75	5.3057 7405	0.1884 7391	191.3677 3536	36.0678 2605	0.0052 2554	0.0277 2554
76	5.4251 5396	0.1843 2657	196.6735 0941	36.2521 5262	0.0050 8457	0.0275 8457
77	5.5472 1993	0.1802 7048	202.0986 6337	36.4324 2310	0.0049 4808	0.0274 4808
78	5.6720 3237	0.1763 0365	207.6458 8329	36.6087 2675	0.0048 1589	0.0273 1589
79	5.7996 5310	0.1724 2411	213.3179 1567	36.7811 5085	0.0046 8784	0.0271 8784
80	5.9301 4530	0.1686 2993	219.1175 6877	36.9497 8079	0.0045 6376	0.0270 6376
81	6.0635 7357	0.1649 1925	225.0477 1407	37.1147 0004	0.0044 4350	0.0269 4350
82	6.2000 0397	0.1612 9022	231.1112 8763	37.2759 9026	0.0043 2692	0.0268 2692
83	6.3395 0406	0.1577 4105	237.3112 9160	37.4337 3130	0.0042 1387	0.0267 1387
84	6.4821 4290	0.1542 6997	243.6507 9567	37.5880 0127	0.0041 0423	0.0266 0423
85	6.6279 9112	0.1508 7528	250.1329 3857	37.7388 7655	0.0039 9787	0.0264 9787
86	6.7771 2092	0.1475 5528	256.7609 2969	37.8864 3183	0.0038 9467	0.0263 9467
87	6.9296 0614	0.1443 0835	263.5380 5060	38.0307 4018	0.0037 9452	0.0262 9452
88	7.0855 2228	0.1411 3286	270.4676 5674	38.1718 7304	0.0036 9730	0.0261 9730
89	7.2449 4653	0.1380 2724	277.5531 7902	38.3099 0028	0.0036 0291	0.0261 0291
90	7.4079 5782	0.1349 8997	284.7981 2555	38.4448 9025	0.0035 1126	0.0260 1126
91	7.5746 3688	0.1320 1953	292.2060 8337	38.5769 0978	0.0034 2224	0.0259 2224
92	7.7450 6621	0.1291 1445	299.7807 2025	38.7060 2423	0.0033 3577	0.0258 3577
93	7.9193 3020	0.1262 7331	307.5257 8645	38.8322 9754	0.0032 5176	0.0257 5176
94	8.0975 1512	0.1234 9468	315.4451 1665	38.9557 9221	0.0031 7012	0.0256 7012
95	8.2797 0921	0.1207 7719	323.5426 3177	39.0765 6940	0.0030 9078	0.0255 9078
96	8.4660 0267	0.1181 1950	331.8223 4099	39.1946 8890	0.0030 1366	0.0255 1366
97	8.6564 8773	0.1155 2029	340.2883 4366	39.3102 0920	0.0029 3868	0.0254 3868
98	8.8512 5871	0.1129 7828	348.9448 3139	39.4231 8748	0.0028 6578	0.0253 6578
99	9.0504 1203	0.1104 9221	357.7960 9010	39.5336 7968	0.0027 9489	0.0252 9489
100	9.2540 4630	0.1080 6084	366.8465 0213	39.6417 4052	0.0027 2594	0.0252 2594

$i = 2\frac{1}{2}\%$

n	$(1 + i)^n$	$(1 + i)^{-n}$	$s_{\overline{n}\rceil i}$	$a_{\overline{n}\rceil i}$	$\frac{1}{s_{\overline{n}\rceil i}}$	$\frac{1}{a_{\overline{n}\rceil i}}$
1	1.0250 0000	0.9756 0976	1.0000 0000	0.9756 0976	1.0000 0000	1.0250 0000
2	1.0506 2500	0.9518 1440	2.0250 0000	1.9274 2415	0.4938 2716	0.5188 2716
3	1.0768 9063	0.9285 9941	3.0756 2500	2.8560 2356	0.3251 3717	0.3501 3717
4	1.1038 1289	0.9059 5064	4.1525 1563	3.7619 7421	0.2408 1788	0.2658 1788
5	1.1314 0821	0.8838 5429	5.2563 2852	4.6458 2850	0.1902 4686	0.2152 4686
6	1.1596 9342	0.8622 9687	6.3877 3673	5.5081 2536	0.1565 4997	0.1815 4997
7	1.1886 8575	0.8412 6524	7.5474 3015	6.3493 9060	0.1324 9543	0.1574 9543
8	1.2184 0290	0.8207 4657	8.7361 1590	7.1701 3717	0.1144 6735	0.1394 6735
9	1.2488 6297	0.8007 2836	9.9545 1880	7.9708 6553	0.1004 5689	0.1254 5689
10	1.2800 8454	0.7811 9840	11.2033 8177	8.7520 6393	0.0892 5876	0.1142 5876
11	1.3120 8666	0.7621 4478	12.4834 6631	9.5142 0871	0.0801 0596	0.1051 0596
12	1.3448 8882	0.7435 5589	13.7955 5297	10.2577 6460	0.0724 8713	0.0974 8713
13	1.3785 1104	0.7254 2038	15.1404 4179	10.9831 8497	0.0660 4827	0.0910 4827
14	1.4129 7382	0.7077 2720	16.5189 5284	11.6909 1217	0.0605 3652	0.0855 3652
15	1.4482 9817	0.6904 6556	17.9319 2666	12.3813 7773	0.0557 6646	0.0807 6646
16	1.4845 0562	0.6736 2493	19.3802 2483	13.0550 0266	0.0515 9899	0.0765 9899
17	1.5216 1826	0.6571 9506	20.8647 3045	13.7121 9772	0.0479 2777	0.0729 2777
18	1.5596 5872	0.6411 6591	22.3863 4871	14.3533 6363	0.0446 7008	0.0696 7008
19	1.5986 5019	0.6255 2772	23.9460 0743	14.9788 9134	0.0417 6062	0.0667 6062
20	1.6386 1644	0.6102 7094	25.5446 5761	15.5891 6229	0.0391 4713	0.0641 4713
21	1.6795 8185	0.5953 8629	27.1832 7405	16.1845 4857	0.0367 8733	0.0617 8733
22	1.7215 7140	0.5808 6467	28.8628 5590	16.7654 1324	0.0346 4661	0.0596 4661
23	1.7646 1068	0.5666 9724	30.5844 2730	17.3321 1048	0.0326 9638	0.0576 9638
24	1.8087 2595	0.5528 7535	32.3490 3798	17.8849 8583	0.0309 1282	0.0559 1282
25	1.8539 4410	0.5393 9059	34.1577 6393	18.4243 7642	0.0292 7592	0.0542 7592
26	1.9002 9270	0.5262 3472	36.0117 0803	18.9506 1114	0.0277 6875	0.0527 6875
27	1.9478 0002	0.5133 9973	37.9120 0073	19.4640 1087	0.0263 7687	0.0513 7687
28	1.9964 9502	0.5008 7778	39.8598 0075	19.9648 8866	0.0250 8793	0.0500 8793
29	2.0464 0739	0.4886 6125	41.8562 9577	20.4535 4991	0.0238 9127	0.0488 9127
30	2.0975 6758	0.4767 4269	43.9027 0316	20.9302 9259	0.0227 7764	0.0477 7764
31	2.1500 0677	0.4651 1481	46.0002 7074	21.3954 0741	0.0217 3900	0.0467 3900
32	2.2037 5694	0.4537 7055	48.1502 7751	21.8491 7796	0.0207 6831	0.0457 6831
33	2.2588 5086	0.4427 0298	50.3540 3445	22.2918 8094	0.0198 5938	0.0448 5938
34	2.3153 2213	0.4319 0534	52.6128 8531	22.7237 8628	0.0190 0675	0.0440 0675
35	2.3732 0519	0.4213 7107	54.9282 0744	23.1451 5734	0.0182 0558	0.0432 0558
36	2.4325 3532	0.4110 9372	57.3014 1263	23.5562 5107	0.0174 5158	0.0424 5158
37	2.4933 4870	0.4010 6705	59.7339 4794	23.9573 1812	0.0167 4090	0.0417 4090
38	2.5556 8242	0.3912 8492	62.2272 9664	24.3486 0304	0.0160 7012	0.0410 7012
39	2.6195 7448	0.3817 4139	64.7829 7906	24.7303 4443	0.0154 3615	0.0404 3615
40	2.6850 6384	0.3724 3062	67.4025 5354	25.1027 7505	0.0148 3623	0.0398 3623
41	2.7521 9043	0.3633 4695	70.0876 1737	25.4661 2200	0.0142 6786	0.0392 6786
42	2.8209 9520	0.3544 8483	72.8398 0781	25.8206 0683	0.0137 2876	0.0387 2876
43	2.8915 2008	0.3458 3886	75.6608 0300	26.1664 4569	0.0132 1688	0.0382 1688
44	2.9638 0808	0.3374 0376	78.5523 2308	26.5038 4945	0.0127 3037	0.0377 3037
45	3.0379 0328	0.3291 7440	81.5161 3116	26.8330 2386	0.0122 6751	0.0372 6751
46	3.1138 5086	0.3211 4576	84.5540 3443	27.1541 6962	0.0118 2676	0.0368 2676
47	3.1916 9713	0.3133 1294	87.6678 8530	27.4674 8255	0.0114 0669	0.0364 0669
48	3.2714 8956	0.3056 7116	90.8595 8243	27.7731 5371	0.0110 0599	0.0360 0599
49	3.3532 7680	0.2982 1576	94.1310 7199	28.0713 6947	0.0106 2348	0.0356 2348
50	3.4371 0872	0.2909 4221	97.4843 4879	28.3623 1168	0.0102 5806	0.0352 5806

$i = 2\frac{1}{2}\%$

n	$(1+i)^n$	$(1+i)^{-n}$	$s_{\overline{n}\rceil i}$	$a_{\overline{n}\rceil i}$	$\frac{1}{s_{\overline{n}\rceil i}}$	$\frac{1}{a_{\overline{n}\rceil i}}$
51	3.5230 3644	0.2838 4606	100.9214 5751	28.6461 5774	0.0099 0870	0.0349 0870
52	3.6111 1235	0.2769 2298	104.4444 9395	28.9230 8072	0.0095 7446	0.0345 7446
53	3.7013 9016	0.2701 6876	108.0556 0629	29.1932 4948	0.0092 5449	0.0342 5449
54	3.7939 2491	0.2635 7928	111.7569 9645	29.4568 2876	0.0089 4799	0.0339 4799
55	3.8887 7303	0.2571 5052	115.5509 2136	29.7139 7928	0.0086 5419	0.0336 5419
56	3.9859 9236	0.2508 7855	119.4396 9440	29.9648 5784	0.0083 7243	0.0333 7243
57	4.0856 4217	0.2447 5956	123.4256 8676	30.2096 1740	0.0081 0204	0.0331 0204
58	4.1877 8322	0.2387 8982	127.5113 2893	30.4484 0722	0.0078 4244	0.0328 4244
59	4.2924 7780	0.2329 6568	131.6991 1215	30.6813 7290	0.0075 9307	0.0325 9307
60	4.3997 8975	0.2272 8359	135.9915 8995	30.9086 5649	0.0073 5340	0.0323 5340
61	4.5097 8449	0.2217 4009	140.3913 7970	31.1303 9657	0.0071 2294	0.0321 2294
62	4.6225 2910	0.2163 3179	144.9011 6419	31.3467 2836	0.0069 0126	0.0319 0126
63	4.7380 9233	0.2110 5541	149.5236 9330	31.5577 8377	0.0066 8790	0.0316 8790
64	4.8565 4464	0.2059 0771	154.2617 8563	31.7636 9148	0.0064 8249	0.0314 8249
65	4.9779 5826	0.2008 8557	159.1183 3027	31.9645 7705	0.0062 8463	0.0312 8463
66	5.1024 0721	0.1959 8593	164.0962 8853	32.1605 6298	0.0060 9398	0.0310 9398
67	5.2299 6739	0.1912 0578	169.1986 9574	32.3517 6876	0.0059 1021	0.0309 1021
68	5.3607 1658	0.1865 4223	174.4286 6314	32.5383 1099	0.0057 3300	0.0307 3300
69	5.4947 3449	0.1819 9241	179.7893 7971	32.7203 0340	0.0055 6206	0.0305 6206
70	5.6321 0286	0.1775 5358	185.2841 1421	32.8978 5698	0.0053 9712	0.0303 9712
71	5.7729 0543	0.1732 2300	190.9162 1706	33.0710 7998	0.0052 3790	0.0302 3790
72	5.9172 2806	0.1689 9805	196.6891 2249	33.2400 7803	0.0050 8417	0.0300 8417
73	6.0651 5876	0.1648 7615	202.6063 5055	33.4049 5417	0.0049 3568	0.0299 3568
74	6.2167 8773	0.1608 5478	208.6715 0931	33.5658 0895	0.0047 9222	0.0297 9222
75	6.3722 0743	0.1569 3149	214.8882 9705	33.7227 4044	0.0046 5358	0.0296 5358
76	6.5315 1261	0.1531 0389	221.2605 0447	33.8758 4433	0.0045 1956	0.0295 1956
77	6.6948 0043	0.1493 6965	227.7920 1709	34.0252 1398	0.0043 8997	0.0293 8997
78	6.8621 7044	0.1457 2649	234.4868 1751	34.1709 4047	0.0042 6463	0.0292 6463
79	7.0337 2470	0.1421 7218	241.3489 8795	34.3131 1265	0.0041 4338	0.0291 4338
80	7.2095 6782	0.1387 0457	248.3827 1265	34.4518 1722	0.0040 2605	0.0290 2605
81	7.3898 0701	0.1353 2153	255.5922 8047	34.5871 3875	0.0039 1248	0.0289 1248
82	7.5745 5219	0.1320 2101	262.9820 8748	34.7191 5976	0.0038 0254	0.0288 0254
83	7.7639 1599	0.1288 0098	270.5566 3966	34.8479 6074	0.0036 9608	0.0286 9608
84	7.9580 1389	0.1256 5949	278.3205 5566	34.9736 2023	0.0035 9298	0.0285 9298
85	8.1569 6424	0.1225 9463	286.2785 6955	35.0962 1486	0.0034 9310	0.0284 9310
86	8.3608 8834	0.1196 0452	294.4355 3379	35.2158 1938	0.0033 9633	0.0283 9633
87	8.5699 1055	0.1166 8733	302.7964 2213	35.3325 0671	0.0033 0255	0.0283 0255
88	8.7841 5832	0.1138 4130	311.3663 3268	35.4463 4801	0.0032 1165	0.0282 1165
89	9.0037 6228	0.1110 6468	320.1504 9100	35.5574 1269	0.0031 2353	0.0281 2353
90	9.2288 5633	0.1083 5579	329.1542 5328	35.6657 6848	0.0030 3809	0.0280 3809
91	9.4595 7774	0.1057 1296	338.3831 0961	35.7714 8144	0.0029 5523	0.0279 5523
92	9.6960 6718	0.1031 3460	347.8426 8735	35.8746 1604	0.0028 7486	0.0278 7486
93	9.9384 6886	0.1006 1912	357.5387 5453	35.9752 3516	0.0027 9690	0.0277 9690
94	10.1869 3058	0.0981 6500	367.4772 2339	36.0734 0016	0.0027 2126	0.0277 2126
95	10.4416 0385	0.0957 7073	377.6641 5398	36.1691 7089	0.0026 4786	0.0276 4786
96	10.7026 4395	0.0934 3486	388.1057 5783	36.2626 0574	0.0025 7662	0.0275 7662
97	10.9702 1004	0.0911 5596	398.8084 0177	36.3537 6170	0.0025 0747	0.0275 0747
98	11.2444 6530	0.0889 3264	409.7786 1182	36.4426 9434	0.0024 4034	0.0274 4034
99	11.5255 7693	0.0867 6355	421.0230 7711	36.5294 5790	0.0023 7517	0.0273 7517
100	11.8137 1635	0.0846 4737	432.5486 5404	36.6141 0526	0.0023 1188	0.0273 1188

$i = 3\%$

n	$(1+i)^n$	$(1+i)^{-n}$	$s_{\overline{n}\rceil i}$	$a_{\overline{n}\rceil i}$	$\frac{1}{s_{\overline{n}\rceil i}}$	$\frac{1}{a_{\overline{n}\rceil i}}$
1	1.0300 0000	0.9708 7379	1.0000 0000	0.9708 7379	1.0000 0000	1.0300 0000
2	1.0609 0000	0.9425 9591	2.0300 0000	1.9134 6970	0.4926 1084	0.5226 1084
3	1.0927 2700	0.9151 4166	3.0909 0000	2.8286 1135	0.3235 3036	0.3535 3036
4	1.1255 0881	0.8884 8705	4.1836 2700	3.7170 9840	0.2390 2705	0.2690 2705
5	1.1592 7407	0.8626 0878	5.3091 3581	4.5797 0719	0.1883 5457	0.2183 5457
6	1.1940 5230	0.8374 8426	6.4684 0988	5.4171 9144	0.1545 9750	0.1845 9750
7	1.2298 7387	0.8130 9151	7.6624 6218	6.2302 8296	0.1305 0635	0.1605 0635
8	1.2667 7008	0.7894 0923	8.8923 3605	7.0196 9219	0.1124 5639	0.1424 5639
9	1.3047 7318	0.7664 1673	10.1591 0613	7.7861 0892	0.0984 3386	0.1284 3386
10	1.3439 1638	0.7440 9391	11.4638 7931	8.5302 0284	0.0872 3051	0.1172 3051
11	1.3842 3387	0.7224 2128	12.8077 9569	9.2526 2411	0.0780 7745	0.1080 7745
12	1.4257 6089	0.7013 7988	14.1920 2956	9.9540 0399	0.0704 6209	0.1004 6209
13	1.4685 3371	0.6809 5134	15.6177 9045	10.6349 5533	0.0640 2954	0.0940 2954
14	1.5125 8972	0.6611 1781	17.0863 2416	11.2960 7314	0.0585 2634	0.0885 2634
15	1.5579 6742	0.6418 6195	18.5989 1389	11.9379 3509	0.0537 6658	0.0837 6658
16	1.6047 0644	0.6231 6694	20.1568 8130	12.5611 0203	0.0496 1085	0.0796 1085
17	1.6528 4763	0.6050 1645	21.7615 8774	13.1661 1847	0.0459 5253	0.0759 5253
18	1.7024 3306	0.5873 9461	23.4144 3537	13.7535 1308	0.0427 0870	0.0727 0870
19	1.7535 0605	0.5702 8603	25.1168 6844	14.3237 9911	0.0398 1388	0.0698 1388
20	1.8061 1123	0.5536 7575	26.8703 7449	14.8774 7486	0.0372 1571	0.0672 1571
21	1.8602 9457	0.5375 4928	28.6764 8572	15.4150 2414	0.0348 7178	0.0648 7178
22	1.9161 0341	0.5218 9250	30.5367 8030	15.9369 1664	0.0327 4739	0.0627 4739
23	1.9735 8651	0.5066 9175	32.4528 8370	16.4436 0839	0.0308 1390	0.0608 1390
24	2.0327 9411	0.4919 3374	34.4264 7022	16.9355 4212	0.0290 4742	0.0590 4742
25	2.0937 7793	0.4776 0557	36.4592 6432	17.4131 4769	0.0274 2787	0.0574 2787
26	2.1565 9127	0.4636 9473	38.5530 4225	17.8768 4242	0.0259 3829	0.0559 3829
27	2.2212 8901	0.4501 8906	40.7096 3352	18.3270 3147	0.0245 6421	0.0545 6421
28	2.2879 2768	0.4370 7675	42.9309 2252	18.7641 0823	0.0232 9323	0.0532 9323
29	2.3565 6551	0.4243 4636	45.2188 5020	19.1884 5459	0.0221 1467	0.0521 1467
30	2.4272 6247	0.4119 8676	47.5754 1571	19.6004 4135	0.0210 1926	0.0510 1926
31	2.5000 8035	0.3999 8715	50.0026 7818	20.0004 2849	0.0199 9893	0.0499 9893
32	2.5750 8276	0.3883 3703	52.5027 5852	20.3887 6553	0.0190 4662	0.0490 4662
33	2.6523 3524	0.3770 2625	55.0778 4128	20.7657 9178	0.0181 5612	0.0481 5612
34	2.7319 0530	0.3660 4490	57.7301 7652	21.1318 3668	0.0173 2196	0.0473 2196
35	2.8138 6245	0.3553 8340	60.4620 8181	21.4872 2007	0.0165 3929	0.0465 3929
36	2.8982 7833	0.3450 3243	63.2759 4427	21.8322 5250	0.0158 0379	0.0458 0379
37	2.9852 2668	0.3349 8294	66.1742 2259	22.1672 3544	0.0151 1162	0.0451 1162
38	3.0747 8348	0.3252 2615	69.1594 4927	22.4924 6159	0.0144 5934	0.0444 5934
39	3.1670 2698	0.3157 5355	72.2342 3275	22.8082 1513	0.0138 4385	0.0438 4385
40	3.2620 3779	0.3065 5684	75.4012 5973	23.1147 7197	0.0132 6238	0.0432 6238
41	3.3598 9893	0.2976 2800	78.6632 9753	23.4123 9997	0.0127 1241	0.0427 1241
42	3.4606 9589	0.2889 5922	82.0231 9645	23.7013 5920	0.0121 9167	0.0421 9167
43	3.5645 1677	0.2805 4294	85.4838 9234	23.9819 0213	0.0116 9811	0.0416 9811
44	3.6714 5227	0.2723 7178	89.0484 0911	24.2542 7392	0.0112 2985	0.0412 2985
45	3.7815 9584	0.2644 3862	92.7198 6139	24.5187 1254	0.0107 8518	0.0407 8518
46	3.8950 4372	0.2567 3653	96.5014 5723	24.7754 4907	0.0103 6254	0.0403 6254
47	4.0118 9503	0.2492 5876	100.3965 0095	25.0247 0783	0.0099 6051	0.0399 6051
48	4.1322 5188	0.2419 9880	104.4083 9598	25.2667 0664	0.0095 7777	0.0395 7777
49	4.2562 1944	0.2349 5029	108.5406 4785	25.5016 5693	0.0092 1314	0.0392 1314
50	4.3839 0602	0.2281 0708	112.7968 6729	25.7297 6401	0.0088 6549	0.0388 6549

$i = 3\%$

n	$(1+i)^n$	$(1+i)^{-n}$	$s_{\overline{n}\rceil i}$	$a_{\overline{n}\rceil i}$	$\frac{1}{s_{\overline{n}\rceil i}}$	$\frac{1}{a_{\overline{n}\rceil i}}$
51	4.5154 2320	0.2214 6318	117.1807 7331	25.9512 2719	0.0085 3382	0.0385 3382
52	4.6508 8590	0.2150 1280	121.6961 9651	26.1662 3999	0.0082 1718	0.0382 1718
53	4.7904 1247	0.2087 5029	126.3470 8240	26.3749 9028	0.0079 1471	0.0379 1471
54	4.9341 2485	0.2026 7019	131.1374 9488	26.5776 6047	0.0076 2558	0.0376 2558
55	5.0821 4859	0.1967 6717	136.0716 1972	26.7744 2764	0.0073 4907	0.0373 4907
56	5.2346 1305	0.1910 3609	141.1537 6831	26.9654 6373	0.0070 8447	0.0370 8447
57	5.3916 5144	0.1854 7193	146.3883 8136	27.1509 3566	0.0068 3114	0.0368 3114
58	5.5534 0098	0.1800 6984	151.7800 3280	27.3310 0549	0.0065 8848	0.0365 8848
59	5.7200 0301	0.1748 2508	157.3334 3379	27.5058 3058	0.0063 5593	0.0363 5593
60	5.8916 0310	0.1697 3309	163.0534 3680	27.6755 6367	0.0061 3296	0.0361 3296
61	6.0683 5120	0.1647 8941	168.9450 3991	27.8403 5307	0.0059 1908	0.0359 1908
62	6.2504 0173	0.1599 8972	175.0133 9110	28.0003 4279	0.0057 1385	0.0357 1385
63	6.4379 1379	0.1553 2982	181.2637 9284	28.1556 7261	0.0055 1682	0.0355 1682
64	6.6310 5120	0.1508 0565	187.7017 0662	28.3064 7826	0.0053 2760	0.0353 2760
65	6.8299 8273	0.1464 1325	194.3327 5782	28.4528 9152	0.0051 4581	0.0351 4581
66	7.0348 8222	0.1421 4879	201.1627 4055	28.5950 4031	0.0049 7110	0.0349 7110
67	7.2459 2868	0.1380 0853	208.1976 2277	28.7330 4884	0.0048 0313	0.0348 0313
68	7.4633 0654	0.1339 8887	215.4435 5145	28.8670 3771	0.0046 4159	0.0346 4159
69	7.6872 0574	0.1300 8628	222.9068 5800	28.9971 2399	0.0044 8618	0.0344 8618
70	7.9178 2191	0.1262 9736	230.5940 6374	29.1234 2135	0.0043 3663	0.0343 3663
71	8.1553 5657	0.1226 1880	238.5118 8565	29.2460 4015	0.0041 9266	0.0341 9266
72	8.4000 1727	0.1190 4737	246.6672 4222	29.3650 8752	0.0040 5404	0.0340 5404
73	8.6520 1778	0.1155 7998	255.0672 5949	29.4806 6750	0.0039 2053	0.0339 2053
74	8.9115 7832	0.1122 1357	263.7192 7727	29.5928 8107	0.0037 9191	0.0337 9191
75	9.1789 2567	0.1089 4521	272.6308 5559	29.7018 2628	0.0036 6796	0.0336 6796
76	9.4542 9344	0.1057 7205	281.8097 8126	29.8075 9833	0.0035 4849	0.0335 4849
77	9.7379 2224	0.1026 9131	291.2640 7469	29.9102 8964	0.0034 3331	0.0334 3331
78	10.0300 5991	0.0997 0030	301.0019 9693	30.0099 8994	0.0033 2224	0.0333 2224
79	10.3309 6171	0.0967 9641	311.0320 5684	30.1067 8635	0.0032 1510	0.0332 1510
80	10.6408 9056	0.0939 7710	321.3630 1855	30.2007 6345	0.0031 1175	0.0331 1175
81	10.9601 1727	0.0912 3990	332.0039 0910	30.2920 0335	0.0030 1201	0.0330 1201
82	11.2889 2079	0.0885 8243	342.9640 2638	30.3805 8577	0.0029 1576	0.0329 1576
83	11.6275 8842	0.0860 0236	354.2529 4717	30.4665 8813	0.0028 2284	0.0328 2284
84	11.9764 1607	0.0834 9743	365.8805 3558	30.5500 8556	0.0027 3313	0.0327 3313
85	12.3357 0855	0.0810 6547	377.8569 5165	30.6311 5103	0.0026 4650	0.0326 4650
86	12.7057 7981	0.0787 0434	390.1926 6020	30.7098 5537	0.0025 6284	0.0325 6284
87	13.0869 5320	0.0764 1198	402.8984 4001	30.7862 6735	0.0024 8202	0.0324 8202
88	13.4795 6180	0.0741 8639	415.9853 9321	30.8604 5374	0.0024 0393	0.0324 0393
89	13.8839 4865	0.0720 2562	429.4649 5500	30.9324 7936	0.0023 2848	0.0323 2848
90	14.3004 6711	0.0699 2779	443.3489 0365	31.0024 0714	0.0022 5556	0.0322 5556
91	14.7294 8112	0.0678 9105	457.6493 7076	31.0702 9820	0.0021 8508	0.0321 8508
92	15.1713 6556	0.0659 1364	472.3788 5189	31.1362 1184	0.0021 1694	0.0321 1694
93	15.6265 0652	0.0639 9383	487.5502 1744	31.2002 0567	0.0020 5107	0.0320 5107
94	16.0953 0172	0.0621 2993	503.1767 2397	31.2623 3560	0.0019 8737	0.0319 8737
95	16.5781 6077	0.0603 2032	519.2720 2568	31.3226 5592	0.0019 2577	0.0319 2577
96	17.0755 0559	0.0585 6342	535.8501 8645	31.3812 1934	0.0018 6619	0.0318 6619
97	17.5877 7076	0.0568 5769	552.9256 9205	31.4380 7703	0.0018 0856	0.0318 0856
98	18.1154 0388	0.0552 0164	570.5134 6281	31.4932 7867	0.0017 5281	0.0317 5281
99	18.6588 6600	0.0535 9383	588.6288 6669	31.5468 7250	0.0016 9886	0.0316 9886
100	19.2186 3198	0.0520 3284	607.2877 3269	31.5989 0534	0.0016 4667	0.0316 4667

$i = 3\frac{1}{2}\%$

n	$(1+i)^n$	$(1+i)^{-n}$	$s_{\overline{n}\vert i}$	$a_{\overline{n}\vert i}$	$\frac{1}{s_{\overline{n}\vert i}}$	$\frac{1}{a_{\overline{n}\vert i}}$
1	1.0350 0000	0.9661 8357	1.0000 0000	0.9661 8357	1.0000 0000	1.0350 0000
2	1.0712 2500	0.9335 1070	2.0350 0000	1.8996 9428	0.4914 0049	0.5264 0049
3	1.1087 1788	0.9019 4271	3.1062 2500	2.8016 3698	0.3219 3418	0.3569 3418
4	1.1475 2300	0.8714 4223	4.2149 4288	3.6730 7921	0.2372 5114	0.2722 5114
5	1.1876 8631	0.8419 7317	5.3624 6588	4.5150 5238	0.1864 8137	0.2214 8137
6	1.2292 5533	0.8135 0064	6.5501 5218	5.3285 5302	0.1526 6821	0.1876 6821
7	1.2722 7926	0.7859 9096	7.7794 0751	6.1145 4398	0.1285 4449	0.1635 4449
8	1.3168 0904	0.7594 1156	9.0516 8677	6.8739 5554	0.1104 7665	0.1454 7665
9	1.3628 9735	0.7337 3097	10.3684 9581	7.6076 8651	0.0964 4601	0.1314 4601
10	1.4105 9876	0.7089 1881	11.7313 9316	8.3166 0532	0.0852 4137	0.1202 4137
11	1.4599 6972	0.6849 4571	13.1419 9192	9.0015 5104	0.0760 9197	0.1110 9197
12	1.5110 6866	0.6617 8330	14.6019 6164	9.6633 3433	0.0684 8395	0.1034 8395
13	1.5639 5606	0.6394 0415	16.1130 3030	10.3027 3849	0.0620 6157	0.0970 6157
14	1.6186 9452	0.6177 8179	17.6769 8636	10.9205 2028	0.0565 7073	0.0915 7073
15	1.6753 4883	0.5968 9062	19.2956 8088	11.5174 1090	0.0518 2507	0.0868 2507
16	1.7339 8604	0.5767 0591	20.9710 2971	12.0941 1681	0.0476 8483	0.0826 8483
17	1.7946 7555	0.5572 0378	22.7050 1575	12.6513 2059	0.0440 4313	0.0790 4313
18	1.8574 8920	0.5383 6114	24.4996 9130	13.1896 8173	0.0408 1684	0.0758 1684
19	1.9225 0132	0.5201 5569	26.3571 8050	13.7098 3742	0.0379 4033	0.0729 4033
20	1.9897 8886	0.5025 6588	28.2796 8181	14.2124 0330	0.0353 6108	0.0703 6108
21	2.0594 3147	0.4855 7090	30.2694 7068	14.6979 7420	0.0330 3659	0.0680 3659
22	2.1315 1158	0.4691 5063	32.3289 0215	15.1671 2484	0.0309 3207	0.0659 3207
23	2.2061 1448	0.4532 8563	34.4604 1373	15.6204 1047	0.0290 1880	0.0640 1880
24	2.2833 2849	0.4379 5713	36.6665 2821	16.0583 6760	0.0272 7283	0.0622 7283
25	2.3632 4498	0.4231 4699	38.9498 5669	16.4815 1459	0.0256 7404	0.0606 7404
26	2.4459 5856	0.4088 3767	41.3131 0168	16.8903 5226	0.0242 0540	0.0592 0540
27	2.5315 6711	0.3950 1224	43.7590 6024	17.2853 6451	0.0228 5241	0.0578 5241
28	2.6201 7196	0.3816 5434	46.2906 2734	17.6670 1885	0.0216 0265	0.0566 0265
29	2.7118 7798	0.3687 4815	48.9107 9930	18.0357 6700	0.0204 4538	0.0554 4538
30	2.8067 9370	0.3562 7841	51.6226 7728	18.3920 4541	0.0193 7133	0.0543 7133
31	2.9050 3148	0.3442 3035	54.4294 7098	18.7362 7576	0.0183 7240	0.0533 7240
32	3.0067 0759	0.3325 8971	57.3345 0247	19.0688 6547	0.0174 4150	0.0524 4150
33	3.1119 4235	0.3213 4271	60.3412 1005	19.3902 0818	0.0165 7242	0.0515 7242
34	3.2208 6033	0.3104 7605	63.4531 5240	19.7006 8423	0.0157 5966	0.0507 5966
35	3.3335 9045	0.2999 7686	66.6740 1274	20.0006 6110	0.0149 9835	0.0499 9835
36	3.4502 6611	0.2898 3272	70.0076 0318	20.2904 9381	0.0142 8416	0.0492 8416
37	3.5710 2543	0.2800 3161	73.4578 6930	20.5705 2542	0.0136 1325	0.0486 1325
38	3.6960 1132	0.2705 6194	77.0288 9472	20.8410 8736	0.0129 8214	0.0479 8214
39	3.8253 7171	0.2614 1250	80.7249 0604	21.1024 9987	0.0123 8775	0.0473 8775
40	3.9592 5972	0.2525 7247	84.5502 7775	21.3550 7234	0.0118 2728	0.0468 2728
41	4.0978 3381	0.2440 3137	88.5095 3747	21.5991 0371	0.0112 9822	0.0462 9822
42	4.2412 5799	0.2357 7910	92.6073 7128	21.8348 8281	0.0107 9828	0.0457 9828
43	4.3897 0202	0.2278 0590	96.8486 2928	22.0626 8870	0.0103 2539	0.0453 2539
44	4.5433 4160	0.2201 0231	101.2383 3130	22.2827 9102	0.0098 7768	0.0448 7768
45	4.7023 5855	0.2126 5924	105.7816 7290	22.4954 5026	0.0094 5343	0.0444 5343
46	4.8669 4110	0.2054 6787	110.4840 3145	22.7009 1813	0.0090 5108	0.0440 5108
47	5.0372 8404	0.1985 1968	115.3509 7255	22.8994 3780	0.0086 6919	0.0436 6919
48	5.2135 8898	0.1918 0645	120.3882 5659	23.0912 4425	0.0083 0646	0.0433 0646
49	5.3960 6459	0.1853 2024	125.6018 4557	23.2765 6450	0.0079 6167	0.0429 6167
50	5.5849 2686	0.1790 5337	130.9979 1016	23.4556 1787	0.0076 3371	0.0426 3371

$i = 3\frac{1}{2}\%$

n	$(1+i)^n$	$(1+i)^{-n}$	$s_{\overline{n}\rvert i}$	$a_{\overline{n}\rvert i}$	$\frac{1}{s_{\overline{n}\rvert i}}$	$\frac{1}{a_{\overline{n}\rvert i}}$
51	5.7803 9930	0.1729 9843	136.5828 3702	23.6286 1630	0.0073 2156	0.0423 2156
52	5.9827 1327	0.1671 4824	142.3632 3631	23.7957 6454	0.0070 2429	0.0420 2429
53	6.1921 0824	0.1614 9589	148.3459 4958	23.9572 6043	0.0067 4100	0.0417 4100
54	6.4088 3202	0.1560 3467	154.5380 5782	24.1132 9510	0.0064 7090	0.0414 7090
55	6.6331 4114	0.1507 5814	160.9468 8984	24.2640 5323	0.0062 1323	0.0412 1323
56	6.8653 0108	0.1456 6004	167.5800 3099	24.4097 1327	0.0059 6730	0.0409 6730
57	7.1055 8662	0.1407 3433	174.4453 3207	24.5504 4760	0.0057 3245	0.0407 3245
58	7.3542 8215	0.1359 7520	181.5509 1869	24.6864 2281	0.0055 0810	0.0405 0810
59	7.6116 8203	0.1313 7701	188.9052 0085	24.8177 9981	0.0052 9366	0.0402 9366
60	7.8780 9090	0.1269 3431	196.5168 8288	24.9447 3412	0.0050 8862	0.0400 8862
61	8.1538 2408	0.1226 4184	204.3949 7378	25.0673 7596	0.0048 9249	0.0398 9249
62	8.4392 0793	0.1184 9453	212.5487 9786	25.1858 7049	0.0047 0480	0.0397 0480
63	8.7345 8020	0.1144 8747	220.9880 0579	25.3003 5796	0.0045 2513	0.0395 2513
64	9.0402 9051	0.1106 1591	229.7225 8599	25.4109 7388	0.0043 5308	0.0393 5308
65	9.3567 0068	0.1068 7528	238.7628 7650	25.5178 4916	0.0041 8826	0.0391 8826
66	9.6841 8520	0.1032 6114	248.1195 7718	25.6211 1030	0.0040 3031	0.0390 3031
67	10.0231 3168	0.0997 6922	257.8037 6238	25.7208 7951	0.0038 7892	0.0388 7892
68	10.3739 4129	0.0963 9538	267.8268 9406	25.8172 7489	0.0037 3375	0.0387 3375
69	10.7370 2924	0.0931 3563	278.2008 3535	25.9104 1052	0.0035 9453	0.0385 9453
70	11.1128 2526	0.0899 8612	288.9378 6459	26.0003 9664	0.0034 6095	0.0384 6095
71	11.5017 7414	0.0869 4311	300.0506 8985	26.0873 3975	0.0033 3277	0.0383 3277
72	11.9043 3624	0.0840 0300	311.5524 6400	26.1713 4275	0.0032 0973	0.0382 0973
73	12.3209 8801	0.0811 6232	323.4568 0024	26.2525 0508	0.0030 9160	0.0380 9160
74	12.7522 2259	0.0784 1770	335.7777 8824	26.3309 2278	0.0029 7816	0.0379 7816
75	13.1985 5038	0.0757 6590	348.5300 1083	26.4066 8868	0.0028 6919	0.0378 6919
76	13.6604 9964	0.0732 0376	361.7285 6121	26.4798 9244	0.0027 6450	0.0377 6450
77	14.1386 1713	0.0707 2827	375.3890 6085	26.5506 2072	0.0026 6390	0.0376 6390
78	14.6334 6873	0.0683 3650	389.5276 7798	26.6189 5721	0.0025 6721	0.0375 6721
79	15.1456 4013	0.0660 2560	404.1611 4671	26.6849 8281	0.0024 7426	0.0374 7426
80	15.6757 3754	0.0637 9285	419.3067 8685	26.7487 7567	0.0023 8489	0.0373 8489
81	16.2243 8835	0.0616 3561	434.9825 2439	26.8104 1127	0.0022 9894	0.0372 9894
82	16.7922 4195	0.0595 5131	451.2069 1274	26.8699 6258	0.0022 1628	0.0372 1628
83	17.3799 7041	0.0575 3750	467.9991 5469	26.9275 0008	0.0021 3676	0.0371 3676
84	17.9882 6938	0.0555 9178	485.3791 2510	26.9830 9186	0.0020 6025	0.0370 6025
85	18.6178 5881	0.0537 1187	503.3673 9448	27.0368 0373	0.0019 8662	0.0369 8662
86	19.2694 8387	0.0518 9553	521.9852 5329	27.0886 9926	0.0019 1576	0.0369 1576
87	19.9439 1580	0.0501 4060	541.2547 3715	27.1388 3986	0.0018 4756	0.0368 4756
88	20.6419 5285	0.0484 4503	561.1986 5295	27.1872 8489	0.0017 8190	0.0367 8190
89	21.3644 2120	0.0468 0679	581.8406 0581	27.2340 9168	0.0017 1868	0.0367 1868
90	22.1121 7595	0.0452 2395	603.2050 2701	27.2793 1564	0.0016 5781	0.0366 5781
91	22.8861 0210	0.0436 9464	625.3172 0295	27.3230 1028	0.0015 9919	0.0365 9919
92	23.6871 1568	0.0422 1704	648.2033 0506	27.3652 2732	0.0015 4273	0.0365 4273
93	24.5161 6473	0.0407 8941	671.8904 2073	27.4060 1673	0.0014 8834	0.0364 8834
94	25.3742 3049	0.0394 1006	696.4065 8546	27.4454 2680	0.0014 3594	0.0364 3594
95	26.2623 2856	0.0380 7735	721.7808 1595	27.4835 0415	0.0013 8546	0.0363 8546
96	27.1815 1006	0.0367 8971	748.0431 4451	27.5202 9387	0.0013 3682	0.0363 3682
97	28.1328 6291	0.0355 4562	775.2246 5457	27.5558 3948	0.0012 8995	0.0362 8995
98	29.1175 1311	0.0343 4359	803.3575 1748	27.5901 8308	0.0012 4478	0.0362 4478
99	30.1366 2607	0.0331 8221	832.4750 3059	27.6233 6529	0.0012 0124	0.0362 0124
100	31.1914 0798	0.0320 6011	862.6116 5666	27.6554 2540	0.0011 5927	0.0361 5927

$i = 4\%$

n	$(1+i)^n$	$(1+i)^{-n}$	$s_{\overline{n}\rvert i}$	$a_{\overline{n}\rvert i}$	$\frac{1}{s_{\overline{n}\rvert i}}$	$\frac{1}{a_{\overline{n}\rvert i}}$
1	1.0400 0000	0.9615 3846	1.0000 0000	0.9615 3846	1.0000 0000	1.0400 0000
2	1.0816 0000	0.9245 5621	2.0400 0000	1.8860 9467	0.4901 9608	0.5301 9608
3	1.1248 6400	0.8889 9636	3.1216 0000	2.7750 9103	0.3203 4854	0.3603 4854
4	1.1698 5856	0.8548 0419	4.2464 6400	3.6298 9522	0.2354 9005	0.2754 9005
5	1.2166 5290	0.8219 2711	5.4163 2256	4.4518 2233	0.1846 2711	0.2246 2711
6	1.2653 1902	0.7903 1453	6.6329 7546	5.2421 3686	0.1507 6190	0.1907 6190
7	1.3159 3178	0.7599 1781	7.8982 9448	6.0020 5467	0.1266 0961	0.1666 0961
8	1.3685 6905	0.7306 9021	9.2142 2626	6.7327 4487	0.1085 2783	0.1485 2783
9	1.4233 1181	0.7025 8674	10.5827 9531	7.4353 3161	0.0944 9299	0.1344 9299
10	1.4802 4428	0.6755 6417	12.0061 0712	8.1108 9578	0.0832 9094	0.1232 9094
11	1.5394 5406	0.6495 8093	13.4863 5141	8.7604 7671	0.0741 4904	0.1141 4904
12	1.6010 3222	0.6245 9705	15.0258 0546	9.3850 7376	0.0665 5217	0.1065 5217
13	1.6650 7351	0.6005 7409	16.6268 3768	9.9856 4785	0.0601 4373	0.1001 4373
14	1.7316 7645	0.5774 7508	18.2919 1119	10.5631 2293	0.0546 6897	0.0946 6897
15	1.8009 4351	0.5552 6450	20.0235 8764	11.1183 8743	0.0499 4110	0.0899 4110
16	1.8729 8125	0.5339 0818	21.8245 3114	11.6522 9561	0.0458 2000	0.0858 2000
17	1.9479 0050	0.5133 7325	23.6975 1239	12.1656 6885	0.0421 9852	0.0821 9852
18	2.0258 1652	0.4936 2812	25.6454 1288	12.6592 9697	0.0389 9333	0.0789 9333
19	2.1068 4918	0.4746 4242	27.6712 2940	13.1339 3940	0.0361 3862	0.0761 3862
20	2.1911 2314	0.4563 8695	29.7780 7858	13.5903 2634	0.0335 8175	0.0735 8175
21	2.2787 6807	0.4388 3360	31.9692 0172	14.0291 5995	0.0312 8011	0.0712 8011
22	2.3699 1879	0.4219 5539	34.2479 6979	14.4511 1533	0.0291 9881	0.0691 9881
23	2.4647 1554	0.4057 2633	36.6178 8858	14.8568 4167	0.0273 0906	0.0673 0906
24	2.5633 0416	0.3901 2147	39.0826 0412	15.2469 6314	0.0255 8683	0.0655 8683
25	2.6658 3633	0.3751 1680	41.6459 0829	15.6220 7994	0.0240 1196	0.0640 1196
26	2.7724 6978	0.3606 8923	44.3117 4462	15.9827 6918	0.0225 6738	0.0625 6738
27	2.8833 6858	0.3468 1657	47.0842 1440	16.3295 8575	0.0212 3854	0.0612 3854
28	2.9987 0332	0.3334 7747	49.9675 8298	16.6630 6322	0.0200 1298	0.0600 1298
29	3.1186 5145	0.3206 5141	52.9662 8630	16.9837 1463	0.0188 7993	0.0588 7993
30	3.2433 9751	0.3083 1867	56.0849 3775	17.2920 3330	0.0178 3010	0.0578 3010
31	3.3731 3341	0.2964 6026	59.3283 3526	17.5884 9356	0.0168 5535	0.0568 5535
32	3.5080 5875	0.2850 5794	62.7014 6867	17.8735 5150	0.0159 4859	0.0559 4859
33	3.6483 8110	0.2740 9417	66.2095 2742	18.1476 4567	0.0151 0357	0.0551 0357
34	3.7943 1634	0.2635 5209	69.8579 0851	18.4111 9776	0.0143 1477	0.0543 1477
35	3.9460 8899	0.2534 1547	73.6522 2486	18.6646 1323	0.0135 7732	0.0535 7732
36	4.1039 3255	0.2436 6872	77.5983 1385	18.9082 8195	0.0128 8688	0.0528 8688
37	4.2680 8986	0.2342 9685	81.7022 4640	19.1425 7880	0.0122 3957	0.0522 3957
38	4.4388 1345	0.2252 8543	85.9703 3626	19.3678 6423	0.0116 3192	0.0516 3192
39	4.6163 6599	0.2166 2061	90.4091 4971	19.5844 8484	0.0110 6083	0.0510 6083
40	4.8010 2063	0.2082 8904	95.0255 1570	19.7927 7388	0.0105 2349	0.0505 2349
41	4.9930 6145	0.2002 7793	99.8265 3633	19.9930 5181	0.0100 1738	0.0500 1738
42	5.1927 8391	0.1925 7493	104.8195 9778	20.1856 2674	0.0095 4020	0.0495 4020
43	5.4004 9527	0.1851 6820	110.0123 8169	20.3707 9494	0.0090 8989	0.0490 8989
44	5.6165 1508	0.1780 4635	115.4128 7696	20.5488 4129	0.0086 6454	0.0486 6454
45	5.8411 7568	0.1711 9841	121.0293 9204	20.7200 3970	0.0082 6246	0.0482 6246
46	6.0748 2271	0.1646 1386	126.8705 6772	20.8846 5356	0.0078 8205	0.0478 8205
47	6.3178 1562	0.1582 8256	132.9453 9043	21.0429 3612	0.0075 2189	0.0475 2189
48	6.5705 2824	0.1521 9476	139.2632 0604	21.1951 3088	0.0071 8065	0.0471 8065
49	6.8333 4937	0.1463 4112	145.8337 3429	21.3414 7200	0.0068 5712	0.0468 5712
50	7.1066 8335	0.1407 1262	152.6670 8366	21.4821 8462	0.0065 5020	0.0465 5020

$i = 4\frac{1}{2}\%$

n	$(1+i)^n$	$(1+i)^{-n}$	$s_{\overline{n}\rvert i}$	$a_{\overline{n}\rvert i}$	$\frac{1}{s_{\overline{n}\rvert i}}$	$\frac{1}{a_{\overline{n}\rvert i}}$
1	1.0450 0000	0.9569 3780	1.0000 0000	0.9569 3780	1.0000 0000	1.0450 0000
2	1.0920 2500	0.9157 2995	2.0450 0000	1.8726 6775	0.4889 9756	0.5339 9756
3	1.1411 6613	0.8762 9660	3.1370 2500	2.7489 6435	0.3187 7336	0.3637 7336
4	1.1925 1860	0.8385 6134	4.2781 9113	3.5875 2570	0.2337 4365	0.2787 4365
5	1.2461 8194	0.8024 5105	5.4707 0973	4.3899 7674	0.1827 9164	0.2277 9164
6	1.3022 6012	0.7678 9574	6.7168 9166	5.1578 7248	0.1488 7839	0.1938 7839
7	1.3608 6183	0.7348 2846	8.0191 5179	5.8927 0094	0.1247 0147	0.1697 0147
8	1.4221 0061	0.7031 8513	9.3800 1362	6.5958 8607	0.1066 0965	0.1516 0965
9	1.4860 9514	0.6729 0443	10.8021 1423	7.2687 9050	0.0925 7447	0.1375 7447
10	1.5529 6942	0.6439 2768	12.2882 0937	7.9127 1818	0.0813 7882	0.1263 7882
11	1.6228 5305	0.6161 9874	13.8411 7879	8.5289 1692	0.0722 4818	0.1172 4818
12	1.6958 8143	0.5896 6386	15.4640 3184	9.1185 8078	0.0646 6619	0.1096 6619
13	1.7721 9610	0.5642 7164	17.1599 1327	9.6828 5242	0.0582 7535	0.1032 7535
14	1.8519 4492	0.5399 7286	18.9321 0937	10.2228 2528	0.0528 2032	0.0978 2032
15	1.9352 8244	0.5167 2044	20.7840 5429	10.7395 4573	0.0481 1381	0.0931 1381
16	2.0223 7015	0.4944 6932	22.7193 3673	11.2340 1505	0.0440 1537	0.0890 1537
17	2.1133 7681	0.4731 7639	24.7417 0689	11.7071 9143	0.0404 1758	0.0854 1758
18	2.2084 7877	0.4528 0037	26.8550 8370	12.1599 9180	0.0372 3690	0.0822 3690
19	2.3078 6031	0.4333 0179	29.0635 6246	12.5932 9359	0.0344 0734	0.0794 0734
20	2.4117 1402	0.4146 4286	31.3714 2277	13.0079 3645	0.0318 7614	0.0768 7614
21	2.5202 4116	0.3967 8743	33.7831 3680	13.4047 2388	0.0296 0057	0.0746 0057
22	2.6336 5201	0.3797 0089	36.3033 7795	13.7844 2476	0.0275 4565	0.0725 4565
23	2.7521 6635	0.3633 5013	38.9370 2996	14.1477 7489	0.0256 8249	0.0706 8249
24	2.8760 1383	0.3477 0347	41.6891 9631	14.4954 7837	0.0239 8703	0.0689 8703
25	3.0054 3446	0.3327 3060	44.5652 1015	14.8282 0896	0.0224 3903	0.0674 3903
26	3.1406 7901	0.3184 0248	47.5706 4460	15.1466 1145	0.0210 2137	0.0660 2137
27	3.2820 0956	0.3046 9137	50.7113 2361	15.4513 0282	0.0197 1946	0.0647 1946
28	3.4296 9999	0.2915 7069	53.9933 3317	15.7428 7351	0.0185 2081	0.0635 2081
29	3.5840 3649	0.2790 1502	57.4230 3316	16.0218 8853	0.0174 1461	0.0624 1461
30	3.7453 1813	0.2670 0002	61.0070 6966	16.2888 8854	0.0163 9154	0.0613 9154
31	3.9138 5745	0.2555 0241	64.7523 8779	16.5443 9095	0.0154 4345	0.0604 4345
32	4.0899 8104	0.2444 9991	68.6662 4524	16.7888 9086	0.0145 6320	0.0595 6320
33	4.2740 3018	0.2339 7121	72.7562 2628	17.0228 6207	0.0137 4453	0.0587 4453
34	4.4663 6154	0.2238 9589	77.0302 5646	17.2467 5796	0.0129 8191	0.0579 8191
35	4.6673 4781	0.2142 5444	81.4966 1800	17.4610 1240	0.0122 7045	0.0572 7045
36	4.8773 7846	0.2050 2817	86.1639 6581	17.6660 4058	0.0116 0578	0.0566 0578
37	5.0968 6049	0.1961 9921	91.0413 4427	17.8622 3979	0.0109 8402	0.0559 8402
38	5.3262 1921	0.1877 5044	96.1382 0476	18.0499 9023	0.0104 0169	0.0554 0169
39	5.5658 9908	0.1796 6549	101.4644 2398	18.2296 5572	0.0098 5567	0.0548 5567
40	5.8163 6454	0.1719 2870	107.0303 2306	18.4015 8442	0.0093 4315	0.0543 4315
41	6.0781 0094	0.1645 2507	112.8466 8760	18.5661 0949	0.0088 6158	0.0538 6158
42	6.3516 1548	0.1574 4026	118.9247 8854	18.7235 4975	0.0084 0868	0.0534 0868
43	6.6374 3818	0.1506 6054	125.2764 0402	18.8742 1029	0.0079 8235	0.0529 8235
44	6.9361 2290	0.1441 7276	131.9138 4220	19.0183 8305	0.0075 8071	0.0525 8071
45	7.2482 4843	0.1379 6437	138.8499 6510	19.1563 4742	0.0072 0202	0.0522 0202
46	7.5744 1961	0.1320 2332	146.0982 1353	19.2883 7074	0.0068 4471	0.0518 4471
47	7.9152 6849	0.1263 3810	153.6726 3314	19.4147 0884	0.0065 0734	0.0515 0734
48	8.2714 5557	0.1208 9771	161.5879 0163	19.5356 0654	0.0061 8858	0.0511 8858
49	8.6436 7107	0.1156 9158	169.8593 5720	19.6512 9813	0.0058 8722	0.0508 8722
50	9.0326 3627	0.1107 0965	178.5030 2828	19.7620 0778	0.0056 0215	0.0506 0215

$i = 5\%$

n	$(1+i)^n$	$(1+i)^{-n}$	$s_{\overline{n}\rceil i}$	$a_{\overline{n}\rceil i}$	$\frac{1}{s_{\overline{n}\rceil i}}$	$\frac{1}{a_{\overline{n}\rceil i}}$
1	1.0500 0000	0.9523 8095	1.0000 0000	0.9523 8095	1.0000 0000	1.0500 0000
2	1.1025 0000	0.9070 2948	2.0500 0000	1.8594 1043	0.4878 0488	0.5378 0488
3	1.1576 2500	0.8638 3760	3.1525 0000	2.7232 4803	0.3172 0856	0.3672 0856
4	1.2155 0625	0.8227 0247	4.3101 2500	3.5459 5050	0.2320 1183	0.2820 1183
5	1.2762 8156	0.7835 2617	5.5256 3125	4.3294 7667	0.1809 7480	0.2309 7480
6	1.3400 9564	0.7462 1540	6.8019 1281	5.0756 9207	0.1470 1747	0.1970 1747
7	1.4071 0042	0.7106 8133	8.1420 0845	5.7863 7340	0.1228 1982	0.1728 1982
8	1.4774 5544	0.6768 3936	9.5491 0888	6.4632 1276	0.1047 2181	0.1547 2181
9	1.5513 2822	0.6446 0892	11.0265 6432	7.1078 2168	0.0906 9008	0.1406 9008
10	1.6288 9463	0.6139 1325	12.5778 9254	7.7217 3493	0.0795 0457	0.1295 0457
11	1.7103 3936	0.5846 7929	14.2067 8716	8.3064 1422	0.0703 8889	0.1203 8889
12	1.7958 5633	0.5568 3742	15.9171 2652	8.8632 5164	0.0628 2541	0.1128 2541
13	1.8856 4914	0.5303 2135	17.7129 8285	9.3935 7299	0.0564 5577	0.1064 5577
14	1.9799 3160	0.5050 6795	19.5986 3199	9.8986 4094	0.0510 2397	0.1010 2397
15	2.0789 2818	0.4810 1710	21.5785 6359	10.3796 5804	0.0463 4229	0.0963 4229
16	2.1828 7459	0.4581 1152	23.6574 9177	10.8377 6956	0.0422 6991	0.0922 6991
17	2.2920 1832	0.4362 9669	25.8403 6636	11.2740 6625	0.0386 9914	0.0886 9914
18	2.4066 1923	0.4155 2065	28.1323 8467	11.6895 8690	0.0355 4622	0.0855 4622
19	2.5269 5020	0.3957 3396	30.5390 0391	12.0853 2086	0.0327 4501	0.0827 4501
20	2.6532 9771	0.3768 8948	33.0659 5410	12.4622 1034	0.0302 4259	0.0802 4259
21	2.7859 6259	0.3589 4236	35.7192 5181	12.8211 5271	0.0279 9611	0.0779 9611
22	2.9252 6072	0.3418 4987	38.5052 1440	13.1630 0258	0.0259 7051	0.0759 7051
23	3.0715 2376	0.3255 7131	41.4304 7512	13.4885 7388	0.0241 3682	0.0741 3682
24	3.2250 9994	0.3100 6791	44.5019 9887	13.7986 4179	0.0224 7090	0.0724 7090
25	3.3863 5494	0.2953 0277	47.7270 9882	14.0939 4457	0.0209 5246	0.0709 5246
26	3.5556 7269	0.2812 4073	51.1134 5376	14.3751 8530	0.0195 6432	0.0695 6432
27	3.7334 5632	0.2678 4832	54.6691 2645	14.6430 3362	0.0182 9186	0.0682 9186
28	3.9201 2914	0.2550 9364	58.4025 8277	14.8981 2726	0.0171 2253	0.0671 2253
29	4.1161 3560	0.2429 4632	62.3227 1191	15.1410 7358	0.0160 4551	0.0660 4551
30	4.3219 4238	0.2313 7745	66.4388 4750	15.3724 5103	0.0150 5144	0.0650 5144
31	4.5380 3949	0.2203 5947	70.7607 8988	15.5928 1050	0.0141 3212	0.0641 3212
32	4.7649 4147	0.2098 6617	75.2988 2937	15.8026 7667	0.0132 8042	0.0632 8042
33	5.0031 8854	0.1998 7254	80.0637 7084	16.0025 4921	0.0124 9004	0.0624 9004
34	5.2533 4797	0.1903 5480	85.0669 5938	16.1929 0401	0.0117 5545	0.0617 5545
35	5.5160 1537	0.1812 9029	90.3203 0735	16.3741 9429	0.0110 7171	0.0610 7171
36	5.7918 1614	0.1726 5741	95.8363 2272	16.5468 5171	0.0104 3446	0.0604 3446
37	6.0814 0694	0.1644 3563	101.6281 3886	16.7112 8734	0.0098 3979	0.0598 3979
38	6.3854 7729	0.1566 0536	107.7095 4580	16.8678 9271	0.0092 8423	0.0592 8423
39	6.7047 5115	0.1491 4797	114.0950 2309	17.0170 4067	0.0087 6462	0.0587 6462
40	7.0399 8871	0.1420 4568	120.7997 7424	17.1590 8635	0.0082 7816	0.0582 7816
41	7.3919 8815	0.1352 8160	127.8397 6295	17.2943 6796	0.0078 2229	0.0578 2229
42	7.7615 8756	0.1288 3962	135.2317 5110	17.4232 0758	0.0073 9471	0.0573 9471
43	8.1496 6693	0.1227 0440	142.9933 3866	17.5459 1198	0.0069 9333	0.0569 9333
44	8.5571 5028	0.1168 6133	151.1430 0559	17.6627 7331	0.0066 1625	0.0566 1625
45	8.9850 0779	0.1112 9651	159.7001 5587	17.7740 6982	0.0062 6173	0.0562 6173
46	9.4342 5818	0.1059 9668	168.6851 6366	17.8800 6650	0.0059 2820	0.0559 2820
47	9.9059 7109	0.1009 4921	178.1194 2185	17.9810 1571	0.0056 1421	0.0556 1421
48	10.4012 6965	0.0961 4211	188.0253 9294	18.0771 5782	0.0053 1843	0.0553 1843
49	10.9213 3313	0.0915 6391	198.4266 6259	18.1687 2173	0.0050 3965	0.0550 3965
50	11.4673 9979	0.0872 0373	209.3479 9572	18.2559 2546	0.0047 7674	0.0547 7674

$i = 6\%$

n	$(1+i)^n$	$(1+i)^{-n}$	$s_{\overline{n}\rvert i}$	$a_{\overline{n}\rvert i}$	$\frac{1}{s_{\overline{n}\rvert i}}$	$\frac{1}{a_{\overline{n}\rvert i}}$
1	1.0600 0000	0.9433 9623	1.0000 0000	0.9433 9623	1.0000 0000	1.0600 0000
2	1.1236 0000	0.8899 9644	2.0600 0000	1.8333 9267	0.4854 3689	0.5454 3689
3	1.1910 1600	0.8396 1928	3.1836 0000	2.6730 1195	0.3141 0981	0.3741 0981
4	1.2624 7696	0.7920 9366	4.3746 1600	3.4651 0561	0.2285 9149	0.2885 9149
5	1.3382 2558	0.7472 5817	5.6370 9296	4.2123 6379	0.1773 9640	0.2373 9640
6	1.4185 1911	0.7049 6054	6.9753 1854	4.9173 2433	0.1433 6263	0.2033 6263
7	1.5036 3026	0.6650 5711	8.3938 3765	5.5823 8144	0.1191 3502	0.1791 3502
8	1.5938 4807	0.6274 1237	9.8974 6791	6.2097 9381	0.1010 3594	0.1610 3594
9	1.6894 7896	0.5918 9846	11.4913 1598	6.8016 9227	0.0870 2224	0.1470 2224
10	1.7908 4770	0.5583 9478	13.1807 9494	7.3600 8705	0.0758 6796	0.1358 6796
11	1.8982 9856	0.5267 8753	14.9716 4264	7.8868 7458	0.0667 9294	0.1267 9294
12	2.0121 9647	0.4969 6936	16.8699 4120	8.3838 4394	0.0592 7703	0.1192 7703
13	2.1329 2826	0.4688 3902	18.8821 3767	8.8526 8296	0.0529 6011	0.1129 6011
14	2.2609 0396	0.4423 0096	21.0150 6593	9.2949 8393	0.0475 8491	0.1075 8491
15	2.3965 5819	0.4172 6506	23.2759 6988	9.7122 4899	0.0429 6276	0.1029 6276
16	2.5403 5168	0.3936 4628	25.6725 2808	10.1058 9527	0.0389 5214	0.0989 5214
17	2.6927 7279	0.3713 6442	28.2128 7976	10.4772 5969	0.0354 4480	0.0954 4480
18	2.8543 3915	0.3503 4379	30.9056 5255	10.8276 0348	0.0323 5654	0.0923 5654
19	3.0255 9950	0.3305 1301	33.7599 9170	11.1581 1649	0.0296 2086	0.0896 2086
20	3.2071 3547	0.3118 0473	36.7855 9120	11.4699 2122	0.0271 8456	0.0871 8456
21	3.3995 6360	0.2941 5540	39.9927 2668	11.7640 7662	0.0250 0455	0.0850 0455
22	3.6035 3742	0.2775 0510	43.3922 9028	12.0415 8172	0.0230 4557	0.0830 4557
23	3.8197 4966	0.2617 9726	46.9958 2769	12.3033 7898	0.0212 7848	0.0812 7848
24	4.0489 3464	0.2469 7855	50.8155 7735	12.5503 5753	0.0196 7900	0.0796 7900
25	4.2918 7072	0.2329 9863	54.8645 1200	12.7833 5616	0.0182 2672	0.0782 2672
26	4.5493 8296	0.2198 1003	59.1563 8272	13.0031 6619	0.0169 0435	0.0769 0435
27	4.8223 4594	0.2073 6795	63.7057 6568	13.2105 3414	0.0156 9717	0.0756 9717
28	5.1116 8670	0.1956 3014	68.5281 1162	13.4061 6428	0.0145 9255	0.0745 9255
29	5.4183 8790	0.1845 5674	73.6397 9832	13.5907 2102	0.0135 7961	0.0735 7961
30	5.7434 9117	0.1741 1013	79.0581 8622	13.7648 3115	0.0126 4891	0.0726 4891
31	6.0881 0064	0.1642 5484	84.8016 7739	13.9290 8599	0.0117 9222	0.0717 9222
32	6.4533 8668	0.1549 5740	90.8897 7803	14.0840 4339	0.0110 0234	0.0710 0234
33	6.8405 8988	0.1461 8622	97.3431 6471	14.2302 2961	0.0102 7293	0.0702 7293
34	7.2510 2528	0.1379 1153	104.1837 5460	14.3681 4114	0.0095 9843	0.0695 9843
35	7.6860 8679	0.1301 0522	111.4347 7987	14.4982 4636	0.0089 7386	0.0689 7386
36	8.1472 5200	0.1227 4077	119.1208 6666	14.6209 8713	0.0083 9483	0.0683 9483
37	8.6360 8712	0.1157 9318	127.2681 1866	14.7367 8031	0.0078 5743	0.0678 5743
38	9.1542 5235	0.1092 3885	135.9042 0578	14.8460 1916	0.0073 5812	0.0673 5812
39	9.7035 0749	0.1030 5552	145.0584 5813	14.9490 7468	0.0068 9377	0.0668 9377
40	10.2857 1794	0.0972 2219	154.7619 6562	15.0462 9687	0.0064 6154	0.0664 6154
41	10.9028 6101	0.0917 1905	165.0476 8356	15.1380 1592	0.0060 5886	0.0660 5886
42	11.5570 3267	0.0865 2740	175.9505 4457	15.2245 4332	0.0056 8342	0.0656 8342
43	12.2504 5463	0.0816 2962	187.5075 7724	15.3061 7294	0.0053 3312	0.0653 3312
44	12.9854 8191	0.0770 0908	199.7580 3188	15.3831 8202	0.0050 0606	0.0650 0606
45	13.7646 1083	0.0726 5007	212.7435 1379	15.4558 3209	0.0047 0050	0.0647 0050
46	14.5904 8748	0.0685 3781	226.5081 2462	15.5243 6990	0.0044 1485	0.0644 1485
47	15.4659 1673	0.0646 5831	241.0986 1210	15.5890 2821	0.0041 4768	0.0641 4768
48	16.3938 7173	0.0609 9840	256.5645 2882	15.6500 2661	0.0038 9765	0.0638 9765
49	17.3775 0403	0.0575 4566	272.9584 0055	15.7075 7227	0.0036 6356	0.0636 6356
50	18.4201 5427	0.0542 8836	290.3359 0458	15.7618 6064	0.0034 4429	0.0634 4429

$i = 7\%$

n	$(1+i)^n$	$(1+i)^{-n}$	$s_{\overline{n}\rceil i}$	$a_{\overline{n}\rceil i}$	$\frac{1}{s_{\overline{n}\rceil i}}$	$\frac{1}{a_{\overline{n}\rceil i}}$
1	1.0700 0000	0.9345 7944	1.0000 0000	0.9345 7944	1.0000 0000	1.0700 0000
2	1.1449 0000	0.8734 3873	2.0700 0000	1.8080 1817	0.4830 9179	0.5530 9179
3	1.2250 4300	0.8162 9788	3.2149 0000	2.6243 1604	0.3110 5167	0.3810 5167
4	1.3107 9601	0.7628 9521	4.4399 4300	3.3872 1126	0.2252 2812	0.2952 2812
5	1.4025 5173	0.7129 8618	5.7507 3901	4.1001 9744	0.1738 9069	0.2438 9069
6	1.5007 3035	0.6663 4222	7.1532 9074	4.7665 3966	0.1397 9580	0.2097 9580
7	1.6057 8148	0.6227 4974	8.6540 2109	5.3892 8940	0.1155 5322	0.1855 5322
8	1.7181 8618	0.5820 0910	10.2598 0257	5.9712 9851	0.0974 6776	0.1674 6776
9	1.8384 5921	0.5439 3374	11.9779 8875	6.5152 3225	0.0834 8647	0.1534 8647
10	1.9671 5136	0.5083 4929	13.8164 4796	7.0235 8154	0.0723 7750	0.1423 7750
11	2.1048 5195	0.4750 9280	15.7835 9932	7.4986 7434	0.0633 5690	0.1333 5690
12	2.2521 9159	0.4440 1196	17.8884 5127	7.9426 8630	0.0559 0199	0.1259 0199
13	2.4098 4500	0.4149 6445	20.1406 4286	8.3576 5074	0.0496 5085	0.1196 5085
14	2.5785 3415	0.3878 1724	22.5504 8786	8.7454 6799	0.0443 4494	0.1143 4494
15	2.7590 3154	0.3624 4602	25.1290 2201	9.1079 1401	0.0397 9462	0.1097 9462
16	2.9521 6375	0.3387 3460	27.8880 5355	9.4466 4860	0.0358 5765	0.1058 5765
17	3.1588 1521	0.3165 7439	30.8402 1730	9.7632 2299	0.0324 2519	0.1024 2519
18	3.3799 3228	0.2958 6392	33.9990 3251	10.0590 8691	0.0294 1260	0.0994 1260
19	3.6165 2754	0.2765 0833	37.3789 6479	10.3355 9524	0.0267 5301	0.0967 5301
20	3.8696 8446	0.2584 1900	40.9954 9232	10.5940 1425	0.0243 9293	0.0943 9293
21	4.1405 6237	0.2415 1309	44.8651 7678	10.8355 2733	0.0222 8900	0.0922 8900
22	4.4304 0174	0.2257 1317	49.0057 3916	11.0612 4050	0.0204 0577	0.0904 0577
23	4.7405 2986	0.2109 4688	53.4361 4090	11.2721 8738	0.0187 1393	0.0887 1393
24	5.0723 6695	0.1971 4662	58.1766 7076	11.4693 3400	0.0171 8902	0.0871 8902
25	5.4274 3264	0.1842 4918	63.2490 3772	11.6535 8318	0.0158 1052	0.0858 1052
26	5.8073 5292	0.1721 9549	68.6764 7036	11.8257 7867	0.0145 6103	0.0845 6103
27	6.2138 6763	0.1609 3037	74.4838 2328	11.9867 0904	0.0134 2573	0.0834 2573
28	6.6488 3836	0.1504 0221	80.6976 9091	12.1371 1125	0.0123 9193	0.0823 9193
29	7.1142 5705	0.1405 6282	87.3465 2927	12.2776 7407	0.0114 4865	0.0814 4865
30	7.6122 5504	0.1313 6712	94.4607 8632	12.4090 4118	0.0105 8640	0.0805 8640
31	8.1451 1290	0.1227 7301	102.0730 4137	12.5318 1419	0.0097 9691	0.0797 9691
32	8.7152 7080	0.1147 4113	110.2181 5426	12.6465 5532	0.0090 7292	0.0790 7292
33	9.3253 3975	0.1072 3470	118.9334 2506	12.7537 9002	0.0084 0807	0.0784 0807
34	9.9781 1354	0.1002 1934	128.2587 6481	12.8540 0936	0.0077 9674	0.0777 9674
35	10.6765 8148	0.0936 6294	138.2368 7835	12.9476 7230	0.0072 3396	0.0772 3396
36	11.4239 4219	0.0875 3546	148.9134 5984	13.0352 0776	0.0067 1531	0.0767 1531
37	12.2236 1814	0.0818 0884	160.3374 0202	13.1170 1660	0.0062 3685	0.0762 3685
38	13.0792 7141	0.0764 5686	172.5610 2017	13.1934 7345	0.0057 9505	0.0757 9505
39	13.9948 2041	0.0714 5501	185.6402 9158	13.2649 2846	0.0053 8676	0.0753 8676
40	14.9744 5784	0.0667 8038	199.6351 1199	13.3317 0884	0.0050 0914	0.0750 0914
41	16.0226 6989	0.0624 1157	214.6095 6983	13.3941 2041	0.0046 5962	0.0746 5962
42	17.1442 5678	0.0583 2857	230.6322 3972	13.4524 4898	0.0043 3591	0.0743 3591
43	18.3443 5475	0.0545 1268	247.7764 9650	13.5069 6167	0.0040 3590	0.0740 3590
44	19.6284 5959	0.0509 4643	266.1208 5125	13.5579 0810	0.0037 5769	0.0737 5769
45	21.0024 5176	0.0476 1349	285.7493 1084	13.6055 2159	0.0034 9957	0.0734 9957
46	22.4726 2338	0.0444 9859	306.7517 6260	13.6500 2018	0.0032 5996	0.0732 5996
47	24.0457 0702	0.0415 8747	329.2243 8598	13.6916 0764	0.0030 3744	0.0730 3744
48	25.7289 0651	0.0388 6679	353.2700 9300	13.7304 7443	0.0028 3070	0.0728 3070
49	27.5299 2997	0.0363 2410	378.9989 9951	13.7667 9853	0.0026 3853	0.0726 3853
50	29.4570 2506	0.0339 4776	406.5289 2947	13.8007 4629	0.0024 5985	0.0724 5985

$i = 8\%$

n	$(1+i)^n$	$(1+i)^{-n}$	$s_{\overline{n}\rvert i}$	$a_{\overline{n}\rvert i}$	$\frac{1}{s_{\overline{n}\rvert i}}$	$\frac{1}{a_{\overline{n}\rvert i}}$
1	1.0800 0000	0.9259 2593	1.0000 0000	0.9259 2593	1.0000 0000	1.0800 0000
2	1.1664 0000	0.8573 3882	2.0800 0000	1.7832 6475	0.4807 6923	0.5607 6923
3	1.2597 1200	0.7938 3224	3.2464 0000	2.5770 9699	0.3080 3351	0.3880 3351
4	1.3604 8896	0.7350 2985	4.5061 1200	3.3121 2684	0.2219 2080	0.3019 2080
5	1.4693 2808	0.6805 8320	5.8666 0096	3.9927 1004	0.1704 5645	0.2504 5645
6	1.5868 7432	0.6301 6963	7.3359 2904	4.6228 7966	0.1363 1539	0.2163 1539
7	1.7138 2427	0.5834 9040	8.9228 0336	5.2063 7006	0.1120 7240	0.1920 7240
8	1.8509 3021	0.5402 6888	10.6366 2763	5.7466 3894	0.0940 1476	0.1740 1476
9	1.9990 0463	0.5002 4897	12.4875 5784	6.2468 8791	0.0800 7971	0.1600 7971
10	2.1589 2500	0.4631 9349	14.4865 6247	6.7100 8140	0.0690 2949	0.1490 2949
11	2.3316 3900	0.4288 8286	16.6454 8746	7.1389 6426	0.0600 7634	0.1400 7634
12	2.5181 7012	0.3971 1376	18.9771 2646	7.5360 7802	0.0526 9502	0.1326 9502
13	2.7196 2373	0.3676 9792	21.4952 9658	7.9037 7594	0.0465 2181	0.1265 2181
14	2.9371 9362	0.3404 6104	24.2149 2030	8.2442 3698	0.0412 9685	0.1212 9685
15	3.1721 6911	0.3152 4170	27.1521 1393	8.5594 7869	0.0368 2954	0.1168 2954
16	3.4259 4264	0.2918 9047	30.3242 8304	8.8513 6916	0.0329 7687	0.1129 7687
17	3.7000 1805	0.2702 6895	33.7502 2569	9.1216 3811	0.0296 2943	0.1096 2943
18	3.9960 1950	0.2502 4903	37.4502 4374	9.3718 8714	0.0267 0210	0.1067 0210
19	4.3157 0106	0.2317 1206	41.4462 6324	9.6035 9920	0.0241 2763	0.1041 2763
20	4.6609 5714	0.2145 4821	45.7619 6430	9.8181 4741	0.0218 5221	0.1018 5221
21	5.0338 3372	0.1986 5575	50.4229 2144	10.0168 0316	0.0198 3225	0.0998 3225
22	5.4365 4041	0.1839 4051	55.4567 5516	10.2007 4366	0.0180 3207	0.0980 3207
23	5.8714 6365	0.1703 1528	60.8932 9557	10.3710 5895	0.0164 2217	0.0964 2217
24	6.3411 8074	0.1576 9934	66.7647 5922	10.5287 5828	0.0149 7796	0.0949 7796
25	6.8484 7520	0.1460 1790	73.1059 3995	10.6747 7619	0.0136 7878	0.0936 7878
26	7.3963 5321	0.1352 0176	79.9544 1515	10.8099 7795	0.0125 0713	0.0925 0713
27	7.9880 6147	0.1251 8682	87.3507 6836	10.9351 6477	0.0114 4810	0.0914 4810
28	8.6271 0639	0.1159 1372	95.3388 2983	11.0510 7849	0.0104 8891	0.0904 8891
29	9.3172 7490	0.1073 2752	103.9659 3622	11.1584 0601	0.0096 1854	0.0896 1854
30	10.0626 5689	0.0993 7733	113.2832 1111	11.2577 8334	0.0088 2743	0.0888 2743
31	10.8676 6944	0.0920 1605	123.3458 6800	11.3497 9939	0.0081 0728	0.0881 0728
32	11.7370 8300	0.0852 0005	134.2135 3744	11.4349 9944	0.0074 5081	0.0874 5081
33	12.6760 4964	0.0788 8893	145.9506 2044	11.5138 8837	0.0068 5163	0.0868 5163
34	13.6901 3361	0.0730 4531	158.6266 7007	11.5869 3367	0.0063 0411	0.0863 0411
35	14.7853 4429	0.0676 3454	172.3168 0368	11.6545 6822	0.0058 0326	0.0858 0326
36	15.9681 7184	0.0626 2458	187.1021 4797	11.7171 9279	0.0053 4467	0.0853 4467
37	17.2456 2558	0.0579 8572	203.0703 1981	11.7751 7851	0.0049 2440	0.0849 2440
38	18.6252 7563	0.0536 9048	220.3159 4540	11.8288 6899	0.0045 3894	0.0845 3894
39	20.1152 9768	0.0497 1341	238.9412 2103	11.8785 8240	0.0041 8513	0.0841 8513
40	21.7245 2150	0.0460 3093	259.0565 1871	11.9246 1333	0.0038 6016	0.0838 6016
41	23.4624 8322	0.0426 2123	280.7810 4021	11.9672 3457	0.0035 6149	0.0835 6149
42	25.3394 8187	0.0394 6411	304.2435 2342	12.0066 9867	0.0032 8684	0.0832 8684
43	27.3666 4042	0.0365 4084	329.5830 0530	12.0432 3951	0.0030 3414	0.0830 3414
44	29.5559 7166	0.0338 3411	356.9496 4572	12.0770 7362	0.0028 0152	0.0828 0152
45	31.9204 4939	0.0313 2788	386.5056 1738	12.1084 0150	0.0025 8728	0.0825 8728
46	34.4740 8534	0.0290 0730	418.4260 6677	12.1374 0880	0.0023 8991	0.0823 8991
47	37.2320 1217	0.0268 5861	452.9001 5211	12.1642 6741	0.0022 0799	0.0822 0799
48	40.2105 7314	0.0248 6908	490.1321 6428	12.1891 3649	0.0020 4027	0.0820 4027
49	43.4274 1899	0.0230 2693	530.3427 3742	12.2121 6341	0.0018 8557	0.0818 8557
50	46.9016 1251	0.0213 2123	573.7701 5642	12.2334 8464	0.0017 4286	0.0817 4286

Table VIII

Table of Integrals.

1. $\int x^n\,dx = \frac{x^{n+1}}{n+1} + C \qquad n \neq -1$

2. $\int (ax+b)^n\,dx = \frac{(ax+b)^{n+1}}{a(n+1)} + C \qquad n \neq -1$

3. $\int \frac{1}{(ax+b)}\,dx = \frac{1}{a}\ln[ax+b] + C$

4. $\int \frac{x}{ax+b}\,dx = \frac{x}{a} - \frac{b}{a^2}\ln(ax+b) + C$

5. $\int \frac{x}{(ax+b)^2}\,dx = \frac{1}{a^2}\left[\ln(ax+b) + \frac{b}{ax+b}\right] + C$

6. $\int \frac{dx}{x(ax+b)} = \frac{1}{b}\ln\left(\frac{x}{ax+b}\right) + C$

7. $\int \frac{dx}{x(ax+b)^2} = \frac{1}{b(ax+b)} + \frac{1}{b^2}\ln\left(\frac{x}{ax+b}\right) + C$

8. $\int \frac{1}{\sqrt{x^2+a^2}}\,dx = \ln(x + \sqrt{x^2+a^2}\,) + C$

9. $\int \frac{1}{\sqrt{x^2-a^2}}\,dx = \ln(x + \sqrt{x^2-a^2}\,) + C$

10. $\int \frac{1}{x^2-a^2}\,dx = \frac{1}{2a}\ln\left(\frac{x-a}{x+a}\right) + C \qquad (\text{if } x^2 > a^2)$

11. $\int \frac{1}{a^2-x^2}\,dx = \frac{1}{2a}\ln\left(\frac{a+x}{a-x}\right) + C \qquad (\text{if } a^2 > x^2)$

12. $\int \frac{1}{x\sqrt{a^2+x^2}}\,dx = -\frac{1}{a}\ln\left(\frac{a+\sqrt{a^2+x^2}}{x}\right) + C$

13. $\int \frac{1}{x\sqrt{a^2-x^2}}\,dx = -\frac{1}{a}\left(\frac{a+\sqrt{a^2-x^2}}{x}\right) + C$

14. $\int e^{ax}\,dx = \frac{1}{a}e^{ax} + C$

15. $\int xe^{ax}\,dx = \frac{e^{ax}}{a^2}(ax-1) + C$

ANSWERS TO ODD-NUMBERED EXERCISES

CHAPTER 1

Exercise Set 1.1, page 5

1. (a) false (b) true (c) true
(d) false (e) true (f) false

3. (a) true (b) false (c) true (d) false

5. (a) $\{A, R, D, V, K\}$ (b) $\{M, I, S, P\}$ (c) $\{T, A, B, L, E\}$

7. (a) $\{x \mid x \text{ is a U.S. citizen}\}$
(b) $\{x \mid x \text{ is a U.S. citizen over 40 years of age}\}$

9. (b)

11. $\varnothing$, $\{2\}$, $\{5\}$, $\{2, 5\}$

13. (a) $\varnothing$, $\{a_1\}$, $\{a_2\}$, $\{a_3\}$, $\{a_1, a_2\}$, $\{a_1, a_3\}$, $\{a_2, a_3\}$, $\{a_1, a_2, a_3\}$
(b) $\varnothing$

15. (a) true (b) false (c) false (d) false

17. (a) $\{1, 2, 3, 4, 5, 6, 7, 9\}$ (b) $\{a, b, c\}$

19. no; yes

Exercise Set 1.2, page 15

1. (a) $\{3, 7\}$ (b) $\{1, 3\}$ (c) $\{2, 3\}$
(d) $\varnothing$ (e) $\{3\}$ (f) $\varnothing$

3. (a) (b)

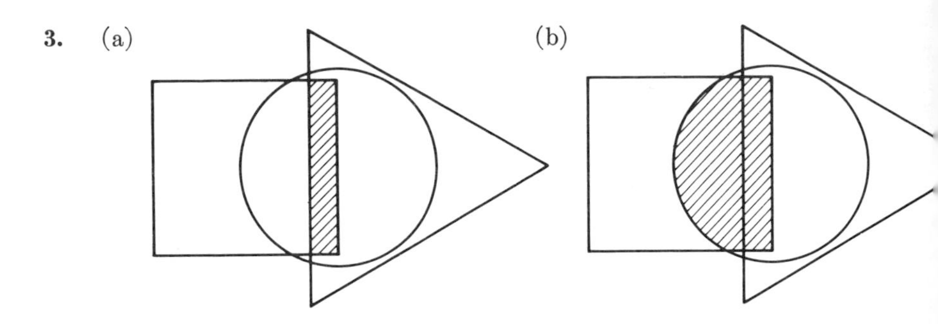

(c)

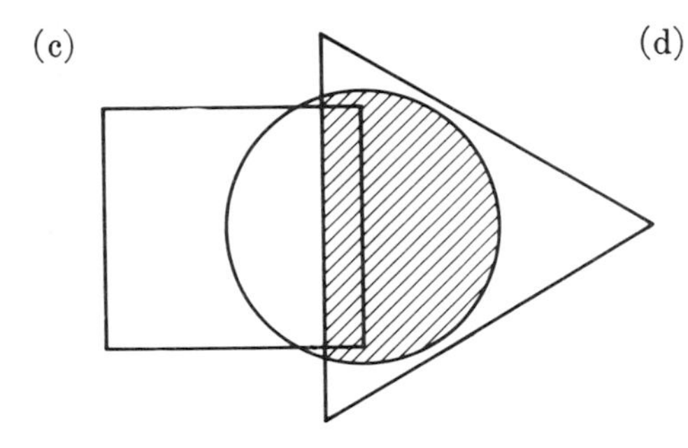

(d)

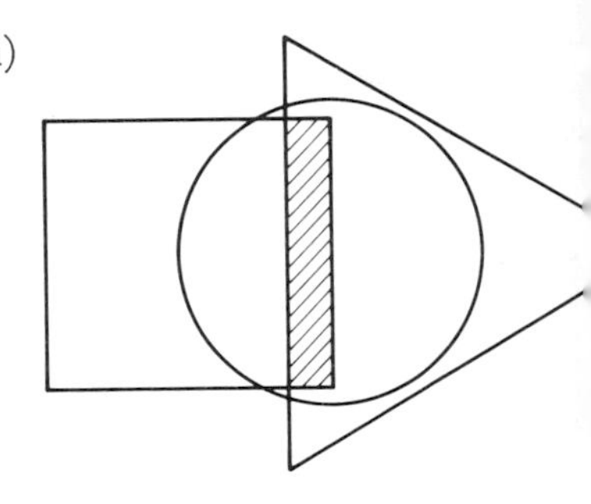

5. (a)

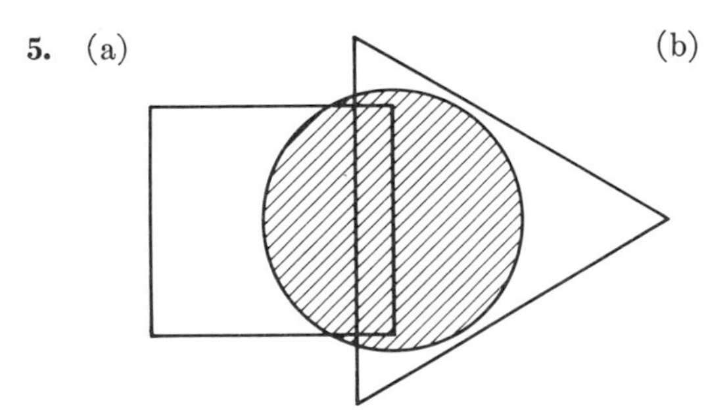

(b)

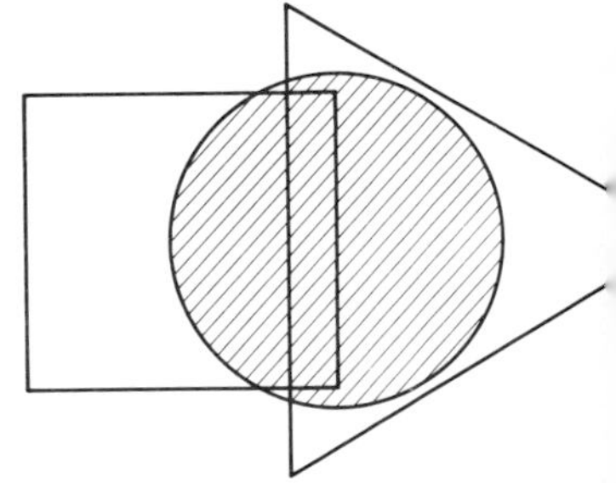

(c)

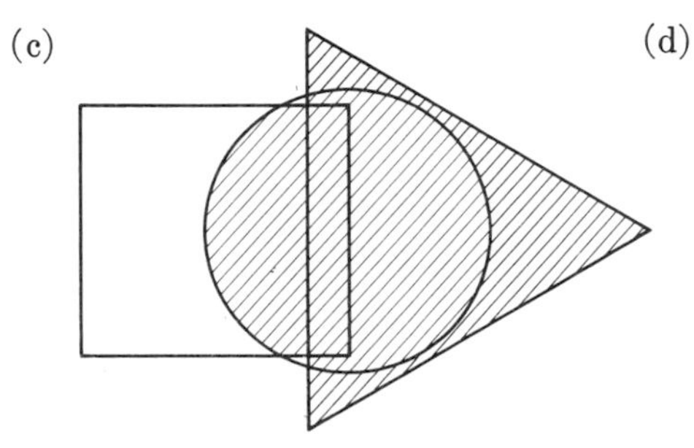

(d)

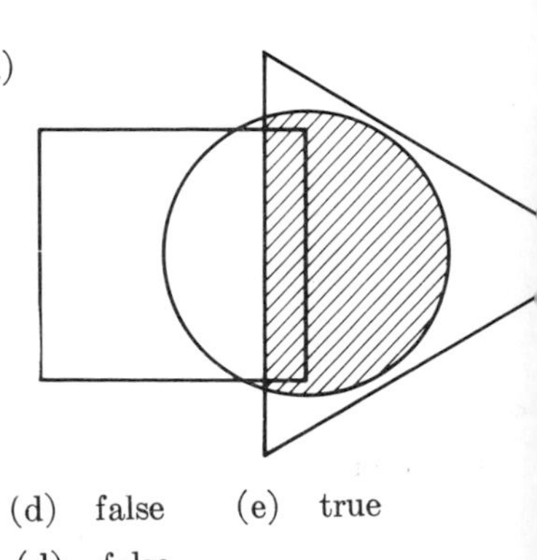

7. (a) true (b) true (c) true (d) false (e) true

9. (a) true (b) true (c) false (d) false

17. (a) all male administrative employees

(b) all employees

(c) all female administrative technical employees

(d) either female or administrative or technical employees

(e) all male administrative employees working for the company at least 5 years

19. (a) those who drive cars with engines that are more than 250 horsepower

(b) those who drive cars with engines that are more than 200 horsepower

(c) those who are male and over 25 years of age and who drive cars with engines that are more than 200 horsepower

(d) those who are female or over 20 years of age or who drive cars with engines that are more than 200 horsepower

(e) those who drive cars with engines that are more than 250 horsepower and are either over 20 years of age or female

Exercise Set 1.3, page 28

1. (a) $\{b, c, e, g\}$ (b) $\{b, c, e, f, g, h\}$

(c) $\{b, c, e, g\}$ (d) U (e) $\varnothing$

3. (a) the set of stocks traded on the New York Stock Exchange that have not paid a dividend for at least one of the past 10 years

(b) the set of stocks traded on the New York Stock Exchange that have a Price-to-Earnings ratio of more than 12

(c) the set of stocks traded on the New York Stock Exchange that have not paid a dividend for at least one of the past 10 years and that have a Price-to-Earnings ratio of more than 12

(d) the set of stocks traded on the New York Stock Exchange that either have not paid a dividend for at least one of the past 10 years or have a Price-to-Earnings ratio of more than 12

5. $U =$ the set of all integers;

$A' = \{x \mid x \text{ is an integer satisfying } x \leq 0\}$

7. $U =$ the set of all letters in the English alphabet;
$C' = \{a, e, i, o, u\}$

9. (a) (b)

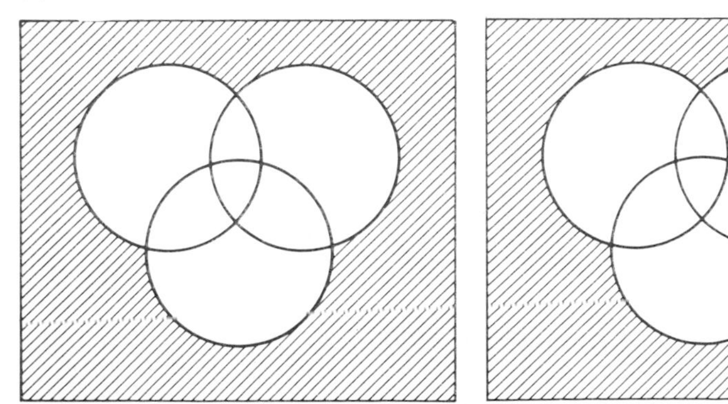

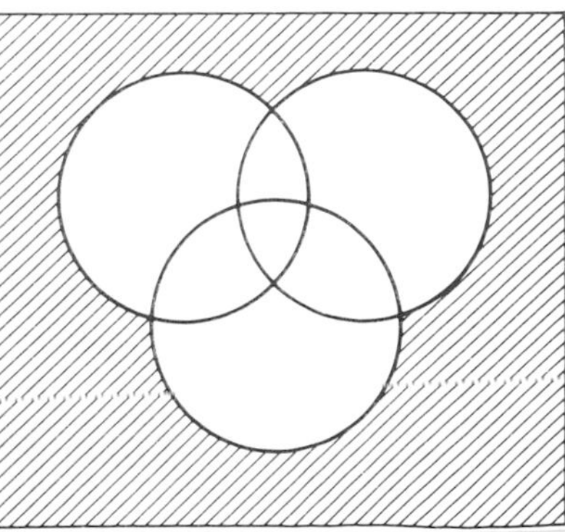

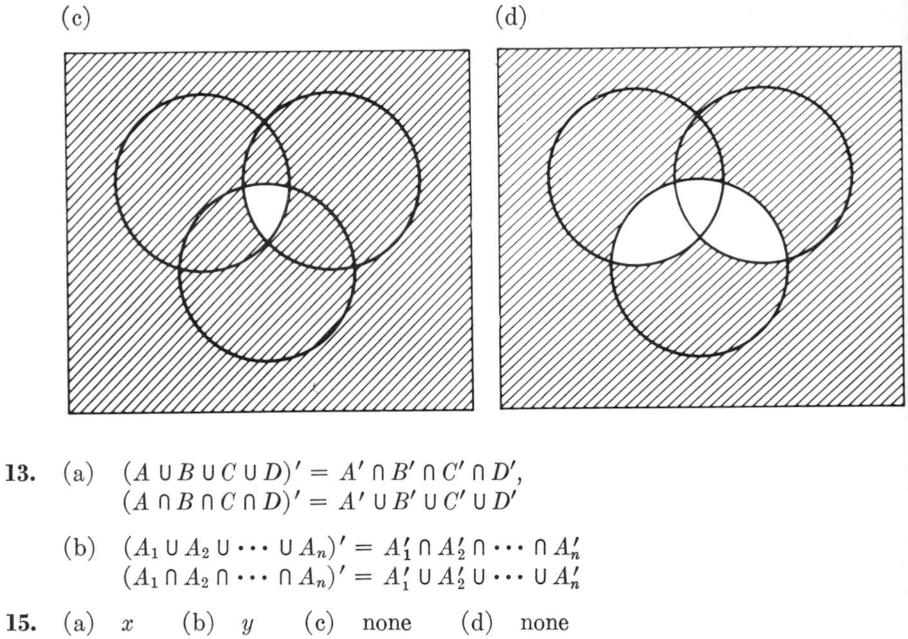

13. (a) $(A \cup B \cup C \cup D)' = A' \cap B' \cap C' \cap D'$,
$(A \cap B \cap C \cap D)' = A' \cup B' \cup C' \cup D'$

(b) $(A_1 \cup A_2 \cup \cdots \cup A_n)' = A_1' \cap A_2' \cap \cdots \cap A_n'$
$(A_1 \cap A_2 \cap \cdots \cap A_n)' = A_1' \cup A_2' \cup \cdots \cup A_n'$

15. (a) x (b) y (c) none (d) none

17. (a) $x = 3$ (b) $x = 3,\ y = 3$

(c) $x = 1,\ y = 7$ (d) $x = 4$ or $x = -4$

19. $(a, 1, x)$ $(a, 1, y)$ $(a, 1, z)$

$(a, 2, x)$ $(a, 2, y)$ $(a, 2, z)$

$(b, 1, x)$ $(b, 1, y)$ $(b, 1, z)$

$(b, 2, x)$ $(b, 2, y)$ $(b, 2, z)$

21. (b) (a, l, l) (a, l, r) (a, r, l) (a, r, r)

(b, l, l) (b, l, r) (b, r, l) (b, r, r)

(c, l, l) (c, l, r) (c, r, l) (c, r, r)

Exercise Set 1.4, page 40

1. (a) 21 (b) 2 (c) 4

(d) 6 (e) undefined (f) 0

3. $B = \varnothing$ **5.** 6 **9.** 450

11. 9 **13.** (b) 10,000,000

15. (a) 51 (b) 17 (c) 73

(d) 7.8% (e) 82 (f) 86

17. 36,000

CHAPTER 2

Exercise Set 2.1, page 50

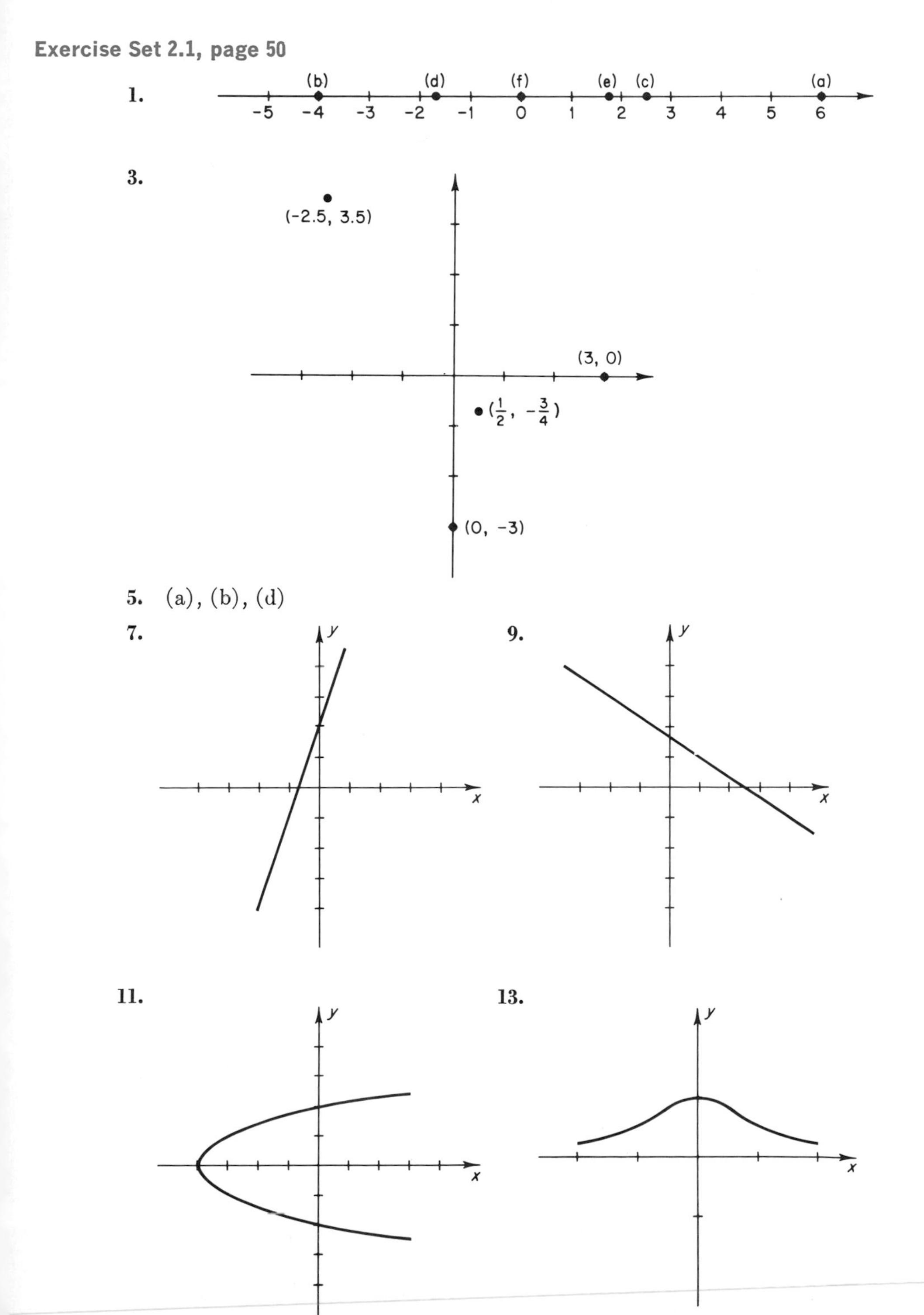

15.

17. $(-2, 9)$ **19.**

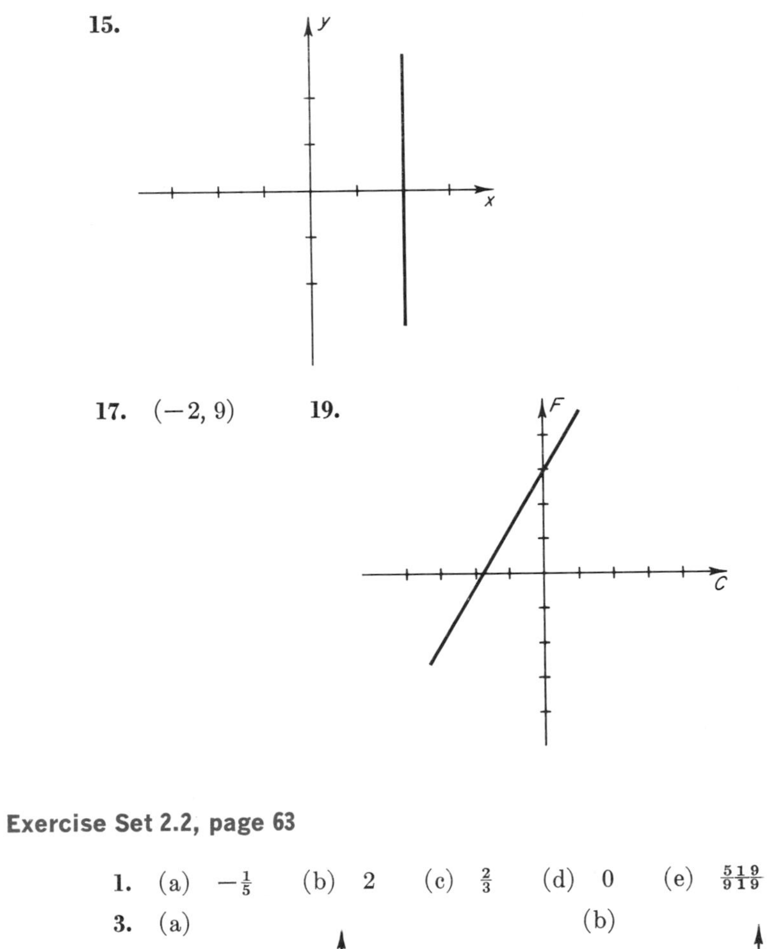

Exercise Set 2.2, page 63

1. (a) $-\frac{1}{5}$ (b) 2 (c) $\frac{2}{3}$ (d) 0 (e) $\frac{519}{919}$

3. (a) (b) (c) (d)

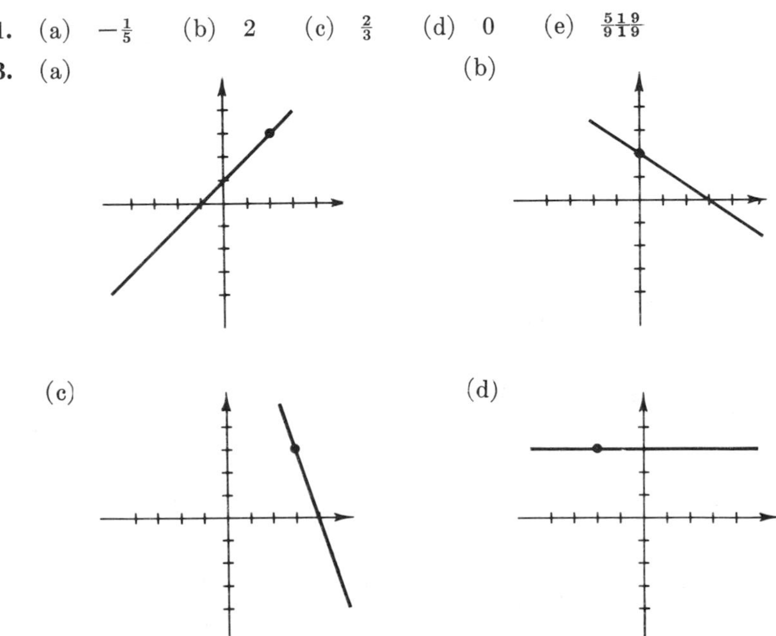

5. (a) 2 (b) $-\frac{3}{2}$ (c) 0 (d) no slope (e) $-\frac{2}{3}$ (f) 2

7. (a) $y = 3 + (x - 2)$ (b) $y = 2 - \frac{2}{3}x$

(c) $y = 3 - 3(x - 3)$ (d) $y = 3$

9. (a) $y = -2$ (b) $x = -3$

11. (c), (d)

13. (a) $x = \frac{46}{5}, y = -\frac{39}{5}$ (b) $x = \frac{18}{25}, y = \frac{26}{25}$

(c) $x = \frac{1}{3}, y = -\frac{1}{2}$ (d) $x = -\frac{17}{26}, y = -\frac{11}{26}$

15. (a) no solution

(b) $x = t, y = \frac{2}{3}t - \frac{5}{3}$, where t is arbitrary

(c) $x = t, y = \frac{3}{4}t + 2$, where t is arbitrary

(d) no solution

17. $y = 2x - 2$

19. (a) $s = 1, t = -1$ (b) $V = 2, W = -1$

Exercise Set 2.3, page 74

1. (a) $S = 12{,}000 + 840t$ (b) \$16,200 (c) \$14,940

3. (a) $V = 24{,}000 - 1920t$ (b) \$10,560 (c) 9.9 (approximately)

(d) 12.5

5. (a) $N = \dfrac{64}{3} + \dfrac{64}{3}t$ (b)

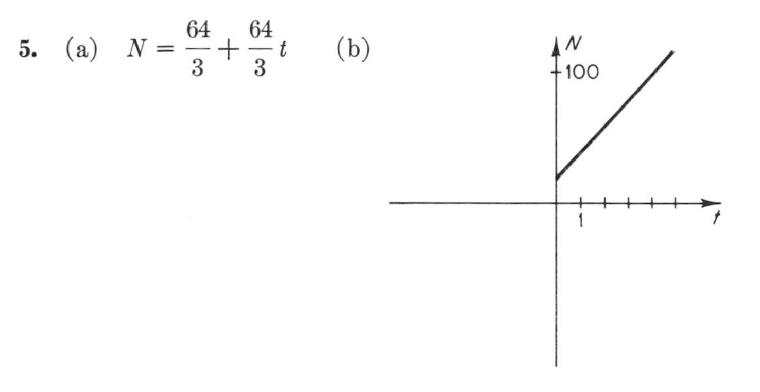

(c) 213,333

9. $y = \dfrac{17}{10}x + \dfrac{3}{10}$

11. (a) $y = 6.177x + 59.190$ (b) 127.1 (c) To simplify the arithmetic

Exercise Set 2.4, page 82

1. (a)

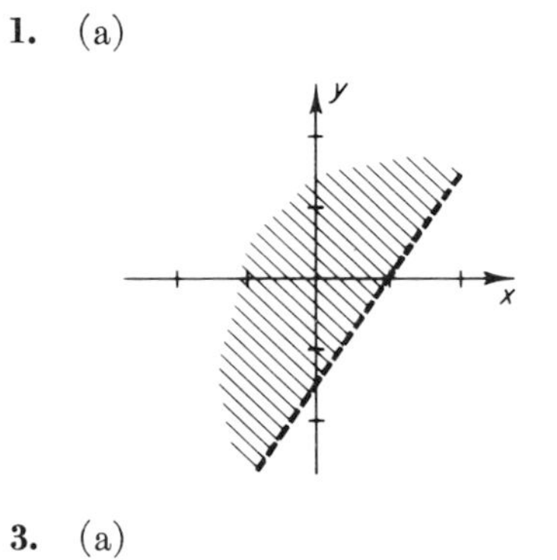

(b)

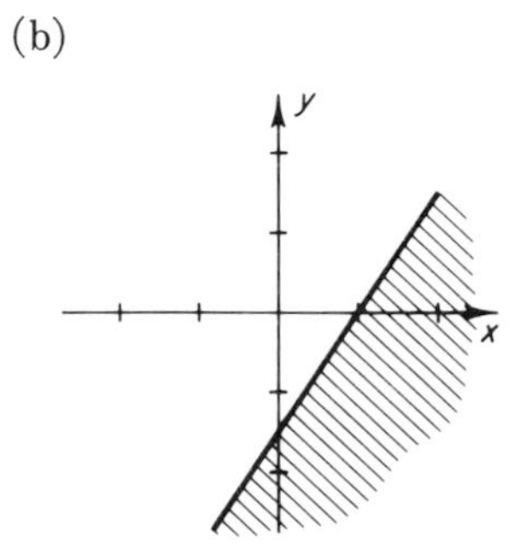

3. (a)

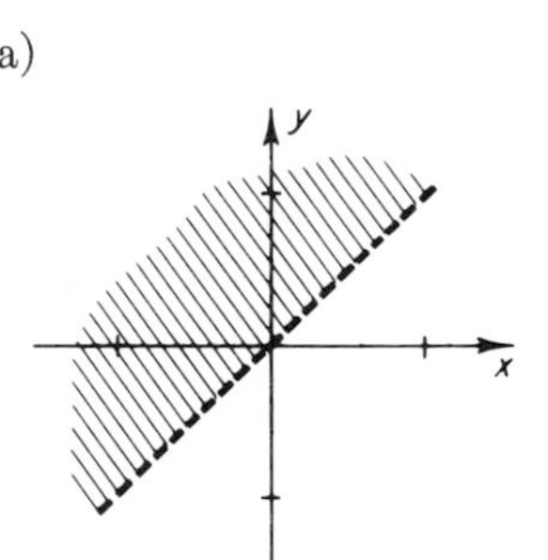

(b)

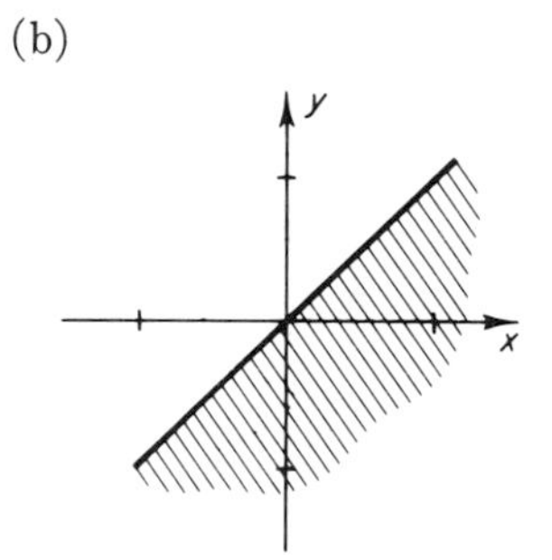

5. (a)

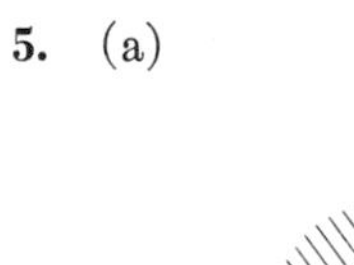

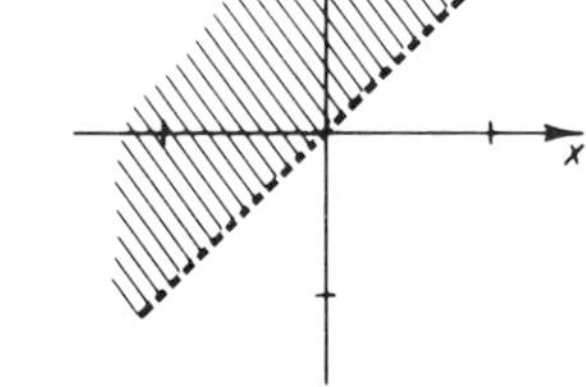

(b)

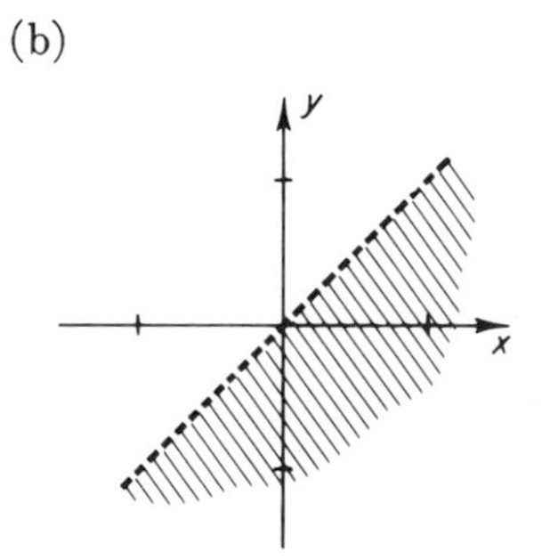

7.

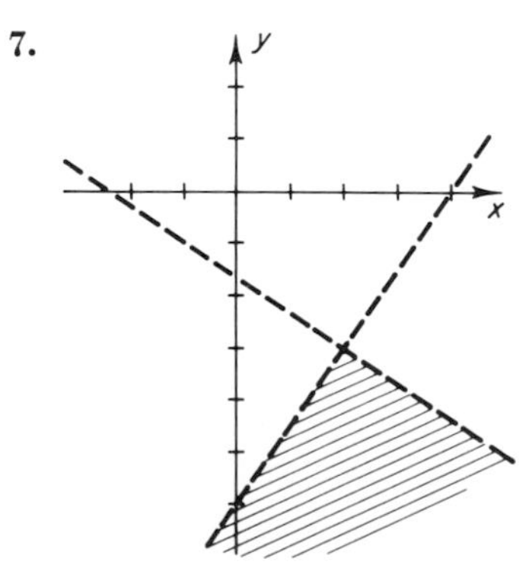

9.

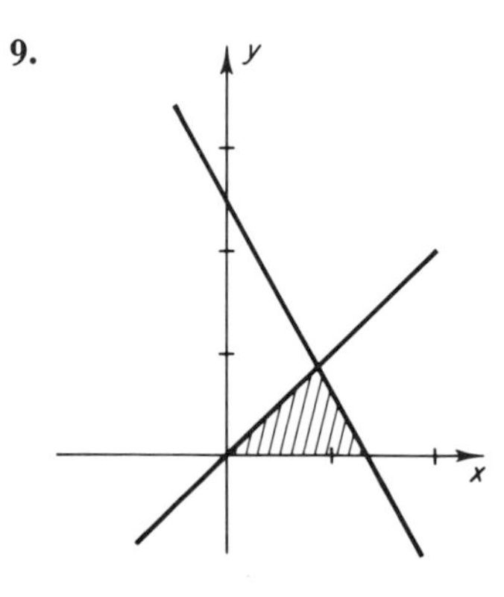

11. no solution

CHAPTER 3

Exercise Set 3.1, page 89

1. make $z = x + 1.5y$ as small as possible, where

$$\begin{aligned} 4x + 10y &\geq 100 \\ 4x + 2y &\geq 60 \\ x &\geq 0 \\ y &\geq 0 \end{aligned}$$

3. make $z = 8x + 12y$ as small as possible, where

$$\begin{aligned} x + y &\geq 7 \\ 2x + y &\geq 10 \\ x &\geq 0 \\ y &\geq 0 \end{aligned}$$

5. make $z = 5x + 3y$ as large as possible, where

$$\begin{aligned} 10x + 5y &\leq 450 \\ 6x + 12y &\leq 480 \\ 9x + 9y &\leq 450 \\ x &\geq 0 \\ y &\geq 0 \end{aligned}$$

7. make $z = 0.3x + 0.4y$ as large as possible (z is in dollars), where

$$\begin{aligned} 5x + 6y &\leq 12{,}000 \\ 5x + 3y &\leq 9{,}000 \\ x &\geq 0 \\ y &\geq 0 \end{aligned}$$

9. make $z = 10{,}000x + 12{,}000y$ as small as possible (z is in dollars), where

$$\begin{aligned} 100x + 200y &\geq 1{,}000 \\ 8x + 8y &\leq 48 \\ x &\geq 0 \\ y &\geq 0 \end{aligned}$$

11. make $z = 25x + 35y$ as large as possible, where

$$\begin{aligned} x + y &\leq 120 \\ 12x + 24y &\leq 1{,}200 \\ 32x + 8y &\leq 160 \\ x &\geq 0 \\ y &\geq 0 \end{aligned}$$

13. make $z = 0.5x + 0.6y$ as small as possible, where

$$\begin{aligned} 0.03x + 0.01y &\leq 6 \\ 0.01x + 0.01y &\leq 4 \\ x + \quad y &\geq 100 \\ x &\geq 0 \\ y &\geq 0 \end{aligned}$$

15. make $z = 1000x + 600y$ as small as possible, where

$$\begin{aligned} x + y &\leq 30 \\ x &\geq 3 \\ x &\leq 10 \\ y &\geq 18 \end{aligned}$$

17. (a) make $z = 3x + y$ as large as possible, where

$$\begin{aligned} 3x + y &\leq 6 \\ x + y &\leq 4 \\ x &\geq 0 \\ y &\geq 0 \end{aligned}$$

(b) make $z = 4x + y$ as large as possible, where

$$\begin{aligned} 3x + y &\leq 6 \\ x + y &\leq 4 \\ x &\geq 0 \\ y &\geq 0 \end{aligned}$$

Exercise Set 3.2, page 100

1. $x = 9$, $y = 0$

3. $x = \frac{7}{3}$, $y = 6$

5. (a) $x = 1$ and $y = 2$, or $x = \frac{7}{3}$ and $y = 6$, or any point on the line segment connecting $(1, 2)$ and $(\frac{7}{3}, 6)$

(b) $x = 0$, $y = 6$

7. 40 tables of type A, 10 tables of type B

9. 0 containers from the Jones Corp.,
2000 containers from the Jackson Corp.

11. 0 planes of type A, 5 planes of type B

13. 20 acres of soybeans

15. 100 gallons of R, 0 gallons of L

17. 3 minutes of H, 18 minutes of L

19. (a) 1 gram of Curine I and 3 grams of Curine II, or
2 grams of Curine I and 0 grams of Curine II, or
any point on the line segment connecting (1, 3) and (2, 0)

(b) 2 grams of Curine 1, 0 grams of Curine II

CHAPTER 4

Exercise Set 4.1, page 108

1. (a), (d) **3.** (b), (d) **5.** (b), (c)

7. (a)
$$\begin{aligned} 4x_1 - 2x_2 &= 7 \\ -2x_1 + 3x_2 + 5x_3 &= -4 \\ 3x_1 + 2x_2 - 5x_3 &= 3 \\ 2x_1 + x_2 + 3x_3 &= 4 \end{aligned}$$

(b)
$$\begin{aligned} 2x_1 + 3x_2 &= 5 \\ 4x_1 + 2x_2 &= 1 \end{aligned}$$

(c)
$$\begin{aligned} 3x_1 + 2x_2 + 4x_3 + 2x_4 &= 5 \\ 2x_1 + 3x_2 + x_3 + 3x_4 &= 1 \end{aligned}$$

(d)
$$\begin{aligned} x_1 &= 1 \\ x_2 &= 2 \\ x_3 &= 3 \end{aligned}$$

11.
$$\begin{aligned} 2x_1 + 2x_2 &= 12 \\ 3x_1 + x_2 &= 16 \\ 2x_1 + 4x_2 &= 14 \end{aligned}$$

Exercise Set 4.2, page 120

1. (a), (c)

3. (a) $x_1 = 3$, $x_2 = -2$, $x_3 = 4$

(b) $x_1 = 3 - 2t$, $x_2 = 6 - 3t$, $x_3 = -2 - 4t$, $x_4 = t$, where t is arbitrary

(c) $x_1 = 1 - 2s - 5t$, $x_2 = s$, $x_3 = -2 - 4t$, $x_4 = 3 - 2t$, $x_5 = t$, where s and t are arbitrary

(d) system is inconsistent

5. (a)
$$\begin{bmatrix} 1 & 0 & -3 & 6 & 0 \\ 0 & 1 & -1 & 3 & 0 \\ 0 & 0 & 0 & 0 & 1 \end{bmatrix}$$

(b)
$$\begin{bmatrix} 1 & 0 & 0 & 0 & 2 \\ 0 & 1 & 2 & 1 & 3 \\ 0 & 0 & 0 & 0 & 0 \end{bmatrix}$$

(c)
$$\begin{bmatrix} 1 & 0 & 0 & 2 & 2 \\ 0 & 0 & 1 & 3 & -3 \\ 0 & 0 & 0 & 0 & 0 \end{bmatrix}$$

7. (a) $x = 6$, $y = 7$, $z = 2$

(b) $x_1 = 0$, $x_2 = 0$, $x_3 = 0$

(c) $x_1 = 3 - s + t$, $x_2 = -5 + s \quad 5t$, $x_3 = -2 - t$, $x_4 = s$, $x_5 = t$, where s and t are arbitrary

(d) $x_1 = 9 - 5s$, $x_2 = 4 - 2s$, $x_3 = 0$, $x_4 = s$, where s is arbitrary

9. (a) $x_1 = -10, x_2 = 2, x_3 = 3, x_4 = 0$

(b) $x_1 = 1, x_2 = 1$

(c) $x_1 = -1, x_2 = 2, x_3 = 2$

(d) $x, = \frac{7}{2} - \frac{1}{2}s, x_2 = s, x_3 = 4$, where s is arbitrary

13. 3 vessels of type A, 4 vessels of type B, 5 vessels of type C

Exercise Set 4.3, page 130

1. (a) 2×3 (b) 3×2 (c) 1×2 (d) 1×1

3. $u = 5, v = 2, w = 0$

5. (a), 3×4; (d), 3×3

9.

	under \$10,000	over \$10,000
Republican	205	192
Democrat	317	128
independent	96	63

Exercise Set 4.4, page 141

1. (a), 3×2; (e), 4×4; (f), 3×2; (g), 3×2

3. (a) $\begin{bmatrix} 40 & -38 & -58 \\ 40 & 32 & 4 \end{bmatrix}$ (c) $\begin{bmatrix} -144 & 157 \\ -84 & 78 \end{bmatrix}$

7.

$$\begin{aligned} 3x_1 + 2x_2 \qquad &= 8 \\ 2x_1 + 4x_2 + 5x_3 &= 0 \\ -3x_1 \qquad + 2x_3 &= -2 \end{aligned}$$

Exercise Set 4.5, page 153

1. yes, (b)

3. (a) $\begin{bmatrix} \frac{1}{2} & -1 & \frac{1}{2} \\ \frac{1}{2} & 0 & -\frac{1}{2} \\ -\frac{1}{2} & 1 & \frac{1}{2} \end{bmatrix}$ (b) $\begin{bmatrix} \frac{1}{3} & \frac{1}{3} & -\frac{2}{3} \\ \frac{1}{6} & -\frac{1}{3} & \frac{7}{6} \\ -\frac{1}{6} & \frac{1}{3} & -\frac{1}{6} \end{bmatrix}$ (c) not invertible

5. (a) $\begin{bmatrix} 3 & 2 & 0 & 6 \\ -2 & -1 & 0 & -3 \\ -2 & -3 & -1 & -6 \\ 2 & 2 & 1 & 4 \end{bmatrix}$ (b) $\begin{bmatrix} 0 & -\frac{1}{3} & \frac{1}{3} \\ 0 & \frac{1}{3} & \frac{2}{3} \\ 1 & -\frac{1}{3} & -\frac{8}{3} \end{bmatrix}$ (c) not invertible

7. $x = 4, y = 2, z = -1$

9. $x_1 = -1, x_2 = 3, x_3 = 2$

11. (a) $x_1 = \frac{27}{25}, x_2 = \frac{17}{25}, x_3 = \frac{13}{25}$

(b) $x_1 = 1, x_2 = 1, x_3 = 2$

(c) $x_1 = -\frac{11}{25}, x_2 = \frac{19}{25}, x_3 = -\frac{34}{25}$

17. (a)

	first bond	second bond
\$180	\$1500	\$1500
\$200	\$500	\$2500
\$160	\$2500	\$500

(b) no

CHAPTER 5

Exercise Set 5.1, page 161

1. (a) **3.** no

5. (a) Maximize $z = 5x + 7y$
subject to

$$4x - 6y + u = 9$$
$$2x + 7y + v = 3$$
$$5x - 8y + w = 2$$
$$x \geq 0, \quad y \geq 0, \quad u \geq 0,$$
$$v \geq 0, \quad w \geq 0$$

(b) Maximize $z = -x_1 + x_2 - x_3 + x_4$
subject to

$$2x_1 + 3x_3 + x_4 + v = 8$$
$$x_1 - 2x_2 + x_4 + w = 6$$
$$x_1 \geq 0, \quad x_2 \geq 0, \quad x_3 \geq 0,$$
$$x_4 \geq 0, \quad v \geq 0, \quad w \geq 0$$

7. (a) Maximize $z = 10x + 12y$
subject to

$$0.2x + 0.4y + v = 30$$
$$0.2x + 0.2y + w = 20$$
$$x \geq 0, \quad y \geq 0, \quad v \geq 0, \quad w \geq 0$$

(b) $x = 50, y = 50, v = 0, w = 0$

9. (a)

$$A = \begin{bmatrix} 1 & 0 & -1 \\ 0 & 1 & -1 \\ 1 & -1 & 0 \end{bmatrix}, \quad B = \begin{bmatrix} 1 \\ 2 \\ 3 \end{bmatrix}, \quad C = \begin{bmatrix} 4 & -2 & 7 \end{bmatrix}, \quad X = \begin{bmatrix} x \\ y \\ t \end{bmatrix}$$

(b)

$$A = \begin{bmatrix} 1 & 3 \end{bmatrix}, \quad B = \begin{bmatrix} 5 \end{bmatrix}, \quad C = \begin{bmatrix} 1 & -1 \end{bmatrix}, \quad X = \begin{bmatrix} x_1 \\ x_2 \end{bmatrix}$$

11. Maximize $z = 6x + 7y + 9t$
subject tc

$$\begin{aligned} 2x - 4y + 5t &\leq 3 \\ 7x - y + 3t &\leq 9 \\ x \geq 0, \quad y \geq 0, \quad t &\geq 0 \end{aligned}$$

Exercise Set 5.2, page 173

1. maximum value of z is 7 for $x = 0$, $y = 1$

3. maximum value of z is 18 for $x = 3$, $y = 0$

5. maximum value of z is 0 for $x = 0$, $y = 0$

9. (a) Maximize $z = 2x + 5y$
subject to

$$\begin{aligned} 3x + 2y + v &= 2 \\ 2x + 5y + w &= 8 \\ x \geq 0, \quad y \geq 0, \quad v \geq 0, \quad w &\geq 0 \end{aligned}$$

(b)
$$\begin{array}{cccccc} x & y & v & w & z & \\ \end{array}$$
$$\begin{bmatrix} -2 & -5 & 0 & 0 & 1 & 0 \\ 3 & 2 & 1 & 0 & 0 & 2 \\ 2 & 5 & 0 & 1 & 0 & 8 \end{bmatrix}$$

(c)
$$\begin{bmatrix} 0 & 0 & 0 & 1 & 1 & 8 \\ \frac{11}{5} & 0 & 1 & -\frac{2}{5} & 0 & -\frac{6}{5} \\ \frac{2}{5} & 1 & 0 & \frac{1}{5} & 0 & \frac{8}{5} \end{bmatrix}$$

(d)
$$\begin{bmatrix} 0 & 0 & 0 & 1 & 1 & 8 \\ 0 & -\frac{11}{2} & 1 & -\frac{3}{2} & 0 & -10 \\ 1 & \frac{5}{2} & 0 & \frac{1}{2} & 0 & 4 \end{bmatrix}$$

11. (a) x, v, z (b) y, v, z (c) x_3, v_1, z (d) x, y, v, z

Exercise Set 5.3, page 189

1. $x = 0$, $y = 0$, $v = 12$, $w = 6$

$$\begin{array}{cccccc} x & y & v & w & z & \\ \end{array}$$
$$\begin{bmatrix} -4 & -3 & 0 & 0 & 1 & 0 \\ 2 & 3 & 1 & 0 & 0 & 12 \\ -3 & 2 & 0 & 1 & 0 & 6 \end{bmatrix}$$

3. $x = 0, y = 0, u = 4, v = 6, w = 1$

$$\begin{array}{ccccccc} x & y & u & v & w & z & \\ \end{array}$$
$$\begin{bmatrix} -8 & -6 & 0 & 0 & 0 & 1 & 0 \\ 1 & 1 & 1 & 0 & 0 & 0 & 4 \\ 1 & 3 & 0 & 1 & 0 & 0 & 6 \\ -1 & 1 & 0 & 0 & 1 & 0 & 1 \end{bmatrix}$$

5. (a) $5x_3 + 7u_3 + 6w + z = 40$

$x_1 - 2x_3 + u + 2w = 6$

$x_2 + 4x_3 + u - w = 14$

$3x_3 + v + 3w = 12$

(b) $x_1 = 6, x_2 = 14, x_3 = 0, u = 0, v = 12, w = 0$

(c) $z = 40$

7. (a), (c) **9.** $y, x, 4$

11.
$$\begin{bmatrix} 0 & 0 & 27 & 4 & 8 & 1 & 540 \\ 1 & 0 & 3 & \frac{1}{2} & 1 & 0 & 50 \\ 0 & 1 & -4 & -\frac{3}{2} & 0 & 0 & 30 \end{bmatrix}$$

13.
$$\begin{bmatrix} 0 & 0 & 0 & 0 & \frac{4}{3} & \frac{23}{3} & 0 & 1 & 360 \\ 0 & 0 & 0 & 1 & -\frac{2}{3} & \frac{5}{3} & -2 & 0 & 80 \\ 0 & 1 & 0 & 0 & \frac{1}{2} & 3 & \frac{5}{2} & 0 & 390 \\ 0 & 0 & 1 & 0 & 0 & -1 & 2 & 0 & 150 \\ 1 & 0 & 0 & 0 & \frac{1}{6} & \frac{1}{3} & \frac{1}{2} & 0 & 30 \end{bmatrix}$$

15. $z = \frac{144}{7}$ for $x = \frac{60}{7}, y = \frac{8}{7}, t = 0$

17. $z = \frac{12}{5}$ for $x_1 = \frac{6}{5}, x_2 = 0, x_3 = 0$

21. 0 gallons of L and E, 125 gallons of R

Exercise Set 5.4, page 199

1. $z = -\frac{26}{3}$ for $x = \frac{8}{3}, y = 1$

3. $z = -\frac{16}{3}$ for $x = 0, y = \frac{8}{3}, t = 0$

5. $z = 8$ for $x_1 = 0, x_2 = 0, x_3 = 8$

7. $z = 6$ for $x_1 = 2, x_2 = 0$

9. $z = \frac{158}{11}$ for $x_1 = \frac{18}{11}, x_2 = \frac{10}{11}$

11. 7 ounces of M, 0 ounces of N

CHAPTER 6

Exercise Set 6.1, page 208

1. (a) $\{0, 1, 2, 3, 4, 5, 6, 7, 8, 9, 10\}$
(b) $\{0, 1, 2, \ldots\}$
(c) $\{x \mid 0 \leq x \leq 100\}$
(d) $\{x \mid x \geq 0\}$
(e) $\{(h, h, h), (h, h, t), (h, t, h), (t, h, h), (h, t, t), (t, h, t), (t, t, h), (t, t, t)\}$

3. $\varnothing$; S, $\{a, b\}$, $\{a, c\}$, $\{b, c\}$, $\{a\}$, $\{b\}$, $\{c\}$

5. (a) the same number is obtained on both tosses
(b) the first number tossed is 3
(c) a 4 is tossed both times.

7. (a) $\{(m, h, d), (m, h, r), (m, h, i), (m, a, d), (m, a, r), (m, a, i), (m, l, d), (m, l, r), (m, l, i), (f, h, d), (f, h, r), (f, h, i), (f, a, d), (f, a, r), (f, a, i), (f, l, d), (f, l, r), (f, l, i)\}$
(b) $\{(m, h, r), (m, a, r), (m, l, r), (f, h, r), (f, a, r), (f, l, r)\}$
(c) $\{(f, h, i), (f, a, i), (f, l, i)\}$
(d) $\{(m, h, d), (m, h, r), (m, h, i), (m, a, d), (m, a, r), (m, a, i), (m, l, d), (m, l, r), (m, l, i), (f, h, d), (f, a, d), (f, l, d)\}$

9. (a) the number tossed is either even or less than 5
(b) the number tossed is divisible by 3 and is less than 5
(c) the number tossed is either divisible by 3 or greater than 4
(d) the number tossed is not divisible by 3 and is less than 5
(e) the number tossed is odd, not divisible by 3, and greater than 4

11. (a) $\{t \mid t \geq 10\}$ (b) $\{t \mid 0 \leq t < 15\}$ (c) $\{t \mid 0 \leq t < 10\}$
(d) $\{t \mid t \geq 15\}$ (e) $\{t \mid t \geq 15\}$ (f) $\{t \mid t \geq 10\}$

13. (a) yes, $E \cap F = \varnothing$, (b) no, $E \cap H \neq \varnothing$
(c) no, $E \cup F \neq S$ (d) yes, $E \cup E' = S$

15. yes, $E \cap F = E \cap G = F \cap G = \varnothing$

Exercise Set 6.2, page 217

1. (a) $\{(h, h)\}, \{(h, t)\}, \{(t, h)\}, \{(t, t)\}$
(b) $\{i\}, \{d\}, \{n\}$
(c) $\{(i, i)\}, \{(i, d)\}, \{(i, n)\}, \{(d, i)\}, \{(d, d)\}, \{(d, n)\}, \{(n, i)\}, \{(n, d)\}, \{(n, n)\}$
(d) $\{a\}, \{b\}, \{c\}, \{d\}, \{e\}$
(e) $\{e\}, \{o\}$

(f) $\{c\}$, $\{d\}$, $\{h\}$, $\{s\}$

(g) $\{0\}$, $\{1\}$, $\{2\}$, $\{3\}$, $\{4\}$, $\{5\}$, $\{6\}$, $\{7\}$, $\{8\}$, $\{9\}$, $\{10\}$

3. (a) $\frac{2}{3}$ (b) $\frac{2}{3}$ (c) 0

5. (a) $\frac{5}{11}$ (b) $\frac{5}{11}$ (c) 1

7. (a) $\{e\}$, $\{o\}$; equally likely

(b) $\{3\}$, $\{\text{not } 3\}$; not equally likely

(c) $\{\text{ace}\}$, $\{\text{not an ace}\}$; not equally likely

(d) $\{\text{black}\}$, $\{\text{red}\}$; equally likely

(e) $\{\{a, b\}\}$, $\{\{a, c\}\}$, $\{\{a, d\}\}$, $\{\{b, c\}\}$, $\{\{b, d\}\}$, $\{\{c, d\}\}$; equally likely

9.

Event	Probability
$\{s_1, s_2, s_3\}$	1
$\{s_1, s_2\}$	$\frac{2}{3}$
$\{s_1, s_3\}$	$\frac{2}{3}$
$\{s_2, s_3\}$	$\frac{2}{3}$
$\{s_1\}$	$\frac{1}{3}$
$\{s_2\}$	$\frac{1}{3}$
$\{s_3\}$	$\frac{1}{3}$
$\varnothing$	0

11. (a) $\frac{3}{8}$ (b) $\frac{7}{8}$ (c) $\frac{1}{2}$ (d) $\frac{1}{2}$

Exercise Set 6.3, page 227

1. (c), (d) **3.** (a) 0.7 (b) 0.9 (c) 0.35

5. (a) 0.5 (b) 0.9 (c) 0.4 (d) 0 (e) 0.2 (f) 0.7

7. 0.4

9. (a) 0.7 (b) 0 (c) 0.6 (d) 0.7

11. (a) 0.3 (b) 0.7 (c) 0.3

Exercise Set 6.4, page 242

1. 350 **3.** 48 **5.** 2^{32} **7.** 7,776; 6^n

9.

abcd	*bacd*	*cabd*	*dabc*
abdc	*badc*	*cadb*	*dacb*
acbd	*bcad*	*cbad*	*dbac*
acdb	*bcda*	*cbda*	*dbca*
adbc	*bdac*	*cdab*	*dcab*
adcb	*bdca*	*cdba*	*dcba*

13. (a) 42 (b) 11,880 (c) 120
(d) 3 (e) 6,720 (f) 20

15.

ab	*ba*	*ca*	*da*	*ea*
ac	*bc*	*cb*	*db*	*eb*
ad	*bd*	*cd*	*dc*	*ec*
ae	*be*	*ce*	*de*	*ed*

17. 720 **19.** 1000 **21.** 720

27. 84 **29.** $C_{25,4} = 303{,}600$ **31.** $\dfrac{21}{128}$

33. (a) $\dfrac{C_{20,6}C_{30,4}}{C_{50,10}}$ (b) $\dfrac{C_{30,10}}{C_{50,10}}$ (c) $1 - \dfrac{C_{30,10}}{C_{50,10}}$

35. 3 **37.** Choose n so $C_{n,3} \geq 52$; $n = 8$

Exercise Set 6.5, page 259

1. (a) $\frac{1}{2}$ (b) $\frac{2}{7}$ **3.** 0.1 **5.** $\frac{2}{33}$

7. $\frac{2}{3}$ **9.** (a) $\frac{1}{5}$ (b) $\frac{1}{3}$ **11.** $\frac{2}{25}$

15. (a) $P(E|H) = \frac{917}{926}$, $P(E) = \frac{951}{1000}$ (b) no

19. (a) 0.3 (b) 0.6 **21.** (a) $\frac{336}{720}$ (b) $\frac{576}{720}$ (c) $\frac{336}{720}$
(d) $\frac{672}{720}$

23. (a) 0.60125 (b) 0.64

Exercise Set 6.6, page 271

1. (a) $\frac{9}{22}$ (b) $\frac{9}{22}$ (c) $\frac{2}{11}$

3. $\frac{12}{13}$ **5.** $\frac{2}{3}$ **7.** 966 **9.** $\frac{1}{3}$ **11.** $\frac{2}{3}$

CHAPTER 7

Exercise Set 7.1, page 280

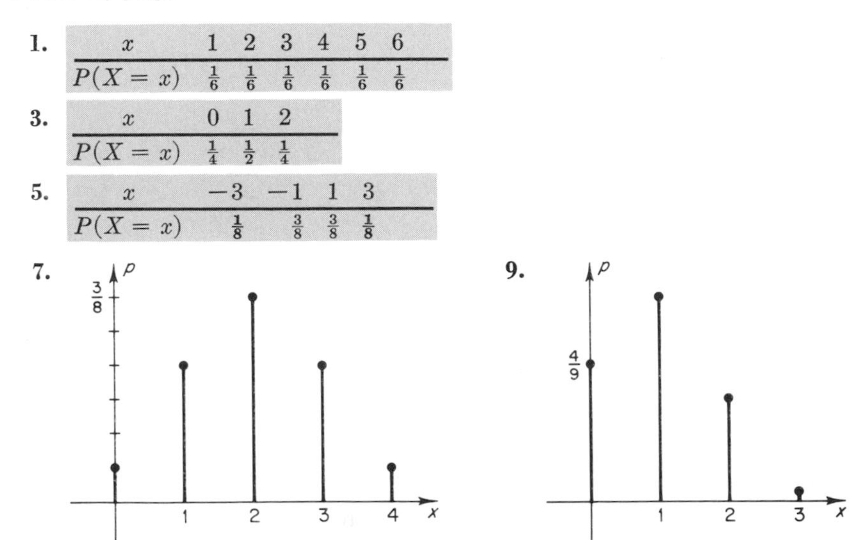

1.

x	1	2	3	4	5	6
$P(X = x)$	$\frac{1}{6}$	$\frac{1}{6}$	$\frac{1}{6}$	$\frac{1}{6}$	$\frac{1}{6}$	$\frac{1}{6}$

3.

x	0	1	2
$P(X = x)$	$\frac{1}{4}$	$\frac{1}{2}$	$\frac{1}{4}$

5.

x	-3	-1	1	3
$P(X = x)$	$\frac{1}{8}$	$\frac{3}{8}$	$\frac{3}{8}$	$\frac{1}{8}$

7. **9.**

11. (a) finite discrete (b) infinite discrete
(c) continuous (d) infinite discrete

Exercise Set 7.2, page 288

1. 3.4 **3.** 2.888 **5.** 2 **7.** 1

9. (a) 3.6 (b) -0.2 **11.** \$4.60 **13.** (a) 3 (b) 3

15. (a) \$240 (b) No; the program costs \$300/wk. and saves only \$500 − \$260 = \$240/wk. on the average

Exercise Set 7.3, page 299

1. 0.81 **3.** $\sigma_X{}^2 \simeq 0.067$, $\sigma_X \simeq 0.26$ **5.** 1

7. $\frac{2}{3}$ **9.** Y

Exercise Set 7.4, page 307

1. (a) 1 (b) 2 (c) 1.5 (d) 2.75

3. (a) 16 (b) -0.5 (c) 24.1 (d) 9.1

5. (a) at least 0 (b) at least 0 (c) at least $\frac{15}{16}$

7. at least 0.91 **9.** at least 15/16 **11.** 0.64

Exercise Set 7.5, page 317

1. (a) 256/625 (b) 256/625 (c) 96/625 (d) 16/625
(e) 1/625 (f) 624/625 (g) 113/625

3. (a) 0.042 (b) 0.017 (c) 0.000
(d) 0.083 (e) 0.157 (f) 0.002

5.

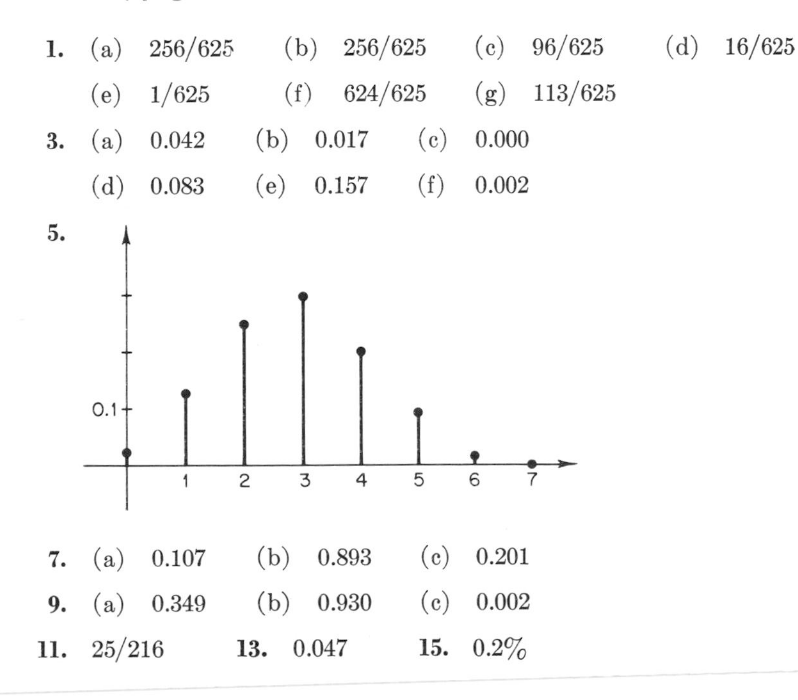

7. (a) 0.107 (b) 0.893 (c) 0.201

9. (a) 0.349 (b) 0.930 (c) 0.002

11. 25/216 **13.** 0.047 **15.** 0.2%

Exercise Set 7.6, page 330

1.

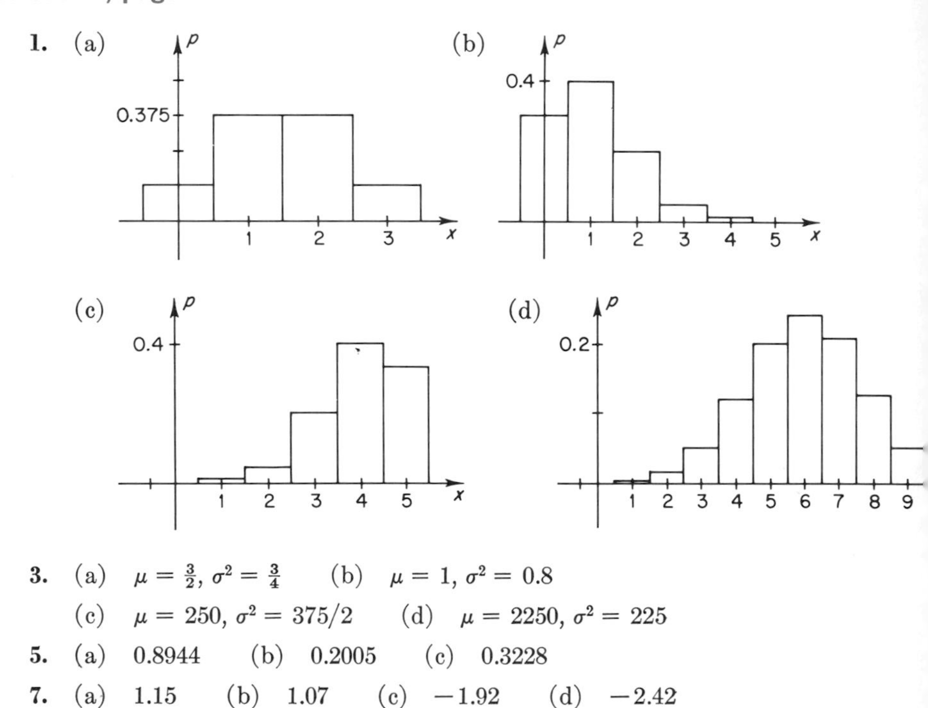

3. (a) $\mu = \frac{3}{2}$, $\sigma^2 = \frac{3}{4}$ (b) $\mu = 1$, $\sigma^2 = 0.8$
(c) $\mu = 250$, $\sigma^2 = 375/2$ (d) $\mu = 2250$, $\sigma^2 = 225$

5. (a) 0.8944 (b) 0.2005 (c) 0.3228

7. (a) 1.15 (b) 1.07 (c) -1.92 (d) -2.42

9. (a) 0.9525 (b) 0.0188 (c) 0.8164

11. 0.0749

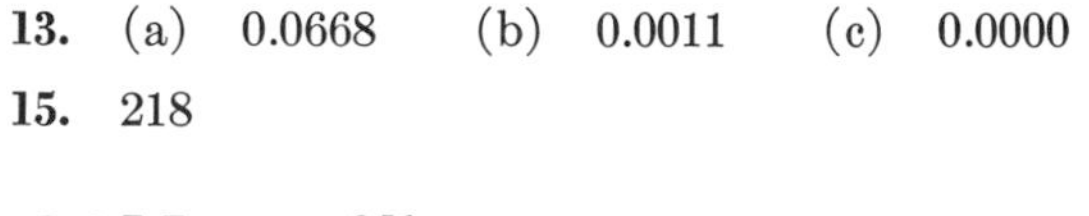

13. (a) 0.0668 (b) 0.0011 (c) 0.0000

15. 218

Exercise Set 7.7, page 341

1. (a) 3.24 (b) 17.64 (c) 4.80

3. the data do not support the assumption

5. the data support the conclusion

7. the data support the contention

CHAPTER 8

Exercise Set 8.1, page 348

1. (a) $P(D_1|A) \simeq 0.438$, $P(D_2|A) \simeq 0.562$ (b) D_2

Exercise Set 8.2, page 357

3. 75% **5.** $\frac{3}{4}$

7. 25% of genotype AA, 50% of genotype Aa, 25% of genotype aa

9. 36% red, 48% pink, 16% white

Exercise Set 8.3, page 364

1. (a) 0.973 (b) 0.933 (c) 0.0273

3. (a) 0.0352 (b) 0.0999 (d) 0.280

5. (a) 0.956 (b) 0.886 (c) 0.993

7. 0.00671

9. at least $346.51

Exercise Set 8.4, page 375

1. (b), (c), (f)

3. (a) row 1, column 2; -5 (b) row 1, column 2; 1

(c) row 1, column 2; -4 (d) row 1, column 2; 4

5.

		Player II	
		3	4
Player I	3	6	−7
	4	−7	8

7.

		Player II		
		stone	scissors	paper
Player I	stone	0	1	−1
	scissors	−1	0	1
	paper	1	−1	0

9. player R shows one finger, player C shows two fingers

13. firms A and B should each use television

Exercise Set 8.5, page 390

1. (a) $\frac{17}{12}$ (b) $\frac{33}{25}$ (c) $\frac{4}{3}$

3. $p_1 = \frac{1}{2}$, $p_2 = \frac{1}{2}$, $q_1 = \frac{3}{4}$, $q_2 = \frac{1}{4}$, $E = \frac{5}{2}$

5. $p_1 = \frac{4}{7}$, $p_2 = \frac{3}{7}$, $q_1 = \frac{1}{14}$, $q_2 = \frac{13}{14}$, $E = \frac{10}{7}$

7. $p_1 = \frac{1}{2}$, $p_2 = \frac{1}{2}$, $q_1 = \frac{1}{2}$, $q_2 = \frac{1}{2}$, $E = \frac{1}{2}$

9. $p_1 = \frac{1}{2}$, $p_2 = \frac{1}{2}$, $p_3 = 0$, $q_1 = \frac{3}{4}$, $q_2 = \frac{1}{4}$, $q_3 = 0$, $E = \frac{1}{2}$

11. $p_1 = 0$, $p_2 = \frac{4}{7}$, $p_3 = \frac{3}{7}$, $q_1 = 0$, $q_2 = \frac{1}{14}$, $q_3 = \frac{13}{14}$, $E = \frac{4}{7}$

13. Columbus should keep going with probability .627

15. $\frac{1}{3}$ male, $\frac{2}{3}$ female

Exercise Set 8.6, page 408

1.
$$\begin{array}{cc|cc} & & \multicolumn{2}{c}{\text{Next state}} \\ & & 1 & 2 \\ \text{Present} & 1 & \frac{2}{5} & \frac{3}{5} \\ \text{state} & 2 & \frac{3}{5} & \frac{2}{5} \end{array}$$

3. (a) If the system is in state 1, the probability that at the next observation the system will be in state 2 is $\frac{3}{4}$.

(b) $\begin{bmatrix}\frac{1}{4} & \frac{3}{4}\end{bmatrix}$ (c) $\begin{bmatrix}\frac{2}{5} & \frac{3}{5}\end{bmatrix}$

5. $\begin{bmatrix}\frac{29}{80} & \frac{51}{80}\end{bmatrix}$

7. $\begin{bmatrix}.722 & .278\end{bmatrix}$

9. $\begin{bmatrix}.320 & .258 & .422\end{bmatrix}$

11. (a) No power of P has all positive entries.

(b) P^2 has all positive entries.

13. (a) .3 (b) .167

15. 279 spaces at Kennedy
115 spaces at LaGuardia
107 spaces at Newark

CHAPTER 9

Exercise Set 9.1, page 417

1. (a) 243 (b) $\frac{1}{256}$ (c) 1 (d) -64 (e) $-\frac{1}{64}$
(f) $\frac{16}{625}$ (g) 16 (h) .2 (i) $\frac{1}{3}$

3. (a) $\dfrac{1}{5^9}$ (b) 2^8 (c) $\dfrac{1}{a^2}$
(d) $\dfrac{1}{a^5}$ (e) t^2 (f) $\dfrac{y^3}{x^3}$

5. (a) 7 (b) $\frac{1}{3}$ (c) $-.6t$ (d) .6

7. (a) .7781 (b) $-.1761$ (c) .9542
(d) .2386 (e) -3.3010 (f) 1.7323

9. (a) -3.4655 (b) 87.664 (c) 4.4817

11.

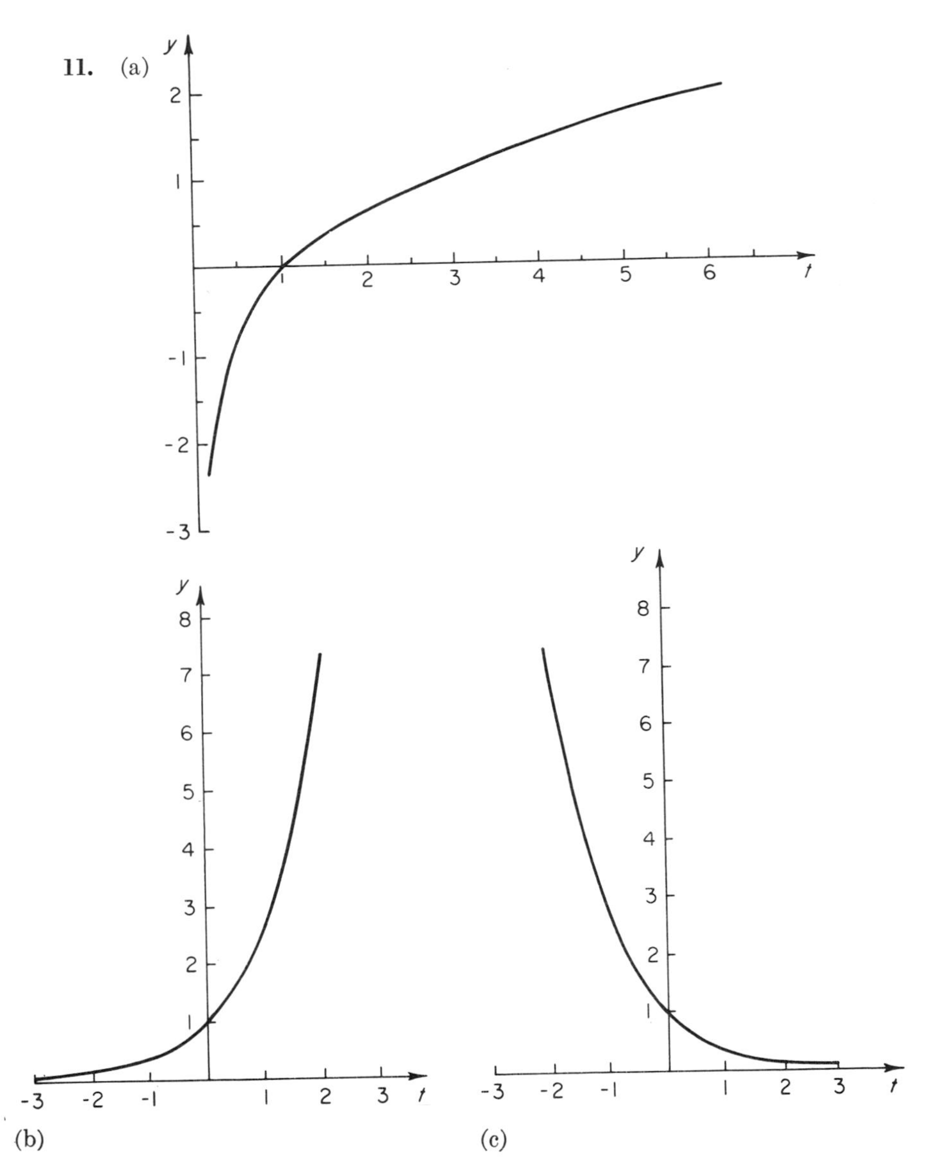

13. (a) $\frac{172}{999}$ (b) $\frac{56}{99}$ (c) $\frac{1293}{1000}$ (d) $\frac{8}{9}$

15. No

17. (a) 3,000 units
(b) 3,738 units

Exercise Set 9.2, page 437

1. (a) $300 (b) $308.39 (c) $306.82 (d) $304.50

3. (a) 6% (b) 6.167781% (c) 6.136355% (d) 6.09%

5. 9% per year compounded quarterly

7. (a) $3150.85 (b) $3108.61 (c) $3122.99

9. $6346.66

11. (a) $5309 (b) $5471 (c) 11.551666 years (d) 6.18%

13. 10.52% **15.** (a) $900 (b) $4100 **17.** 10%

19. (a) $57,507.39 (b) $30,200.99 (c) $175,610.81 (d) $2790.80

21. $219,199.25

23. (a) $4100.20 (b) $13,677.74 (c) $83,721.17 (d) $2069.02

25. $4769.05 **27.** $520.83 **29.** $522.50 **31.** $116,263.17

33. (a) $6181.92 (b) $6450.24 (c) $781.92 (d) $1050.24

CHAPTER 10

Exercise Set 10.2, page 450

1. (a) 5627 (b) 8796 (c) 3579

3. (a) 1 (b) 10 (c) 11 (d) 100
(e) 101 (f) 110 (g) 111

5. ◯ ◯ ◯ ◯

Exercise Set 10.3, page 461

1. LET M = 48.7

3. (a) Increase the number stored in X by 2 and store the new result back into X.
(b) Replace the number stored in X by its square.
(c) Replace the number stored in I by K minus I.

5. a, c, and d.

7. (a) 5 (b) 15.625 (c) 7.5 (d) 12.5

9. (a) `(A + B)/C` (b) `C/(D + E)`

(c) `C/(D * D)` or `C/(D ↑ 2)`

(d) `(C + E)/(D * D)` or `(C + E)/(D ↑ 2)`

(e) `(A + B)/2` (f) `(4/3) * 3.14 * (R ↑ 3)`

11. (a)
```
5
10
```
(b)
```
10
```

13. One possible solution is:

```
10  LET  A = 7.2
20  LET  B = 3.8
30  LET  C = 1.6
40  LET  D = 2.7
50  LET  X = (A + B + C + D)/4
60  PRINT  X
70  END
```

15. One possible solution is:

```
10  LET  A = 1
20  LET  B = 2
30  LET  C = 3
40  LET  D = 4
50  LET  E = 5
60  LET  X = (A ↑ 2) + (B ↑ 2) + (C ↑ 2) + (D ↑ 2) + (E ↑ 2)
70  PRINT X
80  END
```

Another possibility is:

```
10  LET  A = 1 ↑ 2
20  LET  B = 2 ↑ 2
30  LET  C = 3 ↑ 2
40  LET  D = 4 ↑ 2
50  LET  E = 5 ↑ 2
60  LET  F = A + B + C + D + E
70  PRINT  F
80  END
```

Exercise Set 10.4, page 467

1. PRINT "INVENTORY"

3. (a) BACTERIA COUNT = K

(b) BACTERIA COUNT = 600000

(c) BACTERIA COUNT K = 600000

(d) 600000 = BACTERIA COUNT

5. (a)

```
TEMP 1,   TEMP 2,   TEMP 3
98.4      98.6      99.1
```

(b)

```
TEMP T1 98.4
TEMP T2
98.6
T3
```

7.

```
2              4              3
```

9. The computer asks the operator to type in a value for X; the computer calculates Y = X^2 and prints the values of X and Y.

11.

```
10  READ  C1, C2, C3
20  DATA  3, 20, 37
30  LET  F1 = (9/5) * C1 + 32
40  LET  F2 = (9/5) * C2 + 32
50  LET  F3 = (9/5) * C3 + 32
60  PRINT  F1
70  PRINT  F2
80  PRINT  F3
90  END
```

13. One possible solution is:

```
 10  READ  X1, Y1, Z1
 20  DATA  2, 7, 9
 30  LET  X2 = X1 ↑ 2
 40  LET  X3 = X1 ↑ 3
 50  LET  Y2 = Y1 ↑ 2
 60  LET  Y3 = Y1 ↑ 3
 70  LET  Z2 = Z1 ↑ 2
 80  LET  Z3 = Z1 ↑ 3
 90  PRINT  "NUMBER", "SQUARE", "CUBE"
100  PRINT  X1, X2, X3
110  PRINT  Y1, Y2, Y3
120  PRINT  Z1, Z2, Z3
130  END
```

Exercise Set 10.5, page 477

1. It prints the larger of two input values.

3. It reads in seven pairs of numbers; then computes and prints out the average for each pair.

5.
```
 5  REM  PROGRAM OUTPUTS FIRST 200 EVEN INTEGERS
10  FOR  I = 1 TO 200
20  LET  J = 2 * I
30  PRINT  J
40  NEXT I
50  END
```

7.
```
10  REM  INTEREST PROGRAM
20  PRINT  "YEAR", "AMOUNT OWED"
30  FOR  I = 1 TO 25
40  LET  S = 1000 * (1.05 ↑ I)
50  PRINT  I, S
60  NEXT I
70  END
```

9. A possible solution is:

```
 10  REM  TOTAL RECEIPTS COMPUTATION
 20  REM  S = STOCK NO., N = NO. SOLD, P = UNIT PRICE
 30  REM  ENTER S = 0 TO STOP
 40  PRINT  "S, N, P"
 50  INPUT  S, N, P
 60  IF S = 0 THEN 110
 70  LET  T = N * P
 80  PRINT  "STOCK NO.", "TOTAL RECEIPTS"
 90  PRINT  S, T
100  GO TO 40
110  END
```

Exercise 10.6, page 483

1. It prints all odd integers from 5 to 85 inclusive.

3. For an integer K entered by the operator, the program prints a 1 if K is odd and a 2 if K is even.

5. The program prints out the tax to be paid, based on the salary entered by the operator. The tax is computed to be 40% of the salary if the salary is greater than $30,000 and 20% of the salary if the salary is $20,000 or less. Otherwise the tax is 25% of the salary.

7.

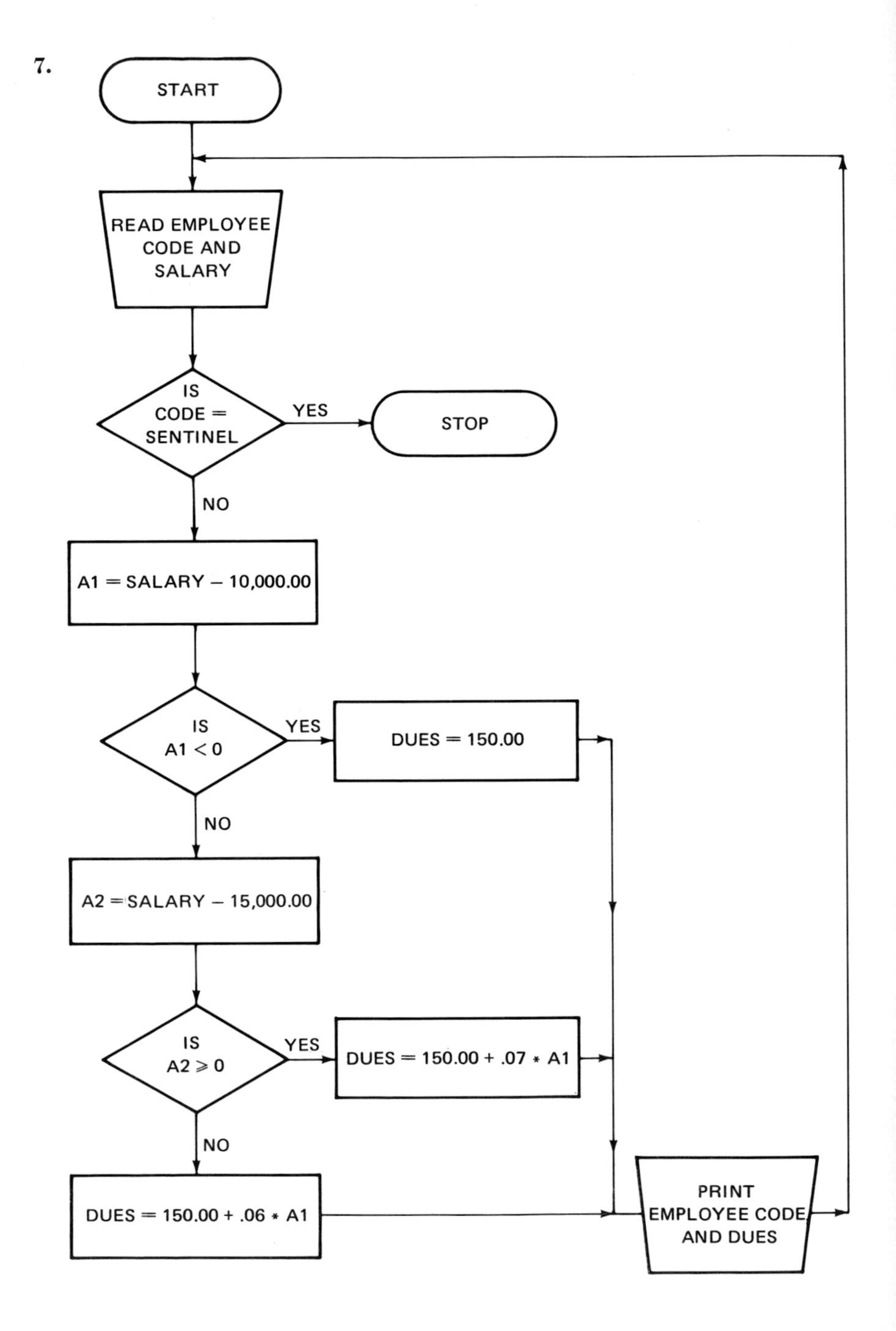
START
READ EMPLOYEE CODE AND SALARY
IS CODE = SENTINEL
YES
STOP
NO
A1 = SALARY – 10,000.00
IS A1 < 0
YES
DUES = 150.00
NO
A2 = SALARY – 15,000.00
IS A2 ⩾ 0
YES
DUES = 150.00 + .07 * A1
NO
DUES = 150.00 + .06 * A1
PRINT EMPLOYEE CODE AND DUES

9.

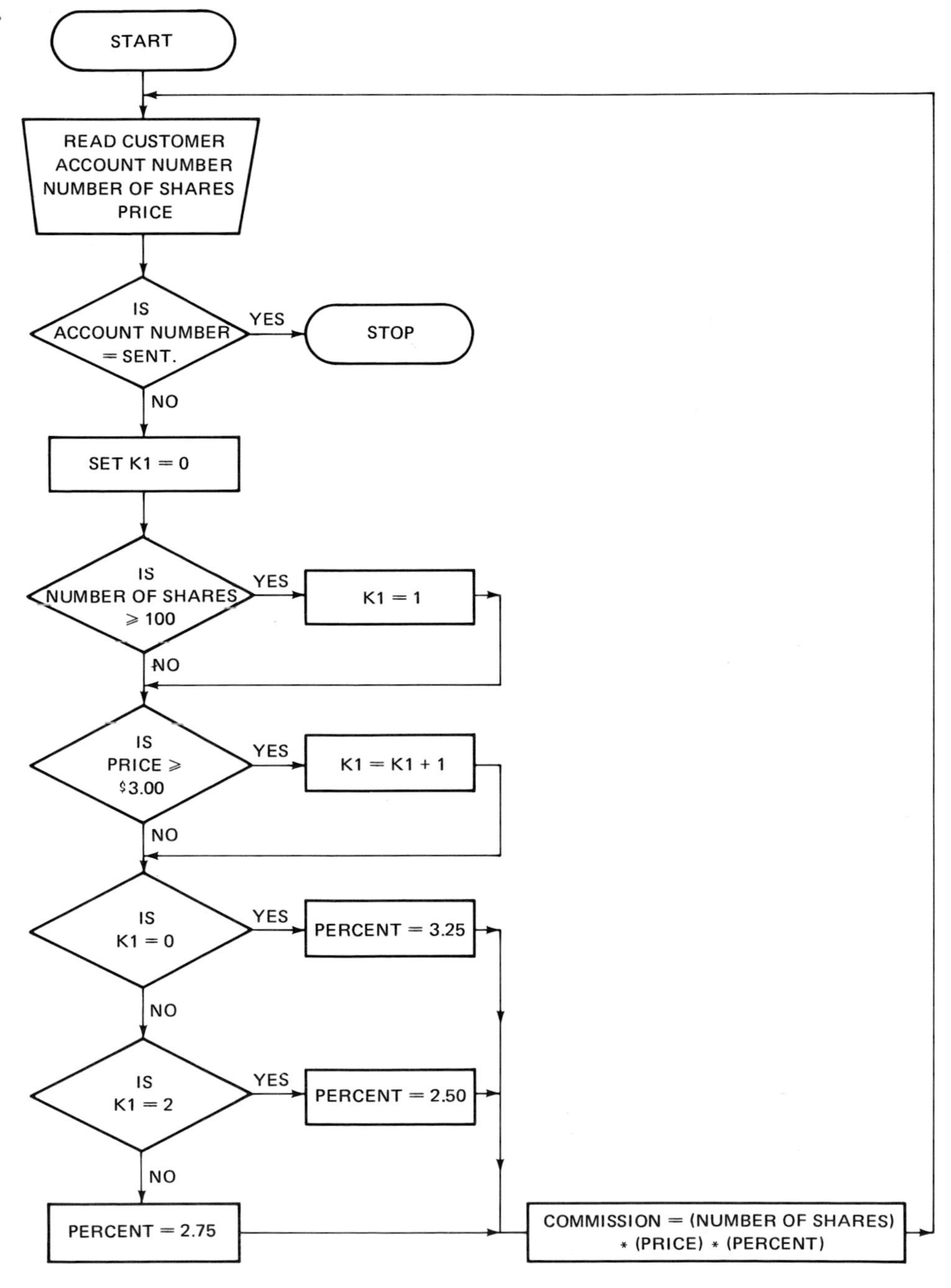
START
READ CUSTOMER ACCOUNT NUMBER NUMBER OF SHARES PRICE
IS ACCOUNT NUMBER = SENT.
YES
STOP
NO
SET K1 = 0
IS NUMBER OF SHARES ≥ 100
YES
K1 = 1
NO
IS PRICE ≥ $3.00
YES
K1 = K1 + 1
NO
IS K1 = 0
YES
PERCENT = 3.25
NO
IS K1 = 2
YES
PERCENT = 2.50
NO
PERCENT = 2.75
COMMISSION = (NUMBER OF SHARES) * (PRICE) * (PERCENT)

CHAPTER 11

Exercise Set 11.2, page 494

1. (a) 3 (b) 7 (c) 3 (d) 8

3. (a) 5 (b) 0 (c) 12 (d) $a^2 - 4$

(e) $x^2 - 4x$ (f) $x^2 + 2hx + h^2 - 4$

5. (a) 1 (b) $-\frac{3}{5}$ (c) $-\frac{3}{2}$

(d) $\dfrac{3}{b-2}$ (e) $\dfrac{3}{b+5}$ (f) $\dfrac{3}{b+h-2}$

7. the set of all real numbers

9. the set of all real numbers except $x = 3$

11. the set of all real numbers except $x = 1$

13.

15.

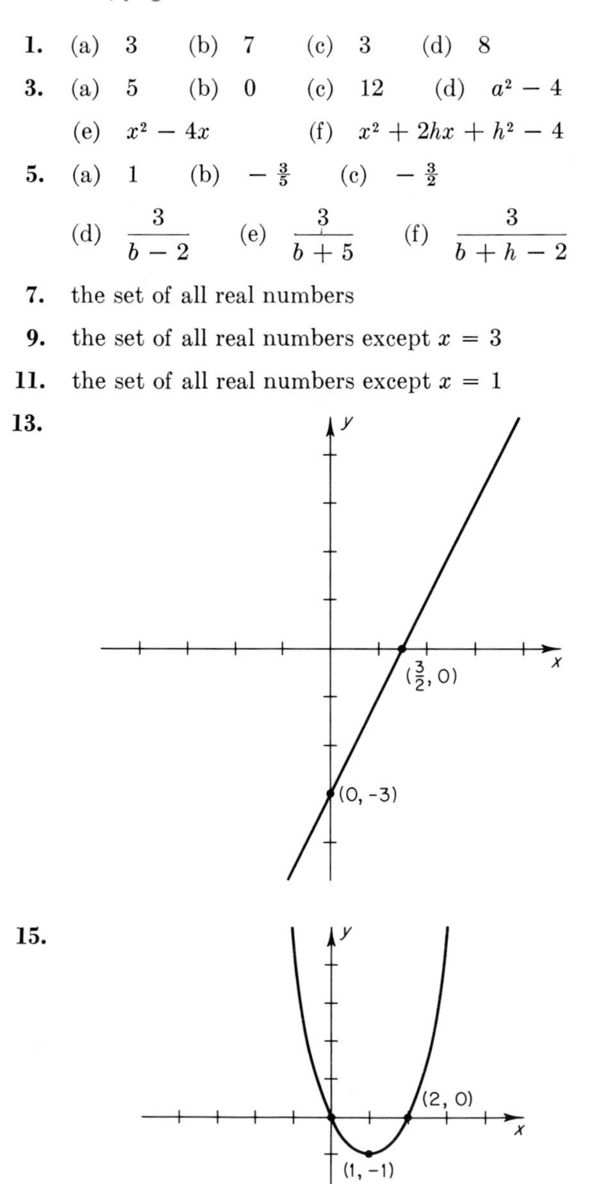

17.

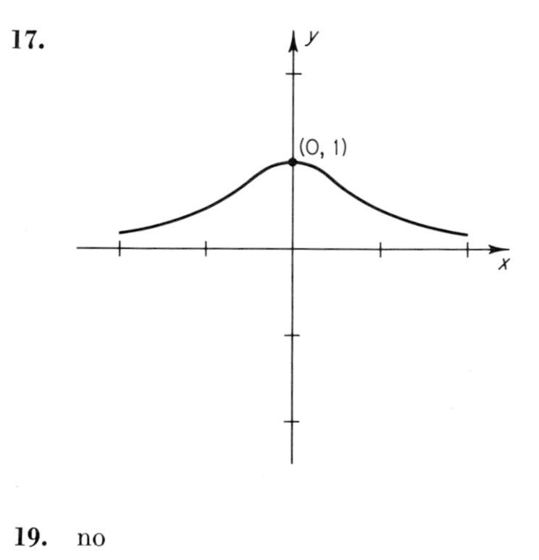

19. no

21. no

23. yes

25. (a) \$1000

(b) \$20,900

(c) \$5950

(d) \$2,000,000

Exercise Set 11.3, page 504

1. (a)

x	2.6	2.7	2.8	2.9	2.99	2.999	3.001	3.01	3.1	3.2	3.3	3.4
$f(x) = 2x - 3$	2.2	2.4	2.6	2.8	2.98	2.998	3.002	3.02	3.2	3.4	3.6	3.8

(b) 3

3. (a)

x	-1	-0.1	-0.01	-0.001	-0.0001	0.0001	0.001	0.01	0.1	1
$f(x) = \frac{\lvert x\rvert}{x}$	-1	-1	-1	-1	-1	1	1	1	1	1

(b) $\lim_{x \to 0} \frac{|x|}{x}$ does not exist

5. 10 **7.** $\frac{11}{5}$ **9.** -2 **11.** 7

13. does not exist

15. $\lim_{x \to 4} \dfrac{x^2 - 16}{x - 4} = 8$

17. $\lim_{x \to 1} \dfrac{x^3 + x^2 - 2x}{x(x + 2)} = 0$

19. $\lim_{x \to 3} \dfrac{|x - 3|}{x - 3}$ does not exist

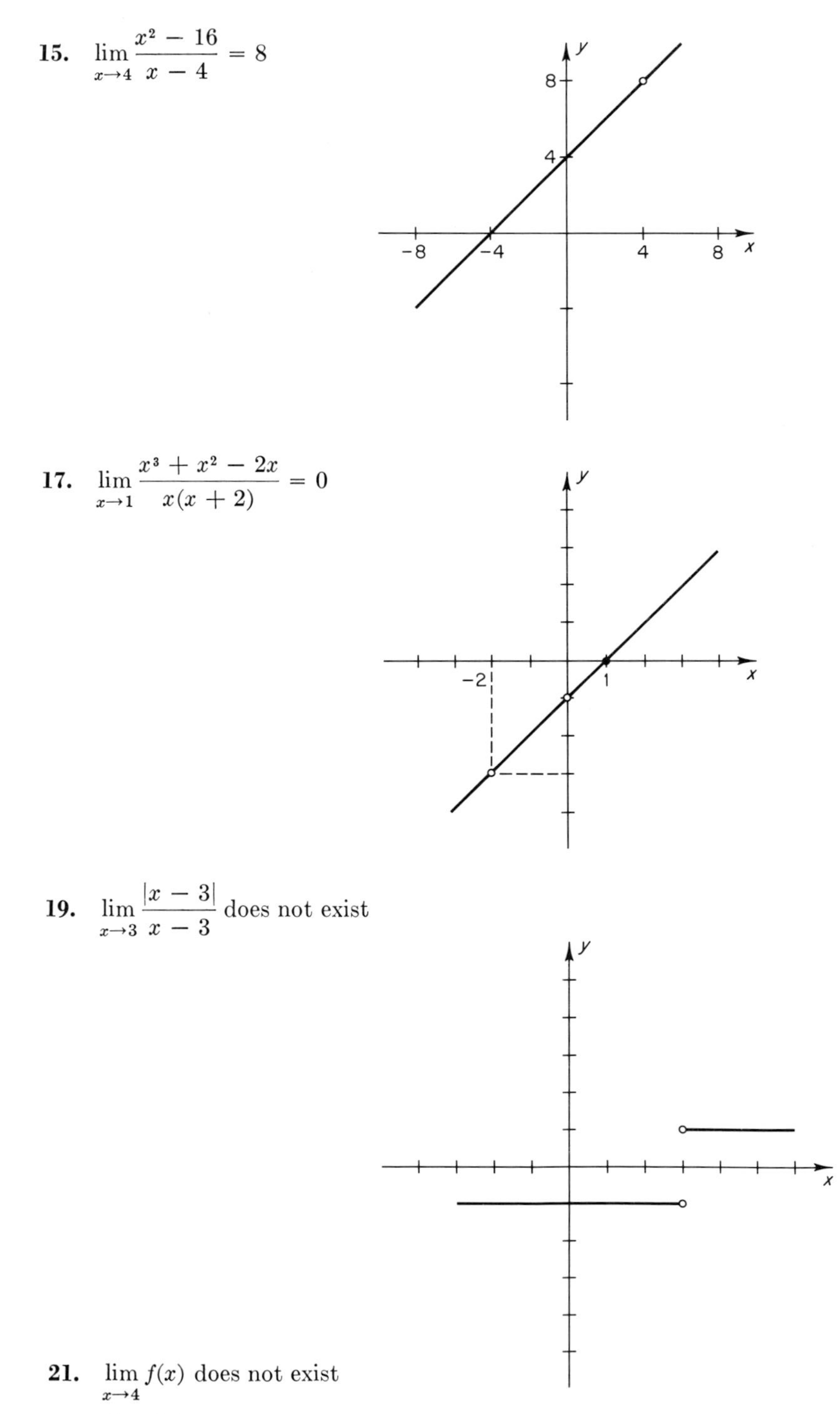

21. $\lim_{x \to 4} f(x)$ does not exist

23. $\lim_{x \to 6} f(x) = 4$

Exercise Set 11.4, page 510

1. (a) yes (b) yes (c) no (d) no
(e) no (f) no (g) no (h) yes

3. Yes, $\lim_{x\to 6} f(x) = 12 = f(6)$

5. Yes **7.** no **9.** no **11.** yes **13.** $x = 3$

15. no points of discontinuity

17. no points of discontinuity

19. $x = 0$

21. No, a jump in the graph occurs whenever a new order of coffee arrives.

Exercise Set 11.5, page 521

1. (a) 3 (b) 3 (c) 3

3. 3 **5.** 0 **7.** $\frac{8}{3}$ **9.** 2 **11.** 0

13. (a) 432 ft (b) 144 ft/sec (c) 96 ft/sec (d) 128 ft/sec

15. (a) 5327.03 people/year
(b) −2464.2 people/year
(c) between 1950 and 1970

17. (a) 1, 0.75, 0.5, 0.25

CHAPTER 12

Exercise Set 12.1, page 534

1. (a) $-\frac{2}{3}$ (b) 1 (c) 0 (d) −1

3. (a) $x + \frac{1}{2}h$ (b) x (c) −2, 0, 4

5. $2(2x + h), f'(x) = 4x$

7. $-\dfrac{1}{x(x+h)}, f'(x) = -\dfrac{1}{x^2}$

9. 2 **11.** 0 **13.** $\frac{5}{2}x^{-1/2}$

15. $\frac{1}{3}x^{-2/3} + \frac{1}{2}x^{-3/2}$ **17.** $-\frac{3}{4}x^{-5/4}$ **19.** $2x - 1$

21. $4x^3 - 6x^2 + 16x - 10$

23. $(x^{1/3} + 7x)(10x^9 - 72x^7) + (x^{10} - 9x^8 + 1)(\frac{1}{3}x^{-2/3} + 7)$

25. $-\dfrac{1}{x^2}$ **27.** $\dfrac{-4x}{(x^2-1)^2}$ **29.** $\dfrac{(3x+2+4\sqrt{x})}{2\sqrt{x}\,(2+3\sqrt{x})^2}$

31. $\frac{17}{25}, \frac{17}{49}, 17$

33. (a) -1 (b) -1 (c) $-\frac{1}{9}$ (d) $-\frac{1}{9}$

35. $(\frac{1}{2}, -\frac{21}{4})$

37. (a) $6r$ (b) $6t$ (c) $6x$ (d) $6s$
(e) $6t$ (f) $2x$ (g) $7x^6$

39. (*a*) $\frac{3}{2}x^{-1/2} + 6x^2$ (b) $-(6/x^3) + 2$

41. (a) not differentiable at $x = 2$

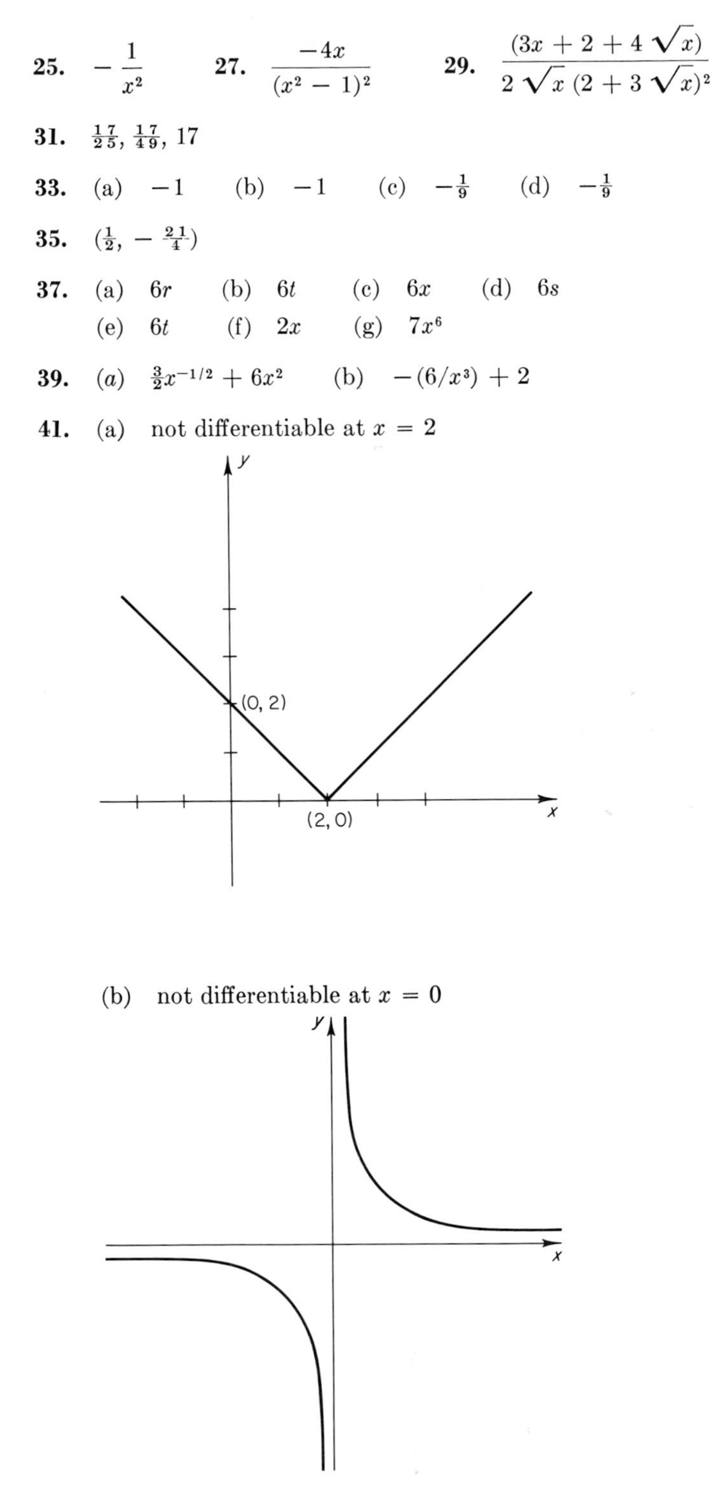

(b) not differentiable at $x = 0$

(c) not differentiable at $x = 0$

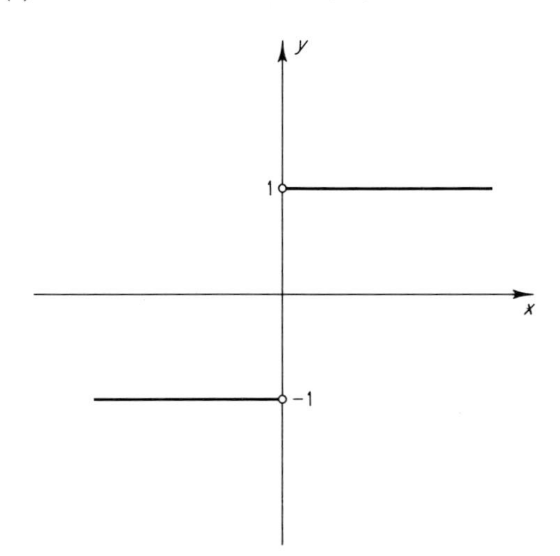

(d) not differentiable at $x = 1$

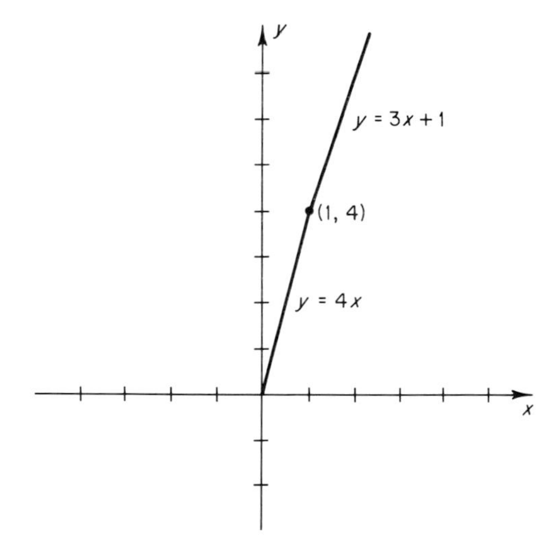

Exercise Set 12.2, page 542

1. $36(1 + 2x)^{17}$ **3.** $\dfrac{-1}{2\sqrt{2 - x}}$ **5.** $\dfrac{6(3 - 4x)}{(2x^2 - 3x)^7}$

7. $(7x^3 - 6x^2 + 2)^{-2/3}(7x^2 - 4x)$

9. $120x^2(3x^4 + 2)^{10}(2x^3 - 3)^{19} + 120x^3(3x^4 + 2)^9(2x^3 - 3)^{20}$

11. $\dfrac{-8x}{(3x^2 + 5)^{5/3}}$

13. (a) $dr/dt = (dr/ds)(ds/dt)$ (b) $dc/dw = (dc/dz)(dz/dw)$

15. 0

Exercise Set 12.3, page 549

1. $y = 4x - 4$

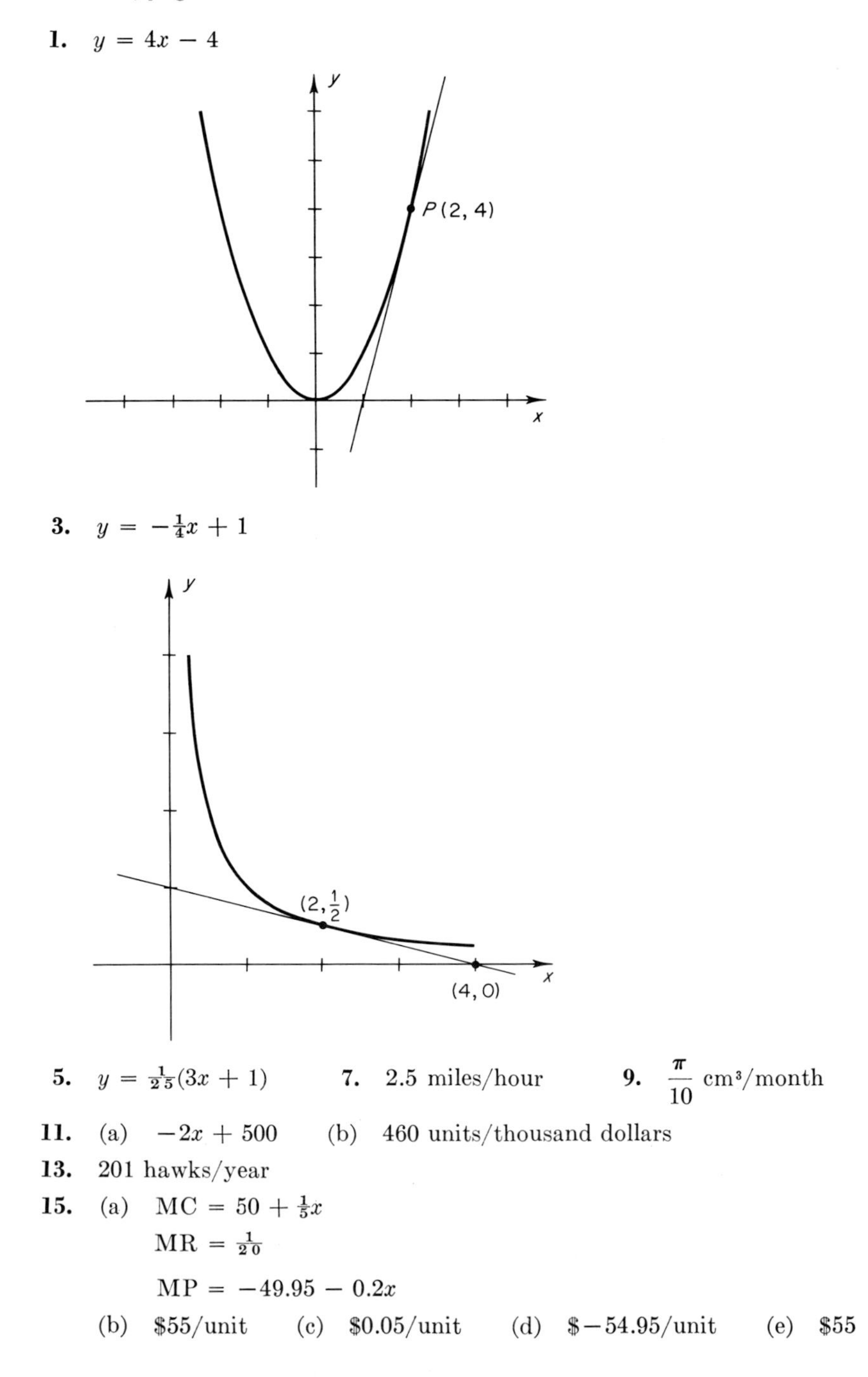

3. $y = -\frac{1}{4}x + 1$

5. $y = \frac{1}{25}(3x + 1)$ **7.** 2.5 miles/hour **9.** $\dfrac{\pi}{10}$ cm³/month

11. (a) $-2x + 500$ (b) 460 units/thousand dollars

13. 201 hawks/year

15. (a) $\text{MC} = 50 + \frac{1}{5}x$

$\text{MR} = \frac{1}{20}$

$\text{MP} = -49.95 - 0.2x$

(b) \$55/unit (c) \$0.05/unit (d) \$$-54.95$/unit (e) \$55

17. (a) $R(x) = 200x - \frac{1}{2}x^2$, $P(x) = -3000 + 200x - \frac{3}{4}x^2$
(b) $\text{MR} = 200 - x$, $\text{MC} = \frac{1}{2}x$, $\text{MP} = 200 - \frac{3}{2}x$
(c) \$175 (d) \$162.50

Exercise Set 12.4, page 556

1. $2e^{2x}$ **3.** $24xe^{4x^2}$ **5.** $2 - 4e^{4x} + 9x^2e^{x^3}$

7. $e^x(x-1)/x^2$ **9.** $-e^{-x^2}\left[\dfrac{2x^2+1}{5x^2}\right]$ **11.** $2e^{2x}$

13. $-7(1-x)^{6,\cdot(1-x)^7}$ **15.** $\dfrac{-3}{x}$ **17.** $\dfrac{1 - 3\ln x}{x^4}$

19. $\dfrac{e^x+1}{e^x+x}$ **21.** $e^x\left(\dfrac{1}{x} + \ln x\right)$ **23.** $y = x + 1$ **25.** 1

29. (a) $k = 0.107$ (b) $\dfrac{dP}{dt} = \dfrac{50{,}000ke^{-kt}}{(1+10e^{-kt})^2}$ (c) $\dfrac{5{,}350e^{-1.07}}{(1+10e^{-1.07})^2}$

CHAPTER 13

Exercise Set 13.1, page 569

1. (a) 0.02 (b) 500
(c)

t	1	5	20	225
Q	510.1	552.6	745.9	45008.5

(d) 10 (e) 10.618

3.

Q	500	1250	2300
t	0	45.815	76.305

5. (a) 1.7328 seconds (b) 3.4655 seconds

7. (a) 13.86% per year (b) 0.6931% per hour **9.** 11 years

11. (a)

Year	1975	1980	1985	1990	1995	2000	2005	2010
World population (billions)	4	4.4208	4.8856	5.3996	5.9672	6.5948	7.2884	8.0552

(b)

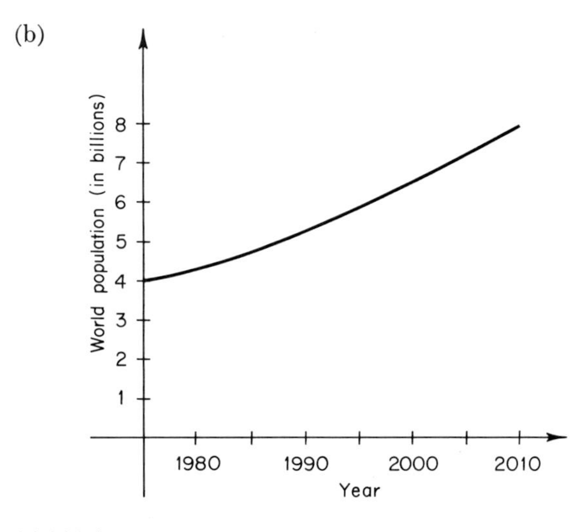

13. 24,964.4 years

15. (a) 10% per year (b) $545,980

Exercise Set 13.2, page 585

1. (a) (b) (c) (d) (e) (f) (g) (h)

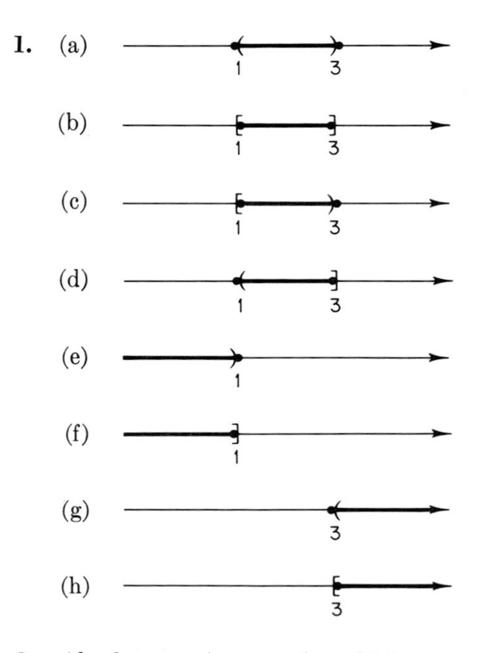

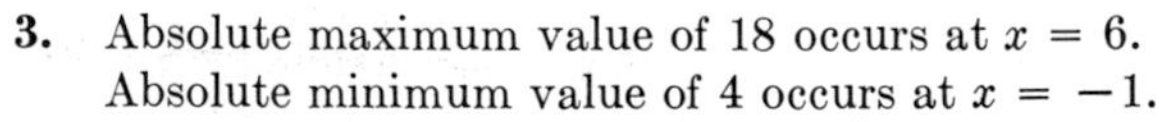

3. Absolute maximum value of 18 occurs at $x = 6$.
Absolute minimum value of 4 occurs at $x = -1$.

5. Absolute maximum value of $\frac{1}{4}$ occurs at $x = \frac{1}{2}$.
Absolute minimum value of -2 occurs at $x = 2$.

7. Absolute maximum value of 0 occurs at $x = 0$ and at $x = 2$.
Absolute minimum value of $-\frac{27}{16}$ occurs at $x = \frac{3}{2}$.

9. Absolute maximum value of 2 occurs at $x = 0$.
Absolute minimum value of $\frac{2}{5}$ occurs at $x = 2$ and at $x = -2$.

11. a square with 60 ft. side

13. 5400 sq yds **15.** $x = 20$ cm **17.** 100 units

19. lot size $= 100$; 25 orders **21.** 20 counters

Exercise Set 13.3, page 597

1. (a) (a, b) and (d, f) (b) (b, d) and (f, g)
(c) (c, e) (d) (a, c) and (e, g)

3. (a) b, f (b) d (c) c, e (d) b, d, f

5. (a) b, d, f (b) (a, b) and (f, g) (c) (b, d) and (d, f)
(d) (c, d) and (e, g) (e) (a, c) and (d, e)

7. decreases for $x < 3$, increases for $x > 3$, relative minimum at $x = 3$

9. decreases for all x

11. decreases for $x < 0$, increases for $x > 0$, relative minimum at $x = 0$

13. decreases for $-3 < x < 1$, increases for $x < -3$ and for $x > 1$; relative maximum at $x = -3$, relative minimum at $x = 1$

15. decreases for $x < 0$, increases for $x > 0$, relative minimum at $x = 0$

17. 6 **19.** 0 **21.** $1/x$

23. concave up for $-\infty < x < \infty$

25. concave up for $x > 2$, concave down for $x < 2$

27. concave up for $x > 3$, concave down for $x < 3$

29. relative maximum at $x = -1$, relative minimum at $x = 1$

31. relative minimum at $x = \frac{3}{2}$

33. no relative maxima or minima exist

35. relative minimum at $x = -1$

37. (a)

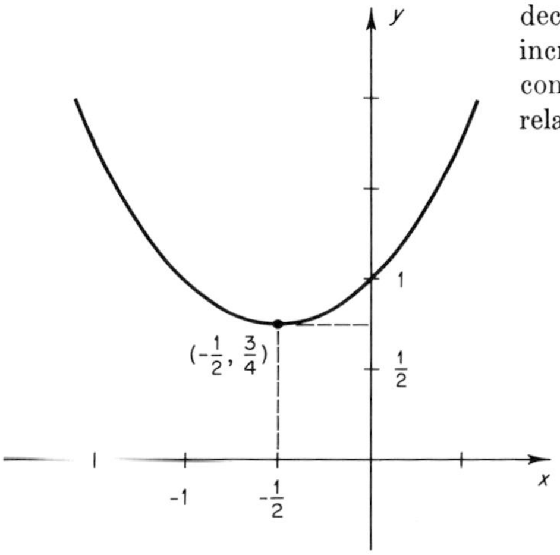

y
1
$(-\frac{1}{2}, \frac{3}{4})$
$\frac{1}{2}$
-1
$-\frac{1}{2}$
x

decreasing for $x < -\frac{1}{2}$
increasing for $x > -\frac{1}{2}$
concave up for $-\infty < x < \infty$
relative minimum at $x = -\frac{1}{2}$

(b)

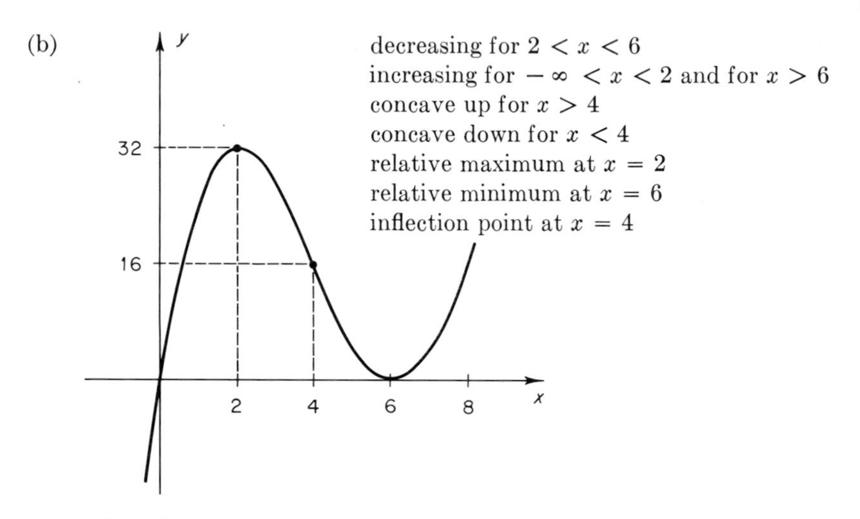

y
32
16
2
4
6
8
x
decreasing for 2 < x < 6
increasing for − ∞ < x < 2 and for x > 6
concave up for x > 4
concave down for x < 4
relative maximum at x = 2
relative minimum at x = 6
inflection point at x = 4

39. at $t = 5$ weeks

Exercise Set 13.4, page 607

1. minimum value of $-\frac{1}{4}$ occurs at $x = \frac{3}{2}$; no maximum value

3. minimum value of 3 occurs at $x = \frac{1}{2}$; no maximum value

5. 2 hours

7. minimum $= 108$ when $x = 6$

9. dimensions 5 in × 5 in × 2.5 in gives the minimum surface area of 75 in^2

11. 30 items **13.** 40 miles per hour **15.** at $t = 49$ hours

17. 110 passengers paying $2200 each maximizes the revenue

19. 55 trees

Exercise Set 13.5, page 617

1. $f_x(x, y) = 3x^2 + y$ $f_y(x, y) = x + 4y$ $f_{yx}(x, y) = 1$

3. $f_x(x, y) = e^y + y^2e^x$ $f_y(x, y) = xe^y + 2ye^x - 3y^2$ $f_{yx}(x, y) = e^y + 2ye^x$

5. $f_x(x, y) = ye^{xy}$ $f_y(x, y) = xe^{xy}$ $f_{yx}(x, y) = e^{xy}[xy + 1]$

7. $f_x(x, y) = \dfrac{1}{x}$ $f_y(x, y) = \dfrac{1}{y}$ $f_{yx}(x, y) = 0$

9. $f_x(-1, 2) = 8$ $f_y(-1, 2) = 6$ $f_{yx}(-1, 2) = 6$

11. $f_x(0, 1) = 1 + e$ $f_y(0, 1) = 2$ $f_{yx}(0, 1) = 2 + e$

13. $f_{xx}(x, y) = 4$ $f_{yy}(x, y) = -2$ $f_{xy}(x, y) = f_{yx}(x, y) = 1$

15. $f_{xx}(x, y) = f_{yy}(x, y) = f_{xy}(x, y) = f_{yx}(x, y) = 0$

17. $f_{xx}(x, y) = (x + 2)e^{x+y}$ $f_{yy}(x, y) = xe^{x+y}$
$f_{xy}(x, y) = f_{yx}(x, y) = (x + 1)e^{x+y}$

19. $f_{xx}(-1, 1) = 6$ $f_{yy}(-1, 1) = -4$ $f_{xy}(-1, 1) = f_{yx}(-1, 1) = -5$

21. $f_{xx}(0, 1) = 4e^3 \quad f_{yy}(0, 1) = 9e^3 \quad f_{xy}(0, 1) = f_{yx}(0, 1) = 6e^3$

23. relative minimum at (1, 2)

25. relative minimum at (0, 0)

27. relative maximum at (1, 0)

29. relative minimum at every point of the form (x, x), $-\infty < x < \infty$

31. relative minimum at $(\frac{2}{3}, -\frac{2}{3})$

33. (a) $\dfrac{\partial C}{\partial x} = 6x, \quad \dfrac{\partial C}{\partial y} = 4$

(b) $\left.\dfrac{\partial C}{\partial x}\right|_{(10,3)} = 60, \quad \left.\dfrac{\partial C}{\partial y}\right|_{(10,3)} = 4$

35. (a) $\dfrac{\partial A}{\partial w} = 0.11, \quad \dfrac{\partial A}{\partial H} = 7.817$

37. A maximum profit of $8000 is obtained when using four salespeople in eight departments.

39. 50 million dollars on research and 4 million on development

CHAPTER 14

Exercise Set 14.1, page 626

1. $5x + C$ **3.** $\frac{1}{6}t^6 + C$ **5.** $\frac{4}{5}x^{5/4} + C$ **7.** $-\dfrac{6}{x} + C$

9. $2e^u + C$ **11.** $50 \ln x + C$ **13.** $\frac{2}{3}x^{3/2} + C$

15. $x^3 - \frac{20}{7}x^{7/4} + 4x + C$ **17.** $\frac{6}{5}x^{5/3} - 3e^x + 2 \ln x + C$

19. $\frac{1}{2}x^2 + 2x - \frac{11}{2}$ **21.** $\frac{1}{3}x^3 + 3x + \frac{16}{3}$ **23.** $2x + 3e^x + 2$

25. $y = x^3 + 9$

27. (a) $50{,}000 + 25{,}000t + \frac{20}{7}t^{7/5}$ (b) $550{,}000 + \frac{20}{7}(20)^{7/5}$

29. (b)

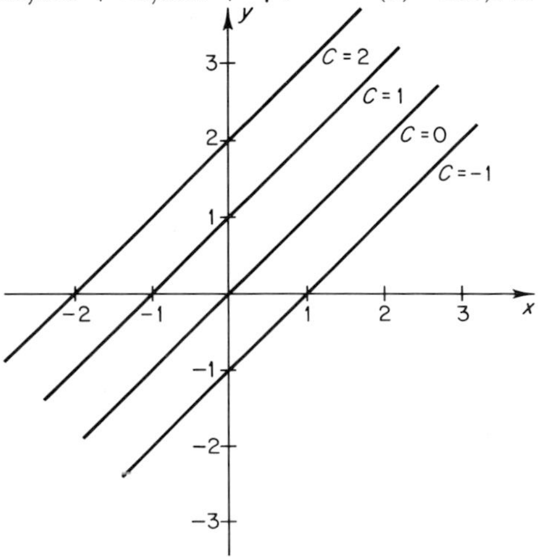

Exercise Set 14.2, page 638

1. 9 **3.** $\frac{45}{2}$ **5.** $\frac{26}{3}$ **7.** $\frac{16}{3}$

9. $\frac{9}{2}$ **11.** $e - 1$ **13.** $2 \ln 2$ **15.** 3

17. 136 **19.** $e^2 - 1$ **21.** $\ln 3$

23. $\int_1^3 x^2\, dx = \frac{26}{3}$ **25.** $\int_{-1}^3 e^s\, ds = e^3 - e^{-1}$

27. $\int_1^9 \sqrt{x}\, dx = \frac{52}{3}$ **29.** $\frac{4}{3}$

Exercise Set 14.3, page 648

1. (a) $\frac{3}{2}$ **3.** (a) $\frac{1}{2}$ **5.** 195 **7.** $-5\frac{1}{12}$

9. $\frac{370}{3}$ **11.** $-\frac{1}{3}$ **13.** 3.5 **15.** 0

17. (a) $e^3 - e^{-1}$ **19.** $\frac{2}{3}$ **21.** $e^2 - \frac{1}{2}$ **23.** 13

25. $\frac{4}{3}$ **27.** $\frac{1}{2} + \ln 3$ **29.** 4.5

Exercise Set 14.4, page 660

1. $\frac{1}{11}(x^3 + 5)^{11} + C$ **3.** $-\frac{1}{14}(x^2 - 1)^{-14} + C$ **5.** $-\dfrac{1}{4(x^2 + 4)^2} + C$

7. $-\ln (x - 2) + C$ **9.** $\frac{1}{6}(t^4 + 2)^{3/2} + C$ **11.** $-\frac{1}{2}e^{-2x} + C$

13. $\frac{1}{3}(\ln x)^3 + C$ **15.** $\frac{1}{3}(e^3 - 1)$ **17.** $e^{3x}[x - \frac{1}{3}] + C$

19. $\frac{2}{15}(x + 1)^{3/2}[3x - 2] + C$ **21.** $\frac{2}{9}x^{3/2}[3 \ln x - 2] + C$

23. $x \ln x - x + C$

25. $\dfrac{x}{3} - \dfrac{2}{9}\ln(3x + 6) + C$ **27.** $\dfrac{1}{3}\ln(3x + \sqrt{9x^2 - 25}) + C$

29. $\dfrac{1}{24}\ln\left(\dfrac{4t - 3}{4t + 3}\right) + C$ **31.** $-\dfrac{1}{2}\ln\left(\dfrac{2 + \sqrt{4 + 9x^2}}{3x}\right) + C$

CHAPTER 15

Exercise Set 15.1, page 670

1. (a) 8 (b) $e^2 - 1$

3. (a) \$8,240 (b) \$14,490 (c) \$14,020

5. $1800(e^{0.4} - 1)$ quadrillion B.T.U.'s

7. (a) 10,000 ft (b) 5 sec

9. $50[e^5 - 1]$ **11.** (b) 0.375 (c) 1

13. (a) $f(x) = \frac{1}{20}$ if $5 \leq x \leq 25$ and $f(x) = 0$ otherwise
(b) $\frac{3}{20}$ (c) 1

15. (a) 0.5 (b) 1 (c) 0.3

17. (a) $1 - e^{-1/2}$ (b) $100e^{1/2}\%$

19. (a) $1 - e^{-1/2}$ (b) $e^{-1/2}$ (c) $100(e^{-1} - e^{-2})\%$

INDEX